PHYSICS-BASED VISION

This is one volume of a three-volume series on Physics-Based Vision. The collection is an important reference source for researchers in the field as well as an educational tool. These volumes are sold as a set and separately. Set of all three volumes: ISBN 0-86720-452-4.

PHYSICS-BASED VISION: Principles and Practice

RADIOMETRY, edited by Lawrence B. Wolff, Steven A. Shafer, and Glenn E. Healey
ISBN 0-86720-294-7

COLOR, edited by Glenn E. Healey, Steven A. Shafer, and Lawrence B. Wolff
ISBN 0-86720-295-5

SHAPE RECOVERY, edited by Lawrence B. Wolff, Steven A. Shafer, and Glenn E. Healey
ISBN 0-86720-296-3

PHYSICS-BASED VISION

Principles and Practice

COLOR

edited by

Glenn E. Healey
Department of Electrical Engineering
University of California
Irvine, California

Steven A. Shafer
School of Computer Science
Carnegie Mellon University
Pittsburgh, Pennsylvania

Lawrence B. Wolff
Computer Science Department
The Johns Hopkins University
Baltimore, Maryland

CRC Press
Taylor & Francis Group
Boca Raton London New York

CRC Press is an imprint of the
Taylor & Francis Group, an **informa** business

First published 1992 by Jones and Bratlett Publishers, Inc.

Published 2018 by CRC Press
Taylor & Francis Group
6000 Broken Sound Parkway NW, Suite 300
Boca Raton, FL 33487-2742

ISBN-13: 978-0-86720-295-3 (hbk)

Visit the Taylor & Francis Web site at
http://www.taylorandfrancis.com

and the CRC Press Web site at
http://www.crcpress.com

Library of Congress Cataloging-in-Publication Data

Color / edited by Glenn E. Healey, Steven A. Shafer, Lawrence B. Wolff.
p. cm. -- (Physics-based vision)
Includes bibliographical references and index.
ISBN 0-86720-295-5
1. Computer vision. 2. Color separation--Digital techniques.
3. Color vision. I. Wolff, Lawrence B. II. Shafer, Steven A.
III. Healey, Glenn E. IV. Series.
TA1632.C63 1992
621.39'9--dc20 92-13370
CIP

Acknowledgments

C.L. Novak and S.A. Shafer. "Color Vision." In Shapiro (ed.), Encyclopedia of Artificial Intelligence, pp. 192-202. © 1992, John Wiley & Sons. All rights reserved. Reprinted with permission of the publisher and author.

C. L. Novak, S.A. Shafer, and R.G. Willson. "Obtaining Accurate Color Images for Machine Vision Research." Proceedings of Perceiving, Measuring and Using Color, SPIE Proceedings Vol. 1250, pp. 54-68 © 1990, Society of Photo Optical Instrumentation Engineers. All rights reserved. Reprinted with permission of the publisher and author.

B.A. Wandell. "The Synthesis and Analysis of Color Images." IEEE Transactions on Pattern Analysis and Machine Intelligence, Vol. 9, No. 1, 1987, pp. 2-13. © 1987, IEEE. All rights reserved. Reprinted with permission of the publisher and author.

S.A. Shafer. "Using Color to Separate Reflection Components." COLOR Research and Application, Vol. 10, No. 4, 1985, pp. 210-218. © 1985, John Wiley & Sons. All rights reserved. Reprinted with permission of the publisher and author.

G. Healey. "Using color for geometry-insensitive segmentation." J. Opt. Soc. Am. A, Vol. 6, No. 6, 1989, pp. 920-937. © 1989, Optical Society of America. All rights reserved. Reprinted with permission of the publisher and author.

S. Tominaga and B. Wandell. "Standard surface-reflectance model and illuminant estimation." J. Opt. Soc. Am. A, Vol. 6, No. 4, pp. 576-584. © 1989, Optical Society of America. All rights reserved. Reprinted with permission ofthe publisher and author.

H.-C. Lee, E.J. Breneman, and C.P. Schulte. "Modeling Light Reflection for Computer Color Vision." IEEE Transactions on Pattern Analysis and Machine Intelligence, Vol. 12, No. 4, 1990, pp. 402-409. © 1990, IEEE. All rights reserved. Reprinted with permission of the publisher and author.

L.T. Maloney. "Evaluation of linear models of surface spectral reflectance with small numbers of parameters." J. Opt. Soc. Am. A, Vol. 3, No. 10, 1986, pp. 1673-1683. © 1986, Optical Society of America. All rights reserved. Reprinted with permission of the publisher and author.

R. Gershon, A.D. Jepson, and J.K. Tsotsos. "Ambient illumination and the determination of material changes." J. Opt. Soc. Am. A, Vol. 3, No. 10, 1986, pp. 1700-1707. © 1986, Optical Society of America. All rights reserved. Reprinted with permission of the publisher and author.

M.H. Brill. "Image segmentation by object color: a unifying framework and connection to color constancy." J. Opt. Soc. Am. A, Vol. 7, No. 10, 1990, pp. 2041-2047. © 1990, Optical Society of America. All rights reserved. Reprinted with permission of the publisher and author.

M.H. Brill. "Photometric Models in Multispectral Machine Vision." SPIE Proceedings, Vol. 1453, 1991, pp. 369-380. © 1991, Society of Photo Optical Instrumentation Engineers. All rights reserved. Reprinted with permission of the publisher and author.

G. Healey. "Color Discrimination by Computer." IEEE Transactions on Systems, Man and Cybernetics, Vol. 19, No. 6, 1989, pp. 1613-1617. © 1989, IEEE. All rights reserved. Reprinted with permission of the publisherand author.

G.J. Klinker, S.A. Shafer, and T. Kanade. "A Physical Approach to Color Image Understanding." International Journal of Computer Vision, 4, 1990, pp. 7-38. © 1990, Kluwer Academic Publishers. All rights reserved. Reprinted with permission of the publisher and author.

G. Healey. "Segmenting Images Using Normalized Color." IEEE Transactions on Systems, Man and Cybernetics, 1992. To appear. © 1991 IEEE. All rights reserved. Reprinted with permission of the publisher and author.

R. Bajcsy, S.W. Lee, and A. Leonardis. "Color Image Segmentation with Detection of Highlights and Local Illumination Induced by Interreflections." Proceedings of ICPR 1990, pp. 785-790. © 1990, IEEE. All rights reserved. Reprinted with permission of the publisher and author.

D.H. Brainard and B.A. Wandell. "Analysis of the retinex theory of color vision." J. Opt. Soc. Am. A, Vol. 3, No. 10, 1986, pp. 1651-1661. © 1986, Optical Society of America. All rights reserved. Reprinted with permission of the publisher and author.

L.T. Maloney and B.A. Wandell. "Color constancy: a method for recovering surface spectral reflectance." J. Opt.Soc. Am. A, Vol. 3, 1986, pp. 29-33. © 1986, Optical Society of America. All rights reserved. Reprinted with permission of the publisher and author.

M. D'Zmura and P. Lennie. "Mechanisms of color constancy." J. Opt. Soc. Am. A, Vol. 3, No. 10, 1986, pp. 1662-1672. © 1986, Optical Society of America. All rights reserved. Reprinted with permission of the publisher and author.

S. Tominaga and B. Wandell. "Component estimation of surface spectral reflectance." J. Opt. Soc. Am. A, Vol. 7, No. 2, 1990, pp. 312-317. © 1990, Optical Society of America. All rights reserved. Reprinted with permission of the publisher and author.

D.A. Forsyth. "A Novel Algorithm for Colour Constancy." International Journalof Computer Vision, Vol. 5, No. 1, 1990, pp. 5-36. © 1990, Kluwer Academic Publishers. All rights reserved. Reprinted with the permission of publisher and author.

J. Ho, B.V. Funt, and M.S. Drew. "Separating a Color Signal into Illumination and Surface Reflectance Components: Theory and Applications." IEEE Transactions on Pattern Analysis and Machine Intelligence, Vol. 12, No. 10, 1990, pp. 966-977. © 1990, IEEE. All rights reserved. Reprinted with permission of the publisher and author.

C.L. Novak and S. A. Shafer. "Supervised Color Constancy for Machine Vision" in Human Vision, Visual Processing, and Digital Display II, SPIE Proceedings Vol. 1453, pp. 353-368 © 1991, Society of Photo Optical Instrumentation Engineers. All rights reserved. Reprinted with permission of the publisher and author.

H.-C. Lee. "Method for computing the scene-illuminant chromaticity from specular highlights." J. Opt. Soc. Am. A, Vol. 3, No. 10, 1986, pp. 1694-1699. © 1986, Optical Society of America. All rights reserved. Reprinted with permission of the publisher and author.

G.J. Klinker, S.A. Shafer, and T. Kanade. "The Measurement of Highlights in Color Images." International Journal of Computer Vision Vol. 2, pp. 7-32. © 1988, Kluwer Academic Publishers. All rights reserved. Reprinted with permission of the publisher and author.

G. Healey. "Estimating spectral reflectance using highlights." Image and Vision Computing, Vol. 9 No. 5, 1991, pp. 333-337. © 1991, Butterworth-Heinemann Ltd. All rights reserved. Reprinted with permission of the publisher and-author.

H.-C. Lee. "Illuminant color from shading." Proceedings of Perceiving Measuring and Using Color, SPIE Proceedings Vol. 1250, 1990, pp. 236-244. © 1990, Society of Photo Optical Instrumentation Engineers. All rights reserved. Reprinted with permission of the publisher and author.

S.A. Shafer, T. Kanade, G.J. Klinker, and C.L. Novak. "Physics-Based Models for Early Vision by Machine." Proceedings of Perceiving, Measuring, and Using Color, SPIE Proceedings Vol. 1250, 1990, pp. 222-235. © 1990, Society of Photo Optical Instrumentation Engineers. All rights reserved. Reprinted with permission of the publisher and author.

B.V. Funt, M.S. Drew, and J. Ho. "Color Constancy from Mutual Reflection." International Journal of Computer Vision Vol. 6, No. 1, 1991, pp. 5-24. © 1991, Kluwer Academic Publishers. All rights reserved. Reprinted with permission of the publisher and author.

B.V. Funt and M.S. Drew. "Color Space Analysis of Mutual Illumination." Simon Fraser University Centre for System Science, Technical Report 91-03, 1991. © 1991, B. V. Funt and M.S. Drew. All rights reserved. Reprinted with permission of the authors.

Contents

Introduction to the Series ix

Introduction to the Color Volume xiii

Color Vision 1
C.L. Novak and S.A. Shafer

Color Image Formation 11

Obtaining Accurate Color Images for Machine Vision Research 13
C.L. Novak, S.A. Shafer, and R.G. Willson

The Synthesis and Analysis of Color Images 28
B.A. Wandell

Color Reflection Models 41

Using Color to Separate Reflection Components 43
S.A. Shafer

Using Color for Geometry-Insensitive Segmentation 52
G.E. Healey

Standard Surface-Reflectance Model and Illuminant Estimation 70
S. Tominaga and B.A. Wandell

Modeling Light Reflection for Computer Color Vision 79
H.-C. Lee, E.J. Breneman, and C.P. Schulte

Evaluation of Linear Models of Surface Spectral Reflectance with Small Numbers of Parameters 87
L.T. Maloney

Color Image Segmentation 99

Ambient Illumination and the Determination of Material Changes 101
R. Gershon, A.D. Jepson, and J.K. Tsotsos

Image Segmentation by Object Color: A Unifying Framework and Connection to Color Constancy 109
M.H. Brill

Photometric Models in Multispectral Machine Vision 116
M.H. Brill

Color Discrimination by Computer 129
G.E. Healey

A Physical Approach to Color Image Understanding 134
G.J. Klinker, S.A. Shafer, and T. Kanade

Segmenting Images Using Normalized Color 166
G.E. Healey

Color Image Segmentation with Detection of Highlights and Local Illumination Induced by Inter-reflections 199
R. Bajcsy, S.W. Lee, and A. Leonardis

Color Constancy 205

Analysis of the Retinex Theory of Color Vision 208
D.H. Brainard and B.A. Wandell

Color Constancy: A Method for Recovering Surface Spectral Reflectance . . . 219
L.T. Maloney and B.A. Wandell

Mechanisms of Color Constancy 224
M. D'Zmura and P. Lennie

Component Estimation of Surface Spectral Reflectance. 235
S. Tominaga and B.A. Wandell

A Novel Algorithm for Color Constancy 241
D.A. Forsyth

Separating a Color Signal into Illumination and Surface Reflectance Components: Theory and Applications 272
J. Ho, B.V. Funt, and M.S. Drew

Supervised Color Constancy for Machine Vision. 284
C.L. Novak and S.A. Shafer

Color Highlight Analysis 301

Method for Computing the Scene-illuminant Chromaticity from Specular Highlights 303
H.-C. Lee

The Measurement of Highlights in Color Images 309
G.J. Klinker, S.A. Shafer, and T. Kanade

Estimating Spectral Reflectance Using Highlights 335
G.E. Healey

Illuminant Color from Shading 340
H.-C. Lee

Color Interreflection 349

Physics-Based Models for Early Vision by Machine 351
S.A. Shafer, T. Kanade, G.J. Klinker, and C.L. Novak

Color Constancy from Mutual Reflection. 365
B.V. Funt, M.S. Drew, and J. Ho

Color Space Analysis of Mutual Illumination 385
B.V. Funt and M.S. Drew

Author Index 411

Introduction to the Series

Physics-Based Vision

When machine vision first became an area of study distinct from pattern classification, in the late 1960s, researchers discovered the power of perspective geometry for recovering 3D shape information from 2D image data. Thus began what continues to be the major trend in machine vision research and application, in which 2D edges and regions are analyzed geometrically. These edges and regions are found by treating the image data as an arbitrary function $I(x,y)$ and differentiating it or modeling it statistically.

However, such traditional analysis fails on real images, because the function $I(x,y)$ is not arbitrary at all. Instead, it is determined by the laws of optics through the processes of illumination, reflection, and image formation. Since the real world is highly structured, so is the intensity or color data in the image. Traditional methods such as stereo or motion analysis routinely fail because they do not account for the systematic nature of the world, such as the movement of specular highlights across a surface as the camera moves.

In the early 1970s, Berthold Horn at MIT made a further striking observation – these optical laws could actually be used as a source of information to aid in the process of visual interpretation, to supplement or even replace traditional geometric analysis. Horn introduced optical models of reflection and imaging to the vision community, and demonstrated over many years that there was power in this approach.

However, few researchers adopted a similar approach to machine vision through the 1970s and early 1980s, largely because Horn's work raised more questions than it answered:

- How to deal with the inherent ambiguity that remains when optical laws are applied to an image;
- How to segment the image, that is, determine the regions with uniform optical properties, so that each can be analyzed and characterized;
- How to obtain reflection models for real materials, or whether there is perhaps some "universal" reflection model;
- How much calibration of cameras is required, and how to do it;
- How much knowledge about the scene is required, and how to get it in realistic scenarios.

Throughout this period of time, there was no sizable "community" of researchers or practitioners in this area, because of skepticism about whether there could be any satisfactory resolution to these issues. What little research was done in this area was typically tested only on relatively simple, synthetic images generated under highly idealized assumptions. The whole approach of modeling optics for machine vision was not widely adopted. Yet, there was an unmistakable promise in the concept, and an apparent potential for great contributions to the fundamental science of computational vision.

In the mid-1980s, an explosion of research in this area began, leading to the adoption of the term "physics-based vision" to distinguish this area from more traditional approaches based entirely on geometry with trivialized models of image data and imaging optics. Several factors contributed to this onslaught of activity in physics-based vision:

- The introduction of differential geometry into the machine vision community, which allowed optical properties to be cleanly related to geometry;
- The spread of photometric stereo, which was for a long time the only technique from physics-based vision that could be easily reproduced in new laboratories and would consistently give good results;
- The development of simple models of reflection and color theory for machine vision, based on work in the computer graphics community;
- The sudden availability of inexpensive, reliable cameras, automated lenses, and digitizers in the mid-1980s;
- The influx of young researchers into machine vision, many of whom have

- more faith in optics than in the traditions of machine vision research.

Thus, in the late 1980s, physics-based vision became accessible to many research sites, and the various sub-problems in the field began to be widely recognized and studied. Since that time, there has been an active community of researchers in physics-based vision.

Current Issues in Physics-Based Vision

Not surprisingly, the current issues in physics-based vision can ultimately be reduced to the issues listed previously, which were raised by Horn's early work. Yet, many specific sub-problems are now recognized and have a distinct literature of their own:

- How to model illumination, particularly shadows and interreflection;
- How to model reflection from smooth and rough surfaces;
- How to model color in illumination and reflection (particularly how to normalize images for illumination color, called "color constancy");
- How to model polarization, refraction, and transparency;
- How to segment an image into regions that correspond to individual surfaces of uniform properties in the scene;
- How to model and control lens optics and camera parameters (particularly the method of "depth from focus");
- How to blend optical analysis with surface smoothness assumptions ("shape from shading");
- How to blend geometric stereo and motion image analysis with optical analysis;
- How to use multiple light sources to gain additional constraint ("photometric stereo");
- How to obtain high-quality image data for research in all these topics.

The State of Research in Physics-Based Vision

As these issues have been raised and addressed in the literature, some questions have been answered with reasonable certainty. For example, the fundamental optics of focusing has been clearly identified and verified experimentally, and several paradigms have been established for exploiting it. Continued research in focusing consists mainly of finding faster and more precise algorithms and hardware, rather than introducing fundamental new concepts.

Unfortunately, the literature of physics-based vision has become scattered throughout many different publications and conferences. Physics-based vision touches on so many areas of optics, geometry, imaging, vision, and robotics that papers tend to appear in conferences in those areas rather than specifically in one journal or conference. Much of the literature is obscure, published in places that are not likely to be within the awareness of the typical reader or researcher in the area. So, while some problems have been "solved," many researchers may not be aware of the true state of the art.

At the same time, while some simple problems have been analyzed fairly well, there are great problems and efforts yet to be addressed. Since they do not appear very clearly in the literature, researchers may not even be aware that these frontiers are waiting to be crossed someday. Two of the most prominent are:

- Physics-Based Image Sensing – The development of new imaging sensors, in the tradition of photometric stereo or "polarization cameras," that exploit strong optical relationships and specially engineered hardware to yield direct measurement of interesting surface properties without the need for symbolic reasoning in software.
- Physics-Based Image Segmentation – Real scenes contain many objects and surfaces. Where specialized sensors cannot be used, a generic color or intensity image must be broken into regions that correspond to each sur-

face, without prior knowledge about the specific objects in the scene.

Physics-based vision is thus at a time of change, as the traditional problems of optical modeling come closer and closer to solution, and the issues of applying these models to vision tasks move to the forefront of research.

This Publication

At this time of rapid progress and change in the field, there is a need to bring together in one place the collected literature of the field of physics-based machine vision. This will give the student or scientist an overview of the state of the art as of mid-1991, in all areas of the field.

This collection was produced because the editors, who are active researchers in the area, felt it was the form of publication most needed to capture the collective wisdom of the community. A textbook might be more concise and clear, but the field is not yet mature enough to support a textbook with a long lifetime. A conference proceedings, or edited collection of newly prepared papers, would be another possible form of publication, but such a collection would of necessity be limited to the few authors and topics for which a new manuscript is available at this moment. It would also fail to provide some papers of great historical interest, such as the NBS and ASTM documents that define standard terminology of radiometry. So, the editors preferred to present the community with a collection of selected papers that were previously published.

Such a collection as this has the great strength that all of the papers here have passed some test of time, and all were selected by the editors through a process of critical comparison with other papers in the literature. However, this kind of collection also suffers from inherent limitations. The editors therefore wish to apologize in advance for omissions of some valuable papers due to limits of space, publication permission, or simply the editors' limited scope of awareness. We hope the collection will be useful nevertheless, and believe it will be so. It cannot present the "final word" in the area, however, since new papers are being produced so rapidly in physics-based vision in 1991 and 1992, even as these volumes are going to press. We therefore present these volumes as a "snapshot" of the current state of the art in the field.

Because of the large number of important papers, we have divided this collection into three volumes: Color, Radiometry, and Shape Recovery. Together, this set of papers should enable researchers in machine vision to gain a current, detailed understanding of what is known in all the exciting areas of research in physics-based vision.

– S.A.S.

Acknowledgments

A publication such as this collection is only made possible by the diligent efforts of many people.

For technical assistance and advice, we thank Shree Nayar, Takeo Kanade, John Krumm, and Reg Willson. They have pointed out important papers and contributed to the overall intellectual structure and scope of the collection.

Our sincere thanks to Alice Peters, of Jones and Bartlett Publishers, for taking on the monumental task of bringing this collection to print, and for coordinating the mountain of paperwork needed to collect these pages and obtain permission to republish them all. We are especially indebted to her for keeping things moving, even when the editors were a bit on the lethargic side!

We also gratefully acknowledge the support of our universities, Carnegie Mellon University, the Johns Hopkins University, and the University of California at Irvine.

Special thanks also to Patty Mackiewicz for handling large amounts of paper, and to Tensie Wenderoth for assistance in the arduous task of preparing permission letters.

– S.A.S., G.E.H., and L.B.W.

Introduction to the Color Volume

For many years, color has been used in image understanding. Early Landsat sensors, for example, provided image measurements for a set of visible and infrared bands. These images were used to classify areas of the earth according to known spectral signatures for region classes such as water or vegetation. In this context, the biggest motivation for using color was that it provided additional features at each pixel in an image to be used by a pattern classification system. In most cases, these additional features improved classification accuracy. For such applications, however, color was treated as a generic feature along with others such as shape or texture measures, and the unique physical attributes of color were not utilized.

In recent years, machine vision researchers have started to decipher and exploit the vast amount of information about a scene that is encoded in color images. The key to progress has been the realization of the importance of physical models that relate a three dimensional scene to its color image. Physical models characterize the relevant components of a scene including illumination, material surfaces, and sensors. To be useful, these models must also describe the color image formation process so that from the arrangement of a scene we can predict the resulting color image. From an analysis of these models, several important uses for color in machine vision have been identified. The articles in this volume describe these physical models and their implications along with algorithms that exploit these models.

The introductory article provides a general introduction to color vision by surveying many of the aspects and uses of color and by establishing terminology. A large set of references is given to guide further reading. Physical models underlying color image formation are presented in the next two sections. The articles in the first section discuss image formation in general and examine issues related to color sensing and the representation of surfaces and light sources. The second section contains a set of articles that focus specifically on color reflection models. These models are the foundation of the algorithms derived in later sections.

The use of color reflection models has significantly improved the capability of image segmentation algorithms. Such algorithms have the critical requirement of locating the different surfaces or objects present in a scene. In the third section, a collection of papers shows how color can be used to segment various optical phenomena that have traditionally caused difficulty for segmentation programs. Such phenomena include shadows, highlights, shading, and interreflection. Several of the articles present segmentation results that demonstrate vividly the power of the physics-based approach.

Color is also a valuable cue for recognizing objects. Unfortunately, the color measurements made by a vision system depend on the ambient illumination as well as on the intrinsic color of an object. In order to recognize an object in different lighting environments, a system must have the ability to discount the effects of the illumination. This ability is called *color constancy.* Most of the recent work on color constancy has been based on explicit physical models describing the lights and surfaces in a scene. This approach has led to several useful color constancy algorithms based on a range of models. A representative collection of these algorithms and their underlying physical assumptions are described in the fourth section.

Until recently, highlights were usually considered a nuisance to machine vision systems. The analysis of reflection models, however, has revealed that highlights contain a wealth of useful spectral and geometric information about a scene. In many situations, for example, the spectral composition of a highlight is approximately the same as the spectral composition of the illuminant. The papers in the fifth section present algorithms for finding highlights and for using them to infer properties of lights and surfaces in a scene. Algorithms are derived, for example, to estimate illuminant chromaticity and to compute illumination invariant color descriptions for a surface.

Another optical process, interreflection, has confounded segmentation and shape estimation algorithms for many years. Interreflection occurs when a visible surface patch is illuminated by light reflected from another surface. In general, this illumination from another surface changes the color and intensity of the light reflected by the surface patch. Thus, interreflection can be regarded as an important special case of spatially varying illumination. Researchers have recently developed models for interreflection and techniques to remove its effects from images. Some authors have observed that, like highlights, interreflection regions can be used to compute additional information about a scene. The articles in the sixth section present models for interreflection and algorithms for interpreting interreflection effects in images.

– G.E.H.

Paper Included in This Collection

[1] Novak, C.L. and Shafer, S.A. "Color Vision". In *Encyclopedia of Artificial Intelligence*. 1992, J.Wiley and Sons. Pages 192-202.

COLOR VISION

Color enriches one's everyday visual experience. In comparison with a monochrome (black-and-white) image, a color picture seems to be alive with detail because of all the additional information in the image. In computer vision, researchers have attempted to harness this additional information. The simplest method for using color is by associating colors with objects, for example "trees are green" and "the sky is blue." If specific object colors are not known, an image can still be chopped into meaningful pieces by finding regions of uniform color. Analyzing color in terms of physical properties of objects and the physics underlying image formation has yielded significant advantages in understanding images. Research has also been conducted into modeling the processing of color information in the human visual system; the goal is to have computers that can mimic the color perception abilities that humans have.

COLOR AND COLOR IMAGING

Color arises from the spectral properties of light. Figure 1 shows the spectrum of electromagnetic energy. Wavelengths of energy are customarily denoted by λ, and the unit of measure is the nanometer (nm). Visible light lies within the range of approximately 380–760 nm, running the gamut from violet and blue at the low end of the visible spectrum, through green, yellow, and orange, to red at the high end.

The wavelengths just higher than the visible spectrum make up the near-infrared (near-IR) portion of the spectrum, which is also frequently used in computer vision. Visible light is normally a mixture of energy at many wavelengths and is characterized by the spectral power distribution (SPD), which tells how much energy is present at each wavelength. An SPD is usually denoted by $S(\lambda)$ (CIE, 1970). At each pixel (point) in an image the SPD of the incident light determines the pixel value.

Monochrome imaging is simpler than color imaging. An imaging sensor is sensitive to the different wavelengths of light to varying degrees, as expressed by the spectral responsivity $s(\lambda)$ of the sensor (Grum and Becherer, 1979). Typical spectral responsivities of the two principal types of sensors, vidicon tubes and silicon CCD chips, are shown in Figure 2. The output pixel value p at any point in the image is defined, for a calibrated camera, by

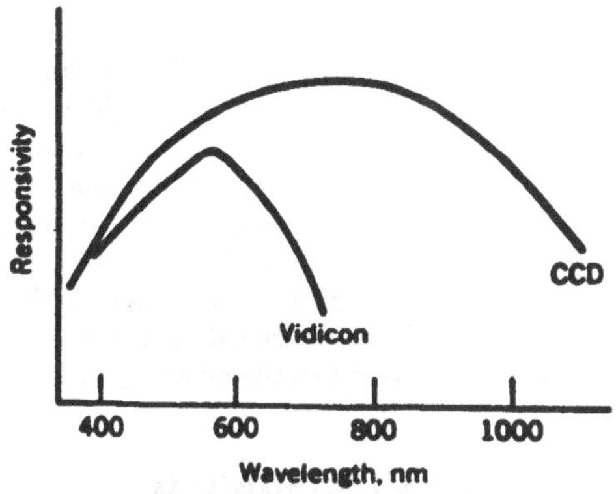

Figure 2. Spectral responsivities of vidicon tubes and CCD chips (Hamamatsu, 1983a, 1983b).

$$p = p_0 \int S(\lambda)s(\lambda)\, d\lambda$$

where p_0 is a scaling factor. Integration is performed over that portion of the spectrum for which $s(\lambda)$ is nonzero. For a vidicon, $s(\lambda)$ is approximately equal to $V(\lambda)$, the spectral luminous efficiency of the human eye, so a monochrome image from a vidicon is similar to the brightness seen by a person. However, CCD sensors are much more sensitive in the near-IR region of the spectrum; for this reason, CCD cameras are frequently fitted with IR cutoff filters or filters to match the responsivity to $V(\lambda)$.

Color imaging is more complex, requiring the assignment of a color to each SPD instead of a single number. However, the set of all perceivable colors, as determined by color-matching experiments, is only a three-dimensional space. Humans cannot distinguish all different SPDs from each other but only those SPDs that correspond to different colors in color space. Since this is a many-to-one correspondence, there can be many SPDs that have the same color; such SPDs are said to be metameric (Kuehni, 1983).

The axes of color space, called primary colors, can be chosen arbitrarily. A convenient set, universally used for color measurement, is the X-Y-Z set of colors adopted by the CIE (International Commission on Illumination). Each distinct point in X-Y-Z space corresponds to a unique color perception. A psychophysical color **C** is defined by

$$\mathbf{C} = \begin{bmatrix} X \\ Y \\ Z \end{bmatrix} = \begin{bmatrix} \int s(\lambda)\bar{x}(\lambda)\, d(\lambda) \\ \int s(\lambda)\bar{y}(\lambda)\, d(\lambda) \\ \int s(\lambda)\bar{z}(\lambda)\, d(\lambda) \end{bmatrix}$$

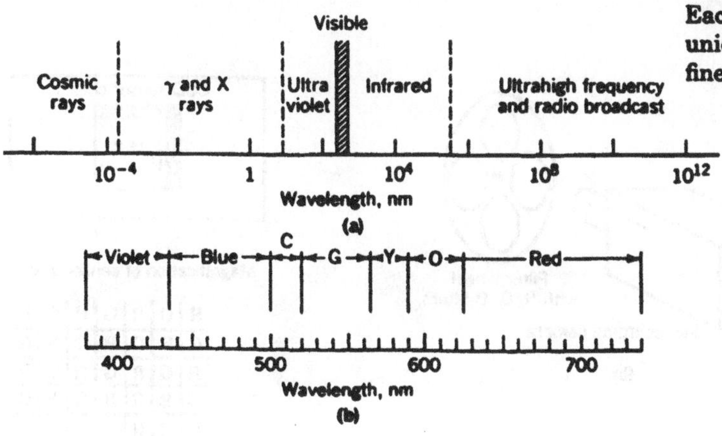

Figure 1. (*a*) Spectrum of electromagnetic radiation (Judd and Wyszecki, 1975); (*b*) magnification of visible portion of spectrum (Williamson and Cummins, 1983): C = cyan, G = green, Y = yellow, and O = orange.

The functions $\bar{x}(\lambda)$, $\bar{y}(\lambda)$, and $\bar{z}(\lambda)$ are CIE tristimulus values that define the primaries X, Y, and Z (Wyszecki and Stiles, 1982). Since $\bar{y}(\lambda) = V(\lambda)$, Y corresponds to luminance (brightness of color as seen by the human eye); the remaining coordinates X and Z determine chromaticity (the aspects of color independent of brightness).

Color imaging for computer vision follows the paradigm of color television, in which colors are measured using color filters. For any filter, the output pixel value p is determined by

$$p = p_0 \int S(\lambda)\tau(\lambda)s(\lambda)\, d\lambda$$

where $\tau(\lambda)$, the transmittance of the filter, is the fraction of light the filter allows to pass through at each wavelength. To yield color values that uniquely correspond to color perceptions, three filters that span the X-Y-Z space must be used. The filters used for color television are red, green, and blue to maximize the gamut of measurable color values. For a particular sensor $s(\lambda)$, the filter transmittances $\tau_r(\lambda)$, $\tau_g(\lambda)$, and $\tau_b(\lambda)$ of the red, green, and blue filters determine tristimulus values $\bar{r}(\lambda) = \tau_r(\lambda)s(\lambda)$, $\bar{g}(\lambda) = \tau_g(\lambda)s(\lambda)$, and $\bar{b}(\lambda) = \tau_b(\lambda)s(\lambda)$. For standard color television, these functions should obey (Wentworth, 1955)

$$\begin{bmatrix} \bar{r}(\lambda) \\ \bar{g}(\lambda) \\ \bar{b}(\lambda) \end{bmatrix} = \begin{bmatrix} 0.587 & -0.164 & -0.089 \\ -0.301 & 0.611 & -0.0087 \\ 0.0178 & -0.0362 & 0.274 \end{bmatrix} \begin{bmatrix} \bar{x}(\lambda) \\ \bar{y}(\lambda) \\ \bar{z}(\lambda) \end{bmatrix}$$

Color pixel values P are then determined by

$$\mathbf{P} = \begin{bmatrix} R \\ G \\ B \end{bmatrix} = \begin{bmatrix} r_0 \int S(\lambda)\bar{r}(\lambda)\, d\lambda \\ g_0 \int S(\lambda)\bar{g}(\lambda)\, d\lambda \\ b_0 \int S(\lambda)\bar{b}(\lambda)\, d\lambda \end{bmatrix}$$

where r_0, g_0, and b_0 are scaling factors. "White" in a color TV image corresponds to CIE Standard Illuminant C (Hunt, 1967). A broadcast color camera contains three separate sensors, each with a color filter, and beam-splitting optics to direct the image simultaneously to all three sensors; for research in computer vision, a single monochrome camera is normally used with a filter wheel to rotate each filter in turn into position (Fig. 3). Three-color CCD sensor chips are also available, offering a cheaper and more compact method of obtaining color images. Unfortunately, the subdivision of the chip into areas with different filters means that the spatial resolution of resulting images is not as good as that obtained from monochrome cameras. In computer vision, most imaging systems do not behave in the ideal way described above. This is because several imaging factors are not controlled or accounted for; these include sensor spectral responsivity, gain within each color band, and nonlinear response to intensity (LeClerc, 1986; Klinker, 1988; Novak and coworkers, 1990). This results in a lack of correspondence between color computer images and National Television System Committee (NTSC) color TV standards. The usual color filters for computer vision are Kodak Wratten filters #25 (R), #58 (G), and #47b (B) (Levine, 1980; Ito, 1975). Infrared filters [$\tau(\lambda)$ highest in near IR] are also used in remote sensing. In this entry, *color* refers to colors measured in a computer vision system using these or similar filters.

COLOR SPACES AND TRANSFORMATIONS

In computer vision, color pixel values usually contain R, G, and B values each measured in 8 bits. The set of image colors is thus a cube called the color space (Fig. 4). Inten-

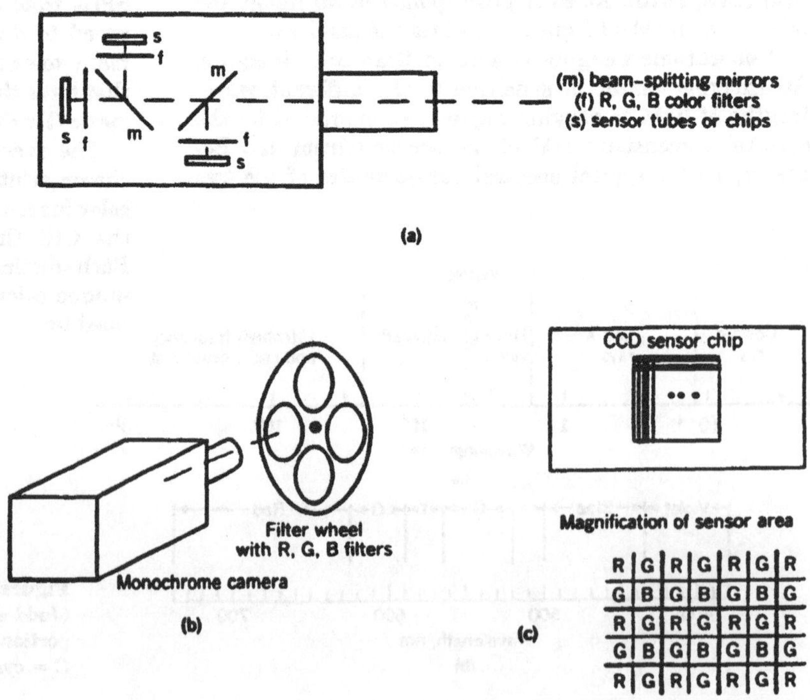

Figure 3. Color camera arrangements. (*a*) Typical broadcast color television camera; (*b*) typical color computer vision camera; (*c*) single-chip color CCD sensor chip with R, G, or B filter on each pixel.

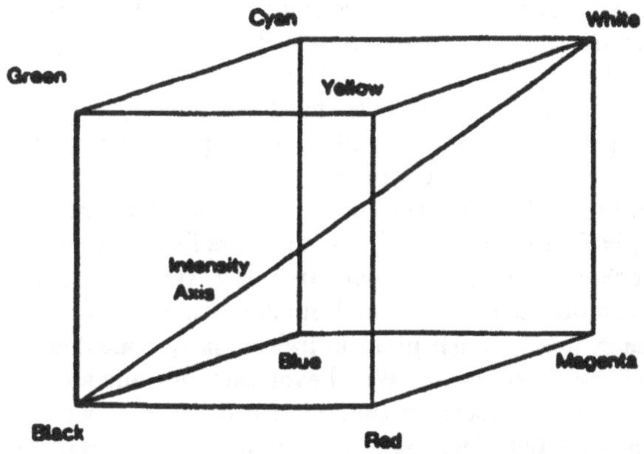

Figure 4. The R-G-B color space.

sity, measured by $I = (R + G + B)/3$, is the main diagonal of this cube from black (0, 0, 0) to white (max, max, max). Researchers in computer vision have frequently used R-G-B coordinates, but have also explored transformations to other coordinate systems that have useful properties. One such system is the CIE X-Y-Z system, a linear transform of R-G-B as defined above. The X-Y-Z system was proposed because of its use as an international standard. However, it is psychophysical rather than psychological and hence does not capture the subjective attributes of color perception in the way that humans talk about color. These are incorporated into systems such as the Munsell color order system (Judd and Wyszecki, 1975). The Munsell system defines three color attributes (value, hue, and chroma) that correspond roughly to the more familiar brightness, hue (color name, eg, blue, purple), and saturation (relative amount of pure hue as opposed to gray). In computer vision, normalized colors r-g-b are first defined by $r = R/I$, $g = G/I$, and $b = B/I$; saturation S and hue H are then defined (Tenenbaum and co-workers, 1974) by

$$S = 1 - 3\min(r,g,b)$$

$$x = \arccos\frac{2r - g - b}{6^{1/2}[(r - 1/3)^2 + (g - 1/3)^2 + (b - 1/3)^2]^{1/2}}$$

$$\text{if } b \le g, \quad \text{then } H = x; \quad \text{otherwise } H = 2\pi - x$$

[See Kender (1977) for a fast algorithm to compute H.] In color space, I corresponds to distance along the intensity axis; on a plane of constant intensity, S and H form a polar coordinate system with S measuring distance from the center (gray) point and H measuring angle from pure red (Fig. 5). It is easily seen that the H-S-I system is analogous to the Munsell system for describing human color perceptions, but numerically the two are quite distinct.

Even the Munsell system, however, has drawbacks. Most work with color in computer vision has been based on the notion of color differences expressed as Euclidean distance in a color space, but Euclidean distance in Munsell coordinates does not correspond well to subjective perceived color difference magnitudes. Spaces with that property have been proposed, however, and are generically called uniform color spaces. The earliest, called U-V-W, has occasionally been proposed for computer vision; it has been replaced in the color science community by a newer system called CIELUV (Grum and Bartleson, 1980). The CIELUV system defines L^*, u^*, and v^* for each color as follows (quantities denoted by subscript *n* refer to the incident illumination color):

$$L^* = 116(Y/Y_n)^{1/3} - 16$$

$$u^* = 13L^*(u - u_n)$$

$$v^* = 13L^*(v - v_n)$$

where

$$u = 4X/(X + 15Y + 3Z)$$

$$v = 9Y/(X + 15Y + 3Z)$$

Color differences ΔE are then expressed by

$$\Delta E = \sqrt{(\Delta L^*)^2 + (\Delta u^*)^2 + (\Delta v^*)^2}$$

However, it seems unlikely that even the use of CIELUV coordinates will solve the fundamental problems of color image segmentation and analysis. Other color spaces frequently used for computer vision include the NTSC television broadcasting encoding system Y-I-Q defined (DeMarsh, 1971; Levine, 1980) by

$$\begin{bmatrix} Y \\ I \\ Q \end{bmatrix} = \begin{bmatrix} 0.299 & 0.587 & 0.114 \\ 0.596 & -0.274 & -0.322 \\ 0.211 & -0.523 & 0.312 \end{bmatrix} \begin{bmatrix} R \\ G \\ B \end{bmatrix}$$

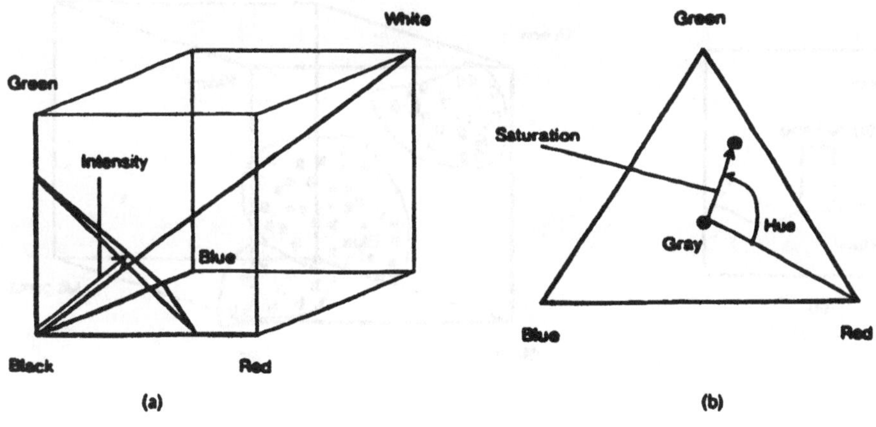

Figure 5. Hue, saturation, and intensity. (*a*) Color space; (*b*) a plane of constant intensity.

Y is the same as the CIE Y coordinate; I and Q measure chromaticity using parameters optimized for the acuity of the human eye. Other researchers have used opponent colors (black/white, blue/yellow, and red/green) or even normalized colors r-g-b themselves.

Kender has pointed out some serious problems in using certain features (Kender, 1977). All features defined by a division (including H-S-I, r-g-b, U-V-W, etc) contain singularities and tend to be unevenly distributed in an image because of the small integer nature of the values used in their computation. This creates problems for algorithms based on histograms, clustering, and edges. Kender recommends a small number of bits per output pixel value for such features, randomizing (adding a random real number from 0 to 1 to each pixel color coordinate), and avoiding these features entirely near their singularities (usually when I is low). It is better to use linear transformations of R-G-B instead, such as Y-I-Q; such linear transformations should themselves be scaled so the maximum value in each row of the transformation matrix is 1.0. Ohta derived statistics for over 100 image regions (using Ohlander's algorithm, described below) and statistically analyzed the color distributions (Ohta, Kanade, and Saki, 1980). He found that the feature that most frequently captured the greatest information was intensity, I. Ohta proposed the use of three features I1-I2-I3 that represent a simple linear transformation from R-G-B and capture the color information he observed very well: $I1 = (R + G + B)/3$; $I2 = R - B$; $I3 = (2G - R - B)/2$. Ohta found these features to perform at least as well as any other set of features (X-Y-Z, R-G-B, Y-I-Q, U-V-W, I-r-g, H-S-I) in his system. However, as he noted, "usefulness of a color feature is greatly influenced by the structure of the color scenes to be [analyzed]."

COLOR AS A STATISTICAL QUANTITY

Much research in color computer vision has been in the area of image segmentation, breaking an image into pieces that have uniform properties. In this work color is usually regarded as a random variable to be analyzed statistically but without regard for the specific physical processes that give rise to color and color variation. The earliest and most obvious technique is called spectral signature analysis, in which prior knowledge about characteristic object colors is used to classify pixels. Spectral signature analysis has been used extensively in remote sensing (satellite and aerial photograph interpretation) and biomedical image analysis; it has been applied occasionally in robotics research. For example, Noguchi classifies pixels in biomedical images of cells (Noguchi and co-workers, 1978). He measures the typical colors of background, cytoplasm, and nucleus in advance; then, for each image, each pixel is individually classified into one of these categories. Whichever category has a characteristic color closest to that pixel's color is assigned as the pixel label. The distance metric used in their work is Euclidean distance in R-G-B space, that is, $\Delta P = \sqrt{\Delta R^2 + \Delta G^2 + \Delta B^2}$. Frequently, nonstandard color filters or features are used that optimize discriminability for the specific task at hand (Akita and Kuga, 1979; Engvall and co-workers, 1981).

If specific object colors are not known in advance, clustering can be used instead. Haralick and many others have applied this standard pattern recognition technique to color image segmentation (qv) (Haralick and Kelly, 1969). A histogram is first created by the color values at all pixels; it tells, for each point in color space, how many pixels exhibit that color (Fig. 6). Typically, the colors tend to form clusters in the histogram, one for each object in the image. By manual or automatic analysis of the histogram, the shape of each cluster is found. Then each pixel in the image is assigned to the cluster that is closest to the pixel color in color space. Clustering differs from spectral signature analysis in that the clusters are found by analysis of the specific image under consideration rather than by prior consideration of the expected data. In some clustering systems, clusters are restricted to be rectangular boxes or ellipsoids (Ali and co-workers, 1979; Connah and Fishbourne, 1981; Sarabi and Aggarwal, 1981). Features that describe texture have been used along with color to create a "feature space" with additional dimensions (Coleman and Andrews, 1979). All clustering techniques suffer from the problem that adjacent clusters frequently overlap in color space, causing incorrect pixel labeling. In conjunction with clustering, a technique called relaxation is sometimes used to improve pixel labeling. In relaxation,

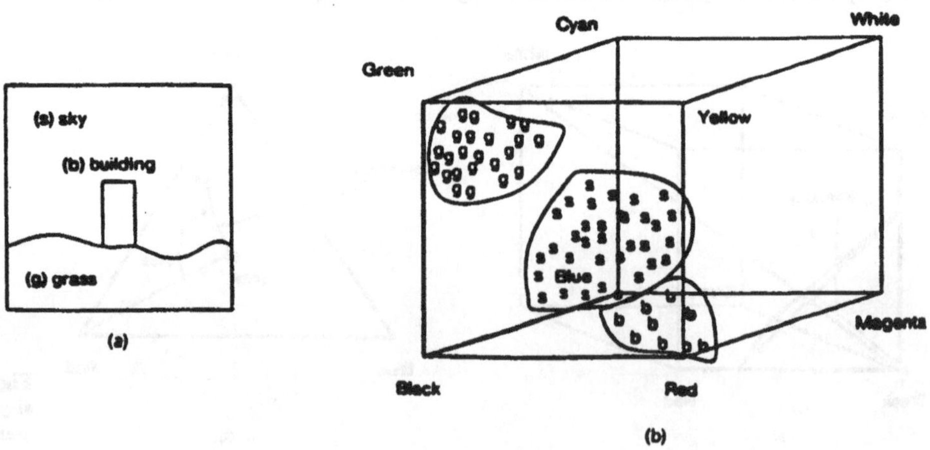

Figure 6. Clustering in color space. (*a*) Input image; (*b*) histogram in color space, wherein g, b, and s indicate pixels in grass, building, and sky; clusters are outlined.

pixel labels are assigned by an iterative method. Each pixel has a probability of belonging to each cluster, and in each iteration step those probabilities are modified. A probability is increased or decreased according to a weighted combination of two factors: the color resemblance of the pixel to the cluster center and the probabilities that the neighboring pixels belong to that same cluster (Rosenfeld, 1978; Nagin and co-workers, 1982).

Another segmentation technique is region splitting, in which the image is broken into successively smaller pieces until each piece has a uniform color (and presumably represents a single object or surface). Ohlander used color to perform this splitting operation (Ohlander and co-workers, 1978). His method begins by computing a variety of color features for each pixel (H, S, I, Y, I, and Q) before the actual segmentation begins. Then the splitting step is applied (the initial region is the entire image). In the splitting step a histogram is created for each color feature (RGBHSIYIQ) within the region. Each histogram is examined independently, and the feature whose histogram exhibits the most prominent peak is selected for use (Fig. 7). The image is then thresholded according to the boundaries of the peak to isolate the pixels that contribute to the selected peak. This splits the original region into smaller regions. The splitting step is then recursively applied to each region, stopping when a region has all flat histograms (ie, it has uniform color). Some of the general problems associated with histogram analysis in the algorithm are discussed in Shafer and Kanade (1982).

Region growing is a complementary segmentation technique in which small regions (initially, individual pixels) are merged together to form larger and larger regions according to color similarity. This method merges the pair of adjacent regions with the greatest similarity of colors, according to a statistical measure. It may take into account both the average color and color variance within each region. Merging continues until no two adjacent regions are sufficiently similar. Variations include simple color difference measures (Yachida and Tsuji, 1971; Bajcsy, 1973; Levine and Shaheen, 1979), the use of color features other than R-G-B (Yachida and Tsuji, 1971; Sloan, 1977; Nagao and co-workers, 1979), and the use of semantic information about object position and relations (Yakimovsky and Feldman, 1973; Tenenbaum and co-workers, 1974). Nagao determined the acceptable color difference limits by finding a valley in local color difference histograms (Nagao and co-workers, 1979).

Edge detection techniques have been adapted for color vision by extending algorithms originally designed for black-and-white (intensity) images to work on R-G-B images (Nevatia, 1977). Researchers in this area have observed that most color features detected by these color edge detectors are also detected by the black-and-white edge detectors. The color detectors offer a limited advantage, however, in that color features are more likely to be significant, since intensities often fluctuate because of lighting variations.

In summary, every major image segmentation method

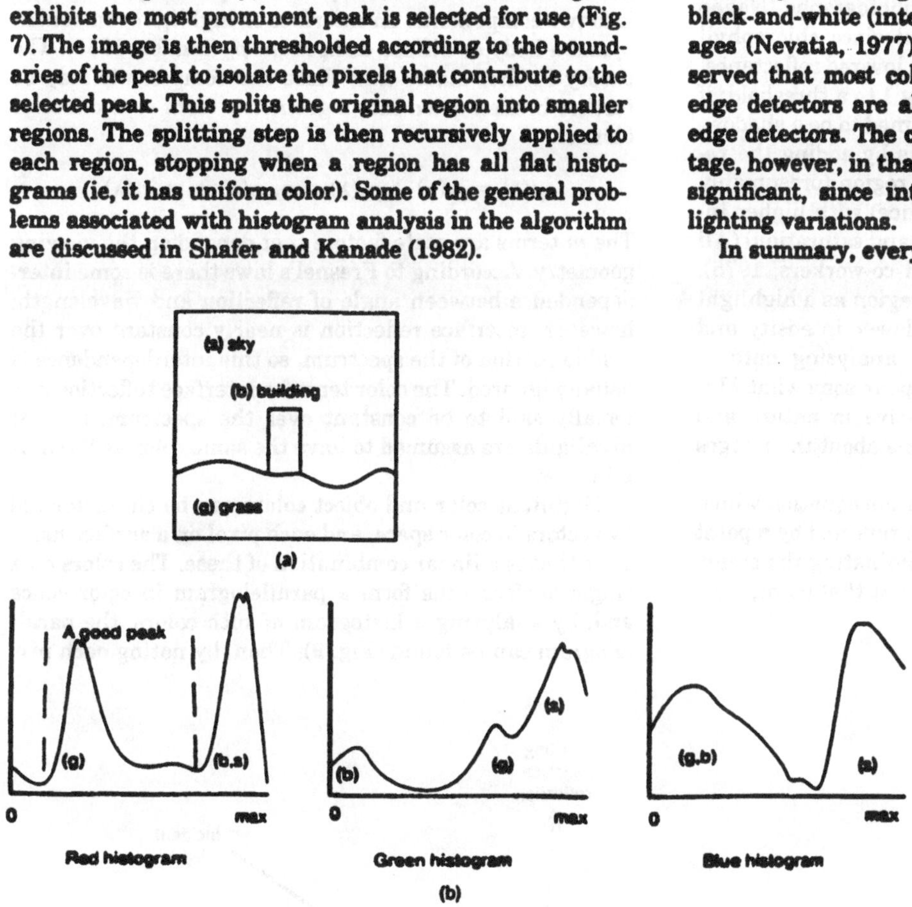

Figure 7. Region splitting by histogram analysis. (*a*) Input image; (*b*) red, green, and blue histograms, wherein g, b, and s indicate pixels of grass, building, and sky; (*c*) after thresholding. Region outlines are indicated by thick lines. Each region will now be split into smaller regions.

has been adapted for color, and some, like relaxation segmentation and region splitting, are almost always performed on color images. These color techniques have been used in robotics applications such as autonomous navigation (Crisman and Thorpe, 1988; Turk and co-workers, 1988).

COLOR AS A PHYSICAL QUANTITY

Most of the early work in color computer vision viewed color as a random variable to be used for image segmentation. Recently many researchers have tried using knowledge about how color is created to analyze a color picture and compute some important three-dimensional facts about the objects being viewed. The most basic avenue of research is this area has used heuristic rules, usually embedded in production systems, to label regions as shadows, highlights, and so on, using knowledge about color behavior. Nagao uses **IR/R**, the ratio of infrared to red at each pixel, to detect vegetation in aerial photographs (Nagao and co-workers, 1979). This is useful since chlorophyll typically has a high IR reflectance but low red reflectance. He also detects shadows by comparing I to a threshold; if the intensity is low, the region is assumed to be a shadow. Several others have detected shadows by adding the requirement that there be an adjacent region (presumably an illuminated part of the same surface) with higher intensity but similar chromaticity (hue and saturation) (Ali and co-workers, 1979; Schachter and co-workers, 1975). Ohlander (1975) similarly labeled a region as a highlight if there is an adjacent region with lower intensity and similar chromaticity. Sloan (1977), analyzing outdoor scenes, noted that distant objects appear somewhat bluish. All these heuristics are qualitative in nature and based on some simplifying assumptions about the images being viewed.

Quantitative analysis of color uses the equation which says that the SPD of light $S(\lambda)$ that is reflected by a point in the scene depends on the light illuminating the scene, $I(\lambda)$, and the reflectance of the surface at that point, $\rho(\lambda)$:

$$S(\lambda) = I(\lambda)\rho(\lambda)$$

The reflectance function $\rho(\lambda)$ may vary widely for different points on the same object because of changes in imaging geometry. In trying to analyze pictures, researchers have looked for models of ρ that contain structure, rather than try to treat it as an arbitrary function. One such model of object reflectance is based on the physical properties of reflection when light strikes an inhomogeneous surface. Inhomogeneous surfaces (including plastics, paints, ceramics, and paper) consist of a medium with particles of colorant suspended in it (Fig. 8). When light hits such a surface there is a change in index of refraction between the air and the medium, so some light is reflected at the interface. This reflection occurs in the "perfect specular direction," where angle of incidence equals angle of reflection, and forms the highlights seen on shiny materials. The light that penetrates through the interface is scattered and selectively absorbed by the colorant. The light that is re-emitted produces body reflection.

The dichromatic reflection model, proposed by Shafer, states that the reflectance ρ of an object may be divided into two components: an interface or highlight component ρ_i and a body or diffuse component ρ_b (Shafer, 1985). The theory further states that these two components may be further broken down into a geometric term m and a color term c:

$$\rho(\lambda) = \rho_i(\lambda) + \rho_b(\lambda) = m_i c_i(\lambda) + m_b c_b(\lambda)$$

The m terms are scale factors that depend on the viewing geometry. According to Fresnel's laws there is some interdependence between angle of reflection and wavelength; however, interface reflection is nearly constant over the visible portion of the spectrum, so this interdependence is usually ignored. The color term for interface reflection c_i is usually said to be constant over the spectrum, so that highlights are assumed to have the same color as the illumination.

Highlight color and object colors can be characterized by vectors in color space, and each pixel on a surface has a color that is a linear combination of these. The colors on a single surface thus form a parallelogram in color space and, by analyzing a histogram of such colors, the parallelogram can be found (Fig. 9). Then, by noting each pix-

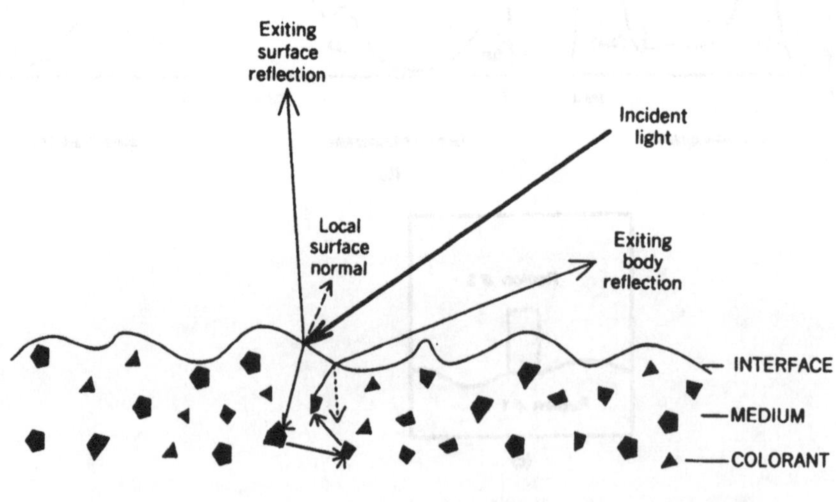

Figure 8. Reflection of light from an inhomogeneous material.

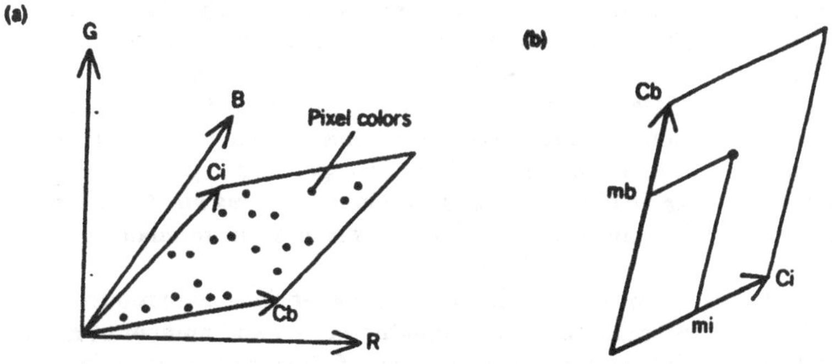

Figure 9. (*a*) Pixel values on a surface lie on a parallelogram in color space; (*b*) position within parallelogram determines reflection magnitudes.

el's color location within this parallelogram, the relative amounts of highlight and object color can be determined at that pixel. This method can then be used to determine the color of the illumination (Klinker and co-workers, 1988a), and to segment an image into regions of uniform color without being fooled by highlights (Klinker and co-workers, 1988b). Other methods of exploiting the dichromatic reflection model have included segmenting according to the normals of the planes defined by the two vectors (Brill, 1989), working in a two-dimensional chromaticity space (Lee, 1986), and finding the vector along which two planes intersect to determine illumination color (Klinker and co-workers, 1987; Tominaga and Wandell, 1989).

Healey has developed a unichromatic reflection model that applies to homogeneous materials such as metals (Healey, 1988). In metals there is no body reflection, and the interface reflection is colored by the surface material. Assuming that there is at least one inhomogeneous surface with highlights in the image, the color of the illumination may be found from the highlight. Metals can be distinguished from inhomogeneous materials, and the object colors of both types of materials can be found (Healey, 1989).

Rubin has presented a method for using color to distinguish material changes from artifacts such as shadows and highlights (Rubin and Richards, 1982). The spectral power distributions of nearby pixels may have a crossover point (wavelength at which the sign of the difference changes) only if the material being viewed at the two pixels is different. By looking for sign changes in the color components of adjacent pixels, material boundaries can be found. Gershon addressed the problem of situations where the ambient illumination color is significantly different from the direct illumination (Gershon, 1987), causing a violation of the crossover assumption.

Funt and Ho proposed a method for determining color based upon a model of lenses in the imaging system (Funt and Ho, 1988). The index of refraction for glass varies as a function of wavelength; this causes entering light to be spread out into its component wavelengths upon exiting, as seen with a prism. When this property is observed in lenses, it is known as chromatic aberration. A single image taken with a black-and-white camera will have some degree of blurring at the edges due to the chromatic aberration. This blurring can be examined to determine the SPD of the incoming illumination, effectively calculating color from a black-and-white image.

Researchers have recently begun to develop physical models further to analyze images containing inter- or mutual reflection. The usual model of illumination for color vision is that of a single point source or ambient illumination that is the same everywhere. There is no interaction between objects in the scene and illumination is usually assumed to be uniform or to vary slowly. However, objects that reflect light into the camera may equally reflect light onto each other, causing a local change in illumination that can cause unexpected color changes in the objects being viewed. New theories are being developed to deal with this difficult problem (Bajcsy and co-workers, 1989; Shafer and co-workers, 1990).

COLOR AS A PERCEPTUAL QUANTITY

Another view of color is to interpret it as a perceptual variable in human vision. Much of this work involves the explicit modeling of color processing within the human visual system in order to adapt it to machine vision. However, the exact mechanisms of human color perception remain unknown, so machine vision researchers concentrate on achieving the same effects. The actual methods used may be biologically plausible or may be suitable for machine computation only.

Researchers are currently studying the phenomenon of color constancy that allows humans to see object colors the same under widely varying illumination. In machine vision the problem is usually to take a picture of some scene under unknown illumination and to calculate automatically some transform that will determine what every object in the scene would look like under standard or known illumination. The solution may involve explicitly calculating the SPD of the illumination and/or the objects in the scene, which gives rise to physics-based methods based on principles similar to the methods described above.

It is not, in principle, possible to perform color constancy for all scenes. A picture consisting of an even orange color may equally well arise from an orange wall lit by white light or from a white wall lit by orange light. In order to perform color constancy, researchers need a reference property or assumption about scene characteristics. The methods of calculation used by researchers may be classified according to the assumptions made and the representation used to describe the SPDs.

One type of reference property is to assume that some kind of statistic holds true for the image. It may be assumed that the different colored objects in the scene are distributed in such a way that the average reflectance value is some known quantity. This quantity may be chosen arbitrarily [typically, that the average color of the scene is gray (D'Zmura and Lennie, 1986)] or may be based on an averaging of color data collected from a variety of materials that are considered likely to be in the scene (Gershon, 1987). The desired average is compared with the observed average of all areas in the picture; then the color bands are scaled to create a new picture with the desired property. Other researchers have looked for reference properties in an area of the scene rather than the whole scene. An object of known reflectance (typically a white test object) may be intentionally put in the scene, or it may be assumed that a white object is likely to occur in the scene naturally. A slight variation is to assume that the brightest object scene in each color band should average out to white (Land, 1977). The appearance of the object or objects yields scale factors to correct for colored incident light. Limitations of scaling methods of achieving color constancy are discussed in Worthey (1986). Hurlbert and Poggio use known correspondences between certain area of the scene under unknown illumination and the values they should have under standard illumination; an algorithm is then used to determine the best transformation that can then be applied to the rest of the scene (Hurlbert and Poggio, 1988).

Other researchers have appealed to the dichromatic reflection model and assumed that highlighted inhomogeneous objects are present in the scene. The highlight areas have the same color as the illumination, since the model assumes that the highlights are not colored by the objects. Once the highlights are found, the color of the illumination can be calculated and factored out of the scene (D'Zmura and Lennie, 1986; Healey and Binford, 1987; Tominaga and Wandell, 1989). Mathematical constraints can also be employed to limit the possible forms of illumination and surface reflectances. This may be done by using more sensors than degrees of freedom in the SPDs (Maloney and Wandell, 1986), by including spectral data in addition to or instead of R-G-B values (Ho and coworkers, 1988; Tominaga and Wandell, 1989), or by using qualitative knowledge about the nature of the SPDs (eg, real-world reflectances are never negative) (Tominaga and Wandell, 1990).

Another issue of research in color constancy is the method of representing the SPDs of the illumination and surface reflectances in the calculations. Researchers need to represent functions $S(\lambda)$ in a finite manner; such a finite representation must necessarily be inexact, since, in principle, SPDs may take any shape. They may be approximated by sampling at regular intervals with a spectroradiometer (eg, every 10 nm within the visible range); by sampling with broad-band physically realizable receptors such as $\bar{r}(\lambda)$, $\bar{g}(\lambda)$, and $\bar{b}(\lambda)$; by representation with a set of abstract basis functions such as Fourier functions (Wandell, 1987) or normalized Legendre polynomials (Healey and Binford, 1988); or by using a set of basis functions derived from a principal components analysis of spectral data taken of natural and synthetic objects (Gershon, 1987). There is also some question of the number of basis functions needed to give an adequate representation of SPDs of real objects and illuminants (Maloney, 1986; Parkkinen and co-workers, 1989). Some researchers have argued that many SPDs corresponding to physically realizable objects are band-limited functions and thus a good fit may be achieved with even a small number of basis functions (Maloney, 1986).

Future work in this area may include more robust algorithms to work on a greater variety and number of materials and under more types of illumination, examination of the band limitation assumption and goodness-of-fit measures (Parkkinen and co-workers, 1989), and greater automation in selecting areas of scenes for analysis.

BIBLIOGRAPHY

K. Akita and H. Kuga, "Towards Understanding Color Ocular Fundus Images," in *Proceedings of the 6th IJCAI*, Tokyo, Morgan-Kaufmann, San Mateo, Calif., 1979, pp. 7–12.

M. Ali, W. N. Martin, and J. K. Aggarwal, "Color-Based Computer Analysis of Aerial Photographs," *Comput. Graph. Image Proc.* **9**, 282–293 (1979).

R. Bajcsy, "Computer Identification of Visual Surfaces," *Comput. Graph. Image Proc.* **2**, 118–130 (1973).

R. Bajcsy, S. W. Lee, and A. Leonardis, *Image Segmentation with Detection of Highlights and Inter-reflections Using Color*, MS-CIS 89-39, University of Pennsylvania, Philadelphia, June 1989.

C. J. Bartleson and F. Grum, *Optical Radiation Measurements*, Vol. 5, Academic Press, New York, 1984.

M. H. Brill, "Object-Based Segmentation and Color Recognition in Multispectral Images," *SPIE/SPSE* **1076**, 97–103 (Jan. 1989).

C. M. Brown, *Color Vision and Computer Vision*, Technical Report 108, University of Rochester Computer Science Dept., Rochester, N.Y., 1982.

International Commission on Illumination, *International Lighting Vocabulary*, CIE, Paris, 1970.

G. B. Coleman and H. C. Andrews, "Image Segmentation by Clustering," *Proc. IEEE* **67**(5), 773–785 (1979).

Committee on Colorimetry, *The Science of Color*, Optical Society of America, Washington, D.C., 1963.

D. M. Connah and C. A. Fishbourne, "The Use of Colour Information in Industrial Scene Analysis," in *Proceedings of the First International Conference on Robot Vision and Sensory Controls*, ICRVSC, Stratford-upon-Avon, UK, April 1981, pp. 340–347.

J. Crisman and C. Thorpe, "Color Vision for Road Following," *Proc. SPIE* **1007**, 175–184 (Nov. 1988).

L. E. Demarsh, "Color Reproduction in Color Television," in *Proceedings of the Inter-Society Color Council 1971 Conference on Optimum Reproduction of Color*, Inter-Society Color Council, Williamsburg, Va., 1971, pp. 69–97.

M. D'Zmura and P. Lennie, "Mechanisms of Color Constancy," *J. Opt. Soc. Am.* **3**(10), 1622–1672 (1986).

J. Engvall and co-workers, "Development of a Mathematical Model to Analyze Color and Density as Discriminant Features for Pulmonary Squamous Epithelial Cells," *Pattern Recognition* **13**(1), 37–47 (1981).

B. Funt and J. Ho, "Color from Black and White," in *International Conference on Computer Vision (ICCV)*, IEEE, New York, 1988, pp. 2–8.

R. Gershon, *Survey on Color: Aspects of Perception and Computation*, RCBV-TR 84-4, University of Toronto Computer Science Dept., Toronto, Canada, 1984.

R. Gershon, *The Use of Color in Computational Vision*, PhD dissertation, University of Toronto, Toronto, Canada, 1987.

F. Grum and C. J. Bartleson, eds., *Optical Radiation Measurements*, Vol. 2, Academic Press, New York, 1980.

F. Grum and R. J. Becherer, *Optical Radiation Measurements*, Vol. 1, Academic Press, New York, 1979.

Hamamatsu Corp., *Vidicons, Catalog SC-5-3*, Hamamatsu Corp., Middlesex, N.J., 1983a.

Hamamatsu Corp., *Silicon Photocells, Catalog SC-3-6*, Hamamatsu Corp., Middlesex, N.J., 1983b.

R. M. Haralick and G. L. Kelly, "Pattern Recognition with Measurement Space and Spatial Clustering for Multiple Images," *Proc. IEEE* 57, 654–665 (April 1969).

G. Healey, "A Color Reflectance Model and Its Use for Segmentation," in *International Conference on Computer Vision (ICCV)*, IEEE, New York, 1988, pp. 460–466.

G. Healey, "Using Color for Geometry-Insensitive Segmentation," *J. Opt. Soc. Am. A* 6(6), 920–937 (1989).

G. Healey and T. O. Binford, "The Role and Use of Color in a General Vision System," in *Proceedings of the ARPA Image Understanding Workshop*, DARPA, Arlington, Va., 1987, pp. 599–613.

G. Healey and T. O. Binford, "A Color Metric for Computer Vision," in *Proceedings of the International Conference on Computer Vision and Pattern Recognition*, IEEE, New York, 1988, pp. 10–17.

J. Ho, B. V. Funt, and M. S. Drew, *Disambiguation of Illumination and Surface Reflectance from Spectral Power Distribution of Color Signal: Theory and Applications*, CSS/LCCR 88-18, Simon Fraser University, Burnaby, B.C., Canada, 1988.

R. W. G. Hunt, *The Reproduction of Color*, Wiley, New York, 1967.

A. C. Hurlbert and T. A. Poggio, "Synthesizing a Color Algorithm from Examples," *Science* 236, 482 (Jan. 1988).

T. Ito, "Color Picture Processing by Computer," in *Proceedings of the Fourth IJCAI*, Tbilisi, USSR, Morgan-Kaufmann, San Mateo, Calif., 1975, pp. 635–642.

D. B. Judd and G. Wyszecki, *Color in Business, Science and Industry*, Wiley, New York, 1975.

J. R. Kender, "Instabilities in Color Transformations," in *PRIP-77*, IEEE Computer Society, Troy, N.Y., June 1977, pp. 266–274.

G. J. Klinker, *A Physical Approach to Color Image Understanding*, PhD dissertation, Carnegie Mellon University, Pittsburgh, Pa., May 1988.

G. J. Klinker, S. A. Shafer, and T. Kanade, "Using a Color Reflection Model to Separate Highlights from Object Color," in *International Conference on Computer Vision (ICCV)*, IEEE, New York, June 1987, pp. 145–150.

G. J. Klinker, S. A. Shafer, and T. Kanade, "The Measurement of Highlights in Color Images," *Int. J. Comp. Vision* 2(1), 7–32 (1988).

G. J. Klinker, S. A. Shafer, and T. Kanade, "Color Image Analysis with an Intrinsic Reflection Model," in *International Conference on Computer Vision (ICCV)*, IEEE, New York, 1988, pp. 292–296.

R. G. Kuehni, *Color: Essence and Logic*, Van Nostrand Reinhold Co., New York, 1983.

E. H. Land, "The Retinex Theory of Color Vision," *Sci. Am.* 237(6), 108–128 (1977).

Y. LeClerc, "A Method for Spectral Linearization," private communication, 1986.

H. C. Lee, "Method for Computing the Scene-Illuminant Chromaticity from Specular Highlights," *J. Opt. Soc. Am. A* 3(10), 1694–1699 (1986).

M. D. Levine, "Region Analysis Using a Pyramid Data Structure," *Structured Computer Vision*, Academic Press, New York, 1980, pp. 57–100.

M. D. Levine and S. I. Shaheen, "A Modular Computer Vision System for Picture Segmentation and Interpretation, Part I," in *PRIP-79*, IEEE Computer Society, Chicago, Ill., 1979, pp. 523–533.

L. T. Maloney, "Evaluation of Linear Models of Surface Spectral Reflectance with Small Numbers of Parameters," *J. Opt. Soc. Am.* 3(10), 1673–1683 (1986).

L. T. Maloney and B. A. Wandell, "Color Consistency: A Method for Recovering Surface Spectral Reflectance," *J. Opt. Soc. Am. A* 3(1), 29–33 (1986).

M. Nagao, T. Matsuyama, and Y. Ikeda, "Region Extraction and Shape Analysis in Aerial Photographs," *Comput. Graph. Image Proc.* 10, 195–223 (1979).

P. A. Nagin, A. R. Hanson, and E. M. Riseman, "Studies in Global and Local Histogram-Guided Relaxation Algorithms," *IEEE Trans. PAMI* PAMI-4(3), 263–276 (1982).

A. Nazif, *A Survey of Color, Boundary Information, and Texture as Features for Low-Level Image Processing*, Technical Report 78-7R, McGill University Electrical Engineering Dept., Montreal, Canada, 1978.

R. Nevatia, "A Color Edge Detector and Its Use in Scene Segmentation," *IEEE Trans. Syst. Man Cybern.* TSMC-7(11), 820–826 (1977).

Y. Noguchi, Y. Tonjin, and T. Sugishita, "A Method for Segmenting a Clump of Cells into Cellular Characteristic Parts Using Multispectral Information," in *IJCPR-4*, Kyoto, 1978, pp. 872–874.

C. L. Novak, S. A. Shafer, and R. G. Willson, "Obtaining Accurate Color Images for Machine Vision Research," *Proc. SPIE* 1250, 54–68 (1990).

R. Ohlander, *Analysis of Natural Scenes*, PhD dissertation, Computer Science Department, Carnegie Mellon University, Pittsburgh, Pa., 1975.

R. Ohlander, K. Price, and D. R. Reddy, "Picture Segmentation Using a Recursive Region Splitting Method," *Comput. Graph. Image Proc.* 8, 313–333 (1978).

Y. Ohta, T. Kanade, and T. Sakai, "Color Information for Region Segmentation," *Comput. Graph. Image Proc.* 13, 222–241 (1980).

J. P. S. Parkkinen, J. Hallikainen, and T. Jaaskelainen, "Characteristic Spectra of Munsell Colors," *J. Opt. Soc. Am. A* 6(2), 318–322 (1989).

A. Rosenfeld, "Some Recent Results Using Relaxation-Like Processes," *Proceedings of the ARPA IU Workshop*, May 1978, pp. 100–101.

J. M. Rubin and W. A. Richards, "Color Vision and Image Intensities: When Are Changes Material?" *Biol. Cybern.* 45, 215–226 (1982).

A. Sarabi and J. K. Aggarwal, "Segmentation of Chromatic Images," *Pattern Recognition* 13(6), 417–427 (1981).

B. Schachter, L. S. Davis, and A. Rosenfeld, *Scene Segmentation by Cluster Detection in Color Space*, Technical Report 424, University of Maryland Computer Science Center, College Park, Md., 1975.

S. A. Shafer, "Optical Phenomena in Computer Vision," in *Proceedings CSCSI-84*, Canadian Society for Computational Studies of Intelligence, London, Ontario, May 1984.

S. A. Shafer, "Using Color to Separate Reflection Components," *Color Res. App.* 10(4), 210–218 (1985).

S. A. Shafer and T. Kanade, "Recursive Region Segmentation by Analysis of Histograms," in *Proc. Intl. Conf. on Acoustics, Speech, and Signal Processing*, Paris, IEEE, May 1982, pp. 1166–1171.

S. A. Shafer, T. Kanade, G. J. Klinker, and C. L. Novak, "Physics-Based Models for Early Vision by Machine," *SPIE/SPSE* 1250, 222–235 (1990).

K. Sloan, *World Model Driven Recognition of Natural Scenes*, PhD dissertation, University of Pennsylvania Moore School of Electrical Engineering, Philadelphia, June 1977.

D. Taenzer, *Physiology and Psychology of Color Vision—A Review*, AIM 369, MIT AI Lab, Cambridge, Mass., 1976.

J. M. Tenenbaum, T. D. Garvey, S. Weyl, and H. C. Wolf, *An Interactive Facility for Scene Analysis Research*, Technical Report TN 87, SRI International, Menlo Park, Calif., Jan. 1974.

S. Tominaga and B. A. Wandell, "Standard Surface-Reflectance Model and Illuminant Estimation," *J. Opt. Soc. Am. A* 6(4), 576–584 (1989).

S. Tominaga and B. A. Wandell, "Component Estimation of Surface Spectral Reflectance," *J. Opt. Soc. Am. A* 7(2), 312–317 (1990).

M. Turk, D. Morgenthaler, K. Gremban, and M. Marra, "VITS: A Vision System for Autonomous Land Vehicle Navigation," *IEEE Trans. Patt. Anal. Mach. Intell.* PAMI-10(3), 342–361 (1988).

B. A. Wandell, "The Synthesis and Analysis of Color Images," *IEEE Trans. Patt. Anal. Mach. Intell.* PAMI-9(1), 2–13 (1987).

J. Wentworth, *Color Television Engineering*, McGraw-Hill, New York, 1955.

S. J. Williamson and H. Z. Cummins, *Light and Color in Nature and Art*, Wiley, New York, 1983.

J. A. Worthey, "Heuristic Analysis of von Kries Color Constancy," *J. Opt. Soc. Am.* 3(10), 1708–1712 (1986).

G. Wyszecki and W. S. Stiles, *Color Science: Concepts and Methods, Quantitative Data and Formulae*, Wiley, New York, 1982.

M. Yachida and S. Tsuji, "Application of Color Information to Visual Perception," *Pattern Recognition* 3, 307–323 (1971).

Y. Yakimovsky and J. A. Feldman, "A Semantics-Based Decision Theory Region Analyzer," in *Proceedings of the Third IJCAI*, Stanford, Calif., Morgan-Kaufmann, San Mateo, Calif., 1973, pp. 580–588.

General References

The reader interested in color measurement is referred to Judd and Wyszecki (1975), a very readable discussion, and the more technical discussion of color standards in Grum and Bartleson (1980). Useful reference handbooks are Wyszecki and Stiles (1982), which contains many tables and formulas, and CIE (1970), which presents formal definitions of terms and units of measure. A discussion of the physiology of human color vision is found in Bartleson and Grum (1984); Committee on Colorimetry (1963) is older but also contains an excellent discussion of color perception. Surveys of color in computer vision are Taenzer (1976), Nazif (1978), Brown (1982), Gershon (1984), and Shafer (1984); however, these tend to survey human color vision more deeply than computer color vision.

Carol L. Novak
Steven A. Shafer
Carnegie Mellon University

Color Image Formation

Color images are obtained for machine vision by using color cameras, which are based on monochrome cameras with color filters added. The complexity of color imaging derives from the composition of all the issues in monochrome imaging, plus the added issues that derive from the placement of color filters. Further, once the image data has been obtained, the representation most useful for image analysis may not be the particular encoding provided by the camera. The two papers in this section describe these matters in some detail.

As background information, the student of color imaging should begin by reading all the papers in the "Camera Calibration" section of the Radiometry volume of this series. These provide a basic description of computer imaging. The issues in *color* imaging are derived from these.

In the world, light energy occurs in many wavelengths (denoted by λ). In general, at any point in the world or in an image, the light arriving at that point contains many wavelengths. Thus, it is common to describe light energy arriving at a point in terms of the *spectral power distribution* (SPD) of the illumination, $I(\lambda)$. When light reflects from a surface, some fraction of light is reflected at each wavelength. This quantity is called the surface's spectral reflectance, $\rho(\lambda)$. (Later sections of this volume explore the nature of $\rho(\lambda)$ in some detail.) The light reflected from the surface at each wavelength is therefore $I(\lambda)\rho(\lambda)$. This light enters the camera and strikes the sensor surface. The sensor has some *spectral responsivity* $s(\lambda)$; thus, the value measured by the sensor is based on the quantity $\int I(\lambda)\rho(\lambda)s(\lambda)$. This is a description of imaging in a monochrome (gray-scale) camera.

The human visual system contains three general classes of color receptor cells (cones) in the eye, each with a characteristic responsivity to different wavelengths of light. This gives humans a much richer perception of light than that afforded by a monochrome camera, which is called *color*. **Color perception** is therefore an attribute of the human perceptual system; while **spectral power** is the corresponding attribute of physical light energy. Color is a three-dimensional quantity; spectral power is a function defined over the range of visible wavelengths of light, from about 380nm (blue) to 760nm (red).

The quantity $\rho(\lambda)$ is particularly interesting to know about a surface because it can encode information about material type, surface roughness, index of refraction, and so on. So, there is a motivation to measure it somehow for machine vision. There exists a vast collection of devices to measure spectral power (thus $I(\lambda)\rho(\lambda)$) directly, called *spectroradiometers*. Unfortunately, these devices generally accept only a single ray of light as input. To create an image of measurements would therefore require mechanically scanning the device across x and y. Since each measurement takes some substantial fraction of a second, to fill a 512x480 image (TV resolution) would require at least several hours, with a potentially cumbersome physical apparatus and a difficult geometric calibration problem.

So, instead, researchers in machine vision generally adopt a less ambitious input: Measurement of *color* rather than *spectral power*. By reducing the amount of spectral information to three samples rather than a continuous spectrum, color imaging can be built on top of monochrome imaging technology to take advantage of the massive parallelism in x and y inherent in cameras.

Color imaging is therefore based on interposing three color filters in the imaging chain, usually fixed to the front of the lens, to provide three independent measurements of spectral information. Each filter has a spectral transmittance $\tau(\lambda)$, which denotes the fraction of light at each wavelength that the filter allows to pass through. If the three filters are denoted τ_i for i=1,2,3, then the camera will make three measurements $\int I(\lambda)\rho(\lambda)\tau_i(\lambda)s(\lambda)$. For convenience, the filter/lens/camera combination is sometimes grouped together as the "color sensor," with $s_i(\lambda)$ used to denote $\tau_i(\lambda)s(\lambda)$.

There are several common methods for attaching a color filter to a monochrome imaging device, as described in [1]. Each has advantages and disadvantages in speed, spatial resolution, and radiometric precision. For the highest-quality color images, however, the method of choice is to use color filters that are individually placed before a high-quality monochrome camera, one at a time. The resulting set of images takes some time to acquire, but the resolution and accuracy of the data will be equal to that of the monochrome camera. This method of color imaging also avoids the problems introduced by conversion of raw color data to an encoded color TV signal.

The first paper of this section describes some of the limitations and problems introduced by this method of color imaging [2]. These problems are of two types: limitations of monochrome imaging that have a more devastating impact on color image quality, and problems that are unique to color imaging.

The first category, limitations of monochrome imaging that are exacerbated in color, includes

- non-linearity of the pixel values, that is, pixel values that are a complicated function of $\int I(\lambda)\rho(\lambda)s(\lambda)$. Since color imaging is based on several such measurements with different color filters, if the pixel values are nonlinearly related to these measurements, then the ratios of the color pixel values to each other may not be the same as the ratios of the raw measurements;
- clipping and blooming pixel values when the sensor becomes saturated. This is also a problem in monochrome imaging, but is more so in color because each of the three measurements may saturate independently.

Problems unique to color imaging include

- spectral sensitivity limitations of the monochrome camera, causing color imbalance and the influence of infrared energy on the color image data;
- chromatic aberration of lenses, which can cause the color bands to be slightly mis-focused and/or mis-registered.

The second paper in this section addresses the issues of color representation that arise once color data is available in the computer [3]. The principal observation of this paper is that, since color data consists of discrete measurements rather than continuous spectra, it is possible to reason about the color image, and even the colors of illumination and objects, in a discrete manner, and thereby reduce problems in continuous mathematics to problems in matrix and vector algebra. This observation underlies nearly all of the work in color constancy (see the "Color Constancy" section of this volume); thus, the discussion in this paper is close to that area. The key concept in this approach is that a continuous function such as $I(\lambda)$ can be modeled as a linear combination of basis functions $I(\lambda) = \Sigma\, i_j I_j(\lambda)$, where i_j are scalars and $I_j(\lambda)$ are the basis functions chosen for the representational system. Illumination, reflectance, and sensor responsivity can all be so represented. Then, many equations from the continuous domain of functions over λ can be reduced to vector manipulations of the coefficients such as $[i_j]$.

However, this paper goes well beyond the literature in color constancy because it shows how such discrete representations are related to physical properties of light in the world, and to color display technology. The key observation about color in the world is that spectral power distributions tend to be fairly smooth, which makes discretization feasible; and that spatial variation of different reflection properties can easily be exploited by linear algebra applied to the discrete representations. For color display, the properties of the display system can be substituted for those of the sensor to derive color transformations for accurate color output.

– S.A.S.

Reference

[1] Novak, C.L., and Shafer, S.A. "Color Vision." In *Encyclopedia of Artificial Intelligence*. 1992, J.Wiley and Sons. Pages 192-202. Reprinted in the "Introduction to the Color Volume."

Papers Included in This Collection

[2] Novak, C.L., Shafer, S.A., and Willson, R.G. "Obtaining Accurate Color Images for Machine Vision Research." In *Proc. SPIE #1250 Perceiving, Measuring, and Using Color*. Santa Clara, CA. SPIE, 1990. Pages 54-68.

[3] Wandell, B.A. "The Synthesis and Analysis of Color Images." *IEEE Trans. on Pattern Analysis and Machine Intelligence* **PAMI-9**(1): 2-13, 1987.

Obtaining Accurate Color Images for Machine Vision Research

Carol L. Novak, Steven A. Shafer, Reg G. Willson

Robotics Institute, Carnegie Mellon University, Pittsburgh PA 15213

ABSTRACT

Algorithms for color machine vision rely upon an explicit or implicit model of how color images are formed. These models usually include an idealized description of how camera systems measure color. Unfortunately, relying on these idealizations can cause serious errors in the performance of algorithms. In this paper, we describe several problems that occur in real color images and how they can be dealt with. A common problem is that many cameras do not have a linear response. Failure to correct this problem will not only cause hue shifts, but will cause problems in any algorithms that rely on the linear assumption. We show how this effect may be measured and corrected. More insidious are the problems of color clipping and blooming, since they can cause sudden changes in the apparent hue. Although it is not possible to correct the problem, we show how it may be detected so that affected measurements may be discounted. Most models assume that integration is performed over the visible spectrum, but many cameras are sensitive to other regions as well. Failure to compensate for this can lead to unexpected results. Also the camera's varying spectral sensitivity within the visible spectrum will cause some measurements to be more prone to noise than others. We show how the use of a computer-controlled lens can compensate for this problem. Chromatic aberration in typical video camera lenses can cause hue changes, and we show how the further use of computer-controlled lens can compensate for this problem. Using these methods to detect and/or correct problems, we can obtain the accuracy needed for applying physics-based methods to actual color images.

1. INTRODUCTION

Algorithms for color machine vision try to relate the physics of what is happening "out there" in the real world, with the resulting measurements made inside the camera. Thus they rely on some sort of model of how color images are formed from light entering the camera. This may be an explicit model, where we talk about how integration is performed over certain wavelengths; or the model may be implicit in an algorithm that merely assumes that there is a linear relationship between intensities of light and pixel values.

These models usually describe the behavior of idealized cameras, based upon how the researcher would like the camera to behave. A model often used for color imaging is

$$m_i = \int L(\lambda)\, s_i(\lambda)\, d\lambda \qquad (1)$$

In equation 1, $L(\lambda)$ refers to the spectral distribution function of the light entering the camera at a particular point, and $s_i(\lambda)$ refers to the spectral sensitivity of the appropriate sensor. In a typical RGB color camera, sensor sensitivities are distributed so that one sensor will measure the long visible wavelengths (red), one will measure the medium-length wavelengths (green), and one will measure the short wavelengths (blue), with some degree of overlap. Integration is performed over the wavelengths for which the sensors are non-zero, usually considered to be those wavelengths that are visible to humans.

Here m_i is the value or values given by the camera for a particular pixel. If there is only one value associated with a particular pixel, then this is a model of a black and white camera. A color imaging system will usually have 3 values associated with each pixel. In our work we usually use a black and white camera as we have found they provide superior spatial resolution to true color cameras. We use our black and white camera to take three successive pictures with a red filter, then a green filter, and then a blue filter. Thus the values obtained from our camera are in the RGB coordinate system, although of course we could use different filters, or even a different number of filters. Throughout this paper we will be referring to the RGB system or the red/green/blue bands as this is the most common way of measuring color with TV cameras, but most of our points apply to other systems as well.

Unfortunately few, if any, cameras behave in the ideal way. Yet vision algorithms which try to recover what is "out there" from the values in the camera depend upon assuming that the camera behaves in some predictable fashion. As long as the camera's limitations are understood and taken into account, or care is taken to operate within the camera's expected range, these algorithms may be expected to succeed. However it is easy to forget these limitations, and the ramifications of some of them are not immediately obvious.

In this paper, we discuss some of these limitations that we have come up against in our work. We explain why they cause hue changes and describe how we are able to detect their effects and the ways in which we have overcome them. These techniques have significantly helped improve our results from applying physics-based vision algorithms to real pictures.

2. LINEARITY

If we model the camera's behavior as in equation 1, we are assuming that our camera behaves in a linear fashion. That is, we assume that as more light enters the camera at a particular pixel point, the resulting measurement will be proportionally larger.

Unfortunately, many TV cameras do not have such a linear response. Many are designed to be used in conjunction with video display devices which typically have a non-linear response. The perceived brightness B of the output of such a display is usually given as

$$B = \alpha V^{\gamma}$$

where V is the input voltage to the device, and α and γ are typically related to the brightness and contrast knobs on the display. The value for γ will therefore vary, but it is usually around 2.[1, 2] Thus the makers of many video cameras design them with a non-linear output, so that when the output values from the camera are input to a display, the resulting display will be roughly linear with respect to the brightness in the scene. This means that circuitry is added so that

$$m_{out} = m^{1/\gamma}$$

where m is the value we would expect from equation 1, and m_{out} is the actual value output by the camera.

2.1. Problem

As long as the values obtained from the camera are used only to drive a display, there is no problem, assuming that the γ value used by the manufacturer of the camera matches the value used by the display.

However, color vision researchers also use those output values as inputs to their algorithms. Many of these algorithms are based upon the physics of reflection and assume a linear response. For example, shape from shading attempts to recover the orientation of a surface based upon the perceived brightness of that surface.

This comes from a physics-based model that relates the amount of light entering the camera to the illumination, object reflectances, and object orientation. One such model might be

$$L_c(\lambda) = E(\lambda)\,\rho(\lambda)\,cos(i)$$

where L_c is the light entering the camera, E describes the light illuminating the scene, ρ describes the object reflectance, and $cos(i)$ is the term that describes the way surface brightness falls off as surfaces are angled away from the light source.[3] Equation 1 assumes that the value measured by the camera is directly proportional to the amount of light entering the camera. The shape from shading algorithm will have serious problems if this assumption is used with a non-linear camera.

Another problem may come with object recognition by perceived hue. The non-linear response of some cameras will cause colors to "look wrong" because many of the hues are shifted. If these hues are used to recognize certain objects, errors will occur. For example a pink object might be associated with a camera RGB triple of (.50, .25, .25) in a camera that ranges from 0 to 1. It might be assumed that as the pink object is viewed at different orientations with respect to the incident light, the overall brightness would change, but that the ratio of red to green to blue would not. However a camera with a non-linear response would not yield triples that have the expected hue. In a camera that had square-root response to compensate for $\gamma = 2$ in the monitor, the measured triple would be (.71, .50, .50). Clearly this triple no longer has the expected ratio of 2:1:1. Thus uncompensated non-linearity can cause hue changes in the image.

2.2. Detection

The responsivity of a camera relative to the intensity of light entering it may be measured with a standard gray scale chart, or with a custom chart prepared for this purpose. Such a chart will have several gray patches of varying intensities. Typically these patches are made from Munsell or other standardized color papers. Since the squares are gray their reflectance is nearly constant for visible wavelengths of light. For example, the Macbeth ColorChecker has six gray squares with reflectances of (2.5%, 9.0%, 21.0%, 36.5%, 59.5%, 88.5%).[4] When the values measured by the camera for these gray squares are plotted against percentage reflectances, they should form a straight line, assuming that the chart is evenly illuminated. If they do not form a straight line, then the camera has a nonlinear response, and the γ correction factor may be estimated by using a semi-log plot.

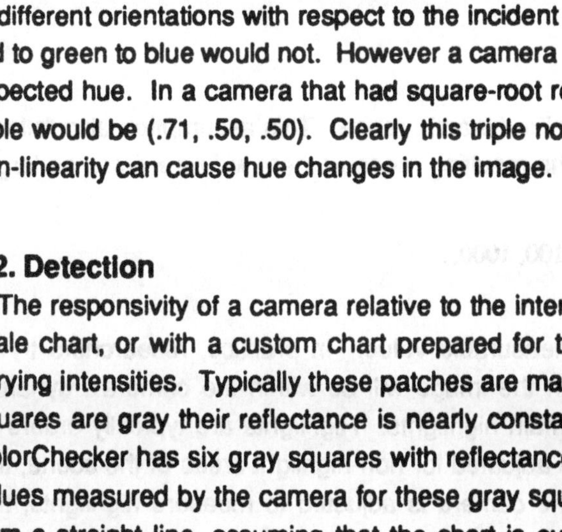

Figure 1: Non-linear camera response

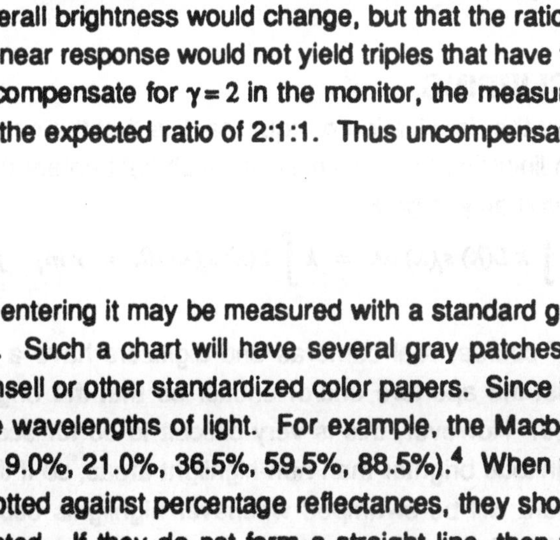

Figure 2: Linear camera response

Figure 1 shows a plot of object reflectance vs. camera response for one typical black and white CCD video camera. Figure 2 shows a graph for a different black and white video camera that does have a linear response.

2.3. Correction

The values measured by a non-linear camera can be corrected by using the same chart. Once the camera values for the gray squares are obtained, a look-up table can be calculated to map non-linear values into linear ones. This table may be calculated by using the γ correction factor calculated, by using linear piecewise interpolation, or by using an interpolated cubic spline.[5, 6]

Figure 18 show a picture of the Macbeth ColorChecker taken by the non-linear camera. Figure 19 shows the same picture after it has been linearized using a look-up table calculated from the camera values for the gray squares. The linearization process can also adjust the slope of the response for linear cameras, making shades of gray appear more distinct, as the blacks become "blacker" and the whites "whiter".

3. CLIPPING

Another implication of equation 1 is that the camera has infinite dynamic range. This assumption also follows from linearity, that if k times as much light enters the camera, the resulting measurement will be k times as large, for arbitrarily large k.

$$\int k\,L(\lambda)\,s_i(\lambda)\,d\lambda \;=\; k\int L(\lambda)\,s_i(\lambda)\,d\lambda \;=\; k\,m_i \qquad \text{for } k = 1, 10, 100, 1000, \ldots$$

Of course, real cameras and digitizers have a maximum measurable value. In practice, researchers try to adjust the aperture and/or shutter so that the brightest point in the image will be within the camera's dynamic range. However, this is very difficult to do for scenes that contain highlights. Highlights are typically orders of magnitude brighter than non-highlight areas, so if the camera is adjusted for non-highlight areas of the scene, the camera will be swamped whenever highlights occur. And if the camera is adjusted to measure highlights, the non-highlighted areas will be so dim as to be right around the noise floor of the camera.

3.1. Problem

In many images there are at least a few regions that are brighter than the camera can handle. In this case, the measured values will be clipped at the camera's maximum value. This can cause hue changes if the values are clipped in some bands but not in others. Consider a light yellow object object with an RGB triple of of (.6, .4, .2) in a camera with a maximum value of 1.0. As the amount of light on the object increases, the values will grow proportionally. As long as all three bands are within the range of the camera, the object will still have the same hue. However, eventually the object can become bright enough that the red value ought to exceed 1.0, but will be held there. The values measured in the green and blue bands can continue to grow however, until they too reach the maximum value. What was a yellowish object will grow whiter and whiter in measured hue.

Figure 3 shows a color cube defined by the measurable limits of the camera. The figure traces the measured color change of the object as its lightness increases. As long as the color remains within the limits defined by the color cube, its measured hue will not change. However, as soon as it reaches one wall of the color cube, the vector will change directions to follow the wall along the path defined by the projection of the desired vector onto that wall. This will continue until the vector hits the next wall, when it will be projected again onto the line where the two saturated bands intersect, until the vector ends at the point of maximum value in all three bands.

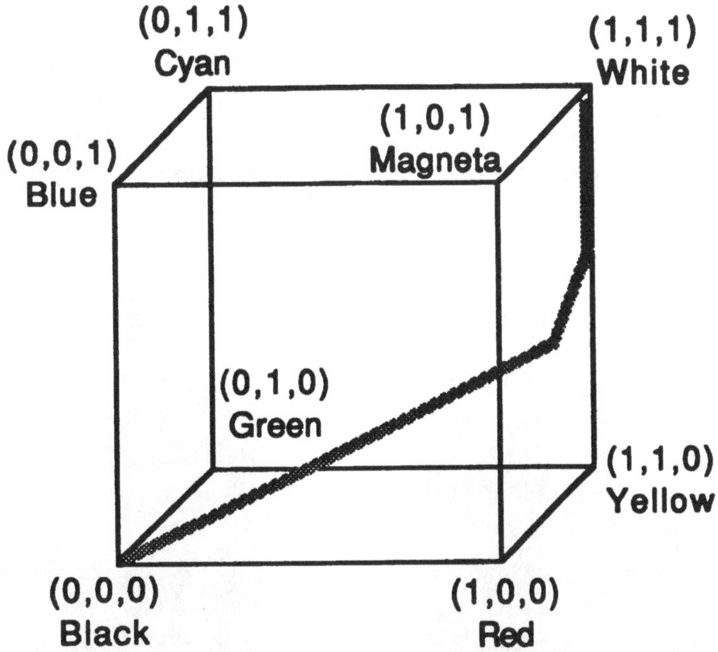

Figure 3: Camera Limits

Clearly, any algorithm that relies on hue to do object segmentation will fail in the presence of clipping. Before the picture is taken, the camera's aperture or shutter may be adjusted to try to make sure the area of interest is within the range of the camera. However, once the picture is taken, clipped values cannot be easily corrected.

3.2. Detection

Fortunately clipped values can be detected. Anytime the measured pixel value is equal to the camera maximum in one or more bands, those values and associated hue should be considered suspect. If the camera is saturated in a particular band, then we can only say that the "true" value at that point is equal to or greater than the camera maximum. We can eliminate the affected pixels from further consideration.

3.3. Correction

Alternatively we can make educated guesses as to what the true hue might be. If we are doing hue segmentation and one of the bands is saturated, we can see if the remaining two bands have values consistent with the hypothesis that we are still looking at the same object. This was done by Klinker et al.[6]

4. BLOOMING

Researchers usually make the implicit assumption that the relation described by equation 1 holds true for each pixel independently. We would like the light entering the camera at one pixel point to have no effect on the measured value at another pixel point.

Unfortunately not all cameras exhibit this property. Some of them fall victim to blooming. When they become saturated at one pixel point, the charge in one well on the CCD chip may spread to neighboring charge wells, affecting the measured values at those pixels.

4.1. Problem

Figure 4 shows a black and white picture taken by a CCD camera that is quite sensitive to blooming. The top half of the scene contains a white card, with all the rest black. This should be the result in the image as well. Unfortunately, the brightness encountered in part of the image has spilled over into the rest of it. Figure 5 shows the same scene imaged by an MOS camera that is more resistant to blooming. The spillover occurs mainly below the saturated pixels.

Figure 4: Severe Blooming

Figure 5: Moderate Blooming

4.2. Detection

If we are using a camera that is sensitive to blooming, then care can be taken to try to keep the brightest value in the scene within the range of the camera. However, as mentioned earlier, this can be very difficult in the presence of highlights. Unfortunately the presence of blooming cannot be directly detected. However, if a particular pixel is saturated in one or more bands, than we can assume that it may have spilled over into the adjoining pixels and added an arbitrary amount to the true values in the saturated band(s). These neighboring pixels may be ignored altogether or only their values in the unaffected bands may be considered. Alternatively, if we are using a camera that only blooms in one direction (usually vertically) we can exploit that knowledge to only ignore pixels below the saturated ones..

5. SPECTRAL SENSITIVITY

Equation 1 explicitly takes into account the spectral sensitivity of the sensors. It says that the measured value is dependent on the spectral characteristics of the camera as well as that of the light entering the camera. However it is sometimes easy to forget the consequences this sensitivity may have.

5.1. Infrared sensitivity

It is commonly assumed that the integration is performed over the wavelengths of light that are sensitive to humans, approximately 400 nanometers (nm) to 700 nm. This way the camera records the same sort of images that we see. However, many video cameras are sensitive to other wavelengths as well. The silicon that is used in CCD chips is particularly sensitive to near infrared (IR) wavelengths (700 nm to 1000 nm). Some cameras may have a built in filter to block these wavelengths, but others do not, or contain a filter that does not completely block infrared. Figure 6 show the spectral sensitivities of two of our cameras. This comparison shows that one of the cameras is fairly sensitive to infrared whereas the other is not.

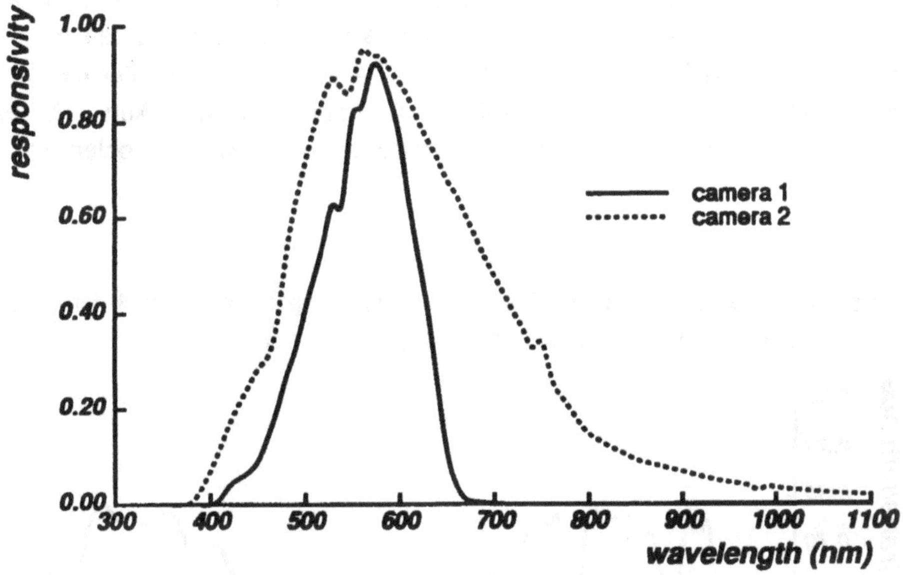

Figure 6: Spectral sensitivities of two cameras

5.1.1. Problem

A camera system that is sensitive to wavelengths to which humans are not can yield some surprising results as the camera will "see" things that we do not. Figure 7 shows an image taken of a black cloth covering a test pattern. The test pattern is clearly visible to the camera, whereas humans would be completely unable to make it out. While the black cloth appears opaque to us since it does not pass "visible" wavelengths of light, the camera shows that the cloth does pass infrared wavelengths. Figure 8 shows the same scene using an IR-block filter, and now the cloth is black as we expected.

Figure 7: "Hidden" Test Pattern

Figure 8: Hidden test pattern, revisited

In color pictures the additional light that enters the camera in the infrared band makes all the colors lighter and less saturated than humans expect to see them. Figure 20 shows a color picture of the chart taken by an IR-sensitive camera. The colors do not look at all the way we expect based upon our own perception of the chart.

5.1.2. Detection

In order to determine whether a camera is sensitive to infrared wavelengths, the spectral sensitivity of the camera may be measured by the researcher, or that information may be requested from the manufacturer. However, a very simple test can reveal if there is a problem. A visible blocking filter (one that blocks only the visible wavelengths of light) may be held in front of the camera and an image taken. The extent to which the scene can be made out in the image gives a simple estimate of how severe a problem will be caused by the infrared sensitivity

5.1.3. Correction

In our research, we always compensate for IR-sensitivity by using a an IR-block filter. Its transmittance is shown in Figure 9. With silicon sensors ultraviolet is generally not a problem.

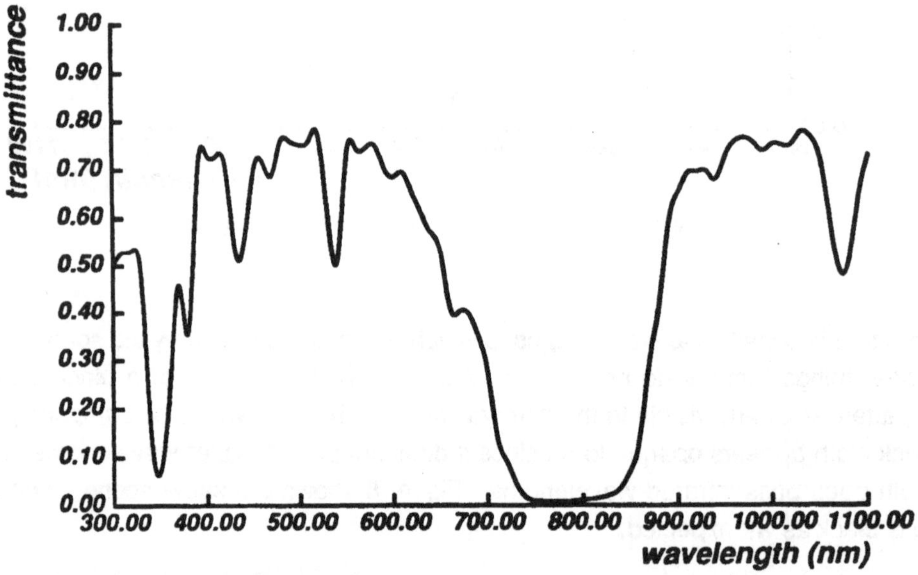

Figure 9: Spectral sensitivity of IR-block

5.2. Color Imbalance

A different artifact of the camera's spectral sensitivity is what we call color imbalance. As Figure 6 shows, both cameras are much less sensitive to the short (or blue) wavelengths than they are to the medium (green) and long (red) wavelengths. Figure 10 shows the spectral sensitivities of the filters we use in conjunction with our black and white camera.

Although the filters have roughly the same magnitude of response in their respective wavelengths of interest, when they are multiplied by the sensitivity of the MOS camera, the blue sensor system (camera with blue filter) is now much less sensitive than the red sensor (camera with red filter) as shown in Figure 11. This problem may be further exacerbated by the choice of illumination, as many light sources put out less energy at blue wavelengths.

5.2.1. Problem

The result is that the blue band of the image will be very dark. White objects will look yellowish and all hues will appear incorrect. Figure 21 shows a picture of the color chart with this problem; the effects are particularly noticeable in the gray squares.

5.2.2. Correction

One simple solution would be to multiply the blue band of the image by some appropriate constant. The problem with this approach is that while it is boosting the blue signal, it is also boosting the noise in the blue band. We would like to increase the blue signal to noise ratio (SNR) so that the blue band values are as reliable as those in the other bands.

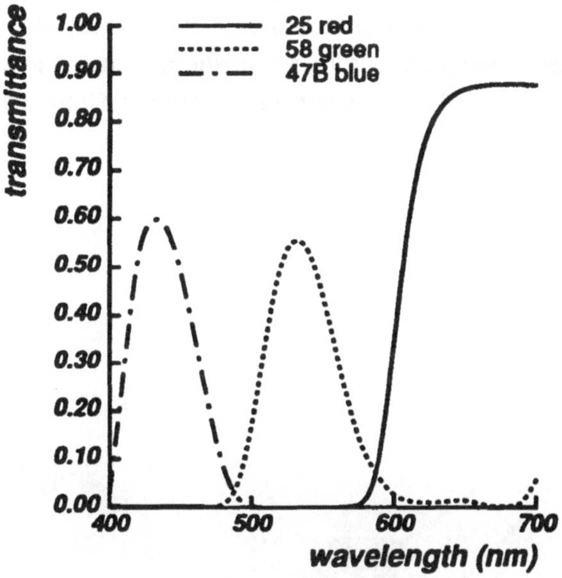

Figure 10: Spectral sensitivity of filters

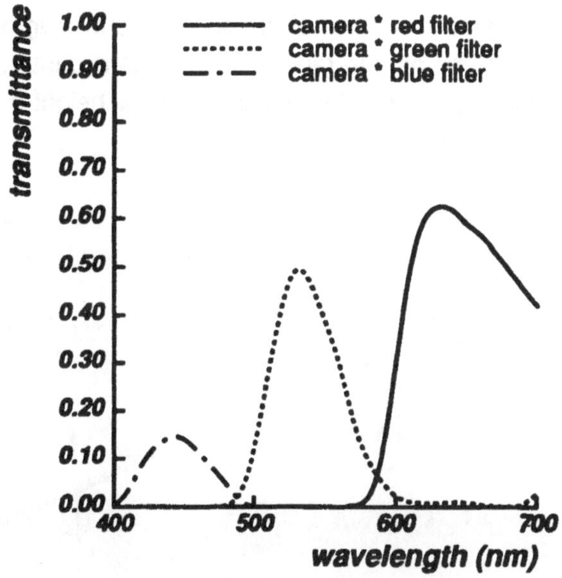

Figure 11: Camera times filters

We achieve this by a method we call aperture balancing: adjusting the aperture of the lens to equalize the measured values in all three bands. We do this under computer control with a motorized lens that has 2700 steps for aperture control. While altering the lens aperture modifies the imaging geometry of the lens affecting the depth of field to some extent, and the focused distance to a much smaller extent, if we use high levels of illumination and correspondingly narrow aperture settings we can neglect most of these effects.

Another method to equalize the signal levels across the three bands would be to use neutral density filters in conjunction with the red and green images. However, this system is not very flexible as the neutral densities would be chosen for a particular camera's sensitivity and a particular illumination. Cross-polarizers would be more flexible, but are difficult to control in a precise, repeatable fashion.

Alternatively, adjustments in the length of exposure may be made in some cameras systems, but most do not have this capability. Furthermore, dark noise is a function of length of exposure, so increasing the exposure time will also increase the amount of noise. Adjusting the illumination has a similar effect to adjusting the aperture, but the former is difficult to do in a controlled fashion.

6. CHROMATIC ABERRATION

An implicit assumption made by researchers is that the lens used with the camera system is doing its job of properly focusing the images on the detector. An ideal lens ensures that the pixel values measured in all three bands correspond to measurements made of light coming from the same point in space. Unfortunately such an ideal lens does not exist.

6.1. Background

Chromatic aberration exists in lens systems because the index of refraction of optical components varies as a function of wavelength. For example, given a simple uncorrected lens with incident white light rays, the blue components of the incident rays will be brought to a focus closer to the lens than the red components, as shown in Figure 12. Effectively the focal length of the lens is a function of the wavelength of the incident light. The change in the point or position of focus along the optical axis resulting from the light's wavelength is called "longitudinal chromatic aberration". If the imaging plane is placed so that the blue image is in focus, the red image will be blurred radially from the lens center, with its apparent height occurring somewhere in the middle of the blurring. This difference in the radial height (magnification) is called "lateral color" or "chromatic difference of magnification".[7]

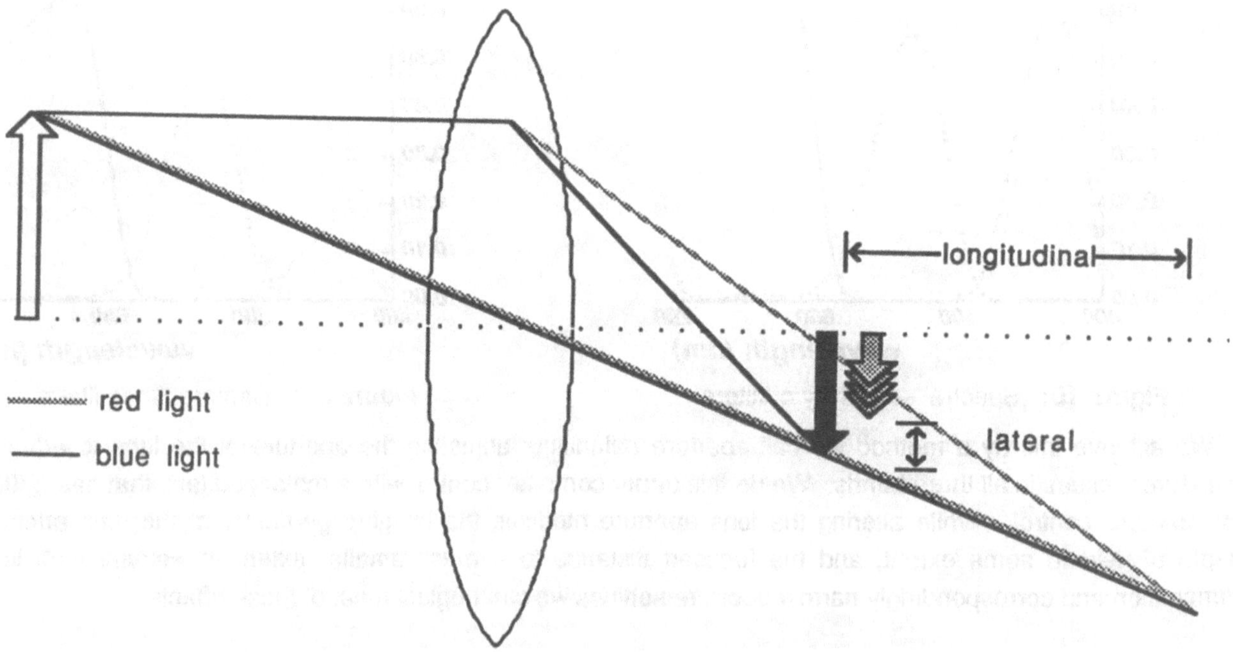

Figure 12: Chromatic aberration in a simple lens

Chromatic aberration is an intrinsic property of a lens. It can be partially compensated for in the lens optics by using pairs of lens elements with offsetting dispersion functions. Lens makers may try to cancel out the chromatic aberration of one element by adding another element with a different dispersion function. These pairs of lens elements are called achromatic doublets. Unfortunately such compensation is usually done only for only two wavelengths (red and blue) and then only at two points in the image field: at the optical axis and some specified radial distance from the optical axis.

6.2. Problem

For color imaging, chromatic aberration results in defocus and magnification differences between the three bands of the image. The magnification differences between the three bands result in color misregistration, particularly in areas of rapid color change such as the edges of regions having different colors. This causes unexpected hue shifts in pixels near the edge as the change in color is registered at different locations for the different bands. With our camera system the difference in the magnification between uncorrected red and blue images was found to be on the order of 0.5%, or 1.2 pixels near the outer edges of images that are 512 pixels wide. Clearly, any algorithm that tries to analyze the colors of objects based upon their hues will have serious problem with the pixels that are affected by chromatic aberration.

Another chromatic phenomenon we have encountered is a color dependent translation of the color images. Whereas a magnification change is radially symmetric with respect to the center of the optical system and has a smaller magnitude nearer the center than at the edge, this translation affects the entire image uniformly. This phenomenon is most likely an extrinsic lens property due to slight misalignments of the optical components of the lens. It may not occur in some lenses. In our work the measured translation between the uncorrected red, green and blue bands ranges from 0.1 to 0.4 pixel widths, depending upon the initial positions of the lens components.

6.3. Detection

Figure 13 shows the test pattern that we use for detecting chromatic aberration. A color image is taken of the grid and regions of interest are examined. The arrows indicate the scanline and columns that are to be used. Figure 14 is a graph of the red, green and blue pixel values verses pixel number for region 1 of the scanline. The vertical black line of the test pattern that falls within region 1 is near the optical axis of the lens and shows no misregistration of the red, green and blue color bands. Figure 15 is a graph of the red, green and blue pixels values verses pixel number for region 2. The vertical black line being examined in this figure is relatively far from the optical axis of the lens and clearly shows the misregistration of the three color bands due to chromatic aberration.

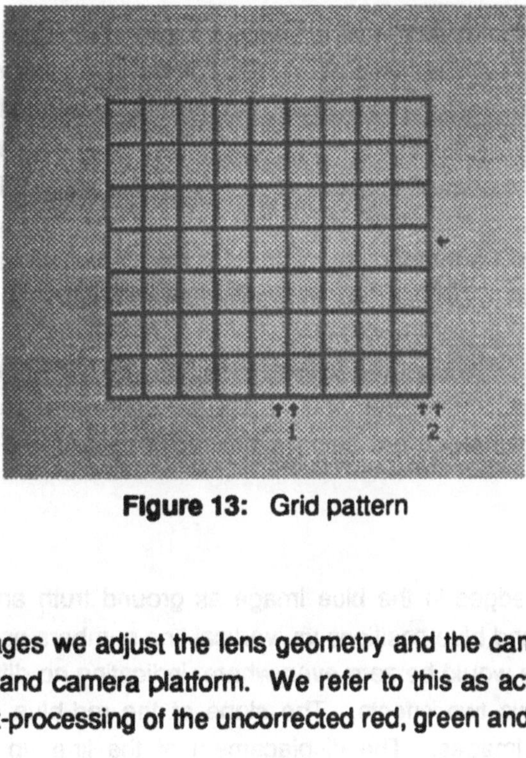

Figure 13: Grid pattern

6.4. Correction

To produce corrected color images we adjust the lens geometry and the camera position for each color image using a computer-controlled lens and camera platform. We refer to this as active compensation to distinguish it from compensation based on post-processing of the uncorrected red, green and blue images.

6.4.1. Camera system control

The camera system consists of a computer controlled filter wheel, a computer controlled lens, a solid state video camera and a six degree of freedom camera platform. The filter wheel houses 6 filters including Wratten #25 (red), 58 (green), and 47B (blue). The filters are selectable under computer control. Lens automation is provided by 3 micro stepping motors providing 3900 steps of resolution for focus distance, 4000 steps for focal length and 2700 steps for aperture. The camera positioning platform has a full six degrees of freedom and can position things to within 0.025 millimeters.

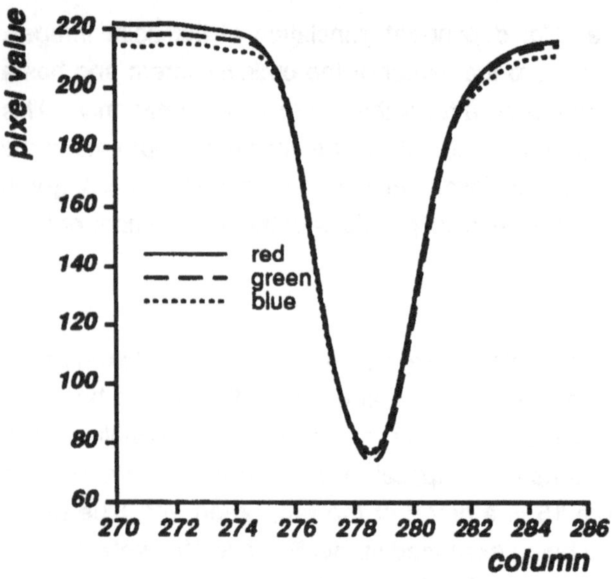

Figure 14: Image profile near center

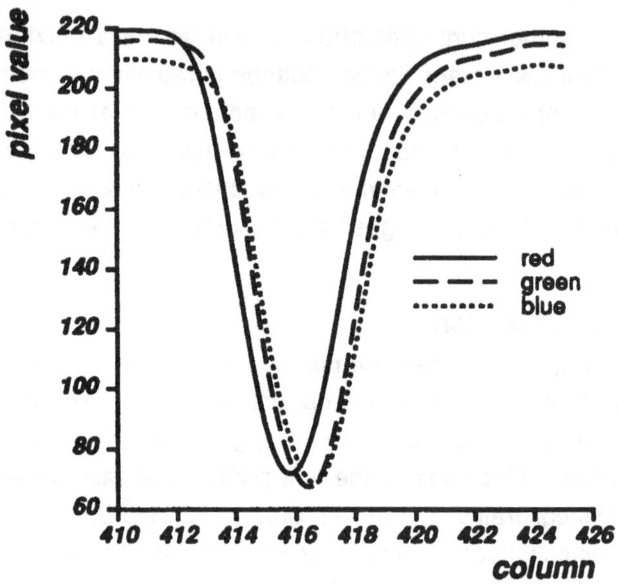

Figure 15: Image profile near edge

6.4.2. Procedure

To actively compensate the lens for chromatic aberration we first change the focused distance of the lens for each color. This is done by an automated focus routine to determine the position of best focus for each band individually. This corrects for the longitudinal chromatic aberration. Next we determine the magnification change between the bands. A calibrated model of the lens relates image magnification to position of the focal length (zoom) motor. We can then change the zoom appropriately for each band to correct the lateral color or change in magnification. To actively compensate the camera system for translation we simply move the camera parallel to the image plane for each color band to compensate for translation introduced by the optics.

In practice we precede these steps by aperture balancing to improve the signal to noise ratio for each image. If a non-linear camera is being used, linearization should also be done, and care must be taken to see that the camera is not saturated in the regions that are being examined, so that the problems of clipping and blooming may be ignored.

6.4.3. Results

Taking the position of the bar edges in the blue image as ground truth and plotting the difference (in pixel widths) between uncorrected red and blue positions vs. vertical line numbers going left to right we obtain the solid line in figure 16. An idealized lens would be zero everywhere, indicating no difference in the positions of the red and blue images. This plot shows two effects. The slope of the red-blue line is a measure of the relative magnification of the red to blue images. The displacement of the line up or down is a measure of the x component of the image translation.

The dotted line shows the red-blue differences after refocusing. The magnification difference has now shifted in the other direction. The reason for this can be visualized in figure 17. Before refocusing, the red image was smeared out, with its apparent magnification falling somewhere within the region of blurring, but at a point less

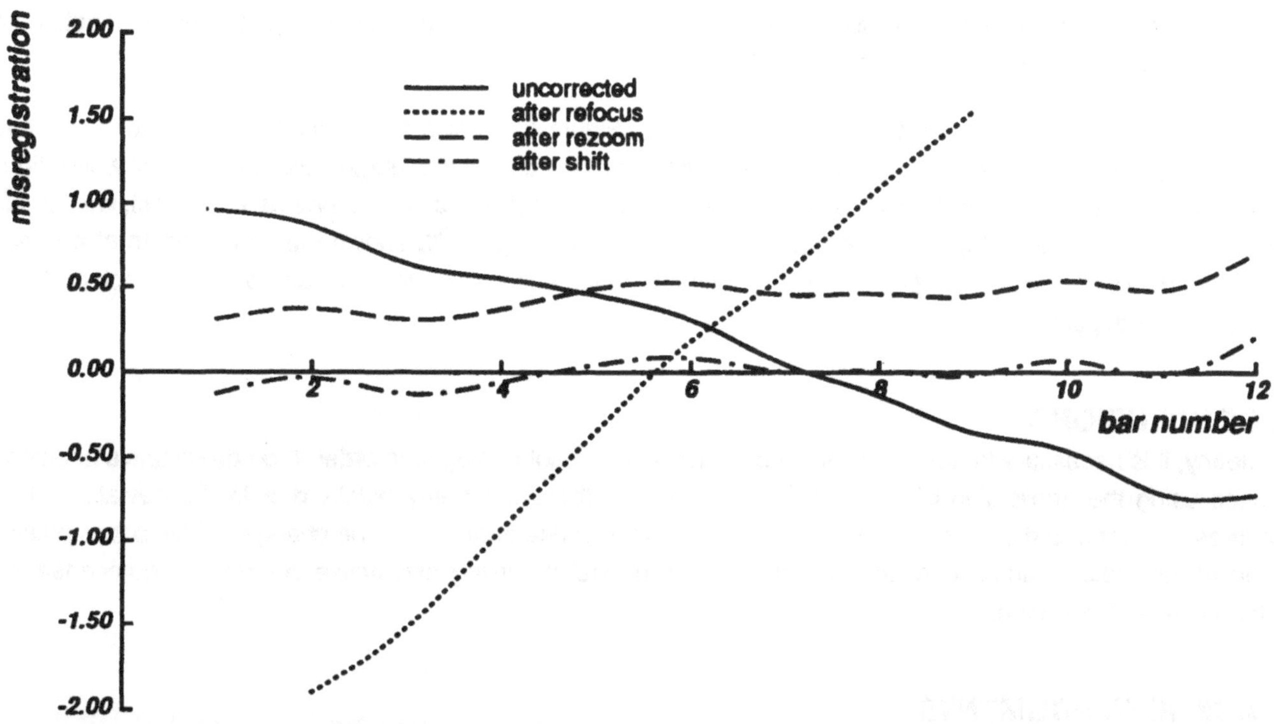

Figure 16: Chromatic aberration correction process

than that for the sharp blue image. After refocusing the red image is in focus, but has a greater magnification than the blue image.

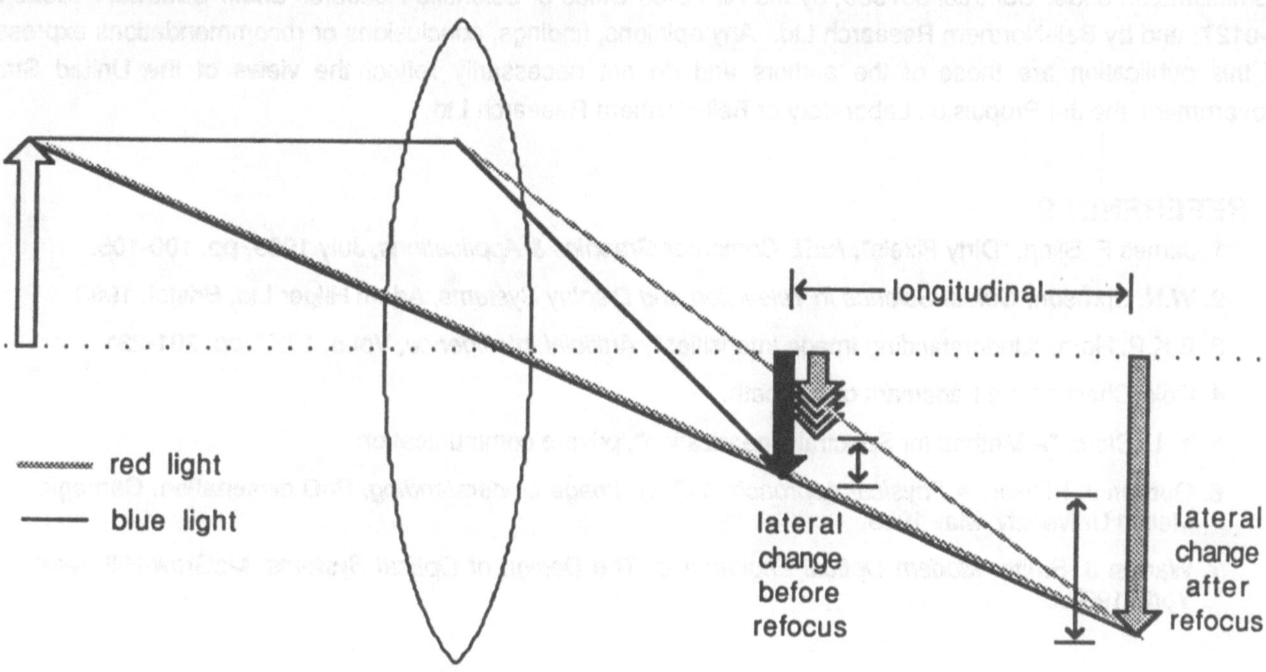

Figure 17: Lateral Color Change

After rezooming, the red-blue differences are shown by the dashed line. Now the slope of the line is approximately zero, so we have succeeded in eliminating the magnification change, but the translation is now obvious. The dash-dot line shows the final result, after the camera has been shifted. The plot is now close to

zero everywhere, emulating the behavior of an idealized lens. At the edge of the image, the misregistration has gone from about one pixel to about one-tenth of a pixel.

The improvement is also noticeable in color images. Figure 22 shows the detailed area of the color image profiled in figure 15. Recall that this vertical line came from the edge of the image, where chromatic aberration is most pronounced. Notice that instead of a simple white background with a black stripe as we expect, we see blue hues on one side of the stripe and red hues on the other side. Figure 23 shows the same region of an image where we performed active compensation to reduce chromatic aberration. Now we see only the black and white colors that we expected.

7. CONCLUSIONS

Clearly, it is necessary for researchers to obtain high-quality color images in order to do quantitative analysis of images using the application of physical laws. However, there are many pitfalls due to the deviation of real cameras from idealized camera behavior. These deviations cause anomalous hue changes. This paper presents some of the major pragmatic issues and shows how careful modeling and active control can compensate for imperfections of real systems.

8. ACKNOWLEDGMENTS

This research was supported by the Defense Advanced Research Projects Agency, DoD, through ARPA Order Number 4976, monitored by the Air Force Avionics Laboratory under Contract F33615-87-C-1499; by the Jet Propulsion Laboratory, California Institute of Technology, sponsored by the National Aeronautics and Space Administration under Contract 957989; by the Air Force Office of Scientific Research under Contract F49620-86-C-0127; and by Bell-Northern Research Ltd. Any opinions, findings, conclusions or recommendations expressed in this publication are those of the authors and do not necessarily reflect the views of the United States Government, the Jet Propulsion Laboratory or Bell-Northern Research Ltd.

9. REFERENCES

1. James F. Blinn, "Dirty Pixels", *IEEE Computer Graphics & Applications*, July 1989, pp. 100-105.
2. W.N. Sproson, *Colour Science in Television and Display Systems*, Adam Hilger Ltd, Bristol, 1983.
3. B.K.P. Horn, "Understanding Image Intensities", *Artificial Intelligence*, Vol.8, 1977, pp. 201-231.
4. ColorChecker is a trademark of Macbeth.
5. Y. LeClerc, "A Method for Spectral Linearization", private communication.
6. Gudrun J. Klinker, *A Physical Approach to Color Image Understanding*, PhD dissertation, Carnegie Mellon University, May 1988.
7. Warren J. Smith, *Modern Optical Engineering, The Design of Optical Systems*, McGraw-Hill, New York, 1966.

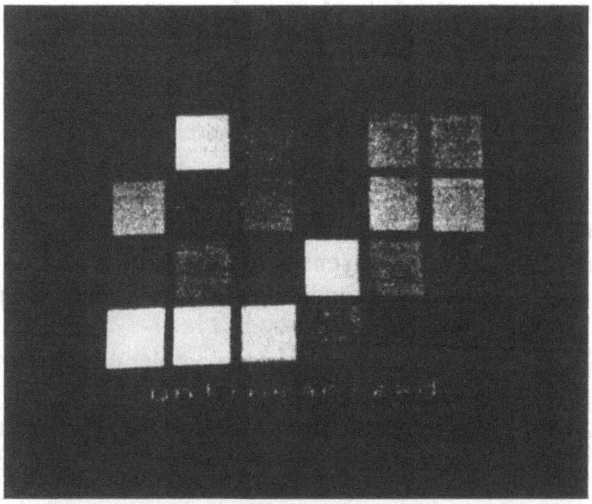

Figure 18: Unlinearized Image

Figure 19: Linearized Image

Figure 20: Image without IR-block

Figure 21: W/o aperture balancing

Figure 22: Before compensation

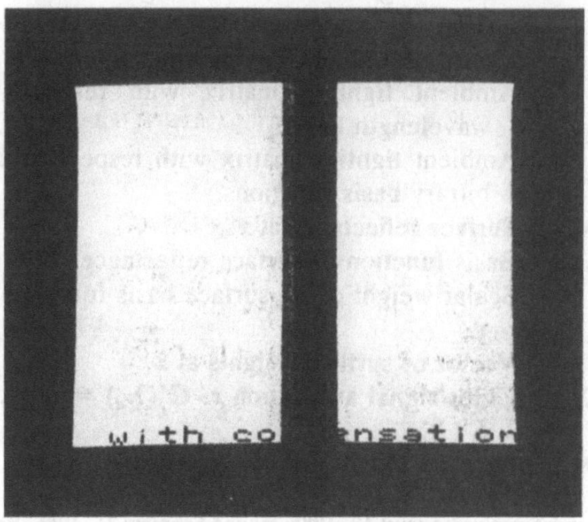

Figure 23: After compensation

See color plate A for Figures 18-23.

The Synthesis and Analysis of Color Images

BRIAN A. WANDELL

Abstract—I describe a method for performing the synthesis and analysis of digital color images. The method is based on two principles. First, image data are represented with respect to the separate physical factors, surface reflectance and the spectral power distribution of the ambient light, that give rise to the perceived color of an object. Second, the encoding is made efficient by using a basis expansion for the surface spectral reflectance and spectral power distribution of the ambient light that takes advantage of the high degree of correlation across the visible wavelengths normally found in such functions.

Within this framework, the same basic methods can be used to synthesize image data for color display monitors and printed materials, and to analyze image data into estimates of the spectral power distribution and surface spectral reflectances. The method can be applied to a variety of tasks. Examples of applications include the color balancing of color images and the identification of material surface spectral reflectance when the lighting cannot be completely controlled.

Index Terms—Color balancing, color constancy, color correction, efficient image encoding, image analysis, image synthesis, subspace methods.

List of Symbols

Symbol	Symbol Definition
N	Number of wavelength samples.
λ_n	Sample wavelength value.
$E^x(\lambda_n)$	Ambient light spectral power distribution at x.
$E_i(\lambda_n)$	Basis function of ambient light representation.
ϵ_i	Scalar weight of *i*th ambient light basis function.
ϵ	Vector of ambient light weights.
Ω_E	Ambient lighting matrix with respect to wavelength basis.
Λ_E	Ambient lighting matrix with respect to arbitrary basis functions.
$S^x(\lambda_n)$	Surface reflectance at x.
$S_j(\lambda_n)$	Basis function of surface reflectance.
σ_j^x	Scalar weight of *i*th surface basis function at x.
σ^x	Vector of surface weights at x.
$C^x(\lambda_n)$	Color signal at location x, $C^x(\lambda_n) = E^x(\lambda_n) S^x(\lambda_n)$.
$R_k(\lambda_n)$	Spectral sensitivity of *k*th sensor.
ρ_k^x	Scalar response of *k*th sensor at x.
ρ^x	Vector of sensor responses at x.
π	Best perpendicular to the sensor response vectors.
$D(E)$	Dimensionality of the ambient light representation.
$D(S)$	Dimensionality of the surface representation.

Manuscript received August 19, 1985; revised February 21, 1986. Recommended for acceptance by S. W. Zucker. This work was supported by NEI Grant 2 R01 EY03164 and NASA Grants NCC 2-307 and 2-332. This work was performed while the author held an NRC Post-Doctoral Fellowship at the NASA Ames Research Center.

The author is with the Department of Psychology, Stanford University, Stanford, CA 94305.

IEEE Log Number 8609223.

Introduction

The two primary physical factors that influence the color appearance of objects in a still image are the spectral power distribution of the ambient light and the surface spectral reflectance of the objects in the image. The information one wishes to display in a synthetic image, and the information available to analyze in during image acquisition, is called the *color signal*. I use the term color signal to refer to the spectral power distribution of the light arriving at the recording device or the human eye. In a natural scene, the color signal is generally computed as the product of the spectral power distribution of the light incident on an object and the surface reflectance of that object. In this paper, I will report on a common representational framework for the synthesis and analysis of color images based upon the synthesis and analysis of the color signal.

In the *synthesis* of color images, we begin with specifications of the surface spectral reflectance function and ambient light spectral power distribution. From these, we calculate the expected color signal. We then seek the intensities of the color display guns such that the display will have the same effect on the human visual system as the actual color signal would have. The application of the synthesis methods is most significant for the problem of accurate rendering of color images.

In the *analysis* of color images, the sensor reponses in the acquired image are determined by the color signal. We then seek a method of inverting the process, that is, recovering the underlying surface spectral reflectance and ambient light spectral distribution functions that gave rise to the color signal. The methods for analysis of color images are useful both for the correct rendering of color images and for the more general vision problem of identifying surface properties of objects from image data. I believe that this method can play an important role in the pattern analysis of color images.

The central issue that unifies the synthesis and analysis of color images is a specification of how the surface and lighting *information* is encoded in the color signal. I characterize the information in the surface and light functions by the number of degrees of freedom in a linear model that is used to approximate these functions. In image synthesis, significant efficiencies in image representation and computation can be obtained by making the number of independent degrees of freedom explicit. In image analysis, the number of independent degrees of freedom in the surface and light functions is the key factor that proscribes how many different classes of color sensors are necessary to decompose the color signal and estimate the surface and light functions. The principal result—due to Maloney and Wandell [12]—is that if the surface reflectance and light functions can be described using N degrees of freedom, then one requires $N + 1$ sensors to unlock the color signal and estimate the surface and light functions. This result has practical significance both for the accurate pattern analysis of color images and for the accurate rendering of color images.

Overview

In the next section, I will develop the basic notation and matrix operations common to the synthesis and analysis of color images. I will then describe the procedures for color synthesis, showing how we may relate the surface spectral reflectance functions, ambient light spectral power distribution, human photoreceptor responses, and display intensities on a monitor. In the third section, I will treat the analysis of color images. I will describe a version of Maloney and Wandell's [12], [20], [11] method for recovering the spectral power distribution of the ambient light and surface spectral reflectance functions of the objects from the sensor responses to the color signal.

The theoretical material in the next two sections is presented using general notation. The basic results apply to a wide class of representations of the ambient light and surface reflectance functions. In the final section, I will turn to the problem of selecting specific representations of these functions that are efficient with respect to a wide class of naturally occurring objects and lights.

Preliminary Considerations

The Asymmetry Between Light and Surface

Suppose that we acquire information at sample wavelengths λ_n for $n = 1, N$. The color signal is obtained by multiplying the spectral power distribution of the ambient light at x, $E^x(\lambda_n)$ times the spectral reflectance function of the surface spectral reflectance at x, $S^x(\lambda_n)$. We denote the color signal at point x as $C^x(\lambda_n) = E^x(\lambda_n)\, S^x(\lambda_n)$.

In developing the mathematical formulation for the synthesis of color images, we can retain symmetry between the role played by the surface and light functions without penalty. This symmetry is evident in the definition of the color signal $C^x(\lambda_n) = E^x(\lambda_n)\, S^x(\lambda_n)$, and can be maintained throughout (cf. Sallstrom [14] and Buchsbaum [2]). In most images, however, there is a physical asymmetry between the surface spectral reflectance and the ambient light spectral power distribution (cf. Horn [8] and Oppenheim and Schafer [13]). The asymmetry arises since the *rate of spatial variation* of the ambient light spectral power distribution $E^x(\lambda_n)$ is generally lower than the rate of spatial variation in the surface spectral reflectance function $S^x(\lambda_n)$. The slow spatial variation of the spectral power distribution makes it possible to represent the ambient light at lower spatial resolution than the surface functions, premitting some additional computational savings.

The most significant consequence of the difference in the rate of spatial variation, however, occurs for the *analysis* of color images. The uniqueness results for the recovery of the surface and light spectral power distributions from the color signal indicate that we cannot uniquely recover the light and surface functions if both vary freely over space. Essentially unique recovery is possible only if one of these functions varies more slowly over space than the other. We adopt the assumption that spatial variation of the ambient light is slower than the spatial variation of the surface reflectance function (cf. Horn [8] and Oppenheim and Schafer [13]).

To evaluate the plausibility of this idea, the reader might notice that in natural scenes in which the ambient light varies more rapidly spatially than the underlying surface—for example, when viewing a slide projected on a screen—the human visual system interprets the spatial variation as if it were due to a surface variation, and not from its true cause, the ambient light variation. The analysis we will use here also assumes that there is an asymmetry between light and surface functions. In any local region of the image, we assume that there is only spatial variation in the surface reflectance function, and no spatial variation in the ambient light spectral power distribution. We will indicate this assumption in our notation by suppressing the superscript x when we refer to the parameters of the spectral power distribution of the ambient light. Thus, we will write $E(\lambda_n)$ instead of $E^x(\lambda_n)$, so that in a small region of the image, $C^x(\lambda_n) = E(\lambda_n)\, S^x(\lambda_n)$.

I emphasize that this assumption does not imply that the ambient light remains constant over the entire image, but only that the spectral power distribution remains constant over a local region of the image in which there is significant spatial variation in the surface spectral reflectance function.

Synthesis of Color Images

Introduction

In generating color images, we wish to determine the appropriate display gun intensities so that the display will have the same *visual effect* as the actual color signal $C^x(\lambda_n)$. In their important work on color standards, the members of the CIE have spent more than 50 years describing various methods and representations for achieving this result. The various methods and representations (using *XYZ* values, *YIQ* values, etc.) are logically equivalent, but have sometimes been adopted because of odd technical limitations in the computation. For purposes of the analysis here, I prefer to use a calculation advocated by Cornsweet [6] based on the photoreceptor absorptions of the human eye. This is the method most closely based on the physical factors governing color perception. Although the method is logically equivalent to the several methods suggested by the CIE and has an excellent physical interpretation, no standard photoreceptor representation has been adopted by the CIE. The estimates by Smith and Pokorny [15] and described by Boynton [1] are widely used at present.

Basic Equations: Wavelength Representation

To determine the display gun intensities that will appear the same as the color signal to the human eye, we first compute the effect that the color signal $C^x(\lambda_n)$ would have on the three types of photoreceptors in the human eye. Suppose that the sensitivity of the receptors in the eye is described by the three sensitivity functions $R_k(\lambda_n)$ for $k = 1, 3$. If we have measurements of the surface reflectance function and spectral power distribution function at N wavelength values for λ_n, $n = 1, N$, then the strength of the kth receptor response at location x is

$$\rho_k^x = \sum_{n=1}^{N} C^x(\lambda_n)\, R_k(\lambda_n). \tag{1}$$

If one begins with the specification of the color signal in terms of the surface and light functions, and if there are N sample points along the wavelength dimension, then the calculation requires $2N$ multiplications and $N - 1$ additions per pixel. To compute the three receptor responses requires $6N$ multiplications and $3(N - 1)$ additions.

The effect of setting the intensity values of the three color guns on the receptors can be calculated as follows. Suppose that the red, green, and blue display guns have spectral emission functions $G_j(\lambda_n)$ for $j = 1, 3$. Denote the intensity of the jth gun as I_j. When the display gun intensities at pixel x are set to the vector of value $I^x = (I_1^x, I_2^x, I_3^x)$, the response of the kth class of photoreceptors will be

$$\rho_k^x = \sum_{j=1}^{j=3} I_j^x \sum_{n=1}^{N} G_j(\lambda_n)\, R_k(\lambda_n). \tag{2}$$

Equation (2) can be written as a matrix equation of the form

$$\rho^x = \Gamma\, I^x \tag{3}$$

where the k, jth entry of the matrix Γ is

$$\sum_{n=1}^{N} G_j(\lambda_n)\, R_k(\lambda_n). \tag{4}$$

The gun intensities that are visually equivalent to the color signal may be determined by first solving (1) for the values of ρ^x and then solving (3) for I^x. Two issues arise immediately in computing the display gun intensities.

First, the solution may have negative values. This will occur when the required color image is outside the gamut of colors that can be displayed by the particular device. I will discuss a method of treating this problem later in the paper when the theory has been more fully developed.

Second, there is the issue of computational efficiency. The entries of the matrix Γ are fixed by the display gun spectral emission functions and the human photoreceptor sensitivity curves. This matrix (or its inverse which generally plays a bigger role) need only be calculated once, and therefore does not represent a significant computational burden. The calculation of the receptor responses ρ^x to a color signal $C^x(\lambda_n)$ must be made for every point in the image. This is the most expensive part of the computation in the synthesis of color images. The cost of this computation is governed by the number of sample points along the wavelength dimension for the functions $E^x(\lambda_n)$ and $S^x(\lambda_n)$. Since the amount of computation is proportional to the sampling rate N, one method of limiting the computation is to reduce the sampling rate N.

A second option is to transform the representation of the surface spectral reflectance and ambient light spectral power distribution functions from a wavelength representation to a more efficient representation. Great savings in efficiency can be obtained because the wavelength functions describing naturally occurring lights and surfaces are highly correlated at nearby wavelengths. To improve computational efficiency, we must seek representations of the spectral power distributions of the lights and spectral reflectance functions of the surfaces that are decorrelated. We will consider specific representations later in this paper. The general effect of applying such transformations is analyzed in the following section.

Basic Equations: Transformed Representation

In order to efficiently encode ambient light and surface vectors, we will seek alternative representations, related to the wavelength representation by a linear transformation. These take the form

$$E(\lambda_n) = \sum_{i=1}^{D(E)} \epsilon_i E_i(\lambda_n) \tag{5}$$

and

$$S^x(\lambda_n) = \sum_{j=1}^{D(S)} \sigma_j^x\, S_j(\lambda_n) \tag{6}$$

where $D(E)$ is the chosen dimensionality of the light representation and $D(S)$ is the dimensionality of the surface representation. In order to span any particular subspace, it is sensible to select functions $E_i(\lambda_n)$ or $S_j(\lambda_n)$ that are orthonormal in the sense that

$$\sum_{n=1}^{N} E_i(\lambda_n)\, E_j(\lambda_n) = 0 \tag{7}$$

when $i \neq j$ and the length of each function is one. This is not necessary—the entire development remains consistent for any choice of linearly independent functions. However, for any subspace one wishes to span, it is always possible to find such a set of functions so that choosing orthonormal functions does not restrict the analysis, and by choosing an orthonormal basis, the recovery of the weights ϵ_i and σ_j^x from the wavelength function is very easy. We refer to these sets of functions $E_i(\lambda_n)$ and $S_j(\lambda_n)$ as *basis functions*.

The usual wavelength representation is equivalent to choosing a vector of weights with respect to the set of functions $E_i(\lambda_n)$ such that

$$E_i(\lambda_n) = 1 \quad \text{if } i = n \text{ and } 0 \text{ otherwise} \tag{8}$$

and $S_j(\lambda_n)$ such that

$$S_j(\lambda_n) = 1 \quad \text{if } j = n \text{ and } 0 \text{ otherwise.} \tag{9}$$

We can express the relationship between the vector of wavelength sample measurements that characterize the ambient light $E(\lambda_n)$ and the column vector of weights that characterize the transformed representation of the ambient light $\epsilon = (\epsilon_1, \epsilon_2, \cdots)$ as

$$E(\lambda_n) = \boldsymbol{E}\epsilon. \tag{10}$$

The columns of the matrix $\boldsymbol{E}$ are equal to the basis vectors $E_i(\lambda_n)$. Similarly, we can express the relationship between the vector of sample measurements of the surface reflectance that characterizes the surface with respect to wavelength and the vector of weights that characterize the surface with respect to the transformed representation $\sigma^x = (\sigma_1^x, \sigma_2^x, \cdots)$ as

$$S^x(\lambda_n) = \boldsymbol{S}\sigma^x. \tag{11}$$

The columns of the matrix $\boldsymbol{S}$ are the basis vectors $S_j(\lambda_n)$. To compute the matrix required to transform from one representation to another, one may use the linear equations (10) and (11) to compute the vector in either wavelength or the transformed domain. When the vectors in the two domains are linearly independent and of the same dimension, a transformation is always possible without loss of information. To improve computational efficiency, we would like to choose the dimensionality in the transform domain to be smaller than the dimensionality in the original representation ($D(E) < N$ and $D(S) < N$), but with little loss of precision. When we choose a smaller dimensionality in the transformed domain than in the wavelength domain, the equations may have no exact solution. In that case, we will always choose *the least squares solution* to (10) and (11) as our representation. When the bases are orthonormal, the least squares fit is found by multiplying by the matrix transpose S' or E'.

Relationship Among Physical Quantities

In order to compute the response of a receptor ρ_k^x from the surface and light functions, we use the equation[1]

$$\rho_k^x = \sum_{n=1}^{N} E(\lambda_n)\, S^x(\lambda_n)\, R_k(\lambda_n). \tag{12}$$

Using matrix notation, we can express the relationship among the sensors, the ambient light, and the surface reflectance function in the form

$$\rho^x = \Omega_E \omega^x \tag{13}$$

where ω^x is the column vector $(S^x(\lambda_1), S^x(\lambda_2) \cdots)'$. When the surface is represented with respect to the wavelength basis terms $S_i(\lambda_n) = 1$ if $i = n$ and 0 otherwise, then the kjth entry of the matrix Ω_E is

$$E(\lambda_j)\, R_k(\lambda_j). \tag{14}$$

I reserve the use of the Greek symbols for omega (Ω and ω) to refer to the special case in which the physical entities are represented with respect to the wavelength basis functions.

The relationship among the surface, the light, and the sensor responses may be characterized as a simple matrix product. The surface vector is mapped by a linear transformation into the vector of sensor responses. The linear transformation is determined by two factors: the ambient light $E(\lambda_n)$ and the receptor sensitivities $R_k(\lambda)$. Across viewing conditions for a particular camera or for the human eye, the receptor sensitivities are unchanging and known. Thus, the only unknown factor that governs the linear transformation Ω_E is the ambient light spectral power distribution. For that reason, we make E explicit when we write down the matrix representing the linear transformation. We will call matrices such as Ω_E *lighting matrices*, although the reader should note that the spectral power distribution of the ambient light and the spectral sensitivities of the sensors both play a role in forming the effective lighting matrix for any particular device.

The relationship among the surface, lights, and sensors takes on a parallel form when we transform the surface representation to a new set of basis vectors. If we use

$$S^x(\lambda_n) = \sum_{n=1}^{N} \sigma_j^x\, S_j(\lambda_n), \tag{15}$$

then the matrix equation becomes

$$\rho^x = \Omega_E S \sigma^x = \Lambda_E \sigma^x \tag{16}$$

where the columns of the matrix S are simply the N vectors $S_j(\lambda_n)$. The matrix S is fixed and we have incorporated the change of basis of the surface representation into the lighting matrix to define a new lighting matrix $\Lambda_E = \Omega_E S$. The kjth entry of Λ_E is now

$$\sum_{n=1}^{N} E(\lambda_n)\, S_j(\lambda_n)\, R_k(\lambda_n). \tag{17}$$

This formula generalizes the formula in (14) for which the surface representation is $S_j(\lambda_n) = 1$ if $j = n$ and 0 otherwise. We will use expression (17) to define the entries of the lighting matrix in general.

The matrix Λ_E can be decomposed into a set of fixed primitive matrices that depend only upon the choice of

[1] In general, the effective surface reflectance at a point in the image may depend upon the geometry of the image. For example, specular surface reflections depend strongly upon the angles among the lighting source, the surface, and the viewer. Although we do not specifically indicate such dependencies, we assume that the surface reflectance function $S^x(\lambda_n)$ will generally depend upon the viewing geometry. Since our analysis here is for a single fixed image, we simply assume that the surface spectral reflectance has been properly specified for the particular viewing geometry. An analysis of the effects of specularity at the physical level may be found in Torrance and Sparrow [18] and Torrance *et al.* [19]. An analysis of the effects of incorporating more sophisticated specular effects in computer-generated images may be found in Cook and Torrance [5]. I will consider the issue of specularity again in the section on the selection of basis functions.

basis functions. If we express the spectral power distribution of the ambient light with respect to the basis vectors $E_i(\lambda_n)$ so that the lighting vector is ϵ, then decomposition of Λ_E is

$$\Lambda_E = \sum_{i=1}^{D(E)} \epsilon_i \Lambda_i \tag{18}$$

where the kjth entry of Λ_i is

$$\sum_{n=1}^{N} E_i(\lambda_n)\, S_j(\lambda_n)\, R_k(\lambda_n). \tag{19}$$

This emphasizes the fact that the effect of the light may be described as the sum of the action of several different linear transformations, one linear transformation per degree of freedom in the lighting model.

Summary of Results

The set of matrix operations that capture the relationship between these various quantities may be summarized as follows. Using the rule that the effect of the display on the receptor responses should equal the effect of the surfaces and light on the receptors, we may write

$$\Gamma I^x = \Lambda_E \sigma^x. \tag{20}$$

Inverting Γ,

$$I^x = \Gamma^{-1} \Lambda_E \sigma^x. \tag{21}$$

If we wish to make the role of the ambient light parameters explicit, we may also write this equation using the expansion of the lighting matrix from (18):

$$I^x = \left[\sum_{n=1}^{D(E)} \epsilon_n \Gamma^{-1} \Lambda_n \right] \sigma^x. \tag{22}$$

Equation (21) expresses the complete relationship among the surface spectral reflectances, light spectral power distribution, and display gun intensities. Once the basis functions are made explicit, and the vector characterizing the lighting ϵ and vectors characterizing the surfaces σ^x are provided with respect to that basis, then all the terms on the right-hand side of the equation are known so that the gun intensities on the left-hand side may be computed.

We may assess the computational burden as follows. In the typical color display, we can control the intensities of three color guns; thus, the vector I^x is three dimensional. The matrix Γ^{-1} is 3 by 3 and its values are fixed by the display phosphors and human photopigment sensitivities. The selection of the basis functions to represent the surfaces and lights completely fixes the matrices Λ_n (19). These matrices are 3 by $D(S)$, so that the product of $\Gamma^{-1}\Lambda_n$ is 3 by $D(S)$ and may be precomputed. They do not represent a significant computational burden.

Thus, we see that the amount of computation depends primarily on the dimensionality of the surface representation $D(S)$. This number plays the same role as the number of sample points on the wavelength dimension N when the equations are written with respect to the wavelength representation. Significant computational savings can be achieved when the representation with respect to an alternative set of basis functions permits us to use a value of $D(S)$ that is significantly smaller than the value of N.

Display Gamut Limitations: When the display monitor cannot generate the desired signal, one of the required gun intensities I^x will be negative. In general, such an outcome cannot be entirely avoided since the span of colors obtainable by additive light mixtures is certainly less than the span of colors obtainable choosing arbitrary surface and light functions. In general, the range of display intensities that may be required is determined by the range spanned by the columns of the matrix

$$\Gamma^{-1}\Lambda_E. \tag{23}$$

Once the surface and lighting basis functions are chosen, the only freedom left in determining the span of the columns of this matrix is the weights of the ambient light spectral power distribution ϵ_i.

By suitable choice of the light and surface basis functions, however, the number of instances in which these gamut errors occur can be greatly limited. For example, Buchsbaum and Gottschalk [3] have considered the spectra of lights that are smoothly varying in the sense that the Fourier transform of these spectra—when treated as functions of wavelength—only have power in the low-frequency region.[2] If these functions are restricted to be bandlimited (i.e., have no power above some frequency of modulation), then the corresponding color signal will be bandlimited and generally fall within a restricted color gamut.

In representing objects outside the color gamut of the display, one may choose two methods. First, one may simply adjust the individual point outside the gamut, for example, by setting the negative value to zero. This method distorts the color representation of that point alone. This solution may be adequate, depending upon the significance of the color of that particular portion of the image.

An alternative method of solution is to adjust all of the display values together, attempting to move all of the colors to within the display gamut. This causes some error at each point in the display, but preserves the relationship among the different objects. A reasonable method of making this type of adjustment is to adjust the value of the lighting vector ϵ_i in order to avoid negative display intensities.

The advantage of this type of correction derives from the psychophysical observation that human observers are reasonably good at discounting the appearance of the ambient light when judging the surface appearance of objects (cf. Helson [7]). This method of color correction in the synthesis of images will be evaluated more fully elsewhere.

[2]*Note:* Frequency in this case refers to the rate of variation of the light and surface functions with respect to the wavelength of the ambient light, a rather unfortunate (multiple) use of the frequency concept.

The Display of Color Camera Data

When the synthesis of the color image on the screen depends upon data acquired from a color camera rather than data generated by an internal model, the relationship between the camera responses and the desired display values may be rather complex. The difficulty arises when the camera sensors are not within a linear transformation of the human receptor sensitivities. In this case (cf. Horn [9]), no simple calculation relating the camera responses to the desired display values is possible.

The methods described in the next section—on image analysis—may be used to decide upon the correct color display values. The idea is the following. Even if the camera sensors are not within a linear transformation of the human photoreceptor responses, their response can still be used to estimate the surface spectral reflectances and light spectral power distributions in the image. These estimated values can then be used to determine the display output. I describe the procedure for using camera data to estimate the surface and the light parameters in the next section. I will return to the problem of correcting camera output at the end of the paper.

THE ANALYSIS OF COLOR IMAGES

In this section, I will treat the analysis of image data. Beginning with the sensor responses acquired in a camera ρ^x, we ask what we can learn about the surface reflectance functions and the ambient light spectral power distribution. The development in this section is based on work done by Maloney and me [11], [20], [12].

The key equation that relates the response of the sensors in a camera system to the ambient light spectral power distribution and surface reflectance functions is (16) which we now write as

$$\rho^x = \Lambda_E \sigma^x. \tag{24}$$

The known values in the acquired image are on the left-hand side of the equation, and the unknown values are on the right-hand side of the equation. An extensive mathematical analysis of the recovery procedure is contained in Maloney's [11] dissertation. The main results are summarized in Maloney and Wandell [12]. I now describe the computational method used in the Stanford color analysis package, written by Brainard and me.[3]

Conditions for Recovery: Spatial Variation

We begin by considering what can be recc 'ered from the color signal when the surface and light functions are both free to vary over space. Recall that the response of the sensors ρ^x is completely determined by the color signal, which is given by $C^x(\lambda_n) = E^x(\lambda_n)\, S^x(\lambda_n)$. To evaluate the uniqueness of recovery, note that we can introduce an arbitrary function of both space and wavelength $G^x(\lambda_n)$ into the equation for the color signal so that

[3] The manual for the package can be obtained by writing the author. The package is written in C on a UNIX BSD 4.2 system.

$$C^x(\lambda_n) = E^x(\lambda_n)\, S^x(\lambda_n) \tag{25}$$

and

$$C^x(\lambda_n) = (E^x(\lambda_n)\, G^x(\lambda_n)) \left(\frac{S^x(\lambda_n)}{G^x(\lambda_n)}\right). \tag{26}$$

The uniqueness of the recovery can be characterized, then, by describing the limitations on the function $G^x(\lambda_n)$. Clearly, without some limitation on the structure of the surface and light functions, unique recovery is impossible. In order to obtain unique recovery, we must introduce restrictions on the variation of the light and surface functions. The consequences of the restrictions I describe here are thoroughly analyzed by Maloney [11].

A Condition on Spatial Variation: Suppose that the scalar intensity of both the ambient light and surface reflectances are free to vary at each point. If we scale the ambient light at x by a factor $g(x)$ and we scale the surface reflectivity by a factor $1/g(x)$, then the color signal remains unchanged. This emphasizes that we cannot discriminate among the alternative decompositions of the signal into $E^x(\lambda_n)$ and $S^x(\lambda_n)$ versus $E^x(\lambda_n)\, g(x)$ and $S^x(\lambda_n)/g(x)$.

Equations (25) and (26) demonstrate that the identical color signal can arise from many different interpretations of the functions describing the ambient light and the surface reflectances. To permit a unique recovery of the light and surface functions, we are forced to adopt some assumption about the spatial variation of the light and surface functions. The general view we will adopt—to be made precise below—is that the rate of variation over space of the surface reflectance is greater than the rate of variation over space of the ambient light.

The basic premise of the analysis is illustrated in Fig. 1. We analyze the image sequentially by dividing it into overlapping spatial areas. Within each of the areas, we analyze the image under the assumption that the ambient light *in that small region* is constant. We choose the image areas with two considerations in mind. First, within each area, there must be significant variation in the sensor responses. As we shall see, this is necessary in order to obtain estimates of the light and surfaces in that area. Second, we choose the areas with a high degree of overlap. This is not necessary, but it is useful so that we can detect slow variations of the spectral power distribution of the ambient light. The complete estimate of the ambient light function is determined by combining the separate estimates of the ambient light into a smoother, interpolated function. The necessary variation in the spectral reflectance function of the surface reflectance functions required within each of the areas is specified in the next section in which I describe the recovery method.

In each local region, then, the spatial variation in the sensor response is assumed to be due entirely to the surface reflectance, and not at all to spatial variation in the ambient light. With this assumption, the estimate of the spectral power distribution of the ambient light is unique

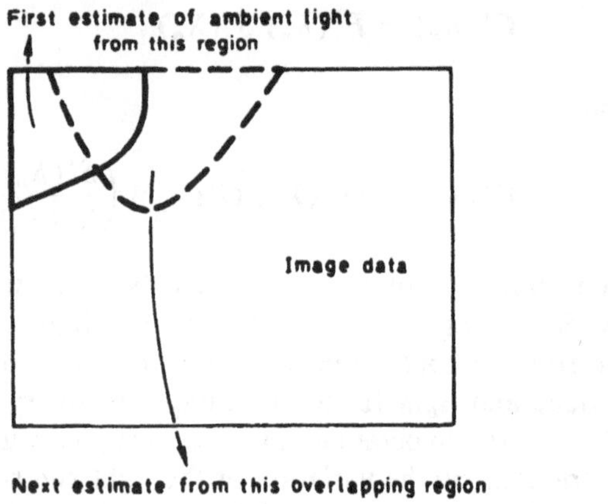

Fig. 1. The analysis of the image data proceeds by selecting regions of the image with enough complexity in order to form independent estimates of the surface and the light. Within each region indicated in the figure, there is enough data to permit an estimate of the surfaces and light functions within that region. As we move from region to region, the light and surface functions estimates are forced to be consistent at the points of overlap.

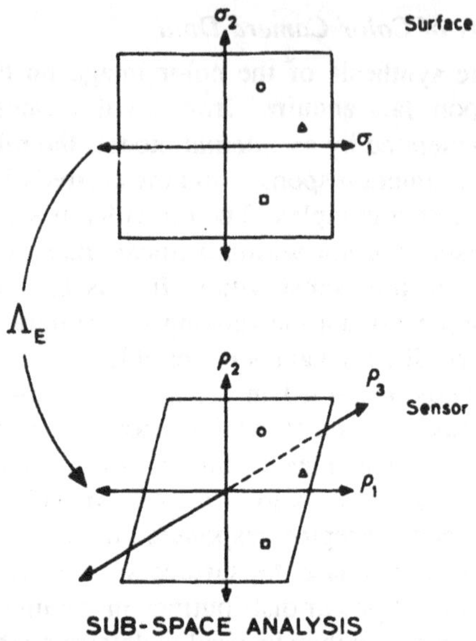

Fig. 2. The surface vectors are represented in a two-dimensional space, shown at the top of the figure. The effect of the light is to act as a linear transformation that maps these two-dimensional vectors into a subspace of the three-dimensional space of receptor responses.

up to a single scalar $\alpha E(\lambda_n)$ and $(1/\alpha)\, S^x(\lambda_n)$ with weak conditions. The value of α is fixed for all points within each region of the analysis. (The final estimate of the ambient light will be a smoothed version of this piecewise constant function.) Further, since the regions may be chosen to overlap, the requirement of consistency at points of overlap leaves us with only a single unknown scalar for the entire image. By convention, we choose α to normalize the total power in the ambient light to unity.

A Condition on the Number of Color Sensors: Let us now consider the second important condition that is necessary for recovery. This is a condition on the number of color sensors necessary for recovery.

Suppose that the dimensionality of the surface representation $D(S)$ is equal to the number of sensor classes, that is, the dimensionality of ρ^x. In that case, again, no unique solution of the equations across space is possible. To see this, let the reader select any vector whatsoever to represent the weights of the vector of the ambient light E. This choice specifies the lighting matrix Λ_E. With this lighting vector chosen and the data ρ^x known, we may invert the lighting matrix and solve for the surface vectors σ^x. This solution provides a completely acceptable account of the image data. For each arbitrary choice of the lighting parameters, we can generate a set of surface vectors that are consistent with the observed sensor responses. It follows that when the number of dimensions in the surface representation equals (or exceeds) the number of sensor classes, we do not have enough data to obtain a unique solution.

This is the first important difference between the synthesis of images and the analysis of images. When we synthesize images, we can use a surface representation with dimensionality equal to or greater than the number of independent display guns. In order to analyze an acquired color image, we must use a surface representation whose dimensionality is strictly *less* than the number of sensor classes. The result emphasizes what is possible—and what is not possible—for systems with different numbers of color sensors. A system with three color sensors, for example, can only recover surfaces and lights that fall within a two-dimensional linear model. Errors will occur should the image data be composed of surfaces and lights requiring a higher dimensional representation. The significance of these errors for a system with three sensors will depend on the nature of the subsequent processing and the severity of the failure of the two-dimensional model.

The Recovery Method

From (16),

$$\rho^x = \Omega_E S \sigma^x = \Lambda_E \sigma^x, \tag{27}$$

we see that when the dimensionality of the surface vector σ^x is less than the dimensionality of the receptor vector ρ^x, the observed data will be restricted to a subspace of the receptor responses. For example, when the set of surface vectors is a two-dimensional space, the image of the surface vectors under a fixed light will span a plane within the three-dimensional space of sensor catches. This is illustrated in Fig. 2. The plane spanned in receptor space is determined by the columns of the matrix Λ_E. The columns of Λ_E are determined entirely by the lighting parameters ϵ. It follows, therefore, that the identity of the subspace is determined by the values of the light parameters, and the position within the subspace is determined by the surface parameters σ^x.

We can use this fact to achieve recovery in a three-step process. First, we identify the subspace which contains the set of receptor responses. Second, we use the identity of the subspace to estimate the lighting vector ϵ. Finally, we use the lighting parameters to determine the pseudoinverse of the lighting matrix Λ_E^{-1}. This is the final step in the problem, since with this matrix, we can determine the estimated surface parameters σ^x from the receptor responses ρ^x.

I will now describe the implementation of each of the surface and light recovery steps used in the Stanford color analysis package. This will permit me to describe the points at which failures can occur.

Identifying the Subspace: The first step in the recovery procedure is to attempt to identify the subspace that contains the set of sensor vectors ρ^x. For image analysis, we will generally wish to use a surface and light representation whose dimensionality is as large as possible in order to minimize the error in the analysis. The best fit can be obtained using the linear models with the largest degrees of freedom. Thus, we will assume that $D(E)$ and $D(S) + 1$ equal the number of sensors since any other assumption will produce an inferior fit. With this assumption, we expect the data to fall in a subspace of the sensor response that can be uniquely characterized by the unit normal vector, perpendicular to all of the vectors in the subspace. Call the unit normal vector, perpendicular to the subspace, π. The dot product of this vector with each of the sensor vectors ρ^x will be zero.

In fact, of course, the data are not likely to fall precisely within a subspace, but rather to span the sensor response space. Our strategy, however, is to find the vector that is "most perpendicular" to the data from the sensor vectors, under the assumption that perturbations outside of the subspace are due to noise. To find the most perpendicular vector, we form the matrix Δ whose rows consists of the sensor values in a region of the image. We then solve the homogeneous linear equation

$$\Delta \pi = 0 \tag{28}$$

where 0 is the zero vector.

In any particular image, the best perpendicular vector π may not be very perpendicular to the data set. That is, the sensor values may not be well confined to a proper subspace. This represents the first potential failure of the recovery method.

The principal difficulty in this case is that the dimensionality of the surfaces and lights that generate the image are actually equal to or greater than the number of sensors in the camera. If this is the case, nothing can be done and accurate recovery is impossible using this particular sensing apparatus. The recovery procedure then has the virtue that it warns the user that accurate recovery cannot take place since the data are not well confined to a subspace. I discuss the selection of good basis functions in the next section.

Notice that in order for the subspace to be properly defined, there must be enough independent surfaces to define the subspace. In general, the required number is $D(E) - 1$ (Maloney [11]). In practice, we require the spatial partitions of the image to be large enough so that each of the overlapping areas in Fig. 1 has enough surface variation to permit a solution.

Finally, notice that if the relative spectral power distribution is constant, but the absolute level of the ambient light varies, the data will continue to fall within a subspace defined by the relative spectral power distribution. Variation in the power of the ambient light may occur due to the geometry of point source illuminations, shadowing, or other factors. This type of spatial variation does not pose a problem for the recovery procedure, but it emphasizes the fact that absolute levels of the surface spectral reflectance and ambient light distributions, in particular for widely separated spatial locations, are not uniquely recovered.

As an alternative to solving the homogeneous linear equation (28) in order to find the perpendicular vector, one may perform a singular value decomposition of the data matrix Δ. The direction corresponding to the smallest singular value would then be selected as the best normal direction to the data.

Recovering the Light Vector: We now wish to recover the light vector ϵ from the vector perpendicular to the sensor data π. We can accomplish this by expanding (28) in the following way.

First, notice that the requirement that π be perpendicular to each observed sensor vector may be put as the set of equations

$$\pi' \rho^x = \pi'(\Lambda_E \sigma^x) = 0 \tag{29}$$

where π' is the transpose of π, and (29) holds for every surface vector σ^x within the model. We may take advantage of the symmetric role played by the light and the surface at each point in the image to rewrite the matrix product $\Lambda_E \sigma^x$ as

$$\Lambda_E \sigma^x = \Lambda_{S^x} \epsilon \tag{30}$$

where the ikth entry of Λ_{S^x} is

$$\sum_{n=0}^{N} E_i(\lambda_n)\, S^x(\lambda_n)\, R_k(\lambda_n). \tag{31}$$

It follows that for any surface within the linear model,

$$\pi' \Lambda_{S^x} \epsilon = 0. \tag{32}$$

For each independent surface in the model, we have a linear equation in ϵ. We may form a matrix Π whose rows are all of the form $\pi' \Lambda_{S^x}$ where the choice of surface function $S^x(\lambda_n)$ varies from row to row. (Note: The surfaces $S^x(\lambda_n)$ need not be in the actual image since the equation must hold for any surface whatsoever.)

Since it is impossible to measure the absolute intensity of the ambient lighting—there is always a free scalar in the estimation of ϵ—it is convenient to choose ϵ so that its first entry is one. By building Π from enough independent surface vectors $S^x(\lambda_n)$, we can solve for epsilon using

$$\Pi \epsilon = 0. \tag{33}$$

Since the first entry of ϵ is normalized to one, we may convert this homogeneous linear equation into a nonhomogeneous equation by moving the first column of Π to the right-hand side. The equation then becomes a nonhomogeneous equation with one less dimension, and we may solve for the remaining entries of ϵ.

Solving for Surfaces Once the Light Is Known: Once the lighting vector is known, the remainder of the problem is straightforward. The dependency of the lighting matrix Λ_E on the lighting vector is made explicit in (17). This matrix may be explicitly computed and its pseudoinverse computed using conventional techniques. The surface vectors can then be computed using

$$\Lambda_E^{-1}\rho^x = \sigma^x. \tag{34}$$

Two difficulties may arise at this stage in the analysis. First, the estimated Λ_E may not be one to one. In this case, there will be no unique solution for the surfaces unless the dimensionality of the surface representation is reduced.

Second, the recovered surface weights σ^x, while consistent with the model, may in fact give rise to estimates of the surface reflectance functions that are not physically realizable, that is, that contain negative values of surface reflectivity. When this happens broadly throughout an image, it is an indication that the linear model for the lights and surfaces is not providing a very good fit to the actual lights and surfaces. This too requires reevaluating the choice and number of basis functions used in the estimation procedure.

Noise Analysis

Two categories of error can enter into the calculation. The first source of error is measurement noise introduced at the level of the sensor responses. The second source of error is failures of the linear models of lights and surfaces to properly characterize the surfaces and lights. The problem of sensor noise is relatively straightforward. If we model sensor noise by a vector η^x added to true sensor vector ρ_0^x

$$\rho^x = \rho_0^x + \eta^x, \tag{35}$$

then the error analysis will depend upon the distribution of the noise vector η^x. Generally, for this source of noise, the typical assumptions of independence of the noise vectors across the image are adequate. The analysis presented in the previous sections is the maximum likelihood calculation under the assumption that the entries of the noise vector are Gaussian and independent.

Model Errors: If the dimensionality of the linear model representations of lights and surfaces $D(E)$ or $D(S)$ are allowed to become arbitrarily large, then model error can be made arbitrarily small. As the dimensionality of the linear representations is restricted to provide increasing efficiency, then an additional error will be introduced by the model approximations to the actual surface and light functions.

Suppose that the dimensionalities of the surface and light functions required to produce an essentially error-free representation of the data are M and N. To reduce the size of the computation, we will seek linear models in

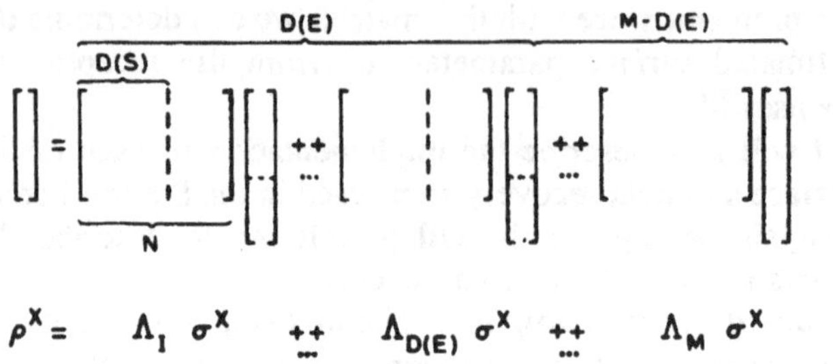

Fig. 3. Figure showing the full model matrix equations with the error terms inside of the reduced model.

which $D(E) < M$ and $D(S) < N$. The sensor data will be correctly modeled by an equation of the form

$$\rho^x = \sum_{n=1}^{M} \epsilon_n \Lambda_n \sigma^x \tag{36}$$

in which σ^x is an N-dimensional vector. The analysis, however, will be based upon the model equation

$$\rho^x = \sum_{n=1}^{D(E)} \epsilon_n \Lambda_n \sigma^x \tag{37}$$

in which σ^x is a $D(S)$-dimensional vector. The difference between the measured sensor data (36) and the best model estimates (37) primarily depends on the ambient light and the surfaces in the image. The error terms that express the difference between these two equations may be described by an additional vector with contributions from various sources, as illustrated in Fig. 3. The distribution of this error vector, then, depends upon the surfaces in the image as well as the ambient light falling on the surfaces. Since we cannot specify the sampling distribution of the actual surfaces without some more knowledge of the specific task of the visual sensor (a robotic inspection system will sample a different class of surfaces from a color correction circuit in a home video recorder), the distribution of the error signal cannot be specified without further information concerning the class of materials and ambient lights that the system will encounter.

The Efficient Representation of Surfaces and Lights

Introduction

The choice of the linear models for the ambient light and surface spectral reflectance terms is crucial in building an efficient computation in the synthesis of images, and it is crucial in permitting one to use a camera with only a few color sensors in the analysis of image data. In this section, I will review the selection of basis functions that permit a more efficient representation and computation of the lighting and surface parameters.

Fourier Basis

The principal means available to us for reducing the computational burden is to reduce the dimensionality of the surface spectral reflectance N. Were the surface spectral reflectance functions and the ambient light spectral power distributions essentially bandlimited functions, then from the sampling theorem, we know that specifying these functions at, say, four sampling points along the wavelength dimension would uniquely determine the identity of the function.

Stiles and Wyszecki [17] and Stiles *et al.* [16] performed an early and important analysis of surfaces reflectance functions in which they suggested that these functions are low-pass functions of wavelength in the visible range. Maloney [11] has put forth the intriguing hypothesis—based on the physical chemistry of surface reflectances—that these functions are generally forced to be low pass in the visible range.

There is wide informal agreement that natural surfaces and lights are generally smoothly varying. There are certain exceptions, such as fluorescent light and the surface reflectance functions of the rare earth metals, but the fact that the overwhelming majority of functions are smoothly varying suggests that in designing a system with only a few color sensors, it is far better to select a representation of the surfaces and lights that is slowly varying across wavelength. Unless there is some specific knowledge of the working environment to override this choice, it seems generally preferable to represent the spectra as the weighted sum of several low-frequency terms across the visual spectrum.

Specular Reflectance: When using a low-frequency representation, the functions $S_j(\lambda_n)$ are illustrated in panel (b) of Fig. 4. The function S_1 is the dc component of the spectral reflectance function, and the functions S_2 and S_3 represent the sinusoidal and cosinusoidal components at one cycle across the visible region. Notice that when this representation is chosen, the weight of the first dimension S_1 is a mixture of the specular reflectance at that point in the image, added together with the dc component of the matte portion of the surface reflectance function. Using such a representation, then, it is relatively straightforward to systematically vary the specularity. In practice, this can be achieved by simply manipulating the value of the first dimension of the surface reflectance function vector.

In Fig. 5, I have analyzed how well the Munsell chips, a widely used set of color surfaces, can be described as bandlimited functions. The figure is derived as follows. First, I used the spectral reflectance functions of the Munsell chips measured by Nickerson (1943)[4] to calculate the least squares fit of Fourier basis terms to the surface spectral reflectance functions of each of the Munsell chips, measured from 400 to 700 nm. I then normalized the root mean-squared error (rmse) of the best fit by the vector length of the surface reflectance function. This is the square root of the proportion of variance accounted for. On the vertical axis, I have plotted the normalized rmse of the median of the 462 Munsell chips available in our data as a function of the number of Fourier basis terms. In Fig. 4(a), I have plotted the worst of the 462 fits to the sample data, along with the three basis functions.

[4] I thank the Munsell Corporation for making these measurements available.

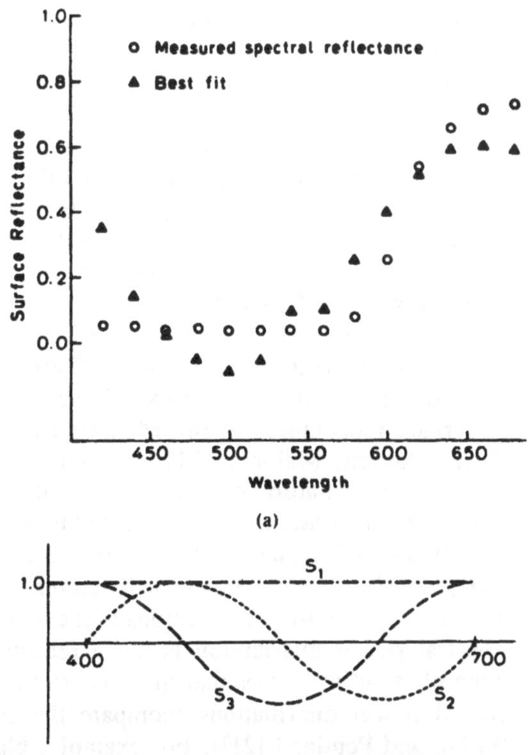

Fig. 4. (a) The surface reflectance function (open symbols) and best fitting estimate (filled symbols) of the Munsell chip with the largest residual error (worst fit) of all 462 chips. (b) The three Fourier basis functions used to fit the surface reflectance functions of the Munsell chip samples.

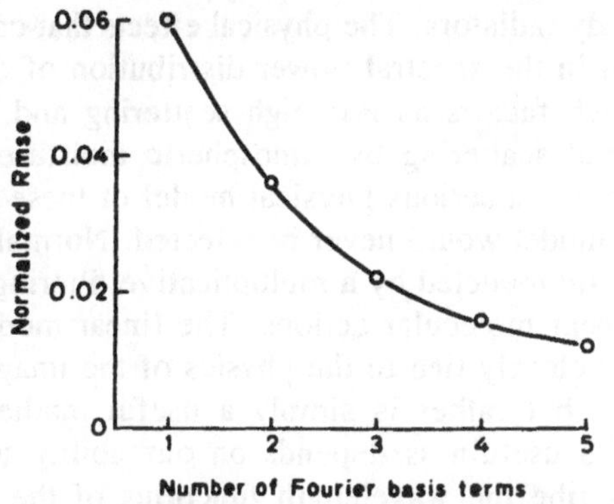

Fig. 5. Each of 462 Munsell chip surface reflectance functions was fit by the sum of sinusoidal and cosinusoidal functions. This figure plots the normalized rmse of the median of 462 Munsell chip fits as a function of the number of Fourier basis terms used in the fit. The percent of variance accounted for may be computed as $\sqrt{(1 - y)}$ where y is the normalized rmse.

Clearly, the fit of the Munsell spectra is rather good. This should not be taken as evidence that *all* surface reflectance functions fall well within a three-dimensional model. Rather, it should be seen as an interesting case of a complex set of objects that can be well described by a linear model with only a few degrees of freedom. The point of the analysis is to demonstrate that the Fourier representation is efficient for representing spectral reflectance functions and ambient light because of the high degree of correlation across the wavelength spectrum. It has the further advantage of identifying specular reflectance with the dc component of the surface reflectance function.

An alternative method for selecting the representation, which is particularly appropriate when the set of surfaces and lights are known in advance, is to perform a principal component analysis of the distribution of surface reflectance functions and spectral power distributions. The principal component analysis—sometimes called the Karhunen–Loeve transformation—explicitly seeks that set of basis functions that simultaneously minimizes the correlation among dimensions and identifies those dimensions that are most descriptive of the data set.

General analyses of the degrees of freedom in spectral data have been applied to the spectral power distribution of ambient daylights by Judd *et al.* [10] and Dixon (1978) and to the surface reflectances of the Munsell chips by Cohen [4]. In addition, Maloney [11] has analyzed the data from a large set of natural objects. All authors have concluded that their data sets are well described using linear models with a small number of degrees of freedom.[5]

Relationship to Physical Processes: It is interesting to note that the linear model for the variations in the ambient daylight spectral power distribution is not a reasonable model for the physical processes that give rise to the observed spectral power distributions (compare the argument in Witkin and Pentland [21]). For example, blackbody radiators form a one-parameter nonlinear family of spectral power distribution curves. Yet Maloney has shown (personal communication) that the set of principal components of the natural daylights also provides a rather good approximation to the spectral power distribution seen in blackbody radiators. The physical effects that cause the fluctuation in the spectral power distribution of daylight include such factors as Rayleigh scattering and the absorption and scattering by atmospheric moisture. Were one to provide a serious physical model of these effects, the linear model would never be selected. Normally such processes are modeled by a multiplicative filtering action and nonlinear molecular actions. The linear model used here is not closely tied to the physics of the imaging environment, but rather is simply a useful mathematical method. Its usefulness depends on our ability to accurately describe the wavelength functions of the desired quantities. The fact that the underlying physical processes are not well modeled by the linear sum of different physical processes is not relevant to the performance of the algorithm.

The Color Camera Revisited: In Fig. 6, I indicate how the analysis and synthesis calculations developed in this paper can be put together in order to obtain a rendering of an image sensed by a color camera. As Horn [9] has pointed out, if the camera sensor spectral sensitivities are within a linear transformation of the spectral sensitivities of the human eye, the problem may be easily solved. The

[5]Judd *et al.* [10] estimated the principal components of daylight at an early stage in the technical development of these methods. The reported functions are not precisely orthogonal. Unfortunately, the original data have been misplaced so that we cannot reanalyze the original data. Maloney (personal communication) is preparing a new analysis of a large set of such data.

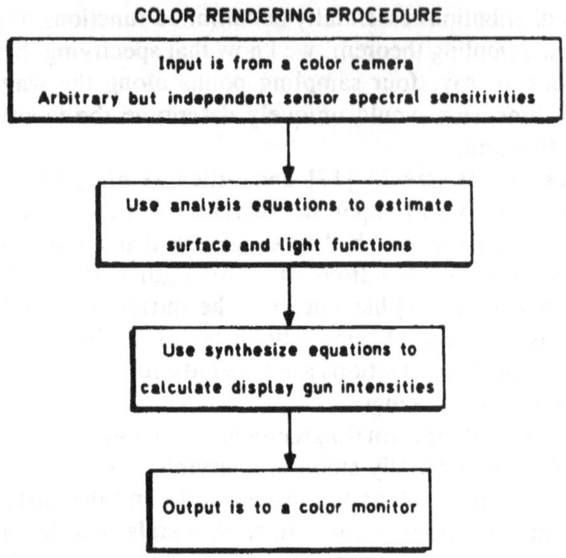

Fig. 6. Flow diagram of how the analyses and synthesis procedures may be used together to convert the outputs of a camera with arbitrary sensor spectral to display an image so that it will appear to an observer as if he were at the camera position.

more difficult—and common—case is when the camera sensors are not related to the visual pigment sensitivities by a linear transformation.

In that case, we may proceed as indicated in Fig. 6. We use the camera sensor responses and the procedure in the analysis section to form an estimate of the ambient light spectral power distribution E. We then calculate estimates of the surface reflectances:

$$\sigma^x = \Lambda_E^{-1} \rho^x. \tag{6}$$

Using these estimates of the surface and ambient light distributions, we set the display intensities using the synthesis calculations in (21), which I repeat here:

$$I^x = \left[\sum_{n=1}^{D(E)} \epsilon_n \Gamma^{-1} \Lambda_n \right] \sigma^x. \tag{7}$$

To the extent that the camera data permit an accurate recovery of the surface and light functions, the synthesized colors will have the same visual effect as if the observer had been at the position of the camera.

Conclusions

The methods described here for the synthesis and analysis of color images are based on the physical factors—lighting, reflectance, and absorption—that give rise to color information. The insistence on referring all calculations of sensor responses back to the underlying physical processes permits us to treat several different problems in the theory of color images from a common framework. In particular, the equations for selecting display gun intensities, and the equations for analyzing color images, are both based on a common formula, (27).

Maloney and Wandell's [12] method for analyzing the color signal into its surface and light components has applications to several different problems in color science.

The analysis of the image data into surface and light components permits us to correct the acquired sensor signal for unwanted effects of ambient lighting. The ability to perform this correction will improve the art of color balancing images. The ability to estimate the surface reflectance of objects in the image from cameras that are not strictly matched to the human eye should also improve the color rendering of images. Finally, the ability to use *object* surface reflectance—rather than gray-level *image* intensity in which confound surface and light information are confounded—should become an important aid in the development of visual inspection systems that are robust with respect to the ambient lighting.

ACKNOWLEDGMENT

The author would like to thank D. Brainard, L. T. Maloney, C. McConnell, and D. Varner for useful discussions and their comments on this paper.

REFERENCES

[1] R. M. Boynton, *Human Color Vision*. New York: Holt, Rinehart and Winston, 1979.
[2] G. Buchsbaum, "A spatial processor model for object color perception," *J. Franklin Inst.*, vol. 310, 1980.
[3] G. Buchsbaum and A. Gottschalk, "Chromaticity coordinates of frequency-limited functions," *J. Opt. Soc. Amer.*, vol. 1, pp. 885-887, 1984.
[4] J. Cohen, "Dependency of the spectral reflectance curves of the Munsell color chips," *Psychon. Sci.*, vol. 1, pp. 369-370, 1964.
[5] R. L. Cook and K. E. Torrance, "A reflectance model for computer graphics," *Comput. Graphics*, vol. 15, pp. 307-316, Aug. 1981.
[6] T. N. Cornsweet, *Visual Perception*. New York: Academic, 1970.
[7] H. Helson, "Fundamental problems in color vision. I. The principle governing changes in hue, saturation and lightness of non-selective samples in chromatic illumination," *J. Exp. Psychol.*, vol. 23, pp. 439-476, 1938.
[8] B. K. P. Horn, "Determining lightness from an image," *Comput. Graphics Image Processing*, vol. 3, pp. 277-299, 1974.
[9] —, "Exact reproduction of colored images," *Comput. Vision, Graphics, Image Processing*, vol. 26, pp. 135-167, 1984.
[10] D. B. Judd, D. L. MacAdam, and G. W. Wyszecki, "Spectral distribution of typical daylight as a function of correlated color temperature," *J. Opt. Soc. Amer.*, vol. 54, p. 1031, 1964.
[11] L. Maloney, "Computational approaches to color vision," Ph.D. dissertation, Stanford Univ., Stanford, CA, 1984.
[12] L. T. Maloney and B. Wandell, "A computational model of color constancy," *J. Opt. Soc. Amer.*, vol. 1, pp. 29-33, 1986.
[13] A. V. Oppenheim and R. W. Schafer, Ed., *Digital Signal Processing*. Englewood Cliffs, NJ: Prentice-Hall, 1975.
[14] P. Sallstrom, "Colour and physics; Some remarks concerning the physical aspects of human colour vision," Inst. Phys. Rep. 73-09, 1973.
[15] V. Smith and J. Pokorny, "Spectral sensitivity of the foveal cone photopigments between 400 and 500 nm," *Vision Res.*, vol. 15, pp. 161-171, 1975.
[16] W. S. Stiles, G. Wyszecki, and N. Ohta, "Counting metameric object-color stimuli using frequency-limited spectral reflectance functions," *J. Opt. Soc. Amer.*, vol. 67, no. 6, pp. 779-784, 1977.
[17] W. S. Stiles and G. W. Wyszecki, "Counting metameric object colors," *J. Opt. Soc. Amer.*, vol. 52, pp. 313-327, 1962.
[18] K. E. Torrance and E. M. Sparrow, "Theory for off-specular reflection from roughened surfaces," *J. Opt. Soc. Amer.*, vol. 57, pp. 1105-1114, 1967.
[19] K. E. Torrance, E. M. Sparrow, and R. C. Birkebak, "Polarization, direction distribution, and off-specular peak phenomena in light reflected from roughened surfaces," *J. Opt. Soc. Amer.*, vol. 56, pp. 916-925, 1966.
[20] B. Wandell and L. Maloney, "Computational methods for color identification," in *Proc. OSA Annu. Meet.*, 1984, p. 21.
[21] A. Witkin and A. Pentland, "Perception and generic modeling," *J. Opt. Soc. Amer.*, vol. 12, p. 1261A, 1984.

Brian A. Wandell was born in New York, NY, on October 6, 1951. He received the B.S. degree in mathematics and psychology from the University of Michigan, Ann Arbor, in 1973, and the Ph.D. degree from the University of California, Irvine, in 1977.

In 1975 he spent a year as a Visiting Scholar at Columbia University, New York, NY, in 1976 a year as a Visiting Scholar at Harvard University, Cambridge, MA, and in 1977 he spent a year and a half as a Post-Doctoral Fellow at the University of Pennsylvania, Philadelphia. In January 1979 he was appointed to the Department of Psychology, Stanford University, Stanford, CA, where he is currently an Associate Professor. Apart from his work on computational vision, he is generally engaged in measuring and analyzing human visual performance.

Dr. Wandell is currently a member of the Editorial Board of *Vision Research* and a member of the Association for Research in Vision and Ophthalmology, the Optical Society of America, and Director of the NASA-Ames-Stanford Center of Excellence in model-based human performance.

Color Reflection Models

The properties of the image of an object critically depend on how light interacts with the object's material. This interaction is dominated by two processes. First, light is reflected when it encounters the surface of an object. Second, light that is not reflected at the interface is scattered and absorbed after it enters the body of the object material. Each of these processes affects the spectral properties of the reflected light.

Accurate color reflection models can significantly improve the capability of a machine vision system. Unfortunately, the interaction of light with real materials is a complicated process, and general reflection models characterize the light reflected by an object as a complicated function of the photometric geometry, the illumination environment, and the optical properties of the object's material. Given the complexity of inverting these models to infer an object's properties from image measurements, it is useful in machine vision to work from reflectance models that are both concise and accurate.

In this section, authors present and analyze a variety of color reflection models. The accuracy of these models is evaluated by experiment and comparison with empirical data sets. These models form the basis of several useful vision algorithms, including highlight separation, segmentation, material classification, and color constancy. The performance of these algorithms demonstrates the effectiveness of using color reflection models in machine vision.

The dichromatic reflection model introduced by Shafer [1] was the first realistic color reflection model used in machine vision and has inspired a large amount of important related work. The dichromatic reflection model applies to inhomogeneous dielectric materials and states that the radiance L of the light reflected by an object is the sum of the radiance L_i reflected at the interface and the radiance L_b of the light scattered from the body of the material. The model further states that each of the components L_i and L_b may be factored into the product of two terms: a spectral composition c_i or c_b that depends on wavelength λ but not on geometry, and a geometric scaling factor m_i or m_b that depends on the photometric angles θ but not on wavelength. The dichromatic reflection model may be summarized by

$$L(\lambda,\theta) = L_i(\lambda,\theta) + L_b(\lambda,\theta)$$

$$= m_i(\theta)\, c_i(\lambda) + m_b(\theta)\, c_b(\lambda).$$

The factors $c_i(\lambda)$ and $c_b(\lambda)$ each depend on the spectral composition of the illumination and the reflecting properties of the object. If the illumination has a spectral composition $E(\lambda)$, then

$$c_i(\lambda) = E(\lambda)\, \rho_i(\lambda) \quad \text{and}$$
$$c_b(\lambda) = E(\lambda)\, \rho_b(\lambda),$$

where $\rho_i(\lambda)$ and $\rho_b(\lambda)$ describe the dependence on wavelength of the interface and body reflectance of the material.

This model predicts that the color pixels corresponding to a single material will lie on a parallelogram in color space. Shafer shows that this constraint can be used to compute intrinsic images that represent separately the amount of interface reflection (highlight) and body reflection at each pixel.

In [2], Healey presents a comprehensive survey of the physics of reflection. Using detailed physical models, he shows that the dichromatic reflection model is an accurate approximation for many inhomogeneous dielectrics. He also shows that a simple model of the form $L(\lambda,\theta) = m(\theta)\, c(\lambda)$ is an accurate color reflection model for metals under a single spectral composition of illumination over most values of the parameters θ. From these models, Healey develops a segmentation algorithm that classifies surfaces according to material independent of surface orientation or the direction of illumination. He also develops an algorithm to distinguish metal and dielectric surfaces. The algorithms are illustrated by experiments.

Tominaga and Wandell experimentally test the dichromatic reflection model under the additional assumption that $\rho_i(\lambda)$ is a constant [3]. Such an assumption predicts that highlights will have the same spectral composition as the illuminant. As with the dichromatic model, this model implies that spectral measurements for a surface will lie on a plane in color space. The additional assumption implies that planes corresponding to different surfaces under a common illuminant will intersect along a line that describes the spectral distribution of the light source. Using principal component analysis and spectroradiometer measurements, the authors show that this reflection model is accurate for a set of plastics and fruits. At the same time, they demonstrate the effectiveness of a procedure that finds the intersection of color space planes to estimate the illuminant spectral distribution.

The constant $\rho_i(\lambda)$ assumption is further investigated by H.C. Lee *et al.* [4]. The measurements presented in this paper show that $\rho_i(\lambda)$ is nearly constant for certain kinds of plastics, leaves, painted surfaces, and fruits. Experiments also show, however, that $\rho_i(\lambda)$ is not approximately constant for other inhomogeneous materials such as colored paper and some ceramics. The authors suggest why such a model might still be useful for these materials when using sensors with broad spectral sensitivity functions such as those used in color cameras.

In [5], Maloney examines the complexity of spectral reflectance functions using large empirical data sets. The analysis shows that the reflectance functions in the data sets can be accurately modeled using a linear model with 5 to 7 parameters. An analysis of the physical structure of absorption bands for materials is used to support this result. The ability to represent reflectance functions using linear models with a small number of parameters has important consequences for several of the algorithms presented in the "Color Constancy" section of this volume.

– G.E.H.

Papers Included in This Collection

[1] Shafer, S.A. "Using Color to Separate Reflection Components." *COLOR Research and Application* **10**(4): 210-218, Winter 1985.

[2] Healey, G.E. "Using Color for Geometry-Insensitive Segmentation." *Journal of the Optical Society of America A* **6**(6): 920-937, June 1989.

[3] Tominaga, S., and Wandell, B.A. "Standard Surface-Reflectance Model and Illuminant Estimation." *Journal of the Optical Society of America A* **6**(4): 576-584, April 1989.

[4] Lee, H.-C., Breneman, E.J., and Schulte, C.P. "Modeling Light Reflection for Color Computer Vision." *IEEE Trans. on Pattern Analysis and Machine Intelligence* **PAMI-12**(4): 402-409, April 1990.

[5] Maloney, L.T. "Evaluation of Linear Models of Surface Spectral Reflectance with Small Numbers of Parameters." *Journal of the Optical Society of America A* **3**(10): 1673-1683, October 1986.

Steven A. Shafer

Computer Science Department
Carnegie–Mellon University
Pittsburgh, Pennsylvania 15213

Using Color to Separate Reflection Components

In computer vision, the goal of which is to identify objects and their positions by examining images, one of the key steps is computing the surface normal of the visible surface at each point ("pixel") in the image. Many sources of information are studied, such as outlines of surfaces, intensity gradients, object motion, and color. This article presents a method for analyzing a standard color image to determine the amount of interface ("specular") and body ("diffuse") reflection at each pixel. The interface reflection represents the highlights from the original image, and the body reflection represents the original image with highlights removed. Such intrinsic images are of interest because the geometric properties of each type of reflection are simpler than the geometric properties of intensity in a black-and-white image. The method is based upon a physical model of reflection which states that two distinct types of reflection—interface and body reflection—occur, and that each type can be decomposed into a relative spectral distribution and a geometric scale factor. This model is far more general than typical models used in computer vision and computer graphics, and includes most such models as special cases. In addition, the model does not assume a point light source or uniform illumination distribution over the scene. The properties of tristimulus integration are used to derive a new model of pixel-value color distribution, and this model is exploited in an algorithm to derive the desired quantities. Suggestions are provided for extending the model to deal with diffuse illumination and for analyzing the two components of reflection.

Introduction

Computer Vision

Computer vision involves the use of computers to analyze images, determining the objects portrayed and their positions in three dimensions. The approaches include both mathematical descriptions of imaging and image analysis, and actually producing working computer programs. Computer vision is aimed at currently usable factory robots, eventual "general-purpose" robots, and understanding human vision by analyzing image information. Computer vision in general is discussed in refs. 1 and 2.

Because computer vision is so complex, it is studied in parts. Such parts include determination of surface orientation at each point ("pixel") in an image, grouping pixels that form smooth connected surfaces, geometrically identifying the solid shape bounded by each visible surface, and recognizing each solid shape as a specific object such as a chair or telephone.

This article is concerned with the first of these parts: determining the surface orientation of the visible surface at each pixel in an image. Much research in computer vision is devoted to this topic, usually producing theories that apply only to idealized "Lambertian" surfaces that are perfectly matte and reflect light equally in all directions. We propose a method for analyzing real surfaces, that exhibit glossiness; to deal with the additional complexity introduced by glossiness, we make use of color information. This article presents both a general discussion of the relationship between gloss and color and a specific method for analyzing a color image to separate the gloss contribution from the matte contribution at each pixel.

The Role of Color and Highlights

When we look around us, the surfaces we see are typically glossy. They may seem to be very shiny, fairly matte, or anywhere in between, but virtually all the surfaces around us exhibit highlights to varying degrees. These highlights are most pronounced when the surface normal bisects the angle between the direction of illumination and the direction of view, making the position and intensity of highlights very sensitive to viewing geometry. This causes problems with many low-level computer-vision methods such as segmentation (which typically assumes uniform or smoothly varying intensity across a surface) or stereo and motion analysis (which attempt to match images taken from different viewpoints).

Highlights are not the only source of intensity variation across a surface. Even with uniform illumination of a matte, uniformly colored surface, there will be smooth shading due to the angle of incidence of the incoming illumination relative to the surface normal.

It would be very useful to be able to separate the effects of shading from those of highlights. This might result in *intrinsic images* telling, at each pixel, the intensity of the shading and the intensity of the highlight at this point. This separation was first suggested by Barrow and Tenenbaum,[3] and would effectively produce an image of the highlights and an image with the highlights removed. These would be useful for analysis since each of these phenomena is more simply related to the angles of illumination and viewing than is their sum (which is measured in a black-and-white image). In addition, the relative insensitivity of shading to viewpoint would make the shading intrinsic image an ideal candidate for stereo or motion analysis.

It is frequently observed that highlights have a different color from the characteristic color of a surface (which is related to shading). In this article, we show how a simple, rather general, model of reflection, called the "Dichromatic Reflection Model," can be used to determine intrinsic images of the two types of reflection from a standard color image (i.e., red–green–blue separation).

The analysis uses the properties of *tristimulus integration*, the process whereby color pixel values are determined from the spectral power distribution (SPD) of incoming light. Combining the Dichromatic Reflection Model with tristimulus integration results in a new model of pixel value distribution in R–G–B color space. The model predicts that pixel values from pixels on a single surface will lie on a parallelogram in color space, and that the position of any pixel's color within that parallelogram yields the coefficients of the two types of reflection.

A simple algorithm is presented that utilizes this model of pixel values to determine the desired intrinsic images. The model and the algorithm are then extended to deal with diffuse illumination and shadows. Additional work for the future includes implementation of the algorithm and verification of the model with real images.

For additional background information, the reader interested in radiometry in general is referred to refs. 4–6; while *appearance measurement* (gloss and color) is discussed in refs. 7–10.

Previous Work in Color-Image Understanding

Color-image understanding has not in the past been based on general models of reflection. Most of the work has been the application to three-dimensional color space of algorithms originally developed and used for analyzing monochrome images. This includes primarily edge detection[11] and clustering.[12–15] In such work, image regions or edges are identified by distances between pixel values in color space, without appeal to any model of color generation in the scene.

Color has also been used for object labelling based on known object colors or object colors measured in the image.[16–21] This approach, known as "spectral signature analysis" in remote sensing, uses the known reflection properties of materials of interest in a particular domain. It depends on having few different types of materials in the scene, and having prior knowledge of their spectral reflectance.

The properties of color-space transformations have also been studied,[22,23] although no such transformations have sufficed to "solve" the image-segmentation or labelling problems.

The only previous work in utilizing general properties of reflected color has been in the form of simple statements such as "far-away outdoor objects look bluish," "outdoor shades are bluish," and "natural colors tend to be desaturated."[16,18] In the psychological field, Rubin and Richards proposed a method for using color to determine changes in material across an image[24]; however, while the method is quite interesting, their assumptions appear to be restrictive.

The approach described in this article is significant because it is applicable to many common types of materials, without prior knowledge about their colors, and because it is based on a very general model of reflection in the real world.

Included is a discussion of previous work in reflectance models for computer graphics and image understanding.

Reflection and the Dichromatic Model

Here we present a brief account of the physics of reflection, followed by the Dichromatic Reflection Model that captures certain aspects of the reflection process. The validity and assumptions of the model are discussed, and it is compared with other reflection models used in computer graphics and image understanding.

Physical Properties of Reflection

The methods of this article deal with materials which are optically *inhomogeneous*, meaning that light interacts both with a *medium* that comprises the bulk of the surface matter and with particles of a *colorant* that produce scattering and coloration. Many common materials can be described this way, including most paints, varnishes, paper, ceramics (including porcelain), plastics, etc. Materials that are homogeneous are not included in this discussion; thus metals and many crystals are not amenable to the analysis presented herein. We also limit this discussion to opaque surfaces, which transmit no light from one side to the other.

Although we sometimes think of a visual surface as a plane, this is an approximation useful only at the macroscopic level. To understand reflection from inhomogeneous materials, we instead view the surface as having a definite thickness (Fig. 1).[8-10]

When light strikes a surface, it must first pass through the interface between the air and the surface medium. Because the medium's index of refraction differs from that of the air, some of the light is reflected at the interface producing *interface reflection* as shown in Fig. 1. The direction of such reflection is in the "perfect specular direction" relative to the local surface normal, i.e., reflected about the surface normal. Note that most materials are optically "rough," with local surface normals that differ from the macroscopic or reference surface normal. The local perfect specular direction is therefore somewhat different from the macroscopic perfect specular direction, so that the interface reflection is somewhat scattered at the macroscopic level.

The amount of light reflected at the interface is governed by Fresnel's laws, which relate interface reflectance to the angle of incidence (relative to the local surface normal), the index of refraction of the material, and the polarization of the incoming illumination.[25] Interface reflectance is thus a function of wavelength of light, since index of refraction generally depends on wavelength, but the amount is typically constant to within a few percent across the visible spectrum; acrylic plastic, for example, has an index of refraction of 1.485 at 400 nm (the blue end of the visible spectrum) and 1.505 at 700 nm (the red end),[9,10,26] producing Fresnel reflection coefficients of 3.8 and 4.1% at the ends of the spectrum. Interface reflection is frequently assumed to be constant with respect to wavelength because of the small magnitude of the difference, and is thus said to have the same color (relative SPD) as the illuminant.[8,9,27] The effect of polarization is more severe, and interface reflection tends to be highly polarized, especially for large angles of incidence ("grazing angles").[10,25]

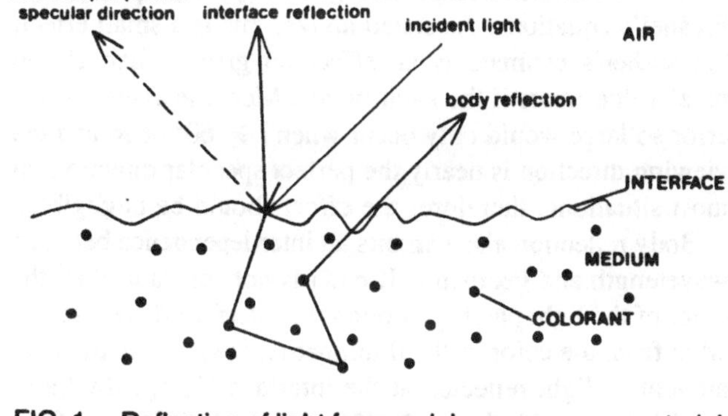

FIG. 1. Reflection of light from an inhomogeneous material.

The light that penetrates through the interface passes through the medium, where it undergoes scattering from the colorant, and eventually is either transmitted through the material (if it is not opaque), absorbed by the colorant, or re-emitted through the same interface by which it entered, producing *body reflection* as shown in Fig. 1. The geometric distribution of this body reflection is sometimes assumed to be isotropic, i.e., independent of viewing direction, although some work is under way to produce improved models of scattering.[7,27] The color of the body reflection is generally different from that of the illumination, since interactions with colorant particles result in absorption with a probability dependent on wavelength. Body reflection is usually considered to be unpolarized.

Real reflection tends to be more complex than described above. However, this explanation is a very useful approximation.

Terminology

This article adopts the terms "interface" and "body" reflection rather than the more common terms "specular" and "diffuse" reflection. While the latter have gained some popularity in the literature of appearance measurement and computer imaging (graphics and vision), they have some severe problems which render them controversial. "Specular reflection" may mean any of three things: interface reflection (the most common usage), interface reflection *in the perfect specular direction at macroscopic level* (the usage most common in gloss measurement), or interface reflection *where the surface is locally optically smooth* (the usage in scattering theory and laser optics[28]). Similarly, the term "diffuse reflection" means "reflected light scattered over a large solid angle," and thus may apply equally well to body and to interface reflection. Having experienced numerous communication problems due to the use of the terms specular and diffuse, I have ultimately adopted the terminology used in this article which is sometimes found in the more technical optics literature.

The terminology used throughout this article for describing reflectance geometry is illustrated in Fig. 2, which defines several angles[29]: the *angle of incidence*, i, between the illumination direction $\boldsymbol{I}$ and the surface normal $\boldsymbol{N}$, the *angle of exitance*, e, between $\boldsymbol{N}$ and the viewing direction $\boldsymbol{V}$, the *phase angle*, g, between $\boldsymbol{I}$ and $\boldsymbol{V}$, and the *off-specular angle*, s, between $\boldsymbol{V}$ and the direction of (macroscopic) perfect specular reflection $\boldsymbol{J}$. (While the name *phase angle* and the symbols i, e, and g are not common in color science, they are quite common in computer vision. They were originally used in astronomical studies of lunar topography using photometry; the *phase angle* g corresponds to the phase of the moon in that context, and the name has been adopted in computer vision as well.)

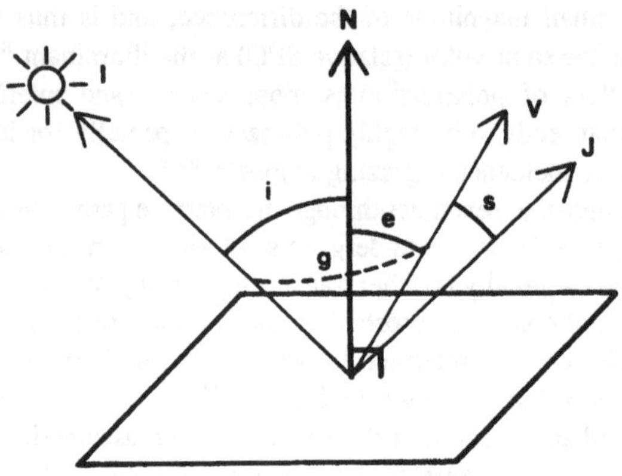

FIG. 2. Photometric angles.

We also use the standard symbol λ to refer to the wavelength of light.

The Dichromatic Reflection Model

We now propose a simple mathematical model of reflectance, based on the above discussion, called the *Dichromatic Reflection Model*.

The Dichromatic Reflection Model states:

$$L(\lambda,i,e,g) = L_i(\lambda,i,e,g) + L_b(\lambda,i,e,g) \quad (1)$$
$$= m_i(i,e,g)\, c_i(\lambda) + m_b(i,e,g)\, c_b(\lambda) \quad (2)$$

This model represents two statements about reflected light, as expressed by the two parts of the equation. Equation (1) says that the total radiance L of the reflected light is the sum of two independent parts: the radiance L_i of the light reflected at the interface and the radiance L_b of the light reflected from the surface body. Equation (2) says that each of these components of the light can be decomposed into two parts: *composition*, a relative spectral power distribution c_i or c_b that depends only on wavelength but is independent of geometry, and *magnitude*, a geometric scale factor m_i or m_b which depends only on geometry and is independent of wavelength. Intuitively, the Dichromatic Reflectance Model says that there are two independent reflection processes, and that each has a characteristic color whose magnitude, but not spectral distribution, varies with the directions of illumination and view.

In the remainder of this section we address the scope of this model and its validity; in later sections we see how this model may be exploited to determine intrinsic images of m_i and m_b, the amount of interface and body reflection at each pixel.

The Dichromatic Reflection Model makes several assumptions. It assumes that the surface is an opaque, inhomogeneous medium with one significant interface, that the surface is not optically active (i.e., has no fluorescence or thin-film properties), and that it is uniformly colored (the colorant is uniformly distributed). Reflection from the surface is assumed to be isotropic with respect to rotation about the surface normal. The model assumes a single light source, i.e., no diffuse ("ambient") illumination, and the relative spectral power distribution $S(\lambda)$ of the illumination must be constant across the scene. There is no account taken of inter-reflection among surfaces.

The assumptions about the surface are typical for reflectance models and not too unrealistic. The assumption of no inter-reflection is also typical, but unfortunately not realistic at all. Finally, the assumption of no ambient light is not at all realistic, but will be relaxed (in fact, eliminated) in a later section.

Equally interesting is a list of assumptions *not* made by the model, that express the scope or generality of the model. The model makes no assumption that orthographic (parallel) or perspective (central) projection is being used. The model applies equally to curved and planar surfaces. It also applies to textured surfaces, i.e., surfaces with macroscopic roughness (but see the note below about analyzing intrinsic images). The model does not assume specific functions m_i, c_i, m_b, or c_b; in particular, there is no specific geometric model of highlights, no assumption that the highlights have the same color as the illumination, and no assumption that the body reflection is isotropic. The model applies equally well to a point light source, an extended light source, or a light source infinitely far away. Finally, the model does not assume that the amount of illumination is the same everywhere in the scene; only that the SPD is the same (see above). This is important, since real (especially extended) light sources produce nonuniform amounts of illumination in different areas of the scene.

It must be pointed out that any complexity of the forms mentioned above will cause great difficulty in *analyzing* the resulting intrinsic images of m_i and m_b. However, the Dichromatic Reflection Model will make it possible to *compute* these intrinsic images regardless of such complexity.

In spite of the apparent simplicity of the Dichromatic Reflectance Model, it suffers from some flaws. In practice, they ought to have only a minimal impact on the usefulness of the model. For example, there is no obvious way to decide how to scale the magnitude functions m_i and m_b against the composition functions c_i and c_b. Therefore, the resulting intrinsic images may be only of relative, rather than absolute, reflection magnitudes. At least, it is reasonable to require that $0 \leqslant m_i,\ m_b \leqslant 1$ for all i, e, and g.

Both types of reflection also have properties that violate the assumptions of the Dichromatic Reflection Model to a small degree. Interface reflection exhibits an interdependence between wavelength and geometry, as expressed by Fresnel's equations. As stated above, this is a small effect; the author's estimate is an effect no greater than 2% in pixel-value errors if the Dichromatic Model is assumed. An error so large would only occur when $i \geqslant 60°$ or so and the viewing direction is nearly the perfect specular direction; in most situations, therefore, the effect should be negligible.

Body reflection also exhibits an interdependence between wavelength and geometry. If $c_i(\lambda)$ is not constant, then the color of the light passing through the interface differs somewhat from the color of the illumination $S(\lambda)$. Since the total amount of light reflected at the interface ($\int_e \int_g L_i[\lambda,i,e,g]\, dg\, de$) varies with the angle of incidence i, the color of the light passing through the interface into the material body also varies with the angle of incidence. Thus, the color of the body reflection should vary with geometry. However, since c_i is generally nearly constant, this should also be a negligible effect in real images.

In summary, the Dichromatic Reflection Model is expressed in a simple equation that should be approximately true over a very wide range of circumstances.

Previously Used Reflection Models

Previous models of reflection used in computer graphics and computer vision have proposed specific reflectance functions to predict the exact amount of reflection.

The most frequently cited model is that of Phong,[30,31] also applied in computer vision by Horn[29]:

$$L(i, e, g) = t\frac{n+1}{2}\cos^n s + (1 - t)\cos i \qquad (3)$$

Here t and n are parameters of the material, with t representing the total amount of light reflected at the interface and n representing the clustering of the interface reflection about the perfect specular direction. The equation shown above is only a monochrome model, but is frequently applied to color images with the color of the interface reflection being white and the color (R–G–B values) of the body reflection being a parameter of the material. Phong's model is easily seen to be an instantiation of the Dichromatic Reflection Model in this article as follows:

$$m_i = t\frac{n+1}{2}\cos^n s \qquad m_b = (1 - t)\cos i$$
$$c_i = 1 \qquad c_b = \text{surface parameter} \qquad (4)$$

Phong's equation, while it produces pleasing-looking images, is not based on any real physical model.

A later model was Blinn's,[32] in which the interface reflection was characterized by a facet model of surface structure (from ref. 33) and Fresnel's equations were used for the color of the interface reflection. Cook and Torrance, in more recent work,[34] used Beckmann's model of reflection from a rough surface[28] for the distribution of the interface reflection, and similarly applied Fresnel's equations for the color of the interface reflection. In both cases, the body reflection was assumed to be isotropic, i.e., $m_b = \cos i$. These models of reflection are also examples of the Dichromatic Reflection Model, with the exception of the use of Fresnel's equations to govern the color of the interface reflection (usually a small effect, as noted above).

In computer vision, Shibata and Frei[35] used the following model to predict reflection in aerial images:

$$L(i,e,g) = t_i \cos^n s + t_2 \cos^m g + (1 - t_1 - t_2)\cos i \qquad (5)$$

Here, the first and last terms are similar to Phong's, and the middle term represents "backscattering" of light in the direction of illumination. The constants t_1, t_2, n, and m are parameters of the material being viewed. Extended to color images, this model would also be an instantiation of the Dichromatic Reflection Model, with the backscattering term added to m_i.

Pixel Values in Color Space

We now examine tristimulus integration, the relationship between an SPD of light and its color coordinates. When this process is applied to the Dichromatic Reflection Model, a new model for pixel values in color space is the result. A simple algorithm is presented for exploiting this model to determine the intrinsic images of m_i and m_b.

Tristimulus Integration

In a monochrome camera, pixel values are computed from the spectral power distribution (SPD) of the measured light very simply. The pixel value p is just a summation of the amount of light at each wavelength $X(\lambda)$, weighted by the responsivity of the camera to the various wavelengths, $s(\lambda)$:

$$p = \int X(\lambda)s(\lambda)d\lambda \qquad (6)$$

The interval of summation is determined by $s(\lambda)$, which is non-zero over a bounded interval of wavelengths λ.

In a color camera, color filters are interposed between the incoming illumination and the camera. Each filter has a transmittance function $\tau(\lambda)$, specifying the fraction of light transmitted at each wavelength; thus, the pixel value computed with a filter is specified by the above integral, with $s(\lambda)$ replaced by $\tau(\lambda)\ s(\lambda)$. Typically, three filters (red, green, and blue) are used with transmittances τ_r, τ_g, and τ_b, resulting in a vector of three color values, $C = [r,g,b]$. If we let $\bar{r}(\lambda)$ be the responsivity of the camera combined with the red filter, $\bar{r}(\lambda) = \tau_r(\lambda)s(\lambda)$, (etc. for $\bar{g}$ and $\bar{b}$), then the color value of any SPD $X(\lambda)$ is given by *tristimulus integration:*

$$C_x = \begin{bmatrix} r_X \\ g_X \\ b_X \end{bmatrix} = \begin{bmatrix} \int X(\lambda)\,\bar{r}(\lambda)\,d\lambda \\ \int X(\lambda)\,\bar{g}(\lambda)d\lambda \\ \int X(\lambda)\,\bar{b}(\lambda)d\lambda \end{bmatrix} \qquad (7)$$

Throughout this article, the "color space" referred to is defined by the responses of a color TV camera rather than the human visual system.

Tristimulus integration is a linear transformation, as shown in ref. 36; in other words, $C_{aX+bY} = aC_X + bC_Y$ where a and b are scalars and $X(\lambda)$ and $Y(\lambda)$ are SPDs. The linearity property of tristimulus integration is important because it says that a mixture of two SPDs of light results in a sum of the corresponding pixel values, taken in the same proportion.

The Dichromatic Model in Color Space

When the linear property of tristimulus integration is combined with the Dichromatic Reflection Model, a powerful new model of pixel color values results.

First, consider a specific point on a surface. At that point, the geometry (angles i, e, and g) is determined, and the magnitudes m_i and m_b may be considered as scalars. So, the Dichromatic Reflection Model may be rewritten at a specific point as:

$$L(\lambda) = m_i c_i(\lambda) + m_b c_b(\lambda) \quad (8)$$

This defines the SPD of the light reflected from the surface. Now, applying the linearity of tristimulus integration, we have:

$$\mathbf{C}_L = m_i \mathbf{C}_i + m_b \mathbf{C}_b \quad (9)$$

where $\mathbf{C}_L$ is the color (pixel value) measured, m_i and m_b are the magnitudes of reflection at the point in question, and $\mathbf{C}_i$ and $\mathbf{C}_b$ are the colors of the interface and body reflection of the material.

Consider the colors $\mathbf{C}_L$ of a set of points on the same uniformly colored surface. Because the geometry is different at each point, the scale factors m_i and m_b vary from point to point. However, the colors $\mathbf{C}_i$ and $\mathbf{C}_b$ of the interface and body reflection are the same at all points on the same surface, because they are simply the results of tristimulus integration of $c_i(\lambda)$ and $c_b(\lambda)$ that do not vary with geometry. In other words, the pixel values are a linear combination of $\mathbf{C}_i$ and $\mathbf{C}_b$, with the coefficients determined by m_i and m_b at each point.

Recalling that we can assume $0 \leq m_i, m_b \leq 1$ without loss of generality, we see that the pixel values $\mathbf{C}_L$ for a set of points on a single surface must lie within a parallelogram in color space, bounded by the colors $\mathbf{C}_i$ and $\mathbf{C}_b$ of the interface and body reflection of the surface (Fig. 3). One corner of this parallelogram is located at the origin, $[r,g,b] = [0,0,0]$. Further, within this parallelogram, the position of any color is determined by the values of m_i and m_b at the corresponding point (Fig. 4).

The Dichromatic Model in Color Space makes two assumptions in addition to those of the Dichromatic Reflection Model. It assumes that there has been a prior segmentation of the image into groups of pixels known (or believed) to lie on a single surface, and that pixel values returned by the camera are linearly related to the irradiance on the sensor plane, i.e., that the camera output is linear with respect to intensity. If the camera is calibrated for monochrome response, it is not necessary to recalibrate it with each color filter separately since the relative SPD of a filter's transmittance is constant with respect to total intensity of illumination.

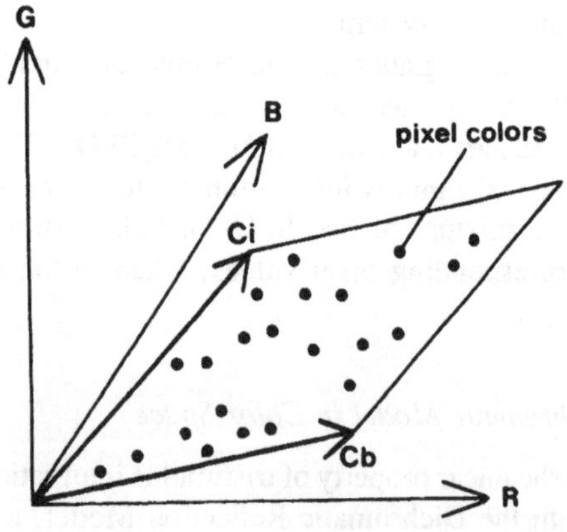

FIG. 3. Pixel values on a surface lie on a parallelogram in color space.

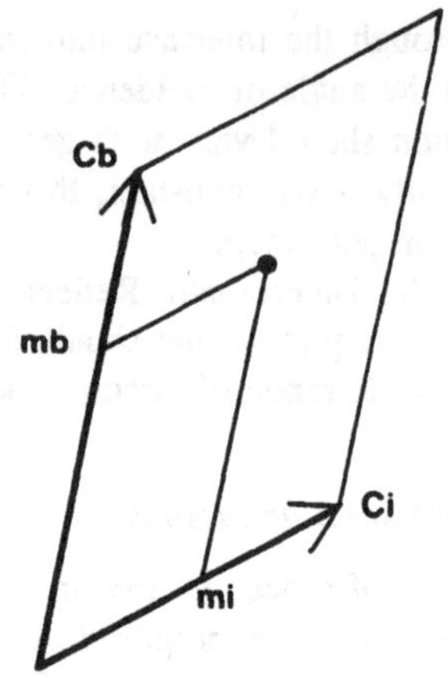

FIG. 4. Position within parallelogram determines reflection magnitudes.

The model of pixel values does not assume that $\mathbf{C}_i$ is achromatic (i.e., $r_i = g_i = b_i$), nor that it is the same for all surfaces in the image. Also, no assumption is made about the color filters $\tau_r(\lambda)$, $\tau_g(\lambda)$, and $\tau_b(\lambda)$, except that they are linearly independent, nor is any assumption made about the camera's spectral responsivity $s(\lambda)$, except that the relative spectral responsivity is constant with respect to the total amount of irradiance and constant across the sensor plane.

Analyzing Color Pixel Values

We can now present a simple algorithm for computing the intrinsic images of m_i and m_b over a set of pixels corresponding to a single surface in the image:

(1) Create a histogram of the pixel values in color space.

(2) Fit a plane in color space to these points, with the restriction that the plane must pass through the origin.

(3) Fit a parallelogram on this plane to the points, with the restriction that one vertex must lie at the origin. The sides of this parallelogram are $\mathbf{C}_i$ and $\mathbf{C}_b$. If it is necessary to distinguish one from the other, either geometric criteria or the fact that $\mathbf{C}_i$ probably lies close to the achromatic axis of the color space may be used.

(4) At each pixel, express its color as a linear combination of $\mathbf{C}_i$ and $\mathbf{C}_b$. This can be performed with a simple linear transformation, i.e., a matrix multiplication of the $[r,g,b]$ color values by a 2×3 matrix (not 3×3 since the direction orthogonal to the plane in color space is of no interest except as an error term).

(5) The resulting values are the magnitudes m_i and m_b.

In addition to the assumptions of the Dichromatic Model in Color Space, this algorithm depends upon a number of additional assumptions for its effectiveness. $\mathbf{C}_i$ and $\mathbf{C}_b$ must be linearly independent, i.e., the interface and body reflection must have different colors. The noise in measuring pixel values must be small enough that plane fitting and parallelogram fitting can proceed. The distribution of pixel values within the parallelogram must not be pathological, i.e., the pixel values must not happen to be colinear in color space, etc. In general, I expect that many points will lie close to the C_b axis, i.e., have m_i close to zero, and the remaining points will lie close to the C_i axis. These latter points may have pixel values "clipped" by the extent of the color cube itself, since highlights may be very bright and thus cause truncation of the measured pixel value at the maximum response of the camera.

It should also be noted that the Dichromatic Analysis Algorithm uses pixel colors to determine m_i and m_b, but it does not interpret m_i and m_b in terms of the photometric angles $i, e,$ and g. Thus, this algorithm does not tell how to analyze the values of m_i and m_b.

Summary

In this section, we have seen a mathematical description of tristimulus integration—the process by which an SPD of light is converted into color values. Tristimulus integration is a linear transformation, a fact that can be applied to the Dichromatic Reflection Model to produce a model of pixel values in color space. The distribution of pixel values is expected to form a parallelogram, bounded by the colors C_i and C_b of the interface and body reflection from the surface. The position of any color within this parallelogram can be used to determine the coefficients m_i and m_b at the corresponding pixel in the image.

Applying the Dichromatic Analysis Algorithm

We now discuss the extension of the Dichromatic Reflection Model for dealing with diffuse illumination, and some possible methods of analyzing the intrinsic images of m_i and m_b.

Extending the Model for Diffuse Illumination

The Dichromatic Reflection Model as presented above depends on the assumption that the illumination at any point comes from a single (point or extended) light source. It is more realistic to model the illumination as consisting of a light source plus "ambient" or "diffuse" light, of lower intensity, coming from all directions in equal amounts, and possibly with a different SPD than the small source. This model is a better approximation of daylight, that contains light from a point source (the yellowish sun) and light from a hemisphere (the bluish sky), and of light in a room, that comes from light fixtures and from inter-reflections off walls and other objects. All the previously mentioned models of reflection in computer graphics presume some sort of diffuse illumination, although the assumption made here (diffuse illumination from all directions equally) is admittedly highly idealized.

The light reflected by diffuse illumination contains a part due to interface reflection and a part due to body reflection. By assuming that this light is incident from and reflected into all directions equally, it can be modelled by adding a single term, $L_a(\lambda)$, to the Dichromatic Reflection Model:

$$L(\lambda,i,e,g) = m_i(i,e,g)c_i(\lambda) + m_b(i,e,g)c_b(\lambda) + L_a(\lambda) \quad (10)$$

In color space, the model becomes:

$$C_L = m_iC_i + m_bC_b + C_a \quad (11)$$

where C_a is the color of the light reflected from diffuse illumination $L_a(\lambda)$. Since $L_a(\lambda)$ does not vary with geometry, the effect of this change is to translate the parallelogram of pixel colors from a single surface away from the origin by the vector C_a, as seen in Fig. 5.

The Dichromatic Analysis Algorithm can still be applied with this change, but note that the plane-fitting and parallelogram-fitting operations specified in the algorithm will be less-constrained and therefore more error-prone. If the relative SPD of the diffuse illumination is the same as that of the small source, then C_a is a linear combination of C_i and C_b[34]; in such a case the *plane* containing the points should pass through the color space origin although the *parallelogram* will not.

It is interesting to note that a point whose color lies exactly at $C_L = C_a$ has $m_i = m_b = 0$, and might therefore be suspected of lying within a shadow. This is a far more precise description of colors within shadows than statements of the form "shadows tend to be bluish," that have been seen in previous work in computer vision. In fact, since for different surfaces C_a is very likely to differ, it may be possible to associate shaded parts of surfaces with illuminated portions, by constructing C_a for each adjacent illuminated area and finding which one matches best with the color of the pixels in the shaded area. Unfortunately, since pixel values in shaded areas of an image tend to be poorly measured by current digitizing cameras, this kind of analysis may prove to be unreliable.

The description presented here is intended to model diffuse illumination; however, it will be a poor substitute for a detailed model of inter-reflection when surfaces are close to each other.

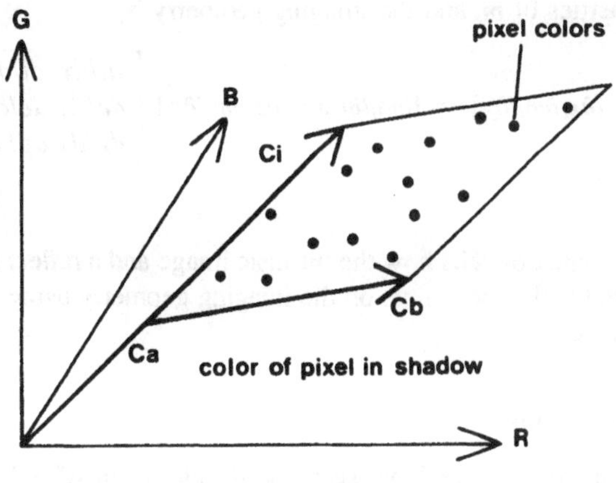

FIG. 5. Color space parallelogram with diffuse illumination.

I continue to call the above equation a "Dichromatic Reflection Model" (rather than a trichromatic model) because the essence of the model is that reflection occurs from two places in a surface: the interface and the surface body.

Analysis of the Intrinsic Reflection Images

The potential utility of the intrinsic images of m_i and m_b stems from two facts: Both m_i and m_b have simpler geometric properties than does intensity in a monochrome image, that represents a weighted sum of the two; and m_b, representing diffusely distributed light, is relatively insensitive to changes in viewpoint from one image to the next.

Here are some possible methods for exploiting these properties of the intrinsic images of m_i and m_b.

In stereo or motion analysis, image matching using m_b might be more reliable because of the elimination of interface reflection, whose position in the image is highly viewpoint-dependent.

If a specific reflectance model such as Phong's is assumed, then the two values m_i and m_b provide two constraints on the surface normal at any point. For example, using Phong's model, the angle s can be calculated from m_i and the angle i can be calculated from m_b. With such constraints, unique surface normals can be established from a color image. (In practice, this kind of analysis would be limited by the applicability of such reflectance models and by the distribution of points on the parallelogram in color space.)

If, as above, a specific model is assumed but the parameters (t, n in Phong's model) are not known, it may be possible to determine or estimate these from the image. At a point with a known orientation, the coefficients m_i and m_b can be determined by color analysis; using these, the surface parameters may be calculated. This might form the basis for a technique of looking at parts of a surface (say, the spots with pronounced highlights) to estimate the nature of the reflectance properties of the surface, then using this estimated-reflectance model to analyze the rest of the surface in detail.

Using differentials, since the photometric angles i, e, and g are functions of x and y in the image, we can relate the differentials of the intrinsic image of m_i to the geometric properties of m_i and the imaging geometry by:

$$[\partial m_i/\partial x \;\; \partial m_i/\partial y] = [\partial m_i/\partial i \;\; \partial m_i/\partial e \;\; \partial m_i/\partial g] \begin{bmatrix} di/dx & di/dy \\ de/dx & de/dy \\ dg/dx & dg/dy \end{bmatrix} \tag{12}$$

This equation tells how the intrinsic image and a reflectance model yield constraints on the imaging geometry using first derivatives.

Conclusions

In this article, we have seen the development of a fairly simple algorithm for determining intrinsic images of reflection from a color image. The presentation has included a fairly general reflection model based on a physical description of the reflection process, a model of pixel-value distribution based on the properties of tristimulus integration, and an algorithm for analyzing color values in an image. This article has also presented a discussion of some related issues such as diffuse illumination, image segmentation, and analysis of the resulting intrinsic images. These intrinsic images promise to be useful for a variety of image-understanding tasks, including stereo matching, geometric surface analysis, and analysis of surface-reflectance parameters.

This work represents a quantum advance over previous efforts in color-image understanding, because it presents a quantitative theory based on the physics of the real world. In this respect, it follows an increasingly important paradigm of three-dimensional image-understanding research, that involves the development of mathematical descriptions of physical and geometric properties of the world, and computer programs to exploit those properties. The current work may help the field of computer vision by demonstrating a theory that depends upon relatively few special assumptions, and by providing "simpler" intrinsic images useful for estimating surface and illumination parameters from examination of the images themselves.

Further efforts in developing this work will include implementation (when a calibrated camera is obtained), pursuing the ideas for intrinsic image analysis suggested above, and developing the theory that will ultimately allow analysis of nonuniform illumination, extended light sources, and perhaps inter-reflection among surfaces. In addition, a segmentation scheme based on the Dichromatic Reflection Model might be investigated.

Acknowledgments

The author would like to thank Fred Billmeyer, of Rensselaer Polytechnic Institute, and David MacAdam, of the Institute for Optics at the University of Rochester, for their comments and suggestions. Takeo Kanade at Carnegie–Mellon University provided much guidance for this work, and Marty Herman and Chris Brown provided comments on the paper. Frank O'Donnell at Rensselaer Polytechnic Institute provided a wealth of papers on gloss measurement.

This research was sponsored by the Defense Advanced Research Projects Agency (DOD), monitored by the Office of Naval Research under Contract N00014–82–K–0193. The views and conclusions contained in this document are those of the author and should not be interpreted as representing the official policies, either expressed or implied, of the Defense Advanced Research Projects Agency or the U.S. Government.

1. D. H. Ballard and C. M. Brown, *Computer Vision*, Prentice–Hall, Englewood Cliffs, NJ, 1982.
2. H. G. Barrow and J. M. Tenenbaum, Computational vision, *Proc. IEEE* **69**, 572–595 (1981).
3. H. G. Barrow and J. M. Tenenbaum, "Recovering Intrinsic Scene Characteristics from Images," in A. R. Hanson and E. M. Riseman,

Eds., *Computer Vision Systems*, Academic Press, New York, 1978, pp. 3–26.

4. F. Grum and R. J. Becherer, *Optical Radiation Measurements, Volume 1: Radiometry*, Academic Press, New York, 1979.
5. B. K. P. Horn and R. W. Sjoberg, Calculating the reflectance map, *Appl. Opt.* **18**, 1770–1779 (1979).
6. F. E. Nicodemus, J. C. Richmond, J. J. Hsia, I. W. Ginsberg, and T. Limperis, *Geometrical Considerations and Nomenclature for Reflectance*, NBS monograph 160, National Bureau of Standards, Washington, D.C., 1977.
7. F. Grum and C. J. Bartleson, Eds., *Optical Radiation Measurements, Volume 2: Colorimetry*, Academic Press, New York, 1980.
8. R. S. Hunter, *The Measurement of Appearance*, John Wiley and Sons, New York, 1975.
9. D. B. Judd and G. Wyszecki, *Color in Business, Science, and Industry*, 3rd ed., John Wiley and Sons, New York, 1975.
10. S. J. Williamson and H. Z. Cummins, *Light and Color in Nature and Art*, John Wiley and Sons, New York, 1983.
11. R. Nevatia, "A Color Edge Detector and its Use in Scene Segmentation," in *IEEE Transactions on Systems, Man, and Cybernetics*, TSMC-7, IEEE, New York, 1977, pp. 820–826.
12. G. B. Coleman and H. C. Andrews, Image segmentation by clustering, *Proc. IEEE* **67**, 773–785 (1979).
13. M. D. Levine and S. I. Shaheen, "A Modular Computer Vision System for Picture Segmentation and Interpretation, Part 1," in *Proceedings of a Conference on Pattern Recognition and Image Processing*, IEEE, New York, 1979, pp. 523–533.
14. P. A. Nagin, A. R. Hanson, and E. M. Riseman, *Region Extraction and Description Through Planning*, COINS TR 77–8, University of Massachusetts, Amherst, MA, 1977.
15. R. Ohlander, *Analysis of Natural Scenes*, Ph.D. thesis, Carnegie–Mellon University Computer Science Dept., Pittsburgh, 1975.
16. M. Nagao, T. Matsuyama, and Y. Ikeda, Region extraction and shape analysis in aerial photographs, *Comp. Graph. Image Proc.* **10**, 195–223 (1979).
17. S. Rubin, *The ARGOS Image Understanding System*, Ph.D. thesis, Carnegie–Mellon University Computer Science Dept., Pittsburgh, 1978.
18. K. Sloan, *World Model Driven Recognition of Natural Scenes*, Ph.D. thesis, University of Pennsylvania, Moore School of Electrical Engineering, Philadelphia, 1977.
19. J. M. Tenenbaum and S. Weyl, A region-analysis subsystem for interactive scene analysis, *Proc. Fourth Intl. Joint Conf. Artificial Intell.* 682–687 (1975).
20. T. E. Weymouth, J. S. Griffith, A. R. Hanson, and E. M. Riseman, Rule based strategies for image interpretation, in *Proc. Natl. Conf. Artificial Intell.*, AAAI, 1983, pp. 429–532.
21. I. T. Young, The classification of white blood cells, *IEEE Trans. Biomed. Eng.* **19**, 291–298 (1972).
22. J. R. Kender, "Instabilities in Color Transformations," in *Proc. Conf. on Pattern Recognition and Image Processing*, IEEE, New York, 1977, pp. 266–274.
23. Y. Ohta, T. Kanade, and T. Sakai, Color information for region segmentation, *Comp. Graph. Image Proc.* **13**, 222–241 (1980).
24. J. M. Rubin and W. A. Richards, *Color Vision and Image Intensities: When are Color Changes Material?*, MIT AI Memo 631, MIT, Cambridge, 1981.
25. F. A. Jenkins and H. E. White, *Fundamentals of Optics*, McGraw-Hill, New York, 1976.
26. F. W. Billmeyer, Jr., private communication, 1984.
27. W. G. Egan and T. W. Hilgeman, *Optical Properties of Inhomogeneous Material*, Academic Press, New York, 1979.
28. P. Beckmann and A. Spizzichino, *The Scattering of Electromagnetic Waves from Rough Surfaces*, MacMillan, New York, 1963, pp. 1–33 and 70–98.
29. B. K. P. Horn, Understanding image intensities, *Artif. Intell.* **8**, 201–231 (1977).
30. B. T. Phong, Illumination for computer generated pictures, *Comm. ACM* **18**, 311–317 (1975).
31. J. D. Foley and A. Van Dam, *Fundamentals of Interactive Computer Graphics*, Addison-Wesley, New York, 1983.
32. J. F. Blinn, Models of light reflection for computer synthesized pictures, *Comp. Graph.* **11**, 192–198 (1977).
33. K. E. Torrance and E. M. Sparrow, Theory for off-specular reflection from roughened surfaces, *J. Opt. Soc. Am.* **57**, 1105–1114 (1967).
34. R. L. Cook and K. E. Torrance, A reflectance model for computer graphics, *Comp. Graph.* **15**, 307–316 (1981).
35. T. Shibata, W. Frei, and M. Sutton, "Digital Correction of Solar Illumination and Viewing Angle Artifacts in Remotely Sensed Images," in *Proc. Seventh Symposium on Machine Processing of Remotely Sensed Data*, 169–177 (1981).
36. S. A. Shafer, Describing light mixtures through linear algebra, *J. Opt. Soc. Am.* **72**, 299–300 (1982).

Received June 26, 1984; accepted October 15, 1984

Using color for geometry-insensitive segmentation

Glenn Healey

Robotics Laboratory, Department of Computer Science, Stanford University, Stanford, California 94305

Received August 24, 1988; accepted February 11, 1989

Physical models indicate that, in general, reflectance is a complicated function of wavelength and geometry. An analysis of general reflectance models, however, shows that approximate reflectance models exist that preserve much of the structure of the more-detailed models. In particular, I show from general models that Shafer's dichromatic reflection model [Color Res. Appl. 10, 210 (1985)] is a reasonable approximation for a large class of inhomogeneous dielectrics. I also show that a unichromatic reflection model is a useful approximation for metals. The approximate color-reflectance model is the basis for two algorithms that use color information. The first algorithm uses normalized color to classify surfaces according to material composition and is insensitive to geometrical variation in the scene. The second algorithm is used to identify metal and dielectric materials from their images. Experimental results are presented.

1. INTRODUCTION

Several computer vision researchers have developed useful algorithms based on models of the image-forming process.[1,2] Modeling the interaction of light with matter is central to these approaches. Unfortunately, selecting appropriate models for light–matter interaction can be difficult. If one desires the most accurate available model, then light–matter interaction can be described in terms of the interaction of photons with atoms or molecules.[3] It would, however, be a formidable task to derive computer vision algorithms from a physical model at such a low level. In computer vision, then, it is often necessary to sacrifice some degree of physical accuracy in order to obtain a model that is useful.

In this paper I present an approximate color-reflectance model (ACRM). The most important property of the ACRM is that the spectral composition of the light reflected by each reflection process is assumed to be independent of the reflection geometry. For the special case of materials that are inhomogeneous dielectrics, the ACRM is equivalent to the dichromatic reflection model suggested by Shafer[1] and used subsequently by many other authors.[4–6] In this paper I use the Reichman body-scattering model[7] to show that the dichromatic reflection model is a reasonable approximation for a large class of inhomogeneous dielectrics. I also show from the Torrance–Sparrow[8] surface-reflection model and the Fresnel equations[9] that a unichromatic reflection model is a reasonable approximation for metals. Thus the ACRM combined the dichromatic reflection model for inhomogeneous dielectrics with a unichromatic reflection model for metals. The analysis includes an estimate of the accuracy of the ACRM for various materials.

Two algorithms have been derived from the ACRM. The first algorithm is used during image segmentation. This algorithm uses normalized color to classify image regions on the basis of the material of the corresponding object surfaces. The structure of the ACRM allows this algorithm to assign stable material descriptions despite changes in the geometry in the scene. The second algorithm is a preliminary method for distinguishing metals and dielectrics in color images. Both algorithms have produced encouraging results on real color images.

2. PRELIMINARIES

A. Optically Homogeneous and Inhomogeneous Materials

It is common to divide materials into two classes on the basis of their optical properties.[1,10,11] Optically homogeneous and optically inhomogeneous materials are defined in this subsection. The methods described in this paper apply to both kinds of material.

Optically homogeneous materials have a constant index of refraction throughout the material. Consequently, for an object composed of a homogeneous material in air, light undergoes reflection and refraction only as it encounters a surface of the object. Metals, glass, and crystals are the most common examples of homogeneous materials.

Optically inhomogeneous materials are composed of a vehicle with many embedded colorant particles that differ optically from the vehicle. While in the body of an inhomogeneous material, light typically interacts with many colorant particles. The net result of many interactions is that the light is diffused. Some examples of inhomogeneous materials are plastics, paper, textiles, and paints.

B. Reflection

When light is incident upon the surface of an object, some fraction of it is reflected. A smooth surface reflects light only in the direction such that the angle of incidence equals the angle of reflection (Fig. 1). The properties of this reflected light are determined by the optical and geometric properties of the surface. For optically homogeneous materials that are not transparent, appearance is determined completely by the properties of light reflected from the surface. For many inhomogeneous materials, such as plastics, effects of surface reflection are also significant.

The fraction of the incident light that is not reflected from the surface enters the body of the material. For inhomogen-

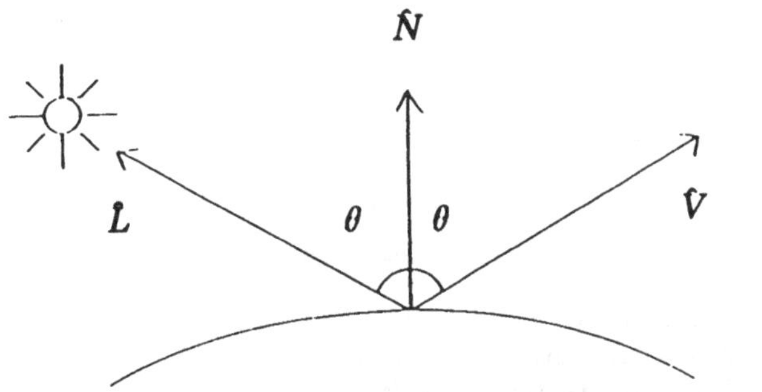

Fig. 1. Surface reflection.

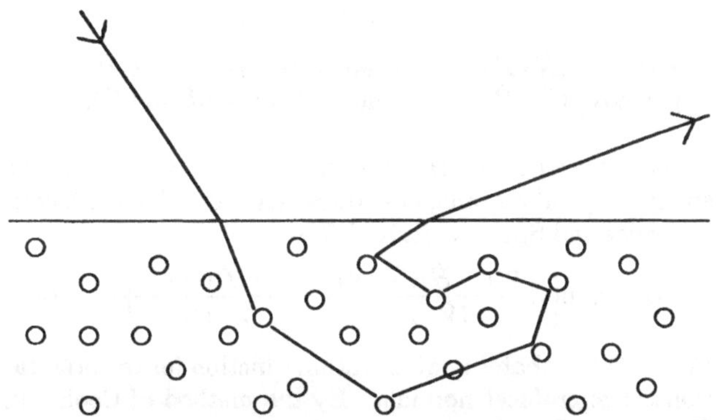

Fig. 2. Scattering by colorant particles.

eous materials, the body is composed of a vehicle and many colorant particles. When light encounters a colorant particle, some portion of it is reflected. After many reflections, the light is diffused, and a significant fraction can exit back through the surface in a wide range of directions (Fig. 2). Shafer[1] gave a detailed discussion of surface reflection (also called interface reflection) and body reflection.

C. Reflectance Model

In this paper opaque samples of matter with one relevant interface are considered. In general, the reflectance R of such a sample depends on four angles.[12] Two angles define the direction of the incident ray, and two angles define the direction into which light is reflected to a viewer. For many real surfaces, reflectance does not depend on the incoming azimuthal angle. The reflectance of such a surface may be quantified by

$$R(\theta_l, \theta_v, \theta_p, \lambda) = R_S(\theta_l, \theta_v, \theta_p, \lambda) + R_B(\theta_l, \theta_v, \theta_p, \lambda), \qquad (2.1)$$

where R_S is the surface-reflectance (or interface-reflectance[1]) term and R_B is the body-reflectance term that is due to colorant-layer scattering. The angles θ_l, θ_v, and θ_p are the photometric angles. θ_l denotes the angle between the surface normal and the illumination direction. θ_v denotes the angle between the surface normal and the viewer. θ_p denotes the angle between the surface projections of the incident ray and the viewing direction. As usual, λ denotes the wavelength.

The power of the light reflected toward a viewer is given by

$$I(\theta_l, \theta_v, \theta_p, \lambda) = R(\theta_l, \theta_v, \theta_p, \lambda)L(\lambda), \qquad (2.2)$$

where $L(\lambda)$ is the spectral power distribution of the light incident upon the surface.

The main result of this paper is that R of Eq. (2.1) can be approximated accurately by a function of the form

$$R(g, \lambda) = \begin{cases} M_S(g)C_S(\lambda) & \text{metal} \\ M_S(g)C_S(\lambda) + M_B(g)C_B(\lambda) & \text{inhomogeneous dielectric} \end{cases} \qquad (2.3)$$

where g indicates dependence on the photometric angles θ_l, θ_v, and θ_p. Standard approximation techniques will be used to construct the approximation of Eq. (2.3). Equation (2.3) is the basis of two image-understanding algorithms that are described in Sections 7 and 8 of this paper.

3. OPTICAL PROPERTIES OF MATTER

The optical properties of a material can be related to the electromagnetic quantities ϵ, μ, and σ (Table 1) and are summarized by the complex index of refraction $M = n - iK_0$, where both n and K_0 are real. The dimensionless quantity n is called the refractive part of M. In a nonattenuating medium ($K_0 = 0$), n is the ratio of the speed of light in vacuum to the speed of light in the material. For absorbing media ($K_0 > 0$), the interpretation of n is more complicated. K_0 is called the absorptive part of M, or the extinction coefficient. If the amplitude of an electromagnetic wave in a material does not decrease as the wave propagates in the material, then $K_0 = 0$. In a homogeneous absorbing material, the amplitude of a light wave is attenuated exponentially. If a plane wave of angular frequency ω traveling in the x direction encounters an interface at $x = 0$, then the irradiance $I(x)$ at a depth x in the material is given by

$$I(x) = I(0)\exp(-2\omega K_0 x/c), \qquad (3.1)$$

where c is the speed of light in vacuum. The distance $x = c/(2\omega K_0)$ is often called the skin depth for a material. It follows from Eq. (3.1) that the irradiance decreases by a factor of e^{-1} as the light wave penetrates to the skin depth.

It is possible to solve for n and K_0 in terms of the fundamental electromagnetic properties of the material by requiring that the incident light wave satisfy Maxwell's equations.[13] The resulting expressions for n and K_0 are given by

$$n^2(\lambda) = \frac{\mu\epsilon c^2}{2}\left\{1 + \left[1 + \left(\frac{\lambda\sigma}{2\pi c\epsilon}\right)^2\right]^{1/2}\right\}, \qquad (3.2)$$

$$K_0^2(\lambda) = \frac{\mu\epsilon c^2}{2}\left\{-1 + \left[1 + \left(\frac{\lambda\sigma}{2\pi c\epsilon}\right)^2\right]^{1/2}\right\}. \qquad (3.3)$$

It is important to note that for a given material ϵ and σ depend on the wavelength, and therefore both n^2 and K_0^2 are functions of the wavelength.

Table 1. Electromagnetic Properties of Matter

Property	Symbol	Units
Electrical permittivity	ϵ	$C^2/(N\text{-}m^2)$
Magnetic permeability	μ	$N\text{-}sec^2/C^2$
Conductivity	σ	$(\Omega\text{-}m)^{-1}$

4. SURFACE REFLECTION

A. Fresnel Reflection

For isotropic homogeneous media, the Fresnel equations describe the light that is reflected specularly from a smooth surface in terms of M and geometry. If unpolarized light is incident at an angle θ_l, then the Fresnel reflectance $F(\theta_l, \lambda)$ is

$$F(\theta_l, \lambda) = 0.5(R_\perp + R_\parallel), \tag{4.1}$$

where $R_\perp$ is the component polarized perpendicular to the plane of incidence and $R_\parallel$ is the component polarized parallel to the plane of incidence. The plane of incidence is the plane that contains the surface normal and the incident ray. From electromagnetic theory, $R_\perp$ and $R_\parallel$ are given by

$$R_\perp(\theta_l, \lambda) = \frac{a^2 + b^2 - 2a\cos\theta_l + \cos^2\theta_l}{a^2 + b^2 + 2a\cos\theta_l + \cos^2\theta_l}, \tag{4.2}$$

$$R_\parallel(\theta_l, \lambda) = R_\perp \frac{a^2 + b^2 - 2a\sin\theta_l\tan\theta_l + \sin^2\theta_l\tan^2\theta_l}{a^2 + b^2 + 2a\sin\theta_l\tan\theta_l + \sin^2\theta_l\tan^2\theta_l}, \tag{4.3}$$

where

$$a = (0.5)^{1/2}(\{[n^2(\lambda) - K_0^2(\lambda) - \sin^2\theta_l]^2 + 4n^2(\lambda)K_0^2(\lambda)\}^{1/2} + [n^2(\lambda) - K_0^2(\lambda) - \sin^2\theta_l])^{1/2}, \tag{4.4}$$

$$b = (0.5)^{1/2}(\{[n^2(\lambda) - K_0^2(\lambda) - \sin^2\theta_l]^2 + 4n^2(\lambda)K_0^2(\lambda)\}^{1/2} - [n^2(\lambda) - K_0^2(\lambda) - \sin^2\theta_l])^{1/2}. \tag{4.5}$$

Equations (4.2) and (4.3) are the Fresnel equations. The Fresnel equations have been derived by many authors, including Born and Wolf[9].

B. Torrance–Sparrow Reflection Model

The Fresnel equations describe reflection from a smooth surface. In practice, most surfaces are not smooth. The Torrance–Sparrow model[8] describes surface reflection from rough surfaces. The Torrance–Sparrow model assumes that a surface is composed of small, randomly oriented, mirrorlike microfacets. Only microfacets with a normal oriented in the perfect specular direction contribute to the surface reflectance R_S. In this paper two modifications are made to the original Torrance–Sparrow model. First, it is assumed that the surface is sufficiently smooth that the effects of multiple scattering between surface facets are negligible. Experimental results[14] suggest that this is a reasonable assumption for many real surfaces. Second, a more general probability distribution function derived by Beckmann[15] is used to represent the orientations of the normals to the planar surface microfacets. This distribution function was used previously in computer graphics.[16] The resulting model for surface reflectance is

$$R_S(\theta_l, \theta_v, \theta_p, \lambda) = \frac{F(\gamma, \lambda)P(\alpha)G(\theta_l, \theta_v, \theta_p)}{\hat{N}\cdot\hat{V}}, \tag{4.6}$$

where $F(\gamma, \lambda)$ is the Fresnel reflectance of Eq. (4.1) and $G(\theta_l, \theta_v, \theta_p)$ and $P(\alpha)$ are explained below. Figure 3 depicts the reflection geometry. Consider a surface point P, and define

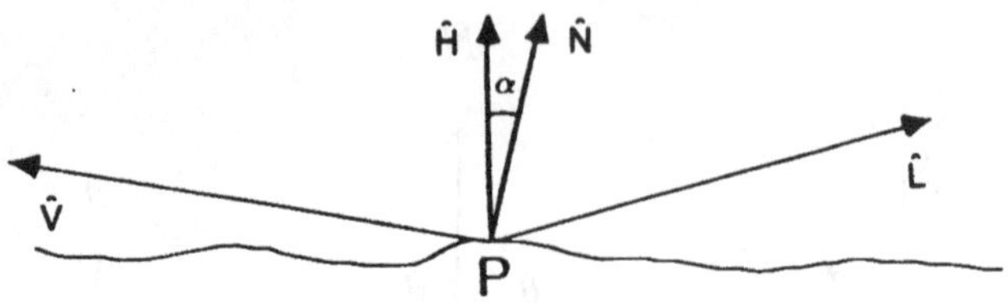

Fig. 3. Reflection geometry.

$\hat{V}$ as a unit vector from P in the direction of the viewer,
$\hat{N}$ as a unit surface normal at P,
$\hat{L}$ as a unit vector from P in the direction of the source,

$\hat{H} = \dfrac{\hat{V} + \hat{L}}{\|\hat{V} + \hat{L}\|}$ (unit angular bisector of $\hat{V}$ and $\hat{L}$; $\|\cdot\|$ is the Euclidean norm),

$\alpha = \cos^{-1}(\hat{N}\cdot\hat{H})$ (the angle between $\hat{N}$ and $\hat{H}$),
$\gamma = \cos^{-1}(\hat{L}\cdot\hat{H})$ (the angle between $\hat{L}$ and $\hat{H}$).

G is a geometrical attenuation factor that accounts for the shadowing and masking of surface facets by adjacent facets. Torrance and Sparrow derived G as

$$G = \min\left\{1, \frac{2(\hat{N}\cdot\hat{H})(\hat{N}\cdot\hat{V})}{(\hat{V}\cdot\hat{H})}, \frac{2(\hat{N}\cdot\hat{H})(\hat{N}\cdot\hat{L})}{(\hat{V}\cdot\hat{H})}\right\}. \tag{4.7}$$

$P(\alpha)$ is the probability distribution function for the orientations of microfacet normals. By the method of Cook and Torrance,[16] $P(\alpha)$ is taken to be the Beckmann distribution function:

$$P(\alpha) = \frac{1}{m^2\cos^4\alpha}\exp\left(-\frac{\tan^2\alpha}{m^2}\right), \tag{4.8}$$

where m is the surface roughness parameter, defined to be the rms slope of microfacet normals. The function $P(\alpha)$ of Eq. (4.8) is more appropriate than the Gaussian distribution function originally proposed by Torrance and Sparrow in that it does not depend on any arbitrary constants.

5. INTERACTION OF LIGHT WITH METALS

A. Intensity of the Reflected Light

The most prominent difference between metals and other materials is the large number of electrons that are free to move throughout a metal, causing metals to have large values of conductivity σ. By contrast, electric charges are not free to move in dielectrics, and thus these materials are poor conductors. Another important property of metals is that they are optically homogeneous. Therefore $R_B \equiv 0$ in Eq. (2.1), and the only relevant reflection process for metals is surface reflection.

The conductivity σ of a material is defined as the ratio of the current density produced by an electric field to the intensity of the electric field. In a perfect dielectric there are no free electrons, and the conductivity is zero. For metals, on the other hand, there are many free electrons, and the conductivity is much greater than zero (Table 2). For a conductor, we have $\sigma = \infty$. Physically this means that a stream of electrons moving in a certain direction would continue to move in that direction without resistance. In most metals, however, electrons experience resistance because of departures of the solid lattice from perfect regularity. These departures from perfect regularity are caused by thermal motion of the atoms and impurities in the solid. When an electron is scattered by the imperfect lattice, electromagnetic energy is converted irreversibly to heat. This process is in part responsible for the attenuation of the incident wave as described by Eq. (3.1).

As a first approximation, metals may be thought of as perfect conductors; i.e., let $\sigma \rightarrow \infty$. If λ is taken to be some visible wavelength, then n and K_0 are functions of only ϵ and σ. It is not possible to measure ϵ directly for metals. It can be assumed from the relevant physics, however, that ϵ for metals is of the same order of magnitude as for dielectrics.[9] Therefore, from Eq. (3.3), we have $K_0 \rightarrow \infty$ as $\sigma \rightarrow \infty$. From the Fresnel equations, this tells us that $R_S \rightarrow 1$ as $\sigma \rightarrow \infty$.

Although the conductivity of most real metals is finite, the approximation $R_S(\theta_i, \lambda) \approx 1$ is reasonable for many metals in the visible wavelength range. At the physical level, this is because the many free electrons present in a metal cause rapid attenuation of the incident light. Thus incident light has little opportunity to penetrate into the metal. Consequently, K_0 is large, and metals have a skin depth that is only a small fraction of a wavelength. Few electrons are able to absorb any energy from the incident light; as a result, most of the incident light is reflected.

B. Spectral Properties of the Reflected Light

The optical properties of a metal are determined largely by how light interacts with the electrons in the metal. We can divide the various possible interactions into two classes. In the first class of interaction, light is scattered by a free conduction electron. This class of interaction is easily the most common in metals because of the large number of free electrons. In the second class of interaction, the light's energy is absorbed by the metal as an electron jumps to a higher energy level.

The most prominent process that contributes to the optical properties of a metal is the interaction of light with free conduction electrons. As is mentioned in Subsection 5.A, these free electrons scatter light effectively, causing metals to have a small skin depth and large reflectance values. Since these free electrons have no natural frequencies, they scatter light of all wavelengths equally well. This is why most metals (e.g., aluminum, silver, sodium) have spectral-reflectance functions that do not vary with wavelength in the visible range and thus appear silver or gray in color.

In addition to being scattered by free electrons, light can also be absorbed during interactions with electrons that are confined to exist in certain energy zones (called Brillouin zones). Each zone is separated from adjacent zones by a gap of disallowed energies. For a certain range of incident light wavelengths, the energy of the incident light will match the energy gap between zones. In this situation, an atom will absorb the light, allowing an electron to jump to a higher-energy zone. This absorbed energy typically is converted rapidly into heat. The important point is that there is a critical wavelength λ_0 such that incident light energy is not absorbed for $\lambda > \lambda_0$, and for decreasing λ there is a dramatic increase in the amount of incident light energy absorbed at λ_0. This absorbed light energy cannot appear in the reflected wave. For decreasing wavelength, therefore, n and K_0 behave in such a way that there is a dramatic decrease in spectral reflectance at λ_0.

Table 2. Conductivity of Materials

Material	Class	Conductivity
Copper	metal	5.9×10^7
Zinc	metal	1.7×10^7
Iron	metal	1.0×10^7
Uranium	metal	3.4×10^6
Graphite	semiconductor	2.0×10^4
Germanium	semiconductor	10^1–10^3
Silicon	semiconductor	10^{-1}–10^1
Glass	ceramic	10^{-11}–10^{-9}
Diamond	ceramic	10^{-11}–10^{-10}
Porcelain	ceramic	10^{-12}–10^{-10}
Mica	ceramic	10^{-11}–10^{-15}
Fused quartz	ceramic	10^{-16}–10^{-18}

For most metals, there is no absorption of light in the visible wavelength range. Thus most metals have spectral-reflectance functions that are constant with respect to wavelength in the visible range. The most-notable exceptions to this rule are copper and gold, both of which have absorption bands in the visible range.[17] The spectral-reflectance functions for both copper and gold are near 1 for long visible wavelengths and decrease sharply below their critical wavelengths λ_0. Copper and gold absorb blue and green light strongly while reflecting red light. This gives these metals their reddish colors. Other metals have absorption bands that do not lie in the visible range. Silver, for example, begins absorbing in the ultraviolet at approximately $\lambda_0 = 310$ nm.

It has been shown[17] that the physical properties of metals provide strong constraints on their spectral-reflectance functions. The scattering of light by free conduction electrons cause most metals (e.g., silver, aluminum, sodium, tin, potassium, niobium, scandium, cesium, yttrium, vanadium, niobium) to have spectral-reflectance functions that are approximately constant over the visible wavelengths. For other metals, absorption bands in the visible range give rise to critical wavelengths λ_0 such that the metal reflects strongly for $\lambda > \lambda_0$ and absorbs strongly for $\lambda < \lambda_0$ (e.g., copper, gold). Note that these are the only possible kinds of spectral-reflectance function for a metal. We would not expect, for example, to see metals with spectral-reflectance functions that decrease sharply with increasing wavelength.

C. Unichromatic Reflection Model

Since metals are optically homogeneous, the only reflection process that contributes to the reflectance of a metal is surface reflection. In Subsection 5.B the surface reflectance R_S for a material was shown to be a complicated function of wavelength and the geometry of reflection. It would be useful to establish a simpler form for R_S that preserves as much of the generality of the full Torrance–Sparrow model as possible. In particular, I examine to what extent the spectral composition of the light reflected by a metal depends on the geometry of reflection.

From Section 4, the power of the light reflected by a metal is given by

$$I(g, \lambda) = \frac{F(g, \lambda)P(g)G(g)L(\lambda)}{\hat{N} \cdot \hat{V}(g)}, \tag{5.1}$$

where g indicates dependence on geometry and λ indicates dependence on wavelength. In this section, I show how $I(g, \lambda)$ can be approximated by a function $\tilde{I}(g, \lambda)$ of the form

$$\tilde{I}(g, \lambda) = I_1(g)I_2(\lambda). \qquad (5.2)$$

The approximation $\tilde{I}(g, \lambda)$ is called a unichromatic reflection model because it predicts that any light reflected from a given metal illuminated by a single spectral power distribution will have a spectral composition given by the single function $I_2(\lambda)$. From Eq. (5.1) it follows that a best approximation $\tilde{I}(g, \lambda)$ will have components of the form

$$I_1(g) = \frac{F_1(g)P(g)G(g)}{\hat{N} \cdot \hat{V}(g)}, \qquad (5.3)$$

$$I_2(\lambda) = F_2(\lambda)L(\lambda), \qquad (5.4)$$

where

$$\tilde{F}(g, \lambda) = F_1(g)F_2(\lambda) \qquad (5.5)$$

is an approximation to $F(g, \lambda)$. Thus it is necessary to examine how well $F(\theta_l, \lambda)$ can be approximated by a function having the form of Eq. (5.5).

Assume that we are given the optical properties of a metal $n(\lambda)$ and $K_0(\lambda)$ over the visible wavelengths. Equation (5.5) then provides a good approximation to the reflecting properties of the metal if functions $F_1(\theta_l)$ and $F_2(\lambda)$ exist such that the sum-squared error,

$$\int_{\theta_l}\int_{\lambda} [\bar{F}(\theta_l, \lambda) - F_1(\theta_l)F_2(\lambda)]^2 d\lambda d\theta_l, \qquad (5.6)$$

is small. $\bar{F}(\theta_l, \lambda)$ represents the function $F(\theta_l, \lambda)$ normalized in the sense that, for a fixed angle $\theta_l = \theta_l'$, $\|\bar{F}(\theta_l', \lambda)\| = 1$, in which $\|\cdot\|$ is the $L^2[\lambda_1, \lambda_2]$ norm over the visible wavelengths $\lambda_1 \le \lambda \le \lambda_2$. The purpose of this normalization is to permit equal representation in the integral of angles θ_l for which $F(\theta_l, \lambda)$ is relatively small.

Since measurements of optical properties typically are reported only for discrete values of λ, a discrete version of this problem in both θ_l and λ is analyzed. We take n values of θ_l in the range $0 \le \theta_l \le \pi/2$ and m values of λ in the visible wavelength range 400 nm $\le \lambda \le$ 700 nm. For the discrete formulation, $\bar{F}(i, j)$ is taken to be the value of $\bar{F}$ at the ith θ_l value and the jth λ value. Thus the discrete version of expression (5.6) is

$$E = \sum_{1 \le i \le n} \sum_{1 \le j \le m} [\bar{F}(i, j) - F_1(i)F_2(j)]^2. \qquad (5.7)$$

For any given $\bar{F}(i, j)$ function, the functions $F_1(i)$ and $F_2(j)$ that minimize Eq. (5.7) may be found by using principal-components analysis.[18] The computational method applied is to perform a singular-value decomposition (SVD)[19] on the $n \times m$ matrix $\bar{F}$. We assume that $n > m$. The SVD produces an $n \times n$ unitary matrix U, an $m \times m$ unitary matrix V, and a diagonal matrix $D = \mathrm{diag}(\sigma_1, \sigma_2, \ldots, \sigma_m)$ with $|\sigma_1| > |\sigma_2| > \ldots > |\sigma_m|$ such that

$$\bar{F} = UDV. \qquad (5.8)$$

Let $F_1^*(i)$ and $F_2^*(j)$ be the functions that minimize Eq. (5.7), for which we choose the Euclidean norm of $F_2^*(j)$ to be equal to 1 to ensure uniqueness. From principal-components analysis,[18] the functions that minimize Eq. (5.7) are given by

$$F_1^*(i) = \sigma_1 u_{i1}, \qquad (5.9)$$

$$F_2^*(j) = v_{1j}. \qquad (5.10)$$

In other words, $F_1^*(i)$ is given by the first column of U multiplied by σ_1, and $F_2^*(j)$ is the first row of V.

Therefore the desired approximation of Eq. (5.2) is given by

$$I_1(g) = \frac{F_1(g)P(g)G(g)}{\hat{N} \cdot \hat{V}(g)}, \qquad (5.11)$$

where

$$F_1(g) = F_1^*(g)\|F(g)\| \qquad (5.12)$$

and

$$I_2(\lambda) = F_2^*(\lambda)L(\lambda), \qquad (5.13)$$

where $F_1^*(g)$ and $F_2^*(\lambda)$ are, respectively, the functions $F_1^*(i)$ and $F_2^*(j)$ in Eqs. (5.9) and (5.10) with the argument θ_l replacing i and the argument λ replacing j. $\|F(g)\|$ is the function of θ_l given by the $L^2[\lambda_1, \lambda_2]$ norm of $F(\theta_l, \lambda)$ at each value of θ_l. This term is necessary to correct the effects of the normalization that is used to get $\bar{F}(\theta_l, \lambda)$ in expression (5.6). Since we have computed F_1^* and F_2^* only for discrete values of θ_l and λ, $I_1(g)$ and $I_2(\lambda)$ of Eqs. (5.11) and (5.13) are defined only at discrete values of θ_l and λ. Since $\bar{F}(\theta_l, \lambda)$ is generally a smooth function of both θ_l and λ for metals,[20–22] and, since it is possible to compute $F_1^*(i)$ and $F_2^*(j)$ for large values of n and m (as we do in Subsection 5.D), little is lost by using a discrete approximation.

D. Testing the Unichromatic Reflection Model

It is natural to ask how good the approximation of Eqs. (5.11) and (5.13) is for real metals. When we examine Eq. (5.7) again the error in the approximaton to the ith row of $\bar{F}(i, j)$ is given by (see Appendix A for mathematical details)

$$\|\bar{F}(i, j) - F_1^*(i)F_2^*(j)\|^2 = \sum_{2 \le j \le m} (\sigma_j u_{ij})^2. \qquad (5.14)$$

Thus the total error in Eq. (5.7) is given by

$$E = \sum_{1 \le i \le n} \sum_{2 \le j \le m} \sigma_j^2 u_{ij}^2 = \sum_{2 \le j \le m} \sigma_j^2, \qquad (5.15)$$

where the last step follows from the fact that U is unitary.

The unichromatic reflection model is tested for several metals by using available optical data. The metals considered include both white metals, e.g., aluminum and silver,

Table 3. Quality of Unichromatic Fit

Metal	E_{45}	F_1	E_{35}
Aluminum	5.973811E-5	0.999940	1.463095E-6
Copper	2.227653E-3	0.997765	4.420242E-5
Gold	6.054643E-3	0.993914	3.372061E-4
Iridium	2.844490E-4	0.999715	8.053443E-6
Molybdenum	1.013760E-4	0.999897	3.336892E-6
Nickel	1.042433E-3	0.998957	6.432593E-5
Osmium	1.996747E-3	0.997990	4.781794E-5
Platinum	5.605146E-4	0.999439	2.189883E-5
Rhodium	1.303187E-4	0.999868	8.165633E-7
Silver	7.070805E-5	0.999929	3.568656E-6
Tungsten	2.091664E-4	0.999791	8.749000E-6

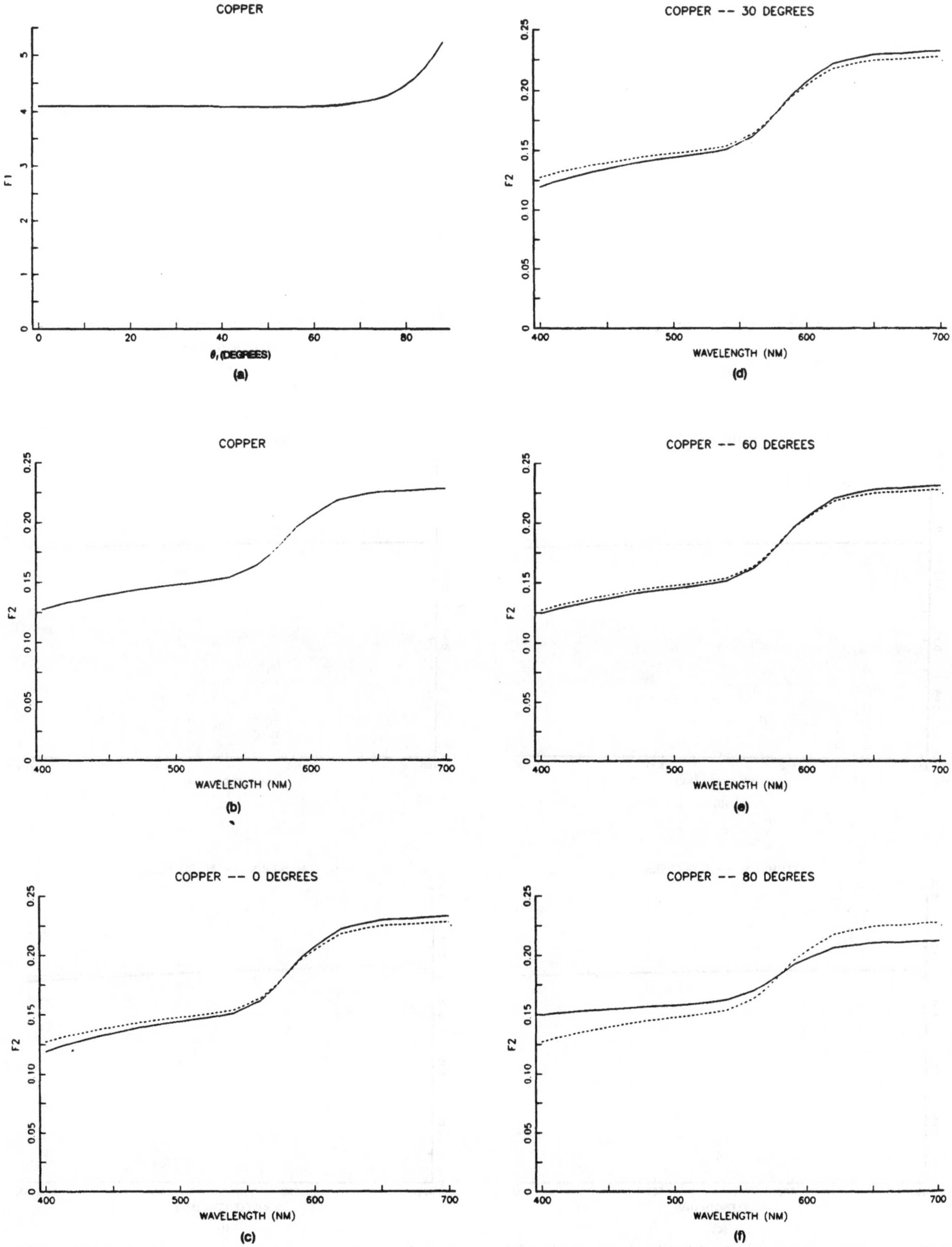

Fig. 4. (a) $F_1(\theta_i)$ for copper. (b) $F_2{}^*(\lambda)$ for copper. (c)–(f) Unichromatic fits for copper for the following angles θ_i: (c) 0°, (d) 30°, (e) 60°, (f) 80°.

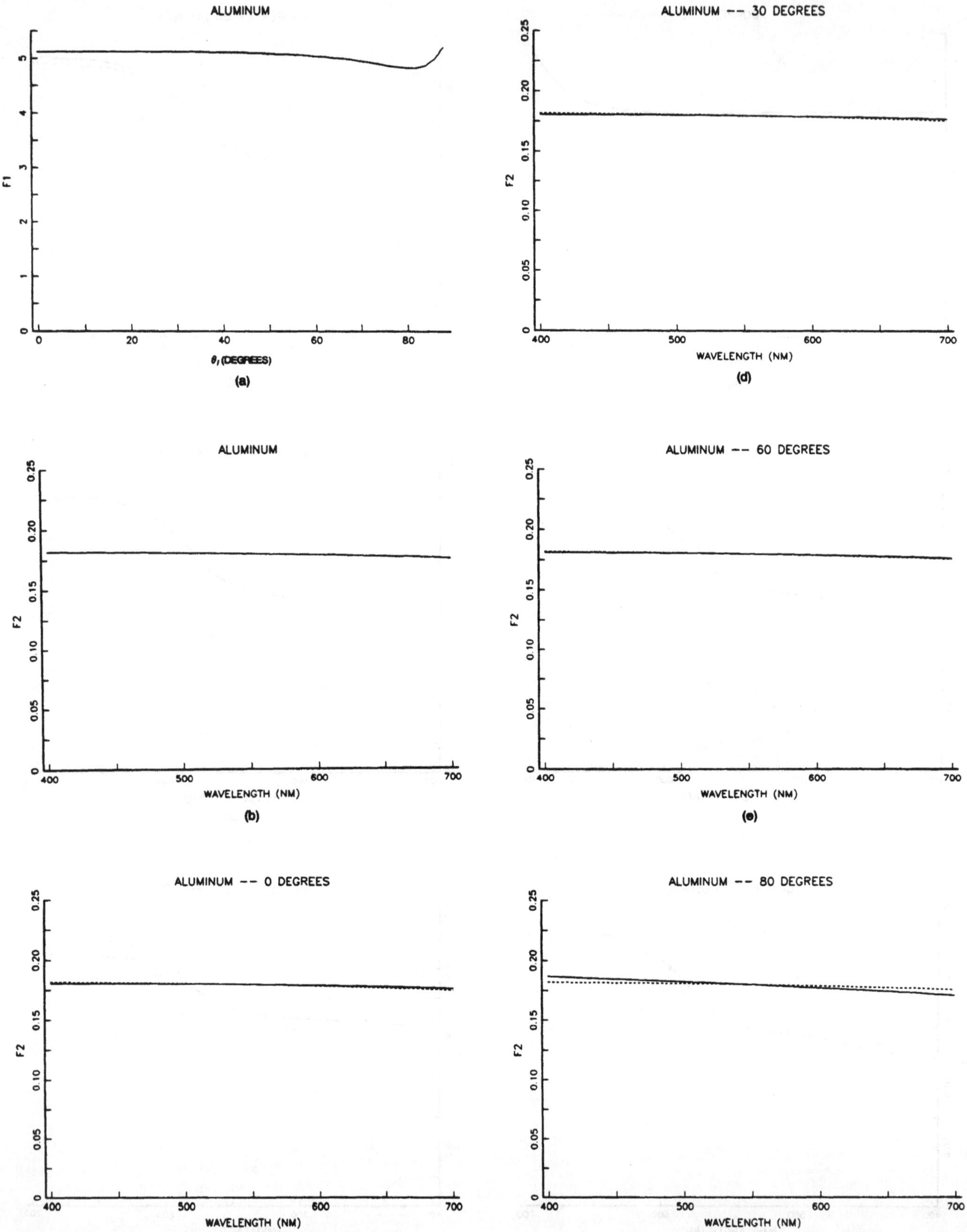

Fig. 5. (a) $F_1(\theta_i)$ for aluminum. (b) $F_2^*(\lambda)$ for aluminum. (c)–(f) Unichromatic fits for aluminum for the following angles θ_i: (c) 0°, (d) 30°, (e) 60°, (f) 80°.

and red metals, e.g. copper and gold. Complex-index-of-refraction data given by $n(\lambda)$ and $K_0(\lambda)$ have been obtained from tables.[21,22] For each metal, the value of $\hat{F}(\theta_i, \lambda)$ is computed at θ_i spacings of 2° in the range $0° \leq \theta_i \leq 90°$ (n = 45) and at λ spacings of 10 nm in the range 400 nm $\leq \lambda \leq$ 700 nm (m = 31). A useful measure for characterizing the quality of the fit of the unichromatic reflection model to the actual reflecting properties of a metal is the average-square error in the approximation to the rows of $\hat{F}(i, j)$. From Eq. (5.15), this average error is given by

$$E_n = \frac{1}{n} \sum_{2 \leq j \leq m} \sigma_j^2. \tag{5.16}$$

Thus E_n is the average-square error obtained when $\hat{F}(j)$ is approximated at a fixed value of i with a unichromatic reflection model.

Another quantity of interest is the fraction of variance accounted for by the unichromatic model. This fraction is defined by

$$F_1 = \sigma_1^2 \Bigg/ \left(\sum_{1 \leq i \leq m} \sigma_i^2 \right). \tag{5.17}$$

Values of E_n and F_1 for 11 metals are shown in the second and third columns of Table 3.

The Fresnel equations have the property that $F(\theta_i, \lambda)$ approaches 1 as θ_i approaches 90° (glancing incidence), independent of the values of $n(\lambda)$ and $K_0(\lambda)$. Therefore, near glancing incidence, all metals reflect nearly all incident light regardless of wavelength. This glancing-incidence effect is one of the largest sources of error in E_{45}, particularly for the red metals, which reflect long visible wavelengths much more efficiently than short visible wavelengths at incidence angles $\theta_i \ll 90°$. Thus it is not surprising that the largest values of E_{45} occur for gold and copper. The fourth column of Table 3, labeled E_{35}, shows E_n when only 35 angles of incidence, in the range $0° \leq \theta_i \leq 70°$, are considered. As can be seen, this produces much smaller values of the error E_n for all the metals.

Despite the behavior of $F(\theta_i, \lambda)$ over a small range of θ_i near glancing incidence, Table 3 shows that a unichromatic reflection model is an excellent approximation for metals. In particular, Table 3 shows that the model accounts for >99% of the variance in the reflectance functions for all 11 metals.

Figure 4(a) shows the function $F_1(\theta_i)$ of Eq. (5.11) for copper. The shape of the function is quite similar to that of the characteristic Fresnel reflectance curve $F(\theta_i)$ for a material at a fixed wavelength.[23] Figure 4(b) shows $F_2{}^*(\lambda)$ for copper and is indicative of copper's reddish color over most illumination geometries. Figures 4(c)–4(f) show the quality of the unichromatic approximation for copper at four angles θ_i. Each plot shows, for a fixed value of θ_i, the approximation $F_2{}^*(\lambda)$ (dashed curve) and the theoretical function $\hat{F}(\theta_i, \lambda)$ (solid curve). In the plots, the function $\hat{F}(\theta_i, \lambda)$ is scaled by dividing by the value $F_1{}^*(\theta_i)$ to facilitate comparison between the curves. Figures 4(c)–4(e) indicate that the unichromatic model provides an excellent approximation over most values of θ_i. At θ_i = 80° [Fig. 4(f)], $\hat{F}(\theta_i, \lambda)$ begins to approach the glancing-incidence limit, and the unichromatic model is not as accurate as for smaller angles of incidence.

Figures 5(a)–5(f) show similar plots for aluminum. Since the reflectance of aluminum varies less with wavelength than does the reflectance of copper, the unichromatic approximation is more accurate for aluminum. Figures 5(c)–5(e) show the nearly perfect fit for most values of θ_i. Figure 5(f) (θ_i = 80°) shows a small error caused by the glancing-incidence effect.

How, then, does the unichromatic reflection model of Eqs. (5.11) and (5.13) compare with the full Torrance–Sparrow model of Eq. (4.6)? The geometrical dependence of the full model and that of the unichromatic approximation are nearly equivalent. P, G, and $\hat{N} \cdot \hat{V}$ in Eq. (4.6) do not depend on wavelength, and thus the exact terms from the Torrance-Sparrow method are used in the unichromatic approximation. F in Eq. (4.6) depends on both geometry and wavelength and is approximated as a function of geometry by Eq. (5.12). $F_1(g)$ in Eq. (5.12) is nearly a constant function of geometry (within approximately 1%) for most metals. At a coarse level, this is because for fixed i in Eq. (5.7) $F_1(i)$ is a scale factor that is used for approximating one unit vector $\hat{F}(i, j)$ with another similar unit vector $\hat{F}_2(j)$. Thus most of the variation of $F_1(g)$ with geometry is caused by variation in $\|F(G)\|$. By replacing $F(g, \lambda)$ in the Torrance–Sparrow model with $\|F(g)\|$ in Eq. (5.12) we are in effect replacing the exact Fresnel reflectance depending on g and λ with an average Fresnel reflectance over the visible wavelengths and leaving variation with λ to be accounted for by the function $I_2(\lambda)$ of Eq. (5.13).

As the examples in this subsection show, $F_2{}^*(\lambda)$ of Eq. (5.13) is useful for approximating the wavelength dependence of reflectance in Eq. (4.6). As explained above, it is impossible for the single function $F_2{}^*(\lambda)$ to give correct approximations near glancing incidence for metals that do not reflect all visible wavelengths equally. However, for angles of incidence in the range $0° \leq \theta_i \leq 70°$, $F_2{}^*(\lambda)$ captures almost exactly the spectral dependence of the reflectance for all metals.

In conclusion, the unichromatic reflection model is nearly equivalent to the Torrance–Sparrow reflection model for metals except for the case of red metals near glancing incidence. For this special case, the Torrence–Sparrow model indicates that the metal will reflect all visible wavelengths equally, whereas the unichromatic model incorrectly predicts that the red metal will still be more efficient at reflecting long visible wavelengths.

6. INTERACTION OF LIGHT WITH DIELECTRICS

Dielectrics differ significantly from metals in that they are poor conductors of electricity (Table 2). Physically this is because in a dielectric all Brillouin zones that contain any electrons are full and there exist large energy gaps separating these full zones from adjacent empty zones. Hence an applied electric field does not produce a current, and the material is an insulator. From Eq. (3.3), an important consequence of having a small conductivity is that K_0 is small and incident light is able to penetrate a significant distance into the body of the dielectric material. Hence, for inhomogeneous dielectrics, both surface reflection and colorant-layer scattering from the body of the material are important optical processes.

A. Surface Reflection

For homogeneous dielectrics, such as glass, the properties of surface reflection are determined by the index of refraction of the material and the surface geometry. For inhomogeneous dielectrics, such as plastic, the properties of surface reflection are determined by the index of refraction of the vehicle material and the surface geometry.

Surface reflection from dielectrics is quantified accurately by the Fresnel equations and the Torrance–Sparrow model explained in Section 4. In certain respects, however, surface reflection from dielectrics is simpler than surface reflection from metals. For a perfect dielectric, the conductivity σ is zero. Therefore, from Eqs. (3.2) and (3.3), it follows that

$$n = (\mu\epsilon c^2)^{1/2}, \tag{6.1}$$

where the units for μ and ϵ are given in Table 1 and

$$K_0 = 0. \tag{6.2}$$

For dielectrics, n defined by Eq. (6.1) is equal to the ratio of the speed of light in vacuum to the speed of light in the material. Since K_0 is zero in Eq. (6.2), a homogeneous dielectric medium does not attenuate the amplitude of an incident light wave [see Eq. (3.1)]. The normal-incidence Fresnel reflectance for such a material is

$$R_S(\theta_l = 0°, \lambda) = \frac{(n-1)^2}{(n+1)^2}. \tag{6.3}$$

For the values of n corresponding to most dielectrics, $R_S(\theta_l = 0°, \lambda)$ is typically less than 0.3 for visible wavelengths.[13] In particular, for angles of incidence that are not near glancing incidence, $R_S(\theta_l, \lambda)$ for metals is typically much larger than $R_S(\theta_l, \lambda)$ for dielectrics.

Although in general $R_S(\theta_l, \lambda)$ depends on wavelength, for dielectrics this dependence is typically weak across the visible wavelengths, because $n(\lambda)$ is a nearly constant function of wavelength in the visible range for most dielectrics.[20] Consequently, light reflected from the surface of a dielectric has approximately the same spectral composition as the incident light. This fact has been verified experimentally for several real materials[24] and is the basis of several color-constancy algorithms.[6,25,26]

B. Colorant-Layer Scattering

For inhomogeneous dielectrics, the most prominent optical process is colorant-layer scattering. Light incident upon an inhomogeneous material that is not reflected at the surface enters the body of the material. In the body of the material the light is reflected by many colorant particles and becomes thoroughly diffused. Scattering refers to this process of diffusion by many reflections. In this subsection a physical model is described for colorant-layer scattering by inhomogeneous dielectrics.

The Kubelka-Munk (K-M) theory[27] is a general mathematical treatment of scattering and absorption in colorant layers. The K-M theory assumes that a colorant layer is composed of a large number of optically identical elementary layers. The thickness of each elementary layer is small compared with the thickness of the entire colorant layer but is large compared with the diameter of an individual colorant particle. Thus it is not necessary to model the optical properties of individual colorant particles. The effects of many colorant particles are modeled by the properties of an elementary layer. An elementary colorant layer is characterized by the parameters α and σ (σ in K-M theory is distinct from the symbol σ for conductivity introduced in Section 3). $\alpha(\lambda)$ is the fraction of light that is absorbed per unit path length. $\sigma(\lambda)$ is the fraction of light that has its direction changed by scattering per unit path length. The color of an inhomogeneous material is due primarily to the scattering and absorbing characteristics of colorant layers as described by $\alpha(\lambda)$ and $\sigma(\lambda)$. Since colorant particles tend to absorb selectively certain wavelengths of light while transmitting others, there is often significant variation of R_B with λ. The K-M model gives rise to simultaneous first-order differential equations. These equations can be solved to give expressions for the reflectance and transmission of a colorant layer.

The original K-M theory makes several limiting assumptions. The original theory assumes the boundary condition of diffusely incident light. This is an unrealistic assumption for most real situations. The original K-M theory also assumes that the vehicle containing the colorant particles has an index of refraction equal to that of air. This assumption eliminates the need to consider internal and external reflections at the air–vehicle interface. Unfortunately, this assumption is also not realistic.

The original K-M theory was extended by Reichman[7] to eliminate the need for these unrealistic assumptions. Reichman derived an expression for the reflectance of an inhomogeneous material that is valid for collimated light at any angle of incidence. Reichman also used a method developed by Orchard[28] to take into account both internal and external reflections at the air–vehicle interface. Experiments have shown that Reichman's model accurately predicts the reflecting properties of real materials.[11,29]

In this paper I consider opaque colorant layers composed of isotropic scatterers. For light incident at an angle θ_l, the extended K-M theory describes the body reflectance R_B as

$$R_B(\theta_l, \lambda) = (1 - R_S')\,\frac{C(\theta_l, \lambda)(1 - r_i)[R_\infty(\lambda) - D(\theta_l)]}{2[1 - r_i R_\infty(\lambda)]\cos\theta_l}, \tag{6.4}$$

where R_S' is the fraction of the incident light that is reflected from the surface. R_S' can be obtained by integrating R_S of Eq. (4.6) over the viewing hemisphere. The term $(1 - R_S')$ describes the fraction of the incident light that is not reflected from the surface and that therefore penetrates into the body of the material. In general, $(1 - R_S')$ depends on both geometry and wavelength. On the basis of the discussion in Subsection 6.A, it is assumed that $(1 - R_S')$ depends on geometry only. Shafer made a similar assumption.[1] r_i is the internal diffuse surface reflectance approximated by Orchard[28] as

$$r_i = 1 - \frac{1.4399 - 0.7099n + 0.3319n^2 - 0.0636n^3}{n^2}, \tag{6.5}$$

where n is the index of refraction of the vehicle. r_i describes the fraction of light that is reflected at the air–vehicle interface as the light attempts to leave the vehicle and return to the air. It is assumed, on the basis of the discussion in Subsection 6.A, that n does not depend on λ for visible wavelengths.

Define

$$w(\lambda) = \frac{\sigma(\lambda)}{[\alpha(\lambda) + \sigma(\lambda)]}. \quad (6.6)$$

$R_\infty(\lambda)$ is the reflectance predicted by the original K-M model for diffusely incident light and is given by

$$R_\infty(\lambda) = \frac{2 - w(\lambda) - 2[1 - w(\lambda)]^{1/2}}{w(\lambda)}. \quad (6.7)$$

$C(\theta_l, \lambda)$ and $D(\theta_l)$ result from the solution of Reichman's differential equations and are

$$C(\theta_l, \lambda) = \frac{w(\lambda)\cos\theta_l(2\cos\theta_l + 1)}{1 - 4[1 - w(\lambda)]\cos^2\theta_l}, \quad (6.8)$$

$$D(\theta_l) = \frac{2\cos\theta_l - 1}{2\cos\theta_l + 1}. \quad (6.9)$$

In contrast to the highly directional properties of surface reflection, colorant-layer scattering produces diffusely reflected light. Consequently R_B in Eq. (6.4) is defined relative to an ideal diffuse surface. One special case of this model is conservative scattering (also called Lambertian scattering), for which $w(\lambda) = 1$ for all visible wavelengths. The Lambertian-scattering model is used frequently in computer vision.

C. Dichromatic Reflection Model

Shafer proposed the dichromatic reflection model originally for inhomogeneous dielectrics.[1] Mathematically, the dichromatic reflection model states that

$$R(g, \lambda) = R_S(g, \lambda) + R_B(g, \lambda) \quad (6.10)$$

$$= M_S(g)C_S(\lambda) + M_B(g)C_B(\lambda), \quad (6.11)$$

where g indicates dependence on geometry and λ indicates dependence on wavelength. Thus M_S and M_B are assumed to depend on geometry and not on wavelength, whereas C_S and C_B are assumed to depend on wavelength and not on geometry.

Equation (6.10) is equivalent to Eq. (2.1) and is accepted generally for opaque objects with one relevant interface. Extending Eq. (6.10) to (6.11), however, requires more justification. One consequence of Eq. (6.11) is that the surface reflectance $R_S(g, \lambda)$ can be written in the form

$$R_S(g, \lambda) = M_S(g)C_S(\lambda). \quad (6.12)$$

The discussion in Subsection 6.A shows that $R_S(g, \lambda)$ is nearly independent of wavelength. Therefore Eq. (6.12) is a reasonable approximation for surface reflectance from inhomogeneous dielectrics.

Equation (6.11) also suggests that the body reflectance $R_B(g, \lambda)$ can be written in the form

$$R_B(g, \lambda) = M_B(g)C_B(\lambda). \quad (6.13)$$

Shafer provides some justification for (6.13).[1] No attempt, however, was made to demonstrate the compatibility of Eq. (6.13) with a physical model for colorant-layer scattering. An inspection of Eq. (6.4) shows that the expression for $R_B(g, \lambda)$ derived by Reichman cannot be written in the form of Eq. (6.13). Therefore it is of interest to determine how well the simple form of Eq. (6.13) can be used to approximate the more general reflectance function in Eq. (6.4).

An approach similar to that described in Subsections 5.C and 5.D is used to determine how well $R_B(g, \lambda)$ of Eq. (6.4) can be approximated by a function in the form of Eq. (6.13). Let $\bar{R}_B(i, j)$ be the discrete normalized version of $R_B(\theta_l, \lambda)$ in Eq. (6.4). $\bar{R}_B(i, j)$ is normalized in the sense that for fixed angle $i' = \theta_l'$, $\|\bar{R}(i', j)\| = 1$. The discretization is done by considering n values of θ_l in the range $0 \le \theta_l \le \pi/2$ and m values of λ in the range 400 nm $\le \lambda \le$ 700 nm. The error E in an approximation of the form Eq. (6.13) is given by

$$E = \sum_{1\le i\le n}\sum_{1\le j\le m}[\bar{R}_B(i, j) - F_1F_2(j)]^2. \quad (6.14)$$

As in Subsection 5.C, a SVD is used to decompose the matrix $\bar{R}_B$ into

$$\bar{R}_B = UDV, \quad (6.15)$$

and the average-square error in the approximation is given as in Eq. (5.16) by

$$E_n = \frac{1}{n}\sum_{2\le j\le m}\sigma_j^2, \quad (6.16)$$

where σ_j is the jth diagonal element of the matrix D. The fraction of the variance in Eq. (6.4) that is accounted for by Eq. (6.13) is

$$F_1 = \sigma_1^2 \bigg/ \left(\sum_{1\le i\le m}\sigma_i^2\right). \quad (6.17)$$

D. Testing the Dichromatic Model

In this subsection the dichromatic reflection model is tested by consideration of a set of inhomogeneous dielectrics with

Table 4. Quality of Dichromatic Fit[a]

Munsell Sample Number	E_{45}	F_1
19	8.684444E-5	0.999913
24	7.493164E-4	0.999251
41	2.065324E-3	0.997934
77	8.175794E-4	0.999182
89	1.272916E-3	0.998725
114	1.775960E-4	0.999822
138	4.365392E-4	0.999563
167	1.084334E-3	0.998912
179	8.534885E-4	0.999146
208	6.494580E-4	0.999350
214	2.453034E-4	0.999755
233	6.412252E-4	0.999359
249	6.937320E-4	0.999306
287	8.371921E-4	0.999163
293	1.185071E-3	0.998814
315	7.986626E-4	0.999201
322	1.006269E-4	0.999899
337	4.328965E-4	0.999567
373	9.892055E-5	0.999901
393	4.849286E-4	0.999515
403	4.617989E-4	0.999538
421	1.852812E-3	0.998146
428	4.702172E-4	0.999528
454	1.327670E-4	0.999867
456	6.477799E-4	0.999352

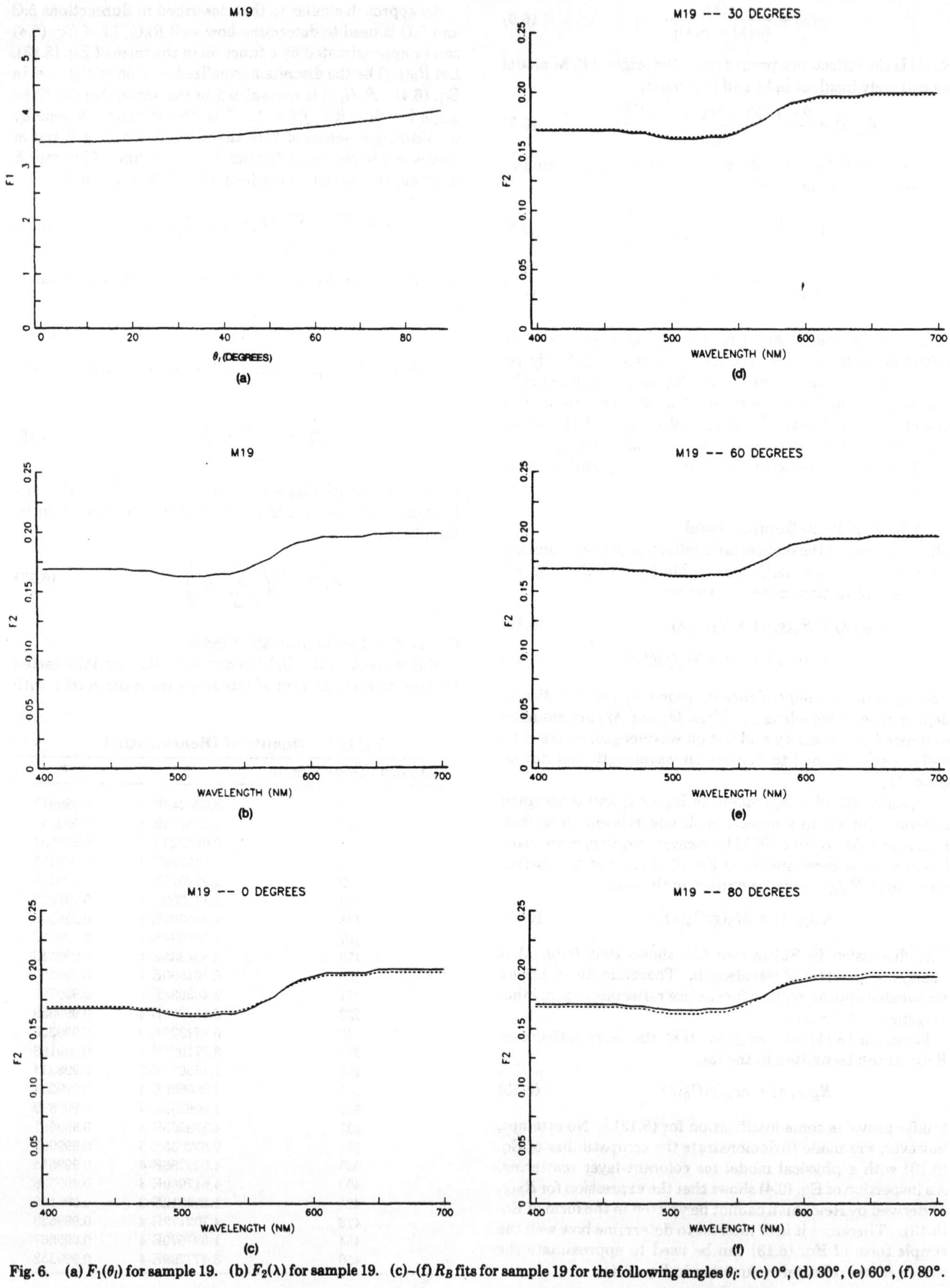

Fig. 6. (a) $F_1(\theta_i)$ for sample 19. (b) $F_2(\lambda)$ for sample 19. (c)–(f) R_B fits for sample 19 for the following angles θ_i: (c) 0°, (d) 30°, (e) 60°, (f) 80°.

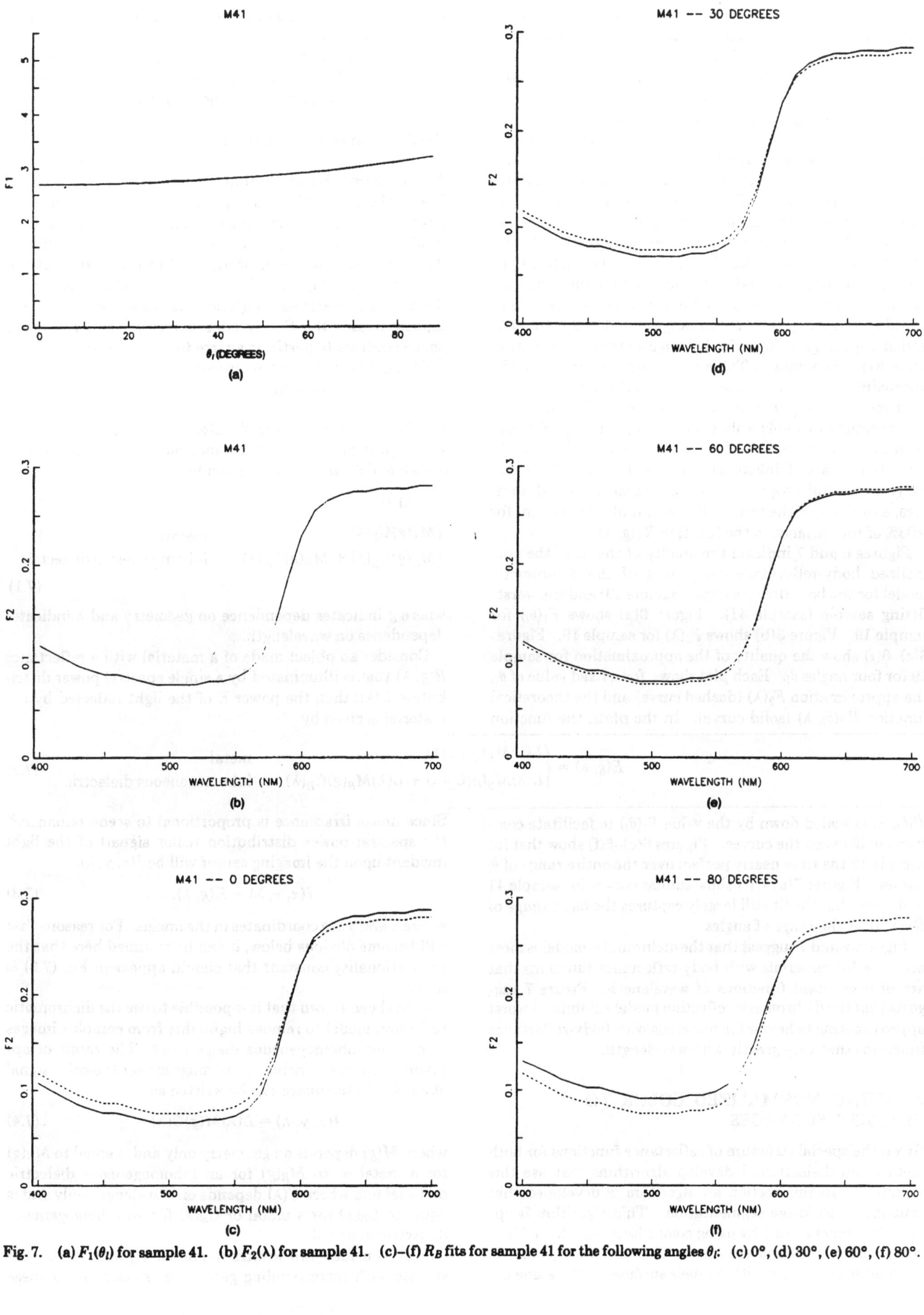

Fig. 7. (a) $F_1(\theta_l)$ for sample 41. (b) $F_2(\lambda)$ for sample 41. (c)–(f) R_B fits for sample 41 for the following angles θ_l: (c) 0°, (d) 30°, (e) 60°, (f) 80°.

widely varying spectral properties. Since the parameters $\alpha(\lambda)$ and $\sigma(\lambda)$ are difficult to measure directly for real materials,[11] optical property data have been constructed with spectral-reflectance data compiled for the Munsell samples.[30] This arrangement permits the simulation of a wide range of scattering and absorbing characteristics.

Twenty-five different inhomogeneous materials are simulated. A vehicle index of refraction of 1.5 across the visible wavelengths is assumed, which is consistent with data for real vehicle materials.[20] For each material, a spectral reflectance function from the Munsell collection is selected at random. This function is taken to be the normal-incidence ($\theta_l = 0$) spectral reflectance of the material. Given this $R_B(\theta_l = 0, \lambda)$ function, it is possible to construct the function $w(\lambda)$ for that material. From $w(\lambda)$, the function $\bar{R}_B(i, j)$ is computed at θ_l spacings of 2° in the range $0° \leq \theta_l \leq 90°$ ($n = 45$) and at λ spacings of 10 nm in the range 400 nm $\leq \lambda \leq$ 700 nm ($m = 31$). The data in Table 4 indicate the quality of the approximation to each material's body reflectance obtained by using an approximation in the form of Eq. (6.13).

The results in Table 4 show that $R_B(g, \lambda)$ in Eq. (6.4) can be approximated well by a function in the form of Eq. (6.13) for a large class of inhomogeneous dielectrics with widely varying spectral properties. In particular, for all 25 samples, a function in the form of Eq. (6.4) is able to account for >99% of the variance in the function $R_B(g, \lambda)$.

Figures 6 and 7 indicate the quality of the fit of the normalized body-reflectance component of the dichromatic model for the best-fitting sample (sample 19) and the worst-fitting sample (sample 41). Figure 6(a) shows $F_1(\theta_l)$ for sample 19. Figure 6(b) shows $F_2(\lambda)$ for sample 19. Figures 6(c)–6(f) show the quality of the approximation for sample 19 for four angles θ_l. Each plot shows, for a fixed value of θ_l, the approximation $F_2(\lambda)$ (dashed curve) and the theoretical function $\bar{R}_B(\theta_l, \lambda)$ (solid curve). In the plots, the function $\bar{F}(\theta_l, \lambda)$ is scaled down by the value $F_1(\theta_l)$ to facilitate comparison between the curves. Figures 6(c)–6(f) show that for sample 19 the fit is nearly perfect over the entire range of θ_l values. Figures 7(a)–7(f) show similar curves for sample 41 and show that the fit still largely captures the basic shape of $\bar{R}_B(\lambda)$ over the range of angles.

Figures 6 and 7 suggest that the dichromatic model is most accurate for materials with body-reflectance functions that are near-constant functions of wavelength. Figure 7 suggests that the dichromatic reflection model exhibits its worst approximating behavior for materials with body-reflectance functions that vary greatly with wavelength.

7. USING NORMALIZED COLOR TO CLASSIFY SURFACES

Given the special structure of reflectance functions for both metals and dielectrics, I develop algorithms that use this structure. In this section an algorithm is developed that contributes to image segmentation. This algorithm is applied to images sensed by using color filters (e.g., R, G, B) as is often done in computer vision. After an initial geometric segmentation, the algorithm labels surfaces in the scene according to material composition. This labeling, in combination with a previously developed color metric,[31] allows surfaces in the scene to be classified on the basis of material composition. This algorithm is relatively insensitive to geometric variations in the scene and is therefore more effective than previous approaches to color segmentation[32,33] when classification based on material is desired.

A. Geometric Segmentation

The surface-classification algorithm is applied to a color image after a geometric segmentation of the corresponding irradiance image. The result of geometric segmentation is that the image has been partitioned into regions that are separated by image irradiance discontinuities. An algorithm for geometric segmentation has been devised.[34] This algorithm uses the Canny edge detector[35] and geometric image relations to partition an irradiance image into regions. This algorithm is used to provide input for the surface-classification algorithm.

B. Stability of Normalized Color

The approximate color-reflectance model discussed in Sections 5 and 6 can be summarized by

$$R(g, \lambda) = \begin{cases} M_S(g)C_S(\lambda) & \text{metal} \\ M_S(g)C_S(\lambda) + M_B(g)C_B(\lambda) & \text{inhomogeneous dielectric} \end{cases}, \tag{7.1}$$

where g indicates dependence on geometry and λ indicates dependence on wavelength.

Consider an object made of a material with a reflectance $R(g, \lambda)$ that is illuminated by a single spectral power distribution $L(\lambda)$; then the power E of the light reflected by the material is given by

$$E(g, \lambda) = \begin{cases} L(\lambda)M_S(g)C_S(\lambda) & \text{metal} \\ L(\lambda)M_S(g)C_S(\lambda) + L(\lambda)M_B(g)C_B(\lambda) & \text{inhomogeneous dielectric} \end{cases}. \tag{7.2}$$

Since image irradiance is proportional to scene radiance,[36] the spectral power distribution (color signal) of the light incident upon the imaging sensor will be $I(x, y, \lambda)$:

$$I(x, y, \lambda) = E(g, \lambda), \tag{7.3}$$

where x and y are coordinates in the image. For reasons that will become obvious below, it can be assumed here that the proportionality constant that should appear in Eq. (7.3) is unity.

It has been shown that it is possible to use the dichromatic reflection model to remove highlights from complex images containing inhomogeneous dielectrics.[5] The result of applying such an algorithm to an image is that the color signal at a point in the image can be written as

$$I(x, y, \lambda) = L(\lambda)M(g)C(\lambda), \tag{7.4}$$

where $M(g)$ depends on geometry only and is equal to $M_S(g)$ for a metal or to $M_B(g)$ for an inhomogeneous dielectric material and where $C(\lambda)$ depends on wavelength only and is equal to $C_S(\lambda)$ for a metal or $C_B(\lambda)$ for an inhomogeneous dielectric material.

Now consider two arbitrary points, P_0 and P_1, on the surface with corresponding geometries g_0 and g_1. If these points are visible in the image, then the corresponding color signals are

$$I_0(x_0, y_0, \lambda) = L(\lambda)M(g_0)C(\lambda), \tag{7.5}$$

$$I_1(x_1, y_1, \lambda) = L(\lambda)M(g_1)C(\lambda), \tag{7.6}$$

where (x_0, y_0) are the coordinates of the image of P_0 and (x_1, y_1) are the coordinates of the image of P_1. Define the normalized color of an image point (x, y) by

$$\bar{I}(x, y, \lambda) = \frac{I(x, y, \lambda)}{\|I(x, y, \lambda)\|}, \tag{7.7}$$

where $\|\cdot\|$ is the $L^1[\lambda_1, \lambda_2]$ norm[37] and $[\lambda_1, \lambda_2]$ is the visible wavelength range. The normalized color of points (x_0, y_0) and (x_1, y_1) is

$$\bar{I}_0(x_0, y_0, \lambda) = \frac{L(\lambda)M(g_0)C(\lambda)}{\|L(\lambda)M(g_0)C(\lambda)\|} = \frac{L(\lambda)C(\lambda)}{\|L(\lambda)C(\lambda)\|}, \tag{7.8}$$

$$\bar{I}_1(x_1, y_1, \lambda) = \frac{L(\lambda)M(g_1)C(\lambda)}{\|L(\lambda)M(g_1)C(\lambda)\|} = \frac{L(\lambda)C(\lambda)}{\|L(\lambda)C(\lambda)\|}. \tag{7.9}$$

Thus the approximate color-reflectance model yields the important result that, given any metal or inhomogeneous dielectric material, the normalized color of all image points of that material illuminated by a single spectral power distribution will be identical, provided that a method is used first to remove highlights from inhomogeneous dielectrics. The key to this result is, of course, the independence of geometry and wavelength in the statement of Eq. (7.1). This independence allows the geometric contribution to the color signal to be factored out, leaving normalized color, which depends on properties of the material and illumination only.

C. Classifying Surfaces

In this subsection an algorithm is described that classifies surfaces in a scene according to material composition. The input to the algorithm is a color image obtained by using three standard color filters and a partition of the irradiance image into regions. In general, the success of the algorithm would require the removal of highlights from inhomogeneous dielectrics. In the images that we have considered so far, however, such highlights are present over only a small part of the image. Thus a highlight-removal step has not been required for inhomogeneous dielectrics. It is assumed that a single spectral power distribution of light illuminates the scene.

The first step in the classification algorithm is to label each image region according to the normalized color of the region. Healey and Binford[31] described a method for estimating the normalized color at a point in an image. This method is based on approximating the incoming color signal from color pixel values by using the spectral properties of the sensors and a finite-dimensional linear approximation to represent the color signal.[38,39] By using such a method, an average normalized color is computed for each image region.

Given the noise properties of real sensors, it is unlikely that two image regions corresponding to surfaces of the same material will have exactly the same average normalized color. In order to manage this situation, a metric for normalized color has been developed[31] that takes into account the noise properties of color image sensors. Thus normalized color space can be considered a metric space. A distance d is defined on the basis of sensor noise properties such that regions within a distance d in normalized color space have a high probability of corresponding to surfaces in the scene with the same reflectance function. By using this distance d, equivalence classes of regions are determined from computed normalized colors.

Each equivalence class of regions corresponds to surfaces in the scene that have similar $R(g, \lambda)$ functions. If an image contains a small number of regions, it is likely that an equivalence class of regions corresponds to a set of surfaces made of the same material. Also, unless there is evidence to the contrary, adjacent image regions that are elements of the same equivalence class probably correspond to surfaces of one material on a single object that are separated by a geometrical discontinuity.

D. Experimental Results

In this subsection the performance of the algorithm of Subsection 7.C is demonstrated with an image of a valve part (Fig. 8). The valve part has both metal and dielectric surfaces made of several different materials. Figure 9 shows the results of an edge detection on the image done by using Canny's edge detector.[5] Figure 10 shows some prominent regions that have been picked by hand, using the edges of Fig. 9. Figure 11 shows the coordinates in normalized color

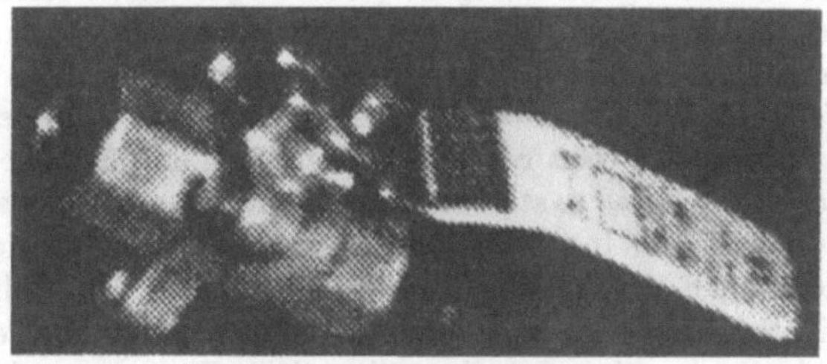

Fig. 8. Image of a valve.

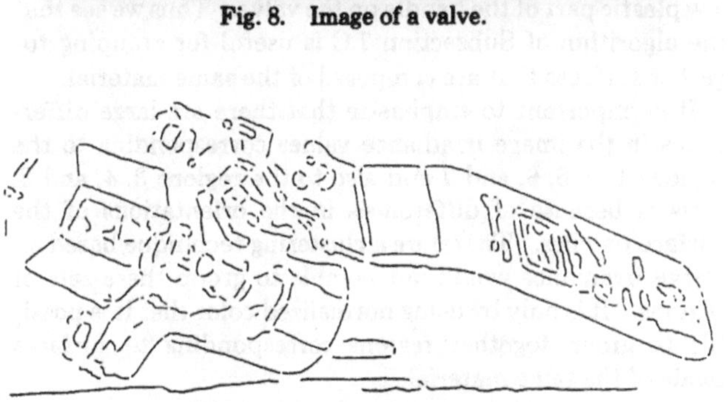

Fig. 9. Result of edge detection.

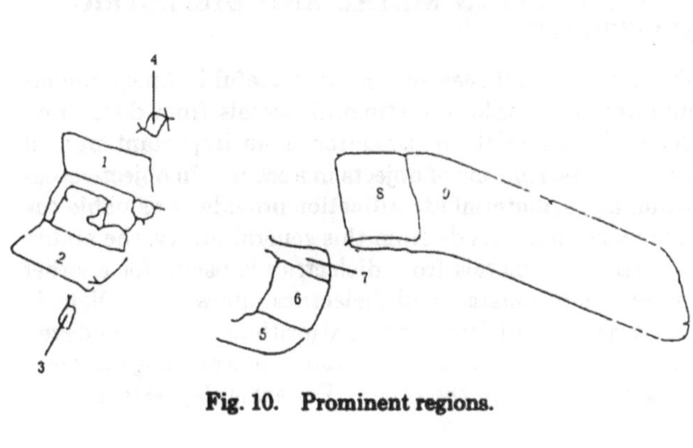

Fig. 10. Prominent regions.

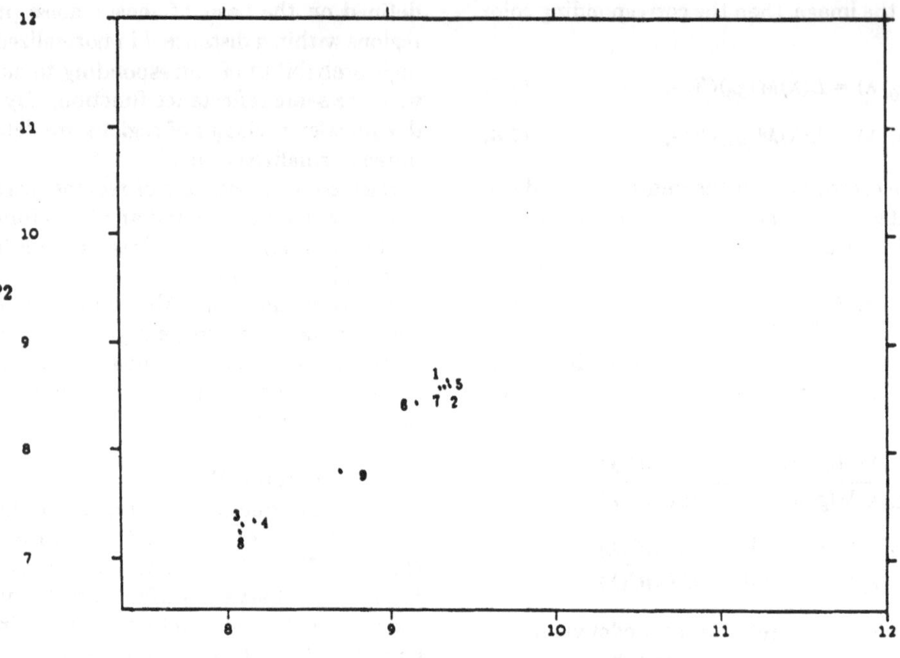

Fig. 11. Regions in normalized color space.

space[31] of the regions in Fig. 10. Three equivalence classes of regions are found. The first equivalence class contains regions 1, 2, 5, 6, and 7. These regions in fact correspond to surfaces of the same brass material on the valve. The second equivalence class contains regions 3, 4, and 8. These regions correspond to steel surfaces on the valve. Regions 3 and 4 correspond to steel bolt heads on the valve, and region 8 corresponds to the steel part of the handle. The third equivalence class contains region 9. This region is the yellow plastic part of the handle on the valve. Thus we see that the algorithm of Subsection 7.C is useful for grouping together surfaces that are composed of the same material.

It is important to emphasize that there are large differences in the image irradiance values corresponding to the regions 1, 2, 5, 6, and 7 and also to the regions 3, 4, and 8. This is because of differences in the orientations of the surface patches. Therefore a clustering technique based on image irradiance would not be able to group these sets of regions. It is only by using normalized color that it is possible to group together regions corresponding to surfaces made of the same material.

8. IDENTIFYING METAL AND DIELECTRIC MATERIALS

There are several reasons why it is useful in image understanding to be able to distinguish metals from dielectrics. Certainly, material identification is an important part of building descriptions of objects in a scene. In object-recognition tasks, material identification provides a valuable cue for recognition. Aside from this general utility, the ability to distinguish metals from dielectrics is useful for another reason. Since metals and dielectrics interact with light in fundamentally different ways, algorithms have been developed in computer vision that apply to one of these classes of material but not to the other. For example, certain vision algorithms require the identification of a dielectric material in a scene.[4,26] Other algorithms are better suited for metals.[40] Before these algorithms can be applied to images, it is necessary to determine whether a material in the scene is a metal or a dielectric.

A. Preliminary Classification Algorithm

Several properties of metals and dielectrics may allow them to be distinguished in color images.[14,41] In this subsection I use the approximate color-reflectance model to derive a preliminary algorithm for distinguishing a metal from an inhomogeneous dielectric. The algorithm assumes that image locations corresponding to a single object have been isolated and that some of the surface-reflected light from inhomogeneous dielectrics reaches the image sensor.

From Eqs. (7.2) and (7.3) it follows that the color signal at a point (x, y) in an image will have the form

$$I(x, y, \lambda) = a_{xy}f(\lambda) \tag{8.1}$$

for a metal or

$$I(x, y, \lambda) = b_{xy}g_1(\lambda) + c_{xy}g_2(\lambda) \tag{8.2}$$

for an inhomogeneous dielectric, in which a_{xy}, b_{xy}, and c_{xy} are constants depending on the image location. Assume that N different spectral sensors are used (e.g., $N = 3$ for R, G, B). At each image location (x, y), N sensor values S_i are measured with

$$S_i = \int_\lambda I(x, y, \lambda)s_i(\lambda)d\lambda \quad (1 \le i \le N), \tag{8.3}$$

where $s_i(\lambda)$ is the spectral sensitivity of the ith sensor.

For a metal, the sensor measurements will lie in a one-dimensional vector space given by

$$[S_1 \cdots S_N] = a_{xy}[\alpha_1 \cdots \alpha_N], \tag{8.4}$$

where

$$\alpha_i = \int_\lambda f(\lambda)s_i(\lambda)\mathrm{d}\lambda. \tag{8.5}$$

For an inhomogeneous dielectric, the sensor measurements will lie in a two-dimensional vector space given by

$$[S_1 \ldots S_N] = b_{xy}[\beta_1 \ldots \beta_N] + c_{xy}[\gamma_1 \ldots \gamma_N], \tag{8.6}$$

where

$$\beta_i = \int_\lambda g_1(\lambda)s_i(\lambda)\mathrm{d}\lambda, \qquad \gamma_i = \int_\lambda g_2(\lambda)s_i(\lambda)\mathrm{d}\lambda. \tag{8.7}$$

Consider sensor measurements made at P locations in an image. Let S be a $P \times N$ matrix for which $S(i, j)$ refers to the jth sensor measurement at the ith image location. The dimensionality of the data in S may be determined by using principal-components analysis.[18] A new matrix $\bar{S}$ is formed by normalizing the rows of S so that, for any row i',

$$\sum_{1 \le j \le N} [\bar{S}(i', j)]^2 = 1. \tag{8.8}$$

This normalization prevents image locations with large sensor values (such as those that occur at specularities) from dominating the dimensionality estimate. $\bar{S}$ can be decomposed by using a SVD into

$$\bar{S} = UDV, \tag{8.9}$$

where U is a $P \times P$ unitary matrix, V is an $N \times N$ unitary matrix, and D is the diagonal matrix $D = \mathrm{diag}(\sigma_1, \sigma_2, \ldots, \sigma_N)$, with $|\sigma_1| > |\sigma_2| > \ldots > |\sigma_n|$.

The rows of V denoted by $\mathbf{v}_1, \mathbf{v}_2, \ldots, \mathbf{v}_N$ can be considered an orthonormal basis for representing the normalized sensor measurements in the rows of $\bar{S}$. From Eq. (8.9), the ith row of $\bar{S}$ can be represented as

$$\bar{S}_i(j) = \sum_{1 \le j \le N} \sigma_k u_{ik} \mathbf{v}_k. \tag{8.10}$$

If we take $D < N$, then we have an approximation to $\bar{S}_i(j)$ given by

$$\bar{S}_i{}^* = \sum_{1 \le k \le D} \sigma_k u_{ik} \mathbf{v}_k. \tag{8.11}$$

The square error in this approximation is

$$\|\bar{S}_i(j) - \bar{S}_i{}^*\|^2 = \sum_{(D+1) \le j \le N} (\sigma_j u_{ij})^2, \tag{8.12}$$

and the average-square error over all rows is

$$E_D = \frac{1}{P} \sum_{1 \le i \le P} \sum_{(D+1) \le j \le N} (\sigma_j u_{ij})^2 = \frac{1}{P} \sum_{(D+1) \le j \le N} \sigma_j{}^2. \tag{8.13}$$

It can be shown that, for any D in $1 \le D \le N$, the vectors $\mathbf{v}_1, \mathbf{v}_2, \ldots, \mathbf{v}_n$ allow a smaller value of E_D than any other set of D vectors.

If Eqs. (8.4) and (8.6) hold exactly and there is no noise in the sensor measurements, then, when considering pixels in the image of a metal, we have $E_1 = 0$. For pixels in the image of an inhomogeneous dielectric, we have $E_1 > 0$ and $E_2 = 0$. These relationships form the basis of a preliminary algorithm for classifying a material as either a metal or an inhomogeneous dielectric.

Table 5. Computed E_D Values for Aluminum and Plastic

Material	$10^5 E_1$	$10^5 E_2$
Aluminum	13.8	3.9
Plastic	108.6	4.1

B. Experimental Results

In this subsection the results are shown for the application of the algorithm described in Subsection 8.A to regions of two images. The first image is of an aluminum cylinder illuminated by a single source. A 100-pixel region on the cylinder is used for analysis. The computed values of E_1 and E_2 are shown in Table 5. The second image is of a blue plastic cup illuminated by a single source. A 100-pixel region on the cup including areas of both surface reflection and body reflection is used for analysis. Table 5 gives the computed values of E_1 and E_2. From Table 5 we see that the value of E_1 is nearly 8 times larger for the plastic cup. It is unlikely that such a large effect is due to noise alone. If we assume that some small value of E_D, say, $10^5 E_D = 20$, can be attributed to sensor noise and modeling approximation error, then the algorithm of Subsection 8.A is able to classify the aluminum cylinder as a metal and the plastic cup as an inhomogeneous dielectric.

9. SUMMARY

In this paper several contributions are made to computer vision. In the first part of this paper a comprehensive description of the physics of reflection is presented. Detailed models are presented for both surface (interface) reflection and body reflection. These models apply to both metal and dielectric materials. A new model for body reflection is introduced from the optics literature,[7] one that has not been used before (to my knowledge) in either computer vision or computer graphics. These physical models are used to show that reflectance functions of geometry and wavelength have a special structure for a large class of materials. Two algorithms have been developed that use the structure of reflectance functions.

The most important contribution of this study is the demonstration of the near independence of the spectral composition and geometrical scaling of the light reflected from a surface. A unichromatic reflectance model for metals is proposed and analyzed. The dichromatic model[1] for inhomogeneous dielectrics is verified rigorously. The independence of wavelength and geometry in reflectance functions allows geometrical effects in a scene to be factored out of color pixel values in an image. This allows algorithms to work on a representation of a color image in which contributions to pixel values that are due to illumination and material properties are preserved but in which factors that are due to scene geometry have been removed.

The analysis of the structure of reflectance functions has led to two algorithms. The first algorithm classifies surfaces according to material composition. The second algorithm is used to identify metals and dielectrics in color images. These algorithms are designed to account for properties of real sensors such as noise and spectral sensitivity. Both algorithms have produced encouraging results on real color images.

APPENDIX A

In this appendix the steps required to obtain the result in Eq. (5.14) are outlined. From Eq. (5.8), the ith row vector of $\bar{F}(i, j)$ is

$$\bar{F}(i, j) = \sum_{1\le j\le m} (\sigma_j u_{ij}) v_j,$$

where v_j denotes the jth row vector of the matrix V. Therefore, from Eqs. (5.9) and (5.10), we have

$$\|\bar{F}(i, j) - F_1{}^*(i)F_2{}^*(j)\|^2 = \left\| \left[\sum_{1\le j\le m} (\sigma_j u_{ij}) v_j \right] - \sigma_1 u_{i1} v_1 \right\|^2$$

$$= \left\| \sum_{2\le j\le m} (\sigma_j u_{ij}) v_j \right\|^2$$

$$= \sum_{2\le j\le m} \|(\sigma_j u_{ij}) v_j\|^2,$$

where the last step follows from the Pythagorean relation in an inner product space[27] and the fact that the rows of V are orthogonal since it is a unitary matrix. Since the rows of V also have unit length, we get the result that

$$\|\bar{F}(i, j) - F_1{}^*(i)F_2{}^*(j)\|^2 = \sum_{2\le j\le m} (\sigma_j u_{ij})^2.$$

ACKNOWLEDGMENTS

I thank Tom Binford, W. E. Blanz, and Brian Wandell. Support for this work was provided in part by the U.S. Air Force Office of Scientific Research under contract F33615-85-C-5106 and by the Advanced Research Projects Agency of the U.S. Department of Defense under Knowledge Based Vision ADS subcontract S1093 S-1.

REFERENCES

1. S. Shafer, "Using color to separate reflection components," Color Research and Application **10**, 210–218 (1985); also available as Tech. Rep. TR 136 (University of Rochester, Rochester, N.Y., 1984).
2. B. K. P. Horn, "Obtaining shape from shading information," in *The Psychology of Computer Vision*, P. Winston, ed. (McGraw-Hill, New York, 1975), pp. 115–155.
3. V. Weisskopf, "How light interacts with matter," Sci. Am. **219**, 60–71 (1968).
4. G. Klinker, S. Shafer, and T. Kanade, "Using a color reflection model to separate highlights from object color," in *Proceedings of the First International Conference on Computer Vision* (Institute of Electrical and Electronics Engineers, New York, 1987), pp. 145–150.
5. G. Klinker, S. Shafer, and T. Kanade, "Image segmentation and reflection analysis through color," in *Proceedings of the Image Understanding Workshop* (Morgan Kaufmann, San Mateo, Calif., 1988), pp. 838–853.
6. H. C. Lee, "Method for computing the scene-illuminant chromaticity from specular highlights," J. Opt. Soc. Am. A **3**, 1694–1699 (1986).
7. J. Reichman, "Determination of absorption and scattering coefficients for nonhomogeneous media. 1: theory," Appl. Opt. **12**, 1811–1815 (1973).
8. K. Torrance and E. Sparrow, "Theory for off-specular reflection from roughened surfaces," J. Opt. Soc. Am. **57**, 1105–1114 (1967).
9. M. Born and E. Wolf, *Principles of Optics* (Pergamon, New York, 1959).
10. R. S. Hunter, *The Measurement of Appearance* (Wiley, New York, 1975).
11. W. Egan and T. Hilgeman, *Optical Properties of Inhomogeneous Materials* (Academic, New York, 1979).
12. B. K. P. Horn, *Robot Vision* (McGraw-Hill, New York, 1986).
13. R. Siegel and J. Howell, *Thermal Radiation Heat Transfer* (McGraw-Hill, New York, 1981).
14. G. Healey and W. E. Blanz, "Identifying metal surfaces in color images," to be published in *Optics, Electro-Optics, and Sensors*, Proc. Soc. Photo-Opt. Instrum. Eng. (1988).
15. P. Beckmann and A. Spizzichino, *The Scattering of Electromagnetic Waves from Rough Surfaces* (Macmillan, New York, 1963).
16. R. Cook and K. Torrance, "A reflectance model for computer graphics," Comput. Graphics **15**, 307–316 (1981).
17. N. Mott and H. Jones, *The Theory of the Properties of Metals and Alloys* (Oxford U. Press, Oxford, 1958).
18. J. Chambers, *Computational Methods for Data Analysis* (Wiley, New York, 1977).
19. K. Atkinson, *An Introduction to Numerical Analysis* (Wiley, New York, 1978).
20. D. Gray, ed., *American Institute of Physics Handbook*, 3rd ed. (McGraw-Hill, New York, 1972).
21. E. Palik, ed., *Handbook of Optical Constants of Solids* (Academic, New York, 1985).
22. J. Weaver and C. Krafka, *Physics Data: Optical Properties of Metals* (Fachinformationszentrum, Karlsruhe 1981).
23. E. Sparrow and R. Cess, *Radiation Heat Transfer* (Hemisphere, Washington, D.C., 1978).
24. H. C. Lee, E. Breneman, and C. Schulte, "An experimental study of a color reflection model," Tech. Rep. 3418-86 (Eastman Kodak, Rochester, N.Y., 1986).
25. M. D'Zmura and P. Lennie, "Mechanisms of color constancy," J. Opt. Soc. Am. A **3**, 1662–1672 (1986).
26. G. Healey and T. O. Binford, "Color algorithms for a general vision system," in *Proceedings of the Tenth International Joint Conference on Artificial Intelligence*, J. McDermott, ed. (Morgan Kaufmann, Los Altos, Calif., 1987), pp. 759–762.
27. P. Kubelka and F. Munk, "Ein Beitrag zur Optik der Farbanstriche," Z. Techn. Physik. **12**, 593 (1931).
28. S. Orchard, "Reflection and transmission of light by diffusing suspensions," J. Opt. Soc. Am. **59**, 1584–1597 (1969).
29. J. Reichman, "Determination of absorption and scattering coefficients for nonhomogeneous media. 2: experiment," Appl. Opt. **12**, 1816–1823 (1973).
30. D. Nickerson, "Spectrophotometric data for a collection of munsell samples," (U.S. Department of Agriculture, Washington, D.C., 1957).
31. G. Healey and T. O. Binford, "A color metric for computer vision," in *Proceedings of the IEEE Conference on Computer Vision and Pattern Recognition*, (Institute of Electrical and Electronics Engineers, New York, 1988), pp. 10–17.
32. R. Ohlander, K. Price, and D. Reddy, "Picture segmentation using a recursive region splitting method," Comput. Graphics Image Process **8**, 313–333 (1978).
33. Y. Ohta, T. Kanade, and T. Sakai, "Color information for region segmentation," Comput. Graphics Image Process. **13**, 222–241 (1980).
34. T. Sumanaweera, G. Healey, B. U. Lee, T. O. Binford, and J. Ponce, "Image segmentation using geometrical and physical constraints," in *Proceedings of the Image Understanding Workshop* (Morgan Kaufmann, San Mateo, Calif., 1988), pp. 1091–1099.
35. J. Canny, "A computational approach to edge detection," IEEE Trans. Pattern Anal. Mach. Intell. **PAMI-8**, 679–697 (1986).

36. B. K. P. Horn and R. Sjoberg, "Calculating the reflectance map," Appl. Opt. **18**, 1770–1779 (1979).
37. E. Kreyszig, *Introductory Functional Analysis with Applications* (Wiley, New York, 1978).
38. B. K. P. Horn, "Exact reproduction of colored images," Comput. Vision Graphics Image Process. **26**, 135–167 (1984).
39. B. Wandell, "The synthesis and analysis of color images," IEEE Trans. Pattern Anal. Mach. Intell. **PAMI-9**, 2–13 (1987).
40. G. Porter and J. Mundy, "Automatic visual inspection of metal surfaces," in *Techniques and Applications of Image Understanding*, J. J. Pearson, ed., Proc. Soc. Photo-Opt. Instrum. Eng. **281**, 176–181 (1981).
41. G. Healey and T. O. Binford, "Predicting material classes," in *Proceedings of the Image Understanding Workshop* (Morgan Kaufmann, San Mateo, Calif., 1988), pp. 1140–1146.

Standard surface-reflectance model and illuminant estimation

Shoji Tominaga

Faculty of Engineering, Osaka Electro-Communication University, 18-8 Hatsu-cho, Neyagawa, Osaka 572, Japan

Brian A. Wandell

Department of Psychology, Stanford University, Stanford, California 94305

Received May 2, 1988; accepted November 17, 1988

In the standard reflectance model for inhomogeneous materials it is assumed that light is reflected by two independent mechanisms. One component is reflected at the interface of the material and air. Light reflected by this mechanism does not interact with surface colorant, and its spectral composition is assumed to equal that of the incident light. The second component is reflected after entering and interacting with the subsurface structure of the material. This interaction substantially changes the spectral composition of the reflected light. We adopt a vector analysis technique for testing the standard reflectance model. Further, we develop a computational method to determine the components of the observed spectra, and we obtain an estimate of the illuminant without using a reference white standard. Finally, we evaluate the accuracy of the standard model and the feasibility of the illuminant spectral estimation by using several test objectives.

1. INTRODUCTION

The visual information that we use to judge the color appearance of an object is derived from the light reflected from the object and the light reflected from nearby objects. The spectral composition of the reflected light depends on the surface-reflectance properties of the objects and the spectral composition of the ambient lighting. For many broadband light sources an object's color appearance often is predicted more closely by the object's surface reflectance than by the spectral composition of the light reflected from the object, which depends on the ambient lighting. This rough approximation, that the effects of the spectral variation of the ambient lighting are discounted and the object appearance is approximately constant, is called color constancy.

The human ability to discount the effect of some changes in the ambient lighting has motivated the construction of a number of algorithms to demonstrate how this feat may be accomplished. In most of these algorithms significant assumptions are made about the surface-reflectance properties of objects. Some of these algorithms are designed to estimate object surface-reflectance functions from the responses recorded by a small number of spectrally distinct sensors even when the ambient lighting is unknown. Within this group of algorithms, some are based on models of surface specularity,[1–3] and others are based on other assumptions concerning the properties of surface-reflectance functions.[4–6] Accurate modeling of surface-reflectance functions has become important in the field of computer graphics,[7] where knowledge of surface-reflectance properties permits more accurate rendering of object appearance.

From these algorithms we can now appreciate the close relationship between the mathematical structure of the surface-reflectance functions and the algorithms that can accurately estimate or render surface properties. As we learn more about the mathematical properties of the surface-reflectance functions, reflectance estimation can be accomplished by using fewer sensors or using more-efficient algorithms.

In the fields of computer graphics and computer vision, a few central ideas are used widely to summarize the mathematical structure of surface-reflectance functions on inhomogeneous substances, which are substances composed of different component materials, such as a vehicle at the surface layer and embedded pigments at the colorant layer. For example, plastics and paints are inhomogeneous. Typical examples of homogeneous objects are metals and crystals. For homogeneous materials, the incident light suffers no scattering; the light is reflected in accordance with Fresnel's law.[8] Homogeneous objects are not represented by a simple reflectance model.

In all reflectance models it is postulated that light reflected from an object is composed of light reflected from two different physical paths. Some light is reflected at the interface between the object's surface and the air. This type of reflection is mirrorlike; light reflected in this way from a point source can be seen only over a narrow range of viewing geometries. This light is said to be the specular (or interface) reflection.

The second reflectance path occurs for light that crosses the object's interface. Once the light enters the pigment colorant layer, significant scattering of the light among the pigment particles in this layer occurs. The light is observed only after it crosses the surface boundary again into the air. Light reflected from this type of path is said to be the diffuse (or subsurface) reflection.

The characterization of the reflected light by the additive component of a small number of terms is implicit in the studies of many authors and was summarized well by Cook and Torrance.[7] Beyond this general formulation, Shafer[9] proposed that, for measurement purposes, specifically for computer vision, several approximations are useful. These approximations are based on the assumption that for inhomogeneous materials the wavelength functions are invariant with respect to the angle of the incident light. Shafer refers to these ideas as the dichromatic reflection model. We believe that the ideas draw on such widely held approximations that the collection of ideas might well be called the standard reflectance model.

First, in the standard reflectance model it is assumed that the ratio between the spectral composition of the light that enters the material and the light that leaves the material is the same for all angles. This ratio of the incident and reflected light is the surface-reflectance function of the diffuse component. Second, in the standard model the specularly reflected light is approximated as being invariant, except for a change in intensity, with respect to viewing geometry. Fresnel's equations for specular reflectance do vary with respect to angle, but this variation is asserted to be sufficiently small (see, e.g., Ref. 9) as to be negligible for many observational conditions.

Although the standard reflectance model is used widely to describe empirical properties of data, there are relatively few experimental results to evaluate the adequacy of the model. Recently Klinker *et al.*[1] used the model to analyze color images including highlights. Their measurements are based on three-color-filter (RGB) camera data. Lee *et al.*[10] presented some spectral wavelength experimental data to show that the standard reflectance model is reasonable for some objects. These models were based on the assumption of the two components of light reflection.

Given the importance of the model for applications, we believed that some further empirical examination of the model and some further examination of algorithms that depend on the model would be of value. In this paper we describe new measurements of the spectral power distributions of lights reflected from objects under a variety of geometric conditions. We use a principal-component decomposition of these observed spectra to test the adequacy of the two-component reflection model.

Several authors have pointed out that, if the standard model is correct and if the spectral composition of the specular component of the light is the same as the spectral composition of the incident light,[11] then some simple algorithms can be designed to estimate the spectral power distribution of the light from sensor data. For example, Lee[2] proposed a method for computing the illuminant chromaticity from the color coordinates of specular highlights, as did D'Zmura and Lennie.[3] In this paper we extend the work of these authors. We define an algorithm that permits one to combine information from two or more surfaces to estimate the spectral composition of the ambient lighting. We show that the problem of estimating the illuminant spectrum is reduced to that of finding the common spectral information from all the measurements from the surfaces.

The illuminant estimation method that we describe can be useful for remote sensing applications. When we measure light reflected from objects at a distance, say, 100 km away, we cannot place a white standard at the remote site. Our method can be used to infer the illuminant spectral power distributions at remote image locations. However, the reader should note that the standard reflectance model applies to inhomogeneous surfaces and is not appropriate for homogeneous materials, such as metallic surfaces. Highlights from metallic surfaces are not equal to the illuminant spectral power distribution.

In Section 2 we give a mathematical formulation of the standard reflectance model. In Section 3 we state the basic principle of our spectral analysis approach to the color reflection problem. The computational method is developed in Section 4. In Section 5 we present our experimental results. The performance and the application are discussed in Section 6.

2. STANDARD REFLECTANCE MODEL

It is generally supposed that light reflected from an object surface can be decomposed into two additive components, the interface reflection and the subsurface reflection (see Fig. 1). In the general case the radiance of the reflected light Y is described as a function of wavelength and geometric parameters,

$$Y(\theta, \lambda) = Y_I(\theta, \lambda) + Y_S(\theta, \lambda), \tag{1}$$

where the wavelength λ ranges over a visible spectrum and the parameters θ include the direction angles needed for describing a reflection geometry, such as the incident angle, the viewing angle, and the phase angle. The subscripts I and S denote the interface and subsurface components, respectively. The key additional assumption of the standard reflectance model is the separability between the wavelength composition and the geometrical conditions.

Additivity and Separability

The spectral composition of each reflected light component is independent of the geometric conditions; that is, the respective components of the spectral radiances Y_I and Y_S are separated into the wavelength factor and the geometric variables as follows:

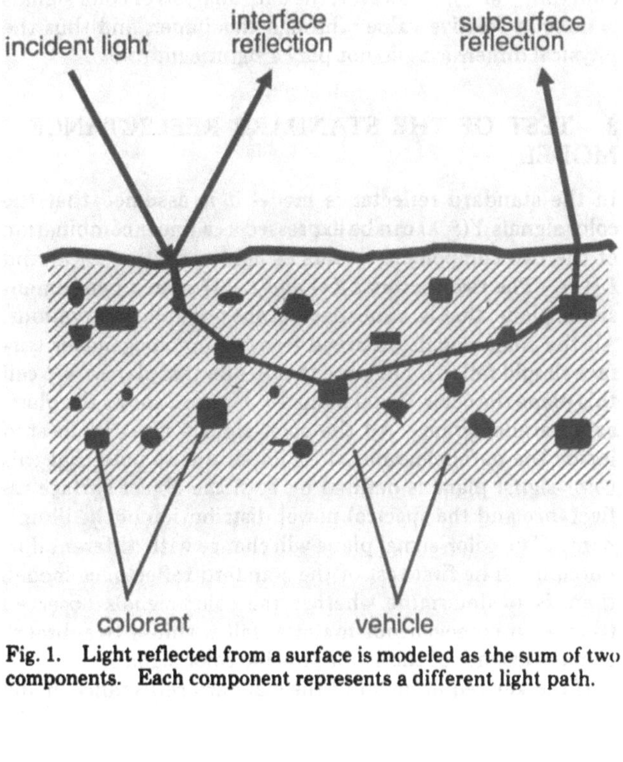

Fig. 1. Light reflected from a surface is modeled as the sum of two components. Each component represents a different light path.

$$Y(\theta, \lambda) = c_I(\theta)L_I(\lambda) + c_S(\theta)L_S(\lambda). \quad (2)$$

The terms $L_I(\lambda)$ and $L_S(\lambda)$ are the spectral power distributions of the interface and subsurface reflection components, respectively. The terms $c_I(\theta)$ and $c_S(\theta)$ are the geometric scale factors.

Equation (2) states that the color signals from an inhomogeneous object are described by the linear combination of two spectral power distributions. The weighting coefficients of each term are geometric scale factors. The important aspect of this assumption is that the spectral composition of each of the reflection components is unchanged as the viewing angle varies.

If we wish to estimate the surface-reflectance functions from the color signal Y, then we must express the reflection model in terms of the surface reflectance of an object and the spectral power distribution of an incident light. Let $S_I(\lambda)$ and $S_S(\lambda)$ be the spectral surface reflectances for the two components. Let $E(\lambda)$ be the spectral power distribution of the incident light. We can represent the reflected light as

$$Y(\theta, \lambda) = c_I(\theta)S_I(\lambda)E(\lambda) + c_S(\theta)S_S(\lambda)E(\lambda). \quad (3)$$

The interface component of reflectance $S_I(\lambda)$ is determined by Fresnel's law. If the refractive index at the surface is constant over the visible wavelength, then the interface component is independent of wavelength. It was reported that many types of material serving as vehicles are some sort of oil and have constant indices of refraction.[2] For these surfaces, the specular reflection appears to have the same color as the illuminant source. A satisfactory approximation for many materials is as follows.

Constant Specular Reflectance

The interface component of surface spectral reflectance is constant over the visible wavelength as $S_I(\lambda)$ = constant.

Note that, from radiometric concepts, the spectral reflectance has the dimension of inverse nanometer inverse steradian ($nm^{-1}\,sr^{-1}$). However, the data analysis of color signals is done in relative values through this paper, and thus the physical dimensions do not play a significant role.

3. TEST OF THE STANDARD REFLECTANCE MODEL

In the standard reflectance model it is assumed that the color signals $Y(\theta, \lambda)$ can be expressed as a linear combination of the two component vectors of light reflection $L_I(\lambda)$ and $L_S(\lambda)$. The two vectors $L_I(\lambda)$ and $L_S(\lambda)$ span a two-dimensional plane, that is, a subspace, of the possible observations. All the color signals observed from an inhomogeneous surface should fall in a subspace that lies on this plane. We call this plane the color-signal plane P. Figure 2 shows the plane as a parallelogram. All the color signals must be located inside the parallelogram. It is important to note that this color-signal plane is defined by both the object surface reflectance and the spectral power distribution of the illuminant. The color-signal plane will change with different illuminants. The first test of the standard reflectance model, then, is to determine whether the color signals observed from an inhomogeneous material fall within a two-dimensional subspace of the set of possible color signals.

If the key requirement of the standard reflectance model

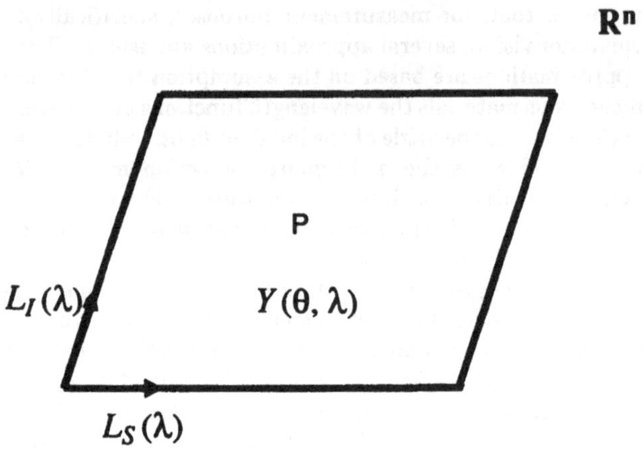

Fig. 2. Color signal plane by a linear combination of the interface reflection component $L_I(\lambda)$ and the subsurface reflection component $L_S(\lambda)$.

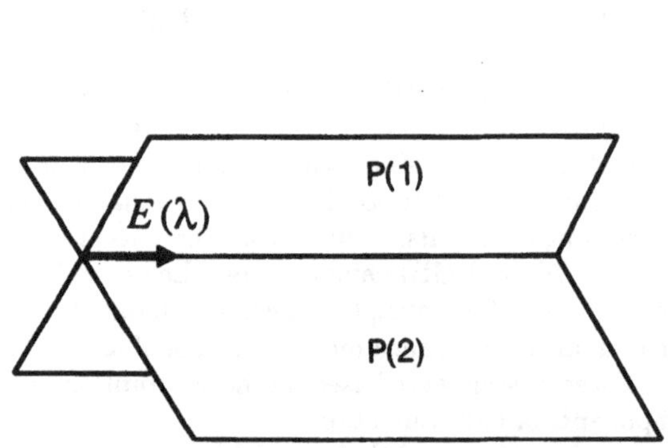

Fig. 3. The intersection of two color-signal planes P(1) and P(2).

is met, namely, additivity and separability, then we may proceed to test the additional assumption of constant specular reflectance. An important practical consequence of this assumption is that the color-signal information can be used to estimate the spectral power distribution of the illuminant. Conversely, the reliability of the estimate may also be used to test the assumption of constant specular reflectance, which is the method that we use here.

Suppose that we measure color signals reflected from two surfaces of inhomogeneous material under the same light source. When we measure the reflected lights with different geometric conditions, then by the first assumption the data will form two different color-signal planes P(1) and P(2) in the space. If the assumption of constant specular reflectance is true, then the spectral power distributions of color signals for the surfaces can be described as follows:

$$Y_1(\theta, \lambda) = c_{I_1}(\theta)E(\lambda) + c_{S_1}(\theta)S_{S_1}(\lambda)E(\lambda), \quad (4)$$

$$Y_2(\theta, \lambda) = c_{I_2}(\theta)E(\lambda) + c_{S_2}(\theta)S_{S_2}(\lambda)E(\lambda). \quad (5)$$

The planes $P(i)(i = 1, 2)$ are constructed by a set of two vectors $[E(\lambda), S_{S_i}(\lambda)E(\lambda)]$. Note that the illuminant vector $E(\lambda)$ is contained in both planes. This vector represents the common spectral information that is due to the light source. The two planes must intersect at this vector, as shown in Fig. 3.

We test the assumption of constant specular reflectance as follows. If the assumption is true, then two color-signal planes from objects under the same illuminant must intersect. Moreover, the vector of their intersection must be in the direction of the illuminant vector. As a useful practical consequence, the standard reflectance model along with the assumption of constant specular reflectance predicts that we can estimate the illuminant spectrum from the color signals reflected from two object surfaces. The estimation problem of an illuminant spectrum can be reduced to a computational problem of finding an intersection of two color-signal planes.

The above-described property can be extended to the case of many surfaces. If there are three or more object surfaces of inhomogeneous material, then more than three color-signal planes are constructed in a vector space of the spectral power distributions. All the planes must intersect at only a common line, which corresponds to the illuminant spectrum.

The points of the present spectral analysis are summarized as follows:

(1) If the assumption of additivity and separability of the reflection model is true, the spectral data must form a two-dimensional subspace.

(2) If the assumption of constant reflectance is true, then the color signal planes from different object surfaces intersect at a common line in the direction of the illuminant.

We evaluate the standard reflectance model by measuring the spectral power distributions of light reflected from inhomogeneous material at various geometric angles. For any single surface material, we test to determine whether the color signals are described well by a plane. We then test to determine whether the subspaces for two different materials intersect and whether the intersection falls in the direction of the illuminant. Notice that strong specular highlights are not required to determine the color-signal plane. In this analysis we require only a wide range of measurements that includes specular and diffuse components. From this pattern of results, we should be able to obtain reliable estimates of the illuminant.

4. COMPUTATIONAL METHODS

A. Determining a Color-Signal Plane

Let us assume that we observe m color signals reflected from an object surface under various geometric conditions but with a constant illuminant. We sample each spectral power distribution at n points in the visible-wavelength region. The m color signals are represented by n-dimensional column vectors denoted by $\mathbf{y}_i$ ($i = 1, 2, \ldots, m$). If a set of data vectors falls in a subspace, their normalized representations fall in the same plane. Thus the absolute radiance values are not necessary to our analysis. We perform all our calculations by using the normalized spectral power distributions; that is, we normalize the measurements to unit power in the sense that $\|y_i\|^2 = 1$. The notation $\|\mathbf{y}\|^2$ is defined as $\mathbf{y}^T\mathbf{y}$, where T denotes a matrix (or vector) transposition. All the normalized spectral power distributions are summarized in an $n \times m$ matrix $\mathbf{Y}$ defined by

$$\mathbf{Y} = [\mathbf{y}_1, \mathbf{y}_2, \ldots, \mathbf{y}_m]. \tag{6}$$

We use the singular-value decomposition (SVD) to determine a color signal plane from the observed light spectra. The SVD is a useful tool for orthogonal decomposition of general rectangular matrices. The application to data analysis is similar to the idea of the well-known principal-component analysis. The SVD of the matrix $\mathbf{Y}$ is given in the following form[12]:

$$\mathbf{Y} = \mathbf{U}\mathbf{\Sigma}\mathbf{V}^T, \tag{7}$$

or, equivalently,

$$\mathbf{Y} = \sigma_1\mathbf{u}_1\mathbf{v}_1{}^T + \sigma_2\mathbf{u}_2\mathbf{v}_2{}^T + \ldots + \sigma_m\mathbf{u}_m\mathbf{v}_m{}^T, \tag{8}$$

where $\mathbf{U}$ (= $[\mathbf{u}_1, \mathbf{u}_2, \ldots, \mathbf{u}_n]$) and $\mathbf{V}$ (= $[\mathbf{v}_1, \mathbf{v}_2, \ldots, \mathbf{v}_n]$) are the $n \times m$ left-hand singular matrix and the $m \times m$ right-hand singular matrix, respectively, and Σ is the $m \times m$ diagonal matrix with elements of the singular values $\sigma_1, \sigma_2, \ldots, \sigma_m$. The n-dimensional left-hand singular vectors $\mathbf{u}_1, \mathbf{u}_2, \ldots, \mathbf{u}_m$ and the m-dimensional right-hand singular vectors $\mathbf{v}_1, \mathbf{v}_2, \ldots, \mathbf{v}_m$ are orthonormal eigenvectors of $\mathbf{YY}^T$ and $\mathbf{Y}^T\mathbf{Y}$, respectively. The singular values σ_i are the square roots of the eigenvalues of $\mathbf{YY}^T$ or $\mathbf{Y}^T\mathbf{Y}$. We assume the descending order $\sigma_i \geq \sigma_{i+1}$.

The dimensionality of the space spanned by the measured color signals can be estimated from the rank of the matrix $\mathbf{Y}$. Let us denote the n-dimensional vector space by the symbol $\mathbf{R}^n$. The measured color signals then fall into a two-dimensional subspace of $\mathbf{R}^n$ when the rank of $\mathbf{Y}$ is 2. This occurs when $\sigma_1 \geq \sigma_2 > 0$ and $\sigma_3 = \sigma_4 = \ldots = \sigma_m = 0$. In practice these strict equations are not satisfied perfectly because nonzero singular values appear as a result of measurement noise and computer roundoff error. A fairly reliable way to estimate the rank is to compare the singular values by using the following performance index:

$$I(K) = \frac{\sum_{i=1}^{K} \sigma_i^2}{\sum_{i=1}^{m} \sigma_i^2} = \frac{\sum_{i=1}^{K} \sigma_i^2}{m}. \tag{9}$$

By hypothesis we expect that $I(2) = 1.0$. If $I(1) = 1.0$, then the measured data are one dimensional. In this case the observation conditions reveal only one component of the interface and subsurface reflections from an object. If $I(2) \ll 1$, then the data have a higher dimensionality than 2. In this case we must consider whether the object might be a homogeneous material or whether the standard reflectance model is in error for this material.

The left-hand singular vectors $\mathbf{u}_1$ and $\mathbf{u}_2$ are a set of basis vectors that span a two-dimensional subspace of $\mathbf{R}^n$. When $I(2) = 1$, all the measured spectral power distributions are described in a linear combination of $\mathbf{u}_1$ and $\mathbf{u}_2$, since

$$\mathbf{y}_i = c_{i1}\mathbf{u}_1 + c_{i2}\mathbf{u}_2 \qquad (i = 1, 2, \ldots, m), \tag{10}$$

where $c_{ij} = \sigma_j v_{ij}$ ($j = 1, 2$). Therefore the color-signal plane of the object surface can be determined from

$$P = \{\mathbf{y} \quad | \mathbf{y} = c_1\mathbf{u}_1 + c_2\mathbf{u}_2, \quad c_i \in R\}. \tag{11}$$

The set R is a set of real numbers such that $\mathbf{y}$ is nonnegative and physically realizable.

B. Estimating an Illuminant Spectral Distribution

Let us assume that two object surfaces are observed under the same light source. Two color-signal planes are determined from the SVD's of the two data sets. The vectors $\mathbf{u}_1$ and $\mathbf{u}_2$ are an orthonormal basis that defines the color-signal plane. [Note that these vectors do not correspond to the physically defined components of the two reflections, $L_I(\lambda)$ and $L_S(\lambda)$.]

Let $[\mathbf{u}_1(1), \mathbf{u}_2(1)]$ and $[\mathbf{u}_1(2), \mathbf{u}_2(2)]$ be the basis vectors of the color-signal planes P(1) and P(2). Since the intersection line must lie in both P(1) and P(2), we have the relation

$$c_1\mathbf{u}_1(1) + c_2\mathbf{u}_2(1) = c_1'\mathbf{u}_1(2) + c_2'\mathbf{u}_2(2). \tag{12}$$

This relationship is equivalent to the homogeneous equation

$$[\mathbf{u}_1(1), \mathbf{u}_2(1), \mathbf{u}_1(2), \mathbf{u}_2(2)]\begin{bmatrix} c_1 \\ c_2 \\ -c_1' \\ -c_2' \end{bmatrix} = 0. \tag{13}$$

A nontrivial solution of Eq. (13) defines the intersection of the planes. A reliable solution method is to apply the SVD algorithm to the above $n \times 4$ matrix again. We then have

$$[\mathbf{u}_1(1), \mathbf{u}_2(1), \mathbf{u}_1(2), \mathbf{u}_2(2)] = [\mathbf{a}_1, \mathbf{a}_2, \mathbf{a}_3, \mathbf{a}_4]\begin{bmatrix} \lambda_1 & 0 & 0 & 0 \\ 0 & \lambda_2 & 0 & 0 \\ 0 & 0 & \lambda_3 & 0 \\ 0 & 0 & 0 & \lambda_4 \\ - & - & - & - \end{bmatrix} \times [\mathbf{b}_1, \mathbf{b}_2, \mathbf{b}_3, \mathbf{b}_4]^T, \tag{14}$$

where the definitions of $\mathbf{a}_i$, $\mathbf{b}_i$, and λ_i (i = 1, 2, 3, 4) correspond to those of $\mathbf{u}_i$, $\mathbf{y}_i$, and σ_i in Eq. (8). If $\lambda_4 = 0$ and $\lambda_3 > 0$ (i.e., rank = 3), the planes P(1) and P(2) intersect in a line in the n-dimensional space $\mathbf{R}^n$. The vector $\mathbf{b}_4$ is a solution for the vector at the intersection. The intersection line is described by the elements of $\mathbf{b}_4$ as

$$\mathbf{e}_1 = \sqrt{2}[b_{41}\mathbf{u}_1(1) + b_{42}\mathbf{u}_2(1)] \tag{15}$$

or

$$\mathbf{e}_2 = -\sqrt{2}[b_{43}\mathbf{u}_1(2) + b_{44}\mathbf{u}_2(2)], \tag{16}$$

where the line vectors are normalized as $\|\mathbf{e}_1\|^2 = \|\mathbf{e}_2\| = 1$. The signs of the coefficients b_{4i} are chosen so that the components of the illuminant vectors $\mathbf{e}_1$ and $\mathbf{e}_2$ are positive, consistent with the requirement that the illuminants be physically realizable. For example, when the right-hand side of Eq. (15) is negative, the signs of both b_{41} and b_{42} are reversed.

In real measurement, the value of λ_4 will not be exactly zero. However, we expect it to be small if the second condition, that of constant specularity reflectance, is satisfied. The same performance index, $I(K)$, may be used to evaluate the hypothesis in this case as well. We can regard $\mathbf{e}_1$ and $\mathbf{e}_2$ as the estimates of the spectral power distribution $E(\lambda)$ in planes P(1) and P(2), respectively. A reliable estimate $\mathbf{e}$ of the illuminant spectrum is obtained from the mean

$$\hat{\mathbf{e}} = \frac{(\mathbf{e}_1 + \mathbf{e}_2)}{2}. \tag{17}$$

The above procedure can be extended straightforwardly to the case of more than three object surfaces under the same light source. The computation to obtain an estimate is based on the repetition of the process presented here for every pair of objects in a scene.

5. EXPERIMENTS

To test the standard model experimentally and to evaluate the precision of the computational methods, we measured the light reflected from plastics and fruits under two different lighting conditions. The light sources were a flood lamp, for a daylight photograph, and a tungsten halogen lamp from a slide projector.

A. Measurement

The objects were placed on a table covered with a black cloth in a black room. The light sources were placed at the same height as the objects for a horizontal projection of the illuminant. Figure 4 shows a monochrome scene of a red plastic cup (left) and a green plastic ashtray (right), ~80 cm apart, illuminated with a flood lamp. Specular highlights are seen clearly at several parts of the surfaces. We used a spectroradiometer to measure the spectral power distributions of the reflected light. The spectroradiometer was placed ~130 cm from the objects and at the same height. The angle between the illumination and the viewing directions was 30 deg.

The surface of each plastic object is uniform and smooth. We measured the spectral radiance reflected from many spots on the surfaces by the angle of the spectroradiometer changing slightly. The power of the reflected light depends greatly on the viewing angle because the intensity in the highlights is several orders of magnitude greater than the intensity in the matte areas. The dynamic range of the spectrometer (and *a fortiori* that of conventional digitizing cameras) does not allow us to make all our measurements with a fixed aperture size. As we measured different positions we increased the aperture size from 6 min. to 1 deg. About half of the measuring spots were selected near the highlight areas and thus contained significant specular contributions. The other spots were from the matte areas and

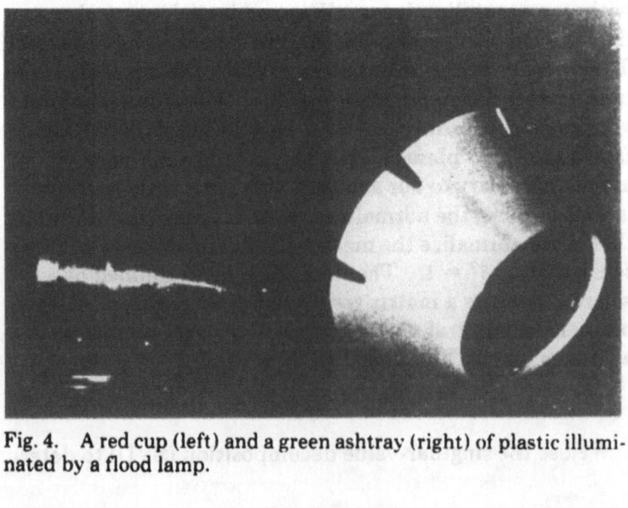

Fig. 4. A red cup (left) and a green ashtray (right) of plastic illuminated by a flood lamp.

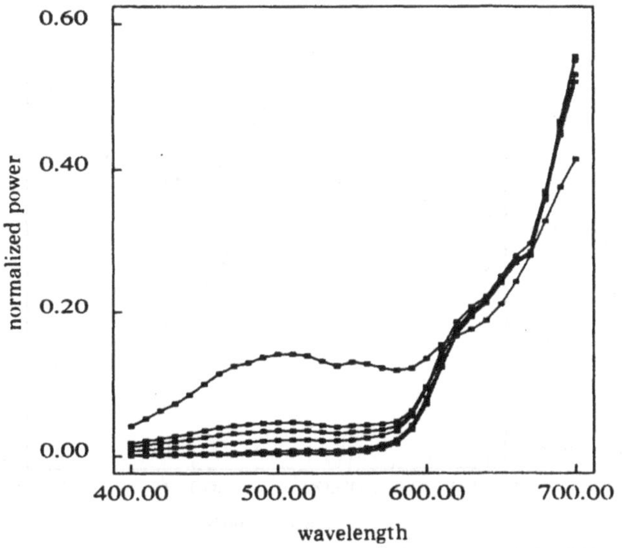

Fig. 5. Normalized curves of the measured spectra of the red plastic cup.

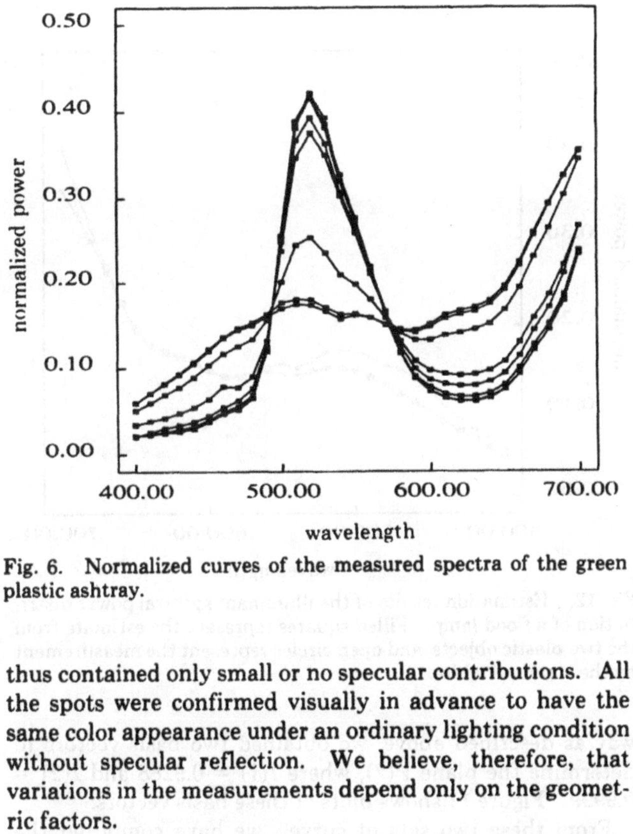

Fig. 6. Normalized curves of the measured spectra of the green plastic ashtray.

thus contained only small or no specular contributions. All the spots were confirmed visually in advance to have the same color appearance under an ordinary lighting condition without specular reflection. We believe, therefore, that variations in the measurements depend only on the geometric factors.

Figure 5 shows the normalized curves of the spectral power distributions measured at nine different spots on the red cup surface. One well-separated curve in the figure corresponds to the strongest highlight spectrum. Figure 6 shows the normalized curves measured at eight spots on the green ashtray surface. These data were obtained separately to eliminate any possible interreflections between the two objects.

Next, we observed fruits with a measuring arrangement similar to that used for the plastics but with a different light source. Figure 7 shows the scene of an apple (left) and a lemon (right), illuminated by a slide projector lamp and ~100 cm apart. It should be noted that the surface conditions of these natural products are quite different from those of the plastic industrial products. The skins of fruits are covered with a waxy tissue, and some of the light reflects at the interface. However, the surfaces are not so hard and smooth that a mirrorlike specular reflection is observed. In Fig. 7 only a weak glossy reflection appears on the lemon surface. A detailed inspection, moreover, shows that the glossy areas consist of fine textures and that the colors are not necessarily the same from a microscopic point of view.

Figures 8 and 9 show the normalized curves of the spectral power distributions measured at 10 different spots on the surfaces of the apple and the lemon, respectively. The spectral curves from the apple are separated into two types of pattern. There is relatively little variation in the spectral curves of the lemon because the glossy points are relatively small compared with the aperture size of the measurement instrument.

B. Results

The measured spectral data have been analyzed according to the computation procedure in Section 4.

1. *Plastics*

First, a 31 × 9 observation matrix **Y** for the red cup was constructed from the normalized spectral curves $\mathbf{y}_1$–$\mathbf{y}_9$, in which each curve was sampled at intervals of 10 nm. The SVD of **Y** gave the singular values of $\sigma_1 = 2.9627$, $\sigma_2 = 0.4698$, $\sigma_3 = 0.0389, \ldots$. The performance index was $I(1) = 0.9753$ for the first component only and $I(2) = 0.9998$ for the first two components. The contributions of the remaining seven components were negligibly small. From this result, we accept the hypothesis that the color-signal space measured from the red cup spans a 2-dimensional subspace of the 31-dimensional measured vector space. Figure 10 shows curves of the basis vectors $\mathbf{u}_1$ and $\mathbf{u}_2$. These curves are the principal-component curves for the measured spectral curves; that is, all the curves in Fig. 5 can be approximated with good accuracy by a linear combination of $\mathbf{u}_1$ and $\mathbf{u}_2$. These vectors determine P(1).

Next, a 31 × 8 observation matrix **Y** for the green ashtray was decomposed into its principal components. In the same

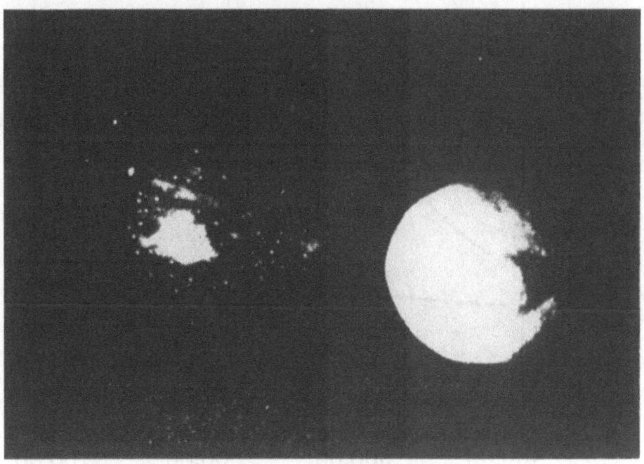

Fig. 7. An apple (left) and a lemon (right) illuminated by a tungsten halogen lamp from a slide projector.

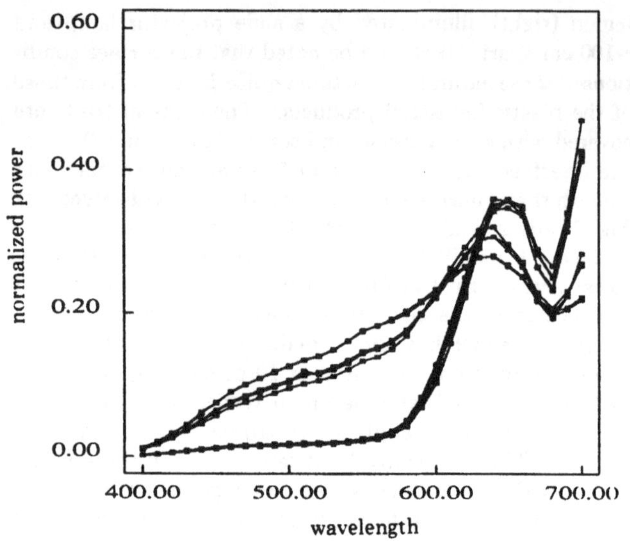

Fig. 8. Normalized curves of the measured spectra of the apple.

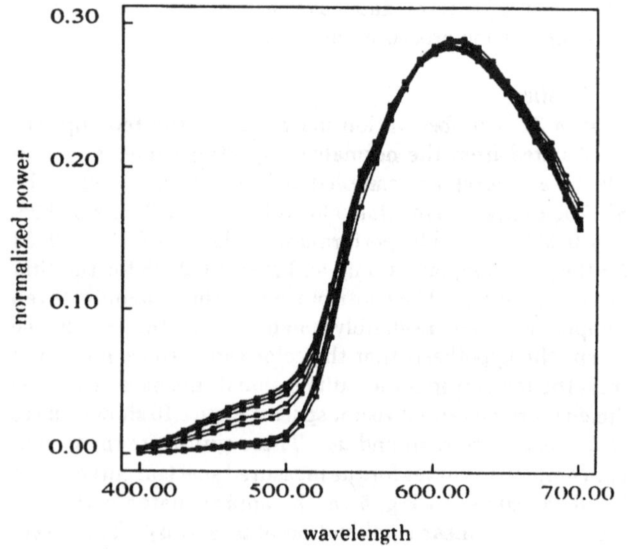

Fig. 9. Normalized curves of the measured spectra of the lemon.

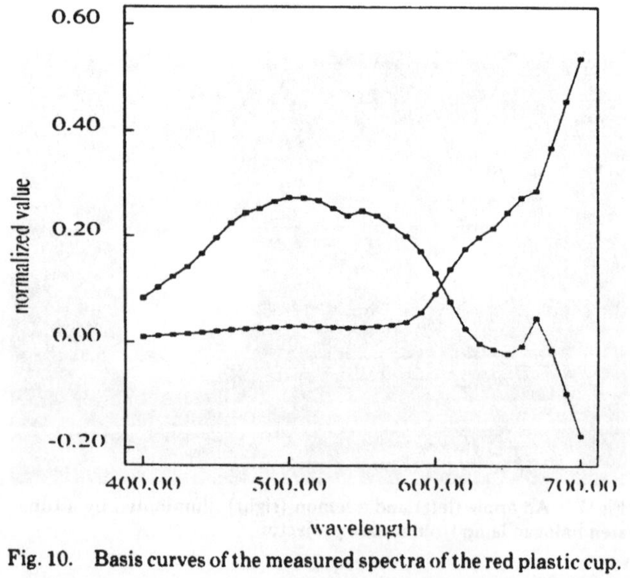

Fig. 10. Basis curves of the measured spectra of the red plastic cup.

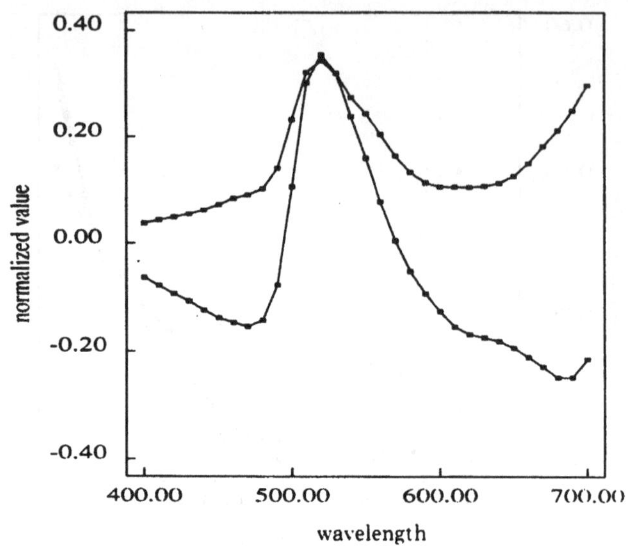

Fig. 11. Basis curves of the measured spectra of the green plastic ashtray.

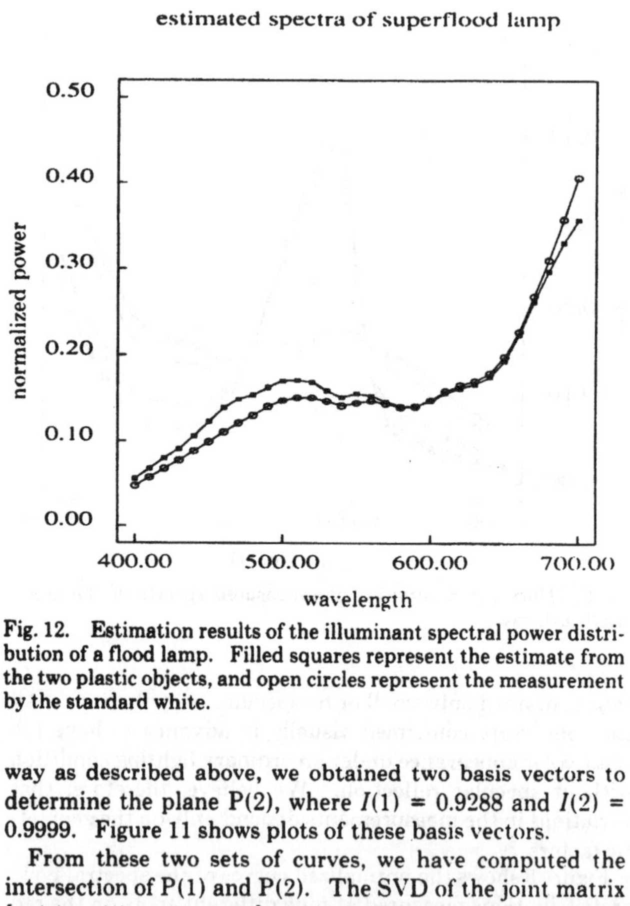

Fig. 12. Estimation results of the illuminant spectral power distribution of a flood lamp. Filled squares represent the estimate from the two plastic objects, and open circles represent the measurement by the standard white.

way as described above, we obtained two basis vectors to determine the plane P(2), where $I(1) = 0.9288$ and $I(2) = 0.9999$. Figure 11 shows plots of these basis vectors.

From these two sets of curves, we have computed the intersection of P(1) and P(2). The SVD of the joint matrix $[\mathbf{u}_1(1), \mathbf{u}_2(1), \mathbf{u}_1(2), \mathbf{u}_2(2)]$ gave the singular values $\lambda_1 = 1.414$, $\lambda_2 = 1.2404$, $\lambda_3 = 0.6792$, and $\lambda_4 = 0.0162$. We conclude that the rank of the joint matrix is 3 and that the planes P(1) and P(2) intersect. The fourth right-hand singular vector $\mathbf{b}_4$ specifies the coordinates of the line. The line vector with a unit length is described as follows:

$$\hat{\mathbf{e}} = 0.4207\mathbf{u}_1(1) + 0.2702\mathbf{u}_2(1) + 0.4567\mathbf{u}_1(2) - 0.2036\mathbf{u}_2(2). \tag{18}$$

Equation (18) expresses an estimate of the spectral power

distribution of the super flood lamp used. The estimate is shown in a solid line marked with black squares in Fig. 12. We compare the estimate with a direct measurement of the spectral power distribution of the light source measured by using a standard white reflectance surface. For the standard white we used a pressed powder of magnesium oxide (MgO). The pressed powder was placed at the same location as the object. The spectral distribution predicted from this measurement is shown by open circles in the figure. A fairly good coincidence between the two curves is seen in the range of middle to long wavelengths.

2. Fruits

We were able to use the same analysis method for the fruits, although the surfaces of the fruits are complicated. The SVD of the 31 × 10 matrix **Y** for the apple gives $\sigma_1 = 3.0513$, $\sigma_2 = 0.8244$, $\sigma_3 = 0.0821, \ldots$, and $I(2) = 0.9990$. Figure 13 shows the basis curves $\mathbf{u}_1$ and $\mathbf{u}_2$. In contrast, the values for the lemon are $\sigma_1 = 3.1592$, $\sigma_2 = 0.1342$, $\sigma_3 = 0.0272, \ldots$, and $I(2) = 0.9999$. Figure 14 shows the basis curves. In this case

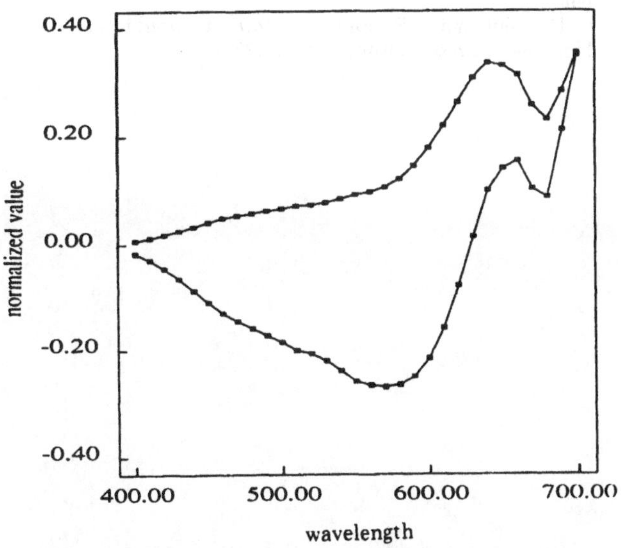

Fig. 13. Basis curves of the measured spectra of the apple.

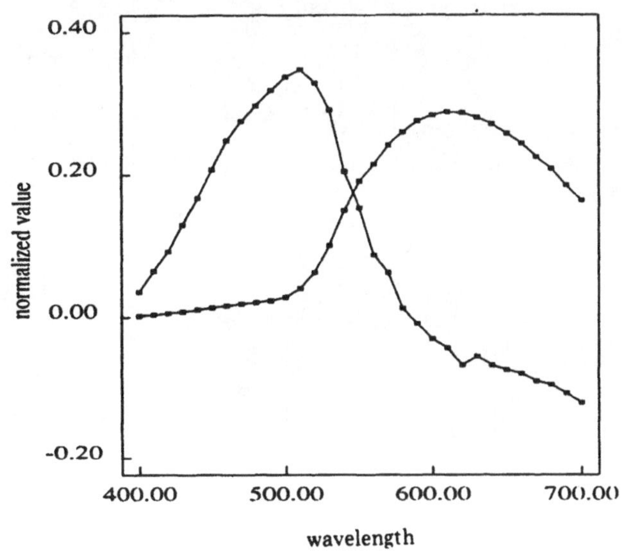

Fig. 14. Basis curves of the measured spectra of the lemon.

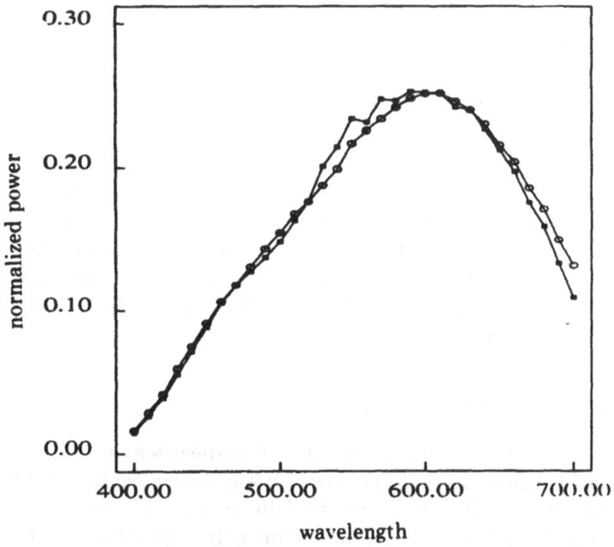

Fig. 15. Estimation results of the illuminant spectral power distribution of a slide projector. Filled squares represent the estimate from the two fruits, and open circles represent the measurement by the standard white.

the joint matrix gives the four values $\lambda_1 = 1.414$, $\lambda_2 = 1.288$, $\lambda_3 = 0.5838$, and $\lambda_4 = 0.0415$. We estimate the spectral power distribution of the slide projector lamp (Fig. 15, filled squares) to be

$$\hat{\mathbf{e}} = 0.4226\mathbf{u}_1(1) + 0.2672\mathbf{u}_2(1) + 0.4655\mathbf{u}_2(2) - 0.1825\mathbf{u}_2(2). \tag{19}$$

The spectral power distribution measured by the reference white is shown by open circles in Fig. 15. A comparison of the two curves shows the reliability of the illuminant spectral estimation for a slide-projector lamp.

These two example measurements are typical of the results that we have obtained with natural products and plastics in our laboratory.

6. DISCUSSION AND CONCLUSION

From our experiments we conclude that the two-component reflection model is adequate for describing color signals from some inhomogeneous materials. For the surfaces that we report here and others that we have measured, the two assumptions of the standard model appear valid. The measured spectral power distributions are described well as falling within a two-dimensional subspace, and the data in the specular highlights are consistent with an interface component that does not affect the spectral composition of the incident light.

We have presented a computational method for estimating the illuminant spectral power distribution based on the two-component reflectance model. An interesting aspect of the computational procedure that we use to measure the illuminant spectrum is that no calibrated reflectance standards are required to measure the light source. Equipped with only the knowledge that the two surfaces are inhomogeneous and with no knowledge of their reflectance functions, the algorithm accurately infers the illuminant spectrum. Since the cost of computational power is often less than that of calibrated, special-purpose materials, this estimation algorithm may prove to be of practical value.

Finally, the algorithm does not require us to establish measurement conditions in which we observe pure specular highlights. Establishing these viewing conditions can be quite difficult. Yet the algorithm estimates the illuminant fairly well under much less elaborate viewing conditions. For example, the fruits are far from being ideal inhomogeneous materials and do not have strong specular reflectance at the spatial resolution of our measurement device. Although the shape of the illuminant spectrum could not be measured directly, since the gloss could never be measured in isolation, an estimate accurate to within a few percent can be extracted from the combined data of the two objects.

ACKNOWLEDGMENTS

This study was performed while S. Tominaga was on leave at Stanford University. The research was supported by NASA grant NCC 2-332 to Stanford University. We thank D. Brainard, H. Lee, and A. Poirson for their helpful comments.

REFERENCES AND NOTES

1. G. J. Klinker, S. A. Shafer, and T. Kanade, "The measurement of highlights in color images," Int. J. Comput. Vision **2**, 7–32 (1988).
2. H.-C. Lee, "Method for computing the scene-illuminant chromaticity from specular highlights," J. Opt. Soc. Am. A **3**, 1694–1699 (1986).
3. M. D'Zmura and P. Lennie, "Mechanisms of color constancy," J. Opt. Soc. Am. A **3**, 1662–1672 (1986).
4. G. Buchsbaum, "A spatial processor model for object color perception," J. Franklin Inst. **310**, 1–26 (1980).
5. M. H. Brill and G. West, "Contributions to the theory of invariance of color under the condition of varying illumination," J. Math. Biol. **11**, 337–350 (1981).
6. L. T. Maloney and B. Wandell, "Color constancy: a method for recovering surface spectral reflectance," J. Opt. Soc. Am. A **3**, 29–33 (1986).
7. R. L. Cook and K. E. Torrance, "A reflectance model for computer graphics," Comput. Graph. **15**, 307–316 (1981).
8. M. Born and E. Wolf, *Principles of Optics* (Pergamon, Oxford, 1983).
9. S. A. Shafer, "Using color to separate reflection components," Color Res. Appl. **10**, 210–218 (1985).
10. H. Lee, E. J. Breneman, and C. P. Schulte, "An experimental study of a color reflection model," Tech. Rep. 229507A (Kodak Research Laboratories, Rochester, N.Y., 1986).
11. All that is required, in fact, is that the specular reflectance function be common and known among the different surfaces. However, it is convenient, and not too inaccurate, to adopt the assumption that the function is constant with respect to wavelength.
12. G. H. Golub and C. F. van Loan, *Matrix Computations* (Johns Hopkins U. Press, Baltimore, Md., 1983).

Modeling Light Reflection for Computer Color Vision

HSIEN-CHE LEE, EDWIN J. BRENEMAN, AND CARL P. SCHULTE

Abstract—In computer vision applications, analysis of shading information requires a proper model of light reflection from object surfaces. The most often used model so far relates the image intensity I to the light source intensity E by a simple relation: $I = E\rho h(\theta, \phi) \cos \theta$, where θ is the incident angle, ϕ is the viewing angle with respect to the surface normal, ρ is the so-called surface reflectance, and $h(\theta, \phi)$, normalized to one at the maximum, is called the reflectivity function, representing the angular dependence of light reflection. Although the terminology is not of standard use, this model seems to be empirically true for a collimated monochromatic incoherent light source. However, the optical geometry and the sensor spectral responsivity of the vision (imaging) system are not explicitly considered in the model. Furthermore, it is not clear that one can always model the light reflection in a product form for a polychromatic light source.

To overcome the above shortcoming and to extend the reflection model for computer color vision, we examine the light reflection problem using the bidirectional spectral-reflectance distribution function (BSRDF) to specify both incident- and reflected-beam geometry. It is shown that the product form can still be retained for a polychromatic light source under two lighting conditions: 1) the light source is collimated; or 2) the spectral factor and the geometric factor can be separated for both the light source and the BSRDF of the surface. The reflection model is then applied to the formulation of a neutral-interface-reflection model, which is tested experimentally. The results show the adequacy of this type of model for surfaces of some material compositions, e.g., plastics, plant leaves, painted surfaces, orange peels, and some glossy cloth, but not for others, e.g., colored paper and some ceramics.

Index Terms—Computer color vision, imaging geometry, reflection measurements, reflection model, spectral reflectance factor.

I. Introduction

One of the visual cues for deriving the three-dimensional shape information from a two-dimensional image is the surface shading. The analysis of this type of cue requires a proper model of light reflection [4], [12], [15], [17], [21], [36]. For color vision, the color variation on a surface also reveals information about the chromaticity of the light source [1], [2], [7], [8], [9], [25], [27], [28]. To make a color vision system see the color of an object to be relatively constant under chromatically different illuminants, one has to compute the light source chromaticity for proper compensation. In order to do this, it is also necessary to formulate a proper color reflection model.

In the past, reflection models used in computer vision applications have been very simple. They generally have the following form:

$$I = E\rho h(\theta_i, \phi_i; \theta_r, \phi_r) \cos \theta_i \tag{1}$$

where I is the image intensity, E is the light source intensity, (θ_i, ϕ_i) and (θ_r, ϕ_r) are the incident angle and the viewing angle with respect to the surface normal, ρ is the so-called surface reflectance, and $h(\theta_i, \phi_i; \theta_r, \phi_i)$, normalized to one at the maximum, is called the reflectivity function, representing the angular dependence of light reflection. There are problems with this simple model because the terminology is not of standard use and the underlying concepts are not clearly defined. Secondly, for algorithms working on the captured images, the model does not consider the imaging geometry and the sensor spectral responsivity explicitly. The model seems to be intuitively true for a collimated monochromatic light source, but its use in a real imaging system needs a more careful formulation [17], [33]. If one wants to study computer color vision, spectral factor in the reflection model also requires special attention [10], [13], [37], [38].

II. Surface and Subsurface Light Reflection

When light illuminates a surface (see Fig. 1), part of it is reflected immediately from the interface between the surface and the air, and part of it penetrates into the surface, is partially absorbed and partially scattered around within the surface material, and then is reflected back to the surface and into the air again. The first component, which is reflected immediately at the interface, is called the interface (surface) reflection, and the second component is called the body (subsurface) reflection (using Shafer's nomenclature [37]). For smooth surfaces, the interface reflection component is very directional and is usually called the specular (or regular) reflection. The body reflection, on the other hand, radiates to all different directions, and is usually called the diffuse reflection.

According to Cook and Torrance's model [6], the interface reflection (Fresnel) component is a function of the index of refraction of the interface material. Since many types of material serving as binders or as overcoats are some sort of oil, and most oils have very constant indexes of refraction for the visible wavelengths [22], the interface reflection component of these surfaces tends to have a relative spectral power distribution close to that of the incident light (i.e., the color of this component is close to illuminate color). This argument is mostly justified by empirical observations, and there have been few measurement data reported in the open literature (see the comments in [33, p. 31]). Most reflection models assume that the body reflection component has a spectral composition different from that of the incident light, while the interface reflection component has the similar composition as the incident light [1], [7], [8], [13], [18], [25], [27]. We will call this type of color reflection model the neutral-interface-reflection (NIR) model. Cook and Torrance [6] have pointed out that this type of color reflection model is not adequate for describing surfaces of homogeneous material, of which the index of refraction cannot be approximated by a constant over the wavelength range of visible light. Metallic surfaces are the most obvious examples, of which the indices of refraction change considerably over the visible light wavelengths [35]. The general light reflection phenomena seem to defy exact analysis, and most practical reflection models used in computer vision and graphics are only empirical approximations.

Although the NIR model is used extensively to generate computer synthesized color images, there are actually very few experimental data available in the literature to show the adequacy of this reflection model. Presumably, the seemingly realistic color appearance of the synthesized images serves as the existence proof of the adequacy of the reflection model. However, subtle color variations are important for designing computer color vision algorithms, creating realistic computer graphics, and studying human color perception. An experimental study on the NIR model is therefore a needed step toward this direction. In the following, we will first formulate a generalized reflection model using the bidirectional spectral-reflectance distribution function (BSRDF) [38] to specify both incident- and reflected-beam geometry. Then we will propose one formulation of the NIR model and present some experimental data to show that the model is quite reasonable for some object surfaces, but not for others.

Manuscript received February 23, 1988; revised September 9, 1989. Recommended for acceptance by O. D. Faugeras.

The authors are with the Imaging Science Laboratory, Eastman Kodak Co., Rochester, NY 14650.

IEEE Log Number 8933079.

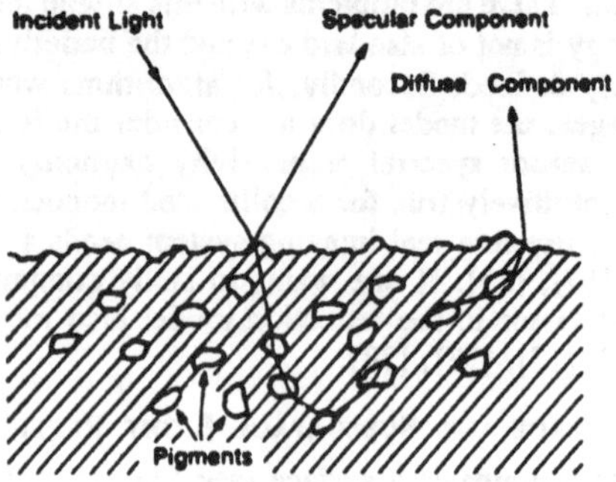

Fig. 1. Reflection from an inhomogeneous surface with two components: specular (interface) reflection and diffuse (body) reflection.

III. Light Reflection and Imaging Geometry

To model the light reflection in a precise term, the terminology and the optical geometry has to be carefully defined. The International Lighting Vocabulary adopted by the Commission Internationale de l'Eclairage (CIE) [5] is widely used. It is the most comprehensive and least controversial, and is recommended by the publications of the U.S. National Bureau of Standards [31, p. 8]. Here, we will only describe some of the concepts in radiometry in order to introduce our reflection model later. Very good descriptions of radiometric terms can be found in various articles and books [3], [11], [16], [33], [38], [42].

Light energy flow per unit time is called "radiant flux," Φ, which is often expressed as joules/second or watts. The radiant output of a point light source in a given direction is called intensity, I, which is measured as radiant flux per solid angle, watts/steradian (see Fig. 2). Light coming from an extended light source or reflected from a surface can be treated as the summation of radiant output from many point sources distributed over the surface. However, we now have to select an infinitely small surface element and treated it as a point source. This means that two other factors have to be considered: the area and the orientation of the surface element. If the elemental area is twice larger, the radiant output of the element point source should be two times more, so that the total surface will deliver the same amount of radiant output. On the other hand, if the surface element is facing away from the viewing direction, the surface area will be foreshortened according to the angle between the surface normal and the viewing direction. To account for this type of geometry, the quantity radiance, L, defined as radiant flux per solid angle per projected surface area, is used to describe the amount of light coming from a surface (see Fig. 3). Radiance is also used to describe radiant flux at a point on the path of a light beam or at a point on the receptor surface. After describing the intuitive concepts, we can now give the formal definitions of some useful quantities in the CIE International Lighting Vocabulary.

Let A denote the area of a surface, ω denote the solid angle of some light beam, and θ denote the angle between the surface normal and the light beam. The quantities which will be used here are summarized as follows [5]:

- radiant flux Φ: Power emitted, transferred, or received in the form of radiation. Unit is watt.
- radiance L (in a given direction, at a point on the surface of a source or a receptor, or at a point on the path of a beam): Quotient of the radiant flux leaving, arriving at, or passing through an element of surface at this point and propagated in directions defined by an elementary cone containing the given direction, and the area of the orthogonal projection of the element of surface on a plane perpendicular to the given direction. $L = d^2\Phi/(d\omega dA \cos\theta)$. Unit is watt per steradian and per square meter.
- irradiance E (at a point of a surface): Quotient of the radiant flux incident on an element of the surface containing the point, by the area of that element. $E = d\Phi/dA$. Unit is watt per square meter.

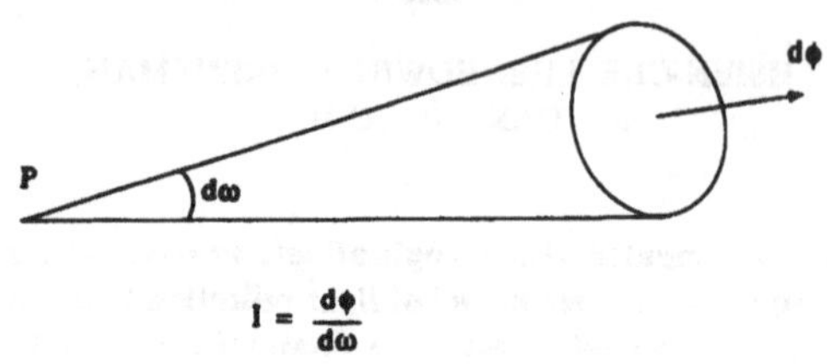

Fig. 2. Geometry of the intensity of a point source.

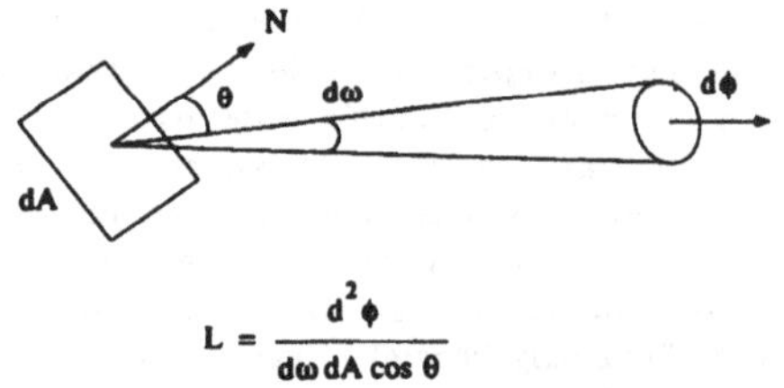

Fig. 3. Geometry of the radiance of a surface element.

- reflectance factor β (at a point of surface, for the part of the reflected radiation contained in a given cone with apex at the point of the surface, and for incident spectral composition and geometric distribution): Ratio of the radiant flux reflected in the directions delimited by the cone to that reflected in the same directions by a perfect reflecting diffuser identically irradiated.
- responsivity R: The responsivity of a detector is defined most generally as the ratio of the output signal to the input power. Its unit depends on the form of the electrical response exhibited by a given detector (see [3, p. 109]).

If the above quantities are specified as a function of wavelength, then the adjective *spectral* is added in front of the terms, and their units become per unit wavelength interval (nanometer). For example, spectral radiance is $L_\lambda = d^3\Phi/(d\omega d\lambda dA \cos\theta)$. It should be noted that reflectance (ratio of the reflected radiant flux to the incident flux) has often been incorrectly used where reflectance factor will be more appropriate. Similarly, the spectral responsivity function has to be distinguished from the spectral sensitivity function, which is the reciprocal of the input power required for the detector to produce a specified level of output signal. (For the human visual system, the spectral responsivity function is difficult to measure, and the spectral sensitivity function is often used instead.) The standard use of the term, intensity, in radiometry and photometry means light flux per solid angle, and thus the use of image intensity is not adopted here. Image irradiance is used instead.

In most applications of interests in modeling the light reflection from solid surfaces, the bidirectional reflectance distribution function (BRDF) proposed by Nicodemus *et al.* [33] is a very general and useful concept to use. It is defined (Fig. 4) as the ratio of the reflected radiance dL_r in the viewing direction to the irradiance dE_i in the direction of the incident light [16], [33]. The original definition of the BRDF f_r can be extended to include the light wavelength as a variable [33, p. 31]. The resulting function is called the bidirectional spectral-reflectance distribution function (BSRDF), which is defined as

$$f_r(\theta_i, \phi_i; \theta_r, \phi_r; \lambda) = dL_r(\theta_i, \phi_i; \theta_r, \phi_r; \lambda; E_i)/dE_i(\theta_i, \phi_i; \lambda), \quad (2)$$

where (θ_i, ϕ_i) and (θ_r, ϕ_r) are the incident and reflected direction angles with respect to the surface normal. It should be noted that dL_r and dE_i in (2) are now the reflected *spectral* radiance and the incident *spectral* irradiance. Let ω_i be the solid angle of the incident light cone, and L_i be the incident spectral radiance (see Fig. 5). Since $dE_i = L_i \cos\theta_i d\omega_i$, where $d\omega_i$ is the differential solid angle in the (θ_i, ϕ_i) direction, the reflected spectral radiance L_r can be

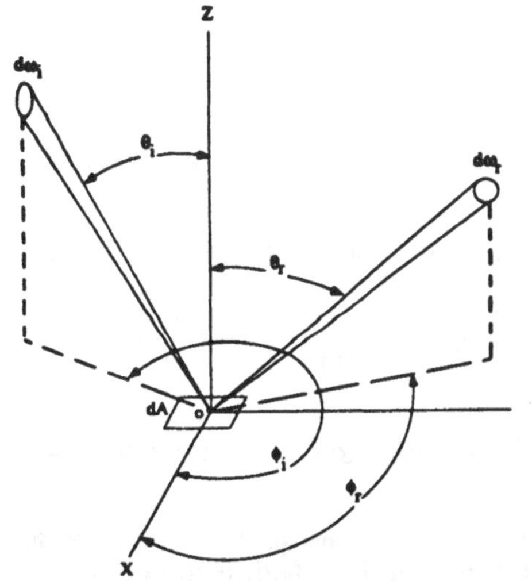

Fig. 4. Geometry of incident and reflected light beam.

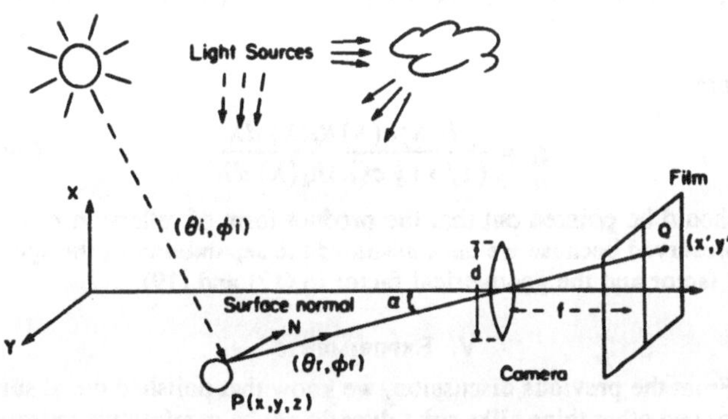

Fig. 5. Simplified imaging geometry.

computed by

$$L_r = \int dL_r = \int f_r \, dE_i = \int_{\omega_i} f_r L_i \cos\theta_i \, d\omega_i. \quad (3)$$

If the object distance is much larger than the focal length and the diameter of the entrance pupil of the imaging system, it can be shown [16] that the image spectral irradiance E_p is proportional to scene spectral radiance L_r and

$$E_p = L_r \frac{\pi}{4}\left(\frac{d}{f}\right)^2 \cos^4\alpha = e(\alpha)\, L_r \quad (4)$$

where d is the diameter of the entrance pupil, f is the focal length of the lens, α is the off-axis angle (see Fig. 5), and

$$e(\alpha) = \frac{\pi}{4}\left(\frac{d}{f}\right)^2 \cos^4\alpha. \quad (5)$$

Combining (3) and (4), we have

$$E_p = e(\alpha) \int_{\omega_i} f_r L_i \cos\theta_i \, d\omega_i. \quad (6)$$

Now let us define the spectral responsivity of the sensor as

$$R(\lambda) = \frac{\text{output signal } V(\lambda)}{\text{input image irradiance } E_p(\lambda)}. \quad (7)$$

Assuming that V is additive over the wavelength, then the output signal V_k from the kth sensor (e.g., red, green, or blue) can be expressed as

$$V_k = \int R_k(\lambda) E_p(\lambda)\, d\lambda \quad (8)$$

$$V_k = e(\alpha) \int_{\omega_i} d\omega_i \cos\theta_i \int d\lambda\, f_r(\theta_i, \phi_i; \theta_r, \phi_r; \lambda) \cdot L_i(\theta_i, \phi_i; \lambda) R_k(\lambda). \quad (9)$$

If the light source is a collimated source incident on the surface from the direction (θ, ϕ), and the surface irradiance on a plane perpendicular to the light source direction is $c(\lambda)$, then it can be shown [16] that

$$L_i(\theta_i, \phi_i, \lambda) = c(\lambda)\delta(\theta_i - \theta)\delta(\phi_i - \phi)/\sin\theta. \quad (10)$$

Since $d\omega_i = \sin\theta_i d\theta_i d\phi_i$, (9) can be simplified to

$$V_k = e(\alpha)\cos\theta \int f_r(\theta, \phi; \theta_r, \phi_r; \lambda) c(\lambda) R_k(\lambda)\, d\lambda. \quad (11)$$

The reflectance factor function $\beta_k(\theta, \phi; \theta_r, \phi_r)$ can be defined as

$$\beta_k(\theta, \phi; \theta_r, \phi_r) = \frac{\int f_r(\theta, \phi; \theta_r, \phi_r; \lambda) c(\lambda) R_k(\lambda)\, d\lambda}{(1/\pi)\int c(\lambda) R_k(\lambda)\, d\lambda} \quad (12)$$

(note that for a perfect diffuser, $f_r = 1/\pi$ [16]); if we let E_k equal the spectral responsivity weighted surface irradiance when the surface is perpendicular to the incident light, i.e., $\cos\theta = 1$, and

$$E_k = \int_{\omega_i} d\omega_i \int d\lambda L_i(\theta_i, \phi_i; \lambda) R_k(\lambda) = \int c(\lambda) R_k(\lambda)\, d\lambda, \quad (13)$$

then we have

$$V_k = (1/\pi) e(\alpha) E_k \beta_k(\theta, \phi; \theta_r, \phi_r)\cos\theta. \quad (14)$$

If we write

$$\beta_k(\theta, \phi; \theta_r, \phi_r) = \hat{\beta}_k h_k(\theta, \phi; \theta_r, \phi_r), \quad (15)$$

where h_k is the function β_k normalized to one at its maximum, then (14) becomes

$$V_k = (1/\pi) e(\alpha) E_k \hat{\beta}_k h_k(\theta, \phi; \theta_r, \phi_r)\cos\theta, \quad (16)$$

which is the same kind of product form as (1). The difference is that the maximum reflectance factor $\hat{\beta}_k$ is a function of the imaging system as well as the optical property of the surface, and the effects of the imaging geometry and sensor spectral responsivity have been made clear in (10), (12), (13), and (15).

IV. The Neutral-Interface-Reflection (NIR) Model

There are many causes of color [32]. To model color shading accurately, one has to know the surface structure and the optical and the physical properties of all the objects in the image. Since the required data are usually not available, empirical models with few parameters are almost always used in light reflection models. Here we are concerned with only the color aspect of the reflected light which does not involve causes like interference, diffraction, polarization, fluorescence, and dispersive refraction as the dominant factors.

Using the notation developed in the previous section, the dichromatic model proposed by Shafer [37] can be reformulated in terms of the BSRDF:

$$f_r(\theta_i, \phi_i; \theta_r, \phi_r; \lambda) = a(\lambda) g(\theta_i, \phi_i; \theta_r, \phi_r) + b(\lambda) h(\theta_i, \phi_i; \theta_r, \phi_r) \quad (17)$$

where functions g and h are normalized to one at their maximum values. The neutral-interface-reflection model can now be formulated by setting $b(\lambda)$ to a constant s and

$$f_r(\theta_i, \phi_i; \theta_r, \phi_r; \lambda) = a(\lambda) g(\theta_i, \phi_i; \theta_r, \phi_r) + s h(\theta_i, \phi_i; \theta_r, \phi_r). \quad (18)$$

The physical meaning of the NIR model as expressed in (18) is that the surface reflects light in two components: the first term represents the wavelength-selective reflection, and the second term, the nonselective reflection. It should be noted that the spectral factors $a(\lambda)$ and s and the geometrical factors g and h are assumed to be separable. As will be seen in the following derivations, this assumption is necessary for the preservation of the reflection models in product form. Empirically this is also a reasonable approximation (see the experimental results presented below and the comments in [33, p. 31]).

To test the adequacy of the model, we will fix the optical geometry for the measurements. Let the solid angles of the incident light source and the sensor reading aperature by ω_i and ω_r, respectively. For a single light source, the incident radiance on a surface can be written as

$$L_i(\theta_i, \phi_i; \lambda) = c(\lambda)L(\theta_i, \phi_i). \quad (19)$$

The spectral reflectance factor is defined as the ratio of the radiant flux at a given wavelength actually reflected by a sample surface to that which would be reflected into the same reflected beam geometry by an ideal (lossless) perfectly diffuse (Lambertian) standard surface irradiated in exactly the same way as the sample [33, p. 8]. The spectral reflectance factor for the NIR model can now be expressed as follows:

$$\begin{aligned}\beta(\omega_i, \omega_r; \lambda) &= \frac{d\Phi}{d\Phi_{\text{ideal diffuser}}}\\ &= \frac{\int_{\omega_i}\int_{\omega_r} f_r(\theta_i, \phi_i; \theta_r, \phi_r; \lambda)L_i(\theta_i, \phi_i; \lambda)\, d\Omega_i\, d\Omega_r}{(1/\pi)\int_{\omega_i}\int_{\omega_r} L_i(\theta_i, \phi_i; \lambda)\, d\Omega_i\, d\Omega_r}\\ &= a(\lambda)U(\omega_i, \omega_r) + sV(\omega_i, \omega_r), \quad (20)\end{aligned}$$

where

$$U(\omega_i, \omega_r) = \frac{\int_{\omega_i}\int_{\omega_r} g(\theta_i, \phi_i; \theta_r, \phi_r)L(\theta_i, \phi_i)\, d\Omega_i\, d\Omega_r}{(1/\pi)\int_{\omega_i}\int_{\omega_r} L(\theta_i, \phi_i)\, d\Omega_i\, d\Omega_r},$$

$$V(\omega_i, \omega_r) = \frac{\int_{\omega_i}\int_{\omega_r} h(\theta_i, \phi_i; \theta_r, \phi_r)L(\theta_i, \phi_i)\, d\Omega_i\, d\Omega_r}{(1/\pi)\int_{\omega_i}\int_{\omega_r} L(\theta_i, \phi_i)\, d\Omega_i\, d\Omega_r},$$

$d\Omega_i = \cos\theta_i d\omega_i$, and $d\Omega_r = \cos\theta_r d\omega_r$.

Since the BSRDF is defined in terms of differentials, dE_i and dL_r, it cannot be measured directly. The operational definition of the NIR model as expressed in (20) will be tested in the following experiments. If we measure the spectral reflectance factor of an object surface at two different angles, then

$$\beta_1(\lambda) = a(\lambda)U_1 + sV_1$$

$$\beta_2(\lambda) = a(\lambda)U_2 + sV_2$$

and

$$\beta_1(\lambda) = \frac{U_1}{U_2}\beta_2(\lambda) + s\left(V_1 - \frac{U_1}{U_2}V_2\right) = C\beta_2(\lambda) + D, \quad (21)$$

where C and D are constants with respect to the wavelength. Since (21) follows directly from (20), we can measure the spectral reflectance factor at two different angles for many wavelengths, and compare them to see if (21) holds. If not, the object does not reflect light according to the NIR model as described by (18).

A simple color reflection model also follows from (9), (18), and (19):

$$\begin{aligned}V_k &= e(\alpha)\int_{\omega_i} d\omega_i \cos\theta_i \int d\lambda f_r(\theta_i, \phi_i; \theta_r, \phi_r; \lambda)\\ &\quad \cdot L_i(\theta_i, \phi_i; \lambda)R_k(\lambda)\\ &= e(\alpha)\int_{\omega_i} d\omega_i \cos\theta_i \int d\lambda(a(\lambda)g + sh)(c(\lambda)L)R_k(\lambda).\end{aligned} \quad (22)$$

Letting

$$E_k = \int c(\lambda)R_k(\lambda)\, d\lambda \quad (23)$$

$$\bar{h}(\theta_i, \phi_i; \theta_r, \phi_r) = \int_{\omega_i} \cos\theta_i h(\theta_i, \phi_i; \theta_r, \phi_r)L(\theta_i, \phi_i)\, d\omega_i \quad (24)$$

$$\bar{g}(\theta_i, \phi_i; \theta_r, \phi_r) = \int_{\omega_i} \cos\theta_i g(\theta_i, \phi_i; \theta_r, \phi_r)L(\theta_i, \phi_i)\, d\omega_i \quad (25)$$

$$\beta_k = \frac{\int a(\lambda)c(\lambda)R_k(\lambda)\, d\lambda}{(1/\pi)\int c(\lambda)R_k(\lambda)\, d\lambda}, \quad (26)$$

we have

$$V_k = (1/\pi)e(\alpha)E_k(\beta_k\bar{g}(\theta_i, \phi_i; \theta_r, \phi_r) + s\bar{h}(\theta_i, \phi_i; \theta_r, \phi_r)), \quad (27)$$

which is in a very convenient product form useful for computer color vision [19], [20], [23], [29], [40]. For the dichromatic reflection model of (17), we have

$$V_k = (1/\pi)e(\alpha)E_k(\beta_k\bar{g}(\theta_i, \phi_i; \theta_r, \phi_r) + s_k\bar{h}(\theta_i, \phi_i; \theta_r, \phi_r)), \quad (28)$$

where

$$s_k = \frac{\int b(\lambda)c(\lambda)R_k(\lambda)\, d\lambda}{(1/\pi)\int c(\lambda)R_k(\lambda)\, d\lambda}. \quad (29)$$

It should be pointed out that the product form of reflection model is preserved because we have assumed the separability of the spectral factor and the geometrical factor in (17) and (19).

V. Experiments

From the previous discussion, we know that polished metal surfaces and other things like solid dyes do not have refractive indexes which are approximately constant in the visible wavelength range, and therefore they cannot satisfy the NIR model (also see [6]). However, for many types of object commonly seen in daily life, it is not clear how good the neutral-interface-reflection model is for describing color reflection from their surfaces. We have picked eight objects for this experiment. They are 1) a yellow plastic cylinder, 2) a green leaf of an indoor plant, 3) a piece of cloth, 4) a painted wood block, 5) an orange, 6) a maroon dish (ceramic), 7) a piece of glossy yellow paper, and 8) a piece of matte blue paper. These objects were chosen because they are made of materials frequently seen, and had been thought to be likely candidates for the NIR models. Obviously there are many others of potential interest, but the scope of the experiment does not permit an exhaustive measurement of all materials. However, it seems that a more complete experiment to measure most of the interested materials will be useful for building up the database for computer color vision and graphics research.

To compute the spectral-reflectance factors, we used pressed barium sulphate powder as the ideal perfectly diffuse standard surface. We measured spectral power distribution curves for light reflected from it three times, one in the beginning, one in the middle, and one at the end of the experiment, to make sure that the instruments were stable during the whole experiment. The three measurements were very close to each other (see Fig. 6, where they have been divided by their average). Their average is used as the reference light source spectral power distribution $c(\lambda)$. The spectral power distribution of the reflected light from a test object surface is then divided by $c(\lambda)$ to give the spectral-reflectance factor curve (multiplied by some constant factor independent of wavelength) for that object. It should be noted that the relative magnitudes are not important in this experiment, because (21) is a linear relation between two spectral reflectance factor curves, and any constant multiplier of a curve will not change the relation we want to verify. Therefore, we can always scale the magnitude of one

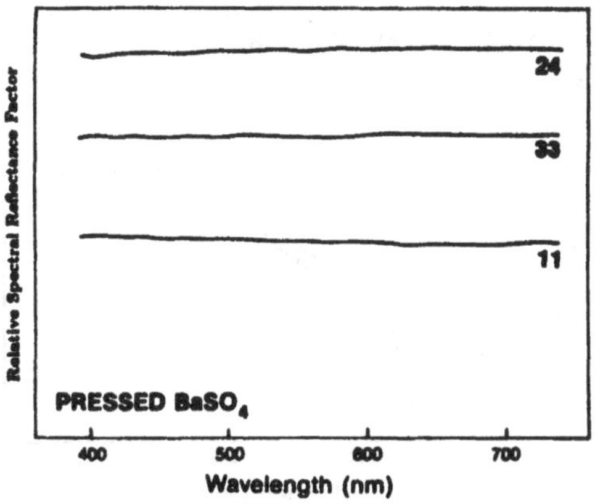

Fig. 6. The relative spectral reflectance factor curves of the pressed barium sulphate powder (arbitrarily scaled).

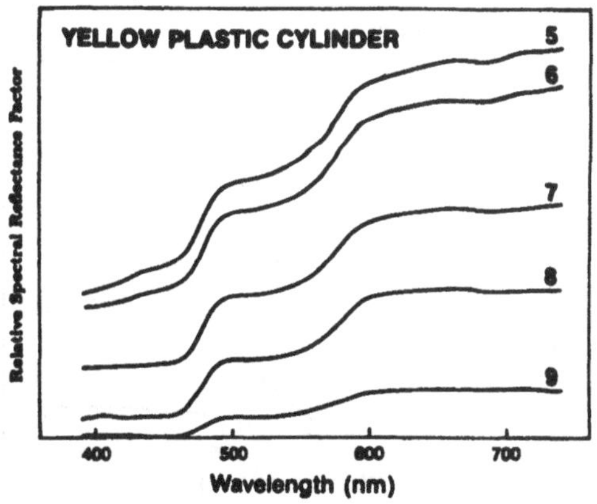

Fig. 7. The relative spectral reflectance factor curves of the yellow plastic cylinder (arbitrarily scaled).

curve relative to another to make the displayed curve plot easier to read.

The object to be measured was placed on a table covered with a black cloth. It was illuminated by a tungsten light source (with a color temperature about 2800°K) about four feet above. A Gamma Scientific tele-spectroradiometer was used to measure the spectral power distribution of the reflected light. The instrument was about five to six feet from the object. The measured surface area was chosen to be larger than any visible texture element of the surface. The spectral power distribution curves were measured at several spots on the object surface, which show different strengths of specular reflection. The reflected radiance differed very much between the angle without visible highlight and that of strong visible highlight. The high voltage on the photomultiplier of the spectroradiometer was constantly adjusted so that the photomultiplier output was kept at a reasonable magnitude (the peak reading from 360 nm to 740 nm in 10 nm increments was maintained at a magnitude of four digits). The selected spots were visually examined (through the telescope attached to the spectroradiometer) and a measuring aperture was selected which would provide an average over any textural fine structure of the sample. Therefore, the measured differences under the experimental condition was mainly from the difference in the illumination angles and viewing angles. The angular apertures used were 20 minutes of arc in all measurements, except in the case of measuring the specular spots on the maroon (ceramic) dish and yellow plastic cylinder, where a 6 minute aperture was used because of the extreme high intensity of the reflected light and because of the small size of the highlight spot.

VI. Results

Figs. 7-12 show the relative spectral reflectance factor curves (multiplied by a constant factor independent of wavelength) computed for the different objects. The curves for the blue paper are shown in Fig. 19. The separation between the curves is mainly due to the strength of the interface reflection and the different gains used for the measurements.

To see if the NIR model of (18) is good for these types of surfaces, pairs of the relative spectral reflectance curves are compared through (21): $\beta_1(\lambda) = C\beta_2(\lambda) + D$. The parameters C and D are determined by the least-squares fit for each pair of curves. The results are shown in Figs. 13-19. The dotted curve is the linear transform of one curve to fit the other by the method of least squares. The specularity of the maroon dish is so high that the readings of certain wavelengths for the off-specular and the specular spots are different by orders of magnitude. Its least-squares fitted curves are far from similar to one another, and therefore need not be considered for (21).

It is clear that except for the yellow paper, the blue paper, and the maroon dish, the other objects are well described by the NIR model. The microstructure of most paper does not seem to have a

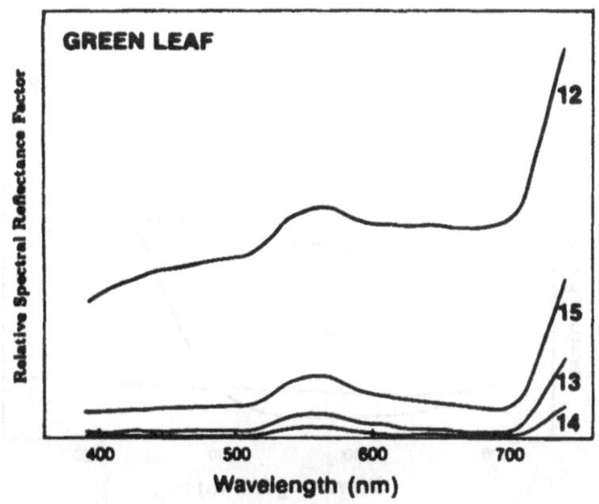

Fig. 8. The relative spectral reflectance factor curves of the green leaf (arbitrarily scaled).

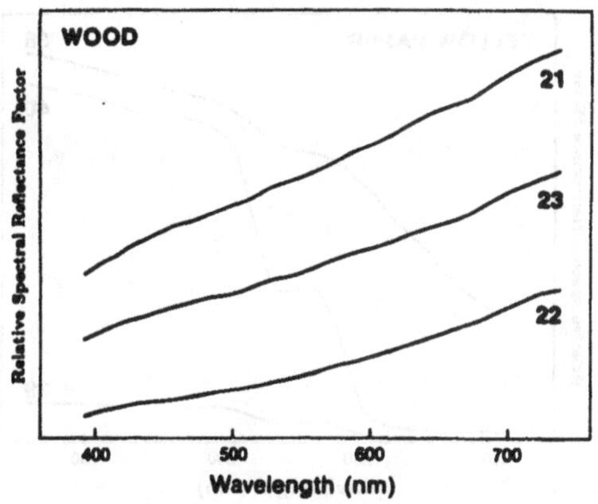

Fig. 9. The relative spectral reflectance factor curves of the painted wood block (arbitrarily scaled).

colorless coating on it to produce the neutral interface reflection component. The fact that the NIR model also failed to describe the glossy yellow paper reflection also reveals the complexity underlying the subtle interaction between the surface material/structure and the incident light. Empirical models like the NIR model are not likely to describe light reflection very accurately (a more detailed analysis can be found in [14]). However, they can be very good approximation for sensors with broadband sensitivity, because of reasons discussed below.

VII. Discussions

The adequacy of the neutral-interface-reflection model for describing the color reflections from such different objects as plastics,

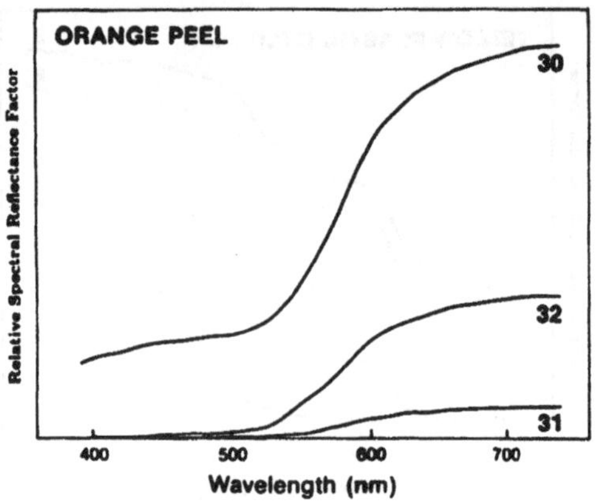

Fig. 10. The relative spectral reflectance factor curves of the orange peel (arbitrarily scaled).

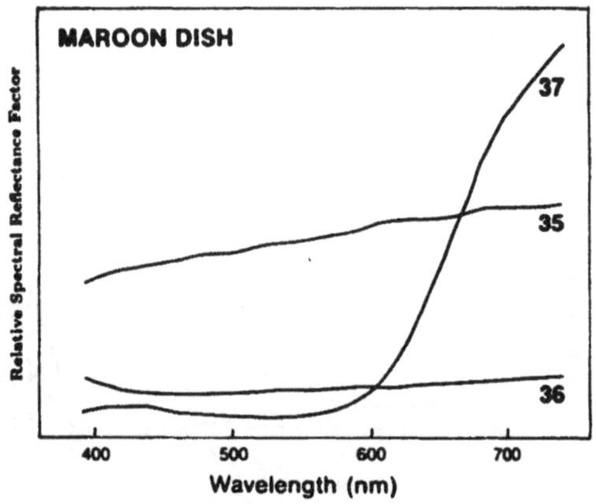

Fig. 11. The relative spectral reflectance factor curves of the maroon ceramic dish (arbitrarily scaled).

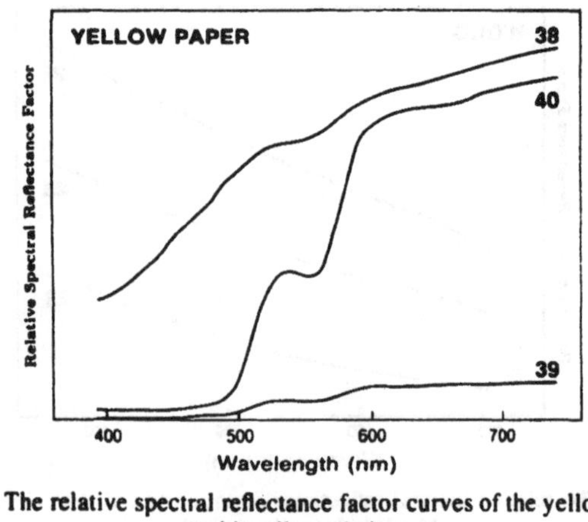

Fig. 12. The relative spectral reflectance factor curves of the yellow paper (arbitrarily scaled).

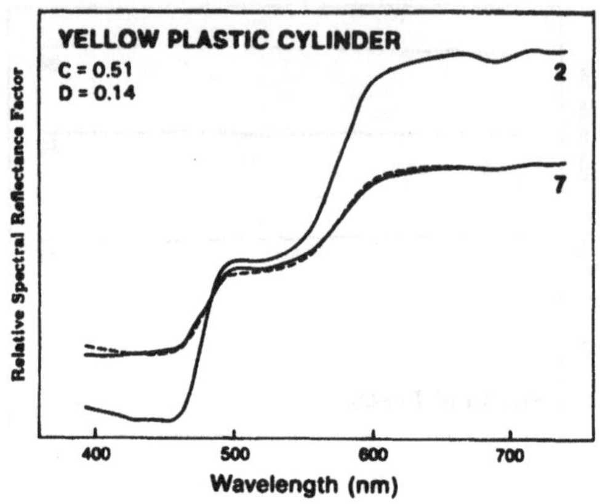

Fig. 13. Comparison of two relative spectral reflectance factor curves of the yellow plastic cylinder (arbitrarily scaled). The dotted curve is the least-squared fit of curve No. 2 to curve No. 7.

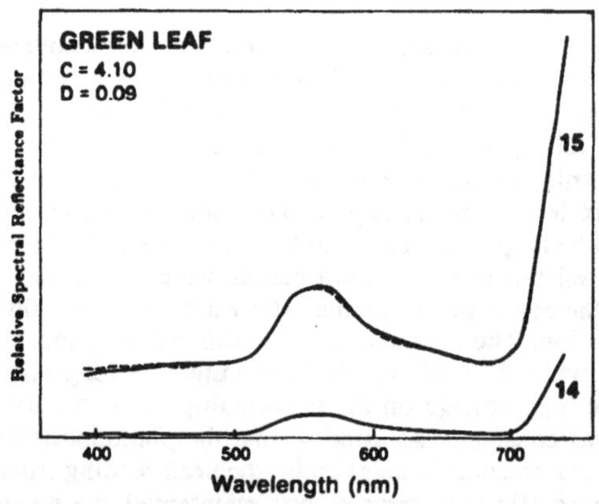

Fig. 14. Comparison of two relative spectral reflectance factor curves of the green leaf (arbitrarily scaled). The dotted curve is the least-squares fit of curve No. 14 to curve No. 15.

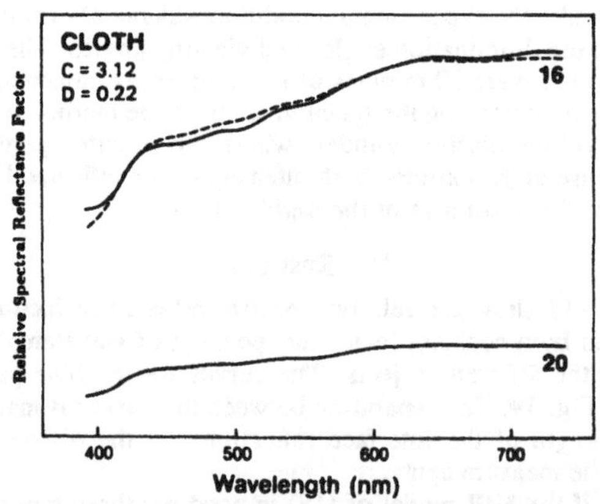

Fig. 15. Comparison to two relative spectral reflectance factor curves of the cloth (arbitrarily scaled). The dotted curve is the least-squares fit of curve No. 20 to curve No. 16.

plant leaves, cloth, painted wood, and orange peels certainly relates to the material composition of these object surfaces. There have been two other experimental studies [14], [30] after our measurements were completed [26]. Both confirm the general usefulness of the NIR model.

It is known [24], [30], [34] that special fatty materials are often secreted to form external surfaces of organisms. For example, the epidermal tissue of plants consists of flat cells with a thick outer wall covered by a heavy waxy cuticle [30, p. 42]. Fatty materials and oils are known to have fairly constant refractive indexes in the visible wavelength region [22]. Probably this is the reason why the color reflection from the surfaces of the plant leaf and the orange peel, as well as the painted wood (the paint vehicle is often a kind of oil, for example, linseed oil), can be so well described by the NIR model. For materials made of polymers, such as plastics, they do not absorb significantly over the range of visible wavelengths, and are naturally more or less colorless [41, p. 269]. Many fabricated articles are artificially colored by adding pigments or dyes to the materials. The material composition and structure of plastic objects, therefore, should be very good for the NIR model, except for the possible complication due to the dependence of light scattering on the pigment particle size and the depth light penetrates

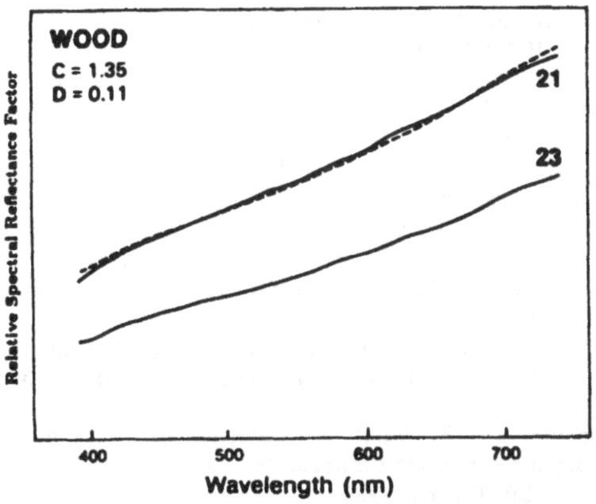

Fig. 16. Comparison of two relative spectral reflectance factor curves of the painted wood block (arbitrarily scaled). The dotted curve is the least-squares fit of curve No. 23 to curve No. 21.

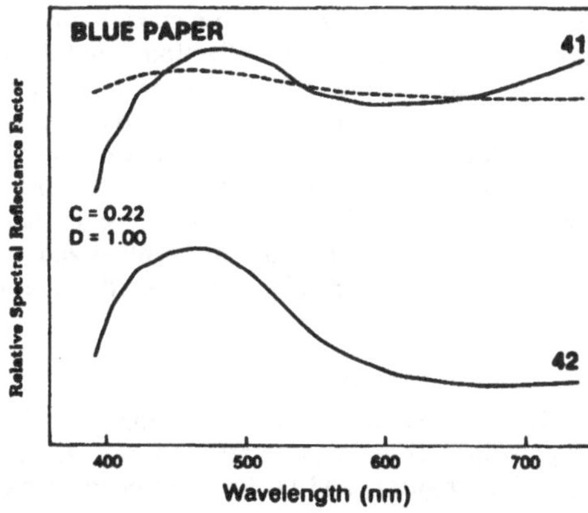

Fig. 19. Comparison of two relative spectral reflectance factor curves of the blue paper (arbitrarily scaled). The dotted curve is the least-squares fit of curve No. 42 to curve No. 41.

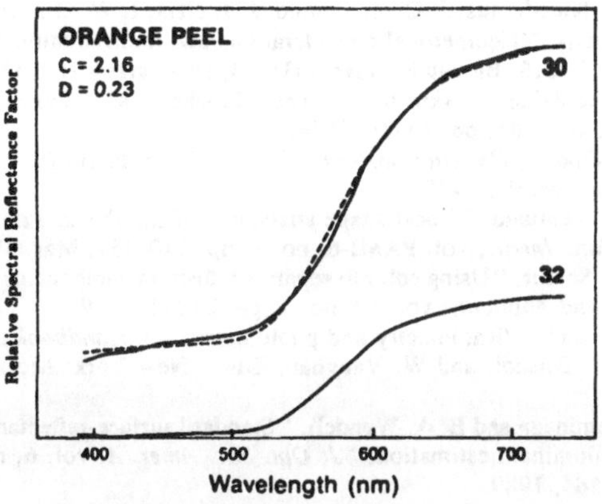

Fig. 17. Comparison of two relative spectral reflectance factor curves of the orange peel (arbitrarily scaled). The dotted curve is the least-squares fit of curve No. 32 to curve No. 30.

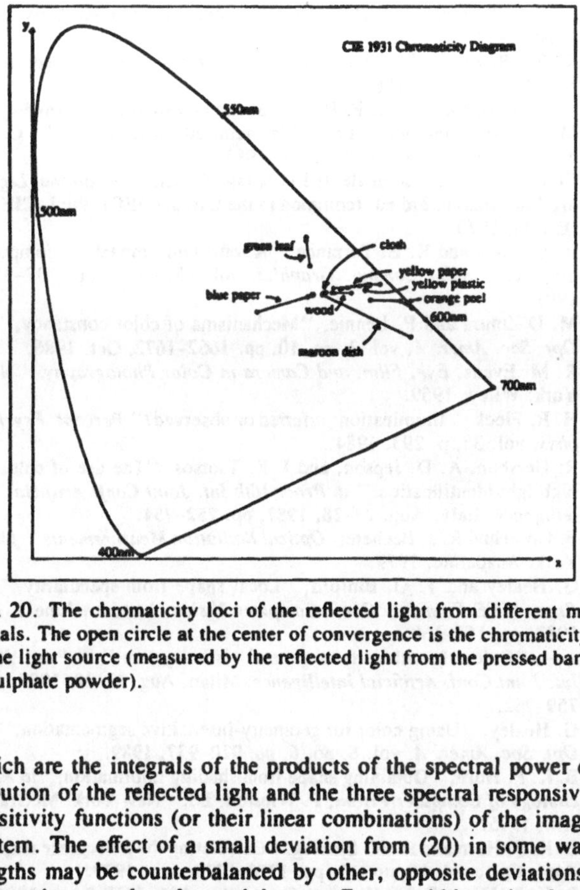

Fig. 20. The chromaticity loci of the reflected light from different materials. The open circle at the center of convergence is the chromaticity of the light source (measured by the reflected light from the pressed barium sulphate powder).

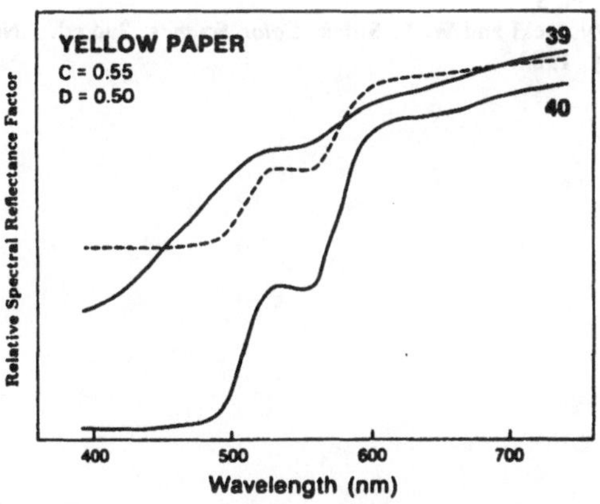

Fig. 18. Comparison of two relative spectral reflectance factor curves of the yellow paper (arbitrarily scaled). The dotted curve is the least-squares fit of curve No. 40 to curve No. 39.

into the surface. Of course, there are many other objects which modify the light spectral composition in completely different ways as described in [32]. The apparent realism of color images synthesized with the NIR model may be an important empirical hint that this model characterizes the nature of reflections from objects in the physical world to a greater extent than we thought. This remains to be seen.

In computer vision and graphics applications as well as human color perception, a color is often represented by three numbers, which are the integrals of the products of the spectral power distribution of the reflected light and the three spectral responsivity/sensitivity functions (or their linear combinations) of the imaging system. The effect of a small deviation from (20) in some wavelengths may be counterbalanced by other, opposite deviations in some other wavelengths, and the net effect would be quite close to what would be produced from the neutral-interface-reflection model. Therefore, an object surface need not strictly follow the model and still has a similar color appearance. This effect can be seen in Fig. 20, where 1931 CIE chromaticity coordinates are computed from the spectral power distributions of the light reflected from different portions of various objects (see [42] for the detailed description of the CIE color specification systems). The CIE color-matching functions used in these computations are related to linear combinations of the three (averaged) spectral sensitivity curves of human cones modified by specular transmittances for the ocular media. If the neutral-interface-reflection model holds for reflection from a given object, then the chromaticity locus of measurements on different spots of a uniform object will be a straight line segment, pointing to the point representing the chromaticity of the illuminant [27]. This is true because the light reflected from any portion of the object will be some additive combinations of interface (surface) and body (subsurface) reflections, and the chromaticities of additive combinations of lights fall on a straight-line locus of points extending between the chromaticities of the individual components. The locus of the yellow paper is close to a straight line segment, although as we have seen in Fig. 18, its spectral reflectance curves do not follow (21). This type of effect may somewhat explain the observation that the synthesized color images appear to be "realistic" in spite of the fact that the model is too simplistic.

Acknowledgment

We would like to thank H. Kwon and S. Fogel for their comments and suggestions about this correspondence, L. Tiefel for drawing some of the figures, and D. Kenworthy for her editorial help.

References

[1] G. A. Agoston, *Color Theory and Its Application in Art and Design.* Berlin: Springer-Verlag, 1979, ch. 5.
[2] J. Beck, *Surface Color Perception.* Ithaca, NY: Cornell University Press, 1972.
[3] R. W. Boyd, *Radiometry and the Detection of Optical Radiation.* New York: Wiley, 1983.
[4] M. J. Brooks and B. K. P. Horn, "Shape and source from shading," M.I.T. Artificial Intell. Lab., Massachusetts Inst. Technol., Cambridge, MA, A.I. Memo 280, Jan. 1985.
[5] Commission Internationale de l'Eclairage (CIE), *International Lighting Vocabulary*, 3rd ed. (common to the CIE and IEC), Publ. CIE 17 (E-1.1), 1970.
[6] R. L. Cook and K. E. Torrance, "A reflectance model for computer graphics," *ACM Comput. Graphics*, vol. 15, no. 3, pp. 307-316, 1981.
[7] M. D'Zmura and P. Lennie, "Mechanisms of color constancy," *J. Opt. Soc. Amer. A*, vol. 3, no. 10, pp. 1662-1672, Oct. 1986.
[8] R. M. Evans, *Eye, Film, and Camera in Color Photography.* New York: Wiley, 1959.
[9] H. R. Flock, "Illumination: inferred or observed?" *Percept. Psychophys*, vol. 35, p. 293, 1984.
[10] R. Gershon, A. D. Jepson, and J. K. Tsotsos, "The use of color in highlight identification," in *Proc. 10th Int. Joint Conf. Artificial Intelligence*, Italy, Aug. 23-28, 1987, pp. 752-754.
[11] F. Grum and R. J. Becherer, *Optical Radiation Measurements.* New York: Academic, 1979.
[12] G. Healey and T. O. Binford, "Local shape from specularity," in *Proc. IEEE First Int. Conf. Computer Vision*, London, June 8-11, 1987, pp. 151-160.
[13] —, "Color algorithms for a general vision system," in *Proc. 10th Int. Joint Conf. Artificial Intelligence*, Milan, Aug. 23-28, 1987, pp. 759-762.
[14] G. Healey, "Using color for geometry-insensitive segmentation," *J. Opt. Soc. Amer. A*, vol. 6, no. 6, pp. 920-937, 1989.
[15] B. K. P. Horn, "Obtaining shape from shading information," in *Psychology of Computer Vision*, P. Winston, Ed. New York: McGraw-Hill, 1975.
[16] B. K. P. Horn and R. W. Sjoberg, "Calculating the reflectance map," *Appl. Opt.*, vol. 18, no. 11, pp. 1770-1779, June 1979.
[17] B. K. P. Horn, *Robot Vision.* Cambridge, MA: MIT Press, 1986.
[18] R. S. Hunter, *The Measurement of Appearance.* New York: Wiley, 1975.
[19] A. Hurlbert, "Formal connections between lightness algorithms," *J. Opt. Soc. Amer. A*, vol. 3, no. 10, pp. 1684-1693, Oct. 1986.
[20] A. Hurlbert and T. Poggio, "Learning a color algorithm from examples," M.I.T. Artificial Intell. Lab., Massachusetts Inst. Technol., Cambridge, MA, A.I. Memo 909, Apr. 1987.
[21] K. Ikeuchi and B.K.P. Horn, "Numerical shape from shading and occluding boundaries," *Artificial Intell.*, vol. 17, pp. 141-185, 1981.
[22] R. Kanthack, *Table of Refractive Indices, Vol. I, Essential Oils.* London; Adam Hilger, 1918.
[23] G. J. Klinker, S. A. Shafer, and T. Kanade, "Using a color reflection model to separate highlights from object color," in *Proc. IEEE First Int. Conf. Computer Vision*, London, June 8-11, 1987, pp. 145-150.
[24] P. E. Kolattukudy, "Biosynthesis of surface lipids," *Science*, vol. 159, pp. 498-505, 1968.
[25] H.-C. Lee, "Method for computing the scene-illuminant chromaticity from specular highlights," *J. Opt. Soc. Amer. A*, vol. 3, no. 10, pp. 1694-1699, Oct. 1986.
[26] H.-C. Lee, E. J. Breneman, and C. P. Schulte, "An experimental study of a color reflection model," Kodak Res. Lab., Tech. Rep. 229507A, July 7, 1986.
[27] H.-C. Lee, "Estimating the illuminant color from the shading of a smooth surface," Massachusetts Inst. Technol., Cambridge, MA, AI Memo. 1068, Aug. 1988.
[28] —, "A physical color balance model for digital image processing," in *Proc. SPSE 42nd Annu. Conf.*, Boston, MA, May 1989, pp. 149-152.
[29] L. Maloney and B. Wandell, "Color constancy: A method for recovering surface spectral reflectance," *J. Opt. Soc. Amer. A*, vol. 3, no. 1, pp. 29-33, Jan. 1986.
[30] D. E. Metzler, *Biochemistry.* New York: Academic, 1977.
[31] U.S. Nat. Bureau Standards, "Self study manual on optical radiation measurements," Part I, NBS Tech. Note 910-1, 1976.
[32] K. Nassau, *The Physics and Chemistry of Color.* New York: Wiley, 1983.
[33] F. E. Nicodemus, J. C. Richmond, J. J. Hsia, I. W. Ginsberg and T. Limperis, "Geometrical considerations and nomenclature for reflectance," U.S. Bureau Standards (U.S.), Monograph 160, Oct. 1977.
[34] N. Nicolaides, "Skin lipids: Their biochemical uniqueness," *Science*, vol. 186, pp. 19-26, 1974.
[35] *Handbook of Optical Constants of Solids*, E. D. Palik, Ed. Orlando, FL: Academic, 1985.
[36] A. P. Pentland, "Local shape analysis," *IEEE Trans. Pattern Anal. Machine Intell.*, vol. PAMI-6, no. 2, pp. 170-187, Mar. 1984.
[37] S. A. Shafer, "Using color to separate reflection components," *Color: Res. and Applicat.*, vol. 10, no. 4, pp. 210-218, 1985.
[38] J. F. Snell, "Radiometry and photometry," in *Handbook of Optics*, W. G. Driscoll and W. Vaughan, Eds. New York: McGraw-Hill, 1978.
[39] S. Tominaga and B. A. Wandell, "Standard surface-reflectance model and illuminant estimation," *J. Opt. Soc. Amer. A*, vol. 6, no. 4, pp. 576-584, 1989.
[40] B. A. Wandell, "The synthesis and analysis of color images," *IEEE Trans. Pattern Anal. Machine Intell.*, vol. PAMI-9, no. 1, pp. 2-13, Jan. 1987.
[41] F. M. Willmouth, "Transparency, translucency, and gloss," in *Optical Properties of Polymers*, G. H. Meeten, Ed. London: Elsevier, 1986, ch.5.
[42] G. Wyszecki and W. S. Stiles, *Color Science*, 2nd ed. New York: Wiley, 1982.

Evaluation of linear models of surface spectral reflectance with small numbers of parameters

Laurence T. Maloney

Human Performance Center, University of Michigan, 330 Packard Road, Ann Arbor, Michigan 48104

Received March 10, 1986; accepted June 27, 1986

Recent computational models of color vision demonstrate that it is possible to achieve exact color constancy over a limited range of lights and surfaces described by linear models. The success of these computational models hinges on whether any sizable range of surface spectral reflectances can be described by a linear model with about three parameters. In the first part of this paper, I analyze two large sets of empirical surface spectral reflectances and examine three conjectures concerning constraints on surface reflectance: (1) that empirical surface reflectances fall within a linear model with a small number of parameters, (2) that empirical surface reflectances fall within a linear model composed of band-limited functions with a small number of parameters, and (3) that the shape of the spectral-sensitivity curves of human vision enhance the fit between empirical surface reflectances and a linear model. I conclude that the first and second conjectures hold for the two sets of spectral reflectances analyzed but that the number of parameters required to model the spectral reflectances is five to seven, not three. A reanalysis of the empirical data that takes human visual sensitivity into account gives more promising results. The linear models derived provide excellent fits to the data with as few as three or four parameters, confirming the third conjecture. The results suggest that constraints on possible surface-reflectance functions and the "filtering" properties of the shapes of the spectral-sensitivity curves of photoreceptors can both contribute to color constancy. In the last part of the paper I derive the relation between the number of photoreceptor classes present in vision and the "filtering" properties of each class. The results of this analysis reverse a conclusion reached by Barlow: the "filtering" properties of human photoreceptors are consistent with a trichromatic visual system that is color constant.

INTRODUCTION

A visual system exhibits perfect or exact color constancy if its estimates of object color are unaffected by changes in ambient lighting. Human color vision is approximately color constant across certain ranges of illumination, although the degree of color constancy exhibited changes with the range of lighting examined.[1] It is not known whether human color vision is exactly color constant over any range of lights; Judd[2] has conjectured that, for illuminants drawn from the range of natural daylight, it is.

Recent computation models of color vision[3–10] demonstrate that it is possible to achieve perfect color constancy over a limited range of lights and surfaces. The algorithms provided by Buchsbaum[8] and by Maloney and Wandell[9] go further: they provide estimates of color appearance for any possible combination of lighting and surface spectral reflectances on a scene. When the lighting and surface spectral reflectances are drawn from the prespecified, limited ranges, the estimates of color are unaffected by changes in the lighting on the scene. When the lighting and surface spectral reflectances in the scene are approximately those of the limited ranges, the color estimates are approximately correct. Both Buchsbaum and Maloney and Wandell provide methods for estimating and bounding the errors in color estimates obtained.[8,10]

Sallstrom[5] shows that, given the kinds of information available to biological visual systems, no better result is obtainable. A visual system with a finite number of different classes of photoreceptors can achieve color constancy only across a limited range of possible lights and surface spectral reflectances. Consequently, any color-vision system that achieves color constancy is making use of constraints on the possible lights and surface spectral reflectances in scenes that it encounters. The constraint takes the form of a specification of the lights and surface spectral reflectances for which the algorithm will operate properly. The algorithms mentioned above all assume the common constraint that surface spectral reflectances are drawn from a linear model with two or three parameters.

TERMINOLOGY

Before the goals and conclusions of this paper are stated, some terminology must be defined. The spectral power distribution of the illuminant in the scene is denoted $E(\lambda)$, where λ is wavelength. The units in which E is measured are not important for the purposes of this paper. The surface reflectance at each retinal location x in the scene is denoted $S^x(\lambda)$. Photoreceptor sensitivities are denoted $R_k(\lambda)$, $k = 1, 2, 3$. The light impinging on photoreceptors is the product of the spectral power distribution of the illuminant with surface reflectance, $E(\lambda)S(\lambda)$, termed the *color signal* (after Buchsbaum and Gottschalk[11]).

A linear model of surface reflectance with n parameters consists of a list of n basis surface reflectances $S_j(\lambda)$. A surface reflectance $S(\lambda)$ is in the model if it can be expressed as the weighted sum of the n basis surface reflectances:

$$S(\lambda) = \sigma_1 S_1(\lambda) + \sigma_2 S_2(\lambda) + \ldots + \sigma_n S_n(\lambda),$$

where the σ_j are weights (real numbers). The n numbers σ_j

are all that is needed to specify a surface reflectance if it is in the linear model. Let $\sigma = (\sigma_1, \ldots, \sigma_n)$ be the vector of weights describing a surface reflectance; then we can denote that surface reflectance as $S_\sigma(\lambda)$.

A surface spectral-reflectance curve viewed simply as a function of wavelength has a Fourier transform, and, if the transform is zero outside a certain range of frequencies, the surface spectral reflectance is band limited. As a consequence of the Shannon–Whittaker sampling theorem, the class of band-limited functions with a fixed band limit forms a linear model with a number of parameters proportional to the band limit. Linear models formed of band-limited surface spectral reflectances will be referred to repeatedly in what follows.

Any function of wavelength across the visible spectrum that ranges between 0 and 1 inclusive is potentially a surface spectral reflectance. The number of parameters required to describe all such functions is basically infinite. The claim that most surface spectral reflectances form a linear model with a small number of parameters is not a trivial constraint.

GOALS AND CONJECTURES

The first part of the paper is concerned with testing three conjectures regarding constraints on surface spectral reflectances: (1) surface spectral reflectances fall within a linear model with a small number of parameters, (2) surface spectral reflectances fall within a linear model composed of band-limited functions with a small number of parameters, and (3) the broad, smooth shape of the spectral sensitivity of human photoreceptors enhances the observed constraints on surface-reflectance functions. Each conjecture is tested by using two large collections of empirical surface spectral reflectances described below.

The conclusions of this analysis are as follows: the first and second conjectures hold for linear models with five to seven parameters but not for linear models having as few as two to four parameters. When the effect of human photoreceptor sensitivities is included, linear models with as few as three to four parameters provide excellent fits to the empirical data sets. The overall shape of human photoreceptors serves to enhance the fit of the models to the available data, confirming the third conjecture.

The second part of this paper examines the role of photoreceptor sensitivities in enhancing color constancy. Barlow[12] suggested that photoreceptors perform two distinct operations on the color signal. The broad, smooth shapes of photoreceptor spectral-sensitivity curves act as low-pass filters on the color signal. The low-pass component of the color signal that passes through the "filter'" is sampled by the three classes of photoreceptors in human vision. If the band limit is low enough, then three samples would serve to reconstruct the color signal. The "filtering" action of the photoreceptors matches the complexity of the effective color signal to the number of samples taken. Barlow terms this result a "low-pass theory of trichromacy."

Barlow analyzes the filtering properties of human photoreceptors and rejects a low-pass theory of trichromacy. While receptors do act as low-pass filters on the color signal, the number of samples required to capture the low-pass component of the color signal passed by the photoreceptors is markedly greater than three. The filtering properties of receptors in human vision do not match the sampling properties.

I reanalyze the relation between the filtering and sampling properties of receptors with the additional assumption that the goal of the filtering and sampling operations is not to reconstruct the color signal but to preserve the information about illuminants and surfaces needed to compute constant color descriptors. The main conclusion is that, with this assumption, the proper band limit for the filtering operation must exceed the limit corresponding to the sampling operation. This conclusion is consistent with Barlow's empirical results.

A natural complement to the analyses of surface reflectances presented here would be an analysis of constraints on common illuminants, including daylight. I know of no large data set of spectral power distributions of illuminants still in existence (although several authors[13] have collected and analyzed samples of daylight). Consequently, no corresponding detailed analysis of illuminants is included.

ANALYSES OF EMPIRICAL SURFACE SPECTRAL REFLECTANCES

In this section I address the first conjecture: surface spectral reflectances form a linear model with a small number of parameters.

Several authors have suggested that naturally occurring spectral-reflectance curves are highly constrained and that many spectral reflectances are well fitted by simple models with only a few parameters. MacAdam[14] notes that the spectral-reflectance curves of many natural substances (including wood, paper, and minerals) are approximately straight lines. Lythgoe[15] asserts that nearly all the biological pigments responsible for yellows, oranges, or reds are approximately cutoff filters (wavelengths longer than a certain value are reflected, shorter wavelengths are absorbed). Neither of the proposed models is a linear model in the sense of the previous section.

Cohen was, to my knowledge, the first to fit a linear model to a set of empirical surface spectral reflectances.[16] He computed the best linear model (through characteristic vector analysis) for a subset of 150 spectral reflectances of Munsell color samples randomly sampled from a subset of 417 of the full set of 433. The first n characteristic vectors form a linear model that provides at least as good a fit to the data points in the least-squares sense as could any n vectors. The first three characteristic vectors in Cohen's analysis accounted for 0.992 of the variance in the overall fit. Adding the fourth characteristic vector improves the fit by only 0.005. Cohen states that the spectral reflectances of individual Munsell chips could be closely approximated by model reflectances, but he does not give quantitative results.

Cohen's analysis suggests that the surface spectral-reflectance curves of the Munsell chips that he analyzed are highly constrained. One limitation of Cohen's analysis is that he was unable to analyze the full set of Munsell spectral reflectances and, consequently, his variance-accounted-for is possibly optimistic. A second limitation is that he reports only the overall goodness-of-fit figure, leaving uncertainty as to how well individual spectral reflectances are fit. Consequently, his results cannot be used to claim that "typical" surface spectral reflectances or even the full set of Munsell color samples is well approximated by a linear model with a small number of basis elements. Third, as Buchsbaum and Gottschalk point out,[12] it is not clear how results concerning the Munsell color samples will generalize to other sets of surface spectral reflectances in the natural environment.

In this section I repeat Cohen's analysis on the full set of Munsell surface spectral reflectances collected by Kelley *et al.*[17] and also on a second set of 337 spectral reflectances of natural formations collected by Krinov[18] and report not only the overall goodness of fit but also the distribution of goodness of fit for individual surface spectral reflectances in the two large samples. I then used Cohen's three characteristic vectors to fit the set of surface spectral reflectances of natural objects collected by Krinov, testing whether the model derived for the Munsell data extends to the data on natural formations. All fitting procedures are given in Appendix A.

The surface spectral reflectance of each of the 462 surface spectral-reflectance curves in the Nickerson–Munsell set[7] had been sampled at 10-nm intervals between 380 and 770 nm. I used only the range between 400 and 700 nm inclusive, the range investigated by Cohen.[19] Each surface spectral reflectance is represented by 31 values. Each surface spectral-reflectance curve in the Krinov set had been sampled at 10-nm intervals from 400 to 650 nm (26 values). Each set was subjected to a modified characteristic vector analysis to determine the best fitting linear models with one through six basis elements (see Appendix A: Procedure 2).

Once the characteristic vector analysis was completed, each of the surface spectral-reflectance vectors in the original set was fitted to linear models based on between two and six characteristic vectors by using multiple regression (see Appendix A: Procedure 1). For each surface reflectance, then, we have a measure of goodness of fit R^2 to the overall best-fit linear model with two parameters, with three parameters, etc.

The overall goodness of fit for two through six characteristic vectors derived from the Nickerson–Munsell data is given in Table 1. Table 2 gives the median, lower, and upper quartiles of the goodness-of-fit values R^2 obtained for the individual fits when linear models based on from two to six characteristic vectors were used. For example, in a model with three characteristic vectors, the R^2 for 75% of the individual fits exceeded 0.9890. The overall goodness of fit for two through six characteristic vectors derived from the Krinov surface spectral-reflectance data is given in Table 3. When the individual Krinov data series were regressed on the first three characteristic vectors derived from the Krinov data set, the R^2 values obtained have median 0.9980 and quartiles 0.9957 and 0.9991.

Table 1. Proportion of Variance Accounted for by a Linear Model with 2–6 Parameters[a]

Parameters				
2	3	4	5	6
0.9583	0.9916	0.9959	0.9985	0.9992

[a] Nickerson's 462 Munsell surface spectral reflectances truncated to 400–700 nm.

Table 2. Distribution of 462 R^2 Multiple Correlation Coefficients for Nickerson's Munsell Surface Spectral-Reflectance Data Fitted to Linear Models Based on 2–6 Characteristic Vectors[a]

Parameters					
2	3	4	5	6	
0.9372	0.9890	0.9948	0.9982	0.9990	1 quartile
0.9758	0.9962	0.9979	0.9992	0.9996	median
0.9923	0.9981	0.9991	0.9996	0.9998	3 quartile

[a] All data truncated to the range 400–700 nm.

Table 3. Proportion of Variance Accounted for by a Linear Model with 2–6 Parameters[a]

Parameters				
2	3	4	5	6
0.9894	0.9960	0.9973	0.9982	0.9987

[a] Krinov's 337 surface spectral reflectances.

It is of interest to compare directly the two data sets. Each of the Krinov spectral-reflectance curves was fitted by Procedure 2 to Cohen's first three characteristic vectors (truncated to the range 400 to 650 nm). The values of R^2 obtained ranged from 0.913 to 0.999 with median 0.993 and quartiles 0.979 and 0.998. The two best and two worst fits are shown in Fig. 1.[20] A model based on Cohen's three characteristic vectors provides a fit to Krinov's data that is not much worse than the fits provided to the Munsell data from which it was derived. The worst fits have noticeable patterned deviations.

As few as six characteristic vectors provide essentially perfect fits to almost all the surface spectral reflectances in both sets. Further, the basis elements of the Munsell set provide excellent fits to the Krinov data. The same cannot be said of the fits with three characteristic vectors: a considerable proportion of the surfaces is not completely captured by the linear models of the two data sets. I postpone discussion of the implications of this result.

The results of this section are subject to a very basic challenge: The precise characteristic vectors derived depend on the sampling of empirical surface spectral reflectances collected. Suppose that we measure an extremely large number of surface spectral reflectances chosen from the natural environment according to some selection procedure. The characteristic vectors of this set of spectral reflectances are then taken as the basis of a finite-dimensional linear model. Stiles *et al.*[21] and Buchsbaum and Gottschalk[11] both outline this approach and emphasize its practical difficulties. However there is a more basic, theoretical, difficulty: In sampling spectral reflectances, we weight each spectral reflectance by its frequency of occurrence under whatever selection procedure we choose. The basis elements of any finite-dimensional linear model chosen to maximize the fit of the finite-dimensional linear model to the data set reflect this (arbitrary) weighting.

An alternative approach, discussed in the following section, is to consider which physical properties are shared by

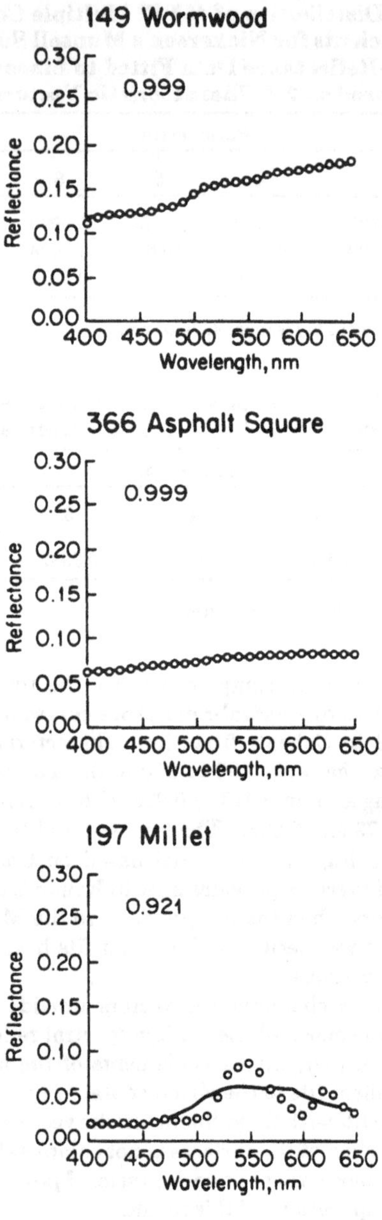

Fig. 1. Naturally occurring surface reflectances. The two best and two worst fits obtained in fitting Krinov's 337 surface spectral reflectances (circles) by Cohen's first three characteristic vectors. The fitted function is shown as a solid line. Each x axis plots wavelength in nanometers; each y axis plots proportion of light reflected. Krinov's identifying information for the surface and the values of R^2 for the multiple-regression fit are given for each surface reflectance.

many naturally occurring spectral reflectances and base models on these properties.

BAND-LIMITED SPECTRAL-REFLECTANCE FUNCTIONS

This section and the following section address the second conjecture: Surface spectral reflectances are band-limited functions. Here, "frequency" refers to the abscissa of the Fourier transform of the spectral-reflectance curve, not the frequency of light, and is measured in cycles/nanometer. If the visible spectrum is taken to be 300 nm wide, ranging from 400 to 700 nm, then a spectral-reflectance curve that is one cycle of a sine wave of approximate frequency 0.0033 cycle/nm will just span the visual spectrum. I will use the term "wave number," which is the reciprocal of wavelength λ and is in units of inverse centimeters, to refer to the frequency of light.

Stiles *et al.*[21] suggest that band-limited functions plotted against wavelength with limits of 0.01–0.02 cycle/nm are adequate models of naturally occurring spectral reflectances for many purposes. As Buchsbaum and Gottschalk[11] point out, the class of essentially band-limited functions with limit 0.01–0.02 cycle/nm can be approximated by a finite-dimensional linear model with 6–12 parameters if the visual spectrum is taken to be $W = 300$ nm in width. Papers by Slepian and Pollak[22] and by Landau and Pollak[23] show that we can find a linear model based on $2WF$ independent functions that approximate any function that has a frequency limit less than F and that has most of its energy in the interval 400–700 nm.

The number of parameters (basis reflectance vectors) across the visible spectrum that is required to capture a band-limited function is proportional to the band limit. For convenience, then, we refer to "the band limit corresponding to three basis vectors," making use of this linear relation.

If the number of parameters in the finite-dimensional linear model is limited to three, then, as Buchsbaum and Gottschalk note, the corresponding frequency limit is approximately 0.005 cycle/nm. Buchsbaum and Gottschalk also show that, under a specific choice of light with equal energy at each wavelength, the set of spectral reflectances with limit 0.005 provides a metamer to every Munsell chip spectral reflectance and covers a large portion of the center of the CIE diagram.

Stiles *et al.* suggest that many spectral reflectances correspond to functions with band limits below 0.01–0.02 cycle/nm. Buchsbaum and Gottschalk advance a stronger conjecture: "the space of practical color signals may . . . be adequately represented by three dimensions . . . ," corresponding to the frequency limit 0.005 cycle/nm. Both the conjecture of Stiles *et al.* and the conjecture of Buchsbaum and Gottschalk (if we replace "color signals" by "surface reflectances," above) are tested in this section. I examine the distribution of power in the power spectra of the Nickerson–Munsell data and Krinov's data by computing modified discrete Fourier transforms (DFT's) (see Appendix A: Procedure 4).

I report all data with 0.0, 0.3, and 1.0 tapering as explained in Appendix A. Table 4 summarizes the results for 462 of the Nickerson–Munsell data series. The full range of the data, 380 to 770 nm (40 samples) was used in estimating

Table 4. Frequency Limits for Nickerson's Munsell Surface-Reflectance Data[a]

0.0033	0.005	0.010	0.015	Frequency Limit
Taper 0.00				
0.9461	0.9584	0.9817	0.9883	1 quartile
0.9605	0.9676	0.9854	0.9907	median
0.9686	0.9724	0.9879	0.9921	3 quartile
Taper 0.30				
0.9454	0.9823	0.9958	0.993	1 quartile
0.9638	0.9884	0.9971	0.9995	median
0.9743	0.9900	0.9978	0.9997	3 quartile
Taper 1.00				
0.9673	0.9881	0.9988	0.9998	1 quartile
0.9840	0.9949	0.9995	0.9999	median
0.9939	0.9980	0.9999	1.0000	3 quartile

[a] Proportion of spectral energy below the frequency limit (cycles/nm) for 462 Munsell surface reflectances with different degrees of tapering.

Table 5. Frequency Limits for Krinov's Surface Reflectances[a]

0.0033	0.005	0.010	0.015	Frequency Limit
Taper 0.00				
0.946	0.960	0.981	0.988	1 quartile
0.954	0.966	0.984	0.990	median
0.959	0.972	0.986	0.993	3 quartile
Taper 0.30				
0.952	0.969	0.990	0.997	1 quartile
0.963	0.972	0.991	0.997	median
0.968	0.976	0.993	0.998	3 quartile
Taper 1.00				
0.917	0.966	0.995	0.999	1 quartile
0.953	0.990	0.999	1.000	median
0.961	0.995	1.000	1.000	3 quartile

[a] Proportion of spectral energy below limit (cycles/nm) for 337 Krinov surface reflectances with different degrees of tapering.

power spectra. The first quartile, median, and third quartile of the power at or below various band limits are given. Table 5 summarizes the results for 337 of Krinov's data series. For the Krinov data, the effect of tapering is more pronounced, reflecting the smaller number of samples in each series. These results must be interpreted with caution because of the nature of the data and the small number of samples in each of Krinov's measured spectral reflectances. Still, the amount of power below the 0.005 cutoff for six basis elements has median 0.9995 for the Munsell data with 0.30 taper. The corresponding value for the Krinov data is 0.997. However, some of the samples have as much as 0.1 of their total power above the three-sample limit of 0.005 cycle/nm. The values of 0.01 and 0.015 are both candidates for a band limit for surface spectral reflectances.

These results are consistent with the conjectured frequency limit of Stiles *et al.* of 0.01–0.02 cycle/nm, which corresponds to a linear model with 6 to 12 parameters. For the Munsell data, a linear model with as few as six parameters again provides an excellent fit to the data.

The results of this section suggest that surface spectral reflectances are characteristically smooth, low-pass functions of wavelength. In the next section I examine the physical processes that generate surface spectral reflectance.

A POSSIBLE PHYSICAL BASIS FOR CONSTRAINTS ON SURFACE SPECTRAL REFLECTANCES

The goal of this section is to formulate a conjecture concerning the physical processes that determine the presence or absence of high-frequency components in any given spectral-reflectance curve. I do not intend to prove that all or even many surface spectral reflectances have a particular, low band limit; I intend, rather, to motivate the use of the class of band-limited functions as a model of surface spectral reflectances. First, I characterize surface spectral-reflectance curves as a superposition of absorption bands associated with electronic transitions within molecules. Second, I describe the physical processes that affect the shape of these bands, and third, I characterize typical band shapes.

Nassau[24] identifies 15 more or less distinct physical causes of color, each a physical mechanism that alters the spectral distribution of the color signal reaching the eye. Of these 15, the phenomena of most interest here concern the absorption of quanta of light by surfaces: molecules in a condensed state (solid, liquid, or in solution). Light reaches the surface of an object, is absorbed, and is reradiated to the viewer, giving a color signal, $E(\lambda)S(\lambda)$, that is a combination of factors intrinsic to the object and the light.

The absorption curves of many organic and inorganic substances are composed of superimposed broad "bands".[24] The broad, smooth shape of these band spectra determines the smooth shape of spectral-reflectance curves.

It is surprising, then, to realize that the processes responsible for energy absorption by molecules across the visible spectrum are all quantized and, in isolation, produce absorption in narrow-wavelength bands such as the Fraunhofer lines observed in the solar spectrum. Each discrete energy transition corresponds to a specific wave number of light through the Einstein relation $E = h\nu$, where h is Planck's constant. A quantized system can absorb light only at discrete frequencies determined by this relation. The processes that permit quantized systems to have smooth absorption curves are described qualitatively in the next few paragraphs.

Three quantized molecular processes—a change in the rotational state of a molecule, a change in its vibrational state, or an electronic transition—absorb energy from incident light. Again, the energy involved in each physical process determines which frequencies of light can be absorbed. Rotational absorptions require little energy and involve frequencies in the far infrared. Vibrational changes require more energy and involve frequencies in the near infrared. Electronic transitions require the most energy and correspond to frequencies in the near and far ultraviolet. Only relatively low-energy electron transitions correspond to visible frequencies. A single electronic transition with an energy difference corresponding to a wave number within the visible spectrum would produce a sharp absorption spike. The visible and ultraviolet Fraunhofer lines of the solar spectrum correspond to electronic transitions.

Two phenomena contribute to the broadband character of spectral absorption bands: (1) interactions among rotational, vibrational, and electronic transitions and (2) molecular interactions. In molecules, the three energy-absorbing processes (vibrational, rotational, electronic) do not occur in isolation: they interact. Even very simple molecules can have hundreds of rotational levels about each vibrational level. The resulting spectrum consists of very fine, closely spaced lines, but,

> "... in fact, the rotational structure fine structure is usually unresolvable, and it affords only a bandwidth to each vibrational subband. Therefore, *the shape of an absorption band due to a single electronic transition may be considered to be determined by the spacing of the vibrational subbands and by the distribution of the total intensity of the electronic transitions among the vibrational subbands*."[25] [Emphasis added.]

When molecules are in a condensed state, as in liquid or solid form or in solution, their interactions further "smear" the observed absorption spectrum.

> "In the dilute vapor, each of the vibrational levels consists of many closely spaced rotational levels.... As the pressure of the vapor is increased, collisions between molecules broaden the rotational levels until a continuum is produced that still shows, however, a group of distinct peaks reflecting the vibrational energy level spacings. In the liquid state the interactions with neighboring molecules causes further broadening, and if this is sufficiently strong the vibrational structure may even disappear."[26]

Kauzmann concludes that "Strong interactions with solvents and with vibrational modes are responsible for the broad absorption bands observed in all organic compounds and many organic compounds in solution."[27]

In summary, an electronic transition whose energy corresponds to a wavelength near the visible spectrum is the center of an absorption peak broadened by other transitions and smoothed by other molecular interactions in the liquid or solid state. A molecule's absorption spectrum can be described as a discrete structure of electronic transitions surrounded by vibrational transitions in turn surrounded by rotational transitions that has been passed through an approximate low-pass filter.

Pringsheim notes that the absorption band has an appreciable width and "an approximately symmetrical intensity distribution around a center" (Ref. 28, p. 300) and, further, that

> "The intensity band within a simple absorption ... band depends, in the first place, on the distribution of vibrational energy in the absorbing or the emitting molecules, and, in the second place, on the fluctuations in the configuration of the surrounding medium. Since both follow the laws of statistics, the spectral intensity curve of each band should also have the shape of a Gaussian error curve.... The nearly symmetric bell shape is lost if two adjacent bands, which may correspond to different electronic transitions, overlap." (See Ref. 28, p. 306.)

The Fourier transform of the Gaussian (also a Gaussian) has tails that drop rapidly to zero with increasing frequency, and convolution with a Gaussian effectively band limits the absorption spectrum and, consequently, the spectral-reflectance curve corresponding to it.

Some quantitative results follow. Recall that energy difference translates directly into wave number ν by the Einstein relation.[29]

> "... the energy separation of different electronic states is of the order of 10 000 cm^{-1}; the separation of different vibrational states associated with the same electronic state is of the order of 1000 cm^{-1}; the separation of different rotational states belonging to the same vibronic state is much smaller, being of the order 1–100 cm^{-1}." (Ref. 25, p. 59).

Kauzmann states that[26] "The absorption bands of most organic and inorganic molecules in solutions have half-widths of the order of 5000 cm^{-1}." The half-width is the width of the spectrum at the points where it falls to half of its peak absorption.

Take the width of the visible spectrum plotted against the wave number to be approximately 10 000 cm^{-1}. Then the half-width of a "typical" absorption band as reported by Kauzmann is approximately one half of the visible spectrum. Further, the spacing between vibrational levels corresponds to frequencies to which human photoreceptors are insensitive, according to Barlow (discussed in the next two sections). This result suggests that human photoreceptors are quite insensitive to whatever vibrational structure remains in typical spectral reflectances.

The conclusions of this section may be stated as a conjecture: many substances in liquid or solid state are band limited by molecular interactions and the superimposed vibrational/rotational patterns. The resultant band limit is consistent with a linear model of band-limited functions with about six parameters. Deviations from a smooth shape with at most two peaks across the visible spectrum reflect primarily the vibrational transitions superimposed on the electronic transitions.

ROLE OF PHOTORECEPTORS

In this section I address the third conjecture: The broad, smooth shape of the spectral sensitivity of human photoreceptors serves to enhance the observed constraints on surface-reflectance functions.

A least-squares criterion ignores potential contributions of the photoreceptor sensitivities to improving the fit of the model. Most obviously, discrepancies between $S(\lambda)$ and $S_\sigma(\lambda)$ over regions of the electromagnetic spectrum where the photoreceptors are completely insensitive can be ignored. Consequently, we measure goodness of fit only over a range (the visible spectrum) where the photoreceptors have some sensitivity to light stimuli. If the photoreceptor sensitivities resemble human photoreceptor sensitivities, however, a least-squares criterion weights discrepancies at

Table 6. Proportion of Variance Accounted for by a Linear Model with 2–6 Parameters[a]

Parameters				
2	3	4	5	6
0.9899	0.9979	0.9993	0.9997	0.9999

[a] Weighted least-squares criterion. See text. Nickerson's 462 Munsell surface spectral reflectances.

Table 7. Distribution of 462 R^2 Multiple Correlation Coefficients for Nickerson's Munsell Surface Spectral-Reflectance Data Fitted to Linear Models Based on 2–6 Characteristic Vectors[a]

Parameters					
2	3	4	5	6	
0.9775	0.9962	0.9986	0.9996	0.9998	1 quartile
0.9940	0.9988	0.9996	0.9999	0.9999	median
0.9983	0.9996	0.9998	0.9999	1.0000	3 quartile

[a] Weighted least-squares criterion.

the ends of the visible spectrum, where all photoreceptors are relatively insensitive, equally with discrepancies in the center of the visible spectrum. In selecting linear models of light or surface spectral reflectance for human vision, we might consider a weighted least-squares criterion across wavelength.

The possibility can be tested by using Procedure 3, as described in Appendix A. A characteristic vector analysis of the Nickerson–Munsell data was done, where the discrepancy between empirical surface reflectance and model reflectance $S_\sigma(\lambda)$ was weighted by the inverse of V_λ, a measure of the sensitivity of the photopic visual system at each wavelength.[30] The full range of Munsell data available, 380 to 770 nm, was used. However, the values of V_λ drop off so rapidly outside the range of 400 to 700 nm that the additional data had little influence. Each data series was then regressed against the second through sixth characteristic vectors obtained, the residual errors again being weighted by V_λ. The results are shown in Tables 6 and 7. The format is that of Tables 1 and 2. Note the marked improvement in fit over the results of Tables 1 and 2. There is no *a priori* reason why weighting by an arbitrary function should not worsen the fit rather than improve it.

PHOTORECEPTOR SPECTRAL SENSITIVITY AND COLOR CONSTANCY

Recall that "frequency" refers to the abscissa of the Fourier transform of the color signal, not to frequency in the electromagnetic spectrum. Recall also that the "color signal" is the product of the illuminant and surface reflectance, not surface reflectance.

The empirical results of the last section lend support to the following argument: Suppose that we could choose the spectral sensitivities of each photoreceptor class so that each photoreceptor completely attenuates frequencies above a certain cutoff. Then photoreceptors act as *ideal low-pass filters* on the color signal. Color signals with many peaks and valleys across the visible spectrum would be effectively smoothed by the filtering action of the photoreceptors. The band limit imposed by the low-pass filter corresponds to a linear model with some number, *p*, of basis vectors. The results of Slepian *et al.* allow us to compute *p* as a function of the band limit. Then, whether or not the color signal is band limited, the low-pass portion of the color signal can be reconstructed from the quantum catches of *p* properly chosen classes of photoreceptors. The quantum catches used to reconstruct the color signal will be referred to as "samples." Since the portion of the color signal above the band limit is removed at the photoreceptors, the *p* samples serve to reconstruct the portion of the color signal that the visual system is capable of sensing.

This line of argument was first advanced by Barlow,[12] who noted that the three broad spectral-sensitivity curves of human photopic vision can act as approximate low-pass filters across the visible spectrum, removing or attenuating high frequencies in the color signal. Barlow examined whether the receptor cutoff frequencies are so low that the filtered color signal is potentially reconstructible from the three samples provided by the three receptor classes at each location ($p \leq 3$). He concludes that color signal is *not* reconstructible and rejects a "low-pass filter theory of trichromacy" for human vision. The cutoff frequency, as estimated from his graph, corresponds to about five to seven classes of photoreceptors at each location.[31]

Barlow's results leave us with a puzzling discrepancy: The shape of human photoreceptor sensitivities permits them to act as low-pass filters on the color signal. Five to seven samples would serve to reconstruct the effective color signal passed by the photoreceptors. The number of different classes of photoreceptors, three, permits reconstruction of a color signal with a band limit corresponding to only three samples. The photoreceptors viewed as filters pass about twice as much information as the photoreceptors viewed as samplers can possibly manage. From the viewpoint of a low-pass theory of trichromacy, the receptors are either "underfiltering" or "undersampling."

In this section I resolve this discrepancy by considering the photoreceptor band limit that is optimal for color-constancy algorithms using linear models of illuminant and surface reflectance and a trichromatic visual system. These algorithms seek to reconstruct information about illuminant and surface reflectances. A proper choice of photoreceptor band limit can serve to enhance the performance of these algorithms by improving the fit between physical environment and linear models of illuminant and surface reflectances. *I will demonstrate that color constancy is possible only if the band limit of the receptors exceeds the band limit that can be reconstructed from three samples.* Barlow's empirical results are a necessary precondition of color-constancy algorithms using linear models. This result is general: In an *n*-chromatic visual system, the band limit of the receptors must exceed the band limit reconstructable from *n* samples.

This result appears paradoxical. The filtered color signal is undersampled, yet a color-constancy algorithm correctly estimates the effective illuminant and surface reflectance that permit us to compute the effective color signal. It seems as if an undersampled signal has been reconstructed contrary to the Shannon–Whittaker sampling theorem.

The paradox is resolved if we note that the color algorithms referred to all assume that the illuminant is constant across several retinal locations and that color constancy is possible only if information from several retinal locations is given. The assumed redundancy in the illuminant permits reconstruction of illuminant and surface reflectance despite undersampling at each retinal location considered in isolation.

Suppose that the illuminants in a scene are band limited with a band limit f_e. Let the number of samples required to reconstruct a light with this band limit be m. Recall that m is proportional to f_e. Suppose that surface reflectances in a scene are also band limited with a band limit f_s. The number of samples required for a surface to be reconstructed with this band limit is n, also proportional to f_s.

The information about lights and surface reflectance is contained in the color signal $E(\lambda)S(\lambda)$ at each retinal location. If illuminant $E(\lambda)$ and surface reflectance $S(\lambda)$ are band limited, then it is not difficult to show that the color signal is also band limited but with a band limit that is the sum of the band limits of illuminant and surface reflectance $f_e + f_s$ requiring $m + n$ samples to reconstruct.[32] If both spectral reflectances and lights were band limited with a limit of 0.005 cycle/nm, then the color signal would be band limited with a limit of 0.010 cycle/nm.

The following figure illustrates that the cutoff for the color signal is the *sum* of the cutoffs of the light and the surface spectral reflectance when both are approximately band limited.

Figures 2(a)–2(c) illustrate the relations among a hypothetical light, a hypothetical surface spectral reflectance, and the resulting color signal. The frequency axis is in arbitary units that may be taken to be the number of samples needed to reconstruct a signal with that band limit. Illuminant and surface reflectance are both band limited with flat spectra within the band −3 to 3. The color signal has a triangular spectrum that is nonzero between −6 and 6. Here $m = n = 3$ and $m + n = 6$. Suppose that light and spectral reflectance are only approximately band limited with limit 0.005 cycle/nm (three samples) and that our intent is to recover only the component of the color signal that corresponds to the band-limited components of light and spectral reflectance. Figures 2(d) and 2(e) illustrate the same illuminant and surface reflectance as Figs. 2(a) and 2(b), with added high-frequency "noise." Figure 2(f) shows the resultant color signal. Applying a filter with a cutoff corresponding to three photoreceptors loses information necessary to reconstruct the band-limited components of light and spectral reflectance. Applying a band-limited filter with a cutoff (0.010 cycle/nm) corresponding to six photoreceptors retains the information necessary to reconstruct the band-limited components [although leakage from higher frequencies (shaded region) may preclude exact reconstuction]. The interesting point here is that the appropriate frequency cutoff for each photoreceptor is dictated *not* by the number of photoreceptor classes (as Barlow suggests) but by the need to preserve information in the color signal corresponding to the light and surface spectral reflectances. We can restate this point in a more general framework.

Suppose that most of the energy in the light $E(\lambda)$ is below a cutoff frequency corresponding to m samples, and we seek to estimate this band-limited component of the light. Assume also that surface spectral reflectances have most of their transform below a cutoff frequency corresponding to n samples and, again, we seek to estimate this band-limited component. Then the transform of the product of the band-limited components of $E(\lambda)$ and $S(\lambda)$ is confined to the region of the transform of the color signal below a cutoff corresponding to $m + n$ samples. A photoreceptor that performs a band-limited filtering of the color signal at a cutoff corresponding to $m + n$ or more samples retains this contribution from the band-limited components of light and spectral reflectance to the transform of the color signal and eliminates (some) contributions due to the high-frequency components. A photoreceptor that performs a band-limited filtering of the color signal at a cutoff corresponding to fewer than $m + n$ samples loses a portion of the contribution of the band-limited components of light and spectral reflectance to the transform of the color signal, precluding reconstruction of the (band-limited components of) light and surface spectral reflectance.

Barlow's analysis of human photoreceptors is carried out when receptor sensitivity is plotted versus both wavelength and wave number. The transforms of receptor sensitivities (wave number) shown in his Fig. 8 are quite similar and, judging from this graph, drop to zero at a cutoff of corresponding to about five to seven samples (photoreceptor classes).[33]

The claim advanced here is that the frequency cutoff of each photoreceptor class must equal or exceed the band limit corresponding to $m + n$ samples in order to preserve information in the color signal on which estimates of light and surface spectral reflectance are to be based. For human vision, Barlow's results suggests that $m + n \leq 5$ to 7. The

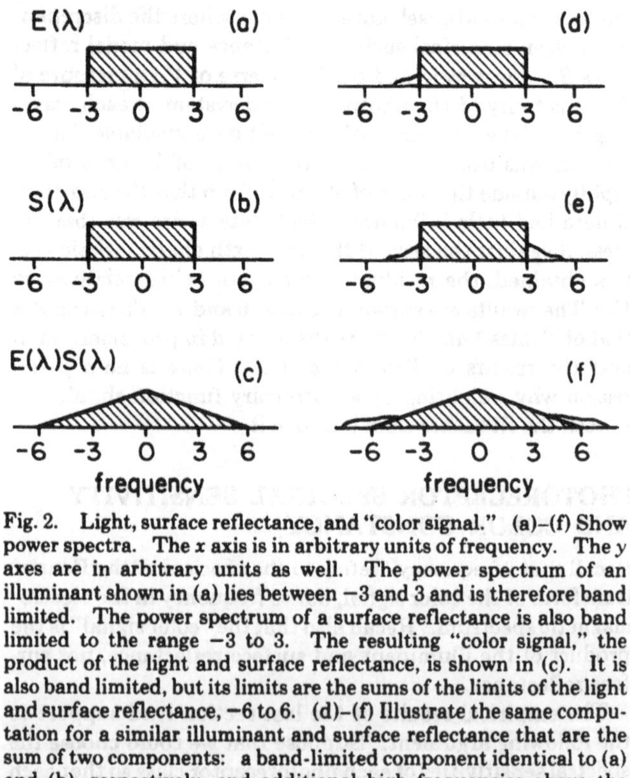

Fig. 2. Light, surface reflectance, and "color signal." (a)–(f) Show power spectra. The x axis is in arbitrary units of frequency. The y axes are in arbitrary units as well. The power spectrum of an illuminant shown in (a) lies between −3 and 3 and is therefore band limited. The power spectrum of a surface reflectance is also band limited to the range −3 to 3. The resulting "color signal," the product of the light and surface reflectance, is shown in (c). It is also band limited, but its limits are the sums of the limits of the light and surface reflectance, −6 to 6. (d)–(f) Illustrate the same computation for a similar illuminant and surface reflectance that are the sum of two components: a band-limited component identical to (a) and (b), respectively, and additional high frequencies above the band limit. Contributions to the power spectrum of the resulting color signal that lie above the −6 to 6 band limit come only from the interaction of the non-band-limited components of light and surface reflectance.

cutoff frequency dictated by the need to retain the information in the color signal (measured in number of samples) corresponds to at least $m + n$ samples; it exceeds the number of samples needed to permit recovery, which varies in accordance with the algorithm (Buchsbaum *et al.*) employed.

CONCLUSIONS

Analyses of two large sets of empirical surface spectral reflectances indicate that a finite-dimensional linear model with five to seven parameters provides an essentially perfect fit to the data examined. Further analyses suggest that empirical surface spectral-reflectance functions fall into a linear model consisting of band-limited functions, but again the number of basis elements required in the space of essentially band-limited functions is about five to seven. Results concerning the surface chemistry of objects suggest a possible physical basis for the observed band limit.

The number of parameters required to fit the two sets of surface reflectances exceeds the number required for color constancy using recent computational theories. It is possible that certain theories are robust enough to tolerate the observed discrepancies between empirical surface spectral reflectances and a two- to four-parameter linear model.

A second route to reconciliation assumes that the shape and sensitivity of the photoreceptors actively emphasized the desired constraints. This claim was first advanced by Barlow. An examination of the effect of the photoreceptors on the fits obtained is promising. The weighted least-squares fits obtained by using weights reflecting the relative sensitivity of human vision across the visible spectrum are better than the corresponding unweighted fits. To a first approximation, the filtering properties of the V_λ function are "worth" about one parameter; the weighted fits with three characteristic vectors are about the same as the unweighted fits with four.

The results suggest that constraints on possible surface-reflectance functions and the "filtering" properties of the shapes of the spectral-sensitivity curves of photoreceptors can both contribute to color constancy. In the last part of this paper I derived the relation between the number of photoreceptor classes present in vision and the filtering properties of each class. The results of this analysis suggest that the filtering properties of human photoreceptors are well matched to a trichromatic visual system that is color constant.

APPENDIX A: FITTING PROCEDURES

An empirical surface spectral reflectance sampled at 10-nm intervals from 300 to 700 nm, inclusive, is essentially a vector of 31 numbers. In the text, surface reflectances and illuminants are notated as continuous functions of wavelength. The analytic procedures described below, however, treat them as finite-dimensional vectors and use statistical procedures such as regression appropriate to vectors.

Procedure 1: Fitting a Surface Reflectance to a Linear Model

Procedure 1 fits a given surface reflectance S sampled at a number of wavelengths across the visible spectrum to a given linear model using a least-squares criterion. After sampling, the surface reflectance is simply a finite vector. The n-parameter linear model is specified by its basis elements: $\{S_1, S_2, \ldots, S_n\}$. The fitting procedure seeks to choose weights σ that minimize

$$|S - S_\sigma|^2 = \sum_l \{S(l) - [\sigma_1 S_1(l) + \sigma_2 S_2(l) + \ldots + \sigma_n S_n(l)]\}^2, \tag{A1}$$

where the summation index l ranges across the wavelengths at which S was sampled. The required minimization is easily computed through multiple regression.[29] The vector of values S is regressed onto the basis elements sampled at the same wavelengths. The regression procedure returns the weights σ corresponding to S_σ, the surface reflectance in the model that best approximates the given surface reflectance S, and a measure of goodness of fit, the multiple correlation coefficient, denoted R^2. S_σ is termed the *fitted surface reflectance*.

Procedure 2: Fitting a Linear Model to a Set of Surface Reflectances

We are given a collection of N surface spectral reflectances $\{S^1, \ldots, S^N\}$, each sampled at a number of wavelengths across the visible spectrum. Procedure 2 computes the linear model with n parameters that best fits the collection. The linear model returned minimizes the sum of the discrepancies between each of the N reflectances and the model computed by using Eq. (A1):

$$\sum_{i=1}^{N} |S^i - S^i_\sigma|^2. \tag{A2}$$

Here S^i_σ denotes the best fitted vector that the model can generate for S^i.

One consequence of the least-squares criterion is that spectral-reflectance curves with entries near zero have less overall weight in determining the choice of characteristic vectors than spectral-reflectance curves with larger entries. The problem encountered here is similar to the problem of influential points in linear regression.[34] Less reflective surfaces are, in effect, given less weight in the fitting procedure, producing a possibly optimistic estimate of variance accounted for. To avoid this problem, all surface spectral-reflectance vectors were normalized to be of length 1 before computing the best-fit linear model.[35]

The computation of the best linear model uses a modified characteristic vector analysis.[36] A characteristic vector analysis determines not the best-fitting linear model for a set of vectors but rather the best-fitting model to a set of vectors once the mean (centroid) of the vectors has been subtracted from each vector. A minor modification in the use of characteristic vector analysis gives us the desired linear model: append to the set of (normalized) vectors S^1, $S^2, \ldots, S^N$ the negations of these vectors: $-S^1, \ldots, -S^N$. The augmented set has mean zero, and the characteristic vector analysis of the augmented set produces precisely the desired best-fitting linear model for the original set.

Procedure 3: Best-Fit Linear Model (Weighted)

Procedure 3 is identical to Procedure 2 except that a weighted-least-squares criterion is used to determine the fit:

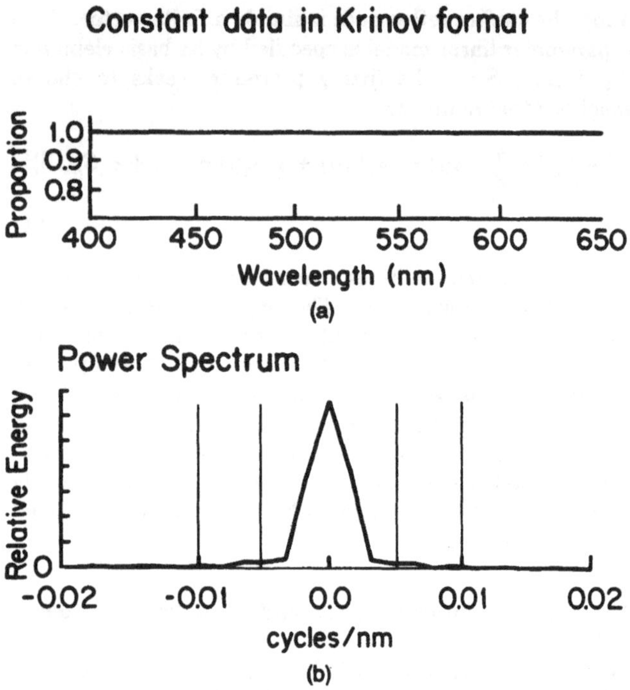

Fig. 3. The effect of windowing and sampling. If the constant function shown in (a) is sampled at 10-nm intervals from 400 to 650 nm, inclusive (the sampling procedure used by Krinov), then the computed power spectrum is that shown in (b). The power spectrum of a constant function that is neither windowed nor sampled is zero, except at the origin. The broadening of the power spectrum is a consequence of windowing and sampling.

$$|S - S_\sigma|_W^2 = \sum_l w_l |S(l) - [\sigma_1 S_1(l) + \sigma_2 S_2(l) + \ldots + \sigma_n S_n(l)]|^2. \qquad (A3)$$

Any discrepancy at the wavelength indexed by l is weighted by a fixed weight w_l. A weight of zero, for example, means that a failure to fit the surface reflectance at a given wavelength carries no penalty. The linear model is now chosen so as to minimize

$$\sum_{i=l}^{N} |S^N - S_\sigma^N|_W^2 \qquad (A4)$$

Procedure 4: Fitting Band-Limited Linear Models

A modified DFT is used to estimate the power spectrum of each surface reflectance. An unmodified DFT misestimates the power spectrum because of the small number of samples available for each spectral-reflectance curve, the effect of sampling the surface reflectance over a finite region ("windowing"), and the abrupt transitions to zero at either end of the sampled region.

Figure 3 illustrates these effects. It shows a constant function across the visible spectrum sampled at the same intervals as Krinov chose. Its power spectrum (the squared norm of its transform) is shown below. The 26 samples of constant value were embedded in an array of length 64 that was filled out with zeros. The theoretical transform of a constant function across the visible spectrum is a Dirac (impulse) function at 0. This Dirac function has been convolved with the transform of a window function that is one across the range where Krinov sampled data and zero elsewhere. The power concentrated at zero in the true transform has been spread out by the windowing operation.

These sources of error are reduced by tapering the ends of the data series by weighting them with a split cosine bell window (a generalization of Hanning).[27] The data are reported with three degrees of Hanning: 0 (no Hanning), 0.3 (partial Hanning), and 1 (full Hanning). The power spectra obtained are of poor quality. However, the intention is to obtain an overall estimate of power in various parts of the power spectrum, which is possible in spite of the obvious smearing.

ACKNOWLEDGMENTS

This work was supported in part by an associateship awarded to the author by the National Research Council of the National Academy of Sciences administered through the NASA Ames Research Center, Life Sciences Division, at Moffett Field, Mountain View, California. I thank William Cowan, David Krantz, Misha Pavel, Roger Shepard, Denise Varner, and Brian Wandell for advice and suggestions concerning this work.

REFERENCES AND NOTES

1. H. Helson, "Fundamental problems in color vision. I. The principle governing changes in hue saturation and lightness of non-selective samples in chromatic illumination," J. Exp. Psychol. **23,** 439–476 (1938).
2. D. B. Judd, "Hue saturation and lightness of surface colors with chromatic illumination," J. Opt. Soc. Am. **30,** 2–32 (1940).
3. H. Yilmaz, "Color vision and a new approach to color perception," in *Biological Prototypes and Synthetic Systems* (Plenum, New York, 1962), pp. 126–141.
4. J. M. Tenenbaum, Ph.D. dissertation (Stanford University, Stanford, Calif., 1971).
5. P. Sallstrom, "Colour and physics: some remarks concerning the physical aspects of human colour vision," Inst. Phys. Rep. no. 73-09 (Institute of Physics, University of Stockholm, 1973).
6. M. H. Brill, "A device performing illuminant-invariant assessment of chromatic relations," J. Theor. Biol. **71** 473–478 (1978).
7. M. H. Brill, "Further features of the illuminant-invariant trichromatic photosensor," J. Theor. Biol. **78,** 305–308 (1979).
8. G. Buchsbaum, "A spatial processor model for object colour perception," J. Franklin Inst. **310,** 1–26 (1980).
9. L. T. Maloney and B. A. Wandell, "Color constancy: A method for recovering surface spectral reflectance," J. Opt. Soc. Am. **3,** 29–33 (1986).
10. L. T. Maloney, "Computational approaches to color constancy," Ph.D. dissertation (Stanford University, Stanford, Calif., 1984).
11. G. Buchsbaum and A. Gottschalk, "Chromaticity coordinates of frequency-limited functions," J. Opt Soc. Am **1,** 885–887 (1984).
12. H. B. Barlow, "What causes trichromacy? A theoretical analysis using comb-filtered spectra," Vision Res. **22,** 635–643 (1982).
13. D. B. Judd, D. L. MacAdam, and G. Wyszecki, "Spectral distribution of typical daylight as a function of correlated color temperature," J. Opt. Soc. Am. **54,** 1031–1040 (1964); S. R. Das and V. D. P. Sastri, "Spectral distribution and color of tropical daylight," J. Opt. Soc. Am. **55,** 319–323 (1965); V. D. P. Sastri and S. R. Das, "Spectral distribution of color of north sky at Delhi," J. Opt. Soc. Am. **56,** 829–830 (1966); "Typical spectra distributions and color for tropical daylight," J. Opt. Soc. Am. **58,** 391–398 (1968); E. R. Dixon, "Spectral distribution of Australian daylight," J. Opt. Soc. Am. **68,** 437–450 (1978).
14. D. L. MacAdam, *Color Measurement; Theme and Variations* (Springer-Verlag, Berlin, 1981), p. 128.

15. J. N. Lythgoe, *The Ecology of Vision* (Clarendon, Oxford, 1979), p. 31.
16. J. Cohen, "Dependency of the spectral reflectance curves of the Munsell color chips," Psychon. Sci. 1, 369–370 (1964).
17. K. L. Kelley, K. S. Gibson, and D. Nickerson, "Tristimulus specification of the *Munsell Book of Color* from spectrophotometric measurements," J. Opt. Soc. Am. **33**, 355–376 (1943); The original spectrophotometric data for Kelley *et al.* were obtained from the Macbeth Corporation, Little Britain Road, P.O. Box 950, Newburgh, N.Y. 12550.
18. E. L. Krinov, *Spectral Reflectance Properties of Natural Formations* (National Research Council of Canada, Ottawa, technical translation: TT-439, 1947). Krinov reports 370 surface spectral reflectances. The 337 analyzed were those that had complete and legible data available from 400 to 650 nm.
19. The analyses reported below were repeated for the full range of Munsell data, 380 to 770 nm, inclusive. The analyses were also repeated after each surface spectral reflectance was transformed from equal-spaced wavelength measurements to equal-spaced wave-number measurements by 4-point Lagrangean interpolation as described by Barlow. The results for both these analyses are qualitatively similar to the results reported here and are not reported.
20. Regression on the first three characteristic vectors derived from the Munsell data directly produced similar results.
21. W. S. Stiles, G. Wyszecki, and N. Ohta, "Counting metameric object-color stimuli using frequency-limited spectral reflectance functions," J. Opt. Soc. Am. **67**, 779–784 (1977).
22. D. Slepian and H. O. Pollak, "Prolate spheroidal wave functions, Fourier analysis and uncertainty. I," Bell Syst. Tech. J. **40**, 43–64 (1961).
23. H. J. Landau and H. O. Pollak, "Prolate spheroidal wave functions, Fourier analysis and uncertainty. II," Bell Syst. Tech. J. **40**, 65–84 (1961); "Prolate spheroidal wave functions, Fourier analysis and uncertainty. III: The dimensions of the space of essentially time and band-limited signals." Bell Syst. Tech. J. **41**, 1295–1336 (1962). The consequences of these results for vision was first set out by Buchsbaum and Gottschalk.[11]
24. K. Nassau, *The Physics and Chemistry of Color: The Fifteen Causes of Color* (Wiley, New York, 1983).
25. H. Suzuki, *Electronic Absorption Spectra and Geometry of Organic Molecules; An Application of Molecular Orbital Theory* (Academic, New York, 1967), p. 79.
26. W. Kauzmann, *Quantum Chemistry* (Academic, New York, 1957), pp. 582 and 669.
27. W. Kauzmann, *ibid.*, pp. 669–670.
28. P. Pringsheim, *Fluorescence and Phosphorescence* (Interscience, New York, 1949).
29. T. H. Wonnacott and R. J. Wonnacott, *Regression: A Second Course in Statistics* (Wiley, New York, 1981).
30. G. Wyszecki and W. S. Stiles, *Color Science; Concepts and Methods, Quantitative Data and Formulas*, 2nd ed. (Wiley, New York, 1982), p. 256.
31. Barlow performs his analysis with color signals plotted against wave number, not wavelength. To avoid confusion, I continue to use wavelength as the independent variable. A change to wave number does not substantially affect the following argument.
32. The spectrum of the light is convolved with the spectrum of the surface reflectance. See R. N. Bracewell, *The Fourier Transform and Its Applications.*, 2nd ed. (McGraw-Hill, New York, 1978), Chap. 3. It is easily shown that the highest frequency present in a convolution is the sum of the two highest frequencies present in the functions being convolved.
33. Barlow uses terahertz rather than inverse centimeters. The conversion factor is 33.33. Barlow assumes that the width of the visible spectrum is $W = 228$ THz. Then, using the result of Slepian *et al.*, $N = 2WF$ becomes $N = 0.456F$, where N is the number of samples required to capture the essentially frequency limited functions with cutoff F in cycles/1000 THz. Note that Barlow's computations use a different formula based on the Shannon–Whittaker Theorem that gives $N = 2WF + 1$.
34. See D. A. Belsey, E. Kuh, and R. W. Welsch, *Regression Diagnostics; Identifying Influential Data and Sources of Collinearity* (Wiley, New York, 1980).
35. The analyses reported in the text were also carried out on the unnormalized data. Comparison of the fits obtained on normalized and unnormalized data showed that normalized data produced worse fits in all cases. In effect, the penalty associated with failing to fit surface reflectances with small entries is reduced by neglecting to normalize.
36. Also known as principal-components analysis or the Karhunen–Loeve decomposition. See K. V. Mardia, J. T. Kent, and J. M. Bibby, *Multivariate Analysis* (Academic, London, 1979). All characteristic vector analyses and multiple regressions were done in AT&T Bell Laboratories' statistical language S. See R. A. Becker and J. M. Chambers, *S: An Interactive Environment for Data Analysis and Graphics* (Wadsworth, Belmont, Calif., 1984).
37. See R. B. Blackman and J. W. Tukey, *The Measurement of Power Spectra from the Point of View of Communications Engineering* (Dover, New York; 1959); P. Bloomfield, *Fourier Analysis of Time Series: An Introduction* (Wiley, New York, 1976).

Color Image Segmentation

Segmentation is the process of partitioning an image into regions that correspond to surfaces or objects in a scene. This process is a fundamental step in the description of scenes and in the recognition of objects. In general, segmentation is a difficult problem since the image measurements corresponding to a single surface can have considerable variation due to effects such as shading, highlights, nonuniform illumination, sensor noise, and surface texture. Traditional approaches to segmentation often implicitly assume simple image models such as constant intensity regions corrupted by additive noise. Unfortunately, algorithms derived using such approaches often produce unacceptable results on real images by, for example, segmenting highlights as separate objects or segmenting a curved surface into more than one surface because of intensity variation caused by shading.

Recent approaches to segmentation have been derived from increasingly accurate physical models for image formation. Such approaches have important advantages over traditional techniques. First, physics-based approaches allow the segmentation of complex images formed according to the modeled processes. Additionally, it is easier to predict the class of images that will be segmented successfully by algorithms based on explicit physical models.

The papers in this section examine ways in which color can be used to distinguish material changes in a scene from other physical events during segmentation. Some of the scene events that have been analyzed successfully using color reflection models include shadows, shading, highlights, and interreflection. This analysis has led to color segmentation algorithms that perform well on complex images.

In the first article, Gershon *et al.* develop a method that can be used to distinguish material changes from shadow boundaries [1]. Using color reflection models, the authors analyze the cases of ideal shadows, for which ambient illumination and direct illumination have the same spectral composition, and nonideal shadows. A *pull factor* is defined that quantifies the deviation of a shadow boundary from the ideal case. Biologically-inspired color operators are discussed whose response is related to a shadow boundary's pull factor. With some knowledge of the scene, a threshold can be used with these operators to distinguish material and shadow boundaries. Properties of the algorithm are demonstrated by applying it to a color image.

In [2], Brill describes computable image quantities that are spatially invariant for a surface using a variety of scene models. He examines in detail the case of rank-2 image fields for which the color pixels measured for a surface lie on a plane in color space. Such rank-2 fields occur for a three dimensional matte object under two different spectral distributions of illumination or for a three dimensional dielectric object with highlights under a single illuminant. For rank-2 fields, invariant quantities are the components of the unit normal to the plane in color space that contains the measurements. Brill generalizes the analysis to higher rank fields that might occur when an object is illuminated by several different illuminants and discusses difficulties inherent in using the segmented image to estimate an object's spectral reflectance. In another paper [3], Brill examines more generally the physical scene models underlying theories of color segmentation, color constancy, and shape from shading. The study shows that the models assumed by each theory are quite different.

Healey develops a color metric that is useful for distinguishing material boundaries from geometric boundaries [4]. The metric is based on the principle that the normalized spectral power distribution reflected by a surface is relatively insensitive to surface orientation except near highlights from inhomogeneous dielectrics. Basis functions over the visible wavelengths are used to estimate normalized reflected spectral power distributions from sensor measurements. The color metric is defined as the distance between these estimates corrected for properties of the sensor noise. A set of experiments with a color chart illustrates the effectiveness of the metric.

A color image segmentation algorithm that takes into account highlights, shading, and camera properties is presented by Klinker, Shafer, and Kanade [5]. Using the dichromatic model for reflection from dielectrics, the authors classify physical events in terms of measured color variation in the image. They show, for example, that a color cluster containing matte and highlight pixels forms a "skewed T" in color space. A hypothesis-based segmentation algorithm is developed using this classification. The algorithm applies physical constraints after computing eigenvalues of local image regions to guide a merging process. The algorithm detects and compensates for cases where camera measurements are not consonant with the dichromatic reflection model such as occurs for blooming or color clipping. Accurate segmentation results are presented for color images of plastic objects.

In [6], Healey presents a segmentation algorithm derived from reflection models and based on Bayesian classification. Using color reflection models for both metals and dielectrics, the author shows that normalized color is a stable surface descriptor over a range of materials and geometries. The algorithm uses region splitting guided by a preliminary edge detection to classify regions based on estimated normalized color probability distributions. By combining the global aspects of region-splitting with the local character of edge detection, the algorithm is able to compute reliable region statistics while obtaining accurate spatial estimates of region boundaries. The algorithm accounts for color variation due to changes in surface orientation and highlights. Experiments on images of metals, plastics, and painted surfaces demonstrate the effectiveness of the approach.

Bajcsy, Lee, and Leonardis describe color image segmentation in the presence of several confounding physical phenomena [7]. The algorithm is derived in hue, saturation, intensity (HSI) space using the dichromatic reflection model under the assumption that highlights have the same spectral composition as the illumination. The algorithm assumes that a reference card of known spectral reflectance is used to generate an image of the scene as it would appear under white illumination. The authors describe the resulting structure of color clusters in HSI space corresponding to scene events such as shading, highlights, shadows, and interreflection. Based on this analysis, the algorithm uses a hue histogram technique to segment individual surfaces and then follows with a local thresholding to identify highlights and interreflection. The algorithm is shown to be effective on color images of glossy objects.

– G.E.H.

Papers Included in This Collection

[1] Gershon, R., Jepson, A.D., and Tsotsos, J.K. "Ambient Illumination and the Determination of Material Changes." *Journal of the Optical Society of America A* **3**(10): 1700-1707, 1986.

[2] Brill, M.H. "Image Segmentation by Object Color: a Unifying Framework and Connection to Color Constancy." *Journal of the Optical Society of America A* **7**(10): 2041-2047, 1990.

[3] Brill, M.H. Adapted from "Photometric Models in Multispectral Machine Vision." In *Proc. SPIE #1453: Human Vision, Visual Processing, and Visual Display II*. Santa Clara, CA. SPIE, 1991. Pages 369-380.

[4] Healey, G. "Color Discrimination by Computer." *IEEE Trans. on Systems, Man, and Cybernetics* **TSMC-19**(6): 1613-1617, Nov/Dec 1989.

[5] Klinker, G.J., Shafer, S.A., and Kanade, T. "A Physical Approach to Color Image Understanding." *International Journal of Computer Vision* **4**(1): 7-38, 1990.

[6] Healey, G. "Segmenting Images Using Normalized Color." *IEEE Trans. on Systems, Man, and Cybernetics*. IEEE, 1992. To appear.

[7] Bajcsy, R., Lee, S.W., and Leonardis, A. "Color Image Segmentation with Detection of Highlights and Local Illumination Induced by Inter-reflections." In *Proc. International Conference on Pattern Recognition*. Atlantic City, NJ. IEEE, 1990. Pages 785-790.

Ambient illumination and the determination of material changes

Ron Gershon, Allan D. Jepson, and John K. Tsotsos

Department of Computer Science, University of Toronto, Toronto, Ontario M5S 1A4, Canada

Received February 3, 1986; accepted May 20, 1986

The task of distinguishing material changes from shadow boundaries in chromatic images is discussed. Although there have been previous attempts at providing solutions to this problem, the assumptions that were adopted were too restrictive. Using a simple reflection model, we show that the ambient illumination cannot be assumed to have the same spectral characteristics as the incident illumination, since it may lead to the classification of shadow boundaries as material changes. In such cases, we show that it is necessary to take into account the spectral properties of the ambient illumination in order to develop a technique that is more robust and stable than previous techniques. This technique uses a biologically motivated model of color vision and, in particular, a set of chromatic-opponent and double-opponent center-surround operators. We apply this technique to simulated test patterns as well as to a chromatic image. It is shown that, given some knowledge about the strength of the ambient illumination, this method provides a better classification of shadow boundaries and material changes.

INTRODUCTION

The perception of color has been a subject of research in many disciplines, including image understanding. One approach to color-image understanding has mainly involved an application of algorithms originally developed for achromatic images to three-dimensional color space. These include edge detection,[1] clustering,[2–4] region splitting,[5] and region growing.[6,7] These schemes regard color as a random variable that is analyzed statistically without regard to any model of specific physical processes of color generation. A second approach involves the explicit use of knowledge about the generation of color. One example is the work of Sloan,[8] which shows that simple statements can be made about outdoor shadows and object colors. Similarly, Rubin and Richards[9,10] demonstrated that by analyzing processes that cause changes in color, one can distinguish material changes from shadow boundaries in some situations. Finally, Shafer[11] presents a theoretical analysis of highlights and object-color reflection that provides a way to remove highlights from portions of images.

In this paper we consider the task, originally proposed by Rubin and Richards, of distinguishing material changes from shadow boundaries in chromatic image. A simple reflectance model is used to illustrate the effect on shadow boundaries of an ambient illumination having a different spectral power distribution from the direct illumination. It is clear from this model that in order to distinguish shadow boundaries reliably from material changes solely on the basis of local measurements of reflected light, some information about the ambient illumination is needed. In particular, it is shown that the technique proposed by Rubin and Richards[10] can incorrectly label a shadow boundary as a material change when the spectral properties of the ambient illumination are sufficiently different from those of the direct illumination. Here we present a new method for labeling shadow boundaries in the presence of ambient illumination. The method is based on the output of biologically motivated operators and on an *a priori* estimate for the strength of the ambient illumination.

A SIMPLE REFLECTANCE MODEL

The appearance of objects is a result of interaction between illumination and reflectance. Thus it is necessary to describe briefly the physics of reflection since it governs the appearance of these regions. In this section we present a simplified model of reflectance that is used in both computer graphics and image understanding[9–12] and that captures the important aspects of the reflection process.

In order to describe the spatial and chromatic properties of the reflected light, we use a reflection model composed of three components: ambient, diffuse, and specular. These reflectance components are properties of any illuminated object. In addition, we assume that there are two types of illumination: ambient (or indirect) and incident (or direct). The ambient illumination represents light incident from the environment, i.e., reflections of the direct light from all the materials in the scene, while the incident illumination is the light originating from specific light sources with no intermediate reflections. The product of illumination and reflectance yields the reflected intensity of the object. Therefore the ambient reflected intensity is the result of the interaction between the ambient illumination and the surface of the object; the diffuse reflected intensity reflects the light that is scattered generally equally in all directions and represents the color of the object itself, since it is the result of incident illumination interacting with the pigments within the object; the specular reflected intensity represents the highlights, the glossy reflection of objects, which is the result of the incident illumination's bouncing off the surface. Therefore, when all three components are taken into account, the image intensity of the reflected light $[I_{ref}(\lambda, \mathbf{x})]$ at a given point, given some illumination and viewing geometry, is formulated in a simple form as

$$I_{ref}(\lambda, \mathbf{x}) = I_{ambient}(\lambda, \mathbf{x})\rho_{ambient}(\lambda, \mathbf{x}) + I_{incident}(\lambda, \mathbf{x}) \times [\delta\rho_{specular}(\lambda, \mathbf{x}) + (1 - \delta)\rho_{diffuse}(\lambda, \mathbf{x})], \quad (1)$$

where $I_{ambient}(\lambda, \mathbf{x})$ and $I_{incident}(\lambda, \mathbf{x})$ are the intensities of the ambient and incident illuminations, respectively; $\rho_{ambient}(\lambda, \mathbf{x})$, $\rho_{specular}(\lambda, \mathbf{x})$, and $\rho_{diffuse}(\lambda, \mathbf{x})$ are the ambient, specular, and diffuse reflectances, respectively; and δ is a fraction of 1 expressing the magnitude of the surface specularity. Because the reflected ambient intensity is a result of interaction of some illumination (ambient) with the surface of the object, it has often been assumed that the ambient reflectance $[\rho_{ambient}(\lambda, \mathbf{x})]$ can be represented as a linear combination of the diffuse and specular reflectances.[11,12] (It is important to emphasize that this assumption is true for the ambient reflectance and not for the ambient illumination.) We can therefore replace the term that represents the ambient reflectance with a combination of the diffuse and specular ones. Furthermore, if we confine our analysis to regions that do not contain highlights, that is, ones in which the proportion of specular reflectance is almost constant (i.e., $\delta \approx$ constant and is usually close to zero), we can represent the reflectance in one term, which will be denoted as $\rho(\lambda, \mathbf{x})$. The formulation for the intensity of the reflected light is therefore

$$I_{ref}(\lambda, \mathbf{x}) = I_{ambient}(\lambda, \mathbf{x})\rho(\lambda, \mathbf{x}) + I_{incident}(\lambda, \mathbf{x})\rho(\lambda, \mathbf{x}) = [I_{ambient}(\lambda, \mathbf{x}) + I_{incident}(\lambda, \mathbf{x})]\rho(\lambda, \mathbf{x}), \quad (2)$$

where $\rho(\lambda, \mathbf{x})$ is the reflectance of the object. From this analysis, it is easy to see that if there is no direct illumination, that is, $I_{incident}(\lambda, \mathbf{x}) = 0$, then the intensity of the reflected light at a given point will be

$$I_{ref}(\lambda, \mathbf{x}) = I_{ambient}(\lambda, \mathbf{x})\rho(\lambda, \mathbf{x}), \quad (3)$$

which represents the intensity of reflected light at a point that is not directly illuminated, hence a point in a shadow region. We therefore see that Eq. (2) represents the intensity at the lit region, while Eq. (3) represents the intensity at the shadow region.

IDEAL AND NONIDEAL SHADOWS

The common factor contributing to the appearance of both the shadow and lit regions of objects is the ambient illumination. This implies that understanding the effects of the ambient illumination on objects may give us important information regarding the disambiguation of these regions, enabling us to distinguish between them. In this section, we examine two conditions under which the relationships between shadow and lit regions change. In the first case, which we call the "ideal" case, we assume that the ambient illumination has the same spectral characteristics as the incident illumination. In the second case, which we call "nonideal," we assume that the ambient illumination differs in its spectral characteristics from the incident illumination. We use the term "chromatic components" to denote the responses of particular chromatic mechanisms. For example, the red component, which is denoted as R, is defined by

$$R(\mathbf{x}) = \int_0^\infty I_{ref}(\lambda, \mathbf{x})S_R(\lambda)d\lambda, \quad (4)$$

where $I_{ref}(\lambda, \mathbf{x})$ is the reflected intensity reaching the sensor (cone) and $S_R(\lambda)$ is the spectral sensitivity function of the red cone. For simplicity, we confine our discussion to the red and green components of the reflected intensities. The extension of the work to include a third chromatic component is straightforward.

In order to show how the ambient illumination affects objects, we first show that conditions hold for the ideal case and then extend the discussion to the nonideal case. We make use of Fig. 1 to illustrate the two cases. Suppose the original setup consists of an illuminant (I) that illuminates the observed object (O). Some of the light does not reach the object O because of a blocking object (B), thus creating a shadow region (S) and a lit region (L) on the observed object O. In the ideal case, the ambient illumination has the same spectral characteristics as the incident illumination. In other words, there is a difference only in the magnitude of the spectral distribution functions representing the incident and ambient illuminations, such that for any given wavelength the ambient illumination will be a fraction of 1 (denoted as α) of the incident illumination. Having assumed that the ambient reflectance is the same as the diffuse reflectance, we find that the ambient reflected intensity has the same spectral characteristics as the diffuse reflected intensity up to the same constant α. In such a case, if the total intensity of the reflected light in the lit regions of object O were the sum of the ambient and diffuse reflected intensities [as given by Eq. (2)], then we can represent the red and green components of this reflected intensity as $(\alpha R + R, \alpha G + G)$. Under ideal conditions, the red and green components of the total reflected intensity from the shadow parts of O (which are illuminated by ambient illumination only) are $(\alpha R, \alpha G)$.

One of the conditions under which an observed object is exposed to ambient illumination that does not have the same spectral characteristics as the incident illumination occurs when there is another object (or objects) casting its color on

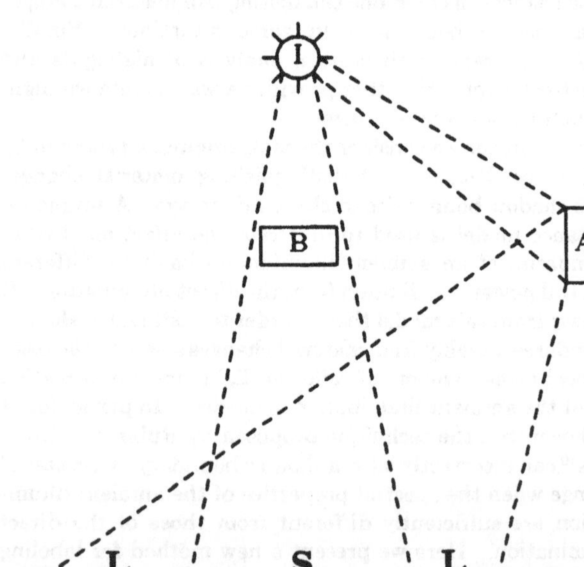

Fig. 1. A case in which the ambient illumination impinging upon an object does not have the same spectral characteristics as the incident illumination.

Table 1. The Response of the Red and Green Mechanisms to the Reflected Intensities of the Shadow and Lit Regions Under Ideal and Nonideal Lighting Conditions

	Reflected Intensity	
Condition	Shadow Region	Lit Region
Ideal	$(\alpha R, \alpha G)$	$(\alpha R + R, \alpha G + G)$
Nonideal	$(r + \alpha R, \alpha G)$	$(r + \alpha R + R, \alpha G + G)$

the observed object. Suppose that there is another object (A) illuminated by I, which casts its color onto O. The object A need not be physically close to O nor necessarily exist in the image. It is used merely to illustrate that the ambient illumination may be biased toward some color other than the incident illumination. Furthermore, let us assume that the color object that A adds is red and that this red color impinges upon both the shadow and the lit regions and is thus reflected off them, since it acts as an illuminant. We therefore have to account for a new factor, which we call the *additional ambient illumination.* Because the regions belong to the same object (O), this additional ambient illumination affects both the shadow and lit regions the same, thus resulting in an additional ambient reflected intensity, of which its red component (the only component in our example) is denoted as r. Therefore the red and green components of the reflected intensity from O are $(r + \alpha R + R, \alpha G + G)$ for the lit region and $(r + \alpha R, \alpha G)$ for the shadow region. A summary of the two cases and their effects on the reflected intensities of the objects is given in Table 1. Note that it is assumed in this paper that *both* the shadow and the lit regions are influenced by the additional ambient illumination. If that were not true, and only one of the regions would have been illuminated by the additional ambient illumination, then the color of the two regions would have been very different, making the task of identifying them as part of the same object complicated.

DETERMINING MATERIAL CHANGES

To this point we have formalized the relationship between shadow and lit regions under different illumination conditions given a simple model of reflection. In this section we will examine the techniques suggesed by Rubin and Richards[9,10] for determining material changes.

Initially, Rubin and Richards[9] suggested the *spectral crosspoint condition,* which stated that the sign of the change in amplitude of a spectral component across a shadow boundary must be the same as the sign for any other spectral component. More formally, this condition states that given two regions X and Y across a discontinuity and intensity samples I taken at two wavelengths λ_1 and λ_2 the following is a test for a material change:

$$(I_{X\lambda_1} - I_{Y\lambda_1})(I_{X\lambda_2} - I_{Y\lambda_2}) < 0. \quad (5)$$

This condition was found to be inadequate as a means of finding material changes since "in a crosspoint, spatial and spectral information are hopelessly intertwined."[10] Therefore a second and independent condition was introduced, the *opposite slope sign condition.*[10] The opposite slope sign condition requires that, given two spectral components measured on both sides of a shadow boundary, on each side the same component must be the maximum. Formally, the opposite slope sign condition is

$$(I_{X\lambda_2} - I_{X\lambda_1})(I_{Y\lambda_2} - I_{Y\lambda_1}) < 0, \quad (6)$$

which again is meant to signal a material change. Indeed, assuming ideal ambient lighting conditions for the image, either expression (5) or (6) can be satisfied at an ideal shadow boundary.

The question arises whether the opposite slope sign condition holds in the nonideal case. As can be observed from Table 1, the treatment of the nonideal case is complicated by the term representing the ambient illumination (in our example, r) that is added to both shadow and lit regions, which are considered to be the two sides of the discontinuity. The complications arise in cases in which the relationship between the shadow and lit regions is not proportional, thus suggesting a material change, although they may in fact be two regions of the same material. To be exact, it may be the case that $(r + \alpha R + R) < (\alpha G + G)$ in the lit region and $(r + \alpha R) > \alpha G$ in the shadow region, or vice versa, therefore causing a change in the sign of the red versus green components, which indicates a material change according to the opposite slope sign condition. Table 2 illustrates this point by using a numerical example taken from an image that will be presented in the section headed Results. Although the opposite slope sign condition indicates that there is a material change, the two samples in Table 2 were obtained from the same object in two different regions: one that was lit directly and the other in a shadow. In summary, we would like to avoid the labeling of such changes as material changes and would rather identify them as changes caused by shadows. This is the nature of the problem that we attempt to solve.

MEASUREMENTS OF THE EFFECTS OF AMBIENT ILLUMINATION

In this section we quantify how the ambient illumination affects the reflected intensity of the object. Without loss of generality, we will limit our discussion to two spectral components of the reflected intensity of the object, namely, red and green. The discussion can be similarly applied to the blue and yellow components without any further assumptions.

Let us define the vector describing the red and green components of the reflected intensity of the observed object as (R, G). This vector describes the reflected intensity that is a result of the direct illumination. If the spectral proper-

Table 2. An Example Taken from an Image Taken Over a Shadow Boundary That Is Classified as a Material Change by the Opposite Slope Sign Condition

Reflected Intensity of Lit Region (X)	Reflected Intensity of Shadow Region (Y)
$I_{X\lambda_1} = 151$	$I_{Y\lambda_1} = 41$
$I_{X\lambda_2} = 176$	$I_{Y\lambda_2} = 16$

Opposite Slope Sign Condition:
$(I_{X\lambda_2} - I_{X\lambda_1})(I_{Y\lambda_2} - I_{Y\lambda_1}) = (25)(-25) < 0$
(Material Change)

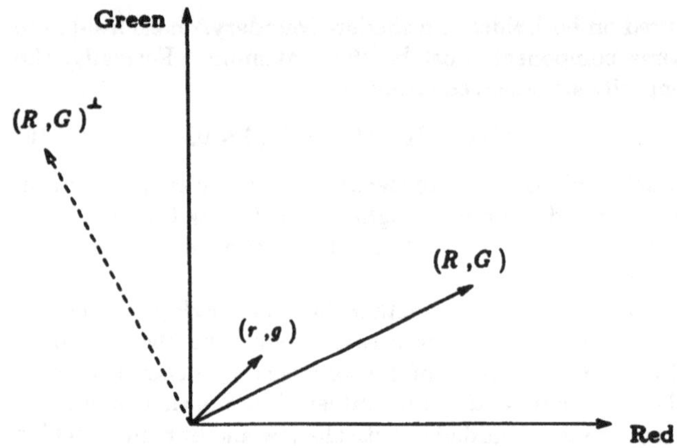

Fig. 2. The red and green components of the vectors describing the diffuse reflected intensity of an object, its normal, and the additional ambient reflected intensity.

ties of the ambient illumination are the same as the direct illumination, then the red and green components of the shadow region are proportional to the vector (R, G). In addition, we define (r, g), the reflected components due to the ambient illumination, as $r = \int_0^\infty I_{\mathrm{ambient}}(\lambda, \mathbf{x})\rho_{\mathrm{ambient}}(\lambda, \mathbf{x}) S_R(\lambda)\mathrm{d}\lambda$ and $g = \int_0^\infty I_{\mathrm{ambient}}(\lambda, \mathbf{x})\rho_{\mathrm{ambient}}(\lambda, \mathbf{x}) S_G(\lambda)\mathrm{d}\lambda$. Note that we assume that the additional ambient reflected intensity is not solely red, as was illustrated in Fig. 1 by r, but rather has two spectral components, red (r) and green (g). The vectors are illustrated in Fig. 2.

In the ideal case the spectral components on either side of a shadow boundary are related by a multiplicative constant (see Table 1). For a general value of (r, g) this will be the case only if (r, g) is in the same direction as (R, G). A measure of the difference from the ideal case is therefore

$$\text{Amount of pull} = (r, g)(R, G)^{\perp}, \tag{7}$$

where $(R, G)^{\perp}$ is orthogonal to (R, G) and therefore $(R, G)^{\perp} \equiv (-G, R)/|(R, G)|$. In order to measure the *relative* amount of pull, we divide by the length of the original vector. Since the sign of the result is not important, but its magnitude is, we take the absolute value of the result. Thus, when put formally, the pull factor is computed as

$$\text{Pull factor} = \frac{(r, g)(-G, R)}{|(R, G)|^2} = \frac{|gR - rG|}{R^2 + G^2}. \tag{8}$$

This quantity is important because the stronger the effect of the additional ambient illumination, the more the shadow region differs from the lit region, thus making the association of both with the same material more unlikely.

PRACTICAL ESTIMATION OF THE CHANGE IN REFLECTED INTENSITY

To this point, we have examined the factors that contribute to the different behavior of ideal and nonideal shadows. We have also suggested a measurement that reflects the relationship between the additional ambient reflected intensity and the original diffuse reflected intensity. We now turn to apply this information to the problem of identifying material changes through the use of a set of operators that respond differently under the two illumination conditions. In this section we introduce the operators suggested for this task and examine their responses under the different conditions, along with their correspondence to the theoretical analysis of the reflection vectors described in the previous section.

Let us first consider a set of operators that have spatial center-surround antagonistic organization and share the same spectral intensity functions in both the center and the surround. Examples of these units, which we will refer to as monochromatic opponent units, are R + center/R − surround, G + center/G − surround, where R and G represent the responses of the mechanisms bearing the R- and G-cone spectral sensitivity functions. A formulation of the response of one of these units, say, R + center/R − surround, is given by

$$\begin{aligned} \mathrm{RESP}(\mathbf{x}; \sigma) &= \mathbf{G}(\mathbf{x}; \sigma_c) * L_R(\mathbf{x}) - \mathbf{G}(\mathbf{x}; \sigma_s) * L_R(\mathbf{x}) \\ &= \int_{-\infty}^{\infty} \mathbf{G}(\mathbf{x} - \mathbf{r}; \sigma_c) L_R(\mathbf{r})\mathrm{d}\mathbf{r} \\ &\quad - \int_{-\infty}^{\infty} \mathbf{G}(\mathbf{x} - \mathbf{r}; \sigma_s) L_R(\mathbf{r})\mathrm{d}\mathbf{r}, \end{aligned} \tag{9}$$

where $*$ is the convolution operator and $L_R(\mathbf{x})$ is the logarithm of the response of the red-sensitive mechanism to the input image, that is, $L_R(\mathbf{x}) = \log R(\mathbf{x})$, and $R(\mathbf{x})$ is defined in Eq. (4). We chose the logarithm since it provides operators with a simple behavior to multiplicative scaling of stimuli. It is also interesting to note that, after adaptation effects have been taken into account, the logarithm provides a rough approximation to photoreceptor responses for low-contrast stimuli.[13,14] $\mathbf{G}(\mathbf{x}; \sigma_i)$ is a Gaussian of the form

$$\mathbf{G}(\mathbf{x}; \sigma_i) = \frac{1}{2\pi\sigma_i^{\,2}} \exp\left(\frac{-|\mathbf{x}|^2}{2\sigma_i^{\,2}}\right). \tag{10}$$

It is evident that these operators are difference-of-Gaussians (DOG's) in space and therefore are bandpass filters with chromatic sensitivity, each one tuned to a different band of wavelengths. If these operators were input the corresponding spectral component of images (e.g., and R + center/R − surround operator with the red component of an image), then they are bound to detect any discontinuity in that component. These discontinuities may be material changes, ideal shadow boundaries, or nonideal shadow boundaries. In terms of the analysis presented in the previous section, the monochromatic opponent units provide information about the *total* change in a given chromatic component. For example, a unit of the type R + center/R − surround will respond to the total change in both r and g along a discontinuity, signaling how much the red component of the reflected intensity varied from one side of the discontinuity to the other. Two examples of responses of such units are given in Figs. 3(a)–3(d).

We now consider double-opponent operators of the same organization and behavior described by physiologists as existing in the cortex of primates.[15,16] These operators, like the monochromatic opponent units, are spectrally sensitive DOG's, although their spectral sensitivity is a combination of two different spectral sensitivity functions. Examples of such operators are $R + G-$ center/$R - G+$ surround, $B + Y-$ center /$B - Y+$ surround. The response of such a unit, say,

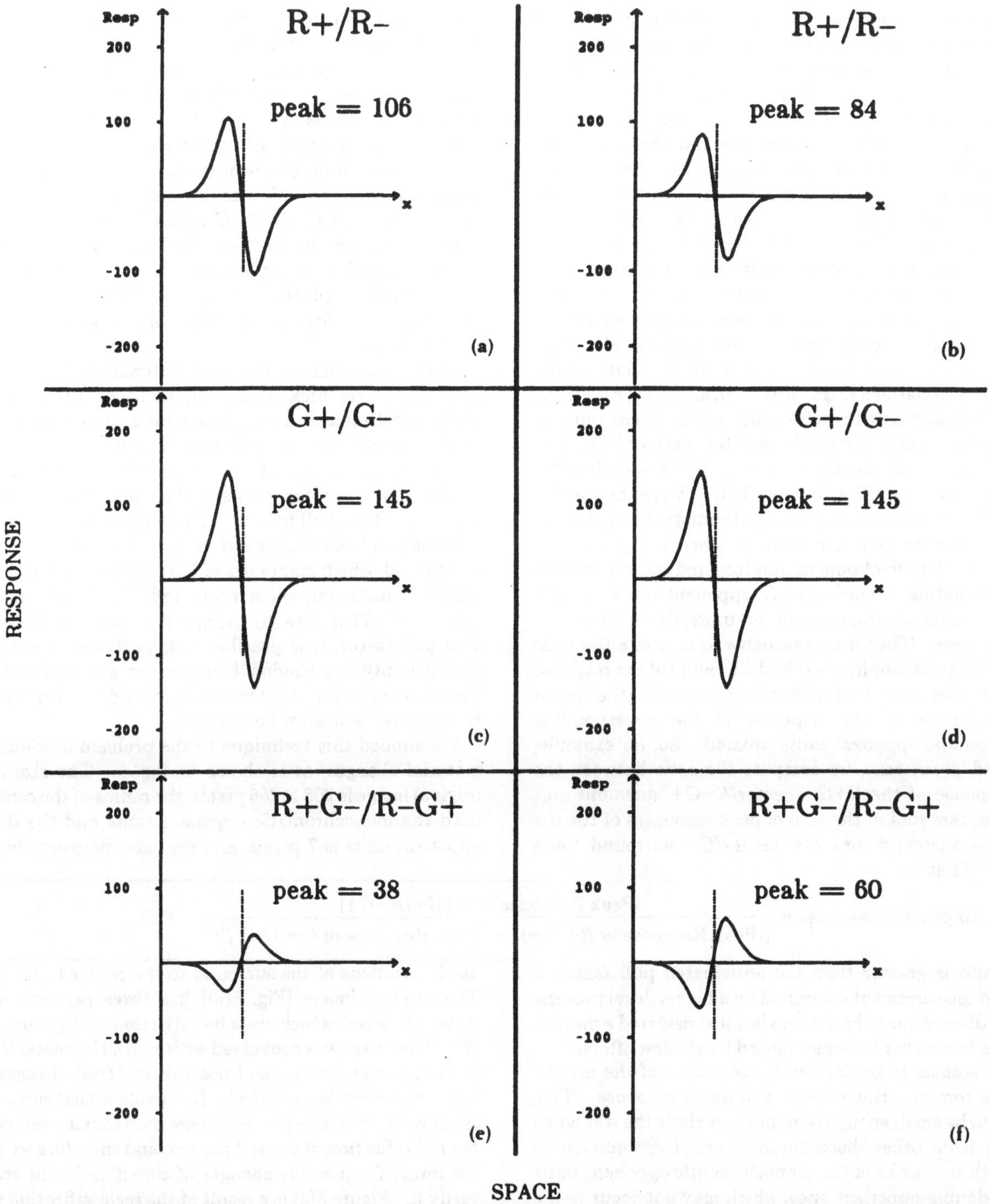

Fig. 3. Responses of monochromatic opponent units and double-opponent operators to two sets of test patterns that include a step-edge in space in which the spectral density changes. The red and green components of the test patterns are (150, 200) followed by (90, 100) for plots (a), (c), and (e) and (90, 100) followed by (60, 50) for plots (b), (d), and (f). The additional ambient reflected intensity was assumed to be 30 units of red, and $\alpha = 0.5$. Plots (a) and (b) are responses of an $R+$ center/$R-$ surround operator, plots (c) and (d) are of a $G+$ center/$G-$ surround operator, and plots (e) and (f) are of an $R+G$ center/$R-G+$ surround operator. The radius of the center of all operators used is 7 pixels, and σ_s/σ_c is 2.5.

$R + G-$ center/$R - G+$ surround, is computed as

$$\begin{aligned}\mathrm{RESP}(\mathbf{x};\sigma) &= \mathrm{G}(\mathbf{x};\sigma_c) * [L_R(\mathbf{x}) - L_G(\mathbf{x})] \\ &\quad - \mathrm{G}(\mathbf{x};\sigma_s) * [L_R(\mathbf{x}) - L_G(\mathbf{x})] \\ &= [L_R(\mathbf{x}) - L_G(\mathbf{x})]\mathrm{DOG}(\mathbf{x}). \qquad (11)\end{aligned}$$

The difference is that double-opponent operators respond to nonproportional changes in their two spectral components. It has already been shown[17] that if the discontinuity represents a proportional change, then other operators, such as achromatic DOG's, respond to the discontinuity while the double-opponent operator does not, thus indicating that although there may have been a discontinuity, it is due to an ideal shadow. By contrast, if there are changes due to shadows that are not proportional, that is, nonideal, and as a result, there exists some additional ambient reflected intensity, then the double-opponent operators *do* respond; this is illustrated in Figs. 3(e) and 3(f).

In summary, the two sets of operators, the monochromatic opponent units and the double-opponent units, signal different aspects of the information needed. The monochromatic opponent units provide information about the total change in each chromatic component, while the double-opponent units provide information about changes in the relative amounts. We anticipate that the stronger the additional ambient reflected intensity becomes (relative to the diffuse reflected intensity), the stronger the response of the double-opponent operator will be. This point is illustrated in Fig. 3, where responses of the units to two test patterns are plotted. Each test pattern is composed of two chromatic stimuli juxtaposed in space (we use in this example only one spatial dimension). In the left-hand test pattern [Figs. 3(a), 3(c), and 3(e)], r is weak relative to R and G, while on the right-hand test pattern [Figs. 3(b), 3(d), and 3(f)], r is strong relative to R and G. Therefore the red-opponent unit responds more strongly to the left-hand test pattern [Figs. 3(a) and 3(b)], while the double-opponent unit responds more strongly on the right [Figs. 3(e) and 3(f)]. Since the double-opponent units are sensitive to two chromatic bands (either red and green or blue and yellow), when comparing the response of a double-opponent unit (say, red and green) with its corresponding monochromatic opponent units, we will have to consider the responses of *two* units, the red ones and the green ones. Thus it is reasonable to compute the ratio between the peak amplitude (absolute value) of the response of a double-opponent unit with the square root of the sum of peak amplitudes of the responses of the corresponding monochromatic opponent units squared. So, for example, in the red–green case, we compute the ratio between the peak response of the $R+G-$ center/$R-G+$ surround unit and the square root of the sum of peak responses of the $R+$ center/$R-$ surround and $G+$ center/$G-$ surround units squared. That is,

$$\text{Relative Amplitude Response} = \frac{|\text{Peak Response of } R{+}G{-}/R{-}G{+}|}{[(\text{Peak Response of } R{+}/R{-})^2 + (\text{Peak Response of } G{+}/G{-})^2]^{1/2}}. \quad (12)$$

If the ratio is greater than the anticipated pull factor, a threshold measurement computed by a higher-level process, then the discontinuity in the area is not considered a material change but rather a change caused by shadow effects.

What remains to be defined is the extent of the area in which we compute this relative amplitude response. This area has to be small enough so as not to include the responses resulting from other discontinuities and large enough to cover both the peaks of the monochromatic opponent units and the double-opponent ones, which may not occur in the same spatial location. The area in which we compare the amplitude responses is the one covered by the center field of both operators, which are of the same size, since this area essentially excludes (with high probability) peaks that are a result of two adjacent discontinuities,[18] yet it is large enough to guarantee the detection of the peak of the responses.[19] The appropriate size of operators relevant to the image (or parts of it) remains an open problem.

RESULTS

First we verify that the relative amplitude response is monotonically related to the pull factor. We have simulated different cases of object colors and different levels of ambient illumination and have computed both the relative amplitude responses of the operators and the corresponding pull factors. The relative amplitude responses were computed as the ratio between the highest peaks of the double-opponent units and the highest peaks of the corresponding monochromatic opponent units, all within an area bounded by the radius of the center of the operators. For example, for the red–green case, the ratio was computed as the one between the highest peak of the $R+G$ center/$R-G+$ surround unit and the sum of the highest peaks of the $R+$ center/$R-$ surround and $G+$ center/$G-$ surround units within a round area of radius 7 pixels. The pull factor was computed by using Eq. (8). The results of the comparisons are summarized in Fig. 4.

In the tests that we ran, four different diffuse reflected intensities were picked, and for each one the amount of ambient reflected intensity was varied. For each test, the relative amplitude response and the pull factor were computed. As can be observed from the results, there is a proportional relationship between the relative amplitude responses and the pull factor. This leads us to the conclusion that the pull factor can be estimated by the practical method introduced, which makes use of the relative responses of the monochromatic opponent units and the double-opponent operators. Therefore, by using a threshold on the acceptable pull factor, it is possible to hypothesize whether the discontinuity is a shadow boundary or a material change. For larger values of this threshold, more discontinuities will be accepted as shadow boundaries.

We applied this technique to the problem of identifying material changes, as is shown in Fig. 5. The size of the original image is 256 × 256 pixels, the radius of the center for both the monochromatic opponent units and the double-opponent units is 7 pixels, and the ratio between the standard deviations of the surround to the center (σ_s/σ_c) is 2.5. The original image [Fig. 5(a)] has three peppers, one of which (1) is red, which casts its reflection on a green pepper (2). The image was convolved with an $R+G-$ center/$R-G+$ surround operator to yield possible material changes [Fig. 5(b)—the white boundaries]. It is evident that some of the shadows in the green pepper (2) are ones that are affected by the red reflection of the red pepper, and therefore we tested the image for possible changes of classification in order to verify it. Figure 5(c) is a result of the reclassification with a predicted pull factor of 0.2, showing that some of the possible material changes are probably just shadows (black boundaries); Fig. 5(d) shows that increasing the pull factor to 0.3 changes the classification of more of the material changes to possible shadows. What we see, then, is that if some global process is able to predict the pull factor, then there is a way of correcting the classification even under complex illumination conditions.

SUMMARY

The relationship between shadow and lit regions in images can be complicated by the existence of spectrally biased ambient illumination that is reflected off objects in the environment. In contrast to previous approaches for inferring

Comparison Results				
Test Number	Diffuse Intensity (R,G)	Ambient Intensity (r,g)	Pull Factor	Relative Amplitude Response
1	(110,160)	(20,0)	0.08	0.13
	(110,160)	(30,0)	0.13	0.17
	(110,160)	(40,0)	0.17	0.20
	(110,160)	(50,0)	0.21	0.23
	(110,160)	(60,0)	0.25	0.26
	(110,160)	(100,0)	0.42	0.33
2	(50,60)	(10,0)	0.10	0.12
	(50,60)	(20,0)	0.20	0.20
	(50,60)	(30,0)	0.29	0.26
	(50,60)	(40,0)	0.39	0.30
	(50,60)	(50,0)	0.49	0.32
	(50,60)	(60,0)	0.59	0.34
3	(100,50)	(15,05)	0.02	0.05
	(100,50)	(30,10)	0.04	0.07
	(100,50)	(45,15)	0.06	0.08
	(100,50)	(60,20)	0.08	0.09
4	(150,100)	(10,20)	0.06	0.08
	(150,100)	(15,30)	0.09	0.11
	(150,100)	(20,40)	0.12	0.13
	(150,100)	(25,50)	0.15	0.15
	(150,100)	(30,60)	0.18	0.19

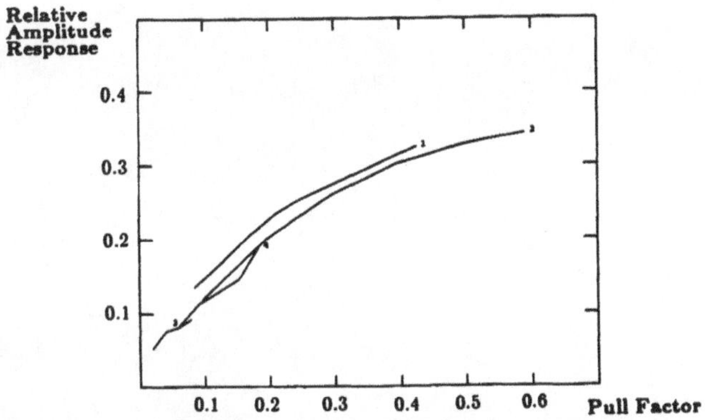

Fig. 4. Comparisons between the pull factor and the relative amplitude responses of the units (red and green opponent units and $R+G-$ center/$R-G+$ surround operator).

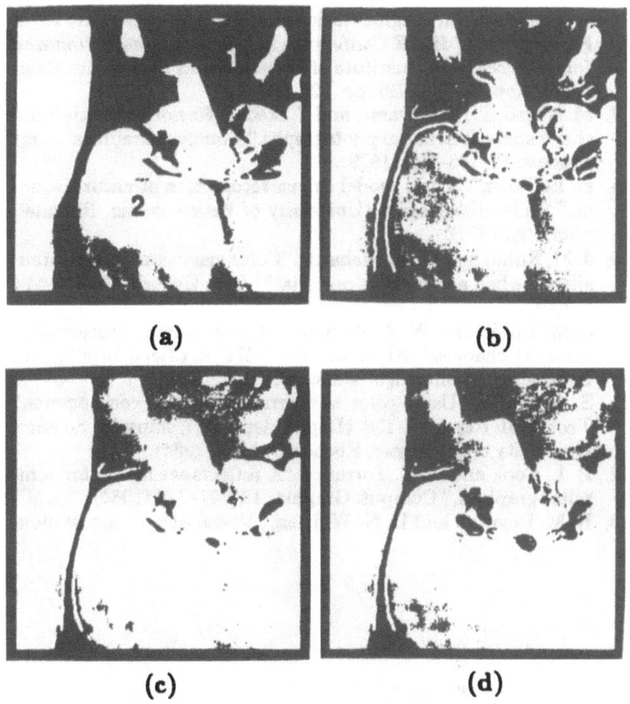

Fig. 5. An application of the algorithm to a chromatic image. (a) Is the original image, with a red pepper (1) casting its color on a green pepper (2). (b) Is the result of the algorithm run with predicted pull factor of zero; the bright lines are the discontinuities hypothesized as material changes. (c) Is the result of the algorithm with predicted pull factor of 0.2, and (d) with a pull factor of 0.3; the darker lines are discontinuities that were reclassified as shadow boundaries. The images in (b), (c), and (d) are shown with reduced contrast so that the discontinuities will be easier to see.

material changes, which assumed that the ambient illumination has the same spectral characteristics as the incident illumination, we have presented a technique that takes into account ambient illumination that is different in its distribution. A measurement that is based on a biologically motivated set of operators has been introduced and found to be related directly to the relative strength of the ambient illumination. It is important to remember that this is an early visual process and as such will not determine perfectly, nor decisively, whether the discontinuities are indeed material changes. Some other process, a higher-level one that involves knowledge about the scene and its structure, may perform the final decision making regarding this classification. Furthermore, it is clear that the threshold that we introduced has to be provided to our technique, since it may vary from one image to another or even from one region to another in the same image. Therefore an additional computation must be carried out by higher-level spatiochromatic mechanisms, ones that should take into account information such as spatial organization and reflectance properties of objects. Nevertheless, the technique that we presented is definitely more flexible, accurate, and robust in its performance.

ACKNOWLEDGMENTS

This work was financially supported in part by the Natural Science and Engineering Research Council of Canada and by the Department of Computer Science at the University of Toronto.

John K. Tsotsos is also a Fellow of the Canadian Institute for Advanced Research.

REFERENCES

1. R. Nevatia, "A color edge detector and its use in scene segmentation," IEEE Trans. Syst. Man Cybern. **SMC-7**, 820–826 (1977).
2. G. B. Coleman and H. C. Andrews, "Imaging segmentation by clustering," Proc. IEEE **67**, 773–785 (1979).
3. P. A. Nagin, A. R. Hanson, and E. M. Riseman, "Studies in global and local histogram-guided relaxation algorithms," IEEE Trans. Pattern Anal. Machine Intell. **PAMI-4**, 263–276 (1982).
4. B. J. Schacter, L. S. Davis, and A. Rosenfeld, "Scene segmentation by cluster detection in color spaces," SIGART Newsl. **58**, 16–17 (1976).
5. R. Ohlander, K. E. Price, and D. R. Reddy, "Picture segmentation using a recursive region splitting method," Comput. Graphics Image Process. **8**, 313–333 (1978).
6. M. D. Levine and S. I. Shaheen, "A modular computer vision

system for picture segmentation and interpretation, Part 1," in *Proceedings of IEEE Conference on Pattern Recognition and Image Processing* (Institute of Electrical and Electronics Engineers, New York, 1979), pp. 523–533.
7. M. Nagao, T. Matsuyama, and Y. Ikeda, "Region extraction and shape analysis in aerial photographs," Comput. Graphics Image Process. **10**, 195–223 (1979).
8. K. E. Sloan, "World model driven recognition of natural scenes," Ph.D. dissertation (University of Pennsylvania, Philadelphia, Pa., 1977).
9. J. M. Rubin and W. A. Richards, "Color vision and image intensities: when are changes material?" Biol. Cybern. **45**, 215–226 (1982).
10. J. M. Rubin and W. A. Richards, "Color vision: representing material changes," AI Memo 764 (MIT Artificial Intelligence Laboratory, Cambridge, Mass., 1984).
11. S. A. Shafer, "Using color to separate reflection components," Technical Rep. TR-136 (Department of Computer Science, University of Rochester, Rochester, N.Y., 1984).
12. R. L. Cook and K. E. Torrance, "A reflectance model for computer graphics," Comput. Graphics **15**, 307–316 (1981).
13. R. M. Boynton and D. N. Whitten, "Visual adaptation in monkey cones: recordings of late receptor potentials," Science **170**, 1423–1426 (1970).
14. W. A. H. Rushton, "Peripheral coding in the nervous system," in *Sensory Communication*, W. A. Rosenblith, ed. (MIT Press, Cambridge, Mass., 1961).
15. M. S. Livingstone and D. H. Hubel, "Anatomy and physiology of a color system in the primate visual cortex," J. Neurosci. **4**, 309–356 (1984).
16. C. R. Michael, "Color vision mechanisms in monkey striate cortex: dual-opponent cells with concentric receptive fields," J. Neurophys. **41**, 557–576 (1978).
17. R. Gershon, "Empirical results with a model of color vision," in *Proceedings of IEEE Conference on Computer Vision and Pattern Recognition* (Institute of Electrical and Electronics Engineers, New York, 1985), pp. 302–305.
18. M. S. Longuet-Higgins, "The distribution of intervals between zeros of a stationary random function," Phil. Trans. R. Soc. London Ser. A **254**, 557–599 (1962).
19. D. J. Fleet, A. D. Jepson, and P. E. Hallett, "A spatio-temporal model for early visual processing," Technical Rep. RBCV-TR-84-1 (Department of Computer Science, University of Toronto, Toronto, 1984).

Image segmentation by object color: a unifying framework and connection to color constancy

Michael H. Brill

Science Applications International Corporation, 1710 Goodridge Drive, McLean, Virginia 22102

Received October 16, 1989; accepted June 8, 1990

A unifying framework is presented for algorithms that use the bands of a multispectral image to segment the image at material (i.e., reflectance) boundaries while ignoring spatial inhomogeneities incurred by accidents of lighting and viewing geometry. The framework assumes that the visual stimulus (image field) from a uniformly colored object is the sum of a small number of terms, each term being the product of a spatial and a spectral part. Based on this assumption, several quantities depending on the reflected light can be computed that are spatially invariant within object boundaries. For an image field either from two light sources on a matte surface or from a single light source on a dielectric surface with highlights, the invariants are the components of the unit normal to the plane in color space spanned by the pixels from the object. In some limited cases the normal to the plane can be used to estimate spectral-reflectance parameters of the object. However, in general the connection of color-constancy theories with image segmentation by object color is a difficult problem. The concomitant constraints on segmentation and color-constancy algorithms are discussed in light of this fact.

1. INTRODUCTION

Fundamental to automatic exploitation of imagery is the localization of objects within the image. An important aspect of localizing objects is delineating them, i.e., segmenting the image so that the segment boundaries correspond to object boundaries. However, there are substantial difficulties in segmenting an image according to object, for the geometry of lighting and viewing of three-dimensional objects incurs spatial inhomogeneities in the image that do not coincide with object boundaries. Some of these inhomogeneities (such as highlights and surface shading) are due to the relationship of the viewer to the plane of incidence of the light on the viewed object. Other inhomogeneities (such as cast shadows and multiply reflected light) depend on the interaction of light with objects other than the viewed object.

For monochrome images the problem is especially acute: One must rely on geometric inferences about objects and illuminations that properly should be made after segmentation and not before it. However, it has long been recognized that, for multispectral images, there is some hope of sorting out object boundaries from illumination boundaries. One popular algorithm involves taking ratios of spectral bands,[1] which discounts the absolute intensity variations over an object that are due to shading of a single light source. The method is not designed to deal with multiple light sources (e.g., sunlight and blue sky) or with object highlights. Another algorithm uses the "spectral crosspoint" operator,[2] which somewhat generalizes the allowed illuminant variations. An algorithm by Klinker,[3] based on a theory of Shafer,[4] assumes that the pixels from a matte object with a highlight span a plane in color space; the identity of the plane (characterized by its normal) permits separation of the object (together with the highlight) from other objects in the image. Independently, a general theory was advanced by Nikolaev[5,6] that subsumes an arbitrary number of illuminants of different spectral compositions and geometries. Nikolaev treats in detail only the case of matte objects without highlights but makes passing reference[6] to an interpretation that involves objects with highlights. The theory can be applied also to lights that undergo multiple reflections between objects; it also has interesting implications for object-color recognition in the presence of an illuminant of unknown spectral composition. The present paper has three objectives pertinent to these issues: (a) to explain an adaptation of Nikolaev's framework in both geometric and algebraic terms, (b) to show how other segmentation algorithms fit as special cases into this framework, and (c) to discuss the relationship of the framework to algorithms for object-color recognition (color constancy) based on the assumption of limited basis-function expansions of reflectances.

2. RANK OF AN IMAGE FIELD

Light reflected from an object and imaged by a retina (or a camera) produces a visual stimulus $F(\lambda, x, y)$, where λ is visible wavelength and x and y are the spatial coordinates on the retina. The quantity F is a spectral power density (SPD) with the units of power per unit wavelength interval per unit retinal area. We define an *image field* as a collection of pixels (retinal locations) x, y that putatively represent light reflected from different parts of the same uniformly colored object.

Nikolaev's framework proceeds from a mathematical model for the way the image field of an object arises from the interaction of lights, the viewer, and the spectral reflectance of the object. To introduce this framework, we imagine a single matte reflecting object [with reflectance $r(\lambda)$] under a single—perhaps spatially distributed—light source with the relative SPD $S_1(\lambda)$. The light received on the retina from the object will have the same relative spectrum at all parts of the image but will be modulated in intensity by shading and shadows, i.e., by a geometrical factor $g_1(x, y)$ that depends on

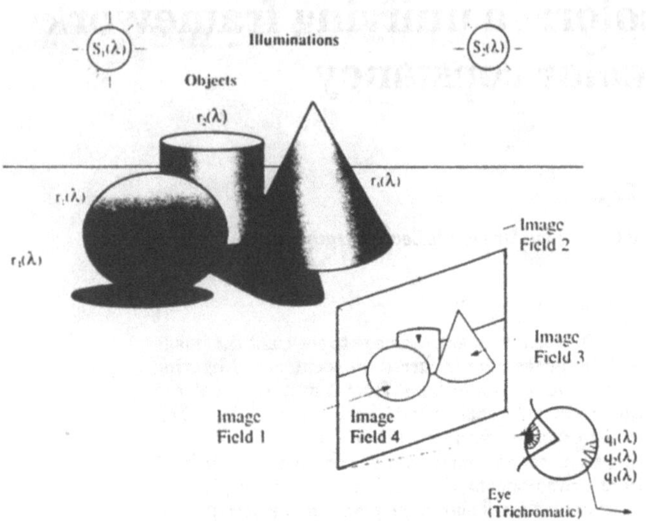

Fig. 1. One origin of rank-2 image fields: matte objects under two spectrally different light sources.

the location and the orientation of the surface of the object with respect to the illumination source and the retina. Hence the image-field SPD $F(\lambda, x, y)$ produced on the viewer's retina from this object has the form $g_1(x, y)S_1(\lambda)r(\lambda)$, where $g(x, y)$ depends on the geometry of lighting and viewing. The function $F(\lambda, x, y)$ is a single spectrum $S_1(\lambda)r(\lambda)$ modulated by a location-dependent factor g_1.

By the same argument, turning off the illumination $S_1(\lambda)$ and turning on another in a different location and with spectrum $S_2(\lambda)$ will produce an image-field SPD $g_2(x, y)S_2(\lambda)r(\lambda)$. Turning on both S_1 and S_2 (see Fig. 1) will produce a SPD that is the sum of the SPD's from having turned on S_1 and S_2 separately:

$$F(\lambda, x, y) = g_1(x, y)S_1(\lambda)r(\lambda) + g_2(x, y)S_2(\lambda)r(\lambda).$$

From this equation it is clear that for two light sources the image-field SPD is a sum of two terms, each one a product of a spatial and a spectral part:

$$F(\lambda, x, y) = g_1(x, y)H_1(\lambda) + g_2(x, y)H_2(\lambda).$$

Now $F(\lambda, x, y)$ is a linear combination of $H_1(\lambda)$ and $H_2(\lambda)$ (weighted by g_1 and g_2).

By extension, if M lights are turned on, the image-field SPD becomes a sum of M terms, each a product of a spatial and a spectral function:

$$F(\lambda, x, y) = g_1(x, y)H_1(\lambda) + \ldots + g_M(x, y)H_M(\lambda). \qquad (1)$$

Now $F(\lambda, x, y)$ is a linear combination of $H_1(\lambda), \ldots, H_M(\lambda)$ with space-dependent coefficients $g_i(x, y)$, $i = 1, M$.

We call the number of terms M in Eq. (1) the *rank* of the image field F.[7] In Section 3 I illustrate, using the rank-2 case of matte objects under two light sources, how an image field with the form of Eq. (1) is amenable to segmentation (automated segregation from other parts of the image). The nomenclature of rank will be shown to correspond to the rank of matrices whose determinants are used in the segmentation. Then, in Section 4, the theory of rank-2 fields is extended by example to image fields that are more general than matte objects under multiple light sources. The straightforward extension of the theory to rank higher than 2 is discussed in Section 5.

3. RANK-2 FIELDS DUE TO TWO LIGHTS

Rank-2 fields are most easily introduced by considering the special case of matte three-dimensional objects under two light sources (see Fig 1). Then the image-field SPD for reflectances $r_k(\lambda)$ under illuminants $S_1(\lambda)$ and $S_2(\lambda)$ is

$$F(\lambda, x, y) = g_1(x, y)S_1(\lambda)r_k(\lambda) + g_2(x, y)S_2(\lambda)r_k(\lambda), \qquad (2)$$

where k enumerates the objects in question and therefore depends discretely on x and y.

Suppose that the rank-2 fields are imaged by a trichromatic eye with spectral sensitivities $q_j(\lambda)$, where $j = 1, 2, 3$ might represent red-sensitive, green-sensitive, and blue-sensitive cones in the eye. The tristimulus values (spectrally integrated sensor inputs) $Q_j(x, y)$ at point (x, y) are then given by

$$Q_j(x, y) = g_1(x, y)\langle S_1 r_k q_j\rangle + g_2(x, y)\langle S_2 r_k q_j\rangle, \qquad (3)$$

where $\langle\ \rangle$ denotes integration over visible wavelength λ.

The problem of object-based segmentation is to find functions of $Q_j(x, y)$ that do not depend on $g_1(x, y)$ or $g_2(x, y)$ but change in x and y only where k changes (i.e., only at the boundaries between imaged objects and not elsewhere). Nikolaev refers to such space-invariant quantities as the codes of an object. The solution for rank-2 fields emerges when one notes that the vectors $\mathbf{Q}(x, y)$ span a plane in the sensor tristimulus space when x and y vary within a single object region k (see Fig. 2). The sought-after invariant is the unit normal $\hat{n}$ to this plane, which can be computed from the vectors $\mathbf{Q}$ for any two points (x, y) and (x', y') in the imaged object. Denoting $\mathbf{Q}$ as $\mathbf{Q}(x, y)$ and $\mathbf{Q}'$ as $\mathbf{Q}(x', y')$, we can generate the unit normal $\hat{n}$ by forming the axial vector $\mathbf{N}$ of determinants,

$$\mathbf{N} = \left[\begin{vmatrix} Q_2 & Q_3 \\ Q_2' & Q_3' \end{vmatrix}, \begin{vmatrix} Q_3 & Q_1 \\ Q_3' & Q_1' \end{vmatrix}, \begin{vmatrix} Q_1 & Q_2 \\ Q_1' & Q_2' \end{vmatrix}\right], \qquad (4)$$

and defining the unit vector $\hat{n} = \mathbf{N}/|\mathbf{N}|$. The components of $\hat{n}$ are the direction cosines of the normal vector with respect to the coordinate axes of the three-dimensional color space (tristimulus space). Clearly $\hat{n}$ has two independent components, and these components are independent of (x, y) and (x', y') if these points lie within image area k.[8]

It is instructive to derive algebraically the spatial invariance of the components of $\hat{n}$ within the boundaries of an image of a uniformly colored object. Rather than dealing algebraically with the invariance of $\hat{n} = \mathbf{N}/|\mathbf{N}|$ (which involves a square root of a sum of squares in the denominator),

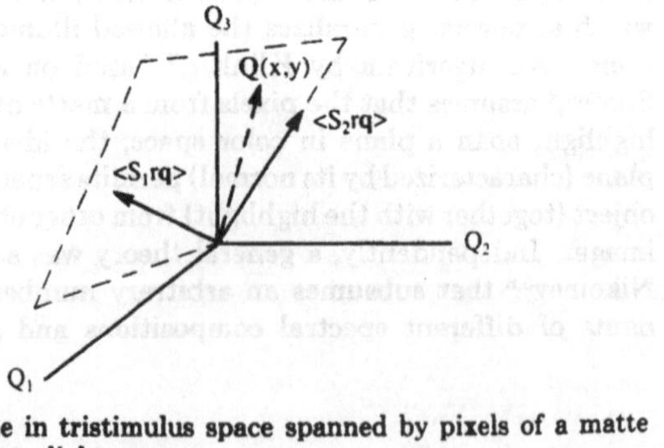

Fig. 2. Plane in tristimulus space spanned by pixels of a matte object under two light sources.

we examine the expressions for $n_i/n_3 = N_i/N_3$ $(i = 1, 2)$, understanding that straightforward modifications must be made in the derivation when $N_3 = 0$. The expressions for n_1/n_3 and n_2/n_3 are

$$\frac{n_1}{n_3} = \begin{vmatrix} Q_2 & Q_3 \\ Q_2' & Q_3' \end{vmatrix} \bigg/ \begin{vmatrix} Q_1 & Q_2 \\ Q_1' & Q_2' \end{vmatrix}$$

$$= \begin{vmatrix} \langle S_1 r q_2 \rangle & \langle S_1 r q_3 \rangle \\ \langle S_2 r q_2 \rangle & \langle S_2 r q_3 \rangle \end{vmatrix} \bigg/ \begin{vmatrix} \langle S_1 r q_1 \rangle & \langle S_1 r q_2 \rangle \\ \langle S_2 r q_1 \rangle & \langle S_2 r q_2 \rangle \end{vmatrix}. \quad (5)$$

To arrive at this result, note that a factor

$$\begin{vmatrix} g_1 & g_2 \\ g_1' & g_2' \end{vmatrix}$$

cancels from the numerator and the denominator.

Similarly,

$$\frac{n_2}{n_3} = \begin{vmatrix} \langle S_1 r q_3 \rangle & \langle S_1 r q_1 \rangle \\ \langle S_2 r q_3 \rangle & \langle S_2 r q_1 \rangle \end{vmatrix} \bigg/ \begin{vmatrix} \langle S_1 r q_1 \rangle & \langle S_1 r q_2 \rangle \\ \langle S_2 r q_1 \rangle & \langle S_2 r q_2 \rangle \end{vmatrix}. \quad (6)$$

At this point the nomenclature of rank becomes self-explanatory, for the rank of 2 is the rank of the matrices whose determinants appear in the above expressions. Note that the spatial parts cancel out these expressions, leaving dependence only on the illuminant spectral power distributions, the reflectance, and the color-matching functions (spectral sensitivities of the device). Note also that the results would not have changed had we interpreted the primes as meaning directional derivative at (x, y) rather than as evaluated at another point (x', y'). The directional-derivative interpretation of $\mathbf{Q}'$, etc. was in fact the first interpretation adopted by Nikolaev. From this starting point, it is then remarkable that a finite-difference approximation to the derivative does not cause the invariant to degrade in any way, so long as both points in the difference are within the same imaged object.

Given computation of the normal $\hat{n}$ to the plane in tristimulus space generated by a reflecting object, two of three of the components of $\hat{n}$ can be used for segmentation of the image. One way of doing this would be to compute $\hat{n}$ as a function of x and y from local triplets of pixels and then segment the image where there is appreciable spatial variation of $\hat{n}$. A more robust technique is to use a clustering algorithm[3] rather than evaluating spatial derivatives as suggested by Nikolaev. Of course, there are residual implementation problems inherent in any multispectral segmentation algorithm, e.g., the problem of saturating nonlinearities in the intensity versus response of the imaging system and the problem of object interreflections.

4. OTHER RANK-2 FIELDS

The physical origins of the image rank are particularly easy to visualize in the case of a matte object under two spectrally different illumination sources with arbitrary spatial arrangements. The functions g_1 and g_2 can then be produced separately by turning on each illuminant separately. When the ith light is turned on, the image field is $g_i(x, y)H_i(\lambda)$, and the spatial variability of the field is due to $g_i(x, y)$. It is less obvious, but nonetheless true, that the rank-2 formalism, and scene segmentation algorithms that derive from it, apply to a more general class of problem.

In particular, the theory applies to dielectric objects with

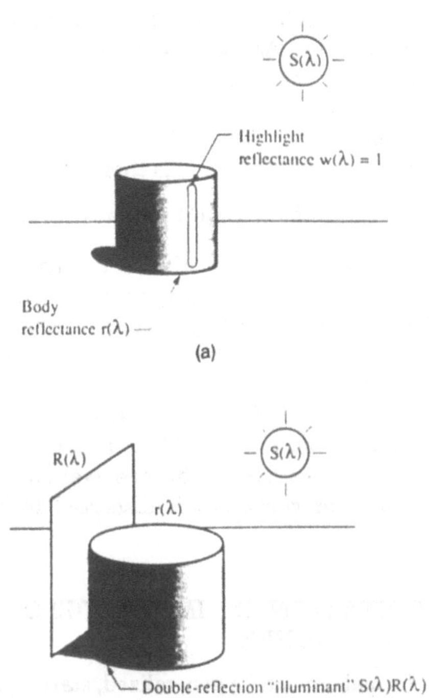

Fig. 3. Rank-2 image fields involving a single light source: (a) a dielectric object with a highlight, (b) a secondary reflection from one matte object to another.

highlights under a single light source [Fig. 3(a)]. In this case, the image-field SPD is given by

$$F_H(\lambda, x, y) = g_1(x, y)S(\lambda)r(\lambda) + g_2(x, y)S(\lambda), \quad (7)$$

where $S(\lambda)$ is the illuminant spectral power distribution and $r(\lambda)$ is the body reflectance. The function $g_1(x, y)$ gives the spatial distribution of the body reflection, and $g_2(x, y)$ gives the spatial distribution of the highlight, which has the color of the illuminant. This is the model of Shafer[4] and of Klinker.[3]

Another rank-2 image field in which there is only a single light source is created by a matte object illuminated not only by the light source itself but by a light reflected from another object [Fig. 3(b)]. For this geometry, the image-field SPD associated with reflectance $r(\lambda)$ is given by

$$F_S(\lambda, x, y) = g_1(x, y)S(\lambda)r(\lambda) + g_2(x, y)S(\lambda)R(\lambda)r(\lambda), \quad (8)$$

where $R(\lambda)$ is the reflectance producing the secondary illumination $S(\lambda)R(\lambda)$ on $r(\lambda)$.

For both F_H and F_S, the rank-2 segmentation would proceed as for the matte-object case discussed in Section 3. It should be noted that Klinker[3] has based a scene-segmentation algorithm on the rank-2 formalism with a single illuminant and highlights. However, instead of generating the normal to the plane spanned by pixels on a single object, Klinker used a cluster-analysis technique to classify pixels as being on or off the plane. This technique is undoubtedly more computationally robust than the method that I suggested in Section 3 of evaluating the normal to the plane based on a few pixels. However, my intent in this paper is to describe the mathematical framework without broaching implementation details, and this objective was also evidently shared by Nikolaev.

As a final example of a rank-2 field, consider a matte

coplanar array of reflectances (such as in a painting) under a single illuminant $S(\lambda)$. Assume that each reflectance in the painting is a linear combination of two basis functions of a wavelength, $\rho_1(\lambda)$ and $\rho_2(\lambda)$. At each point (x, y), the reflectance encountered is $b_1(x, y)\rho_1(\lambda) + b_2(x, y)\rho_2(\lambda)$. The spectrum of light in the image at (x, y) is then

$$F_C(\lambda, x, y) = g(x, y)[b_1(x, y)\rho_1(\lambda) + b_2(x, y)\rho_2(\lambda)]S(\lambda), \quad (9)$$

where $g(x, y)$ represents the spatial variations incurred by lighting and viewing geometry. Although one might want to segment this painting according to reflectance, the rank-2 algorithm will ignore all the interreflectance borders and will see only a blank canvas. Thus sets of reflectances must differ in at least three independent basis-function components in order to be discriminated by a trichromatic rank-2 segmenter. The interaction of this observation with existing theories of color constancy is discussed in Section 7 below.

5. SEGMENTATION OF IMAGE FIELDS WITH RANK 3 AND HIGHER

The rank-2 formalism can be generalized, starting from Eq. (1), to images with matte objects under M spectrally different light sources and also to images with highlights and multiple light sources. For example, matte objects under three light sources produce image fields of rank 3 (see Fig. 4). We can visualize this, as in Section 2, by turning on each light source separately, retrieving the spatial shadowing functions $g(x, y)$, and then turning on all three lights at once. As might be expected, segmentation of rank-3 fields requires four kinds of photosensor, and hence a tetrachromatic eye. Within an object boundary, the reflected-light tetrastimulus vector $\mathbf{Q}$ varies in x and y in such a way as to span a three-dimensional subspace of the four-dimensional sensor space. The normal $\hat{n}$ to this subspace is theoretically invariant to the locations of the points used to compute it (so long as these points stay within the object boundaries). Because there are three independent components of $\hat{n}$, there are three spatially invariant quantities inherent in a rank-3 image field. Ratios of the components of $\hat{n}$ are ratios of determinants

$$\begin{vmatrix} Q_i & Q_j & Q_k \\ Q_i' & Q_j' & Q_k' \\ Q_i'' & Q_j'' & Q_k'' \end{vmatrix},$$

where i, j, and k are distinct triads from the sensor index set 1, 2, 3, 4. These determinants are components of a four-dimensional axial vector, the analog of the cross product in three dimensions (encountered in rank-2 fields).

When there are highlights, the terms in the sum of Eq. (1) are more difficult to visualize than in the matte case, for it is impossible to turn off a highlight except by image processing. However, for a dielectric colored object, the highlight increases the rank of the image field by introducing an independent spatial variation for each nondiffuse illuminant; this field has the relative spectral composition of that illuminant. On the other hand, each body reflection gives rise to a spectrum that is the product of illuminant and reflectance spectra. Highlights and body reflections therefore contribute as separate additive degrees of freedom to the image field.

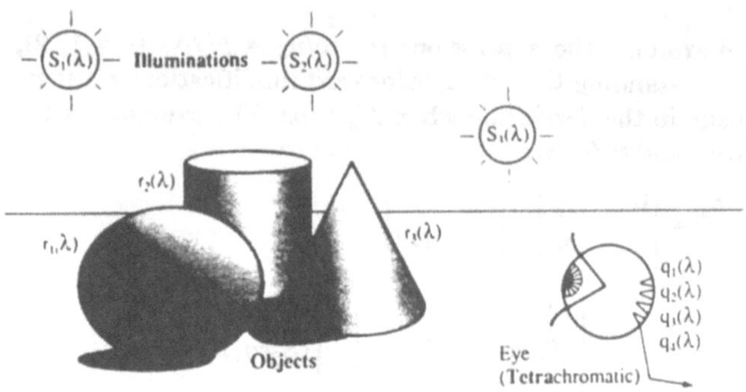

Fig. 4. Rank-3 image fields from three light sources on matte objects.

For a metal, a highlight is not likely to increase the rank of the image field. A metallic highlight does not generally have the color of the illuminant and actually is similar to the body reflectance in relative spectral composition. Therefore the spatially varying image field for a highlight and for the associated body reflection are most likely multiplied by the same function of wavelength.

6. RELATION TO PREVIOUS ALGORITHMS

The rank-M theory of Nikolaev has a great affinity with the theory of Shafer,[4] which originally addressed a problem quite different from object-based segmentation. Shafer's theory determines a single illuminant's tristimulus values from its interactions with a dielectric surface with highlights. The theory then permits separation of the image, pixel by pixel, into highlight and body-reflection components. The image field—including the body reflection and the highlight—has a rank of 2, for the highlight and the body reflections superpose in varying proportions at each pixel to generate a planar region in tristimulus space. Shafer used this planar region to identify the illuminant tristimulus direction (which lies in the plane); the later image-segmentation work of Klinker (which effectively involves the normal to the plane) evolved from this formulation.

As has already been stated, the band-ratio method[1] is the rank-1 case of Nikolaev's framework and serves to null out pure intensity variations that are due to a single illuminant on a matte surface. Of course, two sensor dimensions are needed for the line in distimulus space generated by the pixels in the image field to be orthogonal to the vector direction dictated by the band ratio.

The spectral-crosspoint criterion of Rubin and Richards[2] is also a special case of the rank-M formalism but uses only two kinds of photosensor, so $\mathbf{Q}(x, y) = [Q_1(x, y), Q_2(x, y)]$. In this case, the image field has rank 2, and one of the g functions is assumed to be spatially constant. (For a matte object under two light sources this means that one light source is diffuse, and for a highlighted object under a single light source this means that the body reflection is sufficiently diffuse.) Then, adopting the prime to mean a directional derivative, and taking $g_2(x, y)$ = constant, we can easily see that $Q_1'Q_2' > 0$ so long as the derivatives are not evaluated across an object boundary. [This is true because $Q_1'Q_2'$ equals $(g_1')^2$ times a product of two positive spectrum integrals.] The inequality $Q_1'Q_2' > 0$ is the spectral-crosspoint criterion, in a somewhat different notation from that of Ref. 2.

7. RELATIONSHIP TO ALGORITHMS OF COLOR CONSTANCY

We now address the problem of recognizing the colors of three-dimensional objects whose images have been segmented by using the rank-M formalism. In its most comprehensive form, recognition of object color amounts to illuminant-invariant estimation of the object's spectral reflectance. Existing algorithms for reflectance estimation, or color constancy, were designed for matte coplanar arrays of reflectances. The ability to adapt these algorithms to three-dimensional objects would add a "what" dimension to the "where" designation of reflectances provided by the segmenter.

The basic problem can be phrased formally as follows. Suppose that there has been a successful delineation of a three-dimensional object with a rank-2 algorithm and that it is desired to estimate the spectral reflectance of the object from the spatial invariants $\hat{h}$. These spatial invariants still depend on the unknown illuminant spectra $S_1(\lambda)$ and $S_2(\lambda)$ when the objects are matte with no highlights. When there is only a single illuminant and the objects have highlights, the invariants depend on the single unknown illuminant spectrum $S(\lambda)$. In either case, it is desired to find functions of $\hat{h}$ that depend on $r(\lambda)$ and not on the spectral power distributions of the illumination.

It would seem at first that reflectance estimation ("what") would be a natural extension of the segmentation approach ("where"). However, there are various theoretical difficulties to be overcome in combining the "where" with the "what" algorithms. A few of these problems are outlined here, with possible directions toward their solution.

Reflectance Recognition without Reference Colors

It would be quite desirable to perform illuminant-invariant reflectance estimation without the need for any reference colors (known reflectances such as a matte white object) placed in the visual field to permit independent assessment of the illuminant. However, although there is an algorithm[9] that can be used to recognize matte coplanar reflectances without any reference colors, the recognition algorithm and the rank-2 segmenter make different demands on the dimensionality of color space. Reconciling these demands would be a prerequisite for using the algorithms together in analyzing a colored image.

To understand the interaction of the separate constraints on dimensionality, it is instructive to examine the interaction of the Maloney–Wandell recognition algorithm[9] with the rank-2 segmenter in a three-dimensional color space. Consider the situation (described in Section 4) of a canvas painted with reflectances that are linear combinations of only two basis functions of wavelength. In that case, the rank-2 image field seen by a segmenter will characterize not a single reflectance but rather the whole painting. Significantly, the algorithm of Maloney and Wandell requires, for trichromatic vision, that the reflectances map to a plane in color space. This is done to permit estimation of the illuminant from the normal to that plane without resorting to reference colors. If the rank-2 image segmenter is to separate the reflectances in the painting, these reflectances must span three dimensions in a color space. That would require the full color space to have four dimensions in order to accommodate the normal that permits illuminant estimation in the Maloney–Wandell theory. The rank-2 segmenter might then operate in a three-dimensional subspace of the full color space.

This constraint on dimensionality is not a fatal difficulty but would have to be addressed for any meaningful color recognition to result from the image segmentation. In view of the dimensionality of human color vision, it is also interesting to speculate how a rank-2 segmenter and a Maloney–Wandell color recognizer might be realized by the human visual system.

Reflectance Estimation Using Reference Colors

Even in the presence of reference colors, the estimation of reflectances of three-dimensional objects by using information acquired during image segmentation presents some theoretical difficulties. Consider a segmented rank-2 image due to matte reflectances under two light sources. Suppose (as is common[9–12]) a reflectance $r(\lambda) = a_1\rho_1(\lambda) + a_2\rho_2(\lambda) + a_3\rho_3(\lambda)$, where the basis functions $\rho_m(\lambda)$ are three reference reflectances, which themselves are known and whose tristimulus values are measured under the same lights that are illuminating $r(\lambda)$.[11,12] Estimating absolute reflectance is impossible because reducing the light intensity by a factor cannot be distinguished from reducing the reflectances by the same factor at all wavelengths. Estimating relative reflectance (equivalent to estimating a_1/a_3 and a_2/a_3) is also impossible given just the plane in tristimulus space to which the reflectance maps. To see why this is so, note that a red and a cyan body reflection may lie upon the same plane with a highlight white[3] but will have different values of a_i/a_3 ($i = 1, 2$). Thus using the normal to the plane, even in conjunction with reference colors, cannot in principle yield good estimates of a_i/a_3 for both red and cyan reflectances in the plane.

From this argument, it follows that relative spectral reflectances cannot be estimated from the tristimulus plane spanned by their pixels, even in the presence of three reference colors. This observation applies to all the rank-2 scenarios discussed here. However, the observation does not preclude finding some nontrivial function of $a_1/a_3, a_2/a_3$ that could be used (imperfectly) to recognize objects according to their spectral reflectances independently of illuminant. This would be a weak form of color constancy.

Single Invariant Using Reference Colors

In the special case of a highlighted object with a single illuminant, there is a method to estimate one of the two independent quantities a_i/a_3. The first version of the method was proposed by D'Zmura and Lennie,[13] and we describe it briefly later in this section.

First, we describe a version of the method that is perhaps easier to understand than the original. In this version, three reference colors are arranged one next to another in such a way that a pure body reflection is obtained from one face of each (see Fig. 5). Furthermore, the reference reflectance ρ_1 is chosen to be white, i.e., $\rho_1(\lambda) = 1$. This produces measured (hence known) reference-color tristimulus values $\langle Sq_j \rangle$, $\langle \rho_2 Sq_j \rangle$, and $\langle \rho_3 Sq_j \rangle$. From Eq. (7), the tristimulus values of pixel (x, y) of an object with body reflectance $r(\lambda)$ are then

$$Q_{Hj}(x, y) = g_1(x, y)[a_2\langle \rho_2 Sq_j \rangle + a_3\langle \rho_3 Sq_j \rangle] + [g_2(x, y) + a_1 g_1(x, y)]\langle Sq_j \rangle. \quad (10)$$

Besides the known tristimulus values of the reference colors

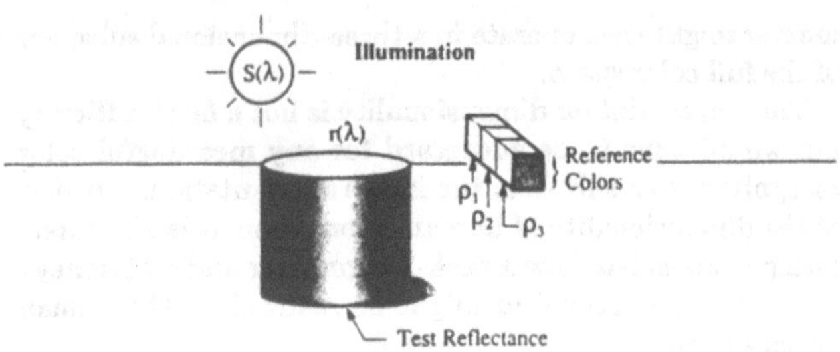

Fig. 5. Role of reference colors in estimating spectral reflectance parameters for a highlighted object under one light source.

in Eq. (10), the measured values $Q_{Hj}(x, y)$ are also known. Without further knowledge, it is possible to determine the ratio a_2/a_3 as follows. Let T be the matrix whose column vectors are the reference-color tristimulus vectors $\langle S\mathbf{q}\rangle$, $\langle \rho_2 S\mathbf{q}\rangle$, and $\langle \rho_3 S\mathbf{q}\rangle$. Then operating on the vector equation (10) with T^{-1} gives

$$T^{-1}\mathbf{Q}_H = \begin{bmatrix} g_2(x, y) + a_1 g_1(x, y) \\ g_1(x, y)a_2 \\ g_1(x, y)a_3 \end{bmatrix}. \tag{11}$$

The ratio of the second and third components of this vector is the desired ratio a_2/a_3:

$$a_2/a_3 = (T^{-1}\mathbf{Q}_H)_2/(T^{-1}\mathbf{Q}_H)_3. \tag{12}$$

Of course, the right-hand side of Eq. (12) is determined completely by known quantities, so a_2/a_3 is computable.

With the three reference colors, the inversion of T is strictly the same as the inversion used in the volumetric theory of color constancy.[11,12]

A geometric interpretation of the above color-constancy algorithm is instructive (see Fig. 6). The reference reflectance ρ_1 is white and therefore generates a pure-highlight tristimulus vector. This vector lies in the plane (call it P_1) in tristimulus space generated by the pixels from a three-dimensional object under the single light. Reflectances ρ_2 and ρ_3 define another plane (P_2) in tristimulus space. The intersection of these planes is a line, L. In a new coordinate system in which the three reference-reflectance tristimulus vectors $\langle S\rho_i\mathbf{q}\rangle$ are orthonormal, line L's direction cosines with these vectors are given by $a_i/[a_1^2 + a_2^2 + a_3^2]^{1/2}$. The color-constancy algorithm transforms the color space to the new coordinate system, finds the intersection of planes P_1 and P_2, and effectively determines the angle ϕ between L and the axis $\langle S\rho_2\mathbf{q}\rangle$. The quantity cot ϕ is the quantity a_2/a_3.

This color-constancy algorithm works in theory because the plane P_1 spanned by the highlight and the body reflection is perpendicular to the plane P_2 in the new coordinate system; hence all the pixels in plane P_1 must enter the computation if it is to yield the same value of a_2/a_3. On the other hand, the method does not extend to other rank-2 fields (such as the fields due to matte objects under two light sources), because not all the points on plane P_1 would project to plane P_2 with the same angle ϕ.

It should be noted that the above color-constancy algorithm is unlike the original formulation of the D'Zmura–Lennie algorithm in requiring the presence of three reference reflectances, ρ_1, ρ_2, and ρ_3, that appear in the same illumination environment as the test object. In contrast, the original method introduced by D'Zmura and Lennie[13] requires only one reference color—the white reflectance ρ_1, which is also required to be one of the three basis functions for the reflectance. In compensation for needing only one reference color, the method requires that the illuminant spectral power distribution be a linear combination of three known functions of wavelength. The first step of the algorithm is to find the expansion coefficients of the illuminant spectrum in terms of the measured tristimulus values from the reference white reflectance. Then the fact that the illuminant basis functions are known means that the coefficients determine an estimate of the illuminant spectrum. In addition to the single reference reflectance ρ_1, the reflectance basis functions ρ_2 and ρ_3 are also known, although they do not appear as reference colors and their tristimulus values are not measured under the test illumination. On the other hand, the estimate for the illuminant spectrum allows the reflected-light tristimulus values of ρ_2 and ρ_3 to be computed from the estimated illuminant spectral power distribution and the known color-matching functions. The reflectances ρ_2 and ρ_3 then assume the role of *virtual* reference colors, because their tristimulus values are known under the test light even though they are not directly measured. The D'Zmura–Lennie algorithm then proceeds by transforming the tristimulus coordinates to a system in which the reference white and the two virtual reference colors form orthonormal vectors in the tristimulus space. Finally, the angle construction of Fig. 6 proceeds exactly as in our version of the method.

The D'Zmura–Lennie algorithm with one reference color has been adopted by Bajcsy *et al.*[14] into a working algorithm, the only one that incorporates both object-based segmentation and color constancy in the same framework.

Test for Situation Amenable to Color Constancy

As we have shown, the D'Zmura–Lennie type of estimate of a_2/a_3 depends on the rank-2 field's having arisen from a highlight and a single light source, not from a matte object with multiple illuminations. Hence, to be fully automatic, an algorithm that segments a scene and then proceeds to perform illuminant-invariant object-color recognition must

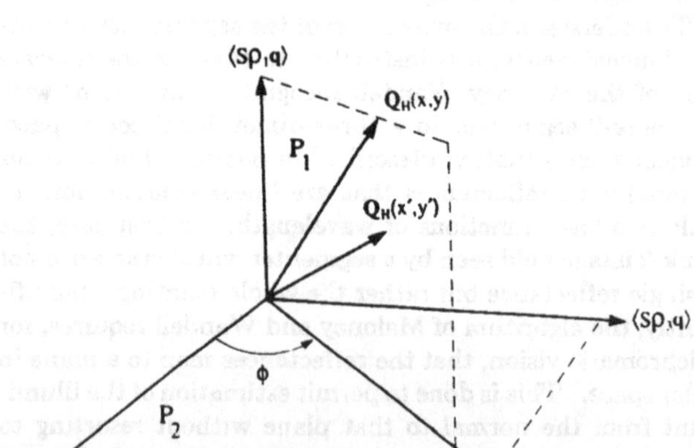

Fig. 6. Geometric representation of the D'Zmura–Lennie method. The quantity a_2/a_3, characteristic of the spectral reflectance and independent of the illumination, is determined from angle ϕ.

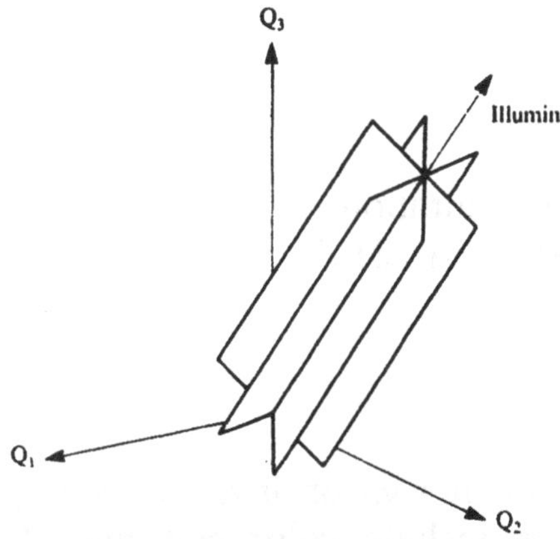

Fig. 7. Method of Tominaga and Wandell for determining the illuminant tristimulus vector. Each depicted plane contains the tristimulus vectors of light reflected from a single object with a highlight.

be able to determine whether a rank-2 field is due to a highlight. This recognition can be done by examining the planes in color space generated by several objects in the now segmented image. If the planes all intersect in nearly a single line (see Fig. 7), this indicates that there is a single illuminant, that the rank of 2 originates with highlights, and that the tristimulus vectors representing the illuminant are in the direction of the line of intersection of all the planes. On the other hand, if the planes intersect in distinct lines, one can infer that there are multiple illuminants and therefore that there are no highlights (for then the image fields would have had rank greater than 2). The use of the intersection of color-space planes to infer the relative tristimulus values of the illuminant is an application of a recent method of Tominaga and Wandell.[15]

8. CONCLUSION

The present paper has incorporated several algorithms for object-based segmentation of multispectral images into a theoretical framework proposed by Nikolaev and has discussed some problems inherent in connecting this framework with color-constancy theories. The key to Nikolaev's theory is not only in his decomposing image fields into spectral and spatial parts but also in requiring the number of sensor dimensions to exceed by one the maximum rank of an image field. The need for an extra sensor to compute invariants is also a feature of the color-constancy theory of Maloney and Wandell,[9] but connecting this—or any—algorithm of reflectance estimation with the algorithms of image segmentation is an intricate problem. The algorithm of D'Zmura and Lennie[13] is a solution to a special subset of the problem: It applies only to rank-2 fields arising from a highlighted object under a single light source, it applies only when there is a reference white reflectance in the visual field, and it computes only one invariant quantity (not enough to permit estimation of the spectral reflectance). Despite these compromises, the algorithm is an exemplary model of what should be expected in future work with the more general problem.

ACKNOWLEDGMENTS

The author thanks Arthur Mansfield at Science Applications International Corporation (SAIC) for his excellent translations of Nikolaev's work from the Russian and for his insightful mathematical discussions of the work. Also, thanks are due to Paulette Rosner and the SAIC Graphics Department for preparation of the illustrations.

REFERENCES AND NOTES

1. R. A. Schowengerdt, *Techniques for Image Processing and Classification in Remote Sensing* (Academic, New York, 1983), pp. 154–159.
2. J. M. Rubin and W. A. Richards, "Color vision and image intensity: when are changes material?" Biol. Cybern. **45**, 215–226 (1982).
3. G. J. Klinker, "A physical approach to color image understanding," Ph.D. dissertation, Doc. CMU-CS-88-161 (Carnegie-Mellon University, Pittsburgh, Pa., May 1988).
4. S. A. Shafer, "Using color to separate reflection components," Color Res. Appl. **10**, 210–218 (1985).
5. P. P. Nikolaev, "Monocular color discrimination of nonplanar objects under various illumination conditions," Biofizika **33**, 140–144 (1988).
6. P. P. Nikolaev, "Algorithms for color discrimination of objects according to the reactions of logarithmic receptors," Biofizika **33**, 517–521 (1988).
7. Nikolaev defines the rank as $2M$ if none of the g functions is constant or linearly dependent on the others; if one of the g functions is constant or linearly dependent on the others, then his rank is $2M - 1$. However, there seems to be no difficulty in adopting the simpler definition that the rank is M.
8. Purists will note that we have defined $\hat{n}$ as perpendicular to all tristimulus vectors $\mathbf{Q}(x'', y'')$ from a particular reflecting object, without having a prior definition of distance in the space. This is permissible because the relation $\hat{n} \cdot \mathbf{Q}(x'', y'') = 0$ that apparently defines perpendicularity actually defines coplanarity of $\mathbf{Q}(x, y)$, $\mathbf{Q}(x', y')$, and $\mathbf{Q}(x'', y'')$ when the substitution [Eq. (4)] is made for $\hat{n}$. The dot product is not between two ordinary vectors in tristimulus space but between an ordinary vector and an axial vector $\hat{n}$ (which in two dimensions is the cross product of ordinary vectors). That no metric structure is needed to define this particular dot product can be verified algebraically by noting that the relation $\hat{n} \cdot \mathbf{Q}(x'', y'') = 0$ is immune to any linear nonsingular transformation on all the $\mathbf{Q}$ vectors.
9. L. T. Maloney and B. A. Wandell, "Color constancy: a method for recovering surface spectral reflectance," J. Opt. Soc. Am. A **3**, 29–33 (1986).
10. G. Buchsbaum, "A spatial processor model for object color perception," J. Franklin Inst. **310**, 1–26 (1980).
11. M. H. Brill, "Computer-simulated object-color recognizer," Progress Rep. 122 (MIT Research Laboratory of Electronics, Cambridge, Mass., 1980), pp. 214–221.
12. M. H. Brill and G. West, "Chromatic adaptation and color constancy: a possible dichotomy," Color Res. Appl. **11**, 196–204 (1986).
13. M. D'Zmura and P. Lennie, "Mechanisms of color constancy," J. Opt. Soc. Am. A **3**, 29–33 (1986).
14. R. Bajcsy, S. W. Lee, and A. Leonardis, "Color image segmentation and color constancy," in *Perceiving, Measuring, and Using Color*, M. H. Brill, ed., Proc. Soc. Photo-Opt. Instrum. Eng. **1250**, 245–255 (1990).
15. S. Tominaga and B. A. Wandell, "Standard surface-reflection model and illuminant estimation," J. Opt. Soc. Am. A **6**, 576–584 (1989).

PHOTOMETRIC MODELS IN MULTISPECTRAL MACHINE VISION

Michael H. Brill

Science Applications International Corporation
1710 Goodridge Drive, 1-11-1, McLean, VA 22102

ABSTRACT

The performance of several tasks in multispectral computer vision involves assumptions about the reflection of light from surfaces. These tasks include color constancy (visual representation of spectral reflectances independent of the illuminant spectrum), object-based image segmentation, and deduction of the shape of a surface from its shading. Most color-constancy theories implicitly assume Lambertian, coplanar reflecting surfaces, a distant viewer, and a distant light source that may have many components that are spatially and spectrally distinct. Object-based-segmentation theories allow curved surfaces, each of whose scattering kernels is the sum of a few separable terms (each of which is the product of a wavelength-dependent part and a geometry-dependent part). There is no restriction on the distances of light sources or observer. However, for these theories the illuminant angular/spectral distribution must consist of only one or two separable terms. Finally, A. Petrov's shape-from-shading theory allows the light source to have arbitrary spectral and spatial composition, but seems to require the surface scattering kernels to have Lambertian dependence on the surface normal. The present paper compares these photometric models.

1. INTRODUCTION

Multispectral images can be used to help solve three problems in computer vision. These problems are (a) to determine, in an image, the boundary of a three-dimensional object with uniform material properties; (b) to recognize the spectral signature of the object apart from the influence of the spectrum of the lighting environment; and (c) to deduce the three-dimensional shape of the object from the shading incurred by the (unknown) light sources. Abbreviated names for the above tasks are, respectively, object-based segmentation, color constancy, and shape-from-shading.

The approach taken in this article, which has been explained in generic terms by W. Richards[1], is to examine for each of the above three tasks a theory of competence for multispectral vision. A theory of competence models the physics of the interaction of the sensed world with a sensory system. On the basis of these models, competence theory derives informational quantities that are accessible to the sensory system and characterize the interesting aspects of the object (e.g., spectral reflectance factor) independently of other, less interesting aspects (e.g., lighting geometry and spectrum). In contrast with competence theory, performance theory deals with algorithms and hardware implementations for extracting the invariants posed by competence theory. A performance theory is by definition "machine dependent", and hence depends on whether one is talking about computer vision or human vision. On the other hand, the theory of competence is, to a large extent, machine-independent. Hence the present discussion, pertaining to competence, may be applied in the contexts of either machine or human vision.

In comparing the assumptions behind the competence theories, I will concentrate on models of the physical interaction between lights, objects, and the photosensors of the visual system. Such models are called photometric. The degree of compatibility of these models will indicate to what extent the three tasks can be done by the same vision system looking at the same scene.

2. FORMAL FRAMEWORK AND DEFINITIONS

It is first appropriate to describe in general terms the formation of an image on a retina given a light source and a uniformly colored reflecting surface.

Let the image be denoted $F(\lambda,\underline{x})$, where $\underline{x}$ is a two-dimensional vector of retinal coordinates and λ is wavelength. The quantity F is a power density of retinal illumination per unit retinal area per unit wavelength interval.

A small patch of reflecting surface can be characterized by the scattering kernel $R(\lambda,\underline{m},\underline{n},\underline{p})$, where $\underline{m}$ is the unit vector from an element of light source to the patch of surface, $\underline{n}$ is the unit vector normal to the surface patch, and $\underline{p}$ is the unit vector from the surface element to the pupil of the eye or machine-vision system. The scattering kernel R conveys the fraction of light at wavelength incident from direction $-\underline{m}$ that gets scattered into direction $\underline{p}$, given a surface normal $\underline{n}$. Of course, the values of $\underline{n}$ and $\underline{p}$ for a patch of surface imaged at point $\underline{x}$ depend on $\underline{x}$. The uniform color of the patch of surface is expressed by the lack of explicit dependence of R on $\underline{x}$.

The incident lighting on the patch of surface is expressed by a spectral power distribution $S(\lambda,\underline{m},\underline{x})$. The dependence on $\underline{x}$ is due to the fact that the surface element imaged at $\underline{x}$ is a certain distance from the source element at $\underline{m}$. It is therefore strictly true that the quantity $S(\lambda,\underline{m},\underline{x})$ depends on the reflecting patch as well as on the light source. However, for a distant light source, S has no $\underline{x}$ dependence.

Note that the $\underline{x}$ dependence of S does not imply that S depends implicitly on the distance of the eye pupil from the surface. The inverse-square distance law that attenuates the received light from a surface patch reduces in the same proportion the image size of the patch, and the intensity is preserved.

The action of the surface, for each separate wavelength (assuming no fluorescence), is to gather light from all directions $\underline{m}$ and return the light with some efficiency to the image point $\underline{x}$, where the light produces $F(\lambda,\underline{x})$. This operation can be expressed mathematically as follows:

$$F(\lambda,\underline{x}) = T(\underline{x}) \int d^2\underline{m}\; S(\lambda,\underline{m},\underline{x})\; R[\underline{m},\underline{n}(\underline{x}),\underline{p}(\underline{x})]/[\underline{n}(\underline{x}).\underline{p}(\underline{x})]. \qquad (1)$$

It is worth a digression to understand the quantities $\underline{n}.\underline{p}$ and T in the above equation. The dot product $\underline{n}(\underline{x}).\underline{p}(\underline{x})$ appears because the more a surface is tilted relative to the line of sight, the more area of the surface fits into a unit solid angle at the pupil; the factor of increase is $1/(\underline{n}.\underline{p})$. The function $T(\underline{x})$ is an efficiency function for carrying light from the pupil of the eye to the retina, and has nothing to do with the world outside the pupil. For a pinhole camera and a planar retina parallel to the pupil, T is proportional to the fourth power of the cosine of the angle from the principal ray (perpendicular to the retina) and the ray through the pupil at point $\underline{x}$ on the retina. This dependence can be understood as follows: one cosine factor comes from the foreshortening of the pupil as seen from $\underline{x}$, one cosine factor comes from the foreshortening of a patch of retina at $\underline{x}$ with respect to the pupil, and two cosine factors come from the distance of the pupil from $\underline{x}$, the inverse square of which modulates the light intensity on the retina. Images from lensed cameras show less drastic falloffs with eccentricity than the pinhole-camera function T, and a spherical retina makes T even closer to a constant in $\underline{x}$ over a significant visual angle. In any event, since T is a function intrinsic to the visual system, it can be calibrated out of all computations. Henceforth it is assumed that $T(\underline{x}) = 1$.

Another property of Eq. (1) should be explained. The integration over $\underline{m}$ is over a hemisphere bounded by the tangent plane to the surface at point $\underline{x}$. Since this tangent plane depends on $\underline{n}(\underline{x})$, it also depends on $\underline{x}$, hence--in all strictness--the integration over $\underline{m}$ depends on $\underline{x}$ as well. A manifestation of this effect is that, as a colored light rises above the horizon, it can change the spectral content of the light incident on the object. However, we will ignore this effect in the present paper, assuming that the surface normal does not change enough to bring substantially new light sources into view.

Given these properties of Eq. (1), now suppose a multispectral image gathers the image field $F(\lambda,\underline{x})$ into a number of spectral bands (which for later convenience we shall call M+1 but which for human photopic vision is 3). The wavelength efficiency of each of these bands is $q_j(\lambda)$, where $j=1,2,...M+1$. Then, the power density per unit retinal area captured by the j-th band at $\underline{x}$ is

$$Q_j(\underline{x}) = \int d\lambda\ F(\lambda,\underline{x})\ q_j(\lambda) \ . \qquad (2)$$

Before discussing the photometric models implicit in competence theories of vision, it will be useful to make several definitions pertaining to surfaces and their defining scattering kernels. First, we define and generalize the <u>Lambertian</u> model for reflecting surfaces. A Lambertian surface can be characterized[2,3] by a scattering kernel

$$R_L(\lambda,\underline{m},\underline{n},\underline{p}) = cr(\lambda)(\underline{n}.\underline{m})(\underline{n}.\underline{p}). \qquad (3)$$

Here, c is a normalization constant, $r(\lambda)$ is reflectance factor as a function of wavelength λ,$\underline{m}$ is the unit vector in the direction of the incident lighting, $\underline{n}$ is the unit vector normal to the surface at the point of incidence, and $\underline{p}$ is the unit vector in the direction from the surface to the viewer. The subscript L of the scattering kernel stands for "Lambertian." [Note: The factor $cr(\lambda)$ in Eq. (3) is usually assumed to be $1/\pi$ in most treatments of Lambertian surfaces, in order to ensure that the scattered power is the same as the incident power. The present definition therefore represents a slight generalization of the usual one.]

It is useful to define a kernel that generalizes Eq. (3) and hence may describe real surfaces more closely. This is the separable kernel, and is defined by the equation

$$R_S(\lambda,\underline{m},\underline{n},\underline{p}) = r(\lambda)\, f(\underline{m},\underline{n},\underline{p}), \tag{4}$$

where f is any nonnegative scalar function and the subscript S stands for "separable." Of course, physical theory requires reciprocity of R_s under interchange of lighting and viewing directions, and this implies that $f(\underline{m},\underline{n},\underline{p}) = f(\underline{p},\underline{n},\underline{m})$. The class of surfaces with separable scattering kernels enters visual competence models in a decisive way.

Finally, it is appropriate to define a "matte" surface as having a separable scattering kernel with the form

$$R_M(\lambda,\underline{m},\underline{n},\underline{p}) = c(\underline{m},\underline{p})\, r(\lambda)\, (\underline{m}\cdot\underline{n})(\underline{n}\cdot\underline{p}), \tag{5}$$

where the subscript M stands for "matte," and c is now an arbitrary symmetric function of $\underline{m}$ and $\underline{p}$ (the symmetry being needed for reciprocity). It can be seen that the variation of R_M with the surface normal $\underline{n}$ is Lambertian.

I and other authors[4,5,6] have used the words "matte" and "diffuse" in our assumptions about surfaces that we knew did not quite have to be Lambertian for our color-constancy models to work. The definition of "matte" set forth here is a generalization of Lambert's law that still works for these models.

3. IMAGE SEGMENTATION

A fundamental question to be answered when interpreting images is, "What image pixels correspond to what objects?" A more particular question is "Where are the boundaries of the object of which the pixel at two-dimensional image point $\underline{x}$ represents a part?" The "where" question is the domain of image segmentation, and multispectral images help answer this question if used in the proper way. The challenge is to segment a multispectral image at object boundaries and not at the edges of highlights and shadows in an image.

The broadest competence theory yet used for multispectral image segmentation is that of Nikolaev[6]. I have discussed the theoretical framework previously.[7] Although Nikolaev's theory is mathematically precise in its assumptions about the interactions of light with surfaces and with viewer, it makes no separate assumptions about the structure of illuminants and reflecting surfaces. However, prior to Nikolaev's work, Shafer[8] presented a sufficient set of assumptions for the theory and applied these assumptions to a restricted version of the theory. Later, Healey[9] provided a detailed physical foundation for the assumptions, for both dielectric and metallic materials. These sufficient assumptions are outlined below:

a.) A few (M) illumination sources, each of which has a particular spectral power distribution $S_k(\lambda,\underline{m},\underline{x})$. The assumption that there is one spectrum for each light source is equivalent to the stipulation that S_k is the product of a spatial and a spectral part:

$$S_k(\lambda,\underline{m},\underline{x}) = s_k(\lambda)\, h_k(\underline{m},\underline{x}) \,, \tag{6}$$

for some scalar functions s_k and h_k, $k=1,...,M$. If the light sources are distant, then S_k and h_k are independent of $\underline{x}$.

b.) There are a few (actually $M+1$) spectral sensitivities $q_j(\lambda)$ in the multispectral visual system.

c.) All reflecting areas are modelled as uniformly colored with separable scattering kernels. Reflecting surfaces are allowed to curve, so the surface normal $\underline{n}$ depends on the part of the surface being imaged, and hence on the image coordinate $\underline{x}$. Furthermore, when the viewer is close to the object, the object-to-viewer vector $\underline{p}$ depends on the image coordinates $\underline{x}$. Hence Eq. (4) can be rewritten

$$R_S(\lambda,\underline{m},\underline{x}) = r(\lambda)\, f[\underline{m},\underline{n}(\underline{x}),\underline{p}(\underline{x})] \,. \tag{7}$$

The usefulness of the independent variables on the left-hand side will be seen presently.

At this point, we note that the allowed curvature of the surfaces may cause interreflections between surfaces; these interreflections count as new illumination components, and demand more spectral bands in the image if the boundaries of interreflected illumination are to be ignored in the segmentation. This is one of the principal problems applying Nikolaev's theory to real-world object scenes.

Given assumptions (a) through (c) about the world and the visual system, the retinal image of a uniformly colored object is represented in band j ($j=1,M+1$) as

$$Q_j(\underline{x}) = \sum_{k=1}^{M} g_k(\underline{x}) \int d\lambda\; q_j(\lambda) r(\lambda) s_k(\lambda) \,, \tag{8}$$

where

$$g_k(\underline{x}) = \int d^2\underline{m}\; h_k(\underline{m},\underline{x})\, f[\underline{m},\underline{n}(\underline{x}),\underline{p}(\underline{x})]/[\underline{n}(\underline{x})\cdot\underline{p}(\underline{x})]. \tag{9}$$

The form in Eq. (8) is the form assumed by Nikolaev, and is sufficient for object-based segmentation. (The segmentation proceeds by clustering together pixels that are in the same M-dimensional hyperplane in the color space, and identifying them as coming from the same uniformly colored object.) The separability of the scattering kernel is surely sufficient for Eq. (8) to hold; we do not know at this time whether the separability condition is absolutely necessary.

The presence of a dielectric highlight demands a slight generalization of Nikolaev's theory. Now the scattering kernel for the surface is the sum of two separable terms, one with wavelength dependence r(λ) and another that is specular and hence nonselective in wavelength. This generalization changes Eq. (8) in a straightforward way, giving rise (for M=1) to the basic equation used by Klinker and others[10-12] to segment an image containing an object with a highlight under a single illuminant.

4. COLOR CONSTANCY

The identification of parameters of a surface scattering kernel from a multispectral image, independent of illuminant spectrum, is called color constancy. Even more so than in image segmentation, the assumptions on the behavior of the scattering kernel are not spelled out in color-constancy theories. Usually[13,14], the assumptions about the spectral behavior of the kernel are made explicit, but the dependences on lighting and viewing geometry are left indefinite.

Here is a review of the assumptions behind a generic class of color constancy models, with an explicit assumption about the geometric dependences of the scattering areas:

a.) The reflecting areas are uniformly colored and coplanar, forming a "Mondrian." Also, the light sources are distant from the scene. Hence each reflecting area receives the same incident light as a function of angle and wavelength. The illuminant spectral power distribution is then $S(\lambda,\underline{m},\underline{x}) = s(\lambda)$, independent of $\underline{m}$ and $\underline{x}$.

b.) The surfaces are Lambertian, and hence the intensity per steradian per unit wavelength at the retina does not depend on the retina's distance from or angle to the surface normal. In conjunction with assumption (a), this assures that the received light intensities are equal from different parts of the same reflecting surface.

When the distance of the viewer from the image is large, the vector $\underline{p}$ is constant over the image, so the Lambertian assumption can be relaxed to the "matte" assumption of Eq. (5) and still will assure color constancy via the usual models. It is not yet known whether the matte assumption is necessary for the models. However, it is clearly the case that not all separable kernels will work.

Now we are in a position to write $Q_j(\underline{x})$ for the color-constancy paradigm, given a matte scattering kernel:

$$Q_j(\underline{x}) = \int d\lambda\ s(\lambda)\ r(\lambda) \int d^2\underline{m}\ c(\underline{m},\underline{p})(\underline{m}.\underline{n})\ . \tag{10}$$

Suppose either that $c(\underline{m},\underline{p})$ is independent of $\underline{p}$ or that $\underline{p}$ does not depend on the image coordinates $\underline{x}$--by dint of a distant receiver. Then, since $\underline{n}$ is independent of $\underline{x}$ for coplanar reflectances, it follows that $Q_j(\underline{x})$ is constant in $\underline{x}$, and color-constancy algorithms can operate directly on these quantities to remove the spectral dependence of the illuminant and arrive at quantities depending only on the reflectance factor r(λ). [The methods of doing so usually involve additional assumptions that s(λ) and r(λ) are decomposable into expansions in finite sets of basis functions. The color-constancy algorithms involve solving for the coefficients of expansion.] In practice, the $\underline{x}$-independence of color-constant operators is obtained approximately by requiring these operators to act locally reflectance boundaries in the Mondrian. However, the exact condition for $\underline{x}$ independence is desirable to know in any event.

5. SHAPE FROM SHADING

For distant light sources and distant viewing, the shading of certain surfaces gives clues as to their three-dimensional shape that can be inferred from a single image.[15] When the light source has a known geometry, the shape of such a reflecting surface can be inferred from its shading in a monochrome image. However, when the light source has an unknown and unconstrained spectral power distribution $S(\lambda,\underline{m})$ as a function of input direction $\underline{m}$, the monochrome image does not provide enough information for a shape-from-shading algorithm. However, Petrov[16] has recently shown a way to obtain shape from shading using a three-band multispectral image. To show how the theory works, we assume a uniformly colored surface with a matte scattering kernel, and substitute Eq. (5) into Eqs. (1) and (2) for $Q_j(\underline{x})$:

$$Q_j(\underline{x}) = \int d\lambda\ q_j(\lambda) \int d^2\underline{m}\ S(\lambda,\underline{m})\ c(\underline{m},\underline{p})\ r(\lambda)\ \underline{m}.\underline{n}(\underline{x}) \ . \tag{11}$$

This expression is equivalent to

$$Q_j(\underline{x}) = \sum_{k=1}^{3} A_{jk}\ n_k(\underline{x}) \ , \tag{12}$$

where

$$A_{jk} = \int d\lambda \int d^2\underline{m}\ q_j(\lambda)\ S(\lambda,\underline{m})\ c(\underline{m},\underline{p})\ r(\lambda)\ m_k \ , \tag{13}$$

and m_k and n_k are respectively the k-th components of the unit vectors $\underline{m}$ and $\underline{n}$.

Because of the distant-lighting assumption, S does not depend on $\underline{x}$; because of the distant-viewing assumption, $\underline{p}$ (and hence c) is independent of $\underline{x}$. Hence the 3-by-3 matrix **A** is constant over all pixels in the image of the uniformly colored surface. For any number of points $\underline{x}_i$ in that image, it is therefore true that

$$\underline{Q}(\underline{x}_i) = \mathbf{A}\ \underline{n}(\underline{x}_i) \ , \tag{14}$$

where the column three-vector $\underline{Q}(\underline{x}_i)$ has components $Q_j(\underline{x}_i)$. Now define 3-by-3 matrix $\mathbf{Q}_{123}$ with column vectors $\underline{Q}(\underline{x}_1)$, $\underline{Q}(\underline{x}_2)$, $\underline{Q}(\underline{x}_3)$; also define 3-by-3 matrix $\mathbf{N}_{123}$ with column vectors $\underline{n}(\underline{x}_1)$, $\underline{n}(\underline{x}_2)$, $\underline{n}(\underline{x}_3)$. Then

$$\det(\mathbf{Q}_{123}) = \det(\mathbf{A})\ \det(\mathbf{N}_{123}). \tag{15a}$$

Similarly, for a point $\underline{x}_i$ not equal to any of the first three $\underline{x}$ points,

$$\det(Q_{i23}) = \det(\mathbf{A}) \det(N_{i23}) \tag{15b}$$

$$\det(Q_{1i3}) = \det(\mathbf{A}) \det(N_{1i3}) \tag{15c}$$

$$\det(Q_{12i}) = \det(\mathbf{A}) \det(N_{12i}). \tag{15d}$$

The ratio of Eqs. (15b) and (15a) yields

$$\det(Q_{i23})/\det(Q_{123}) = \det(N_{i23})/\det(N_{123}) \;. \tag{16}$$

The left-hand side of Eq. (16) is determined by the Q-quantities measured from the image, and the right-hand side is determined only by the surface normals at the points 1, 2, 3, and i. Hence computing the left-hand side of Eq. (16) from the Q-quantities gives a number characterizing the normals to the imaged surface, independent of lighting spectrum, reflectance spectrum, or lighting or viewing geometry. In that sense Eq. (16) represents the extraction of shape from shading information.

By evaluating the ratios of Eqs. (15c) and (15d) to (15a), two more shape invariants are obtained among the points 1, 2, 3, and i. Examining the three determinant-ratio invariants yields the interpretation that they arose from Cramer's rule. That insight leads to the geometric interpretation that the set of three invariants are the components of a vector that represents the surface normal $\underline{n}(\underline{x}_i)$ in the basis spanned by $\underline{n}(\underline{x}_1)$, $\underline{n}(\underline{x}_2)$, and $\underline{n}(\underline{x}_3)$. If somehow the normals at points 1, 2, and 3 are known ahead-of-time, the normals to the surface at all other points are determined. Hence the shape of the surface is completely determined by the shading if three surface normals are known ahead-of-time.

Petrov's result depends on the assumption, embodied in Eq. (14), that an invertible and constant (over the image) linear transformation relates the surface normal at a point to the measured Q-values at the image of the point. Can this assumption be true when the surface scattering kernel departs from the "matte" model in Eq. (5)? It turns out (see Appendix) that the "matte" model can be generalized only to the extent of allowing the quantity c to depend on wavelength as well as on $\underline{m}$ and $\underline{p}$. Hence we define a "generalized matte" surface as having the following scattering kernel:

$$R[\lambda,\underline{m},\underline{n}(\underline{x}),\underline{p}] = c(\lambda,\underline{m},\underline{p})\,(\underline{m}\cdot\underline{n})(\underline{n}\cdot\underline{p}) \;, \tag{17}$$

where reciprocity requires that c be symmetric with respect to interchange of $\underline{m}$ and $\underline{p}$. This model also involves a Lambertian dependence on the surface normal, modulated by other factors that depend on wavelength and on the lighting and viewing directions.

It should be noted that we have been dealing here with a special case of Petrov's theory. Petrov's version of Eq. (14) has $\underline{Q}(\underline{x}_i) = \mathbf{A}\,\underline{n}(\underline{x}_i) = \underline{b}$, where $\underline{b}$ is a constant three-vector. As we have shown here, the matrix $\mathbf{A}$ arises from light sources whose spectral power distributions depend on angular location, but which are guaranteed to be "above the horizon" for all surface elements. Hence the $\underline{m}$ integration is assumed to include all these sources. However, besides these sources, Petrov includes a spatially uniform (perfectly diffuse) light source that produces the constant term $\underline{b}$. This term for this light source does not have the form $\mathbf{A}\,\underline{n}$ because the $\underline{m}$ integration encompasses different parts of the light source for different surface normals $\underline{n}$, and this violates our assumption that the limits of $\underline{m}$ integration are independent of $\underline{n}$. It is easily shown, however, that the scattering kernel of Eq. (17) also preserves Petrov's more general form for $\underline{Q}(\underline{x}_i)$.

Finally, a remark is due about the conditions under which the matrix **A** is nonsingular--a requirement for the computation of any of the ratios of **Q**-determinants in this section. From the definition of **A** in Eq. (13), **A** is singular if $S(\lambda)$ can be expressed as

$$S(\lambda) = s_1(\lambda)\, h_1(\underline{m}) + s_2(\lambda)\, h_2(\underline{m}). \tag{18}$$

Hence Petrov's shape-from-shading algorithm depends on there being more than two light sources such as defined in Nikolaev's theory (i.e., $M > 2$ in Section 3). Since the image segmentation described in Section 3 requires $M < 3$ for a trichromatic visual system[6,7], it follows that--for trichromatic vision--the object-based-segmentation and shape-from-shading theories cannot be applied under the same regime of illuminants. A visual system with more than three kinds of photosensor ($M + 1 > 3$) will support an illuminant variety $M > 2$ for image segmentation, and hence would allow illuminant regimes accessible to the shape-from-shading algorithm. If the number of species of photosensor is greater than three, the shape-from-shading algorithm could operate using only three of the species, or by using all the species of photosensor and performing a pseudoinverse wherever a matrix inverse is required by the three-dimensional version of the model.

6. CONCLUSION

It will be clear by now that the ideal worlds in which object-based image segmentation, color constancy, and shape from shading can be performed are quite different from each other. The color-constancy world has predominantly planar matte surfaces. The image-segmentation world has curved surfaces with more general scattering kernels, but the light sources are restricted in geometry and spectrum. Finally, the shape-from-shading world has curved surfaces and unrestricted (though distant) light sources, but the surfaces must once again be matte.

The coexistence of image segmentation and color constancy in the same world has been demonstrated by D'Zmura and Lennie[17] and by Bajcsy, et al.[12] (see also Ref. 7 for a discussion). The conditions for that merger were (a) a separable illuminant distribution (i.e., S is a product of a wavelength-dependent part and a geometry-dependent part); and (b) a scattering kernel having a separable "body reflection" part and a spectrally nonselective "dielectric highlight" component. Of course, there were also spectral constraints having to do with the expandability of the illuminant and reflectance spectra $s(\lambda)$ and $r(\lambda)$ into basis-function components, and knowledge of these components through "ground-truth" reference colors.

A similar coexistence among the three worlds promises to be more difficult, but may still be possible.

APPENDIX

To show that Petrov's shape-from-shading permits "generalized matte" surfaces, we first simplify Eq. (1) by setting T = 1 and by invoking distant light sources [$S(\lambda,\underline{m},\underline{x}) = S(\lambda,\underline{m})$] and distant viewer [$\underline{p}(\underline{x}) = \underline{p}$]. Then, we substitute Eq. (1) into Eq. (2):

$$Q_j(\underline{x}) = \int d\lambda\ q_j(\lambda) \int d^2\underline{m}\ S(\lambda,\underline{m})\ R[\lambda,\underline{m},\underline{n}(\underline{x}),\underline{p}]/[\underline{n}(\underline{x}).\underline{p}]. \tag{A1}$$

The requirement on $Q_j(\underline{x})$ imposed by Eq. (14) is that

$$Q_j(\underline{x}) = \sum_{k=1}^{3} A_{jk}\ n_k(\underline{x})\ , \tag{A2}$$

where A_{jk} are quantities that do not depend on the image coordinates $\underline{x}$, but may depend on $\underline{p}$ (which is constant over the image). The next step in the proof is to notice that, since Eq. (A2) must hold regardless of the illuminant distribution S, it follows that the integrand in Eq. (A1), divided by $S(\lambda,\underline{m})$, must have the form of Eq. (A2), for any values of λ and $\underline{m}$. Therefore, we fix λ and $\underline{m}$, and write

$$q_j(\lambda)\ R[\lambda,\underline{m},\underline{n}(\underline{x}),\underline{p}]/[\underline{n}(\underline{x}).\underline{p}] = \sum_{k=1}^{3} B_{jk}\ n_k(\underline{x})\ , \tag{A3}$$

where **B** is also a 3-by-3 matrix that does not depend on $\underline{x}$, but now **B** can depend on $\underline{p}$, $\underline{m}$, and λ.

Now, define

$$C_k = B_{jk}/q_j(\lambda).$$

The quantity C_k must be independent of j, but can still depend on $\underline{p}$,$\underline{m}$, and λ. Then

$$R[\lambda,\underline{m},\underline{n}(\underline{x}),\underline{p}]/[\underline{n}(\underline{x}).\underline{p}] = \sum_{k=1}^{3} C_k\ n_k(\underline{x}) = \underline{C}.\underline{n}(\underline{x}). \tag{A4}$$

Now, by reciprocity, $R[\lambda,\underline{m},\underline{n},\underline{p}] = R[\lambda,\underline{p},\underline{n},\underline{m}]$. Hence

$$[\underline{C}(\lambda,\underline{m},\underline{p}).\underline{n}(\underline{x})][\underline{n}(\underline{x}).\underline{p}] = [\underline{C}(\lambda,\underline{p},\underline{m}).\underline{n}(\underline{x})][\underline{n}(\underline{x}).\underline{m}]. \tag{A5}$$

Any function $\underline{C}$ of the form

$$\underline{C}(\lambda,\underline{m},\underline{p}) = c(\lambda,\underline{m},\underline{p})\underline{m} , \tag{A6}$$

where c is symmetric in $\underline{m}$ and $\underline{p}$, will ensure that Eq. (A5) is satisfied. Substituting Eq. (A6) into (A4) gives the "generalized matte" scattering kernel.

To show that Eq. (A6) is the most general form for $\underline{C}(\lambda,\underline{m},\underline{p})$, we first define in Eq. (A5) the parts of $\underline{C}$ that are symmetric ($\underline{C}_S$) and antisymmetric ($\underline{C}_A$) in the quantities $\underline{m}$ and $\underline{p}$:

$$\begin{aligned} \underline{C}_S &= [\underline{C}(\lambda,\underline{m},\underline{p}) + \underline{C}(\lambda,\underline{p},\underline{m})]/2 \\ \underline{C}_A &= [\underline{C}(\lambda,\underline{m},\underline{p}) - \underline{C}(\lambda,\underline{p},\underline{m})]/2 \ . \end{aligned} \tag{A7}$$

Then

$$\underline{C}(\lambda,\underline{n},\underline{p}) = \underline{C}_S + \underline{C}_A \text{ and } \underline{C}(\lambda,\underline{p},\underline{m}) = \underline{C}_S - \underline{C}_A. \tag{A8}$$

Substituting Eq. (A8) into Eq. (A5) and solving for $\underline{C}_A.\underline{n}$ in terms of $\underline{C}_S.\underline{n}$ yields

$$\underline{C}_A.\underline{n} = (\underline{C}_S.\underline{n})[\underline{n}.(\underline{m}-\underline{p})]/[\underline{n}.(\underline{m}+\underline{p})]. \tag{A9}$$

From Eqs. (A8) and (A9),

$$\underline{C}(\lambda,\underline{m},\underline{p}).\underline{n} = 2(\underline{n}.\underline{m})(\underline{C}_S.\underline{n})/[(\underline{m}+\underline{p}).\underline{n}]. \tag{A10}$$

Equation (A10) holds for all $\underline{n}$, and particular holds for $\underline{n}$ being in coordinate direction k:

$$C_k = 2\, m_k\, C_{Sk}/(m_k+p_k). \tag{A11}$$

From Eq. (A11), it is evident that each component of $\underline{C}$ is a symmetric function (call it c_k) of $\underline{m}$ and $\underline{p}$ multiplied by the corresponding component of $\underline{m}$:

$$C_k = c_k\, m_k\ . \tag{A12}$$

It is now possible to show that c_k is independent of k--and hence that Eq. (A12) reduces to (A6)--by substituting (A12) into (A5) for general $\underline{n}$:

$$\sum_{j,k=1}^{3} c_k\, m_k\, p_j\, n_k\, n_j = \sum_{j,k=1}^{3} c_j\, p_j\, m_k\, n_k\, n_j\ . \tag{A13}$$

Equating the coefficients of the quantities $n_k\, n_j$ yields $c_k = c_j$, which completes the proof of necessity of Eq. (A6), and hence of Eq. (17) in the text. I thank Dr. A. P. Stokes of SAIC for helpful discussions in connection with this proof.

7. REFERENCES

1. W. Richards (Ed.), Natural Computation. MIT Press, Cambridge, Massachusetts, 1988, pp. 1-13.

2. G. Wyszecki and W. S. Stiles, Color Science, Wiley, New York, 1967, pp. 387-390.

3. J. W. T. Walsh, Photometry, Dover, New York, 1965, pp. 136-157.

4. E. H. Land and J. J. McCann, "Lightness and Retinex Theory," J. Opt. Soc. Am. 61, 1-11 (1971).

5. M. H. Brill, "A Device Performing Illuminant-Invariant Assessment of Chromatic Relations," J. Theor. Biol. 71, 473-478 (1978).

6. P. P. Nikolaev, "Monocular Color Discrimination of Nonplanar Objects Under Various Illumination Conditions," Biofizika 33, 517-521 (1988) [in Russian].

7. M. H. Brill, "Image Segmentation by Object Color: A Unifying Framework and Connection to Color Constancy," J. Opt. Soc. Am. A 7 (1990), 2040-2047.

8. S. Shafer, "Using Color to Separate Reflection Components," Color Research and Application 10,210-218 (1985).

9. G. Healey, "Using Color for Geometry-Insensitive Segmentation," J. Opt. Soc. Am. A 6, 920-937 (1989).

10. G. Klinker, "A Physical Approach to Color Image Understanding," Ph.D. dissertation, Doc. CMU-CS-88-161 (Carnegie-Mellon University, Pittsburgh, PA, May, 1988).

11. H.-C. Lee, "Illuminant Color from Shading," in Perceiving, Measuring, and Using Color, M. H. Brill, ed., Proc. Soc. Photo-Opt. Instrum. Eng. 1250, 245-255 (1990).

12. R. Bajcsy, S. W. Lee, and A. Leonardis, "Color Image Segmentation and Color Constancy," in Perceiving, Measuring, and Using Color, M. H. Brill, ed., Proc. Soc. Photo-Opt. Instrum. Eng. 1250, 245-255 (1990).

13. M. H. Brill and G. West, "Chromatic Adaptation and Color Constancy: A Possible Dichotomy," Color Research and Application 11, 196-204 (1986).

14. L. T.Maloney and B. A. Wandell, "Color Constancy: A Method for Recovering Surface Spectral Reflectances," J. Opt. Soc. Am. A 3, 29-33 (1986).

15. B. K. P. Horn, "Obtaining Shape from Shading Information," in The Psychology of Computer Vision, P. Winston, ed., (McGraw-Hill, New York, 1975), pp. 115-155.

16. A. P. Petrov, "Color and Grassmann-Cayley Coordinates of Shape," in Human Vision, Visual Processing, and Digital Display II, B. E. Rogowitz, M. H. Brill and J. P. Allebach, eds., Proc. Soc. Photo-Opt. Instrum. Eng. 1453, 342-352 (1991).

17. M. D'Zmura and P. Lennie, "Mechanisms of Color Constancy." J. Opt. Soc. Am. A 3 (1986), 29-33.

Note: The last paragraph of Section 5 was added to this paper after the SPIE meeting at which the paper was delivered, following a conversation with David Forsyth at Oxford University.

This paper was adapted from the following:

Michael H. Brill, "Photometric Models in Multispectral Machine Vision," in Human Vision, Visual Processing, and Digital Display II, B. E. Rogowitz, M. H. Brill and J. P. Allebach, eds., Proc. Soc. Photo-Opt. Instrum. Eng. 1453, 369-380 (1991).

Color Discrimination By Computer

GLENN HEALEY

Abstract—A color space metric that is useful for computer vision is developed. While there is no shortage of proposed methods for quantifying how the human eye perceives color differences, it is difficult to apply these perceptual metrics directly to the color images sensed by a computer vision system. The metric developed can be applied to images sensed using color filters (e.g., R, G, B) as is often done in computer vision. This metric is defined in terms of the spectral characteristics of the filters and camera and accounts for their noise properties. The metric is designed to respond to material changes while remaining insensitive to geometric variation in the scene. The color distance function is derived as an estimate of the distance between normalized spectral power distributions. Components of this distance are weighted to account for sensor noise properties. A color metric has several possible uses in a computer vision system. A color metric is useful for detecting color edges, classifying intensity edges, and estimating color variation within image regions. The usefulness of this color metric is demonstrated by evaluating its performance on real images.

I. Introduction

Until recently most work in computer vision has dealt exclusively with images obtained using a single sensor. It is perhaps for this reason that many of the available algorithms for color images are multidimensional generalizations of algorithms originally intended for intensity images, e.g., [1], [7], [18], [21]. Many of these generalized algorithms have been effectively applied to real images. It is reasonable to expect, however, that better uses for color information can be obtained by first examining the physics that is specific to the formation of color images. In this spirit, computer vision researchers have recently started developing algorithms that are based on the physics of color image formation, e.g., [22]. It is only by examining the relevant physics that it is possible to say whether color images contain useful information that is not present in intensity images and, if so, how this information can be reliably extracted.

Perhaps the most important advantage of having color information in addition to image irradiance information is that the normalized color of a surface of one material is more stable under changes in geometry than the corresponding image irradiance values. It is shown in [8] that if we are not viewing a high-light on an inhomogeneous dielectric, or if an algorithm is used to first remove such highlights [11], then the spectral power distribution $I(x, y, \lambda)$ of the light incident on the imaging sensor can be approximated by

$$I(x, y, \lambda) = L(\lambda) M(g) C(\lambda) \tag{1}$$

where x and y indicate coordinates in the image plane, λ is wavelength, $L(\lambda)$ is the spectral power distribution of the light incident on the imaged surface, and $M(g)$ and $C(\lambda)$ are determined by the optical properties of the imaged surface. The parameter g indicates dependence on the photometric geometry. Define the normalized color at an image point (x, y) by

$$\hat{I}(x,y,\lambda) = \frac{I(x,y,\lambda)}{\int_{\lambda_1}^{\lambda_2} I(x,y,\lambda)\, d\lambda} \tag{2}$$

Manuscript received September 20, 1988; revised March 30, 1989. This work was supported by AFOSR contract F33615-85-C-5106 and KBV Contract AIADS S1093-S-1(Phase II).

The author is with the Robotics Laboratory, Computer Science Dept., Stanford University, Stanford, CA 94305.

IEEE Log Number 8930338.

where $[\lambda_1, \lambda_2]$ is the visible wavelength range. From (1), the normalized color at (x, y) is given by

$$\hat{I}(x, y, \lambda) = \frac{L(\lambda) M(g) C(\lambda)}{\int_{\lambda_1}^{\lambda_2} L(\lambda) M(g) C(\lambda)\, d\lambda} = \frac{L(\lambda) C(\lambda)}{\int_{\lambda_1}^{\lambda_2} L(\lambda) C(\lambda)\, d\lambda} \tag{3}$$

and we see that $\hat{I}(x, y, \lambda)$ does not depend on the geometry g.

In this correspondence, I develop a color metric for computer vision. This metric is intended to quantify the significance of a color change in an image. Some work in computer vision has been done on developing color metrics for the purpose of detecting color edges [16], [17], [19]. None of these proposed metrics, however, consider the underlying properties of the sensors being used. Similarly all of these techniques are sensitive to geometrical variations in the scene. The most significant advancement of this correspondence is the development of a color metric that accounts for the spectral properties of the camera and filters and their noise characteristics and that also is insensitive to geometric variations in the scene unless they are accompanied by material changes.

Based on previous work, it does not seem likely that color edge detection is a particularly useful step to be taken in computer vision. The experiments of Novak and Shafer [17] and Nevatia [16] have demonstrated that a large majority of detected color edges are also detected as irradiance edges. Interestingly human subjects have been unable to bring color edges into focus when the color edge does not coincide with a brightness edge [24].

The color metric of this correspondence is intended to be applied to regions of an image following the detection of image irradiance edges. This is useful for grouping regions corresponding to the same material in the scene and for classifying irradiance edges. For example, given an object of a single material that is illuminated by a single spectral-power distribution, there will be an irradiance edge, but not a normalized color edge across a surface orientation discontinuity. Similarly a specular irradiance edge in an image will indicate the presence of an inhomogeneous dielectric if there is a corresponding normalized color edge [8]. On the other hand a specular irradiance edge without a normalized color edge indicates the presence of a homogeneous material. The color metric is also useful for estimating the color variation within image regions of continuous image irradiance. This variation has a strong relationship to the material composition of an object's surface and may therefore be useful for recognition.

II. Color Sensing

The properties of light impinging on a sensor plane can be characterized by the function $I(x, y, \lambda)$ where I is irradiance, x and y are spatial coordinates in the image plane, and λ is wavelength. Information about the spectral properties of the incideat light can be obtained by using several sensors (e.g., color filters) with differing wavelength responses. For the remainder of this section, I consider a fixed image location (x_0, y_0) and abbreviate $I(x_0, y_0, \lambda)$ by $I(\lambda)$.

It has been standard in computer vision to digitize an image of RGB values using "red," "green," and "blue" filters and to refer to theses RGB values as if they represent something fundamental about the scene. In reality there are an infinite number of combinations of filters and cameras that might be used to obtain RGB values. In general different "red" filters placed in front of different CCD cameras will give us different R values for the same incident light. In this work, I transform the sensor measurements into an approximation to the spectral power distribution of the light entering the camera. This approximation attempts to capture a physical property of the light entering the camera.

Therefore this approximation is somewhat less dependent on the actual sensors used than the sensor measurements themselves.

Following [10] and [23], I approximate $I(\lambda)$ as a linear combination of basis functions. This color recovery method is similar to techniques that express surface reflectance as a linear combination of basis functions. I believe that expressing surface reflectance using such a linear model was first done bySallstrom [20] and subsequently used by many others, e.g., [3], [4], [12]. In this correspondence, I consider only approximations to $I(\lambda)$ and not approximations to surface spectral reflectance.

Consider approximating the function $I(\lambda)$ at a point in an image. At each image point, we measure the outputs s_i of n sensors. Each sensor has a certain spectral response denoted by the function $f_i(\lambda)$. In a typical system, each function $f_i(\lambda)$ will correspond to the product of a color filter transmission function with the spectral response of the camera. Therefore at each image point we have the measured values s_i $(0 \leq i \leq n-1)$ given by

$$s_i = \int_\lambda f_i(\lambda) I(\lambda)\, d\lambda \tag{4}$$

where λ ranges over the entire electromagnetic spectrum. In this work we restrict our interest in the behavior of $I(\lambda)$ to the visible region of the spectrum. Therefore the functions $f_i(\lambda)$ will be nonzero only for values of λ in the visible range. We denote the visible range by $[\lambda_1, \lambda_2]$. Since many cameras respond to light in the near infrared region of the spectrum, it may be necessary to use an infrared (IR) cutoff filter to block light for which $\lambda > \lambda_2$.

Suppose we approximate the function $I(\lambda)$ by a linear combination of m basis functions $b_j(\lambda)$. If we let a_j be the components of $I(\lambda)$ on this basis, then we have

$$I(\lambda) \approx \sum_{0 \leq j \leq m-1} a_j b_j(\lambda). \tag{5}$$

Substituting, (4) becomes

$$s_i = \int_{\lambda_1}^{\lambda_2} f_i(\lambda) \Bigg(\sum_{0 \leq j \leq m-1} a_j b_j(\lambda) \Bigg) d\lambda \tag{6}$$

which may be written

$$s_i = \sum_{0 \leq j \leq m-1} a_j \left(\int_{\lambda_1}^{\lambda_2} f_i(\lambda) b_j(\lambda)\, d\lambda \right). \tag{7}$$

Let

$$k_{ij} = \int_{\lambda_1}^{\lambda_2} f_i(\lambda) b_j(\lambda)\, d\lambda \tag{8}$$

denote the integral in (7). Then k_{ij} is a constant that depends only on the ith filter function and the jth basis function. Let S be the n dimensional vector defined by $S = [s_0, s_1, \cdots, s_{n-1}]^T$. Let K be the $n \times m$ constant matrix defined by $K(i, j) = k_{ij}$. Let A be the m dimensional vector defined by $A = [a_0, a_1, \cdots, a_{m-1}]^T$. Then we have the linear system of n equations

$$S = KA. \tag{9}$$

If we choose our filters $f_i(\lambda)$ and basis functions $b_j(\lambda)$ such that K has maximal rank, then the n sensor outputs $s_0, s_1, \cdots, s_{n-1}$ uniquely determine n components $a_0, a_1, \cdots, a_{n-1}$ of $I(\lambda)$. Therefore by letting $m = n$ in (9) we can recover an approximation to the function $I(\lambda)$ given by

$$\widetilde{I(\lambda)} = B^T(\lambda) A \tag{10a}$$

where

$$A = K^{-1} S \tag{10b}$$

and $B(\lambda)$ is the vector $[b_0(\lambda), b_1(\lambda), \cdots, b_{n-1}(\lambda)]^T$.

In summary, $\widetilde{I(\lambda)}$ is the unique approximation to $I(\lambda)$ that lies in the space spanned by the n basis functions $b_0(\lambda)$, $b_1(\lambda), \cdots, b_{n-1}(\lambda)$ and that also satisfies the n constraints of (4).

There are two sources of error in the approximation $\widetilde{I(\lambda)}$. The first source of error is noisy sensor measurements S. The second source of error is the possibility that $I(\lambda)$ cannot be exactly represented in the space spanned by the basis functions $b_0(\lambda), b_1(\lambda), \cdots, b_{n-1}(\lambda)$. The effects of sensor noise are discussed in Section V. A discussion of approximation error is not central to the development of this work, but is included for completeness in an appendix.

III. A Metric Space for Normalized Color

This section develops a normalization procedure for color. The analysis of this section assumes the recovery of a continuous function $I(\lambda)$ using noiseless sensors. In Section IV, I specialize the analysis to finite dimensional approximations like those recovered using the technique of Section II.

The total power of $I(\lambda)$ is given by the $L^1[\lambda_1, \lambda_2]$ norm

$$\| I(\lambda) \|_1 = \int_{\lambda_1}^{\lambda_2} I(\lambda)\, d\lambda. \tag{11}$$

As in (2), define the normalized color of $I(\lambda)$ by the function $\overline{I(\lambda)}$ having unit total power

$$\overline{I(\lambda)} = \frac{I(\lambda)}{\| I(\lambda) \|_1}. \tag{12}$$

The space of normalized physical colors is the space of all continuous nonnegative functions $\overline{I(\lambda)}$ on $[\lambda_1, \lambda_2]$ having unit total power.

As shown in Section I and analyzed in detail in [8], normalized color is only weakly dependent on reflection geometry. Thus given a technique for removing highlights such as [11], we get the desirable property that for a surface of a single material illuminated by a single spectral-power distribution the normalized color of all image points corresponding to that surface will be identical. In this sense (12) serves to factor out geometric information and preserve information about source color and surface spectral reflectance.

Given any two normalized physical colors $\overline{I_1(\lambda)}$ and $\overline{I_2(\lambda)}$ define the distance from $\overline{I_1(\lambda)}$ to $\overline{I_2(\lambda)}$ by the $L^2[\lambda_1, \lambda_2]$ distance function

$$d\left(\overline{I_1(\lambda)}, \overline{I_2(\lambda)}\right) = \sqrt{\int_{\lambda_1}^{\lambda_2} \left[\overline{I_1(\lambda)} - \overline{I_2(\lambda)}\right]^2 d\lambda}. \tag{13}$$

The $L^2[\lambda_1, \lambda_2]$ distance of (13) is used rather than the distance defined by

$$d_{\max}\left(\overline{I_1(\lambda)}, \overline{I_2(\lambda)}\right) = \max_{\lambda_1 \leq \lambda \leq \lambda_2} \left| \overline{I_1(\lambda)} - \overline{I_2(\lambda)} \right| \tag{14}$$

because inaccuracies over a small range of λ can cause large deviations of $d_{\max}$. On the other hand the distance d of (13) depends on the distance between the functions integrated over the entire spectrum. Therefore the $L^2[\lambda_1, \lambda_2]$ distance will usually give a more reliable characterization of the distance between two measured functions than the distance $d_{\max}$.

Although we have not motivated our decision to compute distances between functions normalized in $L^1[\lambda_1, \lambda_2]$ using the $L^2[\lambda_1, \lambda_2]$ metric, functions treated in this way have several useful properties. These properties are discussed in [9].

IV. A Finite-Dimensional Representation

The functions $I(\lambda)$ in the discussion of Section III were assumed to be arbitrary nonnegative continuous functions on $[\lambda_1, \lambda_2]$, i.e., nonnegative elements of the space $C[\lambda_1, \lambda_2]$. The color recovery method of Section II, however, is only able to recover finite-dimensional approximations to these functions. This section specializes the analysis of Section III to finite-dimensional approximations that can be recovered using the method of Section II.

In general we begin by restricting ourselves to some n dimensional subset S of $C[\lambda_1, \lambda_2]$. If S is specified by a set of basis functions, then the Gram–Schmidt process [5] can be used to construct an orthonormal basis for S. This orthonormal basis may then be taken to be the basis $B(\lambda)$ of Section II.

For the purpose of concreteness, we will take S to be the set of polynomials of degree $n-1$ and the orthonormal basis to be the first n normalized Legendre polynomials. The normalized Legendre polynomials are given by

$$p_i(\lambda) = \sqrt{\frac{2i+1}{2}}\, P_i(\lambda) \tag{15}$$

where $P_i(\lambda)$ is the Legendre polynomial of degree i. The functions $p_i(\lambda)$ are orthonormal in the sense that

$$\int_{\lambda_1}^{\lambda_2} p_i(\lambda)\, p_j(\lambda) = \begin{cases} 1, & \text{if } i = j; \\ 0, & \text{if } i \neq j. \end{cases} \tag{16}$$

We note that the analysis of this section applies more generally to any orthonormal basis for which one of the basis functions is a constant.

A. Computing Normalized Physical Color $\widehat{I(\lambda)}$

Using the normalized Legendre polynomial representation, the technique of Section II recovers an approximation to $I(\lambda)$ of the form

$$\widetilde{I(\lambda)} = \sum_{0 \leqslant i \leqslant n-1} a_i p_i(\lambda). \tag{17}$$

The total power of $I(\lambda)$ is given by

$$\left\|\widetilde{I(\lambda)}\right\|_1 = \int_{\lambda_1}^{\lambda_2} \widetilde{I(\lambda)}\, d\lambda = \int_{\lambda_1}^{\lambda_2} \left[\sum_{0 \leqslant i \leqslant n-1} a_i p_i(\lambda) \right] d\lambda = \sqrt{2}\, a_0 \tag{18}$$

where the last step follows from the orthogonality of the functions $p_i(\lambda)$. Therefore the total power of $I(\lambda)$ may be determined by only considering the first coefficient in the normalized Legendre polynomial representation. From the total power of $\widetilde{I(\lambda)}$, we can determine the normalized physical color $\widehat{I(\lambda)}$ by

$$\widehat{I(\lambda)} = \frac{\widetilde{I(\lambda)}}{\sqrt{2}\, a_0} = \sum_{0 \leqslant i \leqslant n-1} \hat{a}_i p_i(\lambda) \tag{19}$$

where

$$\hat{a}_i = \frac{a_i}{\sqrt{2}\, a_0}. \tag{20}$$

B. Metric Space Properties

For two functions $I_1(\lambda)$ and $I_2(\lambda)$ given by

$$I_1(\lambda) = \sum_{0 \leqslant i \leqslant n-1} r_i p_i(\lambda), \qquad I_2(\lambda) = \sum_{0 \leqslant i \leqslant n-1} s_i p_i(\lambda) \tag{21}$$

the normalized physical colors are

$$\widehat{I_1(\lambda)} = \sum_{0 \leqslant i \leqslant n-1} \hat{r}_i p_i(\lambda), \qquad \widehat{I_2(\lambda)} = \sum_{0 \leqslant i \leqslant n-1} \hat{s}_i p_i(\lambda) \tag{22}$$

where the $\hat{r}_i$ and $\hat{s}_i$ are computed as in (20).

The color space distance between $\widehat{I_1(\lambda)}$ and $\widehat{I_2(\lambda)}$ is given

$$d\left(\widehat{I_1(\lambda)}, \widehat{I_2(\lambda)}\right) = \sqrt{\int_{\lambda_1}^{\lambda_2} \left[\sum_{0 \leqslant i \leqslant n-1} (\hat{r}_i - \hat{s}_i) p_i(\lambda) \right]^2 d\lambda} \tag{23}$$

which simplifies because of orthogonality to

$$d\left(\widehat{I_1(\lambda)}, \widehat{I_2(\lambda)}\right) = \sqrt{\sum_{0 \leqslant i \leqslant n-1} (\hat{r}_i - \hat{s}_i)^2}. \tag{24}$$

C. Color Space as a Subset of R^{n-1}

From (19) the representation for normalized physical colors is

$$\widehat{I(\lambda)} = \sum_{0 \leqslant i \leqslant n-1} \hat{a}_i p_i(\lambda). \tag{25}$$

From (20) we see that all normalized physical colors have $\hat{a}_0 = 1/\sqrt{2}$. We can write

$$\widehat{I(\lambda)} = \frac{1}{2} + \sum_{1 \leqslant i \leqslant n-1} \hat{a}_i p_i(\lambda). \tag{26}$$

We can consider the normalized physical color $\widehat{I(\lambda)}$ to be the point in R^{n-1} with coordinates $\hat{a}_1, \hat{a}_2, \cdots, \hat{a}_{n-1}$. Since we require $\widehat{I(\lambda)} \geqslant 0$ on $\lambda_1 \leqslant \lambda \leqslant \lambda_2$, normalized physical colors form a proper subset of R^{n-1}. To be precise, normalized physical colors are those points of R^{n-1} for which

$$\sum_{1 \leqslant i \leqslant n-1} \hat{a}_i p_i(\lambda) \geqslant -\frac{1}{2} \qquad -1 \leqslant \lambda \leqslant 1. \tag{27}$$

Let C^{n-1} be the set of points in R^{n-1} that are normalized physical colors. Points in C^{n-1} will be referred to as

$$A = (\hat{a}_1, \hat{a}_2, \cdots, \hat{a}_{n-1}) \tag{28}$$

where the coordinates $\hat{a}_1, \hat{a}_2, \cdots, \hat{a}_{n-1}$ have the same significance as in (26).

An important consequence of viewing colors as points in C^{n-1} is that the usual euclidean metric in R^{n-1} is equivalent to the metric we defined for normalized physical colors in Section III. This is seen by examining (24) and recalling that $\hat{r}_0 = \hat{s}_0$. Thus a representation in terms of normalized Legendre polynomials allows intuition about the familiar distance in R^{n-1} to be applied to distance in color space.

V. Modeling Sensor Noise

If we are able to make perfect measurements $s_0, s_1, \cdots, s_{n-1}$, then (10) allows us to compute the approximation to $I(\lambda)$ given by

$$\widetilde{I(\lambda)} = B^T(\lambda) A. \tag{29}$$

In real physical situations each of the measurements $s_0, s_1, \cdots, s_{n-1}$ will have some amount of variability. This section examines how the variance in the measurements S affects the computed approximation of (29).

Let $\bar{A}$ denote $E(A)$ and $\bar{S}$ denote $E(S)$. From (10b), we have

$$\bar{A} = E[K^{-1}S] = K^{-1}\bar{S}. \tag{30}$$

Let Σ_A be the $n \times n$ convariance matrix of A defined by

$$\Sigma_A = E\left[(A - \bar{A})(A - \bar{A})^T\right] \tag{31}$$

$$= E\left[(K^{-1}S - K^{-1}\bar{S})(K^{-1}S - K^{-1}\bar{S})^T\right] \tag{32}$$

$$+ K^{-1}E\left[(S - \bar{S})(S - \bar{S})^T\right](K^{-1})^T \tag{33}$$

which may be written

$$\Sigma_A = K^{-1}\Sigma_S(K^{-1})^T \tag{34}$$

where Σ_S is the $n \times n$ covariance matrix of S.

For most real sensors it is reasonable to assume that A has the multivariate normal density given by

$$p(A) = \frac{1}{(2\pi)^{n/2}\|\Sigma_A\|^{1/2}} e^{-1/2(A-\bar{A})^T\Sigma_A^{-1}(A-\bar{A})}. \quad (35)$$

Contours of constant density are hyperellipsoids satisfying

$$(A-\bar{A})^T\Sigma_A^{-1}(A-\bar{A}) = C \quad (36)$$

where C is a constant. Thus, from Σ_A we can determine the scatter of the data in any direction. It follows that, in general, points in C^{n-1} will tend to exhibit greater dispersion in certain directions than in others. Thus the euclidean metric for C^{n-1} that is appropriate for noiseless measurements (see Section IV) will be inappropriate for real sensor measurements. This is because directions in C^{n-1} along which sensor noise is amplified can dominate the euclidean distance of (24). It is desirable to develop a color metric that can take into account this anisotropic property of C^{n-1} and thereby produce a distance between two normalized physical colors that is relatively independent of scatter due to noise. Define as a color metric the Mahalanobis distance [6] given by

$$d(\widehat{A_1},\widehat{A_2}) = \sqrt{(\widehat{A_1}-\widehat{A_2})^T\Sigma_A^{-1}(\widehat{A_1}-\widehat{A_2})} \quad (37)$$

where $\hat{A}_1$ and $\hat{A}_2$ are normalized colors in C^{n-1}. The metric of (37) has the effect of normalizing by the variance in each direction, thus giving a more representative estimate of the physical distance between A_1 and A_2 than the euclidean distance. We observe that if Σ_A is unitary, then C^{n-1} is isotropic. As expected for this case (37) simplifies to the scaled Euclidean distance.

VI. Experiments

This section gives some experimental results obtained using the metric of (37). The metric has been applied to an image of a color chart [14] illuminated by a lamp intended to simulate daylight. Four color filters (25,47A,57A,96) are used to obtain information over different regions of the visible spectrum. A CCD camera having a linear response and equipped with an infrared blocking filter is used. A set of orthonormal basis functions that span the same space as the sensor functions $f_0(\lambda), f_1(\lambda), f_2(\lambda), f_3(\lambda)$ are used. These basis functions proved to be the most effective in color discrimination experiments. Some theoretical justification for this choice of basis functions is given in the appendix. The matrix Σ_S is estimated from images of uniform regions. The filter transmission functions and camera response function are taken from manufacturer's specifications.

The chart is made up of 24 matte patches characterized by different spectral reflectance functions. Patches 19–24 are different neutral grays; the spectral reflectance functions of these patches differ only by a scalar multiple. From (1), these patches might correspond to a single material illuminated by a single source but viewed under different geometric conditions. We would like a color metric to identify these patches as corresponding to the same material. Ideally the color metric should distinguish the other 18 patches from each other and from the neutral patches.

Average sensor values s_0, s_1, s_2, s_3 for each patch are computed from the image and the distance (37) between each pair of patches is evaluated. Patch 24 (the darkest of the neutral grays) is not considered because the signal-to-noise ratio is too low for the metric to be reliable. Table I lists distances between those pairs of patches for which distance is smallest. All pairs of patches not listed have distances exceeding 5.0×10^{-2}. It is inappropriate to compare the metric of this correspondence with any of the previously mentioned metrics because each of the other techniques respond to intensity variation and therefore do not interpret the neutral patches as instances of a single material.

TABLE I
EXPERIMENTAL RESULTS

Patches	$d(\times10^{-2})$
21–23	0.05
22–23	0.06
21–22	0.09
19–20	0.13
20–21	0.33
19–21	0.27
19–23	0.47
19–22	0.58
20–23	0.64
20–22	0.76
1–2	1.30
9–15	2.82
8–13	3.19

Based on the properties of the Mahalanobis distance, any pair of patches separated by a distance less than $D=1.5\times10^{-2}$ is taken to correspond to patches of the same material. Any pair of patches separated by a distance of D or more is taken to correspond to patches of different materials. Thus the metric assigns all pairs of neutral patches (19–23) to the same material class and also assigns patches 1 (dark skin) and 2 (light skin) to the same material class. Assigning patches 1 and 2 to the same class is not unreasonable since the corresponding spectral reflectance functions differ by only a scalar multiple if one ignores a small wiggle in the spectral reflectance of patch 2. Such small differences are difficult to distinguish from noise when using broad-band color filters like those used in these experiments.

From Table I, we see that the color metric is able to assign to different material classes patch 9 (moderate red) and patch 15 (red) and also patch 8 (purplish blue) and patch 13 (blue). All other pairs of patches are easily distinguished by the color distance function.

Appendix

Given the recovery technique described in Section II, I examine how the choice of the sensors $f_i(\lambda)$ and the basis functions $b_i(\lambda)$ can affect the quality of the recovered approximation to $I(\lambda)$. One result is that given an orthonormal system of n basis functions, it is possible to choose n sensors such that the recovered approximation $\widehat{I(\lambda)}$ is the best least squares approximation to $I(\lambda)$ in the space spanned by the n basis functions. By itself this result is not of great use in practice since given an arbitrary orthonormal system of basis functions it is unlikely that the theoretically optimal sensors will be among those that are readily available. This result however does lead to some insight into the problem of selecting sensors and basis functions in practical situations.

Let $\phi_0(\lambda), \phi_1(\lambda), \cdots, \phi_{n-1}(\lambda)$ be an orthonormal set of basis functions in $L^2[\lambda_1, \lambda_2]$. Then given n sensor measurements S, the procedure of Section II allows us to recover an approximation $\widehat{I(\lambda)}$ of the form

$$\widehat{I(\lambda)} = \sum_{0\le i\le n-1} d_i\phi_i(\lambda). \quad (38)$$

The best approximation $\widehat{I(\lambda)}$ in the least-squares sense is the choice of the d_i that minimizes

$$M = \int_{\lambda_1}^{\lambda_2}\Big[I(\lambda) - \sum_{0\le i\le n-1} d_i\lambda^i\Big]^2 d\lambda. \quad (39)$$

Equation (4) may be written

$$s_i = \langle f_i, I\rangle \quad (40)$$

where $\langle \cdot, \cdot \rangle$ denotes the $L^2[\lambda_1, \lambda_2]$ inner product. Thus if we define

$$f_i(\lambda) = \phi_i(\lambda) \qquad i = 0, 1, \cdots, n-1 \tag{41}$$

then we will have

$$s_i = \langle \phi_i, I \rangle. \tag{42}$$

Since the $\phi_i(\lambda)$ are orthonormal functions, K will be the identity matrix. From (10), the technique of Section II will recover the approximation $\widetilde{I(\lambda)}^*$ given by

$$\widetilde{I(\lambda)}^* = \sum_{0 \leq i \leq n-1} \langle \phi_i, I \rangle \phi_i(\lambda). \tag{43}$$

It is easy to show [5] that the approximation $\widetilde{I(\lambda)}^*$ given by (43) minimizes M and that the error is given by

$$\left\| I(\lambda) - \widetilde{I(\lambda)}^* \right\|^2 = \| I(\lambda) \|^2 - \left\| \widetilde{I(\lambda)}^* \right\|^2. \tag{44}$$

Therefore it is possible to choose sensors $f_i(\lambda)$ such that the procedure of Section V computes the best least-squares approximation to an arbitrary function $I(\lambda)$.

What is perhaps more interesting in practice is that the previous analysis implies that given a set of n linearly independent sensor functions $f_0(\lambda), f_1(\lambda), \cdots, f_{n-1}(\lambda)$, it is possible to find basis functions $b_0(\lambda), b_1(\lambda), \cdots, b_{n-1}(\lambda)$ such that the recovered approximation $\widetilde{I(\lambda)}$ is the best least-squares approximation to $I(\lambda)$ in the space spanned by the n basis functions. These basis functions will be defined such that they span the same space as $f_0(\lambda), f_1(\lambda), \cdots, f_{n-1}(\lambda)$. Selecting basis functions in this way has the disadvantage that these basis functions may not accurately capture the properties of typical $I(\lambda)$ functions. On the other hand in a situation where a set of basis functions $b_0(\lambda), b_1(\lambda), \cdots, b_{n-1}(\lambda)$ is known to provide a good fit to a certain class of $I(\lambda)$ functions that are of interest, it may be advantageous to search the space of available sensors for an n dimensional set that as nearly as possible spans the same space as $b_0(\lambda), b_1(\lambda), \cdots, b_{n-1}(\lambda)$. Selecting sensors in this way can help to minimize the effects of approximation error on the recovered $\widetilde{I(\lambda)}$ functions.

It is natural to ask how large n should be in order that the function $\widetilde{I(\lambda)}$ is a good approximation to $I(\lambda)$. This depends of course on the properties of the function $I(\lambda)$ and how well these properties are captured by the basis functions $B(\lambda)$. Maloney [13] has done a thorough analysis of reflectance functions $R(\lambda)$ for real materials by considering both empirical data and the fundamental physical processes that determine the properties of $R(\lambda)$. He has concluded that at least five basis functions are required to accurately characterize $R(\lambda)$. Unfortunately no corresponding analysis has been done for illuminants.

Acknowledgment

I would like to thank Tom Binford, W. E. Blanz, and Brian Wandell for many helpful suggestions and comments on a previous draft of this correspondence.

References

[1] M. Ali, W. Martin, and J. Aggarwal, "Color-based computer analysis of aerial photographs," *Computer Graphics and Image Processing*, vol. 9, 282–293, 1979.

[2] T. O. Binford, T. Levitt, and W. Mann, "Bayesian inference in model-based machine vision," *Proc. Workshop on Uncertainty in Artificial Intell.*, 1987.

[3] M. Brill, "A device performing illuminant-invariant assessment of chromatic relations," *J. Theoret. Biol.*, vol. 71, pp. 473–478, 1978.

[4] G. Buchsbaum, "A spatial processor model for object colour perception," *J. Franklin Institute*, vol. 310, pp. 1–26, 1980.

[5] R. Courant and D. Hilbert, *Methods of Mathematical Physics*, vol. 1, New York: Wiley, 1953.

[6] R. Duda and P. Hart, *Pattern Classification and Scene Analysis*. New York: Wiley, 1973.

[7] R. Haralick and G. Kelly, "Pattern recognition with measurement space and spatial clustering for multiple images," *Proc. IEEE*, vol. 57, Apr., 1969, pp. 654–665.

[8] G. Healey, "Using color for geometry insensitive segmentation," *J. Opt. Soc. Amer. A*, pp. 920–937, June 1989.

[9] G. Healey and T. O. Binford, "The role and use of color in a general vision system," *Proc. ARPA Image Understanding Workshop*, Univ. of Southern California, Los Angeles, Feb. 1987, pp. 599–613.

[10] B. K. P. Horn, "Exact reproduction of colored images," *Comput. Vision Graphics and Image Processing*, vol. 26, pp. 135–167, 1984.

[11] G. Klinker, S. Shafer, and T. Kanade, "Using a color reflectionmodel to separate highlights from object color," *Proc. First Int. Conf. Computer Vision*, London, England, June 1987, pp. 145–150.

[12] L. Maloney and B. Wandell, "Color constancy: A method for recovering surface spectral reflectance," *J. Opt. Soc. Amer. A*, vol. 3, pp. 1673–1683, Oct. 1986.

[13] L. Maloney, "Evaluation of linear models of surface spectral reflectance with small numbers of parameters," *J. Opt. Soc. Amer. A*, vol. 3, pp. 1673–1683, Oct. 1986.

[14] C. McCamy, H. Marcus, and J. Davidson, "A color-rendition chart," *J. Appl. Photographic Engineering*, vol. 2, no. 3, Summer 1976.

[15] V. Nalwa, "Edge-detector resolution improvement by image interpolation," *IEEE Trans. Pattern Anal. Machine Intell.*, vol. 9, no. 3, May 1987.

[16] R. Nevatia, "A color edge detector and its use in scene segmentation," *IEEE Trans. Syst. Man Cybern.*, vol. SMC-7, no. 11, pp. 820–826, 1977.

[17] C. Novak and S. Shafer, "Color edge detection" in *Image Understanding Research at CMU*, Takeo Kanade, in *Proceedings of ARPA Image Understanding Workshop*, Univ. of Southern California, Los Angeles, pp. 35–37, 1987.

[18] R. Ohlander, K. Price, and D. R. Reddy, "Picture segmentation using a recursive region splitting method," *Comput. Graphics and Image Proces.*, vol. 8, pp. 313–333, 1978.

[19] G. S. Robinson, "Color edge detection," *Optical Engineering*, vol. 16, no. 5, pp. 479–484, Sept.-Oct. 1977.

[20] P. Sallstrom, "Colour and physics: Some remarks concerning the physical aspects of human colour vision," Univ. of Stockholm: Institute of Physics Report 73-09, 1973.

[21] A. Sarabi and J. Aggarwal, "Segmentation of chromatic images," *Pattern Recognition*, vol. 13, no. 6, pp. 417–427, 1981.

[22] S. Shafer, "Using color to separate reflection components," Univ. of Rochester, Rochester, NY, vol. TR 136, Apr. 1984.

[23] B. Wandell, "The synthesis and analysis of color images," *IEEE Trans. Pattern Anal. Machine Intell.*, vol. PAMI-9, no. 1, Jan. 1987.

[24] J. Wolfe, "Hidden visual processes," *Scientific American*, pp. 94–103, Feb. 1983.

A Physical Approach to Color Image Understanding

GUDRUN J. KLINKER
Cambridge Research Lab, Digital Equipment Corporation, One Kendall Square, Bldg. 700, Cambridge, MA 02139
STEVEN A. SHAFER, TAKEO KANADE
Computer Science Department, Carnegie Mellon University, Pittsburgh, PA 15213

Abstract
In this paper, we present an approach to color image understanding that can be used to segment and analyze surfaces with color variations due to highlights and shading. The work is based on a theory—the Dichromatic Reflection Model—which describes the color of the reflected light as a mixture of light from surface reflection (highlights) and body reflection (object color). In the past, we have shown how the dichromatic theory can be used to separate a color image into two intrinsic reflection images: an image of just the highlights, and the original image with the highlights removed. At that time, the algorithm could only be applied to hand-segmented images. This paper shows how the same reflection model can be used to include color image segmentation into the image analysis. The result is a color image understanding system, capable of generating physical descriptions of the reflection processes occurring in the scene. Such descriptions include the intrinsic reflection images, an image segmentation, and symbolic information about the object and highlight colors. This line of research can lead to physics-based image understanding methods that are both more reliable and more useful than traditional methods.

1 Introduction

When humans are asked to describe a picture, they generally give a list of objects and their relative positions in the scene. A closer look at the image reveals that they have omitted a lot of detail in the description. Objects cast shadows, possibly upon other objects. The brightness on the objects varies, appearing much darker at object parts that are tilted away from the light than where surfaces face the light directly. Moreover, some objects have highlights; we may even see a mirror image of one object on another object. These optical effects are caused by various physical processes by which light interacts with matter. If a vision system is expected to generate descriptions similar in quality to the ones given by humans, the system will have to discount the influence of these physical processes.

Physical processes in the scene have not been a strong point of interest in the traditional line of computer vision research. It has been common practice to divide the image understanding problem into two phases, a low-level segmentation or feature extraction phase and a higher-level reasoning phase in which the image features are related to object features described in object models of the scene [9, 17]. Within this line of research, image segmentation has been considered to be a statistical image processing problem where the major concern is to determine statistically significant changes of pixel values under the presence of noise—under the implicit assumption that such significant changes generally correspond to object boundaries in the scene. However, the generated edge or region images outlined not only material boundaries, but also shadows, highlights, and object edges along which the surface orientation changes abruptly. The segmentation was then passed to a higher-level phase which tried to combine regions across highlights or shadows either by matching image features with objects models [9] or by determining the physical cause of edges or regions [6, 28].

Recently, work in image understanding has started to use intrinsic models of physical processes in the scene to analyze intensity or color variations in the image [2, 6, 11, 14, 16, 18, 19, 28, 32, 35]. But such analyses have not been applied to image segmentation. This paper presents an approach to color image understanding that uses an intrinsic reflection model, called the Dichromatic Reflection Model [30], to generate an image segmentation, along with descriptions of object and highlight colors and intrinsic reflection images, showing how shading and highlight reflection vary in the image. The color image analysis system alternates between generating hypotheses about the scene from the image data and verifying whether the hypotheses fit the image. The hypotheses relate object color, shading, highlights, and camera limitations to the color variation in local image areas. They are used to segment the images and to separate them into two intrinsic images, one showing the scene without highlights, and the other one showing only the highlights. In this way, the algorithm is driven by a physical model of light reflection to incrementally identify local and global properties of the scene and to use them in interpreting and segmenting pixels in the images. The system adapts the image interpretation process to local scene characteristics and reacts differently to color and intensity changes at different places in the image. This line of research is aimed at developing physics-based low-level image analysis methods that are both more reliable and more useful than traditional approaches.

2 The Dichromatic Reflection Model

This section provides a short overview of Shafer's Dichromatic Reflection Model [30]. A more detailed description and discussion can be found in [20, 22, 30].

The Dichromatic Reflection Model describes the light, $L(\lambda, i, e, g)$, which is reflected from a point on a dielectric, nonuniform material as a mixture of the light $L_s(\lambda, i, e, g)$ reflected at the material surface and the light $L_b(\lambda, i, e, g)$ reflected from the material body (see figure 1). The parameters i, e, and g describe the angles of the incident and emitted light and the phase angle; λ is the wavelength parameter. L_s is called the *surface reflection component*. It generally has approximately the same spectral power distribution as the illumination and appears as a highlight or as gloss on the object. L_b is called the *body reflection component*. It provides the characteristic object color and exhibits the properties of object shading.

$$L(\lambda, i, e, g) = L_s(\lambda, i, e, g) + L_b(\lambda, i, e, g) \quad (1)$$

The model separates the spectral reflection properties of L_s and L_b from their geometric reflection properties, modeling them as products of spectral power distributions, $c_s(\lambda)$ or $c_b(\lambda)$, and geometric scale factors, $m_s(i, e, g)$ or $m_b(i, e, g)$, which describe the intensity of the reflected light. Substituting these terms into equation (1), we obtain the Dichromatic Reflection Model equation:

$$L(\lambda, i, e, g) = m_s(i, e, g)c_s(\lambda) + m_b(i, e, g)c_b(\lambda) \quad (2)$$

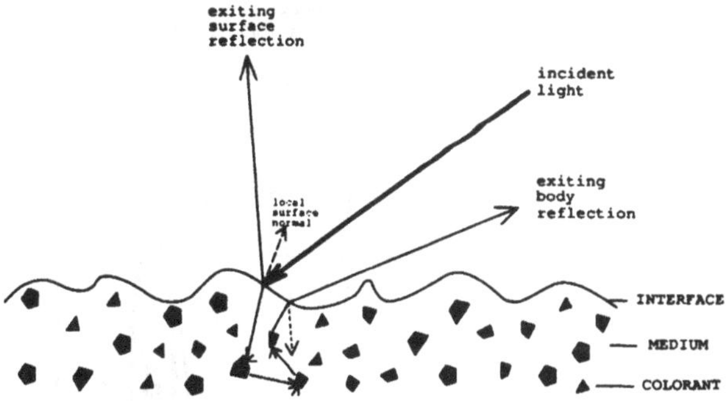

Fig. 1. Light reflection of dielectric materials.

The model thus describes the light that is reflected from an object point as a mixture of two distinct spectral power distributions, $c_s(\lambda)$ and $c_b(\lambda)$, each of which is scaled according to the geometric reflection properties of surface and body reflection. In the infinite-dimensional vector space of spectral power distributions (each wavelength defines an independent dimension in this vector space [15, 29]), the reflected light can be described as a linear combination of the two vectors $c_s(\lambda)$ and $c_b(\lambda)$.

3 Object Shape and Spectral Variation

The Dichromatic Reflection Model describes the spectral properties separately for every single pixel on an object. In itself, this description is not yet very helpful in computer vision since it describes the light at each object point by a set of four (so far unknown) factors. It does not provide a mechanism to uniquely determine these factors from the reflected light beam. This section explores the observation that the light reflected from all points on an object uses the same two spectral vectors and that these two factors are thus constant over an object. We will now show how object shape determines spectral variation on the object and how this spectral variation is related to the body and surface reflection vectors.

3.1 Basic Principles

An investigation of the geometrical properties of surface and body reflection reveals that the light mixtures form a dense spectral cluster in the dichromatic plane. The shape of this cluster is closely related to the shape of the object. Figure 2 shows a sketch of a shiny cylinder. The left part of the figure displays the

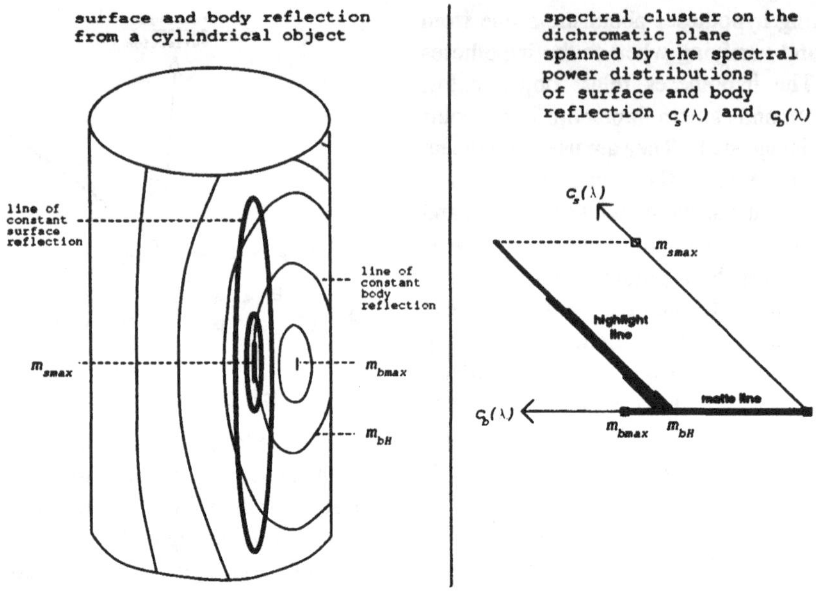

Fig. 2. The shape of the spectral cluster for a cylindrical object.

magnitudes of the body and surface reflection components. The lighter curves show the loci of constant body reflection, assuming perspective viewing and illumination geometry and Lambertian body reflection. The darker curves are the loci of constant surface reflection. Since $m_s(i, e, g)$ decreases sharply around the object point with maximal surface reflection, m_{smax} these curves are shown only in a small object area. We call the points in this area *highlight points*. The remaining object points are *matte points*. The right part of the figure shows the corresponding spectral histogram in the dichromatic plane. As we will describe below, the object points form two linear clusters in the histogram.

Light reflection at **matte points** is primarily determined by the body reflection process. In principle, matte points may contain some surface reflection, as the result of light diffusion at a rough material surface. However, the following analysis assumes that this component of surface reflection can be neglected. The observed light at matte points then depends mainly on $c_b(\lambda)$, scaled by $m_b(i, e, g)$ according to the geometrical relationship between the local surface normal of the object and the viewing and illumination directions. Consequently, the matte points form a *matte line* in the dichromatic plane in the direction of the body reflection vector, $c_b(\lambda)$, as shown in the right part of figure 2.

Highlight points exhibit both body reflection and surface reflection. However, since $m_s(i, e, g)$ is much more sensitive to a small change in the photometric angles than is $m_b(i, e, g)$, the body reflection component is generally approximately constant in a highlight area, as displayed by the curve with label m_{bH} in figure 2. Accordingly, the second term of the Dichromatic Reflection Model equation (2) has a constant value, $m_{bH}c_b(\lambda)$, and all spectral variation within the highlight comes from varying amounts of $m_s(i, e, g)$. The highlight points thus form a *highlight line* in the dichromatic plane in the direction of the surface reflection vector, $c_s(\lambda)$. The line departs from the matte line at position $m_{bH}c_b(\lambda)$, as shown in figure 2. More precisely, the highlight cluster looks like a slim, skewed wedge because of the small variation of the body reflection component over the highlight.

The combined spectral cluster of matte and highlight points looks like a skewed T. The skewing angle of the T depends on the spectral difference between the body and surface reflection vectors while the position of the highlight line depends on the illumination geometry, as described below.

3.2 Relationship Between Illumination Geometry and Spectral Histogram Shape

There exists a close relationship between the illumination geometry and the amounts of body and surface reflection on an object depend. This relationship influences the shape of the clusters in the spectral histogram. It describes constraints that an image analysis program can use when it analyzes spectral variation. We will now derive such a constraint, the *50% heuristic*, for spherical objects. The heuristic states that, under reasonable assumptions, the highlight cluster starts in the upper 50% of the matte cluster. It will be used in section 5 for image segmentation to distinguish spectral changes between matte and highlight points on one object from spectral changes between neighboring objects.

The amount of body reflection under the highlight depends on the phase angle g between the illumination and viewing direction: If g is very small, (i.e., the camera is very close to the light source), the incidence direction of the light at the highlight is close to the surface normal and the underlying amount of body reflection is very high. The highlight line then starts near the tip of the matte line, and the skewed T becomes a skewed L or a "dog-leg" [6, 7]. When the phase angle g increases, the highlight moves away from the area with maximal body reflection, and the amount of body reflection under the highlight decreases. The highlight line moves accordingly away from the tip of the matte line.

The following analysis investigates this relationship more precisely for the case of a spherical object with radius r, viewed and illuminated under perspective projection (see figure 3). The analysis assumes that the sphere is illuminated by a point light source at some distance d from the object center. It further assumes that the camera is positioned at the same distance d from the object center and that an arbitrary phase angle g exists between the viewing and the illumination direction. Although these assumptions on the illumination geometry and the object shape are limiting, the following analysis provides general insight into the relationship between object shape and the shapes of spectral clusters.

Distance d determines the size of the cone that illuminates the sphere (described by γ), as well as the range of phase angles $[-g_{max}, g_{max}]$ under which the camera sees a part of the illuminated area:

$$\gamma = \arc\tan \frac{r}{\sqrt{d^2 - r^2}} \tag{3}$$

$$g_{max} = 180° - 2\gamma = 2 \arc\tan \left(\frac{\sqrt{d^2 - r^2}}{r}\right) \tag{4}$$

Figure 4 describes how the highlight and its underlying body reflection component m_{bH} are related to phase angle g. Since camera and light source are positioned at the same distance d from the object center, the maximal surface reflection component occurs where the surface normal, $\mathbf{n_H}$, bisects the phase angle. Under perspective projection, the illumination angle i_H at the highlight depends on the position of the highlight which, in turn, depends on g. Using the law of sines and the law of cosines, i_H can be described as

$$i_H = \arc\sin \frac{d \sin (g/2)}{\sqrt{d^2 + r^2 - 2dr \cos (g/2)}} \tag{5}$$

Assuming Lambertian body reflection, $m_b(i, e, g) = \cos i$, the underlying body reflection component, m_{bH}, at the highlight is then given by

$$m_{bH} = \cos(i_H) = \sqrt{1 - \frac{d^2 \sin^2(g/2)}{d^2 + r^2 - 2dr\cos(g/2)}} \tag{6}$$

According to equations (5) and (6), m_{bH} approaches 1 when g goes to 0°, confirming our introductory informal statement that the spectral cluster looks like a skewed L when camera and light source are close. When g approaches g_{max}, on the other hand, m_{bH} decreases to 0. When relating m_{bH} to the length of the matte color cluster, we need to take into account that the point with globally maximal body reflection, m_{bMax}, is not always visible from the camera. If it is occluded, the matte line in the spectral histogram does not extend entirely to $m_{bMax}c_b(\lambda)$ but only to the point representing the brightest visible point, $m_{bMaxVisible}$ $c_b(\lambda)$. Figure 5 displays the illumination geometry for the brightest matte point that is visible from the camera. The surface normal, $\mathbf{n_{MaxVisible}}$, at this object point is determined by the angle α, which is given as

$$\alpha = \max (g - (90° - \gamma), 0) = \begin{cases} g - (g_{max}/2) & \text{if } g \geq g_{max}/2 \\ 0 & \text{otherwise} \end{cases} \tag{7}$$

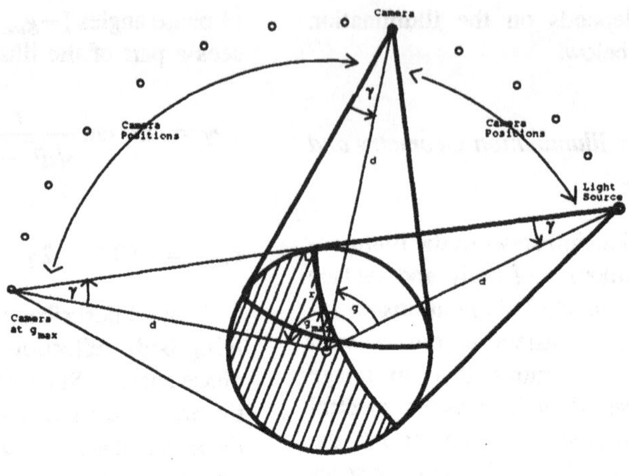

Fig. 3. Assumed illumination geometry.

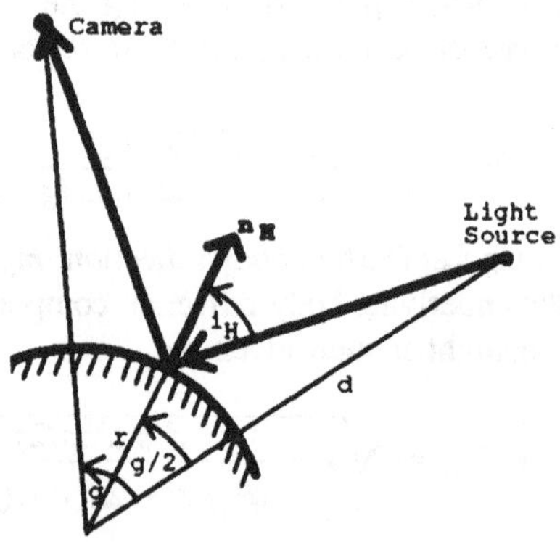

Fig. 4. Photometric angles at highlight.

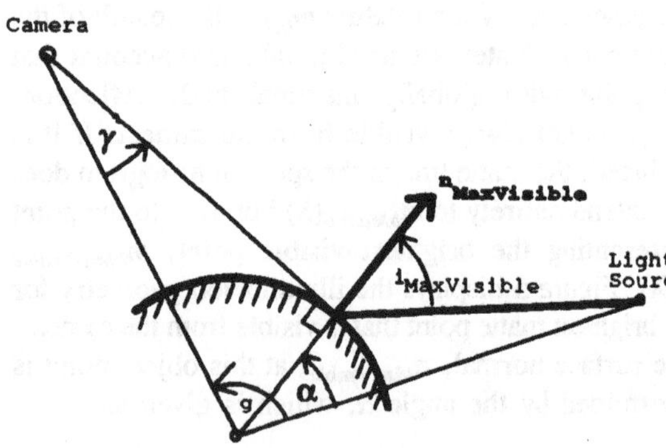

Fig. 5. Brightest matte point that is visible from the camera.

The angle between the illumination vector and the surface normal at the local maximum, as well as the amount of body reflection at that point are given by equations (8) and (9):

$$i_{MaxVisible} = \text{arc sin} \frac{d \sin \alpha}{\sqrt{d^2 + r^2 - 2dr \cos \alpha}} \tag{8}$$

$$m_{bMaxVisible} = \cos (i_{MaxVisible}) = \sqrt{1 - \frac{d^2 \sin^2 \alpha}{d^2 + r^2 - 2dr \cos \alpha}} \tag{9}$$

The starting point of the highlight line on the matte line can now be described relative to the length of the matte line. The ratio q of body reflection at the highlight in relation to the maximally visible amount of body reflection is expressed as

$$q = \frac{m_{bH}}{m_{bMaxVisible}} \tag{10}$$

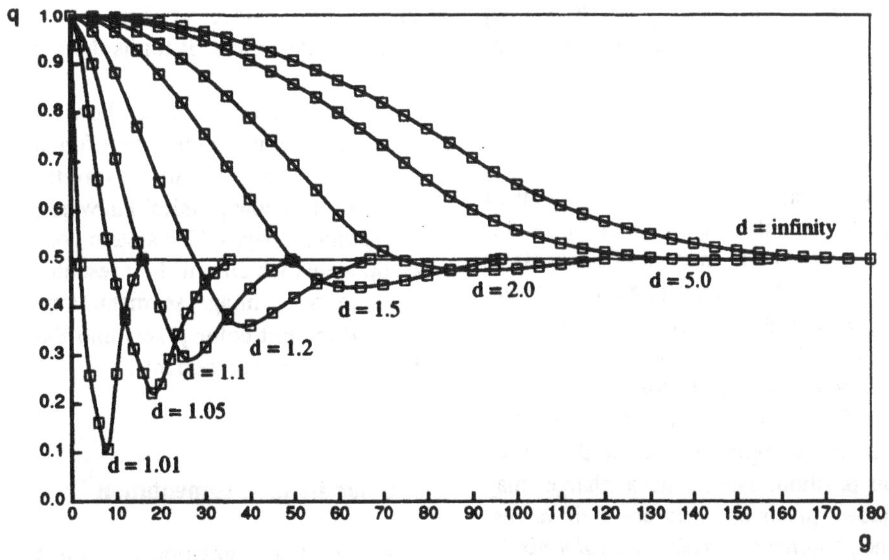

Fig. 6. Starting point q of the highlight line as a function of phase angle g and distance d.

Figure 6 shows how q varies as a function of g and d. Each curve in the figure describes for a different fixed d how q depends on g. In all curves, q approaches 0.5 as g approaches g_{max}. Note, however, that q is not defined for $g = g_{max}$ where $i_{bMaxVisible} = i_{bH} = 90°$ and thus $m_{bMaxVisible} = m_{bH} = 0$. Distance d is varied between the curves. For small values of d, the curves exhibit a minimum in the middle of the curve. With increasing d, this minimum becomes less pronounced and q becomes a monotonically decreasing function with an inflection point. In the extreme, when $d = \infty$, the curve for q assumes the shape of a cosine function for phase angles g under which the globally brightest point is visible ($g \leq 90°$), and the shape of a cosecant function, if the brightest point is invisible. A detailed derivation of the formula is given in [20].

$$q_\infty(g) = \begin{cases} \cos(g/2) & \text{if } g \leq 90° \\ 1/(2\sin(g/2)) & \text{if } g \geq 90° \end{cases} \tag{11}$$

The figure indicates that if d is a moderate or large multiple of r ($d \geq 5r$), q does not drop below 0.5. The body reflection component under the highlight then is always at least half as large as the body reflection component at the brightest visible matte point. Under such illumination geometries, the highlight line always starts in the upper 50% of the matte line. Since laboratory set-ups for scenes with several objects in the field of view generally use camera and illumination distances that are larger than the fivefold object size, we use the above finding as the *50% heuristic* in the segmentation algorithm in section 5.

4 A Camera Model

To account for camera limitations in real images, we combine the Dichromatic Reflection Model with a camera model. This section describes briefly how some characteristics of CCD-cameras influence the pixel values in real images. A more detailed description, including color pictures with color clusters from real images, can be found in [20, 22].

- *Spectral Integration:* Instead of using the infinite-dimensional vector space of spectral color distributions, research in computer color vision generally analyzes a three-dimensional *color-space*. This color space comes from color images obtained by taking three pictures of the scene, through a red, a green, and a blue color filter. The process transforming the spectrum of an incoming light beam into a color triple $\mathbf{C} = [R, G, B]$ is called *spectral integration*. It is a linear transformation from the infinite-dimensional vector space to the three-dimensional color space [8, 29]. The Dichromatic Reflection Model is then describable as a function of two three-dimensional vectors, $\mathbf{C}_s = [R_s, G_s, B_s]$ and $\mathbf{C}_b = [R_b, G_b, B_b]$ which span a dichromatic plane in the three-dimensional color space:

$$\mathbf{C} = m_s\mathbf{C}_s + m_b\mathbf{C}_b \tag{12}$$

- *Color Clipping:* Real cameras have only a limited dynamic range to sense the brightness of the incoming light. This restricts our analysis of light reflection to a *color cube.* If the incoming light is too bright at some pixel position, the camera cannot sense and represent it adequately and the light measurement is *clipped* in one or more color bands.
- *Blooming:* In CCD-cameras, too much incident light at a pixel may completely saturate the sensor element at this position. The electrical charges may spread to adjacent pixels and increase their values. We call such neighboring pixels *bloomed color pixels.*
- *Color Balancing:* CCD-cameras are generally much less sensitive to blue light than to red light. To provide an equal scaling on the three color axes in the color space, we rescale the pixel data separately in the color bands by changing the aperture. We refer to this procedure as *aperture color balancing.* We also use a total IR suppressor in front of our camera to eliminate the CCD-camera's sensitivity to infrared light.
- *Gamma Correction:* The color pixels depend on the camera response to incident light flux. Due to *gamma-correction*, the camera output is generally related by an inverse power law to the incident flux. It introduces curvature into the color space, distorting the linear properties of the Dichromatic Reflection Model. We linearize the color data, reversing the gamma-correction process.
- *Chromatic Aberration:* Since the refraction index of optical glass is a function of wavelength, picture taking suffers from *chromatic aberration:* The focal length, f, of the camera increases with wavelength and, for a fixed focal length, only one color band of an image can be perfectly in focus. The other two image bands show slightly magnified or shrunk images of the scene. Chromatic aberration has a strong influence on color changes in the outer image areas. Since the color bands do not increase or decrease simultaneously, pixels in the rising and falling parts of the profiles generally exhibit color changes that have little relationship to the physical processes in the scene. This effect depends on the quality of the camera lens. It appears quite strongly in some of our images [20]. Since we have not yet been able to eliminate this effect, it limits the performance of our algorithm.

So far, we have discussed how optical and camera properties influence the image function. We will now present an image understanding system that demonstrates how physical knowledge can be used for interpreting images. The system consists of two major components which will be presented in the next sections: a color image segmentation algorithm and a method to split color pixels into their body and surface reflection components.

5 Color Image Segmentation

The goal of segmentation is to identify objects in an image, as delimited by material boundaries. Because most current color segmentation methods are based on a very simple interpretation of color changes in an image, they generally segment images not only along material boundaries but also along other lines exhibiting color or intensity variations, such as highlight and shadow boundaries, or internal object edges with significant shading changes. The Dichromatic Reflection Model provides a more sophisticated interpretation scheme relating the physics of light reflection to color changes in the image and in the color space. We use the Dichromatic Reflection Model in a segmentation algorithm to distinguish color changes at material boundaries from color changes due to shading changes or highlights.

5.1 Using the Dichromatic Reflection Model in Color Image Understanding

The Dichromatic Reflection Model and the sensor model describe processes that occur in the scene and in the camera. The previous sections have projected the influence of these processes from scene properties onto image properties. We will now invert this line of reasoning and design an algorithm that concludes from the color variations in the image which processes have taken part in the image formation process. Unfortunately, the interpretation process is locally ambiguous because the influence of any optical process causes local color variation. The algorithm needs to accumulate local color variations over extended image areas to determine distinguishing characteristics of the processes, such as the T-shape of a color cluster. But how can the optimal extent of an area be determined? There seems to be a circular problem of, on one hand, needing a prior segmentation to relate color variation to physical processes and, on the other hand, needing an understanding of the physical processes to provide a good segmentation.

Conceptually, there are two different ways to approach this problem. The algorithm may start with very large image areas and subsequently shrink or split them until the regions correspond to objects in the scene. Alternatively, it may start with small local areas and then merge or grow them. Within the framework of the Dichromatic Reflection Model, the first method identifies object areas by projecting the entire image into the color space and then distinguishing between several skewed T's in the color space. It may encounter problems when several objects with very similar colors exist in the scene, such that color clusters from different objects overlap. This problem is illustrated in color figure 2. The figure displays the color histogram of the image in color figure 1, showing a scene of eight plastic objects under white light. The various clusters overlap significantly. The second method starts out with small image areas and merges or grows them, including neighboring pixels or areas that fit the same skewed T. Since local color variation on flat or dark objects may be similar in magnitude to camera noise, the local approach has to know how to distinguish camera noise from systematic color variation. This paper presents an algorithm that uses this local approach to color image understanding.

Our segmentation method alternates between generating optical hypotheses from local image data and verifying whether the hypotheses fit the image (see figure 7). The hypotheses relate object color, shading, highlights, and camera limitations to the shapes of color clusters in local image areas. The algorithm looks, in a bottom-up process, for color clusters from local image areas that exhibit the characteristic features of the body and/or surface reflection processes. When it finds a "promising" cluster in an image area, the algorithm generates a hypothesis that describes the object color and/or highlight color in the image area and determines the shading and highlight components of every pixel in the area. The algorithm then applies the new hypothesis to the image, using a region-growing

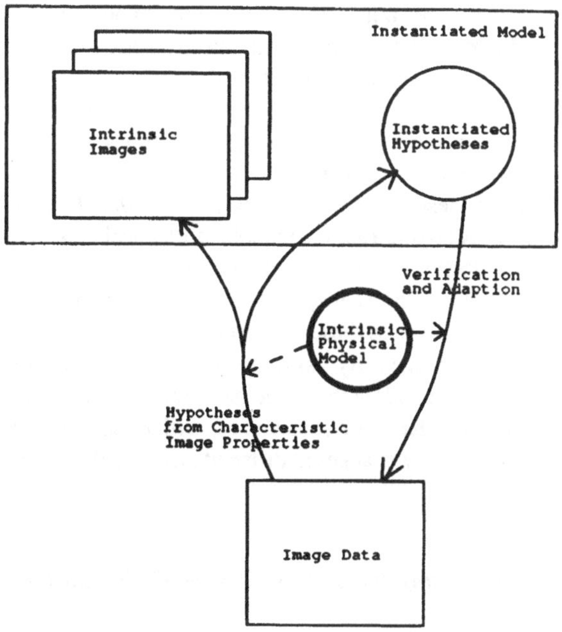

Fig. 7. Using an intrinsic model for image understanding

approach to determine the precise extent of the image area to which the hypothesis applies. This step verifies the applicability of the hypothesis. The physical knowledge embedded in the new hypothesis can be used to split every pixel in the new image area into its intrinsic body and surface reflection components, as will be shown in section 6. The resulting intrinsic images and the hypotheses together instantiate the concepts of shading and highlights of the Dichromatic Reflection Model, describing the specific reflection processes that occur in this part of the scene.

The physical knowledge from all hypotheses instantiated so far can be used to incrementally adapt the steps of the image analysis to the already established knowledge about the scene. Our algorithm performs its generate-and-test analysis in several stages, each of which is related to a particular aspect of the Dichromatic Reflection Model or the sensor model. It starts out with the simplest and most common aspects and then uses the established knowledge to address the more complicated aspects. In this fashion, the algorithm performs the following steps:

1. Compute an initial, rough description of color variation in local image areas.
2. Generate hypotheses on matte color clusters. Exploit these hypotheses for image segmentation.

3. Extend the matte hypotheses into hypotheses on skewed T's in dichromatic planes. Resegment the image, exploiting these hypotheses.
4. Analyze the effects of blooming and color clipping.
5. Exploit the hypotheses to split the pixels into their reflection components.

Each of these steps will be described in detail below.

This control structure exploits a physical model of light reflection to incrementally identify local and global properties of the scene, such as object and illumination colors. It uses these properties to interpret the pixels in the images. By using this control structure, the algorithm can adapt its image interpretation process to local scene characteristics and react differently to color and intensity changes at different places in the image.

5.2 Generating Initial Estimates for Color Clusters

To start the image analysis process, our algorithm generates some initial estimates about the color variations in the image. These initial estimates are too coarse to be useful as hypotheses without further considerations. But they can be used as a starting point for the formulation of hypotheses about linear and planar color variation in the next sections.

5.2.1. A Classification Scheme for Local Color Variation. To estimate and classify local color variation in the image, the algorithm divides the image into small, nonoverlapping windows of fixed size. It projects the color pixels from one window at a time into the color space and finds the principal components of the color distributions, as indicated by the eigenvectors and eigenvalues of the covariance matrix of the cluster [1, 10]. The eigenvalues and eigenvectors determine the orientation and extent of the ellipsoid that optimally fits the data. For convenience, the eigenvalues are sorted by decreasing size such that a reference to the first eigenvalue corresponds to the largest one, the second eigenvalue to the medium one, and the third eigenvalue to the smallest one.

Principal component analysis is not a new concept in color image segmentation. Gershon and Ohta used it to determine optimal color feature axes for a histogram-based, recursive region-splitting approach. They expected the major principal component to indicate color changes between objects—a color feature along which they could best separate objects from one another. Such use of principal component analysis implicitly assumes that the significant color changes in an image occur at the object boundaries while color variation within a single object area is minimal. This is a special case of the dichromatic framework we will develop below: the assumption holds only for matte clusters from dark or flat objects for which the color clusters from single objects form points or small spheres in the color space—no linear or even planar color variation on an object is accounted for. We will now present a more general investigation of the relationship between the principal components of the color clusters and physical scene properties.

The shape of the ellipsoids provides information that relates local color variation to physical interpretations. We distinguish between eight classes of ellipsoids, for each of which exist a few physical interpretations. The classification is based on the number of eigenvalues that are approximately zero, within the limit of pixel value noise, σ_0, in the image. The decision for each eigenvalue is based on a χ^2 test with $(n^2 - 1)$ parameters, where n is the window size. The algorithm then classifies each color cluster according to how many eigenvalues are significantly greater than zero. A summary of the classification is given in Table 1.

Table 1. Summary of color cluster interpretations.

Summary of Color Cluster Interpretations		
$\lambda_1\lambda_2\lambda_3$	Class	Interpretation
s s s	point	a) window on one object with little curvature b) window on one object with dark color, not on highlight
l s s	line	a) window on one object: matte cluster b) window on one object: highlight cluster c) window on two objects: matte clusters, from dark or flat objects
l l s	plane	a) window on one object: dichromatic plane b) window on two objects: matte clusters or highlight clusters
l l l	volume	a) window on one object: noise (clipping or blooming) b) window on two objects: both matte and highlight pixels on one object, matte and/or highlight pixels on the other object c) window on more than two objects

with s = small and l =large

- In zero-dimensional (*pointlike*) clusters, all three eigenvalues of the window are very small. No significant color variation exists in such a window and the body and surface reflection components are nearly constant. This is the case, if the window lies on a very flat object such that the photometric angles are nearly constant within the window. Alternatively, a pointlike cluster can originate from a matte dark object, independently of the surface curvature of such object. In the extreme, a perfectly black object does not exhibit any shading variation.
- One-dimensional (*linear*) clusters are clusters for which only the first eigenvalue is significantly larger than the estimated camera noise. Pixels in such a window may come from a matte object area forming part of a matte cluster. They can also come from the interior of a highlight area such that they form part of a highlight cluster. As a third possibility, the window may overlap the matte object areas of two neighboring objects that are very dark or flat. Such a window consists of two pointlike color clusters or of one dark pointlike cluster and one linear cluster which together fit well into one linear ellipsoid.
- Two-dimensional (*planar*) clusters have large first and second eigenvalues. The local color data fits a plane in the color cube. Such clusters occur at windows that cover some matte and some highlight pixels of one object. In this case, the third eigenvector of the color cluster determines the normal to the dichromatic plane of the object. But planar clusters also arise in windows that overlay matte pixels from two neighboring regions. Such windows have planar clusters because the matte clusters of all objects converge at dark pixels such that matte clusters are pairwise coplanar.
- In three-dimensional (*volumetric*) clusters, all three eigenvalues are large. Such color clusters may arise in the middle of highlights where color clipping and blooming significantly increase the noise in the pixel measurements. Volumetric color clusters also occur along material boundaries when three or more objects in different colors share a window or when a window overlays matte pixels of one object and matte and highlight pixels of another object.

Color figure 3 shows the classification of the color clusters from color figure 1, using windows that each contain 10 × 10 pixels. Pointlike clusters are displayed in yellow, linear clusters in red, planar clusters in green, and volumetric clusters in blue. The image shows that the classifications relate in the expected way to scene properties: most matte object areas are covered by linear (red) windows, while windows at material boundaries and at highlights are planar (green) or volumetric (blue).

5.2.2 Determining Compatible Color Variation. If the window size is chosen appropriately, it is small compared to the sizes of the objects in the scene. Most windows lie, then, inside object areas; and only few windows overlay material boundaries. There will generally be many windows within a single object area. All linear clusters of such windows indicate the direction of the same matte or highlight cluster and all planar clusters support the same hypothesis about a dichromatic plane. This subsection describes how our algorithm merges such windows to gather increased support for a small number of estimates about color clusters in the image.

The algorithm merges neighboring windows that have similar color characteristics. It proceeds in row-major order, testing for all pairwise combinations of neighboring areas whether the two can be merged. In order not to merge windows across material boundaries, it merges windows only if both of them, as well as the resulting larger window, have the same classification. It uses the χ^2 test described above to classify the larger window. Accordingly, it combines windows with pointlike clusters into larger pointlike windows; it merges linear windows into larger linear windows; and it merges planar windows into larger planar windows. It does not merge volumetric color clusters since there is no constraint on the resulting cluster. This process continues until no more areas can be merged. The results are initial hypotheses about the positions and orientations of pointlike, linear, and planar clusters in color space and their respective approximate extents in the image.

Color figure 4 presents the results of merging neighboring windows of the same class from color figure 1. The image shows a few large image areas. Such image areas correspond roughly to the center of matte object areas. They are generally surrounded by many smaller areas close to material boundaries or highlights. Only very few planar windows in highlight areas have been merged, probably due to the increased spread of highlight pixels from color clipping and blooming.

5.3 Linear Hypotheses

In its subsequent steps, our algorithm uses the initial estimates of color clusters to generate and exploit hypotheses about the existing color variation in the image. The above discussion and color figure 4 show that the initial estimates are rather coarse indications of local color variation: there are far more image areas than objects and, due to the coarse initial window width, the area boundaries do not coincide with the material boundaries. We now need to combine neighboring areas at a much finer granularity, resegmenting the image by considering every pixel on an individual basis. The algorithm reanalyzes and combines the estimates of local color variation and uses them to resegment the image. The hypotheses relate the color clusters to properties of the Dichromatic Reflection Model, describing color pixels as matte or highlight pixels or as pixels showing camera problems.

5.3.1 Generating Linear Hypotheses. The algorithm starts by choosing large image areas with linear color clusters as linear hypotheses. Large linear clusters generally correspond to the internal parts of large matte object areas. Such large areas are less influenced by estimation errors in local color variation than small image areas along material boundaries and in highlight centers with increased camera noise from color clipping and blooming. Linear color clusters are also easier to interpret than planar, pointlike, or volumetric color clusters because their first eigenvector describes the orientation of major color variation in the cluster; information that can be related to a physical interpretation. The eigenvectors of planar clusters (e.g., in areas close to highlights) are less useful because the directions of the first and second eigenvectors are linear combinations of the directions of several linear clusters (e.g., of a matte and a highlight cluster). Their direction does not describe the branches of the skewed T of a dichromatic cluster and they cannot be related to the physical properties of body and surface reflection.

5.3.2 Exploiting Linear Hypotheses. The merging algorithm of the previous section used a bottom-up approach to extract information about the scene from the image. In a top-down step, we now use a selected hypothesis to locally resegment the image. The linear hypothesis provides a model of what color variation to expect on the object part on which the image area lies. The mean value and the first eigenvector describe the position and orientation of a linear color cluster, while the second and third eigenvalues determine the extent of the color cluster perpendicular to the major direction of variation. According to the Dichromatic Reflection Model, color variation along the major axis can be attributed to a physical property of the scene—for example, a changing amount of body or surface reflection or a material boundary. At this stage, we attribute color variation perpendicular to the first eigenvector to random noise. Accordingly, we model the linear color cluster as a cylinder (see figure 8). The radius depends on the estimated camera noise. Our algorithm excludes dark-color pixels from its color analysis because all matte clusters merge near the dark corner of the color cube. By this *dark heuristic*, the cylinder is delimited at its dark end by a sphere which is centered at the black corner.

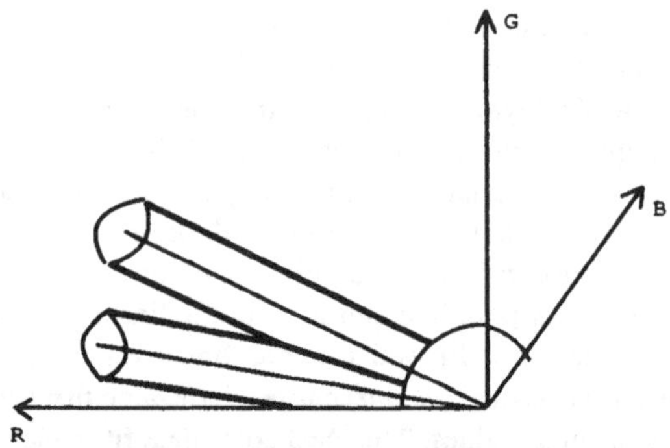

Fig. 8 Proximity heuristic to resolve a color conflict between the color clusters of two neighboring matte object areas.

We use the color cylinder of the current hypothesis to locally resegment the image. The algorithm selects a start pixel from the image area associated with the color cluster. This pixel must have a color that is contained within the color cylinder. The algorithm then grows a four-connected region from this starting point, recursively examining the four neighbors of pixels on the fringe of the region, and including them if their color lies within the color cylinder. The result is an image area of pixels that are consistent with the current linear hypothesis. The boundary of this new region may be very different from the image area that was initially associated with the linear hypothesis. Due to the coarse initial window size, the initial image area may have contained pixels from neighboring objects or highlight pixels with some amount of surface reflection. Such pixels are excluded from the new region. The initial window may also have contained dark pixels in very shaded object areas or—as is the case in figure 1—pixels from a dark background. Due to the dark heuristic, such pixels are also excluded from the new region. On the other hand, the initial segmentation may have had several neighboring image areas on the same matte object part that could not be merged because the characteristics of the color clusters were biased by pixels that came from other objects. Many pixels in such neighboring image areas may lie on the current matte object part and be consistent with the current linear hypothesis. They are now included into the new image area. As a consequence, the new region can be significantly different from the initial image area.

Since there is generally more than one object in the scene, the algorithm iterates the above steps for each large image area with a linear cluster, selecting the areas by decreasing size. It stops when the next selected area is too small. Since all matte clusters converge at dark pixels, there exists a potential conflict between neighboring matte areas. The dark heuristic eliminates the most difficult cases; yet, the cylinders of neighboring clusters may still intersect beyond the selected dark threshold. This depends on the cylinder radius and on the angle between the two cylinders. Neighboring objects with very similar colors have conflicts even at fairly bright colors. Our algorithm assigns pixels with a color conflict to the cluster with the closest axis, as shown in figure 8. This is called the *proximity heuristic*.

In principle, the resulting "linear" regions may be related in any of several different ways to the physical processes in the scene. As discussed in section 5.2.1, a linear color cluster may be a matte cluster or a highlight cluster or even a combination of two clusters across a material boundary. However, linear color clusters from highlights and across material boundaries are generally much smaller than clusters from matte object parts. Since this section considers only linear hypotheses of large image clusters, the following sections assume that all newly grown regions correspond to matte linear clusters.

Color figure 5 shows the results of selecting and applying linear hypotheses to color figure 1. The region boundaries outline the matte object parts in the scene, with the material boundaries being well observed. The highlight areas on the objects have not yet been analyzed. This sometimes divides the matte pixels on an object into several matte areas, as shown on the green donut in the lower right part of the image.

5.4 Planar Hypotheses

The Dichromatic Reflection Model states that color variations on a single object lie in a plane in the color space. The linear hypotheses and the linear segmentation generated so far provide a good starting point to describe color variation in an image. But they do not account for the effects of surface reflection. The hypotheses will now be extended into planar hypotheses that describe dichromatic planes and skewed T's.

5.4.1 Generating Planar Hypotheses. In principle, information on color variation in a dichromatic plane may be obtained from inspecting the characteristics of the initially determined planar color clusters. However, such information is generally not very reliable, due to the small size of most planar image areas. Furthermore, the eigenvectors of such planar clusters do not describe the skewed-T structure of the color cluster because the directions of the first and second eigenvectors are linear combinations of the directions of several linear clusters (e.g., of a matte and a highlight cluster). Their direction does not describe the branches of the skewed T of a dichromatic cluster and they cannot be related to the physical properties of body and surface reflection. Instead of using such planar color information, our algorithm uses the existing linear hypotheses as a basis describing already one branch of skewed T's. It now determines the orientations and positions of the second branch of the skewed T's.

In this process, the algorithm considers all neighbors of a linear region as prospective highlight candidates. Some of such neighboring regions may be matte object areas of neighboring objects while others may be a part of a highlight on the current object. Such types of regions must be distinguished from one another such that the algorithm does not wrongly combine color information from two neighboring matte image areas into a planar hypothesis. To distinguish neighboring matte regions from highlight regions, Gershon has tested whether two clusters are nearly parallel or whether they intersect, suspecting that parallel clusters are neighboring matte clusters [7]. We replace this test by the more precise 50% heuristic of section 3: the color clusters of a matte and a highlight region on the same object intersect in the upper 50% of the matte cluster, while color clusters from neighboring matte regions converge at dark pixels. Accordingly, our algorithm tests whether the two clusters form a skewed T and meet in the upper 50% of the matte cluster. For this purpose, it searches for the brightest matte point in the color cluster to determine the length of the matte line. However, the matte cylinder may contain some highlight points from the foothill of the highlight cluster, as shown in figure 9.

Such highlight points may be brighter than the brightest matte point. To distinguish them from matte points, we exploit the observation that highlight clusters always grow inward into the color cube, due to the additive nature of body and surface reflection. Accordingly, the algorithm chooses the brightest matte point only from pixels with color values on the outside of the matte line.

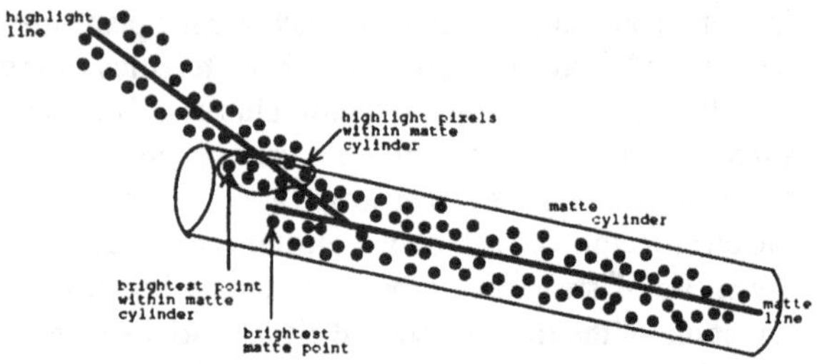

Fig. 9. Finding the brightest matte pixel.

Once the length of the matte line is determined, the algorithm computes the intersection points of the current matte cluster and of the neighboring prospective highlight cluster. It uses their color means and first eigenvectors to describe a matte and a prospective highlight line and to determine the two points on the two lines that are closest to one another. Comparing the distance to the camera noise, the algorithm decides whether the clusters meet in a skewed T. Similarly, the algorithm tests whether the clusters intersect in the upper 50% of the matte cluster. We also check whether the neighboring cluster has a positive direction—a test related to the additive properties of body and surface reflection. If this is not the case, the window of the highlight candidate probably overlays a material boundary, partially covering pixels from the current matte region just outside its color cylinder and partially covering pixels from a neighboring matte region. Candidates that fail any of these tests are discarded.

After all these tests, there may still exist several highlight candidates because the highlight on the object may consist of a series of small windows that had not been merged. Since many such highlight areas are very small and since they may contain some clipped and bloomed pixels, the orientations of their color clusters may vary significantly. To select a good representative of the entire highlight, the algorithm averages the intersection points of all highlight candidates, weighted by the number of pixels in the regions. It then selects the highlight region whose intersection point is closest to the average.

Since the surface reflection color of dielectrics is generally very similar to the illumination color, all highlight clusters are parallel to one another and to the illumination color vector. As a consequence, all dichromatic planes intersect along one line, which is the illumination vector. The algorithm uses this constraint to further reduce the error in estimating the orientations of the highlight clusters. It computes the average direction of all highlight clusters to combine all highlight hypotheses into a single hypothesis on the illumination color vector. The algorithm then generates planar hypotheses, combining the existing linear hypotheses with the direction of the illumination vector, as well as with information on the starting points of the respective highlight clusters. Such hypotheses describe the skewed-T shapes of the color clusters in terms of the positions and orientations of the matte and highlight clusters. The cross-product of the illumination vector and the first eigenvector of a matte cluster determine the normal to the dichromatic plane of a color cluster. The position of the dichromatic plane in the color space is given by the color mean.

5.4.2 Exploiting Planar Hypotheses. Starting from the previously generated linear segmentation, the algorithm now resegments the image to account for the effects of surface reflection on objects. Applying the planar hypotheses one at a time, it proceeds iteratively until no more unprocessed planar hypotheses for large image areas exist.

The chosen planar hypothesis describes the position and orientation of a dichromatic plane in the color space, as well as the shape of the color cluster within the plane. To account for camera noise, the algorithm extends the plane into a slice, as shown in figure 10.

When the algorithm uses the chosen planar slice to locally resegment the image, it starts, in principle, from the selected matte region and expands it until no more pixels at the region boundaries fall into the planar slice. This planar-region growing method is augmented with special provisions to handle coplanar color clusters from neighboring objects. Such coplanar color clusters occur when the illumination vector lies in the plane spanned by the matte vectors of two neighboring objects. The color clusters of such objects lie in the same dichromatic plane and cannot be distinguished in a simple planar-region growing approach. The resulting segmentation would generally be quite counterintuitive since the matte object colors of objects with coplanar

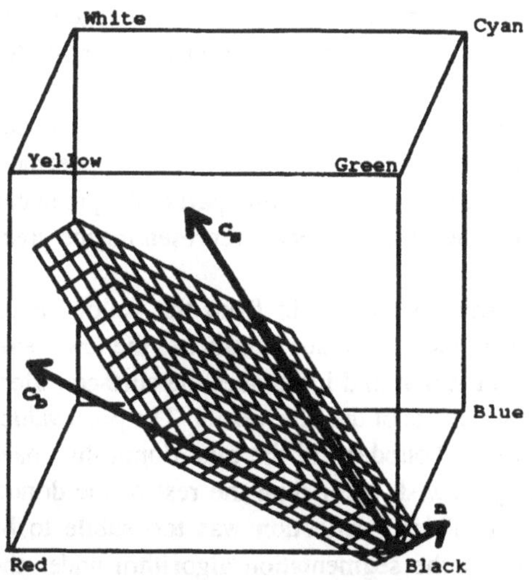

Fig. 10. A planar slice.

clusters may be very different and even complementary. For example, a red-body reflection vector (255,0,0) and a white illumination vector (255,255,255) span a dichromatic plane with normal $(0, -1/\sqrt{2}, 1/\sqrt{2})$. This plane contains all body reflection vectors (r, g, g) for which the *g*-component and the *b*-component are identical. In particular, the plane contains the vector (0,255,255) which describes a cyan object color, as shown in figure 11. In this case, a simple planar region-growing method would not be able to distinguish a red object from a cyan object.

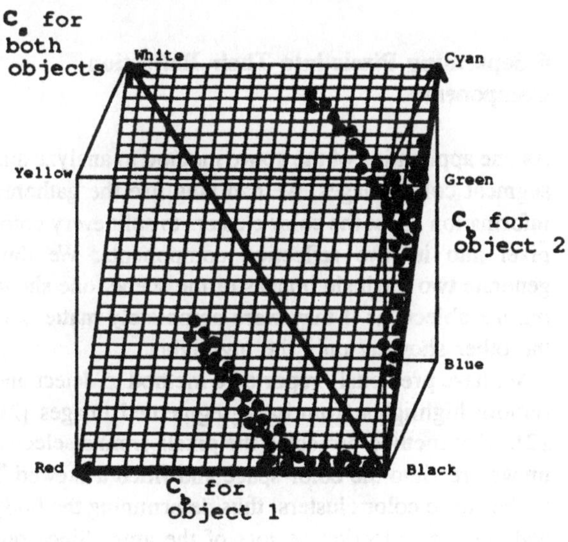

Fig. 11. Coplanar color clusters.

To avoid such segmentation problems, the algorithm exploits the previously gathered knowledge about existing matte color clusters. When the planar-region growing process encounters pixels from a previously grown matte region other than the starting region, it only continues growing if the pixel lies within the matte color cylinder of the starting region. It thus applies the unrestricted planar growing criterion only to pixels that have not been previously recognized as matte pixels, while it falls back to the linear-region growing method when matte pixels are concerned. This reflects the observation that if several matte areas exist in an object area, separated by highlight regions, all such matte areas form a single matte cluster. The algorithm also applies the proximity heuristic described in 5.3.2 to resolve ambiguities for color pixels at the intersection of dichromatic planes.

Color figure 6 displays the results of segmenting the scene using the generated planar hypotheses. In comparison to the linear segmentation in color figure 5, the segmented image areas have grown into the highlight areas. As a result, the two matte image areas on the green donut—previously separated by a highlight—are now united. Due to camera limitations, not all pixels in the centers of the highlights are yet integrated into the object areas. This will be discussed and remedied in the next subsection.

5.5 *Accounting for Camera Limitations*

Unfortunately, real images generally do not fully comply with the Dichromatic Reflection Model. Among other things, color clipping and blooming may significantly distort the color of pixels in and around highlights. As a result, the color pixels in the centers of highlights generally do not fall into the planar slice defined for the dichromatic plane of an object area. The planar segmentation excludes such pixels, as can be observed in color figure 6.

Since color information is so unreliable for these pixels, our algorithm does not use it. Instead, it uses a geometric heuristic to include distorted pixels into the region. Since pixels with distorted colors generally occur in the middle of the highlight areas, the algorithm starts from highlight pixels and expands the planar regions into areas that are next to highlight pixels and contain very bright pixels (i.e., brighter than the intersection point between the matte and highlight cluster).

This heuristic successfully includes highlight centers with clipped colors into the object areas, while leaving other bright image areas (without neighboring highlight pixels) untouched.

Color figure 7 displays the results of segmenting the scene while using the generated planar hypotheses and accounting for camera limitations. Nearly all pixels in the highlight centers have not been integrated into the segmented regions, and the image segments correspond quite well to the objects in the scene. A few pixels on the highlights have been excluded, due to the heuristic of only integrating very bright pixels. Any image processing method for filling holes in image segments, such as a region-expansion operation followed by region shrinking [27], should be able to include these pixels.

5.6 Optical Effects Beyond the Scope of the Segmentation Algorithm

Color figure 7 shows that some pixels at object boundaries have not been included into the object areas. These exclusions are related to effects in the scene that are not modeled in the Dichromatic Reflection Model, such as shadow casting and inter-reflection between objects.

- A small area in the lower right part of the yellow donut was not included into the large image segment covering the donut. The color of these pixels has been significantly altered by inter-reflection from the orange cup, which reflected a part of its body color onto the donut, influencing the reflected body color of the yellow donut.
- A small area at the upper left edge of the yellow donut has wrongly been assigned to the red donut above it. In the original image, this area is covered by a shadow, resulting in very dark pixel values. Since the color clusters of the two donuts are already merged at these color values, the assignment was based on the distance from the two cylinder axes, which happened to be smaller for the red donut. Inter-reflection may also have influenced the color values, biasing them toward the red-body reflection vector.
- There is a mirror image of the yellow cup on the side of the orange cup, resulting in yellow surface reflection. Such yellow surface reflection is added to the orange-body reflection—which, in addition, has also been influenced by the light reflected from the yellow cup. As a consequence, the segmentation does not include these pixels into the region representing the orange cup.
- Similarly, inter-reflection between the orange and green cup causes a few pixels on the uppermost planar surface of the polyhedral part of the green cup to be excluded from the region representing the green cup.
- Finally, some yellow light from the yellow donut underneath the blue donut illuminates the lower side of the blue donut and influences the surface reflection components of the blue pixels. The pixel values at the lower boundary of the blue donut thus have a more yellowish color than the rest of the donut. This case of inter-reflection was too subtle to be detected by the segmentation algorithm under the chosen width of the planar slice. Color figures 5 and 6 show that such yellow pixels have been separated from the blue ones in the linear segmentation step but were merged during the planar segmentation step because the yellow inter-reflection cluster was close to the dichromatic plane of the blue donut.

There seems to be a good mapping between pixel areas in which our algorithm "fails" and object areas with optical properties outside the scope of the model. Such results are quite different from "traditional" segmentation results for which the algorithm performs correctly on some instances of an optical phenomenon (e.g., highlights) but incorrectly on others (as shown in section 7.2).

6 Separating Pixels into Their Reflection Components

As one application of the above method to analyze and segment color images, we can now use the gathered information about the color clusters to split every color pixel into its two reflection components. We thus generate two intrinsic images of the scene, one showing the objects as if they were completely matte, and the other showing only the highlights.

We have previously reported a method to detect and remove highlights from hand-segmented images [21, 22]. That method projected the pixels from a selected image area into the color space and fitted a skewed T to the entire color clusters, thus determining the body and surface reflection vectors of the area. Since our new segmentation algorithm already provides this information as a result of its analysis of local color variations, we can now skip this step. However, due to possible estimation errors in the segmentation process and due to camera problems such as blooming and chromatic aberration, the vectors may not yet fit perfectly. In order to obtain a more precise fit to the data, we retest every pixel in the segmented area and label it as a matte or highlight pixel, depending on whether it is closer to the matte line or to the highlight line. We then refit the matte and highlight line to the matte and highlight pixels by determining the first eigenvectors and the color means of the clusters.

The orientations of the corresponding reflection vectors are shown in table 2. In principle, the length of the reflection vectors, $\mathbf{C}_b$ and $\mathbf{C}_s$, depends on the brightness of the object such that a white and a grey reflection vector have the same orientation but different lengths. However, it is impossible to determine these lengths from the color clusters since the length of a cluster depends both on the magnitude of the reflection curve of the material and on the geometric influence from the photometric angles: a strongly curved, dark object may produce a color cluster of the same length as a brighter, but relatively flat object. For this reason, the vectors in table 2, $\bar{\mathbf{C}}_b$ and $\bar{\mathbf{C}}_s$, are unit vectors, describing only the orientations of the reflection vectors. The table also provides the average surface reflection vector and an independent estimate of the illumination color that has been obtained by taking a color picture of a grey chart under the given illumination. The surface reflection vectors and their average vector are generally close to the illumination vector. This demonstrates that the surface reflection components of dielectric materials have the same color as the illumination and that the surface reflection vectors from real images can be used to determine the illumination color—information which is useful for color constancy algorithms [4, 23, 24].

Our algorithm then uses $\bar{\mathbf{C}}_b$ and $\bar{\mathbf{C}}_s$ of every segmented region to split the pixels into their reflection components. Since the physically correct length of $\bar{\mathbf{C}}_b$ and $\bar{\mathbf{C}}_s$ is not known, the algorithm can determine the reflection components only up to scale factors a_s and a_b that depend on the brightness of the material:

$$\mathbf{C}(x,y) = m_s\mathbf{C}_s + m_b\mathbf{C}_b = \tilde{m}_s\bar{\mathbf{C}}_s + \tilde{m}_b\bar{\mathbf{C}}_b \tag{13}$$

with

$$\tilde{m}_s = m_s \cdot a_s, \qquad \|\bar{\mathbf{C}}_s\| = 1 = \frac{\|\mathbf{C}_s\|}{a_s}$$

$$\tilde{m}_b = m_b \cdot a_b, \qquad \|\bar{\mathbf{C}}_b\| = 1 = \frac{\|\mathbf{C}_b\|}{a_b}$$

Table 2. Body and surface reflection vectors of the eight plastic objects under white light.

Reflection Vectors: Eight Plastic Objects under White Light		
	body reflection vector	surface reflection vector
dark red donut	(0.99,0.11,0.13)	(0.63,0.52,0.58)
orange cup	(0.95,0.26,0.14)	(0.68,0.48,0.55)
yellow cup	(0.84,0.52,0.14)	(0.50,0.63,0.59)
green cup (right half)	(0.27,0.89,0.37)	(0.53,0.58,0.62)
yellow donut	(0.77,0.61,0.19)	(0.70,0.50,0.51)
bright red donut	(0.98,0.17,0.11)	(0.67,0.54,0.51)
green donut	(0.19,0.79,0.58)	(0.45,0.64,0.62)
green cup (left half)	(0.27,0.87,0.42)	(0.61,0.54,0.58)
blue donut	(−0.08,0.19,0.98)	(0.48,0.65,0.59)
illumination vector:		
-computed by alg.		(0.59,0.57,0.58)
-independent meas.		(0.58,0.57,0.58)

To split the pixels into their reflection components, we use $\bar{\mathbf{C}}_b$, $\bar{\mathbf{C}}_s$ and their cross product, $\bar{\mathbf{C}}_b \mathbf{x} \bar{\mathbf{C}}_s$, to define a new (not necessarily orthogonal) coordinate system in the color cube (see figure 12). This coordinate system describes every color in the cube in terms of the amounts of body reflection $\tilde{m}_b$, surface reflection $\tilde{m}_s$ and noise ϵ. There exists an affine transformation, and thus a linear transformation matrix T, which transforms any color vector $[R, G, B]^T$ into a vector $[\tilde{m}_b, \tilde{m}_s, \epsilon]^T$. After computing T from $\bar{\mathbf{C}}_b$ and $\bar{\mathbf{C}}_s$, we can thus transform every color pixel in the image area into its constituent body and surface reflection components, $\tilde{m}_b$ and $\tilde{m}_s$. By selecting the $\tilde{m}_b$-components of all pixels, we generate the *body reflection image* of the region. By selecting the $\tilde{m}_s$-components, we generate the corresponding *surface reflection image*. The ϵ-components provide the *intrinsic noise image*.

We cannot apply this method immediately to bloomed and clipped pixels since their color is altered by the sensing process in the camera. In order to restore the physically correct color of such pixels, we exploit the observation that in many cases, clipping and blooming occurs only in one or two color bands. The pixels may thus have correct data in the other color bands. We assume that the smallest of the three values of a color pixel comes from a color band without clipping and blooming. We then replace the clipped or bloomed pixel

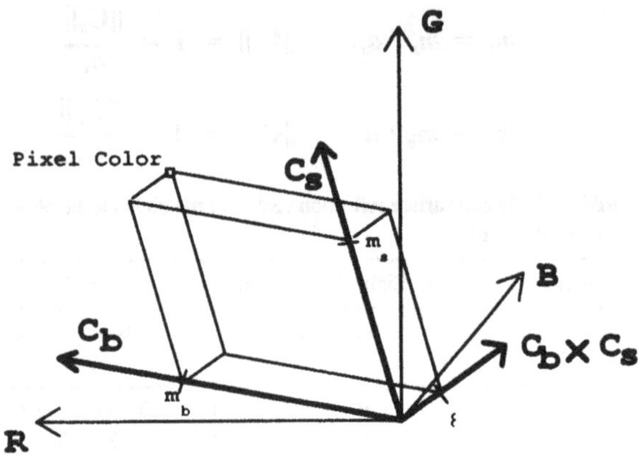

Fig. 12. Decomposing a color pixel into its constituent body and surface reflection components.

by a pixel on the matte or highlight line that has the same value in the undistorted band (see [22] for a detailed description).

Color figures 8–10 display the resulting intrinsic images of the scene with the eight plastic objects. The images demonstrate that we are able to determine the body and surface reflection components of the various objects reasonably well. When evaluated qualitatively, the body reflection image provides smooth shading across most highlight areas. We expect that this image may therefore be a useful tool to determine object shapes from shading information [14]. There exist thin dark rings in the body reflection image around the highlight on the green donut. This error is due to chromatic aberration in the camera [20] which currently limits the performance of our algorithm.

The surface reflection image exhibits gradual changes between areas with no surface reflection and areas with very high amounts of surface reflection, demonstrating that surface reflection is a function of the photometric angles. The surface reflection images may be useful input for methods that determine object shapes from highlights [12]. A careful inspection of the surface reflection image reveals that the surface reflection component also increases at the material boundaries. This effect is related to the aliasing that occurs at pixels that integrate light from two neighboring objects. The colors of such pixels are a linear combination of two matte object colors. Depending on the orientations of the dichromatic planes, they may be included in either of the two object areas in the image, or they may be left unassigned. If they are included into one of the object areas, they generally do not lie close to the matte line, thus resulting in higher surface reflection components.

The intrinsic noise image displays the distance ϵ of every pixel in an image area from the associated dichromatic plane, scaled relative to the width of the planar slices and of the cylinders. Grey pixels represent color pixels that lie on the dichromatic plane. Increased noise values are displayed as increasingly bright or dark pixels, depending on whether the color pixels lie on the front side or on the back side of the dichromatic plane. Very bright or dark areas in the noise image indicate places where the Dichromatic Reflection Model is not satisfactory in explaining color variation as a mere combination of body and surface reflection with added random camera noise. Such areas generally occur at material boundaries where light from two neighboring objects is integrated into the pixel values. The colors of such pixels are a linear combination of the two matte object colors of the neighboring objects. Higher noise values also occur in the centers of the highlights at clipped and bloomed pixels and at internal object edges where the surface orientation changes abruptly. Chromatic aberration causes increased noise values at the highlight on the green donut in the lower right corner [20]. Finally, large noise values exist at places of inter-reflection between objects, such as on the yellow donut, between the orange and yellow cup, and on the right half of the green cup.

To illustrate the changing amounts of body and surface reflection more quantitatively, color figure 11 displays the reflection profiles for a row of pixels in the intrinsic body and surface reflection images of the plastic scene in color figure 1. It also shows the amounts of noise, ϵ, at these pixels. The selected row runs across the yellow donut, the orange cup, and the yellow cup. The body reflection profile shows that the highlights have been removed reasonably well from the objects. Some error was introduced in the body reflection components of restored clipped and bloomed pixels. The surface reflection profile consists of a series of spikes at the positions of the highlights on the objects. Each spike has the form of a peak which falls off to the sides, demonstrating the quantitative nature of the highlight analysis. A few small peaks exist at the object boundaries. These are pixels on material boundaries where light from more than one object is mixed into a single pixel value. Since such pixels generally do not lie close to the matte line, they generate a disproportionately high surface reflection component.

7 Discussion

The previous sections have presented an approach to color image understanding that exploits the Dichromatic Reflection Model. The description of the algorithms was accompanied by a few exemplary results that demonstrated how the algorithm performs on real images. This section now presents more results and discusses strengths, limitations, and extensions of the approach.

7.1 Further Results

We have tested our algorithm on a series of color images of scenes with dielectric objects under varying illumination conditions. The upper left quarters of color figures 12–14 show the images of three plastic cups under yellow, white, and pink light.

The upper right quarters of the figures display the segmentations that were generated by our algorithm. They demonstrate that our segmentation method performs well on dielectric materials under varying illumination conditions. In all images, the segmentations outline the material boundaries of the objects very well. Because the algorithm models matte and highlight color variations as physical reflection processes, it ignores color changes along highlight boundaries and it also tolerates shading changes on the objects.

The lower quarters of color figures 12–14 show the intrinsic reflection images that were generated using the body and surface reflection vectors of tables 3–5. The intrinsic images demonstrate that our algorithm detects and removes highlights well on dielectric materials under varying illumination colors. When evaluated qualitatively, the body reflection vectors have reasonable orientations and the body reflection images generally describe smooth shading across the highlight area. The surface reflection images in the lower right quarters show that the highlights from the original images have been detected well. The surface reflection images exhibit gradual changes between areas with no surface reflection and areas with very high amounts of surface reflection, demonstrating that surface reflection is a function of the photometric angles. The surface reflection vectors in tables 3–5 are generally reasonably close to the estimated illumination colors. The reflection images may be a useful tool to determine object shapes from shading or from highlights while the surface reflection vectors can be used in color constancy algorithms as an indication of the illumination color.

Table 3. Body and surface reflection vectors of the three plastic cups under yellow light.

Normalized Reflection Vectors: Plastic Cups under Yellow Light		
	body reflection vector	surface reflection vector
yellow cup	(0.85,0.52,0.08)	(0.79,0.59,0.17)
orange cup	(0.97,0.23,0.08)	(0.74,0.65,0.18)
green cup	(0.39,0.91,0.15)	(0.70,0.68,0.20)
illumination vector:		
-average surf. refl.		(0.74,0.64,0.18)
-independent meas.		(0.73,0.66,0.19)

Table 4. Body and surface reflection vectors of the three plastic cups under white light.

Normalized Reflection Vectors: Plastic Cups under White Light		
	body reflection vector	surface reflection vector
yellow cup	(0.80,0.58,0.15)	(0.50,0.53,0.68)
orange cup	(0.95,0.27,0.16)	(0.47,0.58,0.66)
green cup	(0.20,0.92,0.33)	(0.43,0.54,0.73)
illumination vector:		
-average surf. refl.		(0.47,0.55,0.69)
-independent meas.		(0.58,0.57,0.58)

Table 5. Body and surface reflection vectors of the three plastic cups under pink light.

Normalized Reflection Vectors: Plastic Cups under Pink Light		
	body reflection vector	surface reflection vector
yellow cup	(0.92,0.38,0.06)	(0.84,0.49,0.23)
orange cup	(0.98,0.19,0.06)	(0.79,0.53,0.30)
green cup	(0.72,0.63,0.29)	(0.89,0.42,0.15)
illumination vector:		
-average surf. refl.		(0.84,0.48,0.23)
-independent meas.		(0.71,0.59,0.38)

7.2 *Comparison with a Traditional Color Image Segmentation Method*

Color figure 15 shows the results of applying a traditional color segmentation method to color figure 1. It has been obtained by using Phoenix [31], a segmentation program that recursively splits color images into smaller regions until the regions are homogeneous. To decide how to split regions, Phoenix uses a set of user-defined color features, such as R, G, B, intensity, hue, saturation, etc., each of which is encoded as a separate image band. When a region is considered for further splitting, Phoenix generates for every feature a histogram of all pixels in the region and looks for distinctive valleys in the histograms. It then splits the region according to the feature with the most prominent valley. If it cannot find a significant valley in any feature histogram, it decides that the region is homogeneous. Phoenix is one of a class of "Ohlander-style" color image segmentation methods which all recursively split the image into segments by seeking for valleys in histograms of some color features. Extensions to the original method [25] centered around determining the color feature axes dynamically through principal-component analysis [6, 26] rather than by user input. We compare our results with Phoenix results since this system is available to us.

Color figure 15 was generated by running Phoenix on three features: intensity, hue, and saturation (IHS). The IHS color features are often used to describe human color perception. We chose this color basis, because our sample scene is composed of objects with relatively few, quite distinct hues, and a comparison on this basis seemed more favorable to Phoenix than a comparison on the RGB-basis. Some of the highlights in color figure 15 are separated from the surrounding matte object parts while other highlights are integrated with the matte areas. In addition, the matte areas are sometimes split into dark and bright areas, as can be seen on the right half of the green cup, as well as on the dark red donut and on the green donut. It is not easy to predict how Phoenix will perform on a given image because its decisions are not directly related to the influences of physical processes. Phoenix uses a predefined, fixed set of color features. It measures color variation only in relationship to these features and segments image areas using only one feature at a time. Projecting the color information from a three-dimensional color space to a one-dimensional feature space, Phoenix loses most of the information about the three-dimensional structure of the color variation.

In comparison to Phoenix, our color cylinders and dichromatic planes capture color variations in a much more flexible way. The color cylinders and planes of the linear and planar hypotheses are determined by analyzing color clusters in their three-dimensional structure. Their position and orientation is independent of a chosen coordinate system in the color space, as well as of a user-defined set of features. Consequently, linear and planar hypotheses can be expected to capture color variation on objects much better and to generate better segmentations of the objects, possibly even under ambient light [34]. This is evident when color figure 15 is compared with color figures 5, 6, and 7. Since the linear, planar, and final segmentations account for shading variation on matte object areas, for highlight reflection and finally for camera limitations, they include all matte and highlight pixels, as well as bloomed and clipped pixels into a single object area.

7.3 *Control Parameters*

We use a few control parameters and simplifying heuristics when we analyze color images. They are related to camera limitations, as well as to the limitations of color information and to the basic assumptions of the Dichromatic Reflection Model. This subsection shows the influence of the parameter settings on the performance of our algorithm; the following subsections discuss the influence of the heuristics and the principal limitations of the dichromatic theory.

The control parameters, as well as the parameter values chosen to generate the results above, are listed in table 6. It seems that most of the parameters are related in quite an intuitive way to the camera limitations and to the reflection model and that they are thus relatively easy to set. The control parameters used were:

- *Clipping Threshold and Noise Density:* The clipping threshold and the noise density parameter are related to the limited dynamic range of cameras. The density parameter expresses a heuristic to discard bloomed pixels (see section 4 and [22]).
- *Camera Noise:* σ_0 describes the estimated camera noise of the camera. We estimate it by repeatedly taking a black-and-white picture of a white board, measuring the intensity of the center pixel, and then computing its statistics. σ_0 corresponds to twice the value of the standard deviation σ of these measurements such that the integral of the Gaussian curve in the range $[\mu - \sigma_0, \mu + \sigma_0]$ describes about 96% of all measurements.

Table 6. Control parameters of the algorithm.

Control Parameters				
	plastic scene	cups yellow light	cups white light	cups pink light
clipping threshold	240	240	240	240
noise density (to detect blooming)	1	1	1	1
camera noise σ_0	2.5	2.5	2.5	2.5
minimal intensity	23	30	30	30
cylinder width, width of planar slice	$4\sigma_0$	$4\sigma_0$	$6\sigma_0$	$4\sigma_0$
window size (for init. grouping)	10 × 10	10 × 10	10 × 10	15 × 15
minimal area size	500	500	500	700

- *Minimal Intensity:* To avoid conflicts in dark image areas where matte clusters overlap, dark pixels are excluded from the color analysis by setting a threshold for minimal intensity. If this threshold is too low, color clusters from neighboring objects overlap beyond the threshold and bleeding may occur between the image areas. If, on the other hand, the threshold in unnecessarily high, color clusters of dark objects may have lost most of their matte cluster such that their body reflection vectors can no longer be estimated reliably.
- *Cylinder Width:* Due to inter-reflection and other optical effects in real scenes for which the Dichromatic Reflection Model does not account, the matte color clusters generally are not perfectly cylindrical. For such reasons and also because of camera noise and possible estimation errors in the direction of the color vector, our algorithm needs to use fairly thick cylinders and planar slices to include most of the pixels on an object into the segmented area. The drawback of unnecessarily thick cylinders and planar slices is a loss of sensitivity to the color properties of the image.
- *Initial Window Size:* The algorithm also depends on the window size in the initialization step of the algorithm. If the windows are too large, many of them overlap several objects, and many local color clusters are classified to be volumetric. If, on the other hand, the windows are too small, color variation on a relatively flat or dark object cannot be detected and the color clusters of such windows are classified as pointlike clusters. The initialization step then does not provide sufficient information to the subsequent steps of generating and applying linear or planar hypotheses.
- *Minimal Area Size:* In order to first concentrate on matte object areas—which are easier to analyze—our algorithm currently inspects only color clusters of image areas beyond a minimal number of pixels. It assumes that linear color clusters of such image areas represent matte branches of skewed T's and it subsequently tries to find the related highlight branches. In principle, however, linear color clusters can also exist in highlight areas or at material boundaries. The size of such highlight areas with linear color properties depends on the surface curvature of the object. Thus, if the selected threshold for the minimal area size is too low, highlight areas of flat objects are selected as candidates for matte branches of skewed T's. If, on the other hand, the threshold is too high, small objects in the scene cannot be detected.

Table 6 shows that most of the parameter settings were fairly constant for all images. Slight adjustments had to be made in a few cases. The changes were triggered by the changing illumination color in the scene and the resulting change in contrast in the color images. A detailed case study showing color pictures of different parameter settings is given in [20]. It may be possible to automatically select—or even to locally adapt—the values of some of the control parameters, such as the window size, the cylinder width, and the area size. We expect that such learning mechanisms fit well into the generate-and-test control structure of our algorithm and into our strategy of starting with the simplest image areas and then progressing into areas where a more complex (or less conservative) analysis is required.

7.4 Simplifying Heuristics

Our algorithm relies on a few simplifying heuristics. They are all related to the problem of determining the structure of a color cluster. Our algorithm currently works with a built-in, very rigid model of what color cluster shapes to expect, assuming that clusters always consist of two linear subclusters that form a skewed T. We now discuss these heuristics.

- *Large Linear Clusters Are Matte Clusters:* As discussed above, the algorithm assumes that linear hypotheses from large image areas describe matte pixels on an object. This heuristic depends on the curvature on the objects in the scene, as well as on their distance from the camera.
- *Color Clusters Have Only One Highlight Branch:* When the algorithm selects a surface reflection vector from a list of highlight candidates, it implicitly assumes that the clusters of all highlight candidates are part of the same highlight cluster. This may not be the case for convex objects. The two planar faces in figure 13 each reflect a highlight into the camera. The incidence angle between the illumination ray and the surface normal is larger at point P_1 than at point P_2. Accordingly, the body reflection component under the highlight at P_1 is smaller than the body reflection component under the highlight at P_2 and the two highlights generate two separate highlight clusters in the dichromatic plane.
- *50% Heuristic:* The 50% heuristic also makes limiting assumptions on the scene. It has only been evaluated so far for spherical objects that were illuminated and viewed from the same distance. Different conditions may exist on cylindrical, elliptical, or planar objects. The heuristic does not apply to convex objects since, on such objects, the brightest visible matte point may lie on one surface while the highlight is on another surface with different illumination conditions. The highlight may then start at an arbitrary place on the matte line.

7.5 Limitations of the Dichromatic Theory and Extensions

The dichromatic theory imposes some limitations on our approach. Since we attribute any linear color variation to the changing illumination geometry of a single material, we are unable to find material boundaries between objects with collinear matte clusters. We will need a geometrical analysis, linking intensity gradients to object shape, to distinguish between such objects. The same will be needed to analyze dark image areas which are currently excluded because their color information is too unreliable.

The model also makes simplifying assumptions about the illumination conditions and the materials in the scene. A color cluster from an object in an unconstrained scene will generally not be a skewed T composed of linear subclusters because the illumination color may vary on different parts on the object surface, and the reflection properties of the object may also change, due to illumination changes and to pigment variations in the material body. The necessary extensions to the model will be the subject of future work.

Furthermore, our method to split color pixels into their reflection components (but not the segmentation) relies on a characteristic color change between the matte object color and the highlight color. There needs to be a certain angle between the orientations of the body and surface reflection vectors of an object. How big the angle needs to be depends on the cylinder width. If the matte and highlight clusters are approximately collinear,

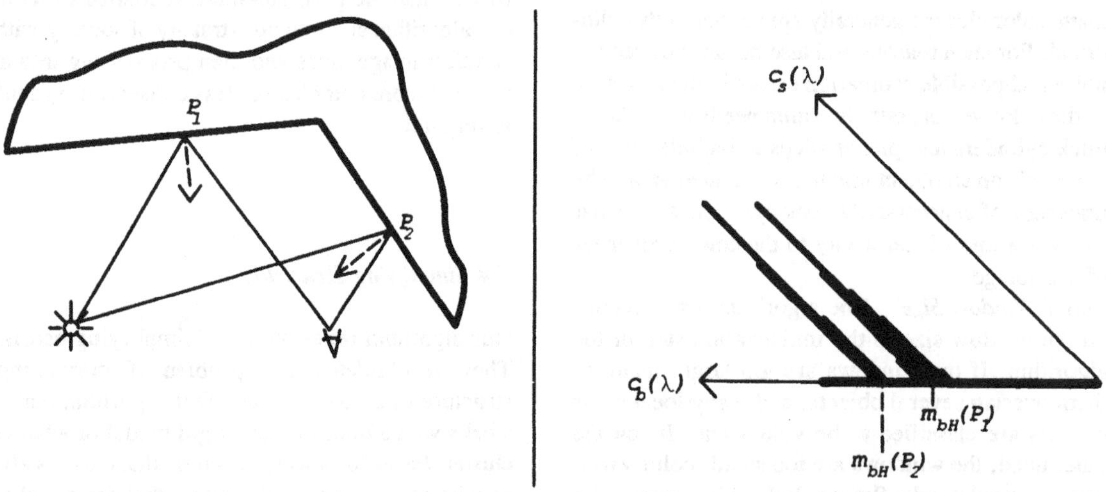

Fig. 13. Cluster shapes for concave objects.

we cannot separate the reflection components. This is the case, if an object has the same color as the light source or if the object is a uniform reflector, such as grey objects or pale pastels. We also have problems when one of the two linear clusters does not exist or is very small for an object. The matte cluster is missing, if the viewed object is very dark, or if the scene is illuminated with a narrow-band illuminant that does not overlap with the wavelengths at which the material reflects light. Matte clusters also do not exist for metallic objects. On the other hand, the highlight cluster may be missing, if an object does not reflect a highlight into the camera, due to its position in the scene and the illumination geometry. As a third case, we need to consider objects with very rough surfaces such that every pixel in the image area has both a significant body and a surface reflection component. The color cluster may then fill out the entire dichromatic plane. A common special case of this is so-called "matte" or "Lambertian" materials—as opposed to glossy materials—which reflect a constant amount of surface reflection in every direction and thus never exhibit a highlight in the common sense of the word. The corresponding color clusters are linear clusters, translated from the origin of the color space according to the constant surface reflection component. Our current method is not capable of distinguishing among all these cases. In combination with exploiting previously determined scene properties, such as the illumination color, we will need to analyze the intensity gradients along the linear axes and relate them to the properties of m_s and m_b, as described in a geometrical model of light reflection.

The Dichromatic Reflection Model also makes several simplifying assumptions which restrict the illumination conditions and the materials in the scene. Because of these assumptions the model can currently only be used to understand scenes taken under very restricted conditions. The results of using the reflection model in analyzing real images, such as the image in color figure 1, have shown that even images of carefully selected scenes often contain areas in which the model does not apply (see color figure 10). To provide a broader applicability of the method, the model needs to be extended to account for more physical processes, especially for more general illumination conditions. Inter-reflection and shadow-casting between objects exist in nearly every moderately complicated scene. Furthermore, many realistic scenes will contain more than a single, point-like light source. There will generally be some amount of ambient light and possibly even several light sources that shed differently colored light onto the scene. A first step in the direction of modeling more complicated illumination conditions has been suggested by Shafer and others [3, 30], adding a constant color term to the reflection model to account for ambient light. Tong and Funt have presented promising first results on analyzing real images with ambient light [34]. Their results indicate that the extended Dichromatic Reflection Model is appropriate for describing color variations under direct and ambient light. We expect our segmentation algorithm to be applicable to images taken under such illumination conditions. Since ambient light adds a constant amount of light to each pixel in an object area, it translates the entire color cluster away from the origin of the color space—without any change to the shape of the cluster. The principal component analysis captures the effect of translational color transformations in the mean color value of the cluster. Our segmentation algorithm uses both the color mean and the principal directions for positioning color cylinders and color planes in the cube, and thus should work as well under ambient light as it does without. However, the intrinsic reflection images are now presenting more complicated body or interface reflection processes since they are influenced by two types of illumination. The ambient light (a constant) may have to be subtracted out before a shape-from-shading or a shape-from-highlights method can be applied.

8 Conclusions

In this paper, we have demonstrated that it is possible to analyze and segment real color images by using a physics-based color reflection model. Our model accounts for highlight reflection and matte shading, as well as for some characteristics of cameras. By developing a physical description of color variation in color images, we have developed a method to automatically segment an image while generating hypotheses about the scene. We then use the knowledge we have gained to separate highlight reflection from matte object reflection. The resulting intrinsic reflection images have a simpler relationship to the illumination geometry than the original image and may thus improve the results of many other computer vision algorithms, such as motion analysis, stereo vision, and shape from shading or highlights [5, 12, 14, 33]. Since the surface reflection component of dielectric materials generally has the same color as the illumination, we can also determine the illumination color from the intrinsic surface reflection image, information which is needed by color constancy algorithms [4, 13, 23, 24].

The key points leading to the success of this work are our modeling of highlights as a linear combination of both body and surface reflection and our modeling of the camera properties. With few exceptions [4, 6, 23, 28], previous work on image segmentation and highlight detection has assumed that the color of highlight pixels is completely unrelated to the object color. This assumption would result in two unconnected clusters in the color space: one line or ellipsoid representing the object color and one point or sphere representing the highlight color. Our model and our color histograms demonstrate that, in real scenes, a transition area exists on the objects from purely matte areas to the spot that is generally considered to be the highlight. This transition area determines the characteristic shapes of the color clusters, which is the information we use to distinguish highlight boundaries from material boundaries and to detect and remove highlights. This view of highlights should open the way for quantitative shape-from-gloss analysis, as opposed to binary methods based on thresholding intensity.

By modeling the camera properties, we are able to obtain high-quality color images (through color balancing and spectral linearization) in which most pixels maintain the linear properties of light reflection, as described in the Dichromatic Reflection Model. We can also detect most distorted color pixels in an image and thus generate an intrinsic error image which then guides our algorithm to separate only undistorted color pixels into their reflection components. We expect that the intrinsic error image will be similarly useful in guiding other computer vision algorithms, such as shape from shading. It may also enable us to automatically control the camera aperture so that we can obtain color images with minimal clipping and blooming.

Our hypothesis-based approach toward image analysis may provide a new paradigm for low-level image understanding. Our method gains its strengths from using an intrinsic model of physical processes that occur in the scene. The result is intrinsic images and hypotheses which are closely related in their interpretation to the intrinsic model, being instantiations of concepts formulated in the model. Our system alternates between a bottom-up step which generates hypotheses and a top-down step which applies the hypotheses to the images. Our analysis thus consists of many small, complete interpretation cycles that combine bottom-up processing with feedback in top-down processing. This approach stands in contrast to traditional image segmentation methods which do not relate their analysis to intrinsic models and also, generally, have a strictly bottom-up control structure. We feel that many low-level image understanding methods such as shape-from-x methods, stereo, and motion analysis may be viewed and approached under this paradigm. We hope to extend our approach into a more complete low-level image analysis system which combines color analysis with a geometrical analysis of the scene, exploiting the body and surface reflection images. Along these lines, we may generate hypotheses about object shapes and about the object materials (see [13]). The highlight image may also provide strong evidence for the position of the light source.

Although the current method has only been applied in a laboratory setting, its success shows the value of modeling the physical nature of the visual environment. Our work and the work of others in this area may lead to methods that will free computer vision from its current dependence on statistical signal-based methods for low-level image analysis.

Acknowledgment

This research was supported by the Jet Propulsion Laboratory, California Institute of Technology, under NASA contract 957989, by the National Science Foundation under Grant DCR-8419990 and by the Defense Advanced Research Projects Agency (DOD), ARPA Order No. 4976, monitored by the Air Force Avionics Laboratory (Air Force Wright Aeronautical Laboratories, Aeronautical Systems Division (AFSC), United States Air Force, Wright-Patterson AFB, OHIO 45433-6543) under contract F33615-87-C-1499 (Arpa Order #4976, Amendment 20). Any opinions, findings, and conclusions or recommendations expressed in this publication are those of the author and do not necessarily reflect the views of the Jet Propulsion Laboratory, the National Science Foundation, the Defense Advanced Research Projects Agency, or the US Government.

References

1. D.H. Ballard and C.M. Brown, *Computer Vision*, Englewood Cliffs, NJ: Prentice-Hall, 1982.
2. H.G. Barrow and J.M. Tenenbaum, "Recovering intrinsic scene characteristics from images," *Computer Vision Systems.* New York: Academic Press, pp. 3–26, 1978.
3. R.L. Cook and K.E. Torrance, "A reflectance model for computer graphics," *ACM Trans. on Graphics* 1 (1):7–24, 1982. Also published in *Computer Graphics* 15(3), SIGGRAPH'81.
4. M. D'Zmura and P. Lennie, "Mechanisms of color constancy," *J. Opt. Soc. Amer.* A 3(10):1662–1672, 1986.
5. L. Dreschler and H.-H. Nagel, "Volumetric model and 3D Trajectory of a moving car derived from monocular TV frame sequences of a street scene," *Comput. Graphics Image Process.* 20:199–228, 1982.
6. R. Gershon, "The use of color in computational vision," Ph.D. thesis, Department of Computer Science, University of Toronto, 1987.
7. R. Gershon, A.D. Jepson, and J.K. Tsotsos, "Highlight identification using chromatic information," *Proc. 1st Intern. Conf. Computer Vision*, pp. 161–171. IEEE, London, June 1987.
8. H. Grassmann, "On the theory of compound colors," *Phil. Mag.*, April 1854.
9. A.R. Hanson and E.M. Riseman, "VISIONS: A computer system for interpreting scenes." In A.R. Hanson and E.M. Riseman (eds.), *Computer Vision Systems*, New York: Academic Press, pp. 303–333, 1978.
10. H.H. Harman, *Modern Factor Analysis*, 2nd. ed. Chicago: University of Chicago Press, 1967.
11. G. Healey and T.O. Binford, "Local shape from specularity." In L.S. Bauman (ed.), *DARPA-Image Understanding (IUS) Workshop*, pp. 874–887. Los Angeles, CA, February 1987.
12. G. Healey and T.O. Binford, Local shape from specularity," *Proc. 1st Intern. Conf. Computer Vision*, pp. 151–161. IEEE, London, June 1987.
13. G. Healey and T.O. Binford, "The role and use of color in a general vision system." In L.S. Bauman (ed.), *DARPA-Image Understanding (IUS) Workshop*, pp. 599–613. Los Angeles, CA, February 1987.
14. B.K.P. Horn, "Understanding image intensities," *Artificial Intelligence* 8(11):201–231, 1977.
15. B.K.P. Horn, "Exact reproduction of colored images," *Comput. Vision, Graphics, Image Process.* 26:135–167, 1984.
16. B.K.P. Horn and B.G. Schunk, "Determining optical flow," *Artificial Intelligence* 17:185–203, 1981.
17. T. Kanade, "Region segmentation: Signal vs. semantics." *Proc. 4th Intern. Joint Conf. Pattern Recog.* pp. 95–105. IEEE, Kyoto, Japan, November 1978.
18. J.R. Kender, "Shape from texture." Ph.D. thesis, Computer Science Department, Carnegie-Mellon University, November 1980. Appeared also as technical report CMU-CS-81-102, Computer Science Department, Carnegie-Mellon University, Pittsburgh, PA.
19. J.R. Kender and E.M. Smith, "Shape from darkness: Deriving surface information from dynamic shadows," *Proc. 1st Intern. Conf. Computer Vision*, pp. 539–546. IEEE, London, June 1987.
20. G.J. Klinker, "A physical approach to color image understanding." Ph.D. thesis, Computer Science Department, Carnegie-Mellon University, May 1988. Available as technical report CMU-CS-88-161.
21. G.J. Klinker, S.A. Shafer, and T. Kanade, "Using a color reflection model to separate highlights from object color," *Proc. 1st Intern. Conf. Comp. Vision*, pp. 145–150. IEEE, London, June 1987.
22. G.J. Klinker, S.A. Shafer, and T. Kanade, "The measurement of highlights in color images," *Intern. J. Comput. Vision* 2(1):7–32, 1988.
23. H.-C. Lee, "Method for computing the scene-illuminant chromaticity from specular highlights," *J. Opt. Soc. Amer.* 3(10):1694–1699, 1986.
24. L.T. Maloney and B.A. Wandell, "Color constancy: A method for recovering surface spectral reflectance," *J. Opt. Soc. Amer.* A 3(1):29–33, 1986.
25. R. Ohlander, K. Price, and D.R. Reddy, "Picture segmentation using a recursive region splitting method," *Comput. Graphics Image Process.* 8:313–333, 1978.
26. Y. Ohta, T. Kanade, and T. Sakai, "Color information for region segmentation," *Comput. Graphics Image Process.* 13:222–231, 1980.
27. A. Rosenfeld and A.C. Kak, *Digital Picture Processing*, 2nd ed. New York: Academic Press, 1982.
28. J.M. Rubin and W.A. Richards, "Color vision and image intensities: When are changes material?" *Biological Cybernetics* 45:215–226, 1982.
29. S.A. Shafer, "Describing light mixtures through linear algebra," *J. Opt. Soc. Amer.* 72(2):299–300, 1982.
30. S.A. Shafer, "Using color to separate reflection components," *COLOR research and application* 10(4):210–218, 1985. Also available as technical report TR-136, Computer Science Department, University of Rochester, NY, April 1984.
31. S.A. Shafer and T. Kanade, "Recursive region segmentation by analysis of histograms," *Proc. Intern. Conf. Acoustics, Speech,* and *Signal Process.*, pp. 1166–1171. IEEE, Paris, France, May, 1982.
32. S.A. Shafer and T. Kanade, "Using shadows in finding surface orientations," *Comput. Vision, Graphics, Image Process.* 22:145–176, 1983.
33. C.E. Thorpe, "FIDO: Vision and navigation for a robot rover." Ph.D. thesis, Computer Science Department, Carnegie-Mellon University, December 1984. Available as technical report CMU-CS-84-168.
34. F. Tong and B.V. Funt, "Specularity removal for shape from shading," *Proc. Conf. Vision Interface*, Edmonton, Alberta, Canada, 1988.
35. S. Ullman, "The interpretation of structure from motion." AI Memo 476, MIT AI Laboratory, Cambridge, MA, October 1976.

COLOR FIGURES

See color plates B-C for Color Figures 1-15.

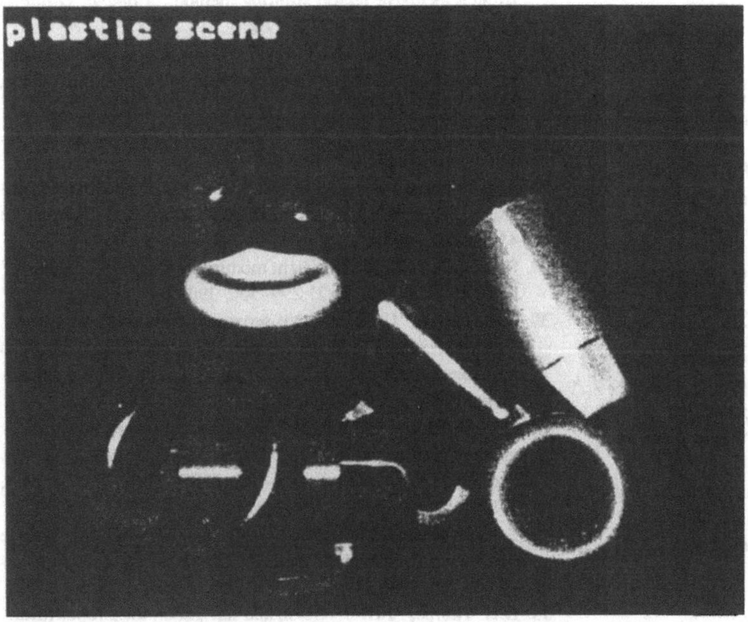

Color Figure 1. Scene with eight plastic objects.

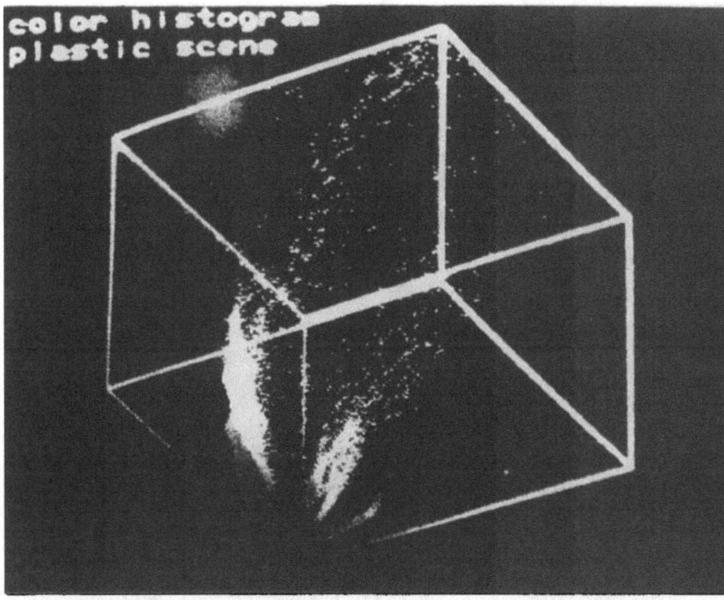

Color Figure 2. Color histogram of the scene with eight plastic objects.

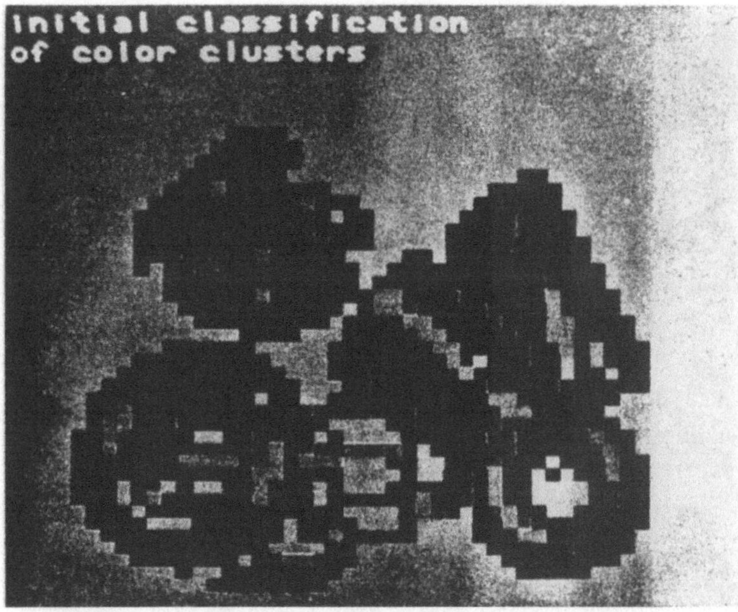

Color Figure 3. Color cluster classification for initial image areas.

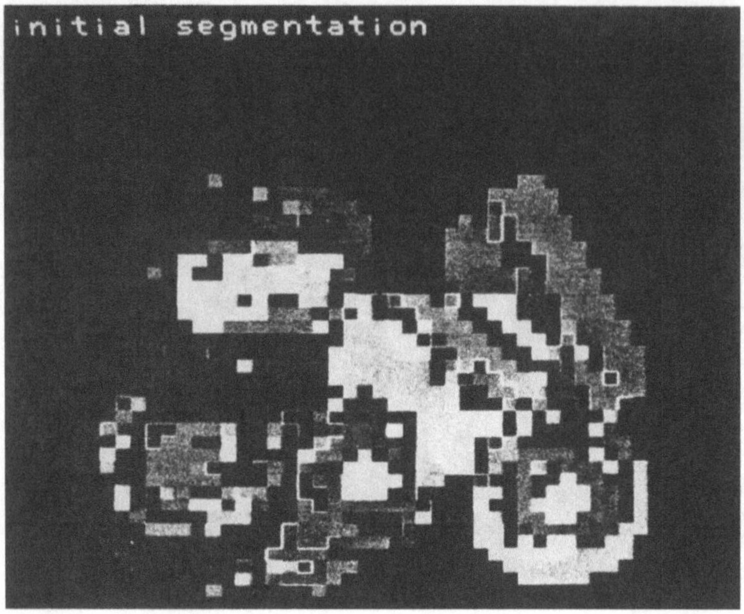

Color Figure 4. Initial grouping into approximate image areas.

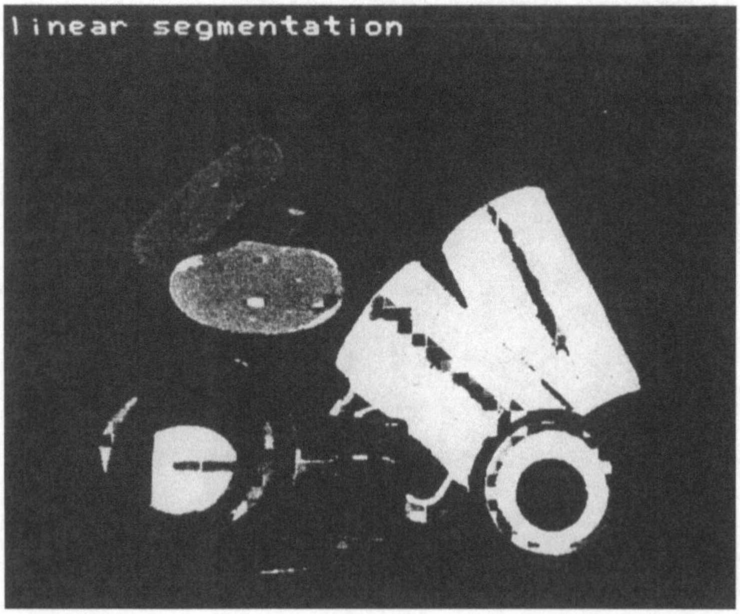

Color Figure 5. Linear segmentation.

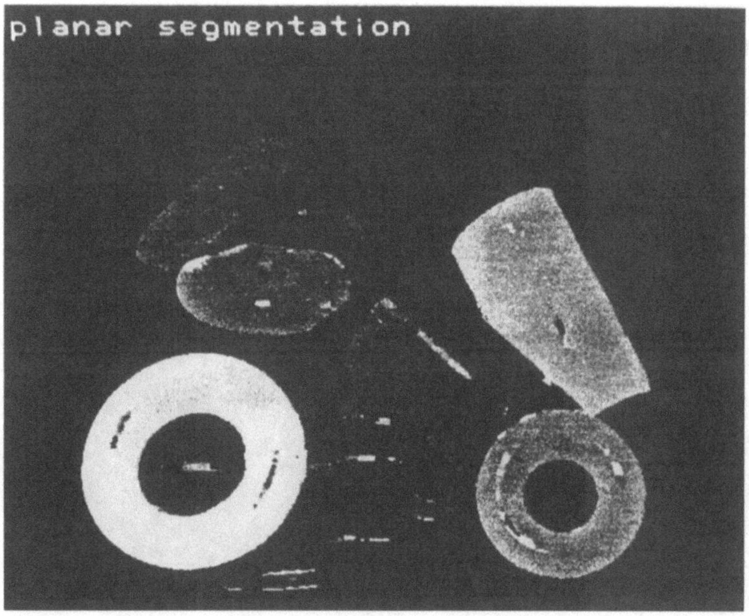

Color Figure 6. Planar segmentation.

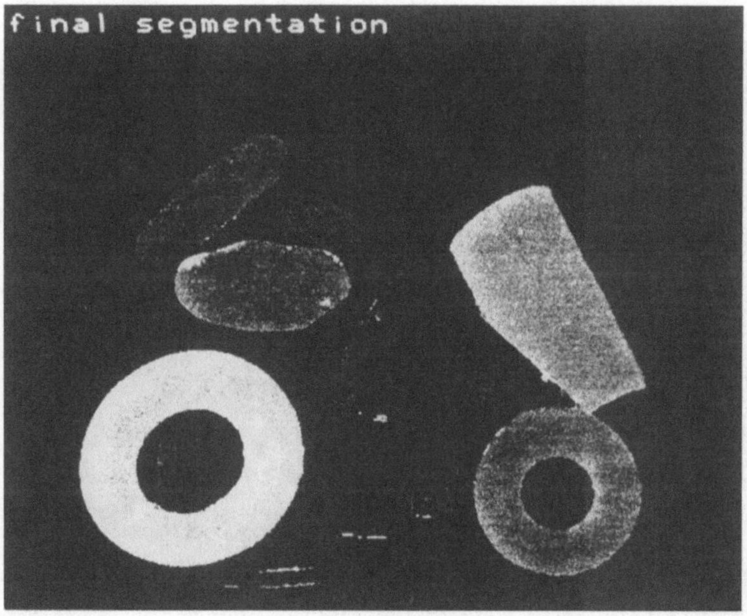

Color Figure 7. Final segmentation.

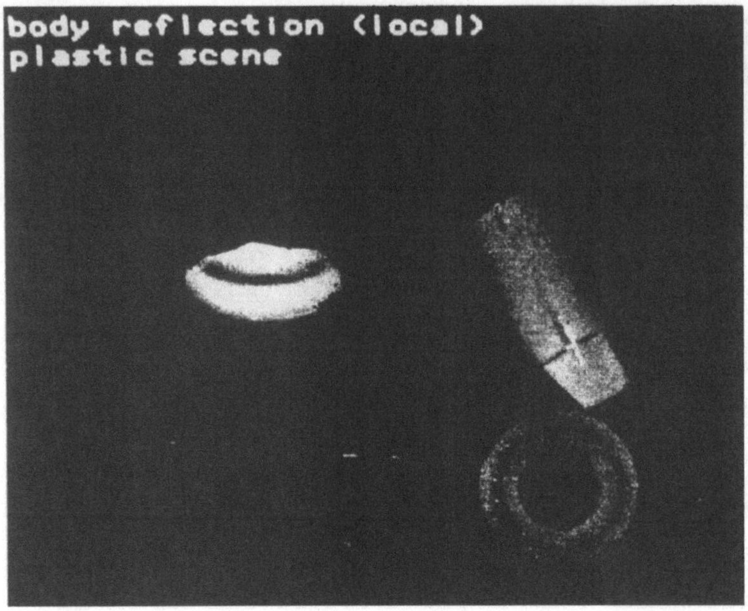

Color Figure 8. Intrinsic body reflection image.

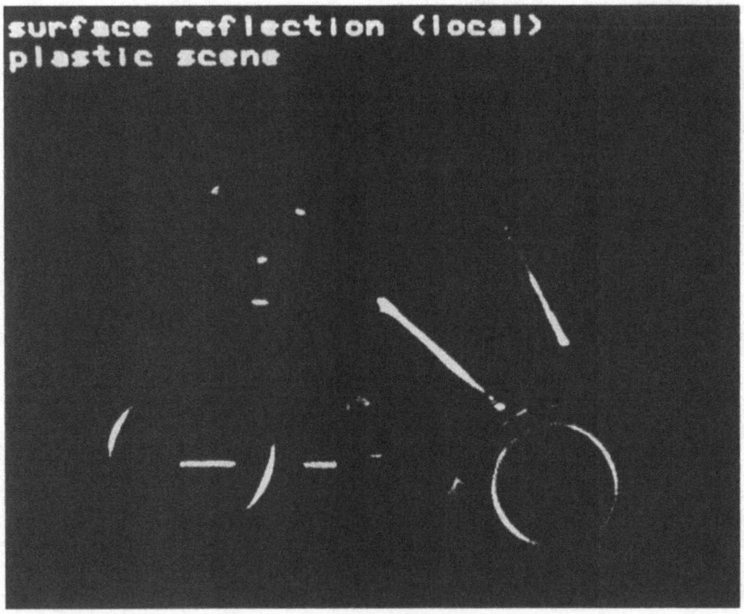

Color Figure 9. Intrinsic surface reflection image.

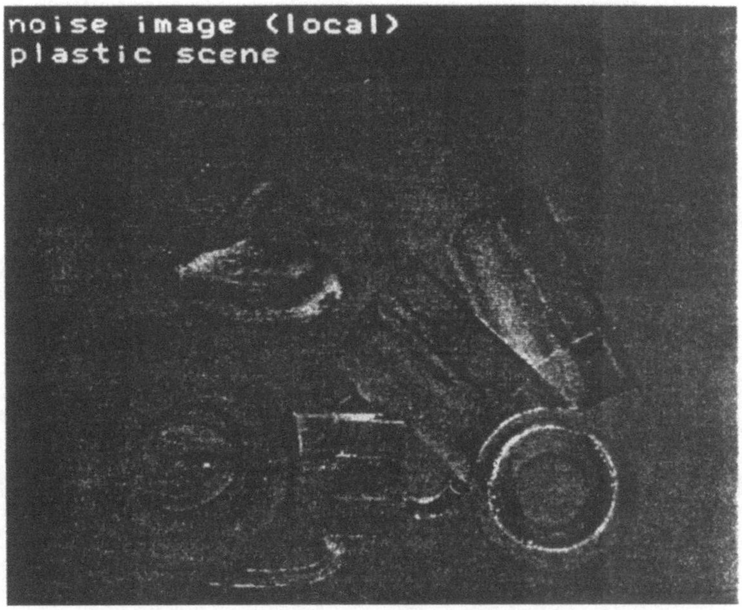

Color Figure 10. Intrinsic noise image.

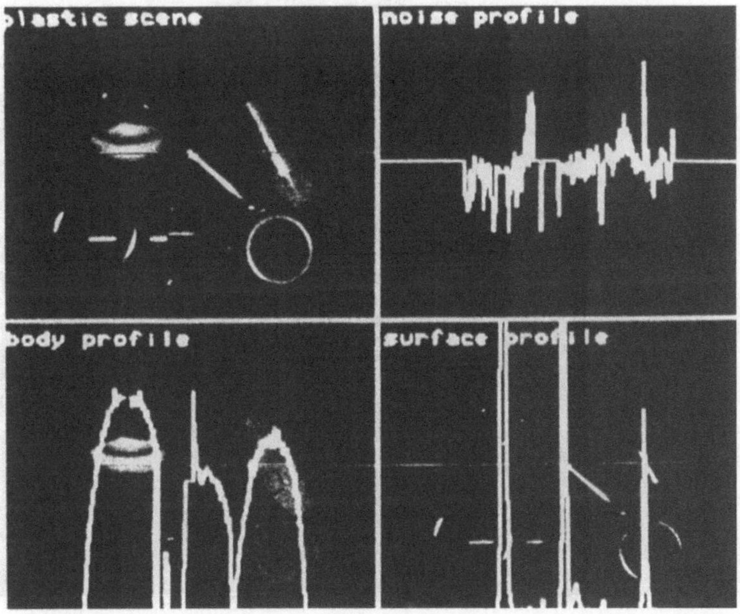

Color Figure 11. Profiles of the intrinsic reflection images along row 226.

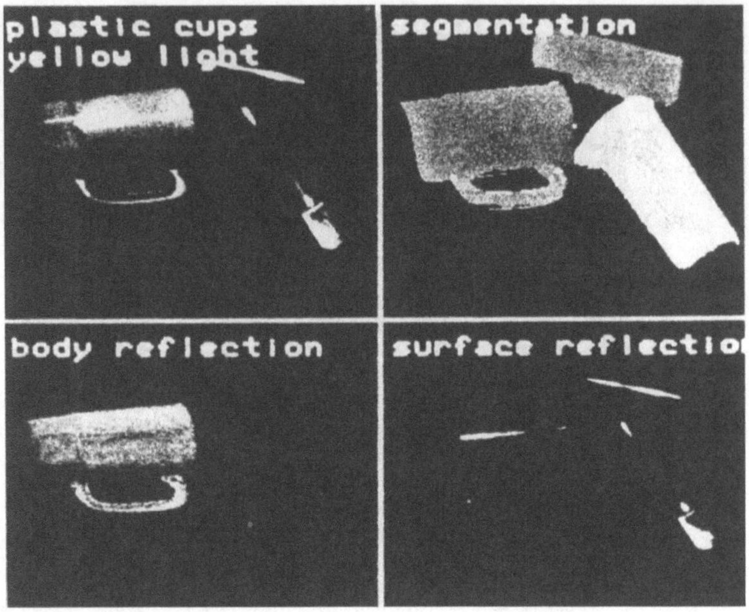

Color Figure 12. Color image segmentation and reflection analysis, scene with three plastic cups under yellow light.

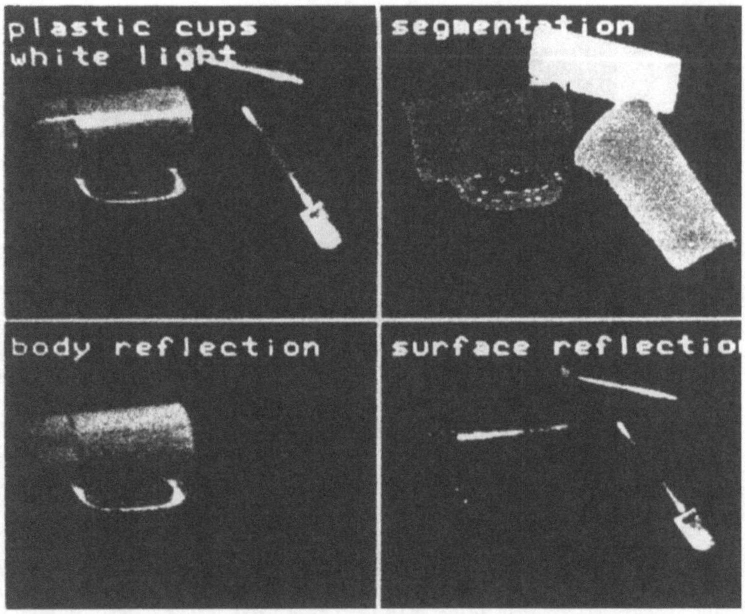

Color Figure 13. Color image segmentation and reflection analysis, scene with three plastic cups under white light.

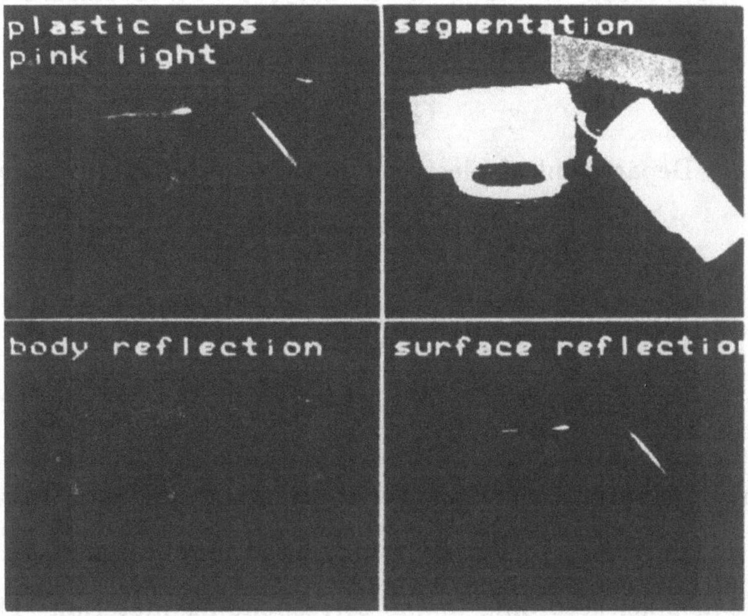

Color Figure 14. Color image segmentation and reflection analysis, scene with three plastic cups under pink light.

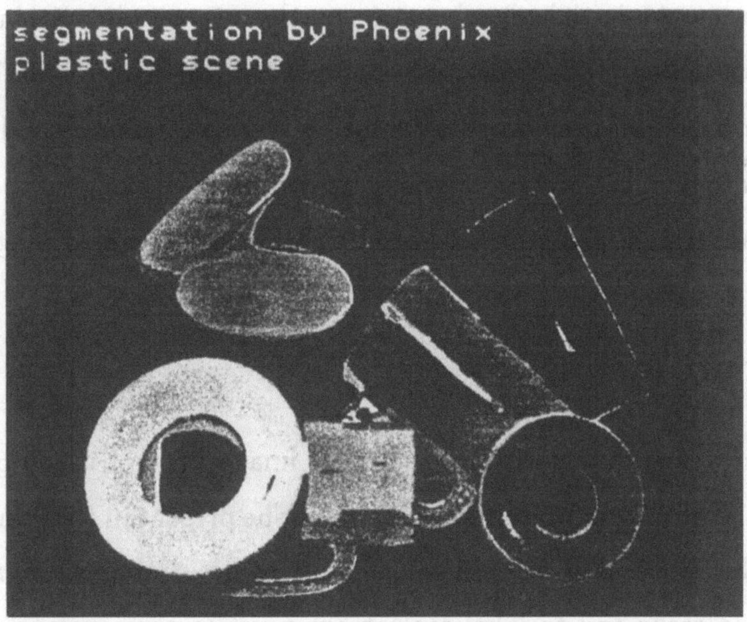

Color Figure 15. Region segmentation by Phoenix, scene with eight plastic objects.

Segmenting Images using Normalized Color

Glenn Healey
Department of Electrical and Computer Engineering
University of California
Irvine, CA 92717

Abstract

This paper presents an algorithm for segmenting images of 3-D scenes. From an input color image, the algorithm determines the number of materials in the scene and labels each pixel according to the corresponding material. This segmentation is useful for many visual tasks including 3-D inspection and 3-D object recognition. The segmentation algorithm is based on a detailed analysis of the physics underlying color image formation and may be applied to images of a wide range of materials and surface textures. An initial edge detection on the intensity image is used to guide the segmentation process and to ensure accurate localization of region boundaries. The algorithm is based on the computation of local image features and can be mapped effectively onto high performance parallel hardware. Issues related to illumination and sensors are addressed. Experimental results obtained for several images are presented.

1 Introduction

This paper develops an algorithm to segment images when little information is available about the objects or materials which might be present in the scene. In particular, I consider the problem of using a color image to determine the number of different materials in a scene and to label each image pixel as corresponding to one of these materials. Color information is essential to the segmentation algorithm because color allows the computation of image statistics which are independent of geometrical variations in the scene [11]. Therefore color, unlike image irradiance, can be used to group together regions of an image which correspond to surfaces of the same material in the scene independent of surface shape and viewing geometry.

Researchers have proposed many different color segmentation techniques. One frequently used approach [5] [10] [24] is to use a clustering procedure [7] on the measured sensor values. Such a procedure attempts to find clusters of pixel values in color space and then assigns each image pixel to one of the clusters. Clustering techniques suffer from several disadvantages. In most real images, color clusters will overlap making pixel misclassification inevitable. Also, it is possible that surfaces of a single material in the scene which are viewed under different geometrical conditions will give rise to different clusters in color space causing these surfaces to be assigned different labels.

Another approach to color segmentation is region splitting. Region splitting involves recursively breaking the input image into smaller and smaller pieces until each piece is uniform in some property. Ohlander [21] used histograms of feature values computed from a color image to guide region splitting. For each region this algorithm computes nine histograms from the input R, G, B values. The best peak in one of the histograms is used to threshold the region to separate out the pixels corresponding to the peak. Since histogram analysis is a global process, this method tends to miss small image regions which will not produce strong peaks in any histogram.

Ohta [22] experimented with Ohlander's algorithm to determine empirically whether a small number of color features might be adequate for segmentation. He applied the Karhunen-Loeve transformation to color images to derive color features with large discriminatory power. Ohta determined from experiments on eight color images that an effective set of color features is given by $I_1 = (R+G+B)/3, I_2 = R-B$, and $I_3 = (2G-R-B)/2$ where I_1 is the most effective for segmentation and I_3 is the least effective. Though Ohta's analysis improves the Ohlander algorithm, it seems doubtful that eight images will be rich enough to lead to the determination of color features which are universally effective for segmentation. Another property of Ohta's color features is that the most significant color feature, I_1, depends on the scene geometry. Applying Ohlander's algorithm with Ohta's color features will cause curved surfaces of a single material to be broken into separate regions as in, for example, the segmentation of the color cylinder presented in [22].

Gershon [8] used a split-and-merge technique [12] for segmentation using color features which are similar to those used in [22] with the exception that a Karhunen-Loeve transformation is used to compute appropriate color features for each image. Since the resulting color features are linear combinations of the sensor measurements, these features will depend on scene geometry. Thus, this algorithm will face the same difficulty faced by Ohta's algorithm in that curved surfaces of a single material will be segmented into separate regions. Gershon presents no experimental results obtained with the segmentation algorithm in [8], but rather presents results obtained in identifying highlights following the segmentation [9]. Therefore, it is difficult to assess the performance of the segmentation algorithm.

More recently, Klinker [15] has developed a color segmentation algorithm based on the dichromatic reflection model [26]. Such an approach is useful because the derived algorithm can account for the processes which contribute to image formation. Furthermore, it is easier to characterize what kinds of images an algorithm derived from a physical model is likely to segment successfully. Klinker's algorithm segments the image into connected regions corresponding to surfaces of a single material. It is more sophisticated than previous segmentation techniques in that it is designed not to break the image because of highlights or because of intensity changes due to geometrical variation in the scene. This algorithm does, however, have several important limitations. Since the algorithm is derived from the dichromatic reflection model, it only applies to inhomogeneous dielectric materials. The algorithm uses a local approach to segmentation which is based on the assumption that the image can be partitioned into small fixed-sized regions that are large enough to provide useful information about a surface, but which are usually small enough to correspond to only one material. Such an assumption is often not reasonable. In situations where the assumption applies, the algorithm requires some way of choosing the size for the elementary regions. The algorithm uses a cylindrical model for clusters in color space which is unable to model accurately clusters having a noncylindrical shape. Thus, these cylinders must be made unnecessarily large making it difficult for the algorithm to discriminate between similar materials. The algorithm also does not attempt to identify separated image regions as instances of the same material in the scene.

2 Normalized Color

In [11], it is shown from general physical models that the reflectance of an opaque sample of material with one relevant interface can be accurately approximated by a function of the form

$$R(g,\lambda) = \begin{cases} M_S(g)C_S(\lambda) & \text{Metal} \\ M_S(g)C_S(\lambda) + M_B(g)C_B(\lambda) & \text{I.D.} \end{cases} \tag{1}$$

where g indicates dependence on the photometric angles and λ indicates dependence on wavelength. Figure 1 defines the photometric angles $\theta_l, \theta_v, \theta_p$ at a point on a surface. Coordinates are chosen so that the surface normal lies in the direction of the Z axis. θ_l is the angle between the source direction and the surface normal. θ_v is the angle between the viewing direction and the surface normal. θ_p is the angle between the surface projections of the source direction and the viewing direction. The subscript S denotes terms associated with surface reflection and the subscript B indicates terms associated with body reflection. I.D. indicates an inhomogeneous dielectric material.

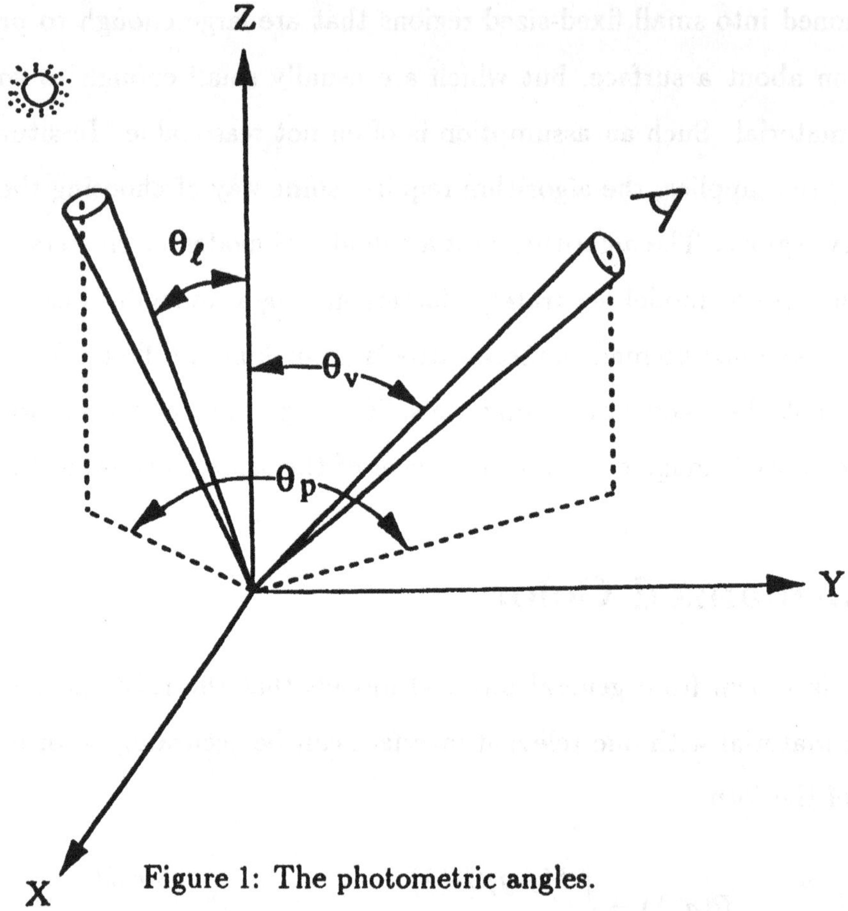

Figure 1: The photometric angles.

Since surface reflection from an inhomogeneous dielectric is concentrated in a single direction [11], the contribution of $M_S(g)$ to measured sensor values for a dielectric surface illuminated from a single direction is zero for most pixels in the image of the surface. The contribution of $M_S(g)$ is, of course, nonzero at highlights. If we temporarily assume that we are not viewing an inhomogeneous dielectric highlight region in the image, then (1) may be approximated by

$$R(g, \lambda) = M(g)C(\lambda) \tag{2}$$

which applies to both metals and inhomogeneous dielectrics.

Consider using N sensors with spectral sensitivities given by $f_i(\lambda)$ $(0 \leq i \leq N-1)$ to obtain an image of a surface illuminated by a spectral power distribution $L(\lambda)$. At each point (x,y) in the image of the surface, the measured sensor values s_i will be given by

$$s_i(x,y) = \int_\lambda f_i(\lambda)L(\lambda)M(g)C(\lambda)d\lambda \quad (3)$$

$$= M(g)\int_\lambda f_i(\lambda)L(\lambda)C(\lambda)d\lambda. \quad (4)$$

where scene geometry g is a function of image location (x, y). Thus for a surface of a material with reflectance given by (2) illuminated by a single spectral power distribution $L(\lambda)$, the measured sensor values for that surface will lie on a line in sensor space which intersects the origin. An important special case in which the assumption of a single illuminating spectral power distribution can be violated occurs when there is inter-reflection among different surfaces in a scene. Inter-reflection is not addressed in this paper but has recently been considered by others [27], [3].

Let $S = (s_0, \cdots, s_{N-1})$ denote the measured sensor vector at a point in the image and let $\widehat{S}$ denote the L^2 normalization of the sensor vector

$$\widehat{S} = \frac{S}{\sqrt{s_0^2 + \cdots + s_{N-1}^2}} = (\widehat{s}_0, \cdots, \widehat{s}_{N-1}). \quad (5)$$

If we let

$$K_i = \int_\lambda f_i(\lambda)L(\lambda)C(\lambda)d\lambda \quad (6)$$

then from (4)

$$\widehat{s}_i(x,y) = \frac{M(g)K_i}{\sqrt{M(g)^2(K_0^2 + \cdots + K_{N-1}^2)}} = \frac{K_i}{\sqrt{K_0^2 + \cdots + K_{N-1}^2}} \quad (7)$$

and $\widehat{s}_i(x, y)$ depends on the sensors, the illuminant, and the reflecting material, but not on the scene geometry g.

Typically normalized color space coordinates (also called chromaticity coordinates) are generated using the L^1 norm [13],[30] so that

$$\widehat{S} = \frac{S}{s_0 + \cdots + s_{N-1}}. \quad (8)$$

It is possible to show from (4) that as with the L^2 normalized coordinates, these L^1 normalized coordinates are independent of scene geometry. One property of this L^1 normalization is that for two distinct material lines in sensor space which are separated by fixed angle, the distance between the normalized coordinates corresponding to these lines depends on the location of the lines in sensor space. To see this in two dimensions, consider two material lines L_0 and L_1 separated by an angle θ as depicted in figure 2. Let θ_1 denote the angle between L_0 and the line $S_1 = 0$. If an L^1 normalization is used (figure 3), then the euclidean distance d_1 between the points where L_0 and L_1 intersect the line $s_0 + s_1 = 1$ is given by

$$d_1 = \sqrt{2}\left[\frac{tan(\theta+\theta_1)}{1+tan(\theta+\theta_1)} - \frac{tan\theta_1}{1+tan\theta_1}\right] \tag{9}$$

From (9), d_1 depends on θ_1.

If an L^2 normalization is used (figure 4), then the euclidean distance d_2 between the points where L_1 and L_2 intersect the circle $s_0^2 + s_1^2 = 1$ is given by

$$d_2 = 2sin\left(\frac{\theta}{2}\right). \tag{10}$$

Therefore, d_2 depends only on θ and not on the location of L_0 and L_1 in sensor space. The situation is similar for higher dimensional sensor spaces. Hence L^1 normalization has this anisotropic property when used for material discrimination. The L^2 normalization is not anisotropic in this sense and will be better suited for our material discrimination task.

It has been suggested on several occasions, e.g. [1], [7], [25], that the normalized color of two points on a surface of the same material will be the same even if one of the points is directly illuminated and the other point is in a shadow. From (7), we see that this will not be generally true unless the light incident on the surface in the shadow has the same spectral power distribution as the direct illumination. This is often not the case. In general, then, the normalized color corresponding to a point in

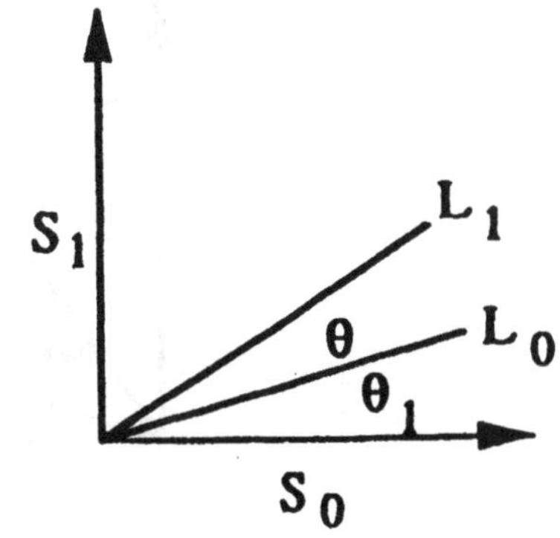

Figure 2: Two lines in sensor space.

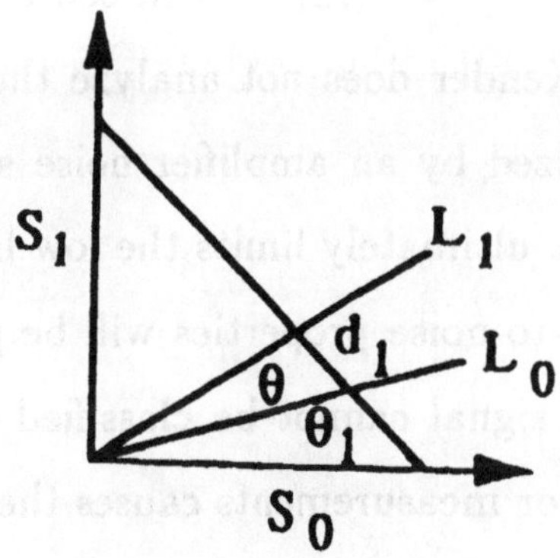

Figure 3: L^1 normalization

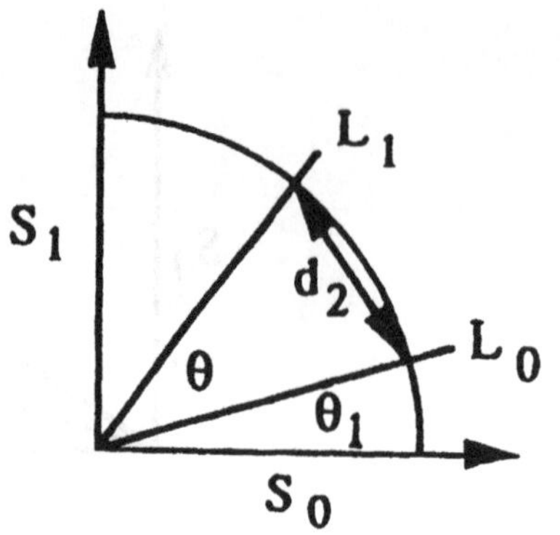

Figure 4: L^2 normalization

a shadow will be different from the normalized color of a directly illuminated point of the same material.

Kender [13] has analyzed two undesirable properties associated with computing normalized color using the L^1 norm. First, he points out that the normalized color transformation has a nonremovable singularity at the zero signal point $s_0 = s_1 = \cdots = s_{N-1} = 0$ in sensor space and is highly unstable near this point. Though Kender does not analyze the effects of noise, solid-state image sensors are characterized by an amplifier noise source which is independent of signal level [6]. This noise ultimately limits the low light level performance of the sensor. Thus sensor signal-to-noise properties will be poor near zero signal. For these reasons, pixels near zero signal cannot be classified reliably. Kender also suggests that using digitized sensor measurements causes the normalized color distribution to have spurious modes and gaps. Both undesirable properties of the normalized color transformation mentioned in [13] will be taken into account by the segmentation algorithm developed

3 The Segmentation Problem

Several things might cause the sensor values measured for points on a surface to lie some distance from the line given by (4). Sensor noise will cause some scatter of points about the line. Small variations in material composition from point to point on a surface described by small variations in the function $C(\lambda)$ will also scatter points about the line. The material discrimination problem is therefore equivalent to the problem of discriminating corrupted lines in N-dimensional sensor space. For most scenes, sensor noise and surface texture due to small variations in material composition will cause lines in sensor space corresponding to different materials to overlap. Therefore it is important to characterize these lines accurately.

Consider a simplified version of the segmentation problem. Suppose that it is known that there are M materials in the scene. We can model the distribution of measured sensor vectors for each material class c_i $(0 \leq i \leq M-1)$ using the multivariate normal density $N(\mu_i, \Sigma_i)$ described by

$$p(S/c_i) = \frac{1}{(2\pi)^{N/2}|\Sigma_i|^{1/2}} e^{\frac{-1}{2}(S-\mu_i)^T \Sigma_i^{-1}(S-\mu_i)} \tag{11}$$

where μ_i is the N-element mean vector for material class c_i and Σ_i is the $N \times N$ covariance matrix

$$\Sigma_i = E\left[(S-\mu_i)(S-\mu_i)^T\right]. \tag{12}$$

Suppose we wish to classify a new sensor vector v as belonging to one of the M material classes. To facilitate comparison which is not influenced by the effects of geometry in the scene as discussed in section 2, each distribution $p(S/c_i)$ is normalized to a multivariate normal distribution $N(\hat{\mu}_i, \hat{\Sigma}_i)$ described by

$$\hat{\mu}_i = \frac{\mu_i}{\|\mu_i\|} \tag{13}$$

$$\hat{\Sigma}_i = E\left[\left(\frac{S}{\|\mu_i\|} - \hat{\mu}_i\right)\left(\frac{S}{\|\mu_i\|} - \hat{\mu}_i\right)^T\right] = \frac{\Sigma_i}{\|\mu_i\|^2} \tag{14}$$

where throughout this paper $\|\cdot\|$ is used to denote the L^2 norm.

Applying a similar normalization to v gives

$$\hat{v} = \frac{v}{\|v\|} \tag{15}$$

so that $\hat{\mu}_i$ and $\hat{v}$ both lie on the unit sphere centered at the origin. The problem is to assign $\hat{v}$ to one of the M material classes described by the distributions $N(\hat{\mu}_i, \hat{\Sigma}_i)$ $(0 \leq i \leq M-1)$.

From Bayes decision theory [7], the set of functions

$$g_i(\hat{v}) = \log p(\hat{v}/c_i) + \log p(c_i) \tag{16}$$

serve as minimum error rate discriminant functions for classification. Thus, if we assign $\hat{v}$ to the class c_i for which $g_i(\hat{v})$ is the largest we will achieve minimum error rate classification. If we assume that all classes are equally likely, then $\log p(c_i)$ is independent of i and can be dropped from each discriminant function. If we model $p(\hat{v}/c_i)$ using the multivariate normal distribution $N(\hat{\mu}_i, \hat{\Sigma}_i)$ described by (13) and (14) and eliminate terms which don't depend on i, then we arrive at the set of discriminant functions given by

$$g_i(\hat{v}) = \frac{-1}{2}(\hat{v} - \hat{\mu}_i)^T \hat{\Sigma}_i^{-1}(\hat{v} - \hat{\mu}_i) - \frac{1}{2}\log|\hat{\Sigma}_i| \tag{17}$$

The algorithm described in the next section will use estimates of the discriminant functions of (17) during segmentation.

4 The Segmentation Algorithm

4.1 Pixel Classification

A segmentation algorithm has been implemented which determines the number of materials in a color image and labels each pixel as an instance of one of the materials. Since, in general, a material discontinuity in the scene will cause an image irradiance discontinuity, the first step of the algorithm is to perform an edge detection on a single band image of the scene. This edge detection guides the color segmentation in the sense that areas of the image which do not contain edge pixels are assumed to correspond to a single material in the scene.

Some work in computer vision has been done on detecting edges in color images [19], [20], [23]. This previous work indicates, however, that the edges obtained from a color edge detection are at best only marginally better than those obtained using a single band. Nevatia [19] and Novak and Shafer [20] have shown experimentally that nearly all edges detected in color images are also detected in the corresponding irradiance images. Since edge detection on a single color band requires significantly less computation than edge detection on a color image, performing edge detection using a single band is a reasonable approach for most problems. In a study of human vision, Wolfe [29] has shown that subjects cannot localize color edges when a corresponding brightness edge is not present.

The segmentation algorithm is similar to region splitting in that it examines increasingly smaller regions of the image. When the algorithm begins, the entire image is placed on a region stack. An initially empty material list is maintained which, for each material, contains the maximum likelihood estimate for μ_i and Σ_i given the pixels which have already been assigned to material class c_i. Thus if at some stage in the processing P sensor vector measurements S_{ij} $(0 \leq j \leq P-1)$ have been assigned to class c_i, then the maximum likelihood estimates for μ_i and Σ_i denoted by μ_i^* and Σ_i^* are given by [2]

$$\mu_i^* = \frac{1}{P} \sum_{0 \leq j \leq P-1} S_{ij} \tag{18}$$

$$\Sigma_i^* = \frac{1}{P} \sum_{0 \leq j \leq P-1} (S_{ij} - \mu_i^*)(S_{ij} - \mu_i^*)^T \tag{19}$$

When an image region from the region stack is processed, the region is first examined for edge pixels. If the region contains edge pixels, it may correspond to more than one material in the scene. Such regions are split into subregions and the subregions are placed on the region stack. Currently, all regions are rectangular and, if necessary, are split into four rectangular subregions of equal size.

If a region contains no edge pixels, the region is assumed to correspond to a single material in the scene. For such a region R, a mean normalized color vector $\hat{v}_R$ is computed

$$\hat{v}_R = \frac{\sum_{S \in R} S}{\| \sum_{S \in R} S \|} \tag{20}$$

where each S is a measured sensor vector in R. Since typically R will contain many pixels and the denominator in (20) will not be an integer, the possibility of encountering any spurious modes and gaps associated with normalized color computed by the division of two integers [13] is eliminated. This normalized color vector $\hat{v}_R$ is analogous to the vector $\hat{v}$ in section 3.

The discriminant functions $g_i(\hat{v}_R)$ of (17) are computed for each class in the material list using the estimates

$$\hat{\mu}_i \approx \frac{\mu_i^*}{\|\mu_i^*\|} \tag{21}$$

$$\hat{\Sigma}_i \approx \frac{\Sigma_i^*}{\|\mu_i^*\|^2}. \tag{22}$$

The pixels in R are assigned the label corresponding to the class c_i for which $g_i(\hat{v}_R)$ is the largest. When the pixels in R are labeled with material class c_i, these pixels are used to update the material list statistics μ_i^* and Σ_i^* for class c_i. If $g_i(\hat{v}_R)$ does not exceed a fixed threshold T for any class c_i, then R corresponds to a new material and a new material class based on R is added to the material list.

To understand the meaning of T, we can show from (11) and (17) that the value of $g_i(\hat{v})$ corresponds to a value of $p(\hat{v}/c_i)$ [7]. Thus a choice of T is equivalent to the choice of a T_p such that if

$$p(\hat{v}/c_i) < T_p \quad \text{for all classes } c_i \tag{23}$$

then $\hat{v}$ is assumed to correspond to a new material. A large choice of T_p will allow for great sensitivity in discriminating similar materials at the risk of giving different labels to regions of the same material. Smaller values of T_p will provide less sensitivity in discriminating similar materials, but will cause fewer instances of a single material receiving more than one label. The value of T_P used in this work is such that a region R is assumed to correspond to a new material if for each class c_i the probability of observing a $\hat{v}$ at least as far from the class mean as $\hat{v}_R$ is less than 0.05. This choice of T has produced high quality segmentation results for relatively complex scenes (see section 6).

When region size becomes very small, e.g. four pixels, regions are merged into spatially adjacent existing classes rather than possibly becoming new classes or forming isolated regions of an existing class. At the final classification level, region size is one pixel and the edge test is eliminated. Edge pixels are merged into the spatially adjacent material class for which $g_i(\hat{v}_R)$ is the largest. Pixel classification terminates when every image pixel has been assigned a label.

4.2 Finding Highlights

The segmentation achieved at this point has been based on the assumption that the $M_S(g)$ term for inhomogeneous dielectrics in (1) is zero. This assumption is violated at highlights generated by inhomogeneous dielectrics (I.D. highlights). Since I.D. highlights can cause a shift in normalized color from regions of pure body reflection, some I.D. highlight pixels might be assigned material labels which differ from the label assigned to other pixels corresponding to the same material. A procedure has been developed to locate image regions corresponding to I.D. highlights and to label these regions according to the material of the surface which generated the highlight.

After the initial pixel classification described in section 4.1, the segmentation algorithm identifies candidate I.D. highlight regions in the image. A candidate I.D. highlight region is a connected image region with a single class label for which measured sensor values are at least as large in all color bands as the measured sensor values for a body reflection region at the same scene geometry. This condition is tested by considering average sensor measurements over narrow (3 pixel wide) strips on either side of the boundary separating a hypothesized highlight region from a hypothesized body reflection region.

It is still possible that a candidate I.D. highlight region is not a highlight, but actually corresponds to a distinct material. The candidate I.D. highlight region might, for example, be a bright surface of some material occluding a background surface of another material. A final test is required to determine the classification of the candidate I.D. highlight region.

If the candidate I.D. highlight region corresponds to a different material from the background region, then the sensor values for the candidate region will lie on a line through the origin. If the candidate region is a highlight region, then the sensor values will lie on a line which intersects the line in sensor space corresponding to the hypothesized body reflection region [14]. For the latter case, the intersection will typically occur some distance from the origin as described by the 50 percent heuristic derived by Klinker in [14]. Thus a technique which estimates where the two lines intersect (i.e. either near the origin or at some distance from the origin) will allow classification of the candidate region. Our implementation uses an approach similar to the technique described in [14].

4.3 Implementation Considerations

Certain precautions are required during the segmentation. Due to considerations discussed in section 2, pixels which fall near the zero signal point in color space cannot be classified reliably. At low incident light levels, the measurements made by a solid-state image sensor are limited by an amplifier noise source with standard deviation σ_N. Sensor measurements that lie within the sphere of radius $4\sigma_N$ centered at the origin in sensor space are labeled as being too dark to classify. These sensor measurements usually correspond to surfaces in the scene which lie in shadows.

Edge detection is the first step in the pixel classification algorithm described in 4.1. Although good edge detectors are available [4] [18], it is not realistic to assume that these detectors will generate perfect edge maps. The output of an edge detector is characterized by two kinds of mistakes. A false positive occurs when an edge is declared, but no edge is present. A false negative occurs when an edge is present, but no edge is declared. Most edge detectors use a threshold at some stage to decide if an edge is present, e.g. [4] [18]. A low value for the threshold leads to a high false positive rate and a low false negative rate. A higher setting of the threshold leads to a lower false positive rate and a higher false negative rate. In the following paragraphs, we consider how each kind of edge detection mistake can affect the pixel classification algorithm.

False positives will cause the pixel classification algorithm to split the image unnecessarily and consequently to compute statistics on smaller regions than might be processed otherwise. Thus, the algorithm will take longer to execute and region statistics will be less reliable than if larger image regions could be processed. A few false positives will not, however, have a profound impact on the final segmentation.

False negatives have more serious consequences. If an image region which is processed during pixel classification corresponds to two materials in the scene yet has no edge pixels, the statistics of this region will be computed and assumed to correspond to a single material in the scene. We can easily test for this situation by examining if the region's distribution in normalized color space has more than one significant mode. This test is not necessary, however, if edge detector thresholds are chosen appropriately. For use with the pixel classification algorithm, edge detector thresholds should be set low enough to minimize the occurrence of false negatives. The associated additional false positives will slow processing, but segmentation results will suffer negligible degradation. The thresholds used for edge detection in this work are based on sensor noise estimates [4].

5 The Experimental Setup

In order to apply the segmentation algorithm to real color images, it is necessary to use illuminants and sensors with properties which are consistent with those assumed by the algorithm. This section describes the setup used for the experiments presented in this paper.

5.1 Illumination

Equation (4) assumes that the imaged scene is illuminated by a single spectral power distribution $L(\lambda)$. This assumption is valid if all surfaces in the scene are illuminated by a single source or by some combination of several sources each of which generates light having the same spectral power distribution. For many real scenes this assumption does not hold. Many scenes, for example, contain surfaces which are illuminated by light which has been reflected onto them by other surfaces. It is unlikely that this light will have the same spectral power distribution as a primary illumination source. The experiments described in section 6 use a single light source with no background illumination.

Many common illuminants, e.g. tungsten, generate light with a spectral power distribution function which is much larger for long visible wavelengths than for short visible wavelengths. Informally, we might say that these illuminants generate reddish light. As discussed in section 2, most solid-state imaging systems contain noise sources which are independent of the amount of incoming light. Therefore, to ensure adequate signal-to-noise for all examined regions of the visible spectrum without exceeding the available dynamic range, it is advisable to use an illuminant with a spectral power distribution function which is approximately constant across the visible wavelength range. The imaged scenes presented in section 6 were illuminated by a lamp with color temperature 4800°. Such a lamp is intended to simulate daylight and produces light which is whiter than ordinary indoor illumination.

5.2 Sensors

Both MOS and CCD sensors [6] have been used for the experiments described in section 6. The principal advantage of a CCD sensor is that it has relatively low noise. As described in 4.3, a low noise sensor is desirable for the segmentation algorithm because it allows for the reliable classification of sensor measurements at low signal levels. For a given sensor, amplifier noise σ_N can be estimated by applying an algorithm such as [17] to a dark uniform image.

Despite their low-noise characteristics, CCD sensors perform poorly when imaging metal surfaces which generate bright specular reflection regions. In these situations, CCD sensors exhibit blooming in which bright light causes charge at a sensor element to overflow into adjacent sensor elements. In many cases, bright incident light can cause blooming to affect a significant portion of the image. MOS sensors do not bloom significantly and are better suited for imaging metal surfaces.

Most imaging systems using a solid-state sensor and color filters respond significantly to radiation in the near infrared (IR) region of the spectrum. There are several reasons why this is a hindrance to the segmentation algorithm. At moderately high temperatures, many materials emit radiation in the IR [28]. This emitted radiation does not follow the approximation of (1) which is derived from models for reflected radiation. In addition, IR radiation is absorbed at a significant distance below the surface of a silicon sensor allowing lateral diffusion of photogenerated electrons. This degrades the spatial resolution of the sensor. For these reasons, an IR blocking filter is placed in front of the camera to prevent infrared radiation from influencing measured sensor values.

Equation (4) assumes that measured sensor values are linear with respect to sensor plane irradiance and that there is no zero offset in digitized sensor values. For many real sensors this is not the case. The sensors used in this work are calibrated using reflectance standards to respond linearly to changes in sensor plane irradiance. In addition, the digitizing board is adjusted so that there is no zero offset in the digitized sensor values.

6 Experimental Results

This section presents experimental results obtained using the algorithm of section 4. Three color filters (25,47A,57) are used to generate the color images. An implementation of the Canny edge detector [4] is used to generate edges using one of the bands as explained in 4.1. The band with the largest average pixel value provides the highest signal to noise ratio of the color bands and is used for the edge detection. A single program with fixed values for T_p and the edge detection thresholds is used to generate the segmentation results for all images. There are no image-dependent parameters which are varied.

Figure 5 is an image of the Macbeth color chart [16]. The chart consists of twenty-four matte painted squares with spectral reflectance constant inside each square but different for any distinct squares. The six squares on the bottom row have spectral reflectance functions which are approximately constant functions of wavelength in the visible range. These squares appear different shades of grey varying from bright white on the left to dark black on the right. The color chart is a simple test case, but it demonstrates some important properties of the segmentation algorithm. Figure 6 shows the edges detected for this image.

Figure 7 is the computer segmentation of figure 5. Each color in figure 7 represents a distinct material class. The darkest neutral square is found to be too dark to classify reliably. Such regions are colored black in the segmented images. Each of the other five neutral squares is segmented into the same material class. This is reasonable since the neutral squares could all correspond to surfaces of a single material viewed under different geometries. The algorithm segments the leftmost two squares of the first row into the same material class. These two squares, colored dark brown and light brown respectively, have spectral reflectance functions which are nearly a scalar multiple of one another. Therefore, they could correspond to a single material viewed under different geometries. Each of the other squares is segmented into a distinct material class.

The color chart image demonstrates some important properties and limitations of the segmentation algorithm. Figure 7 shows that the algorithm is able to discriminate many materials in a single image even when several of the materials have similar spectral reflectance functions. Figure 7 also shows an important situation where the algorithm will label distinct materials as corresponding to the same material class. This occurs when two different materials have spectral reflectance functions which are nearly a scalar multiple of one another even if one of the materials reflects significantly more light than the other. Since the algorithm uses only the local information given by (4), it cannot distinguish this case from the situation where a single material is viewed under different geometries. In other words, in order to ensure that different surfaces of a single material are always given the same class label independent of surface orientation, the algorithm will give the same label to distinct materials with spectral reflectance functions that differ by a scalar multiple. It is relatively unlikely, however, that natural scenes will contain many instances of distinct materials having spectral reflectance functions which are nearly scalar multiples of one another.

Figure 8 is an image of several toys against a blue background. The toy objects represent several geometrical shapes including a polyhedron, a torus, a cube, and a circular cylinder. The objects are made of both plastic and painted wood and exhibit both highlights and body reflection. There are several shadow regions in the image. Figure 9 shows the edges detected for figure 8.

Figure 10 is the segmentation of figure 8. The shadow regions are labeled as being too dark to reliably classify. The vast majority of the remaining pixels are labeled correctly despite the wide range of geometrical variation across objects and the presence of highlights. Some misclassifications occur near occluding contours of the purple polyhedron and the torus. This occurs because a large fraction of the incident light at these locations is not direct illumination from the primary illumination source, but rather is light which has been reflected by other surfaces in the scene. Hence the assumption of the algorithm that all incident light has the same spectral power distribution is violated and misclassifications occur.

Figure 11 is an image of a valve against a blue background. The body of the valve consists of brass faces and several steel bolt heads. The visible parts of the handle of the valve are made of yellow plastic, steel, and brass. An MOS sensor was used to obtain this image to avoid the blooming which occurs when imaging metal surfaces with a CCD sensor. Figure 12 shows the edges detected for figure 11. Since the image is relatively complex, there are many broken edges and many edgels in locations where there are not geometric, material, or illumination discontinuities.

Figure 13 is the segmentation of figure 11. Shadow regions and some parts of the body of the valve are labeled as too dark to classify reliably. The algorithm correctly labels nearly all of the remaining image pixels in the presence of varying scene geometry.

Figure 13 demonstrates several properties of the segmentation algorithm. It shows that the algorithm can be successfully applied to both metal and dielectric materials in the same scene. This image also shows that the algorithm can be applied to images with some texture and less than perfect edge maps. The algorithm correctly labels small regions such as those corresponding to the steel bolt heads on the body of the valve.

Figure 14 is an image of a plastic part composed of two plastic materials on a blue background. There is significant geometric detail on several of the surfaces of the part and both plastic materials generate highlights. Since the scene is relatively dark and contains no metal surfaces, a CCD sensor has been used to provide improved signal-to-noise properties over an MOS sensor. Figure 15 shows the edges detected for figure 14.

Figure 16 is the segmentation of figure 14. Shadow regions and some poorly illuminated areas on the part are labeled as too dark to classify. Nearly all of the other pixels are labeled correctly as one of the plastic materials or as background. A few misclassifications occur near an occluding contour of the grayish plastic where a significant fraction of the incident light is not direct illumination from the primary source.

7 Discussion

This paper has presented an efficient segmentation algorithm which is more general in several respects than previous segmentation techniques. The algorithm may be applied to images of both metals and dielectrics and accounts for surface texture due to small local variations in surface geometry and material composition. Although the algorithm is based on region splitting, edges are used to localize region boundaries accurately. The algorithm is based on a detailed analysis of the physics of color image formation and has been used successfully to segment color images of 3-D scenes containing a wide variety of materials and surface geometries. Such an algorithm is a useful component of systems designed to inspect or recognize three dimensional

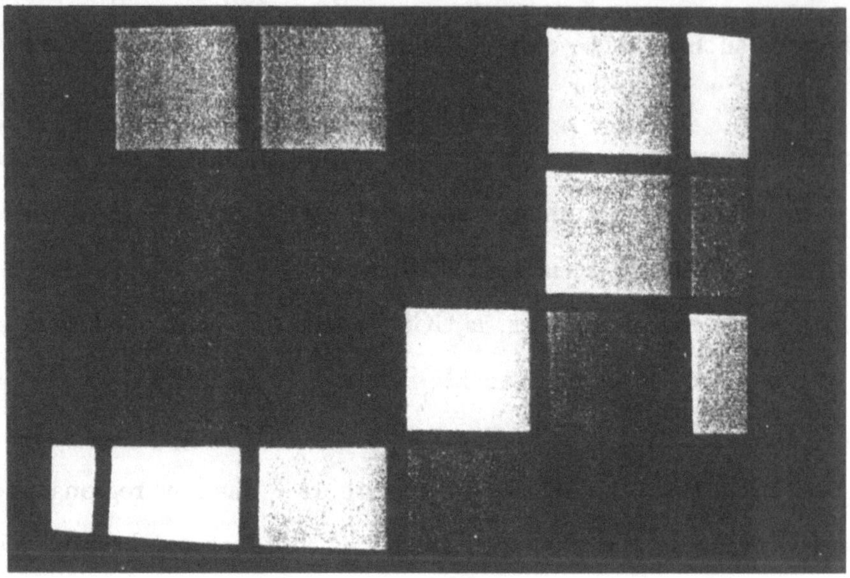

Figure 5: Image of a color chart.

Figure 6: Chart image edges.

See color plate D for Figures 5-6.

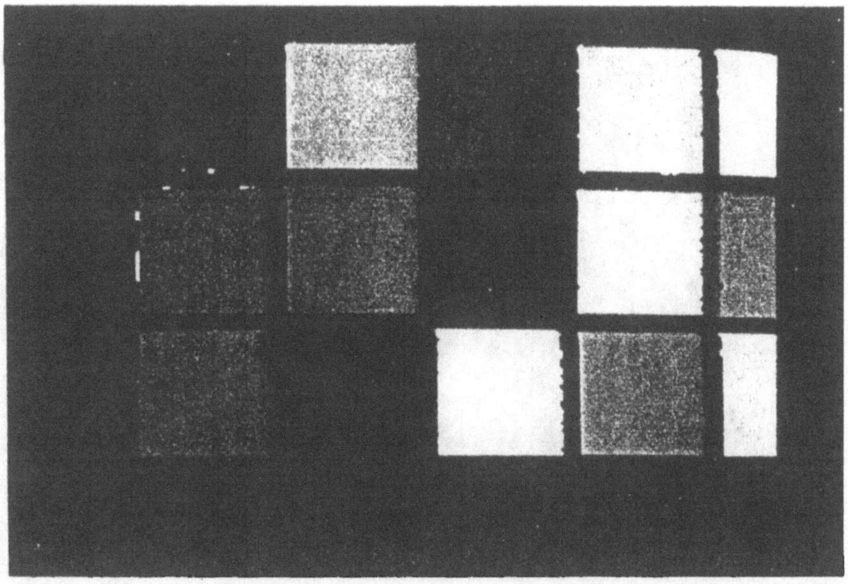

Figure 7: Chart image segmentation.

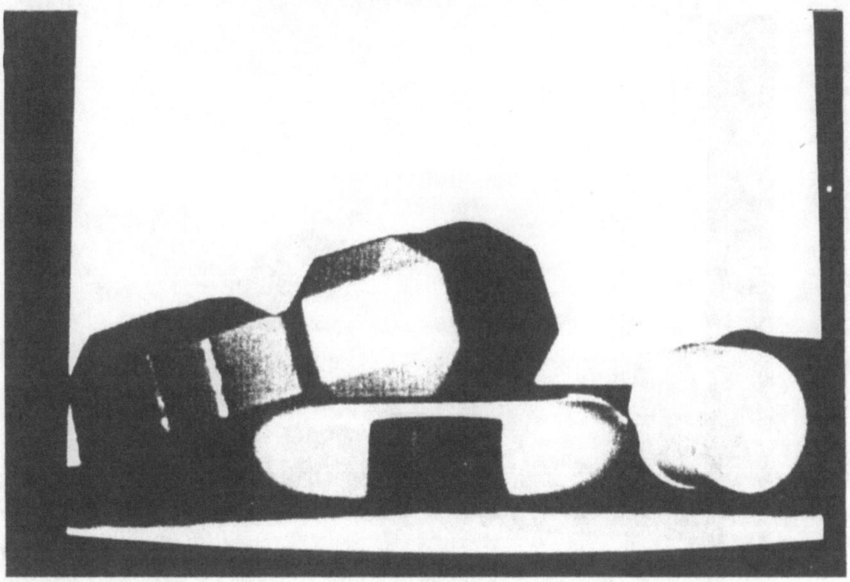

Figure 8: Image of toys scene.

See color plate D for Figures 7-8.

Figure 9: Toys image edges.

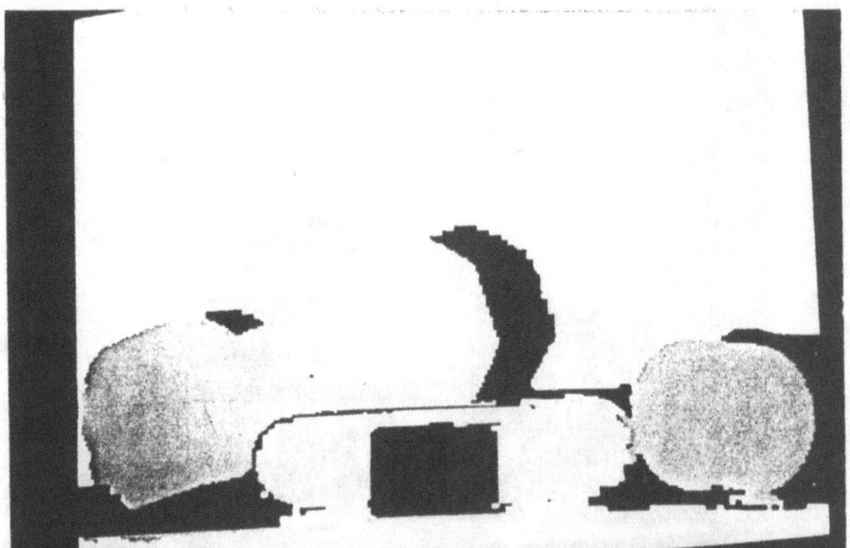

Figure 10: Toys image segmentation.

See color plate D for Figures 9-10.

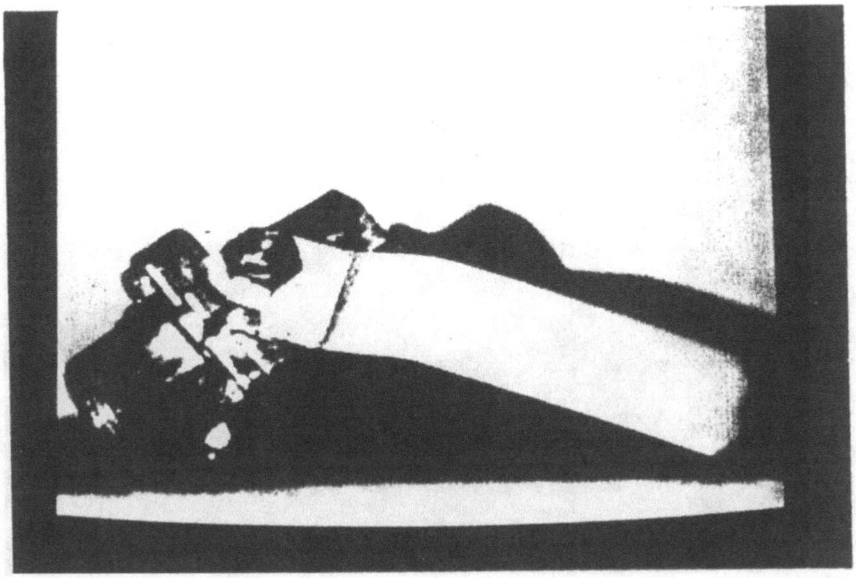

Figure 11: Image of a valve.

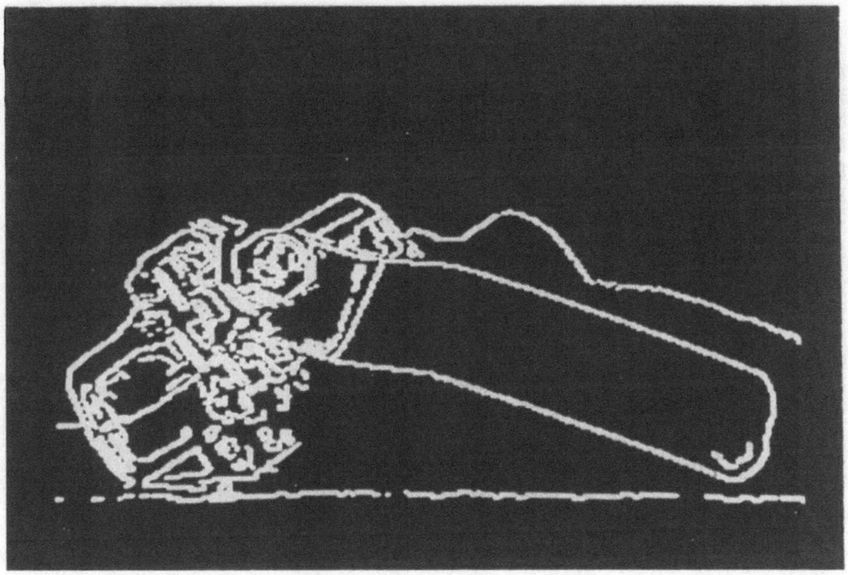

Figure 12: Valve image edges.

See color plate D for Figures 11-12.

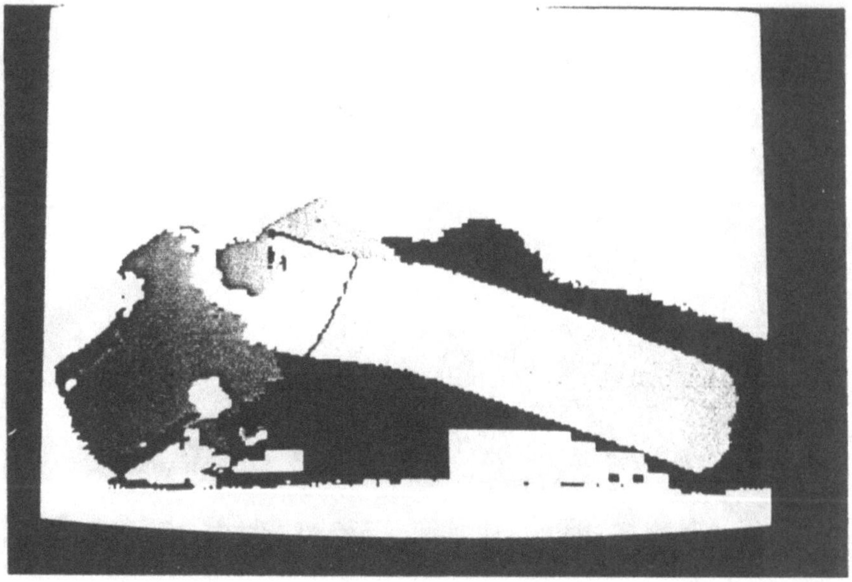

Figure 13: Valve image segmentation.

Figure 14: Image of a plastic part

See color plate E for Figures 13-14.

Figure 15: Plastic part edges.

Figure 16: Plastic part image segmentation

See color plate E for Figures 15-16.

objects.

An important aspect of the segmentation algorithm is that it combines the advantages of a global region-based approach like region-splitting with the advantages of a local edge-based approach. This differs from previous approaches to color segmentation which either are based exclusively on processing regions [8], [21], [22] or are based exclusively on analyzing local information [14]. Region-based approaches process large connected areas of an image. This ensures that reliable statistics will be computed for a region and that segmented regions will be spatially connected. On the other hand, since region-based methods are relatively global, they are less likely to segment successfully small regions or to localize region boundaries accurately. Local edge-based approaches, however, are suitable for detecting small objects and localizing region boundaries. By using edges to guide region-splitting, we achieve an algorithm which can compute reliable statistics from large image regions but which also finds small regions and accurately localizes region boundaries.

The segmentation algorithm can be adapted for high-speed 3-D applications because it is based on the computation of local image features which are independent of scene geometry. Thus, the computation of statistics for an image region can be done in parallel with the computation of statistics for any other image region. In addition, the computation required by the segmentation algorithm is scene dependent. In contrast to purely local approaches, e.g. [14], the segmentation algorithm will be particularly efficient for images containing few edges.

8 Acknowledgements

I thank Tom Binford for his help. This work was begun while the author was at the Robotics Laboratory of the Computer Science Department at Stanford University. This work has been supported by the Air Force Office of Scientific Research under contract F33615-85-C-5106, by the Advanced Research Projects Agency of the Department of Defense under Knowledge Based Vision ADS subcontract S1093 S-1, and by the National Science Foundation under grant IRI-9009811.

References

[1] M. Ali, W. Martin, and J. Aggarwal. Color-based computer analysis of aerial photographs. *Computer Graphics and Image Processing*, 9:282–293, 1979.

[2] T. Anderson. *An Introduction to Multivariate Statistical Analysis.* Wiley, New York, 1958.

[3] R. Bajcsy, S.W. Lee, and A. Leonardis. Color image segmentation with detection of highlights and local illumination induced by inter-reflections. In *Proceedings of the International Conference on Pattern Recognition*, pages 785–790, Atlantic City, 1990.

[4] J. Canny. A computational approach to edge detection. *IEEE Transactions on Pattern Analysis and Machine Intelligence*, PAMI-8(6):679–697, November 1986.

[5] G.B. Coleman and H.C. Andrews. Image segmentation by clustering. *Proceedings of the IEEE*, 67(5):773–785, May 1979.

[6] M. Collet. Solid-state image sensors. *Sensors and Actuators*, 10:287–302, 1986.

[7] R. Duda and P. Hart. *Pattern Classification and Scene Analysis.* Wiley-Interscience, New York, 1973.

[8] R. Gershon. The use of color in computational vision. Technical report, University of Toronto, Computer Science Ph.D. Thesis, 1987.

[9] R. Gershon, A. Jepson, and J. Tsotsos. Highlight identification using chromatic information. In *Proceedings of the First International Conference on Computer Vision*, pages 161–170, London, June 1987.

[10] R. Haralick and G. Kelly. Pattern recognition with measurement space and spatial clustering for multiple images. *Proceedings IEEE*, 57:654–665, April 1969.

[11] G. Healey. Using color for geometry insensitive segmentation. *J. Opt. Soc. Am. A*, 6:920–937, June 1989.

[12] S. Horowitz and T. Pavlidis. Picture segmentation by a tree traversal algorithm. *Journal of the Association for Computing Machinery*, 23(2):368–388, April 1976.

[13] J. Kender. Saturation, hue, and normalized color: Calculation, digitization effects, and use. Technical report, CMU Computer Science Department, November 1976.

[14] G. Klinker. A physical approach to color image understanding. Technical Report CMU-CS-88-161, Carnegie-Mellon University, Computer Science Ph.D. Thesis, May 1988.

[15] G. Klinker, S. Shafer, and T. Kanade. Color image analysis with an intrinsic reflection model. In *Proceedings of Second International Conference on Computer Vision*, pages 292–296, Tampa, 1988.

[16] C. McCamy, H. Marcus, and J. Davidson. A color-rendition chart. *Journal of Applied Photographic Engineering*, 2(3), Summer 1976.

[17] P. Meer. A fast parallel algorithm for blind estimation of noise variance. Technical Report CAR-TR-373, Center for Automation Research, University of Maryland, June 1988.

[18] V.S. Nalwa and T.O. Binford. On detecting edges. *IEEE Transactions on Pattern Analysis and Machine Intelligence*, 8(6):699–714, November 1986.

[19] R. Nevatia. A color edge detector and its use in scene segmentation. *IEEE Transactions on Systems, Man, and Cybernetics*, SMC-7(11):820–826, 1977.

[20] C. Novak and S. Shafer. Color edge detection. In *Proceedings of the Image Understanding Workshop*, pages 35–37, USC, 1987.

[21] R. Ohlander, K. Price, and D.R. Reddy. Picture segmentation using a recursive region splitting method. *Computer Graphics and Image Processing*, 8:313–333, 1978.

[22] Y. Ohta, T. Kanade, and T. Sakai. Color information for region segmentation. *Computer Graphics and Image Processing*, 13:222–241, 1980.

[23] G Robinson. Color edge detection. *Optical Engineering*, 16(5):479–484, September-October 1977.

[24] A. Sarabi and J. Aggarwal. Segmentation of chromatic images. *Pattern Recognition*, 13(6):417–427, 1981.

[25] B. Schachter, L. Davis, and A. Rosenfeld. Scene segmentation by cluster detection in color space. Technical Report 424, University of Maryland Computer Science Department, 1975.

[26] S. Shafer. Using color to separate reflection components. Technical Report TR 136, University of Rochester, April 1984.

[27] S. Shafer, T. Kanade, G. Klinker, and C. Novak. Physics-based models for early vision by machine. In *SPIE Volume 1250 on Perceiving, Measuring, and Using Color*, pages 222–235, Santa Clara, February 1990.

[28] R. Siegel and J. Howell. *Thermal Radiation Heat Transfer.* McGraw-Hill, 1981.

[29] J. Wolfe. Hidden visual processes. *Scientific American*, pages 94–103, February 1983.

[30] G. Wyszecki and W. Stiles. *Color Science: Concepts and Methods, Quantitative Data and Formulae.* Wiley, New York, 1982.

Color Image Segmentation with Detection of Highlights and Local Illumination Induced by Inter-reflections

Ruzena Bajcsy, Sang Wook Lee and Aleš Leonardis

Department of Computer and Information Science
University of Pennsylvania, Philadelphia, PA 19104

Abstract

A computational model for color image segmentation is proposed based on the physical properties of sensors, illumination lights and surface reflectances. For image segmentation to depend only on the change of material surface, variations of surface reflection due to illumination, shading, shadows, and highlights should be discounted from a measured image. We use the dichromatic model for dielectric materials and develop a metric space of intensity, hue and saturation in order to better interpret various light-surface interactions. Using the established model for surface reflection, we perform color image segmentation based on the material change with detection of highlights and small inter-reflections between adjacent objects. A reference plate is used to whiten global illumination. The detected inter-reflections represent local variation of illumination.

1 Introduction

Color is one of the major features in identifying objects, and color vision has been the most intensively studied sensory process in human vision. The purpose of color vision is to extract aspects of the spectral property of object surfaces in order to provide useful information in visual process. The most notable applications in machine vision include recognition and identification of colored objects and image segmentation.

There have been approaches to color image segmentation based on histograms of red (R), green (G) and blue (B) values or on transformed values such as YIQ [13], or by using color edges in addition to intensity edges [12]. These algorithms are primarily extensions of intensity segmentation to multidimensional processing with extra information provided by color. Thus they cannot distinguish between different regions due to material changes, highlights, shading, and shadow.

Also, there has been an approach to detection of highlights by combining the traditional segmentation with physics-based post-processing [4]. Since traditional color image segmentation algorithms generally have not accounted for the influence of optical effects on object colors, the method suffers from many erroneous regions. If the physical models were included at the segmentation stage, many of these regions could be properly classified. The first physical approaches to color image understanding have been introduced only recently [1] [2]. Detection of highlights is achieved by properly clustering color points in RGB space using the dichromatic property of dielectric materials. Since the spectral curves of the camera and filters are not included in these models, the metric in the RGB space varies with the choice of the sensors..

Image segmentation should be based on changes in material surfaces. In addition to material changes, however, a measured image contains variations due to shading, shadows and highlights. Besides there are other impediments to observing real material changes such as variation of illumination and multiple reflections between objects. Objects receive not only the lights directly from illumination sources but also the reflected lights from other objects. The latter induces a change of the perceived surface color which we call inter-reflection. The process of discounting illuminations from the surface reflectance, namely color constancy, is a heavily studied subject in color vision [8] [6] [5], though general applications in machine vision are yet limited.

In this paper we present an approach to the construction of a computational model for color image segmentation, not only with detection of highlights but with detection of small color changes induced by inter-reflections as well. We use the dichromatic model [1] for dielectric materials and develop a metric space based on the physical properties of sensors in order to better manipulate light-surface interactions. We use a reference plate to discount global illumination, and the detected inter-reflections represent local variation in illumination due to adjacent objects.

We begin in Section 2 with sensing and metric representation of color in a machine vision system. In the following section, we describe a calibration method of illumination whitening using a reference plate. In Section 4 a surface reflectance model is established from reflection mechanisms. Various optical phenomena such as shading, shadow, highlights and inter-reflections between objects are explained in Section 5, in terms of the established surface reflectance model. In Section 6, we propose a method of segmentation and highlights and inter-reflection detection. We present some of our experimental results in Section 7.

2 Representation and Sensing of Color

Representation of color with finite-dimensional linear models has been a topic of many studies [14] [15] [7] [16]. There have been some approaches to obtaining characteristic basis functions by investigating many samples of daylight and surface reflectances [15] [14]. It has been suggested that although the number of basis functions required to completely describe full spectra is essentially infinite, a small number of basis functions can provide good spectral approximations of most natural illuminants and surface reflectances.

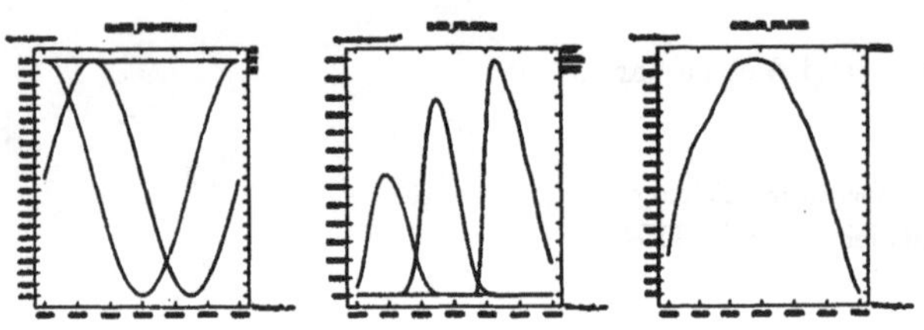

Figure 1: (a) Fourier basis functions (b)Spectral responses of 3 filters (c) Spectral response of CCD and IR filter

Our approach is to represent measured colors in a metric space in order to better manipulate physical properties of surface colors for image segmentation. Both for surface reflectance and for illumination, we here use the first three Fourier bases for their mathematical simplicity and for their sufficient effectiveness in resolving natural colors. Within our framework developed here, however, better models of basis functions (such as the first three of Cohen's reflectance basis functions [12] and those of Judd's illuminant basis functions [13]) can be used.

For the three-dimensional linear model, surface reflectance and illumination can be represented as a weighted sum of basis functions expressed as:

$$S(x,y,\lambda) = \sum_{i=0}^{2} \sigma_i(x,y) S_i(\lambda) \quad E(x,y,\lambda) = \sum_{j=0}^{2} \varepsilon_j(x,y) E_j(\lambda), \tag{1}$$

where $S_i(\lambda)$ and $E_j(\lambda)$ are the basis functions, and $\sigma_i(x,y)$ and $\varepsilon_j(x,y)$ are the scalar weighting factors at (x,y). The basis functions $S_0(\lambda) = 1$, $S_1(\lambda) = sin\lambda$ and $S_2(\lambda) = cos\lambda$ are shown in Figure 1 (a).

The color image sensing is performed with a CCD camera using filters of different spectral responses. The measured color signal $I(x,y,\lambda)$ is obtained as a product of the spectral power distribution (*SPD*) of illumination $E(x,y,\lambda)$ and the spectral response of the surface reflectance function $S(x,y,\lambda)$, i.e., $I(x,y,\lambda) = E(x,y,\lambda)S(x,y,\lambda)$. With 3 filters (usually R, G and B), the quantum catch or the measured signal from the camera is given by

$$\rho_k(x,y) = \int_{\lambda_1}^{\lambda_2} I(x,y,\lambda) \cdot Q_k(\lambda) d\lambda, \tag{2}$$

where $Q_k(\lambda)$ and $\rho_k(x,y)$ for k = 0, 1, 2 are the spectral response of the $k-$th filter and the camera output through the $k-$th filter at (x,y), respectively. The wavelengths $\lambda_1 = 400$ nm and $\lambda_2 = 700$ nm cover the range of the visible spectrum. Figure 1 (b) shows the spectral response of the 3 filters: $Q_0(\lambda)$, $Q_1(\lambda)$ and $Q_2(\lambda)$. The curves are obtained from filter measurements with a spectro-photometer and are multiplied by the product of the spectral sensitivity of the CCD camera and the infra-red cut-off (*IR*) filter which is shown in Figure 1 (c). The filters integrate the spectral responses over the given range, and consequently act as an anti-aliasing filter in sampling the color signal.

With 3 basis functions, the relationship between the sensor response, illumination and reflectance is given as

$$\rho = \Omega^\sigma \varepsilon = \Omega^\varepsilon \sigma \tag{3}$$

where the elements of Ω^σ and Ω^ε in the k-th row and j-th (or i-th) column are respectively

$$\Omega^\sigma_{kj} = \int_{\lambda_1}^{\lambda_2} \sum_{i=0}^{2} \sigma_i S_i E_j Q_k d\lambda, \quad \Omega^\varepsilon_{ki} = \int_{\lambda_1}^{\lambda_2} \sum_{j=0}^{2} \varepsilon_j E_j S_i Q_k d\lambda. \tag{4}$$

3 Illumination Whitening

To obtain the surface reflectance of objects, which is the primary information for image analysis, the influence of illumination should be discounted from the measured color in the image. This process is called color constancy. Computationally, the color constancy problem is to solve Equation (3) for ε's and σ's with the measurements ρ's. While the general solution for color constancy with unknown illumination demands complex algorithms and some spectral and/or spatial constraints [15] [4] [7] [9] [8], it is simple to remove the known illumination measured with a reference object. One assumption is that the objects of interest are illuminated by light sources of the same *SPD*, and the reference object is applicable. Since we are given the product $E(\lambda)S(\lambda)$, $E(\lambda)$ and $S(\lambda)$ can be recovered only up to a multiplicative constant. This scaling ambiguity cannot be generally resolved, and thus we only solve for the spectral components of surface reflectance and illumination up to a multiplicative factor.

If we use a reference plate with known reflectance σ^{ref}, the *SPD* of illumination obtained from

$$\varepsilon^{ref} = (\Omega^{\sigma^{ref}})^{-1} \rho \tag{5}$$

represents the spectral composition of the global illumination throughout the image area. With the normalized ε^{ref}_{norm}, the calculation of

$$\sigma = (\Omega^{\varepsilon^{ref}_{norm}})^{-1} \rho \tag{6}$$

leads to the color signal under whitened illumination.

4 Surface Reflectance Model

Once the effects of color illumination have been eliminated, the color image $I(x,y,\lambda)$ represents only the surface reflectance properties multiplied by the illumination intensity ε_0.

The analysis of the color image demands proper models for various optical phenomena of surface reflections. We use the dichromatic model for dielectric materials such as plastic, paper and paint [1]. The surface reflectance of dielectric materials has two different optical properties. Highlights are from specular reflection and depend on the refractive indices of the vehicle materials. On ther other hand, body or diffuse reflection occurs inside the material due to the pigments. The distribution of pigments determines the color of the exiting light by selectively absorbing certain wavelengths.

The magnitude of the image signal is also affected by geometrical factors such as surface orientation, illumination and viewing directions [16]. So far, we have assumed that the geometrical factors are implicitly included either in $E(\lambda)$ or in $S(\lambda)$. More precisely, the geometrical weighting can be regarded as the local variation of illumination density on a surface patch. However, since the geometrical weighting has different mechanisms for specular and body reflections, it would be convenient to include it in the surface reflectance. If we consider the factors as part of the surface reflectance of the dichromatic model, the surface reflectance can be modeled as

$$S(\lambda) = g_S(\mathrm{G})S_S(\lambda) + g_B(\mathrm{G})S_B(\lambda), \tag{7}$$

where $S_S(\lambda)$ and $S_B(\lambda)$ are the specular and the body reflectance, respectively; g_S and g_B are the geometrical factors for the specular and the body reflection, respectively; and G is the geometrical variable that accounts for the viewer direction and the source direction, where each is relative to the surface normal. Now $S_S(\lambda)$ and $S_B(\lambda)$ represent the real surface reflectance of object, i. e., *albedos*. For most vehicle materials, the refractive index is weakly dependent on λ. Thus, $S_S(\lambda)$ is almost constant over the visible range of light, while the body color is determined by $S_B(\lambda)$.

Given that the specular reflection is flat over the spectrum, we can rewrite the color image signal as

$$\begin{aligned} I &= (\varepsilon_0 E_0 + \varepsilon_1 E_1 + \varepsilon_2 E_2)\cdot \\ &\quad [(g_S\hat{\sigma}_{OS} + g_B\hat{\sigma}_{OB})S_0 + g_B\hat{\sigma}_{1S}S_1 + g_B\hat{\sigma}_{2B}S_2], \end{aligned} \tag{8}$$

where

$$\sigma_0 = g_S\hat{\sigma}_{OS} + g_B\hat{\sigma}_{OB}, \quad \sigma_1 = g_B\hat{\sigma}_{1B}, \quad \sigma_2 = g_B\hat{\sigma}_{2B}. \tag{9}$$

By whitening the illumination, we can remove the terms $\varepsilon_1 E_1$ and $\varepsilon_2 E_2$.

5 Image Analysis on (S_0, S_1, S_2)

For identification of the surface reflection mechanisms explained above, we represent the color on the orthogonal 3-dimensional S space. The S space is similar to the opponent space which is convenient for describing intensity, hue and saturation. The definitions of intensity, saturation and hue for our work are defined as:

$$\begin{aligned} \text{intensity} &= \sigma_0, \\ \text{hue} &= \arctan\frac{\sigma_2}{\sigma_1}, \\ \text{saturation} &= \frac{|\sigma_{12}|}{\sigma_0}, \\ &\quad \text{where } |\sigma_{12}| = \sqrt{(\sigma_1^2 + \sigma_2^2)}. \end{aligned} \tag{10}$$

5.1 Shading

Shading on surfaces of uniform color is due to variations in geometry G in $g_B(\mathrm{G})$ in Equation (7) As can be seen in Equation (8), the shading can be modeled as the weighting of $\hat{\sigma}_{OB}$, $\hat{\sigma}_{1B}$ and $\hat{\sigma}_{2B}$ by the scaling factor g_B. In the absence of highlights, this corresponds to a linear cluster of color points through the origin in S space, as shown in Figure 2 (a).

Since the saturation is the slope of this line with respect to the S_0-axis, the saturation values of the color points on the linear cluster are constant. The slope of the line can be changed by colored illumination. On the normalized S space in which S_1 and S_2 are normalized with respect to S_0, the line due to shading is parallel to the S_0 axis, so that the saturation values (the distance between the line and the S_0 axis) are kept constant as shown in Figure 2 (a).

5.2 Highlight

Since the spectral response of the specular reflection is flat for dielectrics, the illuminating light is reflected from the surface without any change of color. Under white (or whitened) illumination, highlights add whiteness (σ_{OS}) to the body colors by $g_S(\mathrm{G})$. As the result, the color points in S space are shifted upward along the S_0 direction due to the highlights. In general, the distribution of highlights can spatially vary over the wide area of shaded matte surface as shown in Figure 2 (a), so that they form planar clusters which include the linear clusters made by shading on the S space as shown in Figure 2 (a). The planes are perpendicular to the $S_1 - S_2$ plane under white or whitened illumination. Since the points are shifted upward, the saturation values are always decreased by highlights compared to those of the points without highlights.

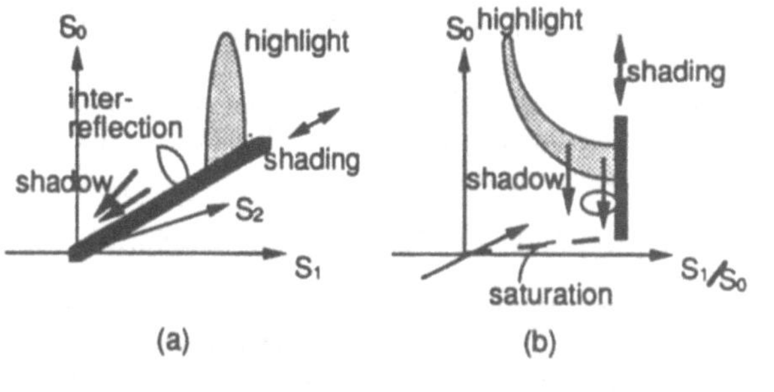

Figure 2: (a)Color clusters on S space, and (b) on normalized S space

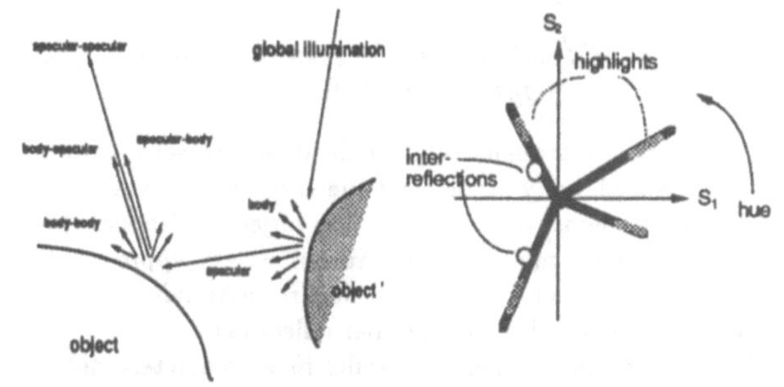

Figure 3: (a) Inter-reflection (b) Color clusters in S space (top view)

5.3 Shadow

Shadows are due to the screening of illumination by other objects, which can be interpreted as the variation of ε_0, ε_1, and ε_2 (only ε_0 under whitened illumination). Thus the shadow does not change the shape and the orientation of the linear cluster, but moves the color points toward the origin along the line passing through the origin as illustrated in Figure 2 (a).

5.4 Inter-reflection

So far we have been concerned only about the first reflection of light. As there are many objects, the object surface of interest receives not only the light from the illumination sources but also the reflected light from the other objects. The latter causes the local change of illumination, as illustrated in Figure 3 (a). The change of surface color due to the multiple reflections between objects is named inter-reflection. The local influence of the reflected light from other objects is in fact the local variation of illumination. Since we discount only the global illumination, the inter-reflection remains as a local change of surface reflectance.

Under whitened global illumination, the color signal of a dielectric surface influenced by another dielectric surface (denoted by ′) can be written as

$$I(\lambda) = \varepsilon_0 E_0[g'_S S'_S(\lambda) + g'_B S'_B(\lambda)][g_S S_S(\lambda) + g_B S_B(\lambda)]. \quad (11)$$

Considering that the specular reflectances are much greater than the body reflectances for most dielectric materials ($S_B(\lambda) << S_S(\lambda)$, $S'_B(\lambda) << S'_S(\lambda)$), we can approximate $I(\lambda)$ as a linear sum of the specular-specular, the specular-body and the body-specular reflections as (neglecting the body-body reflection)

$$I(\lambda) \simeq \varepsilon_0 E_0[g'_S g_S S'_S(\lambda) S_S(\lambda) + g'_S g_B S'_S(\lambda) S_B(\lambda) + g'_B g_S S_S(\lambda) S'_B(\lambda)]. \quad (12)$$

The first term represents white reflection, the second term shows increased body color, and the last term is the influence from the body color of the other object. Although the inter-reflection is a multiplicative process, we can approximate it as a additive process for most dielectirc materials since the dominant reflections involve specular reflections.

While large inter-reflection results in a completely different cluster of points, small inter-reflection can be identified as a small deviation from the linear cluster as illustrated in Figure 2 (a). In general, the color points of inter-reflections do not fall in the planar region formed by highlights and body reflections, since inter-reflection can change the perceived hue as well as saturation of the surface.

6 Image Segmentation

Our segmentation method is based on the assumption that the image consists of patches of object surfaces which have uniform color, i.e., the image can be divided into many regions of uniform hue and saturation regardless of surface structure and finishing.

Hue segmentation can be done by hue-histograming or region-splitting based on hue values. Under white (or whitened) illumination, the planes formed by highlights and body reflections are perpendicular to the $S_1 - S_2$ plane as shown in Figure 3 (b). Because the highlights are in the same plane as the body color, the specular reflection and the body reflection from the same object are segmented together by the hue segmentation.

Further segmentation and detection of highlights and inter-reflections can be achieved using saturation values based on the observation that

- Shading, highlights, shadow and inter-reflections change the intensity,
- Shading and shadow do not change the hue or the saturation value,
- Highlight decreases the saturation value, and
- The inter-reflections in general cause the change in hue and saturation. Since reflections from neighboring objects are dominantly specular, in most cases, the purity of the surface color is usually decreased.

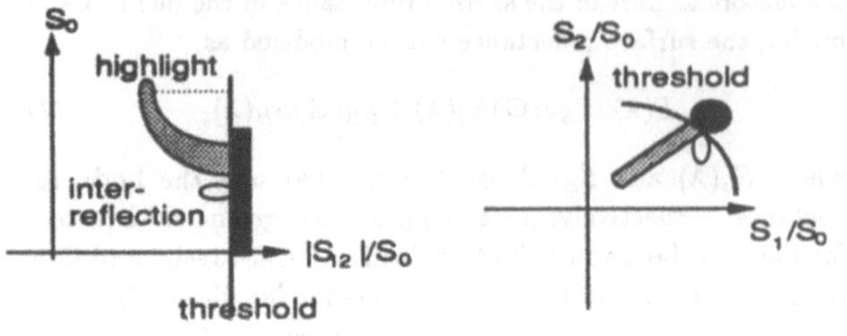

Figure 4: (a) Intensity-saturation plane, and (b) Hue-saturation plane

Therefore application of local thresholding on saturation values enables us to separate the color clusters of highlights and most of the small inter-reflections from those of body reflections.

Figure 4 (a) and (b) illustrate the color clusters in the $S_0 - saturation$ plane within a given range of hue values after hue segmentation, and in the projected $S_1/S_0 - S_2/S_0$ plane, respectively. As explained earlier, shading and shadow do not change the saturation value of an object region with a given color. On the other hand, highlights increase only σ_0 under white illumination, thereby always decreasing the saturation. The inter-reflections may affect σ_0, σ_1 and σ_2, which implies that hue and saturation change. Although inter-reflections can possibly increase the saturation or change the hue values of the surface without affecting the saturation value, in most cases they decrease the saturation. This is because the neighboring objects usually decrease the purity of the surface color since dominant inter-reflections involve specular reflections. Therefore application of local thresholding on saturation values enables us to separate the color clusters of highlights and most of the small inter-reflections.

7 Experimental Results

In order to test the algorithm, we perform an experiment on plastic color balls. Four tungsten lights are used as illumination sources in two directions (right and left sides). They are spatially diffused in order to reduce the contrast between highlights and matte colors. All the measurements are in the linear range of our 8-bit image digitizer, and the camera output is linearized by setting the *gamma* correction factor of the CCD camera to 1. Illumination is whitened using a reference plate of known spectral reflectance. Figure 5 (a), (b) and (c) show the intensity (S_0) image, and the images projected on S_1, and S_2 axes, respectively, after illumination whitening.

Figures 6 (a) shows the saturation image and Figures 6 (b) shows the color clusters in S space after illumination whitening. As expected, the color points from highlights form planar clusters while shading makes linear clusters. The deviations from the lines and/or planes are due to the errors and noise in measurements as well as the errors in modeling and nonuniformity the surface color. Hue segmentation is achieved using the hue-histogram shown in Figures 6 (b). The results are partially shown in Figure 7 (a). After the hue segmentation, application of local thresholding in each segment results in the separation of body reflections from highlights and inter-reflections. Figure 7 (c) shows the intensity (S_0) image of separated highlights and inter-reflections which are the deviations from the linear clusters of body reflections. Most of the strong inter-reflections are due to the specular-specular reflections. The intensity (S_0) image of the separated body reflections is shown in Figure 7 (b).

Another experiment is carried out for a scene which consists of a glossy red plate and glossy yellow blocks. Again four diffused tungsten lights and a reference plate are used. Figure 8 (a), (b) and (c) show the intensity (S_0) image and the images projected on S_1, and S_2 axes, respectively, after illumination whitening. The red plate is partly highlighted, and each side of the yellow blocks except the shadowed area is evenly highlighted depending on the surface orientations.

Figure 9 (a), (b) and (c) show S_0, S_1 and S_2 images of the separated matte part, respectively, and Figure 10 (a), (b) and (c) show S_0, S_1 and S_2 images of the separated highlights/inter-reflections part, respectively. In Figure 10, it can be seen that multiple reflections are strong in the wedge region where the red plate and the long block meet. It is not clearly seen in the original image before separation. The slightly colored highlights in the plate and the blocks result from the small errors in whitening the global illumination. With body reflections removed, the objects look like metals having only specular reflections.

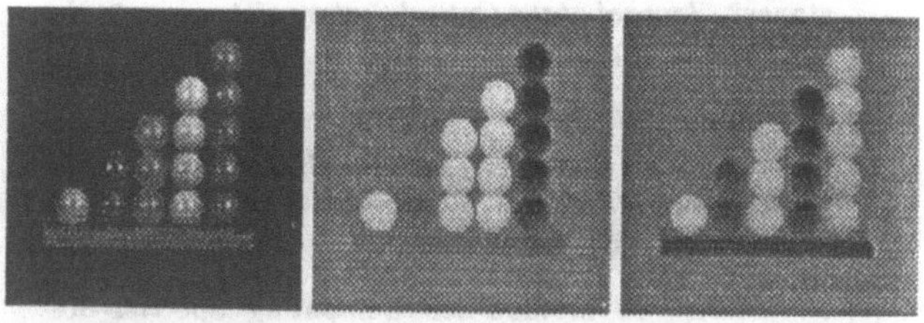

Figure 5: (a) S_0 (b) S_1 (c) S_2

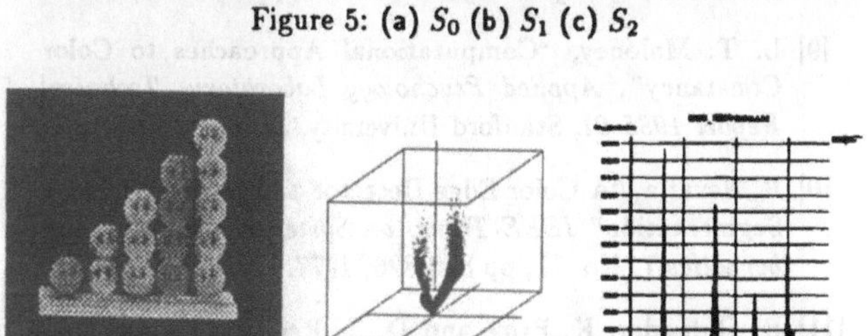

Figure 6: (a) Saturation (b) Color clusters in S space (c) Hue histogram

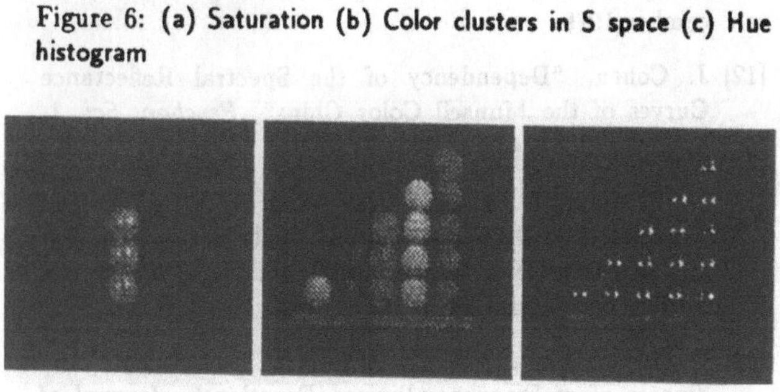

Figure 7: (a) Hue-segmented red balls (S_0) (b) S_0 Matte (c) S_0 Highlights and inter-reflections

Figure 8: (a) S_0 (b) S_1 (c) S_2

Figure 9: Matte (a) S_0 (b) S_1 (c) S_2

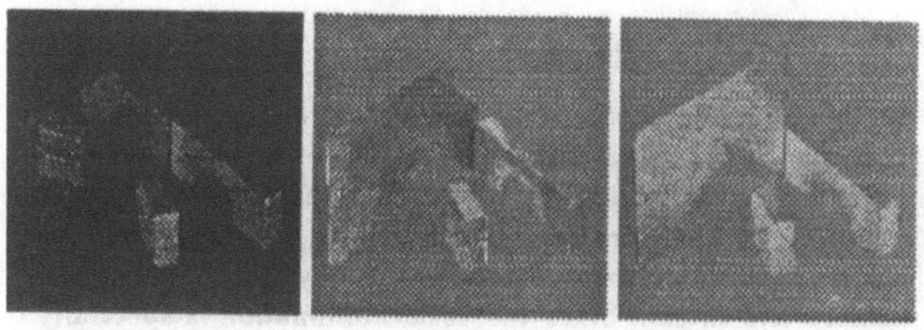

Figure 10: Highlights and inter-reflections (a) S_0 (b) S_1 (c) S_2

8 Discussion

We have carried out experiments for image segmentation with objects of saturated colors in order to demonstrate the efficacy of the algorithm. The experimental results agree with our interpretations on the surface reflections, and the detection of highlights and inter-reflections is successful under the condition that the scene consists of small patches of uniform hue and saturation (not intensity).

The segmentation is performed only in S space without using any spatial information. Since separated regions of similar color can be segmented together, segmentation of more complex scenes may require partitioning an image in the spatial domain. Therefore, the use of spatial information is desirable in order to further improve the segmentation. More extensive use of spatial processes may involve the use of color edge detectors. Color edge detection can be based on the distance in the S space between adjacent pixels, and from our interpretation of various reflection mechanisms we can distinguish between different types of edges. For example, the edge detected due to shading should not show the change in hue and saturation while the one due to highlights should indicate the change in saturation.

An important issue is the proper determination of the local thresholds on the saturation values, since this is the key to the successful detection of highlights and inter-reflections. For our sample image, we obtained the threshold values by examining the deviation in the saturation values in uniform color regions. We found that the size of the deviations across different parts of the image is about the same. Therefore, the observed deviation values can be used for determining thresholds for any measurements from the same settings for the sensors. The detection of highlights and inter-reflections can be further improved by locally adaptive adjustment of thresholding. With proper local thresholds, the algorithm is successful in detecting highlights and small inter-reflections which cannot be easily clustered in RGB space.

We obtained the global illumination using a reference object, and the detection of approximate colors of local illumination by inter-reflections completes the color constancy. The detected local illumination is highly varying in space, and therefore it cannot be easily obtained with the color constancy algorithms proposed so far. Most of the color constancy algorithms require severe spatial or spectral constraints such as smoothly varying illumination in space, many image points of different surface reflectances under same illumination [7].

9 Conclusion

In this paper we present an approach to color image segmentation with the detection and separation of highlights and local illumination induced by inter-reflections between adjacent objects. To better represent and process the image color, a color space is developed based on the physical model of the camera and filters. Since the usual illumination is spectrally colored, a white object of reference is used to whiten the illumination. While the spectral distribution of the object surface is not changed by shading and shadow under the white illumination, it is affected by highlights and inter-reflections between the objects. Image segmentation is done by hue and saturation values. As the highlights add whiteness to the object color, they can be detected by observing the change in saturation of the uniformly colored objects. The inter-reflections are also detected with the change in saturation, as well as in hue values. The experimental results show that within the framework, our model of color image interpretation performs well.

Acknowledgement

This work was supported by E. I. du Pont de Nemours and Company, Inc. and by the following grants: Air Force AFOSR F49620-85-K-0018, ARPA N00014-88-K-0630. We are grateful to P.M. Tannenbaum at DuPont for many discussions, and to S.A. Shafer and L.T. Maloney for their helpful suggestions when we were initiating our color work. We would like to thank F. Fuma and H. Anderson for their help in the experimental setup, and A. Garito, C. Grossman and J.W. Wu in the Department of Physics for their help in the spectral measurements of camera and filters.

References

[1] S. A. Shafer, "Using Color to Separate Reflection Components", *COLOR Research and Application*, **10**, No. 4, pp 210-218, Winter, 1985.

[2] G. J. Klinker, S. A. Shafer and T. Kanade, "Image Segmentation and Reflection Analysis Through Color", *Proceedings of the DARPA Image Understanding Workshop*, Pittsburgh, pp. 838-853, 1988.

[3] R. Gershon, *The Use of Color in Computational Vision*, PhD dissertation, Department of Computer Science, University of Toronto, 1987.

[4] L. T. Maloney and B. A. Wandell, "A Computational Model of Color Constancy", *Journal of the Optical Society of America* **1**, No. 1, pp 29-33, April 1986.

[5] L. T. Maloney, "Evaluation of Linear Models of Surface Reflectance with Small Number of Parameters", *Journal of the Optical Society of America* **3**, No. 10, pp 1673-1683, October 1986.

[6] M. D'Zmura and P. Lennie, "Mechanisms of Color Constancy", *Journal of the Optical Society of America* **3**, No. 10, pp 1662-1672, October, 1986.

[7] R. Bajcsy, S.W. Lee, A. Leonardis, "Computational Aspects of Color Constancy", *GRASP Laboratory Technical Report MS-CIS-89-50*, University of Pennsylvania, September, 1989.

[8] A. Yuille, "A Method for Computing Spectral Reflectance", *Biological Cybernetics* **56**, pp 195-201, 1987.

[9] L. T. Maloney, "Computational Approaches to Color Constancy", *Applied Psychology Laboratory, Technical Report 1985-01*, Stanford University, January, 1985.

[10] R. Nevatia, "A Color Edge Detector and Its Use in Scene Segmentation" *IEEE Trans. on Systems, Man, and Cybernetics* **7**, No. 11, pp 820-826, 1977.

[11] R. Ohlander, K. Price and D. R. Reddy, "Picture Segmentation Using a Recursive Region Splitting Method", *Computer Graphics and Image Processing* **8**, No. 3, December 1979.

[12] J. Cohen, "Dependency of the Spectral Reflectance Curves of the Munsell Color Chips", *Psychon. Sci.* **1**, pp 369-370, 1964.

[13] D. B. Judd, D. L. MacAdam, and G. W. Wyszecki, "Spectral Distribution of Typical Daylight as a Function of Correlated Color Temperature", *Journal of the Optical Society of America* **54**, p 1031, 1964.

[14] G. Buchsbaum and A. Gottschalk, "Chromaticity Coordinates of of Frequency-limited Functions", *Journal of the Optical Society of America* **1**, p 885, 1984.

[15] G. Buchsbaum, "A Spatial Processor Model for Object Colour Perception", *J. Franklin Inst.* **310**, pp 1-26, 1980.

[16] B.K.P. Horn, **Robot Vision**, The MIT Press, 1986.

Color Constancy

Surface color is an important characteristic of an object that can be used for its identification. When the illumination environment changes, however, the spectral properties of the light reflected by an object to a sensing device also change. Color constancy is the ability to recover surface descriptors from color images that do not change as the illumination changes. The human visual system is approximately color constant over a range of surfaces and illumination conditions. Color constancy is also highly desirable for machine vision systems that must operate in environments where illumination cannot be controlled.

Achieving color constancy is difficult because the color signal that reaches a vision system depends on the illumination, the spectral reflectance of the observed object, and the scene geometry. These properties of the imaging environment are, in general, unknown. In machine vision, color constancy is often formulated as the problem of recovering an estimate of the spectral reflectance of each surface in the scene. Alternatively, the problem can be posed as one of determining the image of a scene as it would appear under some known illuminant.

Without constraint on the structure of a scene, color constancy is not possible. Most of the color constancy algorithms presented in this section are derived from specific assumptions about the properties of illuminants and surfaces. Several authors, for example, work from the assumption that surface spectral reflectance can be represented as a linear combination of a fixed set of basis functions and that the illuminant spectral power distribution can be represented as a linear combination of another set of basis functions. Such model-based approaches are useful because they allow characterization of the set of scenes for which a given algorithm will be color constant.

The first algorithm for color constancy was the retinex theory originally proposed by Land and McCann to model human color vision [1]. The retinex algorithm computes a vector color descriptor at each image location using a normalization rule. The normalization applied to a measured photoreceptor value depends on other measurements across the image for the same photoreceptor class. Several authors, e.g. [2], have verified that the retinex algorithm computes approximately constant color descriptors for certain classes of scenes as the spectral power distribution of the illumination changes.

The properties of the retinex algorithm are examined by Brainard and Wandell [4]. In consistency with the original formulation of the retinex theory, the study considers the domain of Mondrian scenes that consist of a collage of matte colored rectangles. The authors assert that an effective color constancy algorithm should generate constant surface descriptors in the presence of changes in illumination and also in the presence of changes in scene composition. Using simulations, they show that although the retinex algorithm is relatively insensitive to changes in illumination, it often produces markedly different color descriptors for a surface as the colors of surrounding surfaces are varied.

In [5], Maloney and Wandell develop an algorithm for estimating the spectral reflectance of surfaces in a scene. The algorithm is based on the assumption that a single (or slowly changing) spectral power distribution illuminates the scene and that both the spectral reflectance of surfaces and the illuminant spectral power distribution can be represented using finite dimensional linear models. Under the condition that there are more classes of photoreceptor than degrees of freedom in the surface reflectance model, the algorithm recovers the spectral power distribution of the illuminant and the spectral reflectance at each location in the scene.

D'Zmura and Lennie propose methods for the recovery of color constant surface descriptors in the presence of varying scene geometry and while also allowing the possibility that surfaces exhibit both specular and body (diffuse) reflection [6]. After assuming that the specular component of the reflectance is constant with wavelength, the authors formulate a three-dimensional linear model for surface spectral reflectance that contains terms accounting for changing scene geometry and the corresponding changes in the amount of specular and body reflection measured at each image location. Significantly, they show that the hue of a surface is constant with respect to changing geometry. In order to recover the three surface descriptors from which hue can be computed across a scene, some mechanism must be available to discount the effects of the illuminant. The authors survey available methods for discounting the illuminant and discuss possible physiological implementations.

Tominaga and Wandell present a method for estimating the body (subsurface) component of the spectral reflectance of a surface [7]. The algorithm makes use of a previously developed technique [3] for estimating the spectral power distribution of the illumination. The previous technique was derived from the dichromatic reflection model with the assumption that the interface (specular) reflectance is a constant function of wavelength. To estimate body reflectance, two physical constraints on reflectance functions are used: 1) reflectance is a non-negative function of wavelength, and 2) reflectance is a linear combination of an interface and a body component where both coefficients are non-negative. The authors show that using these constraints, measurements of the light reflected by a surface restrict the body reflectance to a range of functions. The method is demonstrated using spectroradiometer measurements on a set of plastics and fruits.

Forsyth [8] develops two color constancy algorithms based on different scene assumptions. The algorithms assume Mondrian scenes and endeavor to recover the photoreceptor values that would be measured for each surface under a canonical light. The first step is to estimate the illuminant, which Forsyth shows is constrained by the color measurements across the image. The first algorithm, Crule, scales each of the photoreceptor measurements to obtain the color descriptors. Crule makes no assumptions about the surface reflectance, but requires sufficiently constrained illuminants for color constancy. The second algorithm, Mwext, is an extension of Maloney and Wandell's algorithm [5] and represents spectral reflectance functions and the spectral power distribution of illuminants using linear models. Forsyth shows that Mwext can be used to recover more parameters in the illuminant model than in the original Maloney-Wandell algorithm for the same number of photoreceptors. Crule is demonstrated on a set of real Mondrian images and performs well. In comparison with the retinex algorithm, Crule generates significantly more stable color descriptors for a surface as the composition of the surrounding Mondrian changes.

Ho, Funt, and Drew present an algorithm for recovering both surface spectral reflectance and illuminant spectral power distribution [9]. As with other techniques, the method models reflectance and illumination using finite dimensional linear models. Critical to the success of the algorithm, the authors use large sets of published data to show that the structure of illuminant functions is significantly different from the structure of reflectance functions. This allows the recovery of three parameters describing the illuminant and three parameters describing the surface from a sampled reflected color signal. The authors suggest how a sampled reflected color signal can be obtained from a small set of photoreceptor measurements using chromatic aberration. Simulations are used to test the separation algorithm and yield good results.

Novak and Shafer describe a reliable technique that can be used for color constancy [10]. The technique makes use of a color chart that is placed in the scene having several patches of known spectral reflectance. Color measurements across the chart provide a large number of constraints on the spectral power distribution of the illumination. The authors show how these constraints can be used to estimate the illumination using a least squares fit to basis functions and also using a neural network. The algorithm has performed well on simulation data. Using a similar approach, Novak and Shafer also show how the spectral reflectance of an unknown surface can be estimated using several images obtained using known illuminants.

– G.E.H.

References

[1] Land, E.H., and McCann, J.J. "Lightness and Retinex Theory." *Journal of the Optical Society of America* **61**: 1-11, 1971.

[2] McCann, J.J., McKee, S.P., and Taylor, T.H. "Quantitative Studies in Retinex Theory: A Comparison Between Theoretical Predictions and Observer Responses to the Color Mondrian Experiments." *Vision Research* **16**: 445-448, 1976.

[3] Tominaga, S., and Wandell, B.A. "Standard Surface-reflectance Model and Illuminant Estimation." *Journal of the Optical Society of America A* **6**(4): 576-584, April 1989.

Papers Included in This Collection

[4] Brainard, D.H., and Wandell, B.A. "Analysis of the Retinex Theory of Color Vision." *Journal of the Optical Society of America A* **3**(10): 1651-1661, October 1986.

[5] Maloney, L.T. and Wandell, B.A. "Color Constancy: A Method for Recovering Surface Spectral Reflectance." *Journal of the Optical Society of America A* **3**(1): 29-33, January 1986.

[6] D'Zmura, M., and Lennie, P. "Mechanisms of Color Constancy." *Journal of the Optical Society of America A* **3**(10): 1662-1672, October 1986.

[7] Tominaga, S., and Wandell, B.A. "Component Estimation of Surface Spectral Reflectance." *Journal of the Optical Society of America A* **7**(2): 312-317, February 1990.

[8] Forsyth, D.A. "A Novel Algorithm for Color Constancy." *International Journal of Computer Vision* **5**(1): 5-36, 1990.

[9] Ho, J., Funt, B.V., and Drew, M.S. "Separating a Color Signal Into Illumination and Surface Reflectance Components: Theory and Applications." *IEEE Transactions on Pattern Analysis and Machine Intelligence*, **PAMI-12**(10): 966-977, October 1990.

[10] Novak, C.L., and Shafer, S.A. "Supervised Color Constancy for Machine Vision." In *Proc. SPIE Vol. 1453: Human Vision, Visual Processing, and Digital Display II*. Santa Clara, CA. SPIE, March 1991. Pages 353-368.

Analysis of the retinex theory of color vision

David H. Brainard and Brian A. Wandell

Department of Psychology, Stanford University, Stanford, California 94305

Received August 16, 1985; accepted June 11, 1986

If color appearance is to be a useful feature in identifying an object, then color appearance must remain roughly constant when the object is viewed in different contexts. People maintain approximate color constancy despite variation in the color of nearby objects *and* despite variation in the spectral power distribution of the ambient light. Land's retinex algorithm is a model of human color constancy. We analyze the retinex algorithm and discuss its general properties. We show that the algorithm is too sensitive to changes in the color of nearby objects to serve as an adequate model of human color constancy.

INTRODUCTION

Introduction to Color Constancy

If color appearance is to be a useful feature in identifying an object, then color appearance must remain roughly constant when the object is viewed in different contexts. People maintain object color appearance despite variation in the color of nearby objects *and* despite variation in the spectral power distribution of the ambient light.[1–4]

Historically, changes in the color appearance of an object caused by variation in the surface reflectance functions of surrounding objects have been called simultaneous contrast, whereas changes in color appearance caused by variation in the spectral power distribution of the ambient light have been called failures of color constancy. *Both* of these effects reduce the usefulness of color appearance as a feature for identifying objects. In our view, *color constancy* should be defined as the maintenance of color appearance despite variation in the color of nearby objects *and* despite variation in the spectral power distribution of the ambient light. Although human color vision does not maintain perfect color constancy, human performance is better than that of any currently available man-made systems.

The fundamental difficulty in designing a color constant system arises becase the light in the visual image confounds two factors: the spectral power distribution of the ambient light and the surface reflectance of the objects in the scene. The problem of separating these two factors cannot be solved for all possible viewing conditions. For example, if there is only a single unknown object in the image illuminated by an unknown light source, no algorithm can correctly determine the surface reflectance of the object. It follows that all color constant algorithms must use information obtained from light reflected from several different objects in the scene. We therefore think it natural to broaden the definition of color constancy to include preservation of appearance across variation in the surface reflectance of nearby objects.

A number of algorithms have been proposed for the purpose of modeling human color constancy or to achieve approximate color constancy.[3–7] Only the more recent algorithms[5–7] make explicit the class of viewing contexts for which the algorithm will be color constant.

The retinex algorithm proposed by Land and McCann[4] and Land[8] is important because it was the first attempt at developing a computational model for human color constancy. There is widespread current interest in the retinex algorithm.[8–17] The experiments that lead to the development of the retinex algorithm and the algorithm itself have been widely discussed in the literature.[6,18–23] In spite of this, the retinex algorithm is not generally well understood.

In this paper we quantitatively describe the problem of color constancy. We then describe Land's most recent retinex algorithm.[8,17] We present some simple results that describe the general behavior of the retinex algorithm. We extend our analysis by describing a computer simulation of the algorithm. The simulation demonstrates that the retinex algorithm is a poor model of human color constancy. The dramatic failures of retinex as a model for human vision arise because the assignment of a color value to a surface is far too dependent on the composition of the other surfaces in the image. In Appendix A we discuss other versions of retinex, suggested by McCann and Houston[11] and Horn.[24]

Formulation of Color Constancy

The retinex algorithm has been proposed for and tested on only a limited class of viewing contexts. We will formulate the problem of color constancy with respect to this simplified model of the natural environment. The viewing context is illustrated in Fig. 1. An observer looks at a flat two-dimensional surface. The materials on the surface are matte, and they reflect the ambient light toward a normal color observer, who has three classes of photoreceptors. In our formulation we describe the values of all functions of wavelength by their values at a discrete number of sample wavelengths λ_n for $n = 1, N$. In particular, we characterize the spectral power distribution of the ambient light by the function $E(\lambda_n)$ and the reflectance at a point x on the surface by the function $S^x(\lambda_n)$. The light arriving at the eye is called the *color signal*. It is equal to the product of the spectral power distribution of the ambient light and the surface reflectance function, $C^x(\lambda_n) = E(\lambda_n)S^x(\lambda_n)$. The color signal defines all the image information that is available at the eye to make judgments concerning the surface reflectance at different points on the surface.

The observer has three photoreceptor arrays that spatially

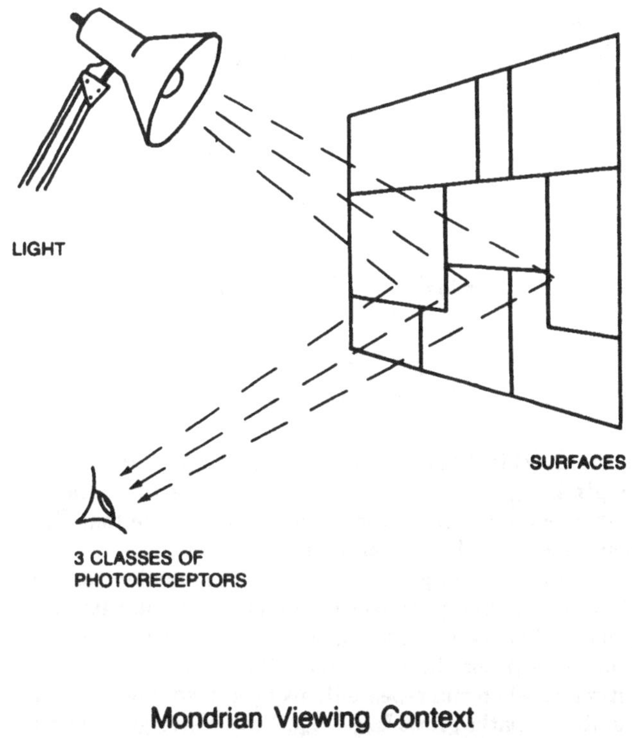

Fig. 1. A flat surface of colored materials is illuminated by a light source.

sample the color signal. The response of the photoreceptors is computed from the color signal and the spectral sensitivity of the photopigment in the kth receptor class, $R_k(\lambda_n)$:

$$\rho_k^x = \sum_{n=1}^{N} C^x(\lambda_n)R_k(\lambda_n)$$

$$= \sum_{n=1}^{N} E(\lambda_n)S^x(\lambda_n)R_k(\lambda_n). \quad (1)$$

This equation can be written as a matrix product of the form

$$\rho^x = \Lambda_E\sigma^x \quad (2)$$

in which the nth entry of the column vector σ^x is the surface reflectance at λ_n, $S^x(\lambda_n)$. The matrix Λ_E is $3 \times N$, and its k, nth entry is $E(\lambda_n)R_k(\lambda_n)$. The entries of the matrix depend only on the spectral power distribution of the ambient light (a physical variable) and the receptor spectral sensitivities (a fixed value). We emphasize the fact that the matrix depends only on the spectral power distribution of the ambient light by calling the matrix the *lighting matrix*.

The problem of color constancy can be expressed as the following challenge: Beginning with the spatial array of data on the left-hand side of Eq. (2), recover the spatial array of surface reflectance functions on the right-hand side of the equation. An algorithm with perfect color constancy uses the photoreceptor values on the left-hand side to calculate color estimates at each pixel that are independent of *both* the spectral power distribution of the ambient light (i.e., the lighting matrix Λ_E) and the other surfaces comprising the scene (i.e., the vectors σ^x).

The class of viewing contexts used to evaluate the retinex is limited in various ways. The spatial distribution of the surface reflectances on the two-dimensional surface is restricted to be composed of an array of rectangular, matte, colored papers. Land refers to this spatial distribution of surfaces as a Mondrian. This means that there is generally a clear signal in the image data at each of the boundaries between the differently colored papers. Other limitations of this viewing context are that there are no secondary reflections among surfaces, there is no specular reflectance, and the surfaces are restricted to lie in a plane.

In this paper, we study the performance of the retinex algorithm for the simple case in which the intensity of the light is spatially uniform across the surfaces. This means that it is not necessary to discount spatial variation in the intensity of the ambient light. The problem of discounting such spatial variation across a single scene is logically distinct from the problem of discounting variation in the spectral power distribution of the ambient light from one scene to another. We discuss this point more fully in Appendix A in connection with our analysis of Ref. 24.

THE RETINEX ALGORITHM

There are several published variants of the retinex theory.[4,8,11,25] Here we consider the algorithm first described in Ref. 8. All the published versions of retinex share basic underlying principles, and in Appendix A we analyze the McCann–Houston algorithm[11] and an algorithm presented by Horn.[24]

The input into the retinex algorithm is an array of photoreceptor responses for each location in the image. We can think of this input as being three separate arrays of data, one for each receptor class. Each of these spatial arrays contains the receptor responses for a single receptor class for each location in the image. We call each separate spatial location in the image a *pixel* and use the superscript x to denote a particular pixel in the spatial array. The algorithm transforms the spatial array of photoreceptor responses in the kth class, ρ_k^x, into a corresponding spatial array of values, l_k^x. Land calls the individual values, l_k^x, *lightness values*. At each pixel, one lightness value is computed for each photoreceptor class. A central principle of the retinex algorithm is that the lightness values at pixel x are calculated independently for each receptor class. Thus we need describe the computation of only one of the spatial arrays of lightness values. This permits us to supress the subscript k that indicates the photoreceptor class.

The Lightness Computation

The algorithm estimates the spatial array of lightness values for a single receptor class by computing a series of *paths*. Each path is computed as follows. We select a starting pixel x_1. We then randomly select a neighboring pixel x_2. We calculate the difference of the logarithms of the sensor responses at the two positions. This value is added into an accumulator register for position x_2 such that

$$A(x_2) \leftarrow A(x_2) + \log(\rho^{x_2}) - \log(\rho^{x_1}). \quad (3)$$

In addition, a counter register $N(x_2)$ for position x_2 is incremented to indicate that a path has crossed this position. At the start of the computation all accumulators and counters are initialized to 0.

The path calculation proceeds iteratively with the random selection of a neighbor of pixel x_2. In general, the accumulator of position x_i on this path is updated by

$$A(x_i) \leftarrow A(x_i) + \log(\rho^{x_i}) - \log(\rho^{x_1}), \tag{4}$$

and the corresponding counter register $N(x_i)$ is incremented. Note that the sensor response of the first element of the path plays a special role in the accumulation for that path calculation: It is used as a normalizing term at every point on the path. The path starting at x_1 need not pass through all the positions in the image, and it may pass through some positions more than once. The number of positions traversed by a path is a parameter N_{pl} of the algorithm. (The subscript pl stands for path length.)

After the first path has been computed, the procedure is repeated for a new path that starts at another randomly chosen position. The number of paths in a complete calculation of the algorithm followed is also a parameter, N_{np}. (The subscript np stands for number of paths.)

After all paths have been completed, the lightness value for each pixel x is computed by simply dividing the accumulated values in $A(x)$ by the contents of the corresponding counter register, $N(x)$.

The full retinex algorithm described in Ref. 8 includes an additional thresholding operation that occurs during the path calculation. The thresholding operation compares the sensor response at adjacent pixels along the path. If the value at pixel x_{i+1} is sufficiently close to the sensor response at the previous pixel, x_i, then the accumulator value at x_{i+1} is updated by using the sensor response at x_i. The intention of the thresholding operation is to "remove the effects of nonuniform illumination over the scene" (Ref. 8, p. 5165). In this paper we will analyze retinex only for images with spatially uniform ambient light distributions. Thus the thresholding operation has no effect on any of our calculations, and we will not address this aspect of the algorithm (but see Ref. 23).

The Meaning of the Lightness Values

The purpose of the retinex algorithm is to compute lightness values that will be invariant under changes of viewing context, much as human performance is roughly invariant under similar changes. At each pixel x, the lightness triplet should depend only on the surface reflectance $S^x(\lambda_n)$ and not on the spectral power distribution of the ambient light or the surface reflectance functions of the other papers comprising the Mondrian. In the following sections we investigate both the extent to which the lightness values are color constant and whether the retinex algorithm in fact models human vision.

ANALYSIS OF THE RETINEX ALGORITHM

The retinex algorithm is stochastic, as both the starting pixels of the paths and the path routes are chosen randomly. To understand the retinex algorithm, we examine its convergence properties.

Consider the kth receptor class. The lightness value l_k^x for pixel x is the average of the contributions to $A(x)$ from the paths that passed through pixel x. Each time a path passes through pixel x, $A(x)$ is incremented by the log of the ratio of the receptor response at x to the receptor response at the starting pixel of the path. Let the total number of times a path has crossed pixel x be N. If we denote the starting points for the paths passing through pixel x as $x_1, \ldots, x_N$, then

$$l_k^x = \frac{\log\left(\dfrac{\rho_k^x}{\rho_k^{x_1}}\right) + \ldots + \log\left(\dfrac{\rho_k^x}{\rho_k^{x_N}}\right)}{N}$$

$$= \log\left[\frac{\rho_k^x}{(\rho_k^{x_1} \ldots \rho_k^{x_N})^{1/N}}\right]$$

$$= \log\left[\frac{\rho_k^x}{\hat{G}_k(x)}\right], \tag{5}$$

where $\hat{G}_k(x)$ is the geometric mean of the receptor responses of pixels $x_1, \ldots, x_N$ that are the path starting locations corresponding to each crossing of pixel x. Note that $\hat{G}_k(x)$ may depend on the spatial position of pixel x.

If the number of paths N_{np} is small, the random variable $\hat{G}_k(x)$ will have large variance, as will the computed lightness values. This is an undesirable property since for a small number of paths the algorithm will not return stable color estimates when run repeatedly using a fixed input. As the number of paths grows large, $\hat{G}_k(x)$ will converge to a limiting value, $G_k(x)$, and hence the lightness values will be stable. Therefore we may supress the parameter N_{np} and examine the properties of the algorithm under the assumption that N_{np} is large enough so that the lightness values may be computed by their value in the limit as N_{np} grows without bound.

The number of times that a path starting at any pixel x_i crosses pixel x is a random variable, as is the total number of paths N that cross pixel x. Denote the expected number of crossings of pixel x by a path starting at location x_i by $E(x|x_i)$. If the path starting points are selected with equal probability among all pixels, then, as N_{np} tends to infinity, the value of $\hat{G}_k(x)$ is given by the equation

$$\log G_k(x) = \frac{1}{E(N)}\left[\sum_{\text{all pixels}} E(x|x_i)\log(\rho_k^{x_i})\right], \tag{6}$$

where $E(N)$ is the expected number of crossings through pixel x, $E(N) = \Sigma_{\text{all pixels}} E(x|x_i)$. In this case, we can compute the lightness values as

$$l_k^x = \log\left[\frac{\rho_k^x}{G_k(x)}\right], \tag{7}$$

where $G_k(x)$ is given by Eq. (6).

Now we must address the question of how to compute the values of the $E(x|x_i)$ as a function of the parameter N_{pl}. The authors of the retinex algorithm have not provided a precise description of the statistical properties of the path-generation process.[26] We have chosen to model the path-generation process by using a Markov process. On each step the path is equally likely to move from its current position to any of the immediately adjacent pixels. There are eight adjacent pixels for points in the center of the image, five adjacent pixels for points at the edge, and three adjacent pixels at the corners. The choice of a Markov process simplifies the

Figure 18. (Novak et al., p. 27) Unlinearized image.

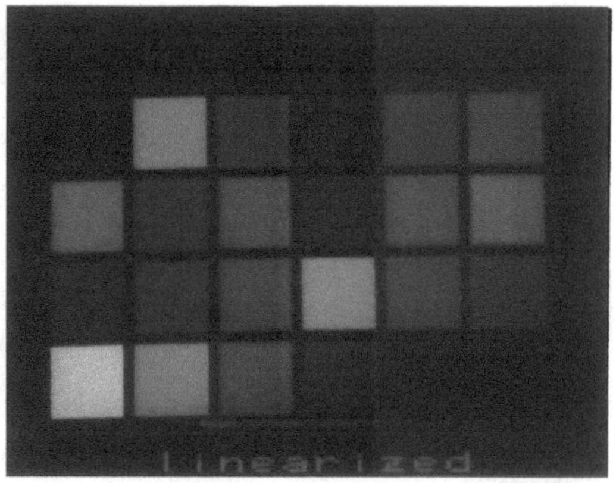

Figure 19. (Novak et al., p. 27) Linearized image.

Figure 20. (Novak et al., p. 27) Image without IR-block.

Figure 21. (Novak et al., p. 27) W/o aperture balancing.

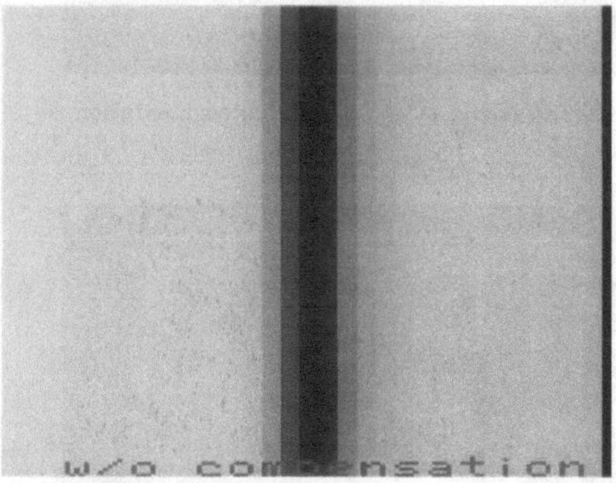

Figure 22. (Novak et al., p. 27) Before compensation.

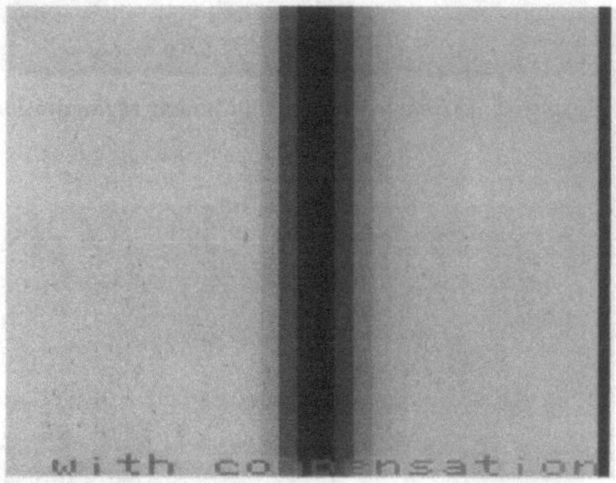

Figure 23. (Novak et al., p. 27) After compensation.

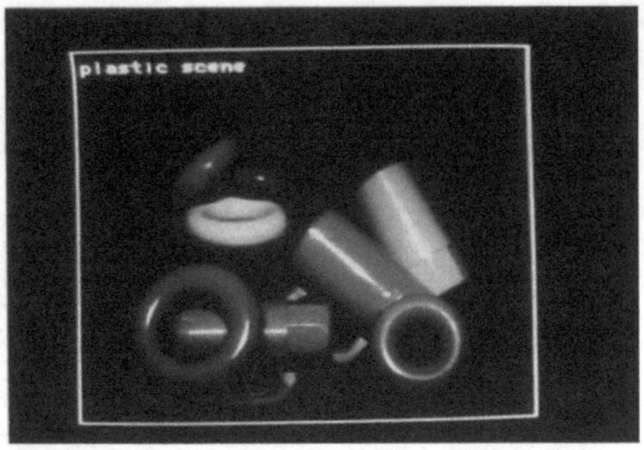

Figure 1. (Klinker et al., p. 158) Scene with eight plastic objects.

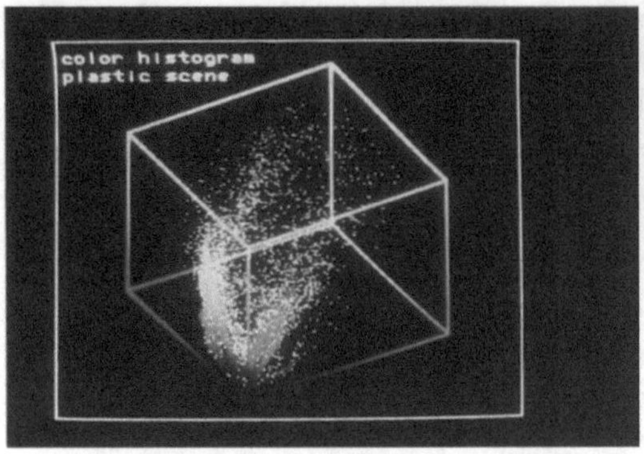

Figure 2. (Klinker et al., p. 159) Color histogram of the scene with eight plastic objects.

Figure 3. (Klinker et al., p. 159) Color cluster classification for initial image areas.

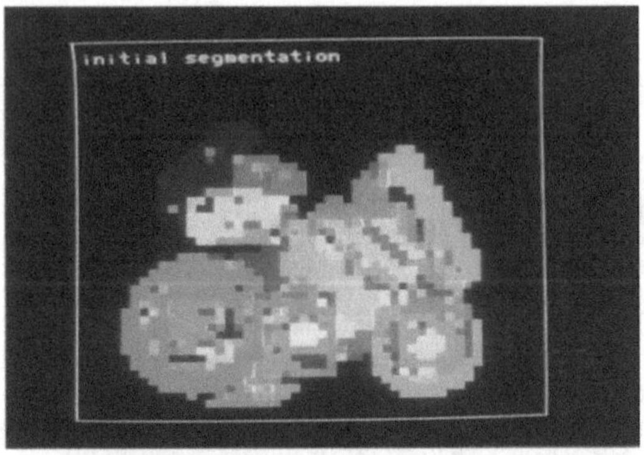

Figure 4. (Klinker et al., p. 160) Initial grouping into approximate image areas.

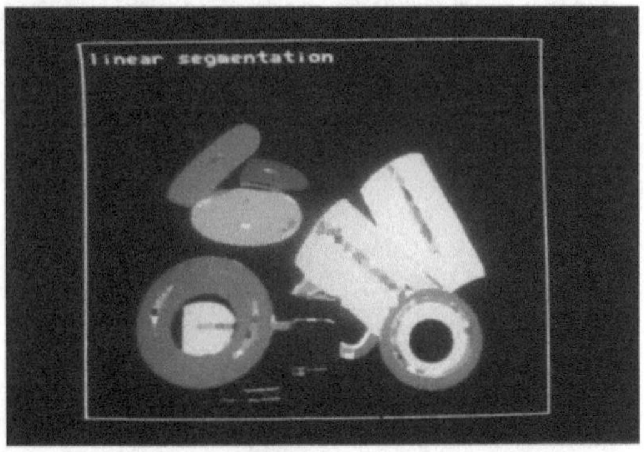

Figure 5. (Klinker et al., p. 160) Linear segmentation.

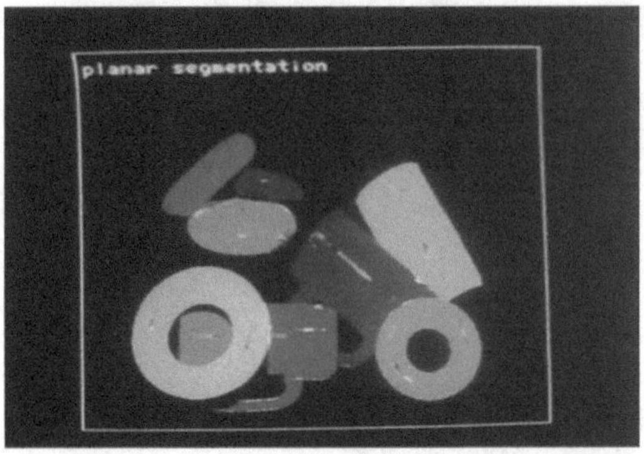

Figure 6. (Klinker et al., p. 161) Planar segmentation.

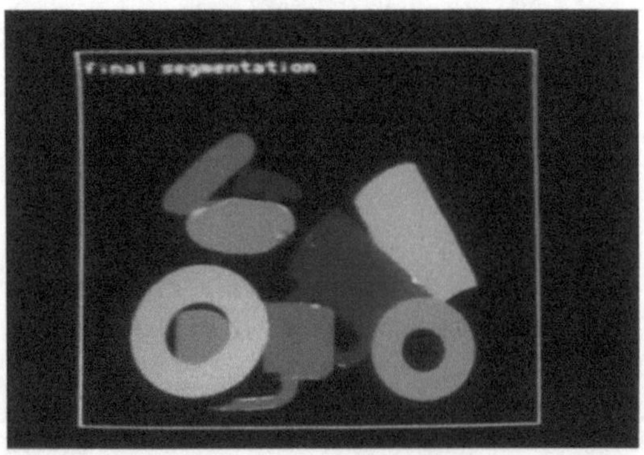

Figure 7. (Klinker et al., p. 161) Final segmentation.

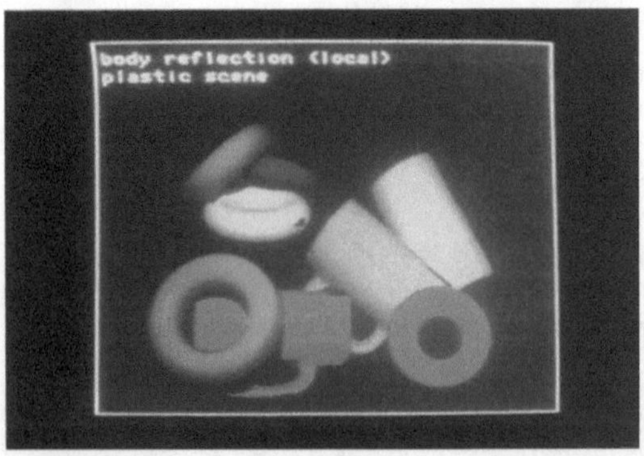

Figure 8. (Klinker et al., p. 162) Intrinsic body reflection image.

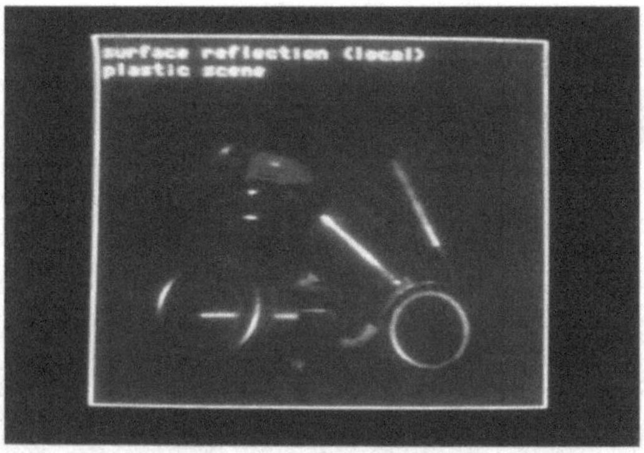

Figure 9. (Klinker et al., p. 162) **Intrinsic surface reflection image.**

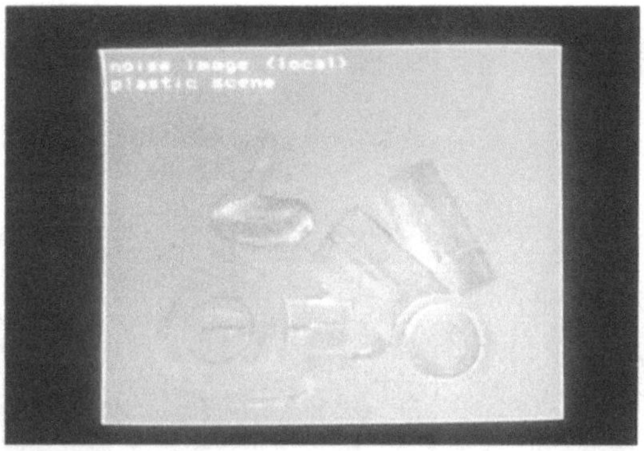

Figure 10. (Klinker et al., p. 163) **Intrinsic noise image.**

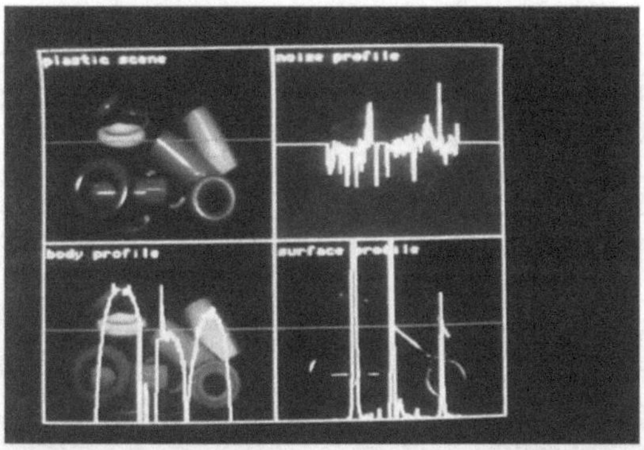

Figure 11. (Klinker et al., p. 163) **Profiles of the intrinsic reflection images along row 226.**

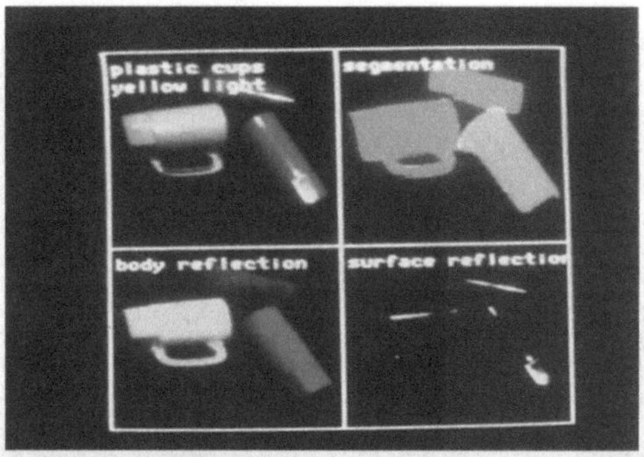

Figure 12. (Klinker et al., p. 164) **Color image segmentation and reflection analysis, scene with three plastic cups under yellow light.**

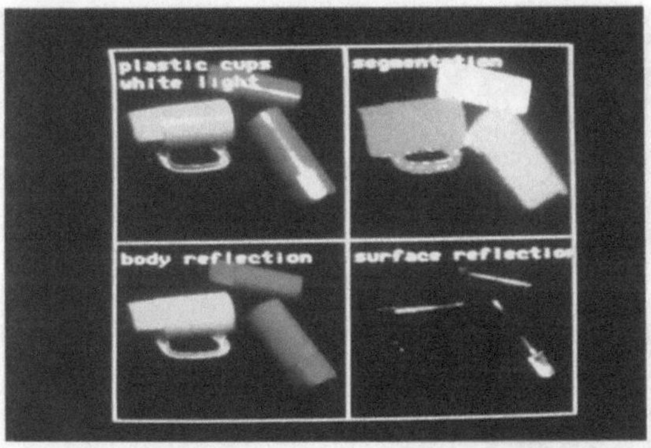

Figure 13. (Klinker et al., p. 164) **Color image segmentation and reflection analysis, scene with three plastic cups under white light.**

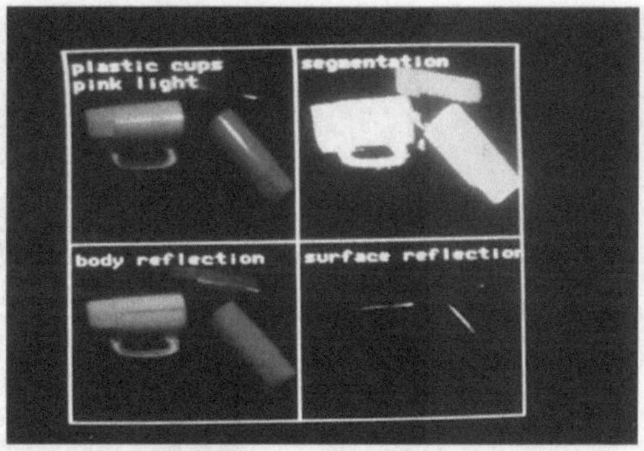

Figure 14. (Klinker et al., p. 165) **Color image segmentation and reflection analysis, scene with three plastic cups under pink light.**

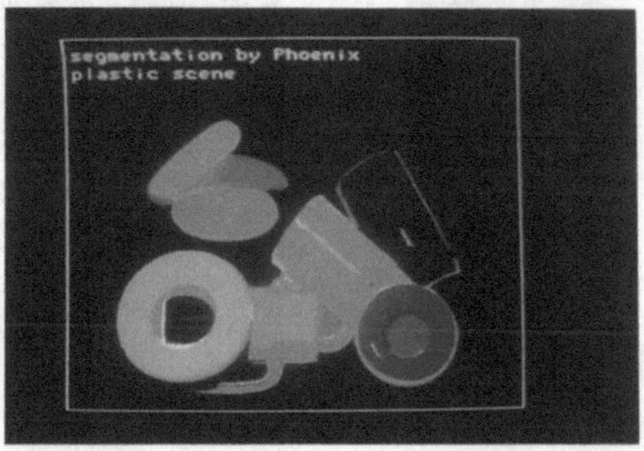

Figure 15. (Klinker et al., p. 165) **Region segmentation by Phoenix, scene with eight plastic objects.**

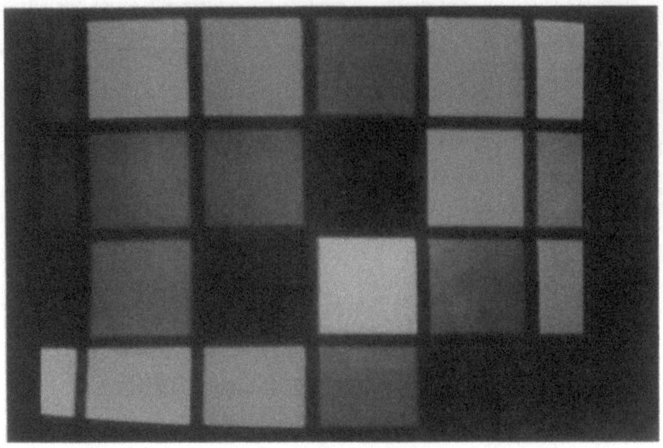

Figure 5. (Healey, p. 188) **Image of a color chart.**

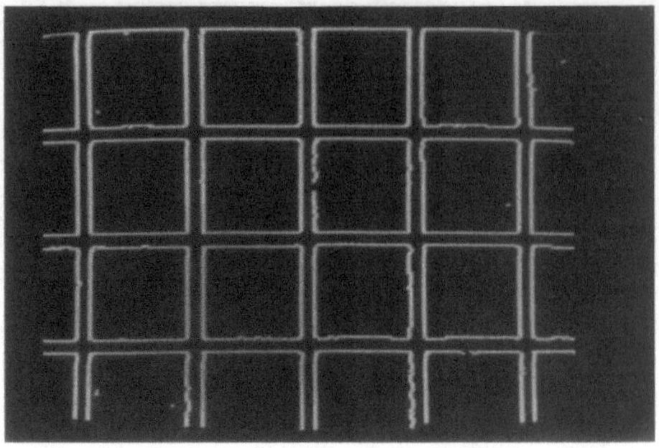

Figure 6. (Healey, p. 188) **Chart image edges.**

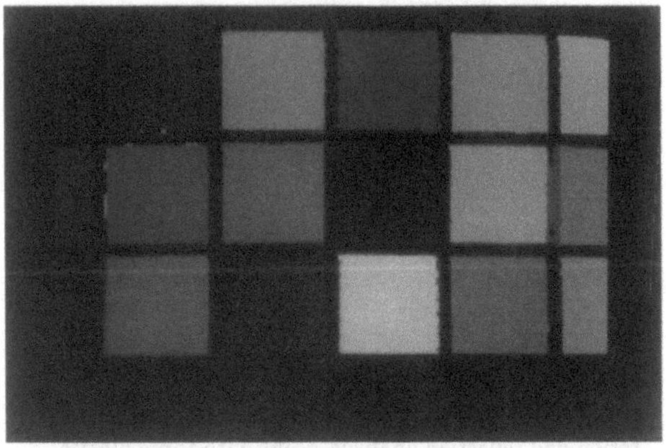

Figure 7. (Healey, p. 189) **Chart image segmentation.**

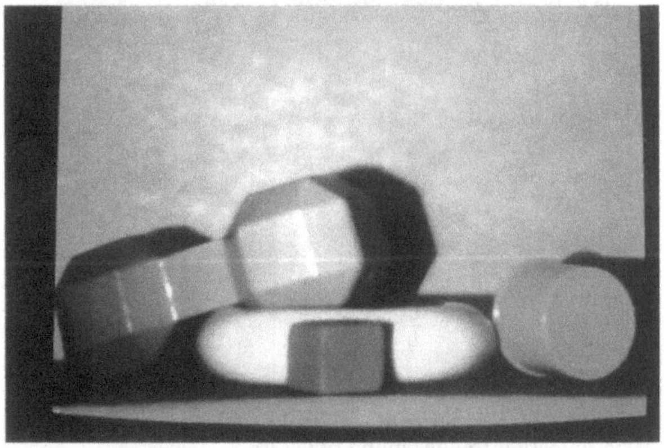

Figure 8. (Healey, p. 189) **Image of toys scene.**

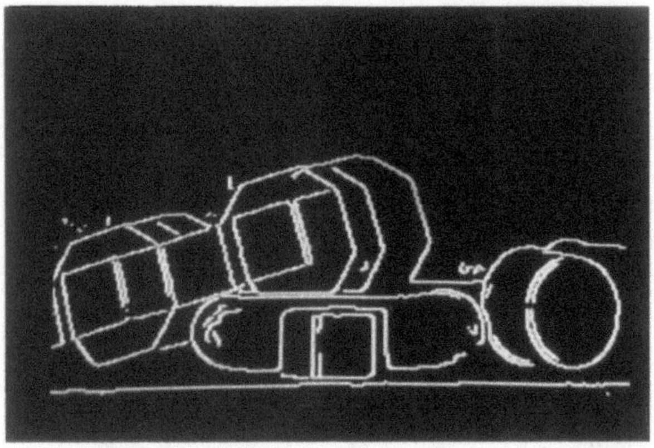

Figure 9. (Healey, p. 190) **Toys image edges.**

Figure 10. (Healey, p. 190) **Toys image segmentation.**

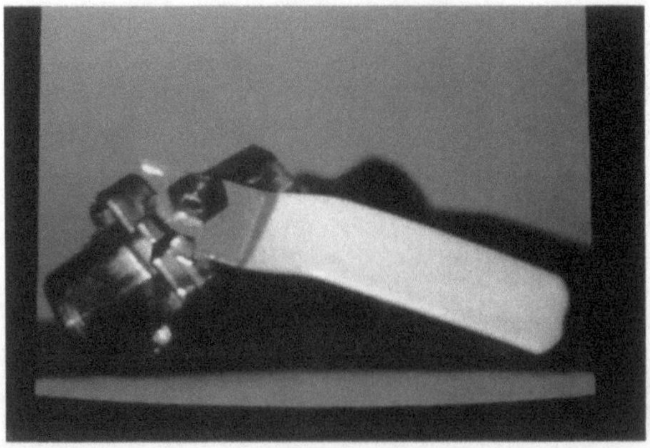

Figure 11. (Healey, p. 191) **Image of a valve.**

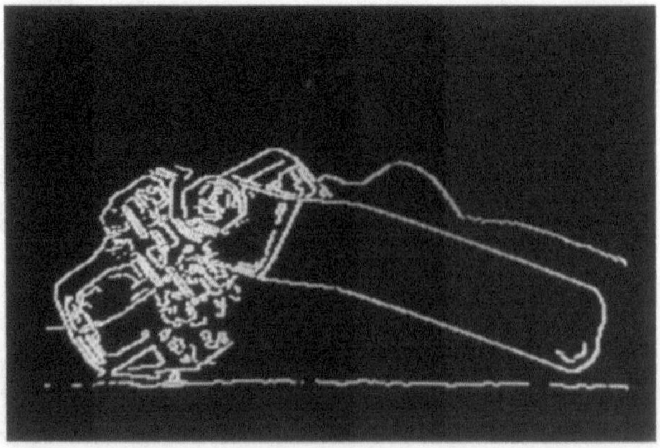

Figure 12. (Healey, p. 191) **Valve image edges.**

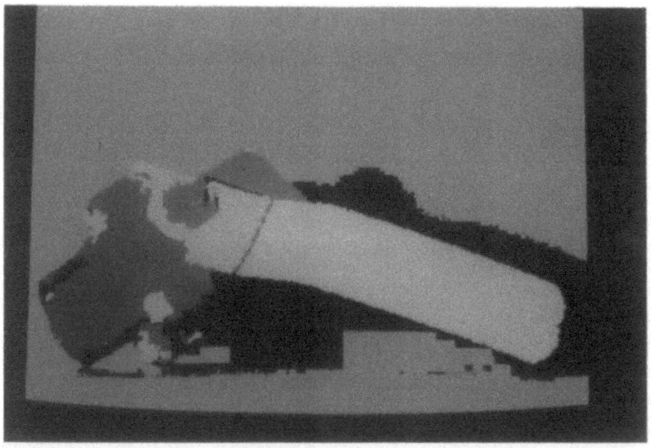

Figure 13. (Healey, p. 192) **Valve image segmentation.**

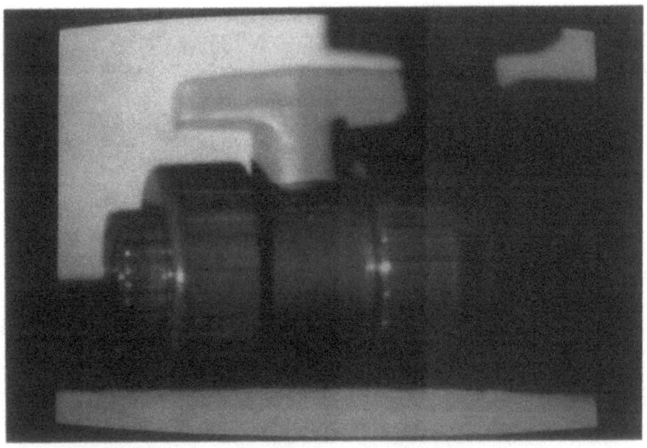

Figure 14. (Healey, p. 192) **Image of a plastic part.**

Figure 15. (Healey, p. 193) **Plastic part edges.**

Figure 16. (Healey, p. 193) **Plastic part image segmentation.**

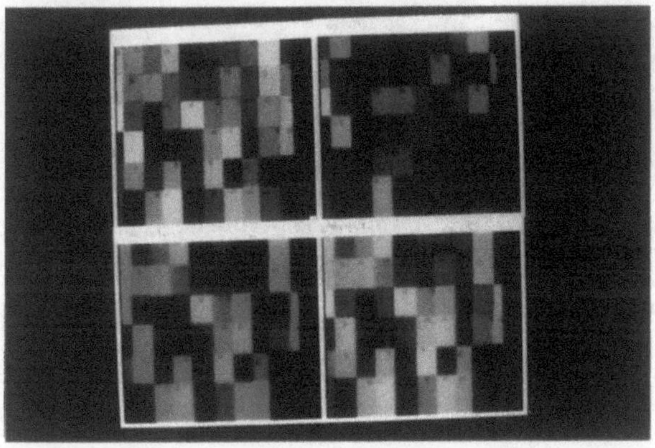

Figure 1. (Forsyth, p. 261) **Images of first Mondriaan, under (clockwise, from top left) white, green, purple, and red lights. Note enormous differences. Images show receptor responses. In each case, a human observer named the colors of the Mondriaan patches.**

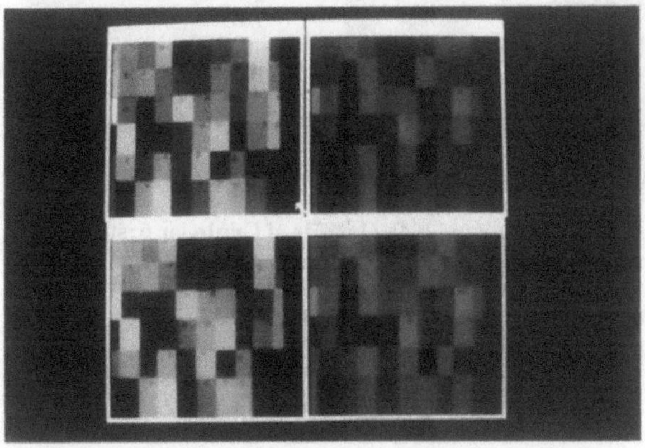

Figure 2. (Forsyth, p. 262) **Images of first Mondriaan, under (clockwise, from top left) white, blue-green, blue, and yellow lights.**

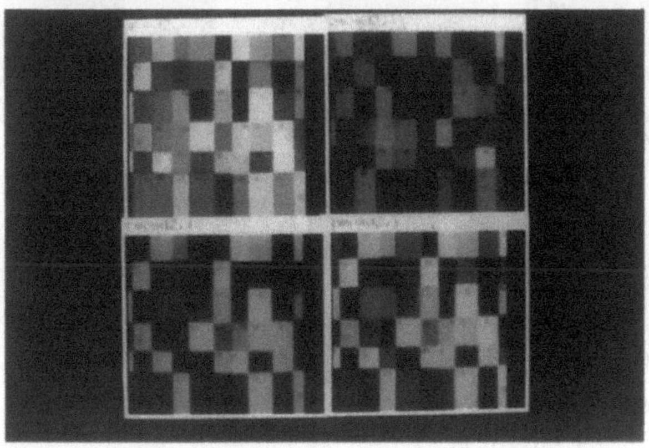

Figure 3. (Forsyth, p. 262) **Images of second Mondriaan, under (clockwise, from top left) white, green, purple, and red lights.**

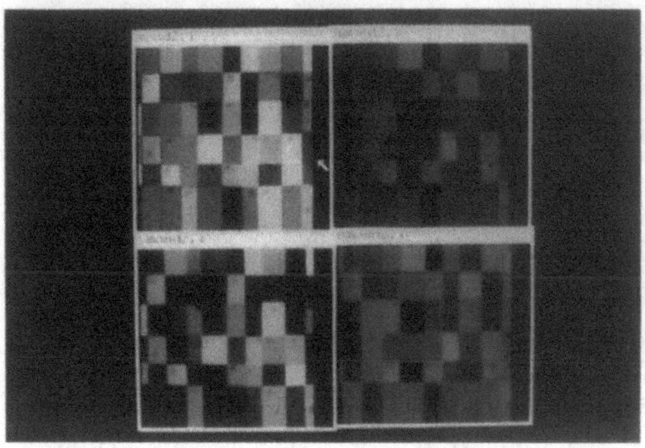

Figure 4. (Forsyth, p. 263) **Images of second Mondriaan, under (clockwise, from top left) white, blue-green, blue, and yellow lights.**

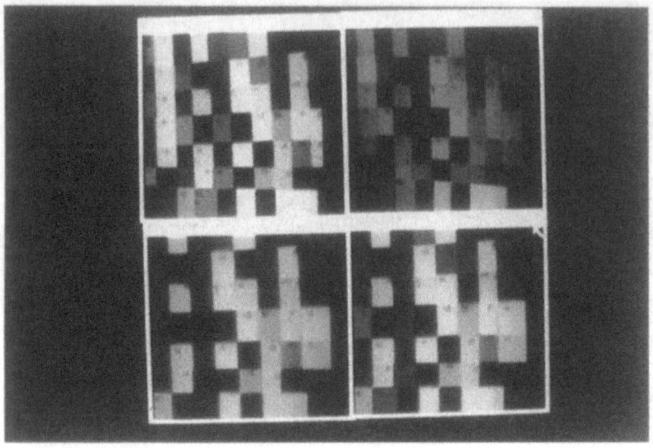

Figure 5. (Forsyth, p. 263) Images of third Mondriaan, under (clockwise, from top left) white, green, purple, and red lights.

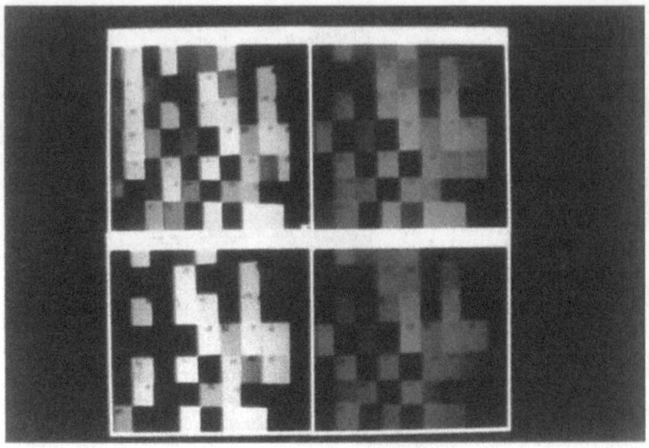

Figure 6. (Forsyth, p. 264) Images of third Mondriaan, under (clockwise, from top left) white, blue-green, blue, and yellow lights.

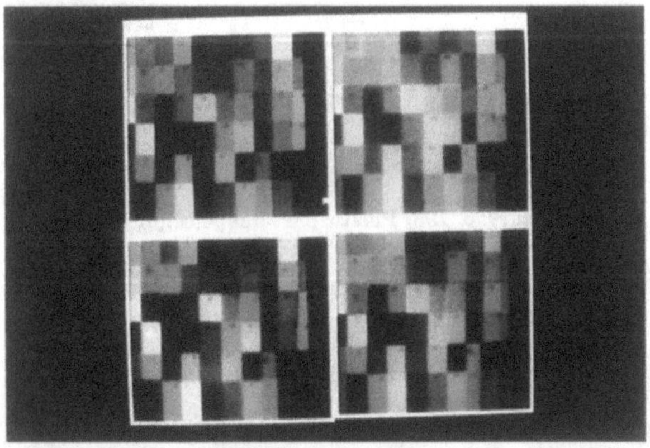

Figure 7. (Forsyth, p. 264) Outputs of **Crule** for first Mondriaan, from images under (clockwise, from top left) white, blue-green, blue, and yellow lights. **Crule** performs well, because these images look similar.

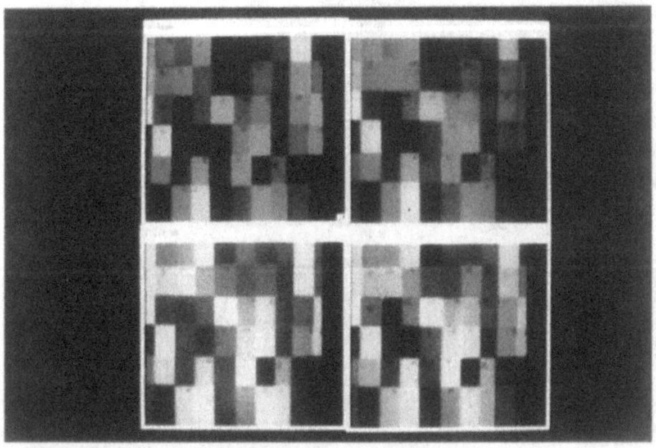

Figure 8. (Forsyth, p. 265) Outputs of **Crule** for first Mondriaan, from images under (clockwise, from top left white, green, purple, and red lights. Note that **Crule** performs poorly under red light.

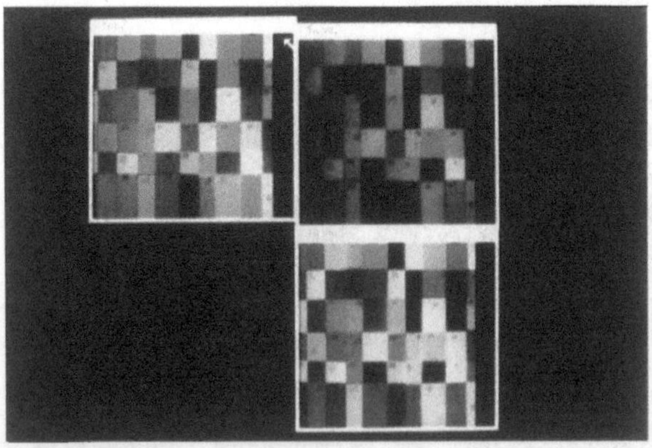

Figure 9. (Forsyth, p.265) Outputs of **Crule** for second Mondriaan, under (clockwise, from top left) white, green, and purple lights. **Crule** performed poorly under red light, which was omitted for this reason.

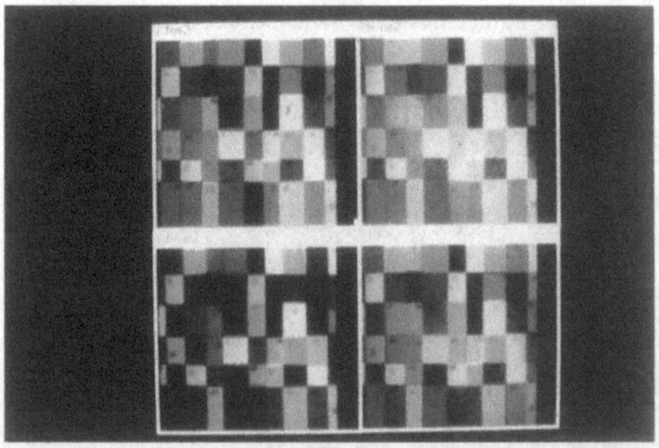

Figure 10. (Forsyth, p. 266) Outputs of **Crule** for second Mondriaan, under (clockwise, from top left) white, blue-green, blue, and yellow lights.

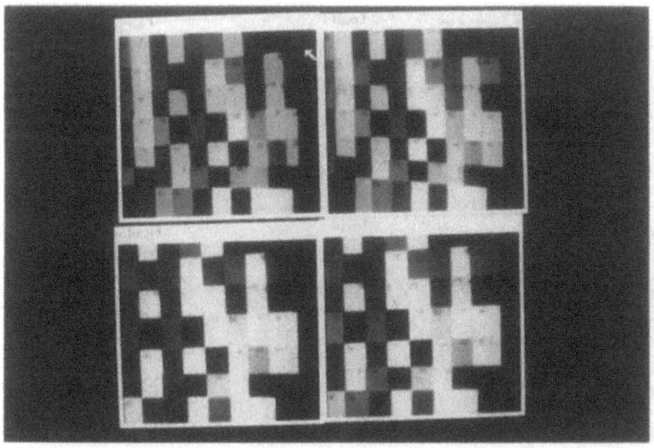

Figure 11. (Forsyth, p. 266) Outputs of **Crule** for third Mondriaan, under (clockwise, from top left) white, green, purple, and red lights. **Crule** performs poorly on these images because they do not contain a wide range of colors.

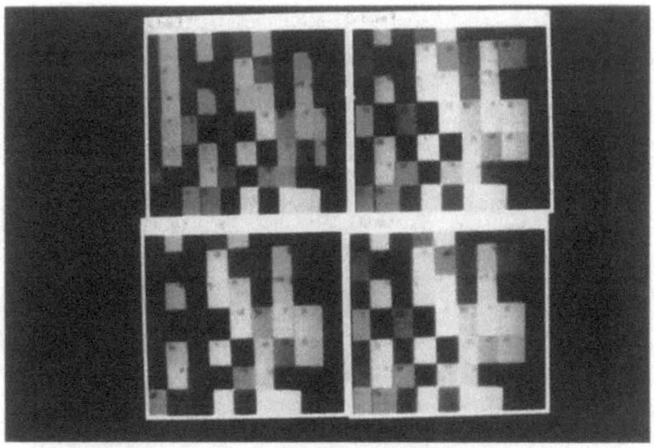

Figure 12. (Forsyth, p. 267) Outputs of **Crule** for third Mondriaan, under (clockwise, from top left) white, blue-green, blue, and yellow lights. **Crule** performs poorly on the images because they do not contain a wide range of colors.

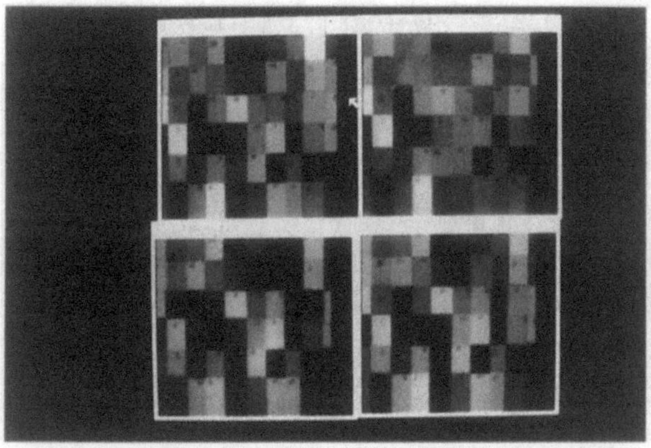

Figure 13. (Forsyth, p. 267) Outputs of Brainard and Wandell (1986) model of Retinex algorithm for Mondriaan, under (clockwise, from top left) white, green, purple, and red lights.

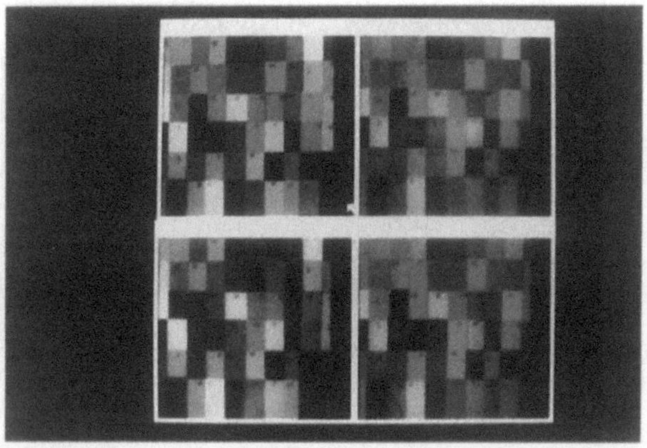

Figure 14. (Forsyth, p. 268) Outputs of Brainard and Wandell (1986) model of Retinex algorithm for first Mondriaan, under (clockwise, from top left) white, blue-green, blue, and yellow lights.

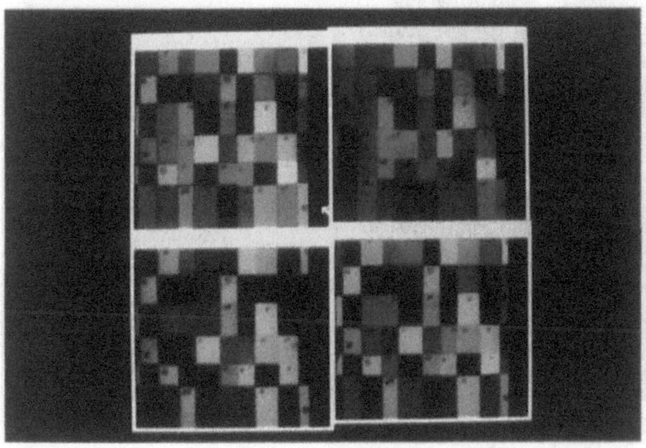

Figure 15. (Forsyth, p. 268) Outputs of Brainard and Wandell (1986) model of the Retinex algorithm for second Mondriaan, under (clockwise, from top left) white, green, purple, and red lights.

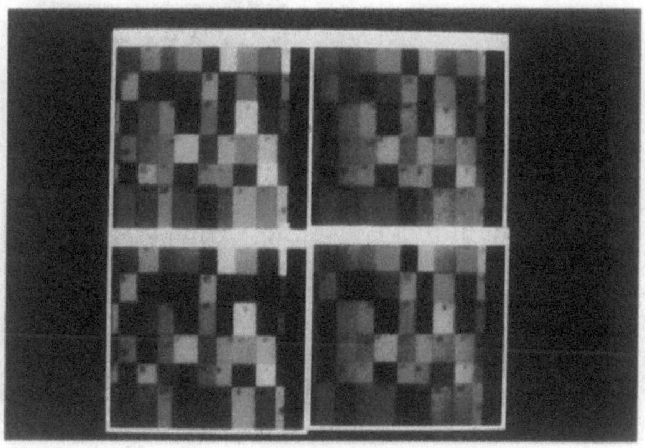

Figure 16. (Forsyth, p. 269) Outputs of Brainard and Wandell (1986) model of Retinex algorithm for second Mondriaan, under (clockwise, from top left) white, blue-green, blue, and yellow lights.

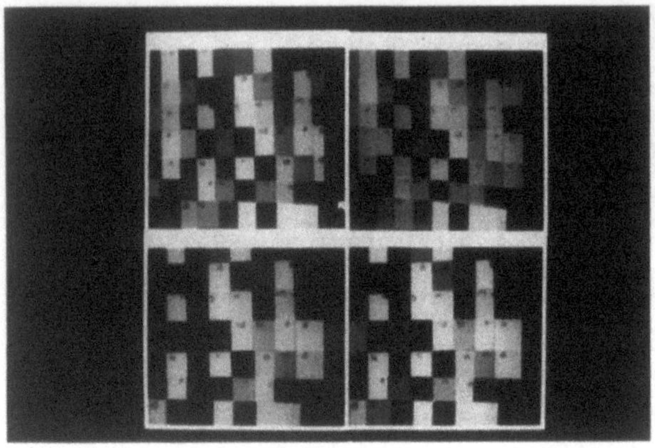

Figure 17. (Forsyth, p. 269) Outputs of Brainard and Wandell (1986) model of the Retinex algorithm for the third Mondriaan derived from images under (clockwise, from top left) white, green, purple, and red lights.

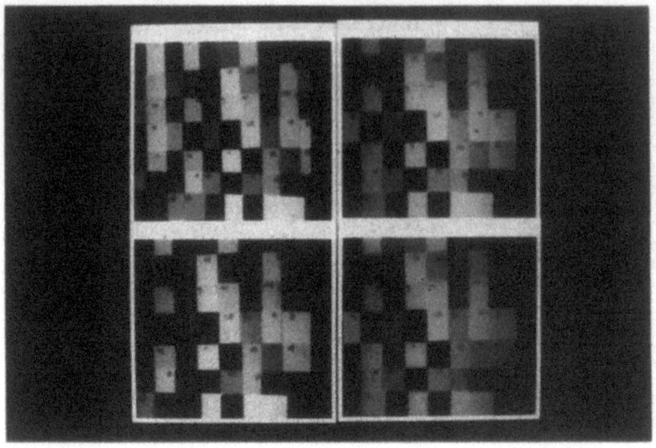

Figure 18. (Forsyth, p. 270) Outputs of Brainard and Wandell (1986) model of the Retinex algorithm for the third Mondriaan derived from images under (clockwise, from top left) white, blue-green, blue, and yellow lights.

Figure 19. (Forsyth, p. 270) Photograph of the first Mondriaan with red borders attached.

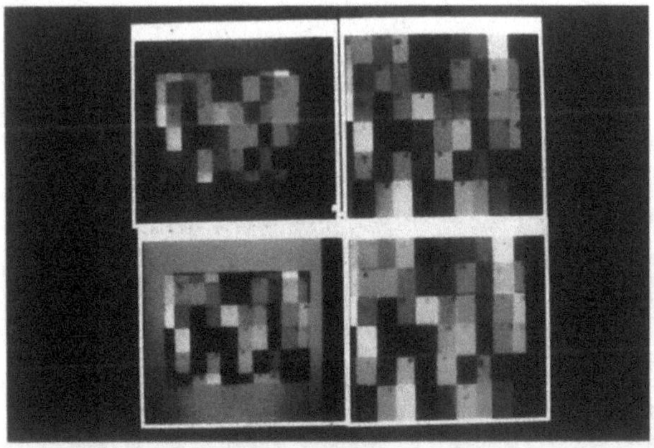

Figure 20. (Forsyth, p. 271) Top: The outputs computed by the Retinex algorithm, for an image of the first Mondriaan under white light (right), and for an image of the first Mondriaan with red borders (left), again under white light. The greenish appearance of the left-hand image is caused by the Retinex algorithm confusing a change in average surface color with a colored illuminant. Bottom: Outputs computed by **Crule** for an image of the first Mondriaan under white light (right), and for an image of the first Mondriaan with red borders (left), again under white light. Because **Crule** does not use average surface color, appending borders to the Mondriaan does affect the color descriptors it computes.

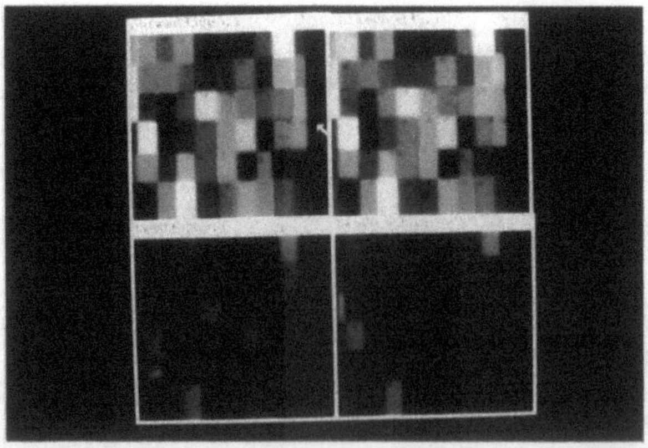

Figure 21. (Forsyth, p. 271) **Crule** outputs for two separate color images. The bottom two images are the input images, the top the output images. The image on the left is for a red filter which passed very much more light than that on the right. Details appear in section 7.

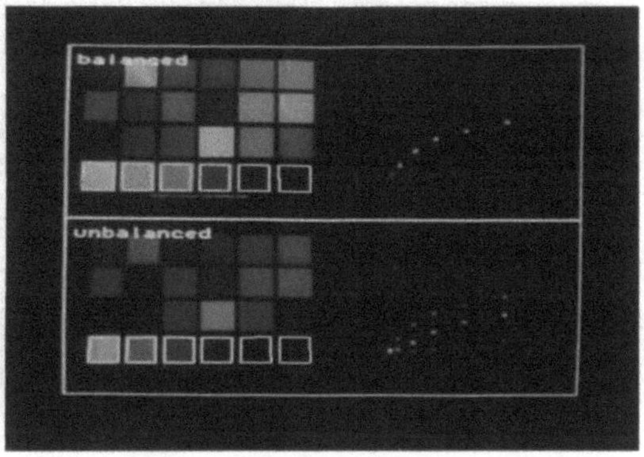

Figure 1. (Klinker et al., p. 329) **The effect of color balancing and gamma correction on color images.**

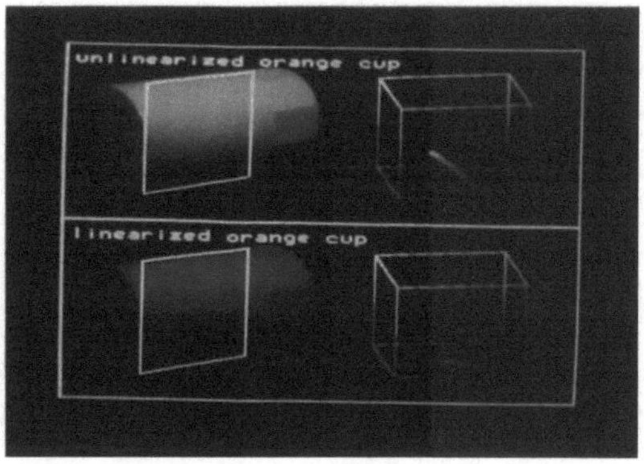

Figure 2. (Klinker et al., p. 329) **Curvature in the color space, caused by gamma-corrected cameras.**

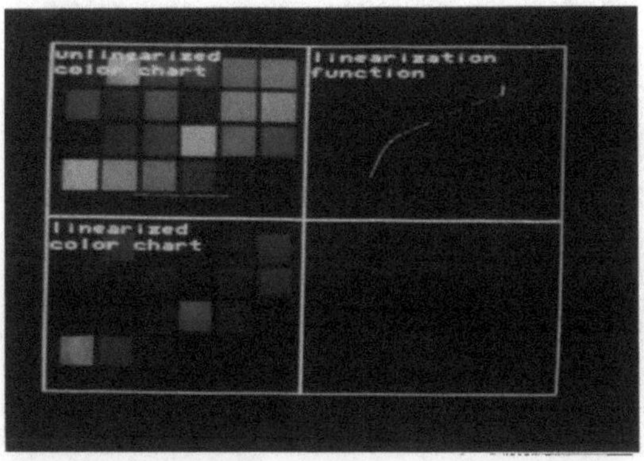

Figure 3. (Klinker et al., p. 330) **Spectral linearization with cubic spline functions.**

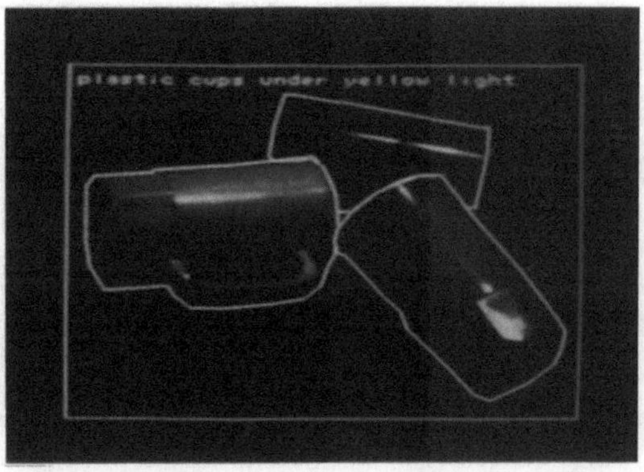

Figure 4. (Klinker et al., p. 330) **Three plastic cups under yellow light.**

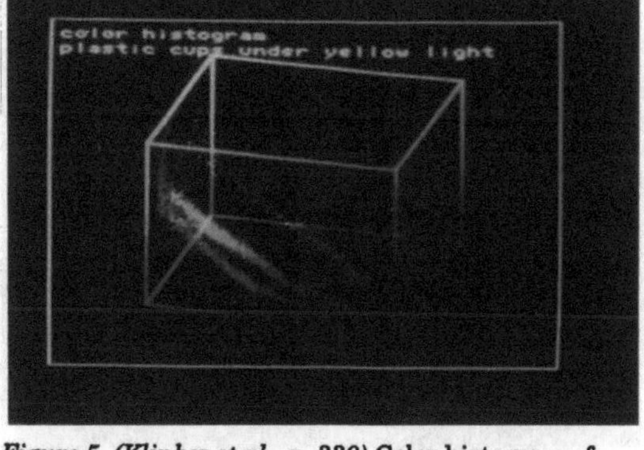

Figure 5. (Klinker et al., p. 330) **Color histogram of three plastic cups under yellow light.**

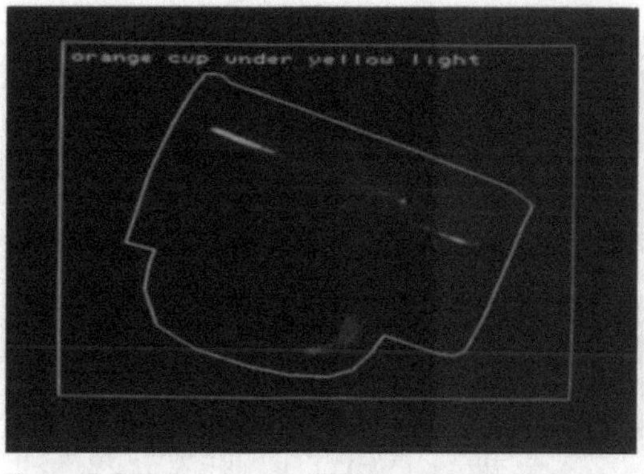

Figure 6. (Klinker et al., p. 331) **Orange cup under yellow light.**

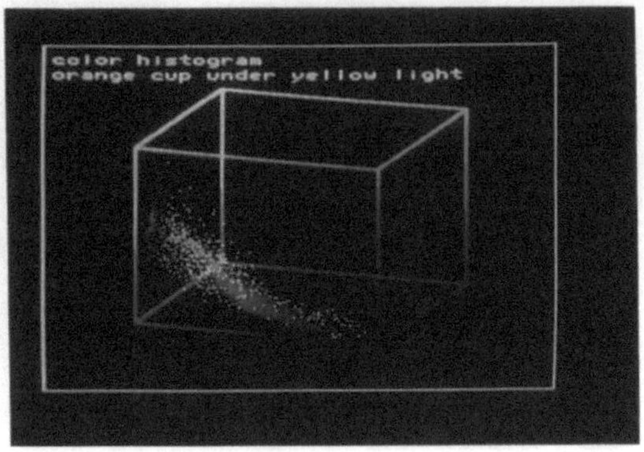

Figure 7. (Klinker et al., p. 331) Color histogram of the orange cup under yellow light.

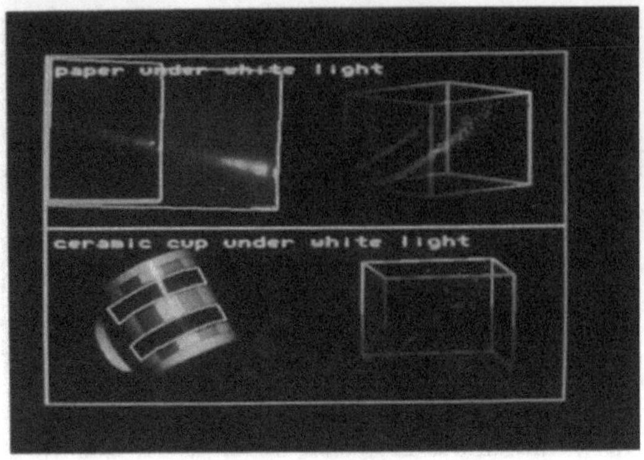

Figure 8. (Klinker et al., p. 331) Images and color clusters of glossy paper and a ceramic cup under white light.

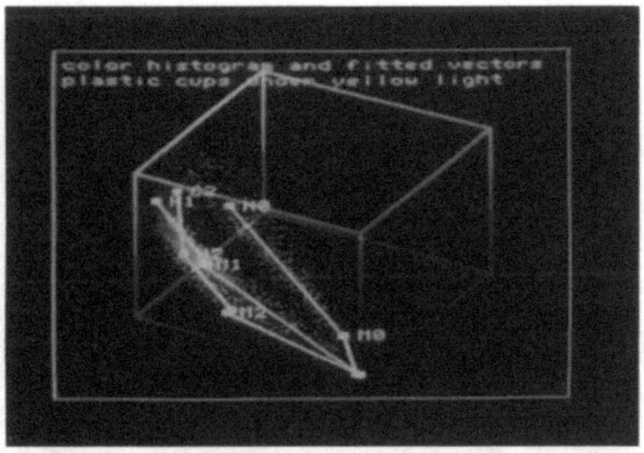

Figure 9. (Klinker et al., p. 332) Color histogram and fitted vectors of the three plastic cups under yellow light.

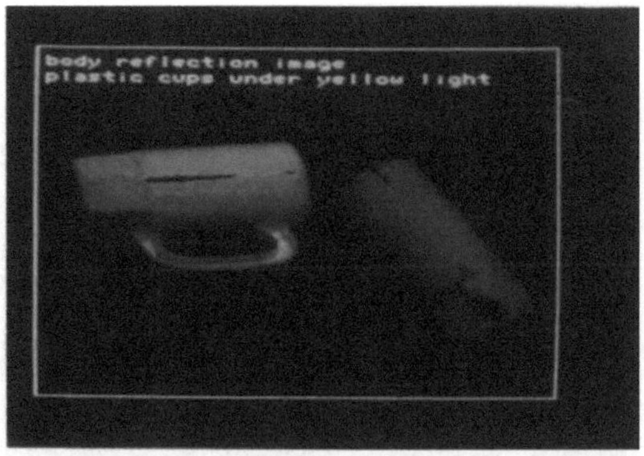

Figure 10. (Klinker et al., p. 332) Intrinsic body reflection image of the three plastic cups under yellow light; clipped and bloomed pixels are shown in black.

Figure 11. (Klinker et al., p. 332) Intrinsic surface reflection image of the three plastic cups under yellow light; clipped and bloomed pixels are shown in black.

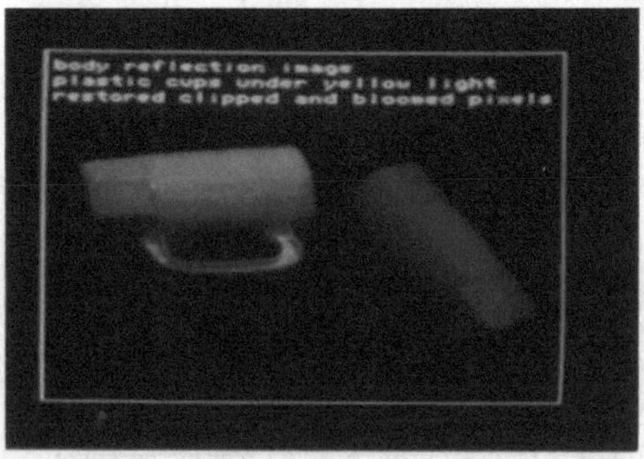

Figure 12. (Klinker et al., p. 333) Intrinsic body reflection image of the three plastic cups under yellow light; with restored clipped and bloomed pixels.

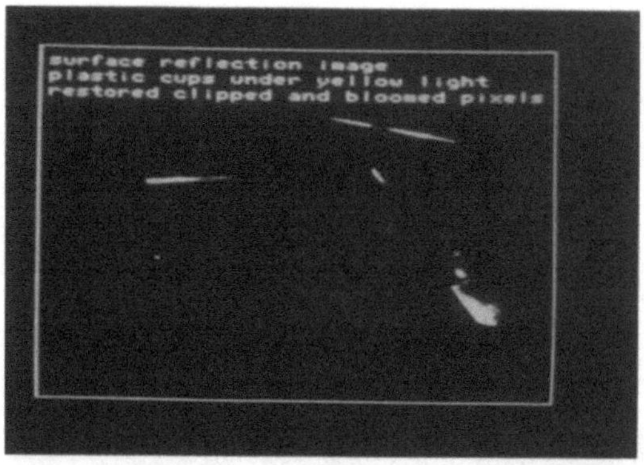

Figure 13. (Klinker et al., p. 333) **Intrinsic surface reflection image of the three plastic cups under yellow light; with restored clipped and bloomed pixels.**

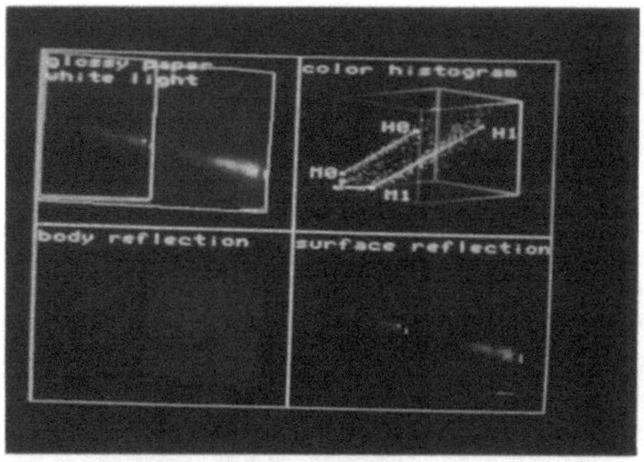

Figure 14. (Klinker et al., p. 333) **Intrinsic reflection images of the glossy paper folders under white light.**

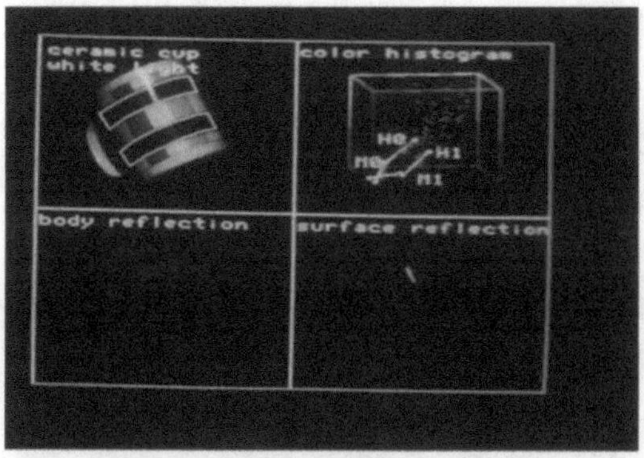

Figure 15. (Klinker et al., p. 334) **Intrinsic reflection images of the ceramic cup under white light.**

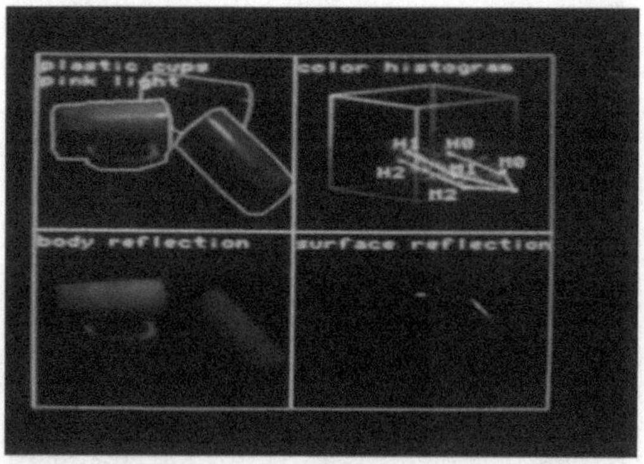

Figure 16. (Klinker et al., p. 334) **Intrinsic reflection images of the three plastic cups under pink light.**

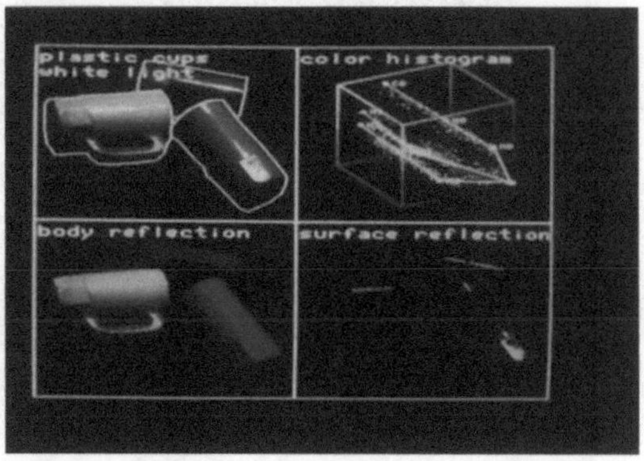

Figure 17. (Klinker et al., p. 334) **Intrinsic reflection images of the three plastic cups under white light.**

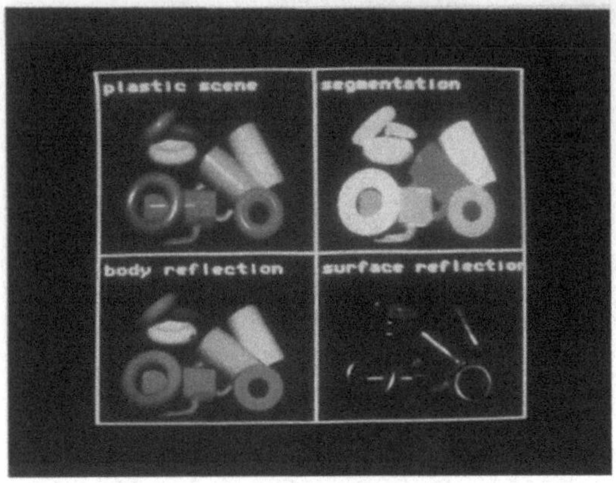

Figure 8. (Shafer et al., p. 364) **Image analysis using the dichromatic reflection model.**

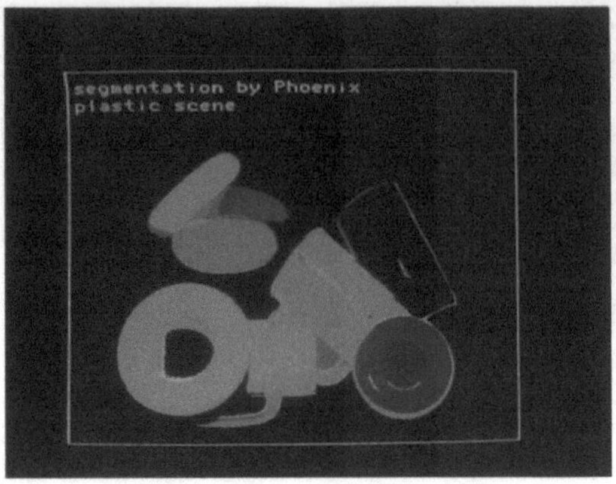

Figure 9. (Shafer et al., p. 364) **Segmentation by an older, traditional method.**

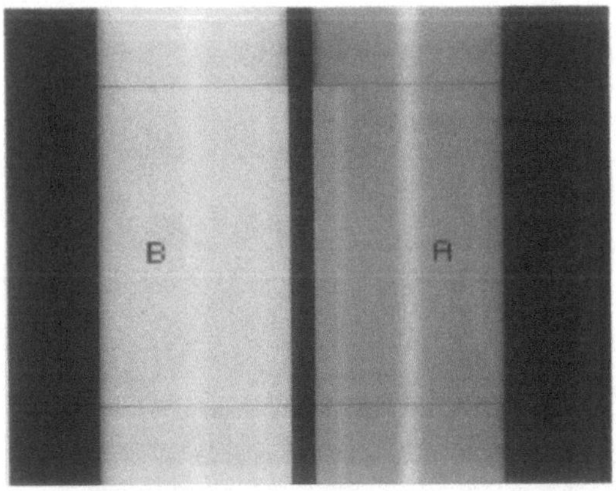

Figure 10. (Shafer et al., p. 364) **Two cylinders showing interreflection (synthetic image).**

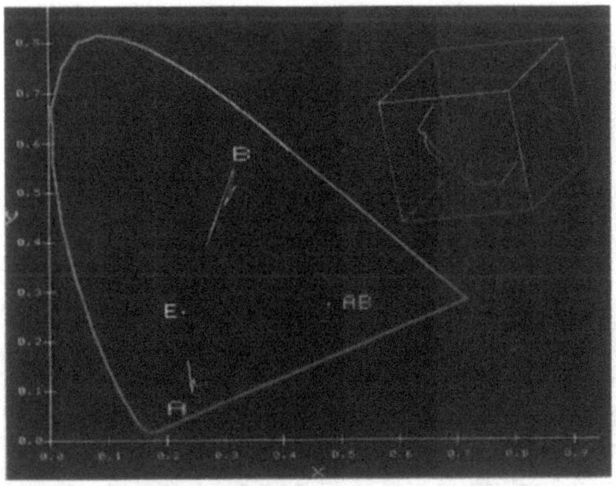

Figure 11. (Shafer et al., p. 364) **Colors of interreflection in chromaticity diagram and color space.**

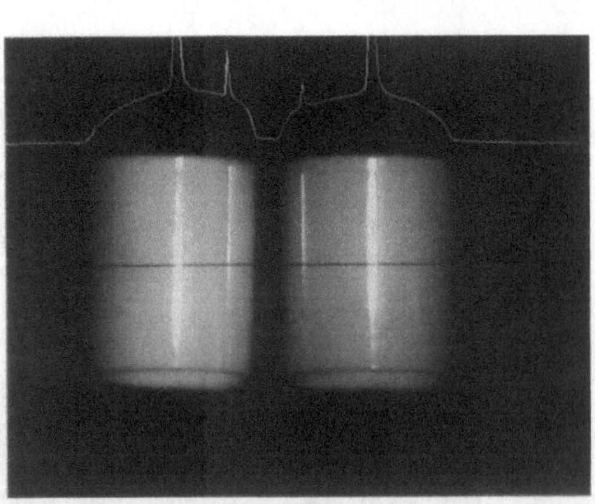

Figure 12. (Shafer et al., p. 364) **Interreflection in a real image.**

analysis of the dependence of the algorithm's behavior on the path length. This choice seems consistent with the intention of the authors, though different in detail.[27] When the Markov path-generation procedure is used, the probability that the path is at pixel x on on step $n + 1$ depends only on the path location on step n. We can calculate the probability, $P(n = x|1 = x_i)$, that a path will be at pixel x on the nth step, given that the path started at pixel x_i. For a given path length N_{pl}, the number of times that we expect a path starting at x_i to pass through pixel x is given by

$$E(x|x_i) = \sum_{n=1}^{N_{pl}} P(n = x|1 = x_i).$$

We have computed the $E(x|x_i)$ for a pixel in the sixth row and sixth column of a Mondrian of spatial dimensions 10 × 10, for several values of N_{pl}. For very small path length ($N_{pl} < 5$), only the neighbors of the pixel contribute to $G_k(x)$. For $N_{pl} = 25$, the neighbors of the pixel contribute six times as frequently as remote pixels. For $N_{pl} = 200$, the neighbors contribute 1.25 times as frequently as remote pixels. When the path length is large ($N_{pl} \gg 200$), the contributions from all pixels are approximately equal. In this case $G_k(x)$ is simply the geometric mean of the receptor responses from all pixels in the image and is independent of x. The lightness of any pixel is given by

$$l_k^x = \log\left(\frac{\rho_k^x}{G_k}\right), \tag{8}$$

where G_k is the geometric mean of all receptor responses for the kth receptor class. Other authors have noted that the retinex algorithms perform some kind of normalization.[4,17,21,22,25] Equation (8) provides the exact expression for this normalization for the Land[8,17] version of the retinex algorithm in the limiting case of large N_{np} and long N_{pl}. In Appendix A, we provide a similar result for the McCann–Houston[11] version of the retinex algorithm.

In order to evaluate the performance of retinex, we must know the appropriate path length. The published versions of retinex are not explicit about how to choose N_{pl}. Land (Ref. 8, p. 5166) writes that contributions to the lightness of a single point in the image are "taken over areas from the entire visual field and not just those nearby; experiments indicate there may be nearly as much contribution from distant areas as nearby ones." This suggests that Land's implementation uses a rather long path length.

In the next section we investigate the performance of the algorithm in the case when N_{pl} is large enough so that we may calculate the lightness values with Eq. (8). Our investigation will show that in this case the lightness values are too dependent on the composition of surfaces in the Mondrian to predict human color sensation.

In earlier work with another varient of the retinex algorithm, McCann *et al.* (Ref. 25, p. 454) used a path length of 200 with a 480-point image array. With this choice of parameter values, all points in the image will not contribute equally to the lightness value calculated for each point. For shorter path lengths, the lightness at each pixel depends to a larger extent on nearby pixels than remote pixels. We take this point up in more detail in the section headed Discussion.

PERFORMANCE OF RETINEX

Introduction

Even within the framework of the Mondrian viewing context with uniform illumination, no algorithm that uses the responses of a small number of receptor classes can be color constant if the lights and surfaces are allowed to have arbitrary spectral power distributions and surface reflectance functions.[21,28] More recent computational models of color constancy[6,7,28,29] specify the classes of surfaces and lights that are within the model's domain of application. The retinex algorithm—formulated much earlier—has not been updated to include such a specification. It is, however, possible to compute the lightness values for any set of lights and surfaces. In this section we discuss the performance of the retinex algorithm for light spectral power distributions and surface reflectance functions that are frequently encountered in the natural environment.

As we have emphasized in our definition of color constancy, there are two ways in which an algorithm can fail. It can fail to return color constant descriptors when the spectral power distribution of the ambient light is varied and it can fail to be color constant when the composition of the surfaces in the Mondrian is varied. We discuss the retinex algorithm's performance with respect to both of these types of image variation.

Dependence of Color Value on Ambient Light

McCann *et al.*[25] have investigated the performance of an earlier version of retinex algorithm when only the spectral power distribution of the ambient light is varied. They used the mixture of three narrow-bandwidth lights as the illuminant in their experiments. Their results indicate that, although the lightness values are not perfectly constant across illuminations, the failures of the algorithm are not dramatic and are of the same order of magnitude as human failures. We have extended their simulations to include broadband illuminants using the Land[8] algorithm. The results of our simulations, not presented here, are consistent with the conclusions drawn by McCann *et al.*[25] The computed lightness values remained roughly constant as the illuminant was varied.

When the composition of the Mondrian is held constant and when the path length is sufficiently long, the retinex algorithm acts by scaling the separate receptor classes. This method of color correction is similar to one widely used in industry. A gray-card reference surface is inserted into the scene, and each receptor class is scaled by its response to the gray card. We pursue this point in more detail in the section headed Discussion.

Dependence of Color Values on Surface Composition

The dependence of the lightness values computed by the retinex algorithm under changes of the composition of the Mondrian has not been addressed in the literature. In order to explore the performance of the retinex algorithm when the composition of the Mondrian is varied, we simulated the algorithm by using a number of different Mondrians.

Simulation Framework

The input to the simulation of retinex is a list of receptor responses for each pixel. We calculated the receptor re-

Blue Mondrian

R8/4	B8/2	G8/2
10YR5/6	Y8/2	P6/8
B2/2	B3/2	B4/2

Green Mondrian

R8/4	B8/2	G8/2
10YR5/6	Y8/2	P6/8
G6/6	G7/6	G8/6

Yellow Mondrian

R8/4	B8/2	G8/2
10YR5/6	Y8/2	P6/8
Y6/8	Y7/8	Y8/8

Purple Mondrian

R8/4	B8/2	G8/2
10YR5/6	Y8/2	P6/8
P7/4	P6/6	P7/6

Test Mondrian Composition

Fig. 2. The spatial composition of the blue, green, yellow, and purple Mondrians used in the simulations of the Land[8] retinex algorithm.

sponses for a Mondrian composed of matte surfaces taken from the Munsell chips, uniformly illuminated by CIE standard daylight D_{65}. (The Munsell chips are a set of colored papers spanning a broad range of natural colors. The chips are widely used for color measurement in science and industry.) To calculate the receptor responses we used the Smith–Pokorny estimates of the three human cone photoreceptors as reported by Boynton.[30] The receptor responses are given by

$$\rho_k^x = \sum_{n=1}^{N} E(\lambda_n) S^x(\lambda_n) R_k(\lambda_n). \tag{9}$$

We used the spectral reflectance data of Nickerson[31] and represented these data as the best-fitting linear combination of Cohen's[32] basis functions for the Munsell chips.[22,28] Judd and Wyszecki[33] described the spectral power distribution of illuminant D_{65}.

From Lightness Triplets to Color Names

We used the following method (see Land[8]) to evaluate the dependence of lightness values on changes in the viewing context. We assigned each Munsell chip to a lightness triplet calculated under standard viewing conditions. The standard viewing conditions consisted of a Mondrian of 462 Munsell color papers illuminated by CIE D_{65}. We will refer to this Mondrian as the standard Mondrian. To evaluate the color appearance predicted by the retinex algorithm of a test chip in some different viewing context, we first comput-

Mondrian	STANDARD	BLUE	GREEN	YELLOW	PURPLE
Predicted Color	R8/4	Y8/2	YR6/2	P6/6	10R6/4
Lightness Triplet	0.9640 0.9476 1.0435	0.9077 0.8250 0.7905	0.3166 0.2310 0.3940	0.2968 0.2824 0.8885	0.3248 0.2904 0.2150

Test Chip: R8/4 Light: D_{65}

Mondrian	STANDARD	BLUE	GREEN	YELLOW	PURPLE
Predicted Color	B8/2	N8/	RP6/2	10PB6/8	R6/2
Lightness Triplet	0.9762 1.0844 1.2631	0.9199 0.9618 1.0102	0.3287 0.3679 0.6137	0.3090 0.4193 1.1081	0.3370 0.4272 0.4346

Test Chip: B8/4 Light: D_{65}

Change of Context Simulation Results

Fig. 3. Predicted color values and lightness triplets computed by the Land[8] retinex algorithm for two chips in the standard, blue, green, yellow, and purple Mondrians. The lightness values were computed according to Eq. (8).

ed the lightness triplet in the new viewing context by using Eq. (8):

$$l_k^x = \log\left(\frac{\rho_k^x}{G_k}\right), \quad (10)$$

where G_k is the geometric mean of receptor responses for the kth receptor class. (This is the limiting lightness value for the algorithm when the number of paths is large and the path length is long.) To find the color value for the test chip, we compare the computed lightness triplet with all the lightness triplets calculated under the standard viewing context. We then find the Munsell chip in the standard viewing context whose lightness value is closest (in a Euclidean metric sense) to the test chip's lightness triplet. The color value of this Munsell chip is used as the predicted color value for the test chip.

The 462 Munsell papers sparsely populate the set of perceptually distinct color values. It is possible, therefore, that our method will assign the same color value to two lightness triplets that produce different color sensations. We can be confident, when the method assigns different color values to two lightness triplets, that the corresponding predicted color sensations are noticeably different.

Simulation Results

To evaluate the dependence of the lightness values on composition of the Mondrian, we computed the lightness triplets for six test chips in each of four Mondrians illuminated by CIE D_{65}. The composition of these Mondrians is shown in Fig. 2. These 3 × 3 Mondrians differ only in the composition of the lowest row: Their upper two rows are identical. We will refer to these four Mondrians as the blue, green, yellow, and purple Mondrians, respectively, as these names indicate the hues of the lower chips. We chose these Mondrians because the variation in the composition of the lower row significantly shifts the geometric mean of the receptor responses when they are illuminated by D_{65}. A human observer viewing these Mondrians on a black background illuminated by daylight perceives virtually no change in the appearance of the upper two rows of chips across the four Mondrians. Figure 3 shows the lightness triplets and corresponding predicted color values for two of the test chips, as computed for each of the Mondrians. Both the computed lightness triplets and the corresponding color values change markedly as a function of the composition of the Mondrian. For example, the retinex algorithm predicts that test chip R8/4, which in fact appears pink in the viewing conditions simulated, will range in appearance from beige (Y8/2) in the blue Mondrian to purple (P6/6) in the yellow Mondrian. This range of predicted color values is typical of all the test chips and is not characteristic of human performance. Although human color vision does exhibit simultaneous contrast effects, these effects are small compared with the magnitude of color shift predicted by the retinex algorithm. We conclude that the retinex algorithm does not correctly predict color appearance when the composition of the Mondrian is varied.

DISCUSSION

The goal of an ideal color constancy algorithm is to use the information in the receptor responses to compute color values that depend only on the surface reflectance spectra. Equation (2),

$$\rho^x = \Lambda_E \sigma^x,$$

expresses the relation between the surface reflectance σ^x and the responses ρ^x. To recover σ^x, we multiply the response vector ρ^x by the pseudoinverse of Λ_E[6,7,28,29]:

$$\sigma^x = \Lambda_E^{-1}\rho^x. \quad (11)$$

The matrix Λ_E^{-1} describes the proper correction for the spectral power distribution of the ambient light. The entries of Λ_E^{-1} are independent of the surfaces in the image. The retinex algorithm always attempts to correct for the illuminant by an equation of the form

$$l^x = \Gamma^x \rho^x, \quad (12)$$

where Γ^x is a diagonal matrix. The entries of Γ^x depend on the spectral power distribution of the ambient light, the position x in the image, *and* the other surfaces in the image. As the path length increases, the algorithm converges toward using a single matrix Γ, for all positions x, whose kth diagonal entry approaches $1/G_k$. Recall that the value G_k is the geometric mean of the receptor responses for the kth receptor class.

The simulation results show that, in the case of long path length, the dependence on the surfaces in the image is too strong. The predicted color values of the retinex algorithm vary too widely with changes in the composition of the Mondrian to serve as an adequate model of human color vision.

For shorter path lengths, the lightness triplet calculated for a pixel x depends on the location of the pixel. In this case, the diagonal entries of the matrix Γ^x are a weighted geometric mean of the responses near location x. While reducing the path length will change the set of points responsible for the overdependence of predicted color names on the surfaces in the Mondrian, the qualitative effect will not change. Rather, the same dependence on changes of surfaces as was exhibited in the long-path-length case will be replayed repeatedly across the image on a local scale. We have not discovered a choice of path length that allows the retinex algorithm to predict color names that are color constant.

Another way to understand the retinex algorithm is to compare it with the following simple color correction procedure. To perform a color correction, we place a reference surface in the image. We divide each of the responses of the kth receptor class by the response of the kth receptor class to the reference surface. (The normalized receptor responses parallel the retinex algorithm's lightness triplets.) This method is currently used in industry for correcting the color output of video cameras. Brill and West[21] (see also Ref. 18) have analyzed this method of color correction. They show that the method performs correctly only for a severely limited class of surface reflectance functions and spectral power distributions of the ambient light. When a reference surface is available, the algorithm presented by Buchsbaum[6] will properly correct for changes in the illuminant for a much broader class of surface reflectance functions and spectral power distributions of the ambient light.

The retinex algorithm is equivalent to normalization with

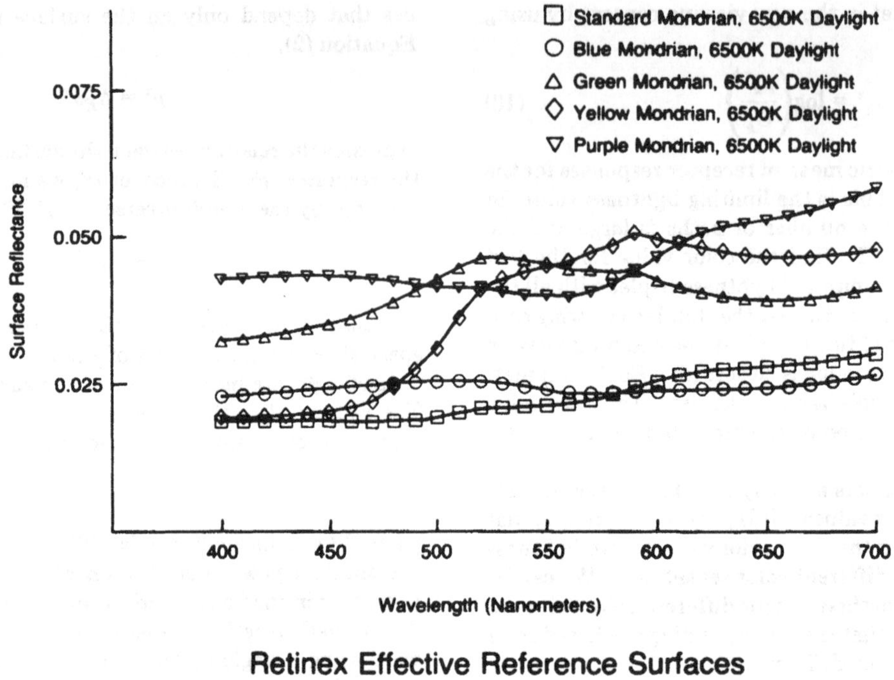

Fig. 4. The effective reference surface for the standard, blue, green, yellow, and purple Mondrians under CIE illuminant D_{65}.

respect to an effective reference surface. Indeed, the reflectance function $\sigma_r{}^x$ for this surface can be found simply:

$$\sigma_r{}^x = \Lambda_E{}^{-1}\gamma, \tag{13}$$

where γ is the column vector consisting of the G_k. The effective reference surface used by the retinex varies both with the illuminant and with the surfaces in the scene. We have calculated how the effective reference surface depends on both the spectral power distribution of the ambient light and the surfaces in the scene. The effective reference surface is almost constant under changes of illuminant. Figure 4 shows the effective reference surface for the standard, blue, green, yellow, and purple Mondrians under D_{65}. The effective reference surface is markedly different for Mondrians containing different surfaces.

The difficulty with the retinex algorithm is that it normalizes to different reference surfaces for different scenes. It is as if, rather than using a standard gray card for color balancing as is done in industry, we were to use a different reference card for every scene. It is not surprising that the retinex algorithm fails to be color constant with changes in scene composition.

Finally, we note that it is possible to perform an empirical test of the retinex algorithm as a model of human performance without specifying the detailed algorithm parameters. To do this, begin with an arbitrary first image. Calculate the three photoreceptor response arrays for this image. Construct a second image so that its red photoreceptor array is a scaled version of the red photoreceptor array of the first image, so that its green photoreceptor array is a scaled version of the green photoreceptor array of the first image, and so that its blue photoreceptor array is a scaled version of the blue photoreceptor array of the first image. No matter how the paths are constructed, the retinex algorithm will assign identical lightness values to these two images. This is because, within each receptor class, the algorithm's normalization will discount the scale factors that distinguish the two. Thus the retinex algorithm predicts that the two images will have identical color appearance. We are unaware of any empirical tests of this prediction.

CONCLUSION

Historically, work on color constancy has emphasized the need to correct for the spectral power distribution of the ambient light. This emphasis has made it easy to forget that the information reflected from a single surface is insufficient to separate ambient light and surface reflectance and that color constancy algorithms require information from multiple surfaces. It is important to evaluate the performance of such algorithms when the surfaces that comprise the scene are varied. Human vision maintains approximate color constancy despite variation in the spectral reflectance functions of nearby surfaces and despite variation in the spectral power distribution of the ambient light.

The complex set of calculations that define the retinex algorithm is equivalent to a simple normalization. This normalization is not color constant: A color constant algorithm must correct for the ambient light independent of the surfaces in the scene. The retinex algorithm corrects for the light in a manner that depends strongly on the surfaces. This dependence is not characteristic of human color vision. We conclude that retinex is not a color constant algorithm and that it is not an adequate model of human performance.

APPENDIX A: RELATED IMPLEMENTATIONS OF THE RETINEX ALGORITHM

Introduction

In this appendix we review two additional methods proposed for calculating the lightness values for an image. The first method is described by McCann and Houston[11] and Frankle and McCann.[10] It is very similar to the calculation described by Land. The method differs because it includes an additional nonlinear operation—the reset—that was also used in the original retinex calculations.[4] The second method is described by Horn[24] and cited by Land[8] as being equivalent to Land's method (but see Ref. 15).

Retinex with Reset

In the original retinex algorithm,[4] a special step is present in the path calculation. Consider the path starting at location x_1. If at any point during the path computation the value to be added to the accumulator $A(x)$ at location x exceeds 0, then the path calculation is *reset*. The reset operation has two parts. First, the value added to $A(x)$ is set to 0. Second, for the remainder of the calculations along the path, the value $\rho_k^{\,x}$ is substituted for the receptor response $\rho_k^{\,x_1}$. A sophisticated computational procedure for the retinex algorithm with reset is described by McCann and Houston[11] and Frankle and McCann.[10] It is not presented in terms of a set of path calculations, although they state that "the lightness calculation is in principle the same as in earlier work: ratio, product, reset, average" (Ref. 10, p. 1005). McCann and Houston compute the value of $\exp(l_k^{\,x})$ rather than $l_k^{\,x}$ itself. In this appendix we use the notation $l_k^{\,x}$ to refer to the quantity calculated by their procedure rather than to the lightness, as defined by Land.[8] Whether the lightness is expressed in linear or logarithmic units has no bearing on the conclusions of this paper.

The computation is defined by a simple algorithm. For a given receptor class, the algorithm is initialized by setting the lightness value at all locations to 1. The algorithm determines the lightness at location x by using an interactive comparison procedure. The lightness at location x is compared with the lightness at a sequence of comparison locations, c_n. Since we wish to study the convergence properties of the algorithm, we assume that the comparison list is infinitely long and that all locations are repeatedly compared with location x. The lightness at all locations is computed concurrently. The lightness value at x is given by the following recursive equation, which defines the lightness after $n+1$ iterations in terms of the lightness after n iterations:

$$l^x(n+1) = \left\{ l^x(n) f^* \left[\frac{\rho^x}{\rho^{c_n}} l^{c_n}(n) \right] \right\}^{1/2}, \tag{A1}$$

where $f^*[\ ]$ is a reset function defined by $f^*(a) = a$ for $a < 1$ and $f^*[a] = 1$ otherwise. In the limit, as the algorithm continues through many iterations, the lightness at pixel x is given by

$$l^x = \lim_{n\to\infty} \left\{ l^x(n) f^* \left[\frac{\rho^x}{\rho^{c_n}} l^{c_n}(n) \right] \right\}^{1/2}. \tag{A2}$$

This limit will converge only if

$$(l^x)^2 = \lim_{n\to\infty} l^x(n) f^* \left[\frac{\rho^x}{\rho^{c_n}} l^{c_n}(n) \right]. \tag{A3}$$

Further, since we are assuming that $\lim_{n\to\infty} l^x(n) = l^x$, it follows that convergence occurs if and only if

$$l^x = \lim_{n\to\infty} f^* \left[\frac{\rho^x}{\rho^{c_n}} l^{c_n}(n) \right]. \tag{A4}$$

To understand the convergence of the lightness l^x in general, we begin by analyzing the lightness at the location at which the receptor response is smallest, x_s. For location x_s, it must be that $(\rho^{x_s}/\rho_{c_n}) \le 1$. Moreover, it may be seen from Eq. (A1) that the lightness at any location is always bounded above by 1. This is because the lightness value is initialized to 1, and at each step in the iteration the lightness is multiplied by a value that is less than or equal to 1. It follows that the location with the smallest sensor response will never encounter a reset operation during the calculation of its lightness value because the argument to the reset function is always less than 1. In studying the convergence of the lightness at x_s, we may remove the reset function so that we have

$$l^{x_s} = \lim_{n\to\infty} \frac{\rho^{x_s}}{\rho^{c_n}} l^{c_n}(n). \tag{A5}$$

Since all locations x are repeatedly compared with location x_s, then, if l^{x_s} converges, Eq. (A5) must be true for any comparison location x. If the computation converges, the lightness of any location can be expressed in terms of the lightness at x_s by the equation

$$l^x = \left(\frac{\rho^x}{\rho^{x_s}} \right) l^{x_s}. \tag{A6}$$

Now we consider the location with the largest receptor response, x_l. We will show that for each location x, $l^x \le (\rho^x/\rho^{x_l})$ and $l^x \ge (\rho^x/\rho^{x_l})$. This forces the conclusion that $l^x = (\rho^x/\rho^{x_l})$.

To show $l^x \le (\rho^x/\rho^{x_l})$, we rewrite Eq. (A6) for location x_l:

$$l^{x_s} = l^{x_l} \left(\frac{\rho^{x_s}}{\rho^{x_l}} \right). \tag{A7}$$

Since l^{x_l} is bounded by 1, it follows that $l^{x_s} \le (\rho^{x_s}/\rho^{x_l})$. Using this inequality in Eq. (A6), we conclude that

$$l^x \le \frac{\rho^x}{\rho^{x_l}} \tag{A8}$$

for all possible comparison locations x.

To show $l^x \ge (\rho^x/\rho^{x_l})$, we use a proof by induction with respect to the steps of the iteration. On the first step of the iteration, $l^x(1) = 1 \ge (\rho^x/\rho^{x_l})$ for every location x. Suppose that at step n, $l^x(n) \ge (\rho^x/\rho^{x_l})$ for every location x. Then we have

$$\frac{\rho^x}{\rho^{c_n}} l^{c_n}(n) \ge \left(\frac{\rho^x}{\rho^{c_n}} \right) \left(\frac{\rho^{c_n}}{\rho^{x_l}} \right) \ge \frac{\rho^x}{\rho^{x_l}}. \tag{A9}$$

Since the reset function is monotonic, we have

$$f^* \left[\frac{\rho^x}{\rho^{c_n}} l^{c_n}(n) \right] \ge f^* \left(\frac{\rho^x}{\rho^{x_l}} \right) = \frac{\rho^x}{\rho^{x_l}}. \tag{A10}$$

We substitute for the reset function in Eq. (A1) to get

$$l^x(n+1) \ge \left[l^x(n) f^* \left(\frac{\rho^x}{\rho^{x_l}} \right) \right]^{1/2} = \left[l^x(n) \frac{\rho^x}{\rho^{x_l}} \right]^{1/2}. \tag{A11}$$

Since $l^x(n) \ge (\rho^x/\rho^{x_l})$ by the induction hypothesis, it follows that $l^x(n+1) \ge (\rho^x/\rho^{x_l})$. We conclude by induction that $l^x \ge (\rho^x/\rho^{x_l})$. It follows that $l^x = (\rho^x/\rho^{x_l})$. Thus, if the computation converges, which we have not shown, the lightness values approach a simple normalization of the receptor response with respect to the response at the largest location for

Mondrian	STANDARD	RED	BLUE	GRAY	GRAY-RED
Predicted Color	R4/4	Y7/6	YR6/8	10YR7/8	GY9/5
Lightness Triplet	−2.1123 −2.2586 −2.1942	−0.9906 −1.0808 −2.0028	−1.0028 −1.2176 −2.2225	−0.6581 −0.9327 −1.9953	0.0000 0.0000 −0.9079

Test Chip: R4/4 Light: D_{65}

Mondrian	STANDARD	RED	BLUE	GRAY	GRAY-RED
Predicted Color	Y4/4	10YR5/6	10Y5/6	Y6/6	GY8/8
Lightness Triplet	−2.1974 −2.1609 −2.9954	−1.5781 −1.5527 −2.5189	−1.5903 −1.6895 −2.7385	−1.2456 −1.4047 −2.5113	−0.5875 −0.4719 −1.4239

Test Chip: Y4/4 Light: D_{65}

Change of Context Simulation Results

Fig. 5. The spatial composition of the red, blue, gray, and gray-red Mondrians used in the simulations of retinex algorithm with reset.

the receptor class. This parallels the computation by Land.[8] Land's calculation is equivalent to normalization with respect to the geometric mean of receptor responses for the receptor class, whereas the retinex with reset calculation is equivalent to normalization with respect to the response at the largest location. Retinex with reset will also be sensitive to changes in the viewing context; in particular, it will be sensitive to any change in the reflectance function of the surface causing the largest receptor response.

We simulated the performance of the McCann–Houston algorithm in the limitihg case that the lightness values are given by

$$\exp(l_k^x) = \frac{\rho_k^x}{\rho_k^{x_l}}. \tag{A12}$$

Figure 5 shows the four Mondrians that we used in this simulation. These 3 × 3 Mondrians differ only in the composition of the lowest row: Their upper two rows are identical. We will refer to these four Mondrians as the red, blue, gray, and gray-red Mondrians, respectively. We chose these Mondrians because the variation in the composition of the lower row significantly alters the largest receptor responses when they are illuminated by D_{65}. Figure 6 shows the lightness triplets and corresponding predicted color values for two of the test chips, as computed for each of the Mondrians. For the retinex algorithm with reset, the predicted color values also change dramatically with small changes in the Mondrian. We conclude that the reset operation does not significantly reduce the retinex algorithm's overdependence on the composition of the Mondrian. On the contrary, the

Red Mondrian

R4/4	Y4/4	BG4/4
B2/2	B3/2	B4/2
10YR5/6	R8/4	P6/8

Blue Mondrian

R4/4	Y4/4	BG4/4
B2/2	B3/2	B4/2
10YR5/6	B8/2	P6/8

Gray Mondrian

R4/4	Y4/4	BG4/4
B2/2	B3/2	B4/2
10YR5/6	N2/	P6/8

Gray-red Mondrian

R4/4	Y4/4	BG4/4
B2/2	B3/2	B4/2
10YR5/6	N2/	RP4/4

Test Mondrian Composition

Fig. 6. Predicted color values and lightness triplets computed by the McCann–Houston retinex algorithm for two chips in the standard, red, blue, gray, and gray-red Mondrians. The lightness values were computed according to Eq. (A12).

reset operation localizes the dependence to just a few pixels: the pixels with the largest response in each receptor class.

It is possible to run the McCann–Houston[11] algorithm without choosing the comparison pixels from the entire image. In this case, the limiting lightness value for a pixel at location x will be given by normalization to the pixel with the largest receptor response among the set of comparison pixels chosen for location x. We do not believe that limiting the set of comparison pixels can eliminate the overdependence on the composition of the Mondrian. The predicted color for a particular location will remain sensitive to variation in the surface reflectance function at pixels in the comparison set for that location.

Determining Lightness from an Image

The second method that we consider is presented in Horn's[24] paper entitled "Determining lightness from an image." Land[8] cites this as another method of implementing the retinex calculation. Recently, it has been developed further by Blake[15] and Terzopoulos.[16] The central premise of Horn's[24] paper is that the surface reflectance and spectral power distribution of the ambient light may be separated in the image data. The physical property of images that permits this, it is argued, is that the spatial variation in the ambient lighting occurs at a lower rate than the spatial variation of the surface reflectance of the objects.

Horn bases his calculations on two steps. First, he assumes that the image may be written as the product of two functions, object reflectance at a point, which we call $r(x, y)$, and ambient light intensity, $s(x, y)$. He writes the recorded image intensity $p(x, y)$ as

$$p(x, y) = s(x, y)r(x, y) \tag{A13}$$

(Ref. 24, p. 281).

In the case of the uniformly illuminated Mondrian, the spatial distribution of the ambient light has no step edges. The boundaries of the colored papers, on the other hand, define many different step edges of the surface reflectance. Horn assumes that, in general, boundaries between objects will give rise to step edges and changes in the ambient light will give rise only to smoothly varying changes in intensity. Based on this assumption, Horn argues that we may recover the surface reflectance function as follows. First, compute the logarithm of the image-intensity record

$$\log[p(x, y)] = \log[s(x, y)] + \log[r(x, y)]. \tag{A14}$$

Next compute the Laplacian of the resulting image. Since the Laplacian is a bandpass spatial operator, it will take on a relatively larger values when approaching step discontinuities, and it will be small or zero over regions of nearly constant value. By assumption, the larger values of the Laplacian arise from surface reflectance changes, owing to the term $\log[r(x, y)]$. Horn applies a threshold operator to the Laplacian of the logarithm of the image data. The threshold operator sets all values smaller than some criterion level to zero and passes unchanged all values larger than criterion. Horn's claim is that, following thresholding, the effects of the ambient light have been removed.

Horn's method does not make the role of wavelength explicit anywhere in the calculation. By doing so, it is easy to see why the method does not recover the surface reflectance function. We rewrite Eq. (A13) with the spatial coordinates made explicit only in the surface reflectance function:

$$\rho(x, y) = \sum_{n=1}^{N} E(\lambda_n)S(x, y, \lambda_n)R_k(\lambda_n). \tag{A15}$$

When the role of wavelength is made explicit, we find that applying the logarithm to both sides of Eq. (A15) does not yield the additive separation of the surface and light functions required for the use of the Laplacian and thresholding operations. Thus Horn's method cannot eliminate the effects of changing the spectral power distribution of the ambient light.

The method may remove some of the effects of variation in the intensity of the ambient illumination. For example, as noted by a referee, if the ambient light function is separable in space and wavelength so that $E(\lambda_n) = E_1(\lambda)E_2(x, y)$, then

$$\rho(x, y) = E_2(x, y) \sum_{n=1}^{N} E_1(\lambda)S(x, y, \lambda_n)R_k(\lambda_n). \tag{A16}$$

Under this assumption, applying the logarithm to both sides of Eq. (A16) permits an additive separation of the *spatial variation* of the ambient lighting from the spatial variation of the surface reflectance function. However, the *spectral power distribution* of the lighting remains confounded with the surface reflectance functions.

ACKNOWLEDGMENTS

This work was supported by grant no. 2 RO1 EY03164 from the National Eye Institute and grant NCC 2-307 from the NASA-Ames Research Center. Some of the material in this paper is based on work supported under a National Science Foundation Graduate Fellowship. We thank W. Brainard, M. Huang, L. Maloney, J. Nachmias, M. Pavel, R. N. Shepard, and D. Varner for useful discussions.

REFERENCES AND NOTES

1. H. von Helmholtz, *Handbuch der Physiologischen Optik*, 2nd ed. (Voss, Hamburg, 1896).
2. H. Helson, "Fundamental problems in color vision. I. The principle governing changes in hue, saturation and lightness of nonselective samples in chromatic illumination," J. Exp. Psychol. **23**, 439–476 (1938).
3. D. B. Judd, "Hue saturation and lightness of surface colors with chromatic illumination," J. Opt. Soc. Am. **30**, 2–32 (1940).
4. E. H. Land and J. J. McCann, "Lightness and retinex theory," J. Opt. Soc. Am. **61**, 1–11 (1971).
5. M. H. Brill, "A device performing illuminant-invariant assessment of chromatic relations," J. Theor. Biol. **71**, 473–478 (1978).
6. G. Buchsbaum, "A spatial processor model for object color perception," J. Franklin Inst. **310**, 1–26 (1980).
7. L. T. Maloney and B. A. Wandell, "Color constancy: a method for recovering surface spectral reflectance," J. Opt. Soc. Am. A **3**, 29–33 (1986).
8. E. H. Land, "Recent advances in retinex theory and some implications for cortical computations: color vision and the natural image," Proc. Nat. Acad. Sci. USA **80**, 5163–5169 (1983).
9. E. H. Land, D. H. Hubel, M. S. Livingston, S. H. Hollis, and M. M. Burns, "Colour-generating interactions across the corpus callosum," Nature **303**, 616–618 (1983).
10. J. Frankle and J. J. McCann, "Method and apparatus for lightness imaging," United States Patent No. 4,384,336, 1983.
11. J. J. McCann and K. Houston, "Calculating color sensations from arrays of physical stimuli," IEEE Trans. Systems Man Cybern. **SMC-13**, 1000–1007 (1983).

12. M. Livingston and D. Hubel, "Anatomy and physiology of a color system in the primate visual cortex," J. Neurosci. 4, 309–356 (1984).
13. D. Ingle, "The goldfish as a retinex animal," Science 227, 651–654 (1985).
14. S. Grossberg and E. Mingolla, "Neural dynamics of form perception: boundary completions, illusory figures, and neon color spreading," Psychol. Rev. 92, 173–207 (1985).
15. A. Blake, "Boundary conditions for lightness computation in Mondrian world," Comput. Vis. Graphics Image Process. 32, 314–327 (1985).
16. D. Terzopoulos, "Image analysis using multigrid relaxation methods," IEEE Trans. Pattern Anal. Mach. Intell. PAMI-8, 129–139 (1986).
17. E. H. Land, "Recent advances in retinex theory," Vision Res. 26, 7–22 (1986).
18. M. M. Woolfson, "Some new aspects of color perception," IBM J. Res. Dev. 3, 313 (1959).
19. G. Walls, "Land! Land!" Psychol. Bull. 57, 29–48 (1960).
20. D. B. Judd, "Appraisal of Land's work on two-primary color projections," J. Opt. Soc. Am. 50, 254–268 (1960).
21. M. Brill, and G. West, "Contributions to the theory of color," J. Math. Biol. 11, 337–350 (1981).
22. J. Worthey, "Limitations of color constancy," J. Opt. Soc. Am. A 2, 1014–1026 (1985).
23. R. M. Shapley, "The importance of contrast for the activity of single neurons, the VEP and perception," Vision Res. 26, 45–62 (1986).
24. B. K. P. Horn, "Determining lightness from an image," Comput. Graphics Image Process. 3, 277–299 (1974).
25. J. J. McCann, S. P. McKee, and T. H. Taylor, "Quantitative studies in retinex theory: a comparison between theoretical predictions and observer responses to the color Mondrian experiments," Vision Res. 16, 445–448 (1976).
26. Partial descriptions of the path-generation process can be found in McCann *et al.* (Ref. 25, p. 454), where they write: "The origin of the path and its direction in the 480-point array were determined by a random-number generator. The paths traveled straight ahead until they reached the perimeter of the target where they either reflected back across the target or traveled along the perimeter. The direction of the reflection from the perimeter was also chosen by the random-number generator." Land (Ref. 8, p. 5165) writes "The signal will proceed ... with freedom to branch within the rule of implied directionality inherent in the concept of a signal radiating from a single source," whereas in Land (Ref. 17, p. 12) the paths are generated by an "optical mouse system [which] is swept along the random pathways" of the image.
27. McCann *et al.*[25] emphasize that the details of the path-generation process are not critical when they write (Ref. 25, p. 453): "The model described in this paper is one of many embodiments using the ratio-multiplication process."
28. L. Maloney, "Computational approaches to color vision," Ph.D. dissertation (Stanford University, Stanford, Calif., 1984).
29. B. A. Wandell, "The synthesis and analysis of color images," IEEE Trans. Pattern Anal. Mach. Vis. (to be published).
30. R. M. Boynton, *Human Color Vision* (Holt, New York, 1979).
31. D. Nickerson, "Spectrophotometric data for a collection of Munsell samples," Report, Cotton Division, U.S. Department of Agriculture, Washington, D.C., 1957.
32. J. Cohen, "Dependency of the spectral reflectance curves of the Munsell color chips," Psychonanal. Sci. 1, 369–370 (1964).
33. D. Judd and G. Wyszecki, *Color in Business, Science, and Industry* (Wiley, New York, 1975).

Color constancy: a method for recovering surface spectral reflectance

Laurence T. Maloney* and Brian A. Wandell

Department of Psychology, Stanford University, Stanford, California 94305

Received July 18, 1985; accepted August 9, 1985

Human and machine visual sensing is enhanced when surface properties of objects in scenes, including color, can be reliably estimated despite changes in the ambient lighting conditions. We describe a computational method for estimating surface spectral reflectance when the spectral power distribution of the ambient light is not known.

INTRODUCTION

When evening approaches, and daylight gives way to artificial light, we notice little change in the colors of objects around us. The perceptual ability that permits us to discount spectral variation in the ambient light and assign stable colors to objects is called color constancy. Much of human color-vision research focuses on the adaptational mechanisms underlying color constancy. Yet, given the kinds of information available in the initial stages of biological vision, it is not known how color constancy is even possible. Indeed, without restrictions on the range of lights and surfaces that the visual system will encounter, color constancy is not, in general, possible.[1]

In this paper we describe an algorithm for estimating the surface reflectance functions of objects in a scene with incomplete knowledge of the spectral power distribution of the ambient light. We assume that lights and surfaces present in the environment are constrained in a way that we make explicit below. An image-processing system using this algorithm can assign colors that are constant despite changes in the lighting on the scene. This capability is essential to correct color rendering in photography, in television, and in the construction of artificial visual systems for robotics. We describe how constraints on lights and surfaces in the environment make color constancy possible for a visual system and discuss the implications of the algorithm and these constraints for human color vision.

PRELIMINARY DEFINITIONS

Consider a visual sensing device consisting of a lens that focuses light from a scene onto a planar array of sensors, analogous to a retina. We introduce the following definitions and assumptions. We begin by restricting attention to a region of the scene where the spectral power distribution of the light is constant.[2] At any location in the scene, the ambient light is specified by its spectral power distribution, $E(\lambda)$, which describes the energy per second at each wavelength, λ. The ambient light is reflected from surfaces and focused onto the sensor array. The proportion of light of wavelength λ reflected from an object toward location x on the sensor array is determined by the surface spectral reflectance, $S^x(\lambda)$. The superscript x denotes the spatial position on the two-dimensional sensor array at which the object is imaged.[3] The light arriving at each location x on the sensor array is described by the function $E(\lambda)S^x(\lambda)$.

We assume that there are p distinct classes of sensors at each location x. In human vision, there are four photoreceptor classes (rods and cones), of which three (cones) are known to be active in daylight vision. We denote the relative wavelength sensitivity of the visual color sensors of the kth class as $R_k(\lambda)$

Each of the p sensors at location x records a sensor quantum catch

$$\rho_k^x = \int E(\lambda)S^x(\lambda)R_k(\lambda)d\lambda, \qquad k = 1, 2, \ldots, p, \qquad (1)$$

where the integral is taken over the entire spectrum. The information about the scene available to the visual system is contained in the p sensor quantum catches at each location x. The spectral reflectance at each location $S^x(\lambda)$ is assumed to be unknown.

Given only the sensor responses ρ_k^x, we show how to recover the surface spectral reflectances $S^x(\lambda)$ over a range of possible ambient lights $E(\lambda)$. Knowledge of $S^x(\lambda)$ permits us to compute color descriptors that are independent of the ambient light $E(\lambda)$.

PREVIOUS WORK

In their early and important work on color constancy, Helson and Judd[4] studied and formally modeled the ability of human observers to achieve this goal. Land and McCann[5] proposed a theory of color vision (the retinex theory) and a method for computing color-constant color descriptors given only the kinds of information available in the sensor responses. The retinex algorithm performs this task only under a limited set of physical conditions.[6] Buchsbaum[7] demonstrated that it is possible to compute color descriptors that are completely independent of the ambient light in an image if the average spectral reflectance of the objects in the image is known. Buchsbaum's result is the strongest that has been obtained until now. His result is useful for many applications, but it is of limited utility in visual sensing applications such as photography and satellite remote sensing in which it is not possible to know in advance the average spectral reflectance function of the objects in the image.

Our purpose here is to improve on Buchsbaum's result. We describe how to recover surface spectral reflectance from an image without knowledge of the average spectral reflectance function.[8]

MODELS OF LIGHTS AND SURFACES REFLECTANCES

We express surface reflectance as a weighted sum of basis spectral reflectance functions $S_j(\lambda)$ as suggested by Sällström,[1] Buchsbaum,[7] and Brill,[9]

$$S^x(\lambda) = \sum_{j=1}^{n} \sigma_j^x S_j(\lambda) \tag{2}$$

and term this representation a linear model of surface reflectance. The basis reflectances are fixed. They do not vary with location in the scene and are assumed known. The number of basis elements, n, is referred to as the number of degrees of freedom in the model. Knowledge of the weights σ_j^x corresponding to a surface reflectance $S^x(\lambda)$ described by the finite linear model amounts to complete knowledge of $S^x(\lambda)$.

Any finite set of surface spectral reflectances can be reproduced by a linear model of this kind if n is large enough. What is surprising is that models with only a few basis reflectances provide excellent approximations to many naturally occurring spectral reflectances. Stiles *et al.*[10] suggest that spectral reflectances may be treated as band-limited functions. A collection of band-limited functions is perfectly captured by a linear model. The number of basis elements is proportional to the band limit.[11] The range of limiting frequencies that they suggest corresponds to three to five basis reflectances in Eq. (2). Buchsbaum and Gottschalk[12] demonstrate that band-limited reflectances generated using as few as three basis reflectances provide metamers to most naturally occurring spectral reflectances.

The Munsell collection includes color chips spanning a wide range of colors. Cohen[13] used a characteristic vector decomposition of the spectral reflectances of 150 Munsell color chips selected at random from a full set of 433 chips to compute the linear models with from 1 to 5 basis reflectances that best approximated the surface reflectances of the Munsell chips.[14] He found that a linear model using as few as three properly chosen basis reflectances captured 99.2% of the overall variance. He states that model reflectances generated by these three basis reflectances provided a good approximation to the spectral reflectances of the Munsell chips. Reflectances generated by these same three basis reflectances also provide good approximations to 337 surface spectral reflectances of naturally occurring objects measured by Krinov.[15]

We also represent the ambient light by a linear model,

$$E(\lambda) = \sum_{i=1}^{m} \epsilon_i E_i(\lambda), \tag{3}$$

with fixed, known basis lights $E_i(\lambda)$. It is natural to inquire how well such a model captures the range of spectral variation of natural lights such as daylight. Judd *et al.*[16] performed a characteristic vector analysis of 622 functions describing the spectral distribution of natural daylight measured over a range of weather conditions and times of day. They determined that, for practical purposes, three to four basis lights provide essentially perfect matches to the spectral distributions measured. The measurements by Judd *et al.* suggest that the number of parameters required to have an adequate linear model of the ambient light may often be small. Dixon and others have independently measured and analyzed spectral power distributions of daylight and drawn similar conclusions.[17]

REFORMULATION OF THE PROBLEM

Next we describe the consequences of assuming linear models of the ambient light and surface reflectances for the problem of color constancy. The m values of ϵ_i in Eq. (3) form a (column) vector ϵ that specifies the light $E(\lambda)$. The n values of σ_j in Eq. (2) form a (column) vector σ that specifies the surface reflectance $S^x(\lambda)$. Substituting Eqs. (2) and (3) into Eq. (1) permits us to express the relationship between the daylight surface reflectances and sensor responses by the matrix equation

$$\rho^x = \Lambda_\epsilon \sigma^x, \tag{4}$$

where ρ^x is a (column) vector formed from the quantum catches of the p sensors at location x. The matrix Λ_ϵ is p by n, and its kjth entry is of the form $\int E(\lambda)S_j(\lambda)R_k(\lambda)d\lambda$. The matrix Λ_ϵ captures the role of the light in transforming surface reflectances at each location x into sensor quantum catches.

Various limits on surface reflectance recovery are dictated by Eq. (4). We consider the limits on recovery (1) when the light on the scene is assumed to be known and (2) when the light on the scene is unknown.

In the simple case in which the ambient light and (therefore) the lighting matrix Λ_ϵ is known, we see that to recover the n weights that determine the surface reflectance we need merely solve a set of simultaneous linear equations. The recovery procedure reduces to matrix inversion when $p = n$. If p is less that n Eq. (4) is underdetermined and there is no unique solution.

If the ambient light is unknown then it easy to show that we cannot do so well: we cannot in general recover the ambient light vector ϵ or the spectral reflectances even when $p = n$. The matrix Λ_ϵ is square. For any ϵ such that Λ_ϵ is nonsingular there is a set of surface reflectances that satisfy Eq. (4). Any such choice of a light vector ϵ and corresponding surface reflectances σ^x could have produced the observed surface reflectances. No unique solution is possible without additional information concerning lights and surfaces in the scene.

Any solution method must therefore resolve the unfavorable ratio of unknown parameters to observed data points. We do so by assuming that there are more classes of sensors than degrees of freedom in surface reflectances: $p > n$. Suppose that there are $p = n + 1$ linearly independent sensors to sample the image at each location spectrally. In this case from s different spatial locations $s(n + 1)$ data values are obtained on the left-hand side of the equation. The number of unknown parameters is only sn unknowns from the different surface vectors and m unknowns from the light vector. After sampling at $s > m$ locations we finally obtain more data values than unknowns.

In principle, then, the response of the $n + 1$ sensors can

contain enough information to permit exact recovery of both the lighting parameter ϵ and the surface reflectance σ^x at each location. Next we outline a method for computing the light ϵ and the n-dimensional surface reflectance vector σ^x given the $n+1$ dimensional sensor response vector ρ^x at each location.

COMPUTATIONAL METHOD

The ambient light vector ϵ and the surface vectors σ^x contribute to the value of the sensor vectors in different ways. The ambient light vector specifies the value of the light transformation matrix, Λ_ϵ. The matrix Λ_ϵ is a linear transformation from the n-dimensional space of surface reflectances σ^x into the $n+1$-dimensional space of sensor quantum catches ρ^x. The sensor response to any particular surface, σ^x, is the weighted sum of the n column vectors of Λ_ϵ. Consequently, the sensor responses must fall in a proper subspace of the sensor space determined by Λ_ϵ and therefore by the lighting parameter ϵ.

Figure 1 illustrates the situation when there are three classes of sensors ($p = 3$) and two degrees of freedom in the surface reflectances ($n = 2$). In the particular example shown in Fig. 1, the two-dimensional surface vectors span a plane (passing through the origin) in the three-dimensional sensor space. The light ϵ determines the plane.

We propose a two-step procedure to estimate light and surface reflectances. First, we determine the plane spanning the sensor quantum catches: knowledge of the plane in Fig. 1 permits us to recover the ambient light vector ϵ. Second, once we know the light vector ϵ, we determine the lighting matrix Λ_ϵ and achieve our goal of recovering the surface vectors simply by inverting this transformation. The exact mathematical conditions that must obtain in order for our procedure to yield the unique correct result are analyzed by Maloney.[18]

We have shown that the formal problem of estimating ambient light and surface spectral reflectance from image data may be reduced to a simple computational procedure. Our procedure is also applicable to the construction of automatic sensor systems capable of discounting fluctuations in the ambient light in unusual working environments.

In some natural scenes, the spectral composition of the ambient light varies with spatial location. The computation above can be extended in a straightforward manner to the problem of estimating and discounting a slowly varying (spatial-low-pass) ambient light. We will describe the details of how to do so elsewhere.

The recovery procedure that we developed is exact when the actual physical conditions fall within the bounds defined by the finite-dimensional models of lights and of surface reflectances. The method is readily extended to the case in which the finite-dimensional models only approximate actual lights and surface reflectances. We can no longer guarantee, for example, that the sensor quantum catches in Fig. 1 will lie exactly in the plane determined by the light. Under these circumstances, we estimate the plane that best fits the sensor quantum catches in the least-squares sense and derive an estimate of the light vector $\hat{\epsilon}$. Given $\Lambda_{\hat{\epsilon}}$, we continue by computing the best estimates of surface reflectance σ^x in the least-squares sense.[19] Small deviations from the assumptions of the models produce small errors in estimation.

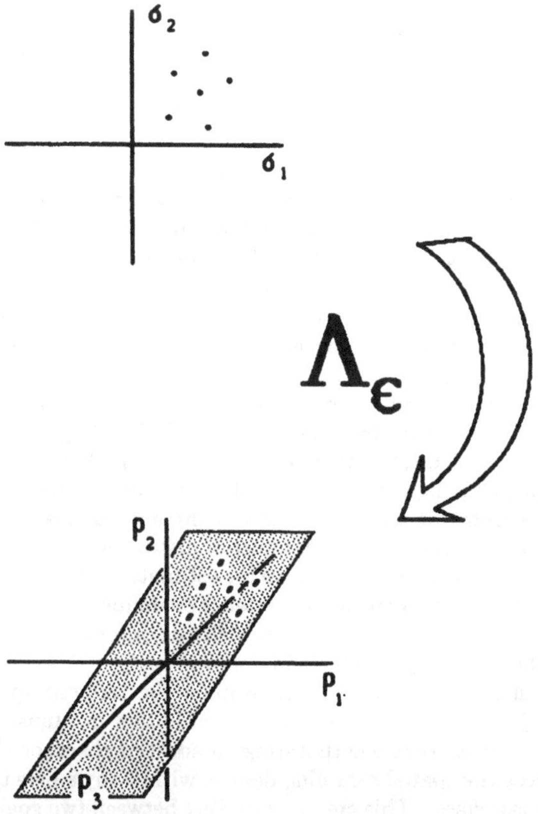

Fig. 1. Outline of the solution method in the case when there are three classes of sensors ($p = 3$) and two degrees of freedom in the model of surface reflectances ($n = 2$). The sensor quantum catches lie on a plane through the origin in the three-dimensional space of sensor quantum catches below. Knowledge of the plane determines the light ϵ and the matrix Λ_ϵ. The matrix Λ_ϵ is inverted to recover the surface reflectances σ^x above.

By increasing the dimensionality of the finite-dimensional approximation, exact solutions may be approached to within any desired degree of precision. A trichromatic visual system can approximate color constancy when reflectances in the visual environment are predominantly captured by a linear model with two degrees of freedom.

IMPLICATIONS

Our calculation has several implications for human color vision. First, it suggests that perfect color constancy is possible only for ranges of lights and reflectances that can be described by a small number of parameters. The human visual system is known to have better color constancy over some ranges of lights than others. Our formulation provides a framework that may be used to determine, experimentally, the range of ambient lights and surfaces over which human color constancy succeeds and over which it fails.

We can test whether a linear model characterizes the range of lights permitting essentially perfect color constancy in human vision as follows. Suppose that the color appearance of a set of surfaces is preserved when measured with two ambient lights, say $E(\lambda)$ and $E'(\lambda)$. If the lights for which human color constancy succeeds form a linear model, then color appearance should also be preserved when the surfaces are viewed under weighted mixtures of the ambient lights $E(\lambda)$ and $E'(\lambda)$.

Second, our results show how the number of classes of photoreceptors active in color vision limits the number of degrees of freedom in the surface reflectances that can be recovered. With three classes of photoreceptors, we can exactly recover surface reflectances drawn from a fixed model of surface reflectance with at most two degrees of freedom. Additional degrees of freedom in the surface reflectance will, in general, introduce error into the estimates of surface reflectance obtained, precluding perfect color constancy.

Third, the parameter counting argument indicates that a minimum number of distinct surface reflectances must be present in the scene to permit recovery. In the illustration of the solution above, at least two distinct surface reflectances are needed to specify the plane (which must pass through the origin) that determines the light. In general, at least $p - 1$ distinct surfaces are needed in order to determine uniquely the light vector ϵ. In the presence of small deviations from the linear models of light and surface reflectances, an increase in the number of distinct surface reflectances will, in general, improve the estimate of the light and the corresponding surface reflectance estimates. Our analysis suggests that color constancy should improve with the number of distinct surfaces in a scene.[20]

Fourth, our algorithm may explain the reduced spatial sampling of the short-wavelength receptors in human vision.[21] Note that incorporating an additional sensor class reduces the spatial sampling density within any single type of sensor class. This creates a conflict between two goals of the visual system. Better color correction for the ambient light can be obtained by including more sensor classes. But including additional sensor classes reduces the spatial sampling density within individual sensor classes. A reasonable trade-off between these two goals may be obtained by the following observation.

Once the light vector ϵ has been estimated, we may calculate the inverse of the lighting matrix Λ_ϵ. Once this matrix is known, the $n + 1$ quantum catches at each location are redundant: only n sensor values are needed to compute the value of σ^x. It follows that if the ambient light varies slowly across the scene, the $n + 1$th sensor class need not be present at as high a sampling density as the other sensor classes.

In a three-sensor system there is no need to place the third sensor with as high a spatial sampling density as the first two sensors. It follows that reducing the retinal space occupied by the short-wavelength sensor class permits the system to obtain a higher spatial resolution of the estimated surface spectral reflectance function with virtually no deterioration in its ability to correct for variation in the spectral power distribution of the ambient light.

ACKNOWLEDGMENTS

This research was supported by grant no. 2 RO1 EY03164 from the National Eye Institute and NASA grant NCC-2-44. We thank M. Pavel, R. N. Shepard, and D. Varner for their advice and suggestions. Stanford University has applied for a patent directed to the matter described herein.

* Lawrence T. Maloney is now at the Human Performance Center, University of Michigan, 330 Packard Road, Ann Arbor, Michigan 48104.

REFERENCES AND NOTES

1. See P. Sällström, "Colour and physics: some remarks concerning the physical aspects of human colour vision," Institute of Physics Rep. 73-09 (University of Stockholm, Stockholm, 1973).
2. The computational method that we develop requires only that the ambient light be approximately constant over small local patches of the image. The method is easier to explain if we restrict attention to a region of the image across which the ambient light does not change.
3. In general the surface reflectance function may depend on the geometry of the scene, the angle of incidence of the light on the surface, and the angle between the surface and the line of sight. We are concerned here with the analysis of a single image drawn from a scene with fixed geometric relations among objects, light sources, and the visual sensor array. $S^x(\lambda)$ refers to the proportion of light returned from the object toward the sensor array within that fixed geometrical framework.
4. D. B. Judd, "Hue saturation and lightness of surface colors with chromatic illumination," J. Opt. Soc. Am. **30**, 2 (1940); H. Helson, "Fundamental problems in color vision. I. The principle governing changes in hue saturation and lightness of non-selective samples in chromatic illumination," J. Exp. Psychol. **23**, 439 (1938).
5. E. H. Land and J. J. McCann, "Lightness and retinex theory," J. Opt. Soc. Am. **61**, 1 (1971); E. H. Land, "Recent advances in retinex theory and some implications for cortical computations: color vision and the natural image," Proc. Nat. Acad. Sci. U.S. **80**, 5163 (1983); E. H. Land, D. H. Hubel, M. Livingston, S. Perry, and M. Burns, "Colour-generating interactions across the corpus callosum," Nature **303**, 616 (1983).
6. D. Brainard and B. Wandell, "An analysis of the retinex theory of color vision," Stanford Applied Psychology Lab. Tech. Rep. 1985-04 (Stanford University, Stanford, Calif., 1985).
7. G. Buchsbaum, "A spatial processor model for object colour perception," J. Franklin Inst. **310**, 1 (1980).
8. It is not possible to recover $E(\lambda)$ better than to within a multiplicative constant given only the sensor quantum catches. If, for example, the intensity of the light is doubled to $2E$ but all reflectances are halved to $\frac{1}{2}S^x(\lambda)$, it is easy to verify that the sensor quantum catches in Eq. (1) are unchanged. When we speak of recovering the ambient light and surface reflectances, we mean recovery up to this unknown mutiplicative constant.
9. M. H. Brill, "A device performing illuminant-invariant assessment of chromatic relations," J. Theor. Biol. **71**, 473 (1978).
10. W. S. Stiles, G. Wyszecki, and N. Ohta, "Counting metameric object-color stimuli using frequency-limited spectral reflectance functions," J. Opt. Soc. Am. **67**, 779 (1977).
11. For a discussion of band-limited functions, see R. Bracewell, *The Fourier Transform and Its Application*, 2nd ed. (McGraw-Hill, New York, 1978), Chap. 10.
12. G. Buchsbaum and A. Gottschalk, "Chromaticity coordinates of frequency-limited functions," J. Opt. Soc. Am. A **1**, 885 (1984).
13. J. Cohen, "Dependency of the spectral reflectance curves of the Munsell color chips," Psychonomic Sci. **1**, 369 (1964).
14. See K. V. Mardia, J. T. Kent, and J. M. Bibby, *Multivariate Analysis* (Academic, London, 1979), Chap. 8, for a discussion of characteristic vector analysis (also known as principal-components analysis or the Karhunen–Loève decomposition).
15. E. Krinov, *Spectral Reflectance Properties of Natural Formations*, Technical translation TT-439 (National Research Council of Canada, Ottawa, 1947); details of the fit of Cohen's characteristic vectors to the Munsell surface reflectances are given in L. Maloney, "Computational approaches to color constancy," Stanford Applied Psychology Lab. Tech. Rep. 1985-01 (Stanford University, Stanford, Calif., 1985).
16. D. B. Judd, D. L. MacAdam, and G. Wyszecki, "Spectral distribution of typical daylight as a function of correlated color temperature," J. Opt. Soc. Am. **54**, 1031 (1964).
17. E. R. Dixon, "Spectral distribution of Australian daylight," J. Opt. Soc. Am. **68**, 437 (1978); G. T. Winch, M. C. Boshoff, C. J. Kok, and A. G. du Toit, "Spectroradiometric and colorimetric characteristics of daylight in the southern hemisphere: Pre-

toria, South Africa," J. Opt. Soc. Am. **56**, 456 (1966); S. R. Das and V. D. P. Sastri, "Spectral distribution and color of tropical daylight," J. Opt. Soc. Am. **55**, 319 (1965); V. D. P. Sastri and S. R. Das, "Spectral distribution and color of north sky at Delhi," J. Opt. Soc. Am. **56**, 829 (1966); "Typical spectral distributions and color for tropical daylight," J. Opt. Soc. Am. **58**, 391 (1968).

18. L. Maloney, "Computational approaches to color constancy," Stanford Applied Psychology Lab. Tech. Rep. 1985-01 (Stanford University, Stanford, Calif., 1985).
19. See Ref. 18, Chap. 4, for details.
20. R. B. MacLeod, "An experimental investigation of brightness constancy," Arch. Psychol. **135**, 1 (1932), reports that human brightness constancy does.
21. E. N. Willmer and W. D. Wright, "Colour sensitivity of the fovea centralis," Nature **156**, 119 (1945); D. R. Williams, D. I. A. MacLeod, and M. Hayhoe, "Punctate sensitivity of the blue-sensitive mechanism," Vision Res. **21**, 1357 (1981); F. M. de Monasterio, S. J. Schein, and E. P. McCrane, "Staining of blue-sensitive cones of the macaque retina by a fluorescent dye," Science **213**, 1278 (1981).

Mechanisms of color constancy

M. D'Zmura and P. Lennie

Center for Visual Science and Department of Psychology, University of Rochester, Rochester, New York 14627

Received April 21, 1986; accepted June 27, 1986

We develop a model of how the visual system finds the colors of objects that have unknown shapes and positions. The model relies on mechanisms of light adaptation, coupled with eye movements, to recover three descriptors of surface reflectance that are represented in the signals of an achromatic mechanism and two color-opponent mechanisms. These descriptors are transformed to yield estimates of hue, the dimension of surface color that is independent of object shape and viewing geometry.

INTRODUCTION

The intrinsic chromatic properties of objects are given by surface reflectance functions, which specify the proportion of incident illumination reflected by a surface as a function of wavelength. Recovery of surface reflectance functions from the lights that reach our eyes is a complex task, because the reflected light is a composite signal: it depends on both surface reflectance and the spectral power distribution (SPD) of the illuminant. Mechanisms that provide color constancy must decompose this signal and discount the spectral properties of the illuminant in order to establish the colors of objects in terms of their surface reflectance functions.

A trichromatic visual system can represent a surface reflectance function using at most three parameters. Similarly, it must represent illuminants by at most three parameters, so unless surface reflectances and illuminants can be represented adequately by only three variables, color constancy will be impossible. Fortunately, three-dimensional descriptions of surfaces and illuminants are adequate because there is an underlying simplicity in natural chromatic signals. Principal components analyses of daylight[1,2] and of surface reflectances[3] suggest that much of the variance in each of these domains may be accounted for by a small set of components: only three components added together in appropriate proportions are required to represent any phase of daylight, and three to five appropriately weighted components are required for the reconstruction of most surface reflectances.[4] This simplicity is the key to color constancy and has been exploited in the formal solutions offered by Sällström,[5] Brill,[6,7] Buchsbaum,[8] and Maloney and Wandell.[4,9] The purpose of this paper is to extend these formal treatments in two ways.

For adequate color constancy the perceived color of an object should be stable not only through changes in the SPD of the light that falls on a scene but also through changes in the position of the object, the illuminant, and the viewer. Because surface reflectance functions depend not only on material properties but also on geometric variables, the lights that reach the eyes from an object change when viewing geometry is changed. Even if the viewing geometry is held fixed, the spectral distribution of the light from a solid object generally varies from point to point across its surface.

Our first aim is to show how the hue of an object can be rendered independently of its shape and position. We show formally how a trichromatic system that can find the SPD of an illuminant can discount the illuminant and find the completely stable hues of objects without discarding the geometric information that is carried by the dimensions of lightness and saturation.

The second aim of the paper is to develop a physiologically plausible account of the mechanisms that might undertake the computation. We show first how mechanisms of light adaptation, both in the separate cone pathways and at opponent sites, can find and discount an illuminant. We then show how the signals from second-stage mechanisms can be transformed to yield a stable hue descriptor.

SURFACE REFLECTANCE AND THE STABILITY OF HUE

We follow Yilmaz,[10] Sällström,[5] Brill,[6,7] Buchsbaum,[8] and Maloney and Wandell[4,9] in using a small set of basis functions to provide an approximate representation of surface reflectance. For expository purposes we use the three basis functions shown in Fig. 1A. These are functions of wavelength; when added together in the correct proportions they describe the fraction of incident light reflected by a surface. The function $S_1(\lambda)$ is a constant function of wavelength; variations in the contribution of $S_1(\lambda)$ to surface reflectance represent variations along a "light–dark" dimension. The function $S_2(\lambda)$ represents variation along a "red–green" dimension, while the function $S_3(\lambda)$ represents variation along a "yellow–blue" dimension. These functions are the first three terms in a Fourier series expansion and have been chosen for their simplicity. The functions that best describe any particular class of surfaces will be the set that accounts for the greatest proportion of the variance in the collection of surface reflectances. For natural surfaces these are probably not the functions shown in Fig. 1A, but they would share the same general features.[4]

The two "color" functions $S_2(\lambda)$ and $S_3(\lambda)$ have negative as well as positive values and cannot, by themselves, describe a real surface reflectance; they are used only in combination with the "light–dark" function $S_1(\lambda)$. Figure 2 shows how any surface reflectance function can be approximately reconstituted by adding together appropriate amounts of

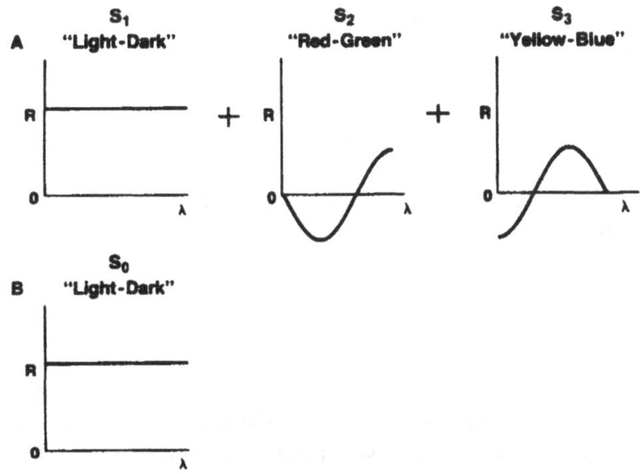

Fig. 1. A, Three basis functions $S_1(\lambda)$, $S_2(\lambda)$, and $S_3(\lambda)$ used to model surface reflectance functions. The absolute values of reflectance (R) for S_1, S_2, and S_3 are not indicated; their contributions to a given surface reflectance are set by scalars in the model. B, The basis function $S_0(\lambda)$, identical to $S_1(\lambda)$, that describes the spectral properties of the specular component of reflectance. Part A of this figure may be reinterpreted as a description of the spectral properties of the diffuse component.

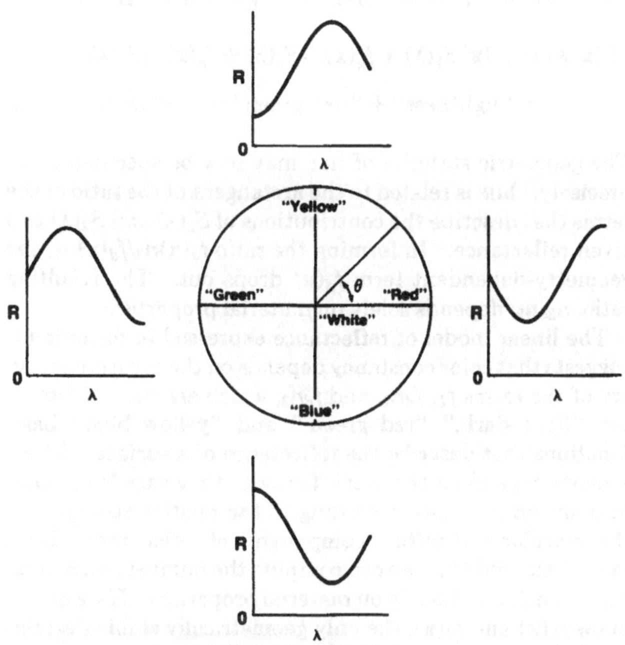

Fig. 2. Color circle showing variation in contributions of S_1, S_2, and S_3 to surface reflectance functions. The basis functions are summed in appropriate ways to generate the surface reflectance functions shown here. The ratio of contributions to reflectance made by S_3 and S_2 determines the azimuth θ, which represents hue.

these functions. For instance, by adding the "red–green" function $S_2(\lambda)$ to the "light–dark" function $S_1(\lambda)$, the result is a "red" reflectance function, while by subtracting $S_2(\lambda)$ from $S_1(\lambda)$ we arrive at a "green" reflectance function.

Color at a Point

We do not know how the psychological dimensions of color are linked to surface reflectance, but a consistent account is suggested by the color circle of Fig. 2.[10] Saturation is related to the proportion of "colored" light to "white" light and so is related to the amounts n_2 and n_3 contributed by $S_2(\lambda)$ and $S_3(\lambda)$ relative to the amount n_1 contributed by $S_1(\lambda)$. Bearing in mind that the values of n_2 and n_3 may be negative; saturation is related to the ratio $(n_2{}^2 + n_3{}^2)^{1/2}/n_1$. The hue of a surface is related to the ratio n_3/n_2 of the two "color" terms: the amount of "yellow–blue" relative to the amount of "red–green." If hue is taken to be the azimuthal angle measured from the "red" half-axis, as shown in Fig. 2, then hue is given by $\arctan(n_3/n_2)$. The lightness of a surface is related to the amount n_1 of the flat component $S_1(\lambda)$ of a given reflectance function.

Formally, a surface reflectance function at a point on the surface of an object is given as the following linear combination:

$$S(\lambda) \simeq \sum_{i=1}^{3} n_i S_i(\lambda). \tag{1}$$

Surface Reflectances of Solid Objects

The problem with the model of reflectance given by formula (1), which is used with minor variations in most formal approaches to color constancy,[4–9] is that it does not adequately describe variations in reflectance that depend on geometric variables. The model may be extended to cover the dependence on geometry by recognizing that the "light–dark" contribution to surface reflectance can come from two physically distinct components. The first of these is the specular component, which arises at the optical interface between air and the surface[11–14]; the second is the diffuse component, which arises from the absorption and scattering of light by molecules underlying the surface.[11,14–16]

Figure 1B indicates that the specular component $S_0(\lambda)$ of most objects is essentially flat across the visible spectrum.[17] The specular reflection therefore has the relative SPD of the illuminant.

At any point on the surface of a uniformly colored object, the total reflectance is the sum of the specular and diffuse components. For such an object (e.g., a green pepper) the contributions of these two terms to total reflectance vary from point to point. Figure 3A depicts this variation in a space that represents the quantum catches in the three classes of cones. Responses to the lights from different points on the surface of the object fall along the locus L. Quantum catches from the specular component alone lie along the axis I that describes the illuminant, while quantum catches from the diffuse component alone lie along the axis D.

Where the specular component dominates the total reflectance, for instance in a "highlight" region, the surface appears light and desaturated and the trichromatic response lies close to the axis I defined by the illuminant. The surface appears desaturated at such points because the color contribution of the diffuse component is small relative to the lightness contribution from the specular component. Points lying outside highlights are typically darker and more saturated and give rise to responses that lie closer to the axis D, which describes the diffuse component.

Since the response to the light from any point on a uniformly colored surface is a sum of responses to specular and diffuse components, the locus L of responses to different points lies in the plane defined by the axes D and I, as noted

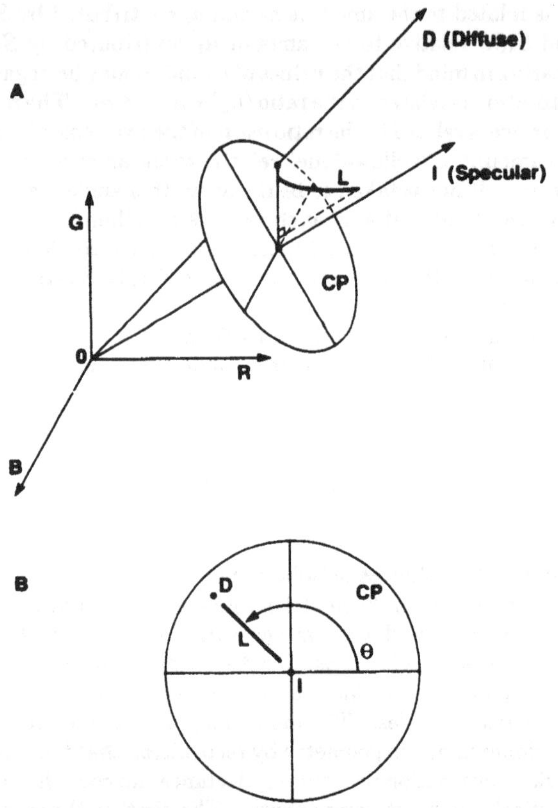

Fig. 3. A, A three-dimensional space of photoreceptor quantum catches (R, G, B) in which lies the locus L of responses to lights from different points on a single, uniformly colored object; both specular and diffuse components of reflectance vary across the surface of the object. L lies within a single plane determined by the response to the specular component, which has the chromaticity of the illuminant and varies along the axis labeled I, and the response to the diffuse component, which varies along the axis labeled D. A chromaticity plane CP is centered on and perpendicular to the axis I. Variations in L perpendicular to CP represent changes in lightness. B, Projection of L onto CP, showing that saturation may also change across a single uniformly colored surface. The azimuth θ that represents hue is a geometrically stable color descriptor.

by Shafer.[17] We use this fact later to show how the chromaticity of the illuminant can be found by using highlights.

The plane CP of Fig. 3, which is centered on and perpendicular to the axis that describes the illuminant, is a chromaticity plane. This is easiest to imagine if the illuminant is white. Figure 3B shows that responses to all the lights from a uniformly colored surface lie along a line segment in the plane CP that joins the chromaticity of the illuminant to the chromaticity of the diffuse component. This line spans the range of possible color signals from the surface: at one extreme (I) the light received from a point on a surface arises entirely from the specular component, while at the other extreme (D) it arises entirely from the diffuse component.

The variation in specular and diffuse components of reflectance from point to point on the surface results in corresponding variations in lightness and saturation. Only the ratio of contributions from "yellow–blue" and "red–green" basis functions is constant across a uniformly colored surface: although the contributions of these two chromatic terms to reflectance may vary across a solid object, the contributions vary identically in a way that is determined by the intensity of the diffuse component of reflectance. This stable hue is represented in Fig. 3B by the azimuth θ.

Formal Expression of Surface Reflectance

The concepts that have been outlined graphically are formalized by expressing the surface reflectance $S(\mathbf{x}, \lambda)$ at a point $\mathbf{x} = (x_1 x_2)$ on a uniformly colored object as the following sum:

$$S(\mathbf{x}, \lambda) \simeq f_s(\mathbf{x}) S_0(\lambda) + f_d(\mathbf{x}) \left[\sum_{i=1}^{3} n_i S_i(\lambda) \right]$$

$$= \text{“specular”} + \text{“diffuse”}. \quad (2)$$

The function $f_s(\mathbf{x})$ describes the variation in the intensity of the specular component with location on the surface, while the function $f_d(\mathbf{x})$ describes the corresponding variation in the intensity of the diffuse component.

Because the basis functions $S_0(\lambda)$ and $S_1(\lambda)$ are identical, the two terms that govern their intensities may be combined into a single term $f_L(\mathbf{x})$ that represents "lightness":

$$f_L(\mathbf{x}) = f_s(\mathbf{x}) + f_d(\mathbf{x}) n_1. \quad (3)$$

Formula (2) may be recast to express surface reflectance as the sum of a "lightness" term and two "color" terms:

$$S(\mathbf{x}, \lambda) \simeq f_L(\mathbf{x}) S_1(\lambda) + f_d(\mathbf{x}) n_2 S_2(\lambda) + f_d(\mathbf{x}) n_3 S_3(\lambda)$$

$$= \text{“lightness”} + \text{“red–green”} + \text{“yellow–blue”}. \quad (4)$$

The geometric stability of hue may now be specified more precisely: hue is related to the arctangent of the ratio of the terms that describe the contributions of $S_3(\lambda)$ and $S_2(\lambda)$ to a given reflectance. In forming the ratio $f_d(\mathbf{x}) n_3 / f_d(\mathbf{x}) n_2$, the geometry-dependent term $f_d(\mathbf{x})$ drops out. The resulting ratio, n_3/n_2, depends solely on material properties.

The linear model of reflectance expressed in formula (4) suggests that color constancy depends on the accurate recovery of the terms f_L, $f_d n_2$, and $f_d n_3$, which are the weights on the "light–dark," "red–green," and "yellow–blue" basis functions that describe the reflectance of a surface. These weights depend on geometric factors: they vary from point to point on a surface according to the relative strengths of the specular and diffuse components of reflectance. If we know $f_L(\mathbf{x})$ and $f_d(\mathbf{x})$ we can compute the numbers n_1, n_2, and n_3, which depend solely on material properties. If we do not know $f_L(\mathbf{x})$ and $f_d(\mathbf{x})$, the only geometrically stable descriptor that can be extracted is hue.

DISCOUNTING A KNOWN ILLUMINANT

The description of surface reflectance given by f_L, $f_d n_2$, and $f_d n_3$ can be extracted only if the spectral properties of the illuminant can be discounted. There are well-developed formal solutions to the problem of discounting a known illuminant[4,8]; these rely on our ability to represent the SPD's of illuminants using a small set of basis functions. As a result, an illuminant may be discounted by applying an appropriate linear transformation to quantum catch signals. In what follows, we show that Buchsbaum's[8] method for discounting the illuminant can be applied to scenes in which surfaces are described by formula (4).

Approximate Representation of Illuminants

By analogy to formula (1), which deals with surface reflectance, the SPD $I(\lambda)$ of a given illuminant is expressed as a sum of appropriate amounts of three basis functions $I_1(\lambda)$, $I_2(\lambda)$, and $I_3(\lambda)$:

$$I(\lambda) \simeq \sum_{i=1}^{3} m_i I_i(\lambda). \tag{5}$$

If the basis functions $I_i(\lambda)$ are known, then illuminants that can be characterized by formula (5) are described fully by the numbers m_1, m_2, and m_3. Formula (5) is known to provide an adequate description of the phases of daylight.[1,2]

Interaction of Illuminants and Reflectances

Illuminants interact multiplicatively with surface reflectances to produce the reflected light. Because the specular component of reflectance is flat, specular reflection has the relative SPD of the illuminant, and so therefore does the light that penetrates the surface to interact with colorants.

The light $L(\mathbf{x}, \lambda)$ reflected by a uniformly colored surface is thus found by multiplying the reflectance $S(\mathbf{x}, \lambda)$ expressed in formula (4) by the illuminant $I(\lambda)$ of formula (5); suppressing dependence on wavelength, the result is

$$L(\mathbf{x}) \simeq f_L(\mathbf{x})\left(\sum_{i=1}^{3} m_i I_i S_1\right) + f_d(\mathbf{x})n_2\left(\sum_{i=1}^{3} m_i I_i S_2\right) + f_d(\mathbf{x})n_3\left(\sum_{i=1}^{3} m_i I_i S_3\right). \tag{6}$$

Formula (6) makes evident the problem of discounting the illuminant. Consider the case in which the illuminant $I(\lambda)$ is spectrally flat. The three chromatic components in the reflected light L will represent precisely the contributions made by each of the basis functions for surface reflectance; it is a straightforward matter to form three mechanisms, e.g., an achromatic mechanism and two color-opponent mechanisms, that combine photoreceptoral signals linearly so as to signal f_L, $f_d n_2$, and $f_d n_3$, respectively. However, when we change the SPD of the illuminant, we change the spectral distributions of the three components of the light L that correspond to the three basis functions for reflectance, so mechanisms set up to discount "white" illumination will no longer abstract the terms f_L, $f_d n_2$, and $f_d n_3$.

Decomposing the Color Signal

The computational solution to this problem is evident in formula (6). If one can find the three numbers m_i that describe the illuminant, then one can find the components in the light L that correspond to the three basis functions for surface reflectance. One can then calculate how the sensitivities of mechanisms that abstract reflectance descriptors must change in response to a change in illumination.

The spectral sensitivities of the three photoreceptors are termed $Q_1(\lambda)$, $Q_2(\lambda)$, and $Q_3(\lambda)$, and their respective quantum catches are q_1, q_2, and q_3. We have in mind broad sensitivities of the sort derived by Smith and Pokorny for human cones.[18] The quantum catch q_k of the kth photoreceptor type with spectral sensitivity $Q_k(\lambda)$ from light $L(\lambda)$ is given by the integral of the product of $L(\lambda)$ and $Q_k(\lambda)$ over the visible spectrum. We employ the "inner-product" notation $\langle\, ,\, \rangle$ as a shorthand for these integrals of products over the visible spectrum and continue to suppress dependence on wavelength:

$$q_k = \langle L, Q_k \rangle = \int_{400}^{700} L(\lambda) Q_k(\lambda) d\lambda. \tag{7}$$

If the expression for reflected light L in formula (6) is now substituted for L in Eq. (7), then the result is that

$$q_k(\mathbf{x}) \simeq f_L(\mathbf{x})\left\langle \sum_{i=1}^{3} m_i I_i S_1, Q_k \right\rangle + f_d(\mathbf{x})n_2\left\langle \sum_{i=1}^{3} m_i I_i S_2, Q_k \right\rangle + f_d(\mathbf{x})n_3\left\langle \sum_{i=1}^{3} m_i I_i S_3, Q_k \right\rangle, \quad \text{for } 1 \le k \le 3. \tag{8}$$

The terms in the brackets are integrals that specify the extent to which each basis function for surface reflectance contributes to the quantum catch of a given photoreceptor, given the illuminant described by the parameters m_i. There are three integrals per photoreceptor type, giving a total of nine. The numbers f_L, $f_d n_2$, and $f_d n_3$ remain the same for each photoreceptor type, and so expression (8) may be reformulated by using vectors and matrices. By writing the illuminant as a function of the vector $\mathbf{m} = (m_1 m_2 m_3)$ we find that

$$\begin{bmatrix} q_1(\mathbf{x}) \\ q_2(\mathbf{x}) \\ q_3(\mathbf{x}) \end{bmatrix} \simeq \begin{bmatrix} \langle I(\mathbf{m})S_1, Q_1\rangle & \langle I(\mathbf{m})S_2, Q_1\rangle & \langle I(\mathbf{m})S_3, Q_1\rangle \\ \langle I(\mathbf{m})S_1, Q_2\rangle & \langle I(\mathbf{m})S_2, Q_2\rangle & \langle I(\mathbf{m})S_3, Q_2\rangle \\ \langle I(\mathbf{m})S_1, Q_3\rangle & \langle I(\mathbf{m})S_2, Q_3\rangle & \langle I(\mathbf{m})S_3, Q_3\rangle \end{bmatrix} \begin{bmatrix} f_L(\mathbf{x}) \\ f_d(\mathbf{x})n_2 \\ f_d(\mathbf{x})n_3 \end{bmatrix}. \tag{9a}$$

The result is that the surface reflectance descriptors f_L, $f_d n_2$, and $f_d n_3$ are related to quantum catches by a square matrix $G(\mathbf{m})$ that depends on the SPD of the illuminant. Formula (9a) may be written more briefly as

$$\mathbf{q}(\mathbf{x}) \simeq G(\mathbf{m})\mathbf{p}(\mathbf{x}), \tag{9b}$$

where $\mathbf{q}(\mathbf{x}) = [q_1(\mathbf{x})\ q_2(\mathbf{x})\ q_3(\mathbf{x})]$ describes the vector of quantum catches as a function of position and where $\mathbf{p}(\mathbf{x}) = [f_L(\mathbf{x})\ f_d(\mathbf{x})\ n_2 f_d(\mathbf{x}) n_3]$ describes surface reflectance at $\mathbf{x}$.

By applying the inverse $H(\mathbf{m})$ of the matrix $G(\mathbf{m})$ to both sides of formula (9b), the illuminant is effectively discounted: if

$$H(\mathbf{m}) = G(\mathbf{m})^{-1}, \tag{10}$$

then

$$H(\mathbf{m}) \simeq \mathbf{p}(\mathbf{x}). \quad (11)$$

The principal implication of formula (11) is that information about surface reflectance can be abstracted by mechanisms whose sensitivities change appropriately in response to changes in the SPD of the illuminant. Such adaptive mechanisms cope best with surfaces and illuminants that are accurately represented as combinations of the three members of their respective basis sets. Moreover, the appropriate set of basis functions must be known by (or in some form built into) any mechanism that discounts the illuminant. Finally, the illuminant can be perfectly discounted only if its description **m** is known, a problem to which we now turn.

FINDING AN UNKNOWN ILLUMINANT

The simplest way to find the chromatic properties of a light source is to look directly at it. Another easy way is to look at its reflection in a mirror. Although the absolute intensity of the illuminant is not available in the reflected light, its chromaticity can be determined by a trichromatic system if the mirror is spectrally flat. For illuminants like daylight, which have SPD's that are fully described by the three parameters of intensity and chromaticity, finding the chromaticity alone is tantamount to finding the *relative* SPD. Several different methods can be used to estimate the chromaticity of the illuminant from reflections.

Brightest Surface

In the early retinex scheme of Land and McCann, the brightest region in a Mondrian display fulfills the function of a mirror.[19,20] If the paper has a flat reflectance function one can obtain the chromaticity of the illuminant. Color constancy would therefore fail if there is no bright, white region in a scene.

Specular Reflection (Highlights)

Information from highlights can be used to estimate the chromaticity of the illuminant. Figure 3 shows that the lights from a uniformly colored object give rise to trichromatic responses that lie in the plane defined by axis I, which describes the specular component of reflectance, and the axis D, which describes the diffuse component. This plane is well defined only if both specular and diffuse components of reflectance contribute significant signals and vary across space in different ways. Such is generally the case for curved, glossy surfaces but not for flat, matte surfaces such as the papers in Mondrians.

If a scene contains two (or more) objects that are lit by the same illuminant and whose surfaces exhibit different variations in the two components of reflectance, then the chromaticity of the illuminant may be found by the means illustrated in Fig. 4. Because the specular component of reflection is common to both objects, their response planes intersect along the axis of the illuminant, as shown in Fig. 4A, and thus yield its chromaticity. This is easier to see in Fig. 4B, in which the response loci L and L' have been projected onto the plane CP. The intersection of these lines in the chromaticity plane defines the chromaticity of the illuminant.

There are two substantial reasons to doubt that the visual system depends on such a method. First, to find the loci of responses that correspond to different objects, one must already have segmented the scene to establish which lights come from which objects. This begs the question of the purpose of color vision, which we believe plays an important role in the discrimination among objects and in their identification. Second, there is evidence that, at least under some circumstances, the visual system provides some degree of color constancy for scenes in which surfaces have constant saturations, e.g., in Mondrian displays in which the specular component of reflectance does not vary across an element.[20]

Space Average

If the average light reflected from the field of view has the chromaticity of the illuminant, then any mechanism that can sample the range of lights in a scene has access to the illuminant. In a uniformly illuminated scene, the space-averaged light bears the chromaticity of the illuminant if the space-averaged surface reflectance is flat.[8,21–23]

Does the average light in natural scenes bear the chromaticity of the illuminant? We know of no measurements of the space-averaged chromaticity in scenes illuminated by daylight. Although there are scenes in which the chromaticity of the average light departs significantly from that of the

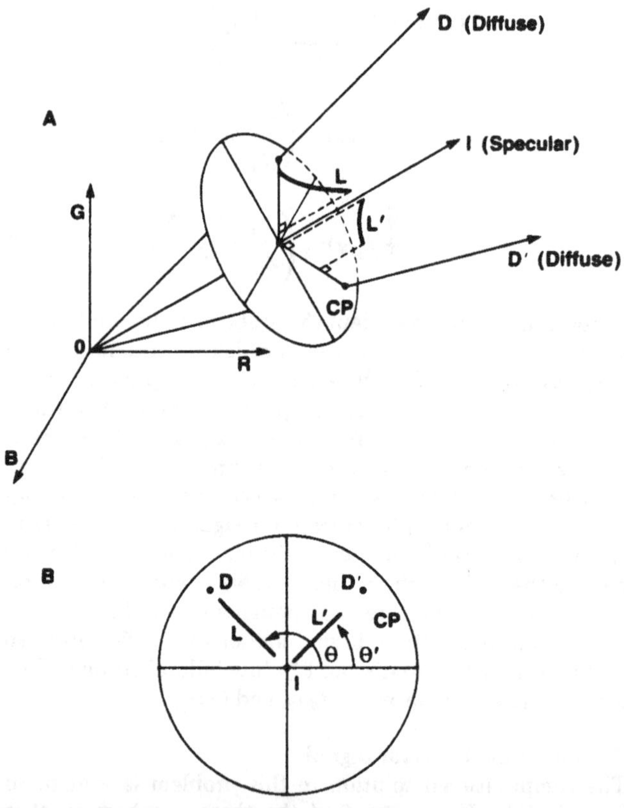

Fig. 4. A, Representation in the (R, G, B) space of responses L and L' to two objects, lit by the same illuminant I, that have different diffuse components of reflectance (D and D'). The two planes described by L and L' intersect along the axis I, which describes the chromaticity of the illuminant, because the specular component of reflectance is common to both objects. The responses from two or more objects that define distinct planes can thus be used to find the axis I that describes the chromaticity of the illuminant. B, Projection of L and L' onto the chromaticity plane CP. The lines described by the responses intersect at the point I marking the chromaticity of the illuminant.

illuminant, we expect that the space-averaged light from most natural scenes will bear a chromaticity that closely approximates that of the illuminant.

The more interesting question, to which we now turn, is whether the human visual system relies on the average chromaticity's being that of the illuminant.

MECHANISMS FOR DISCOUNTING THE ILLUMINANT

Visual signals from a large number of distinct surfaces must be made available to mechanisms that recover the chromaticity of the illuminant.[8,9,24–26] This does not mean that these mechanisms must be spatially extensive; the important requirement is that any point on the retina be able to sample the signals from a large part of the visual field. Mechanisms that are spatially localized have access to the space-averaged light from a scene through eye movements. Retinal mechanisms of adaptation can discount the spectral effects of the illuminant, provided that the space-averaged light, sampled through eye movements, has the chromaticity of the illuminant. The analysis proceeds by showing how the adjustments of retinal sensitivity made by known mechanisms of adaptation can bring about a transformation of the kind required by formulas (9)–(11).

Scaled Cone Signals

There is ample evidence from psychophysical experiments,[27,28] now bolstered by physiological observations,[29] that a multiplicative process regulates sensitivity independently in the three cone mechanisms, generally according to Weber's law.

Figure 5 shows the effects of this earliest stage of adaptation on the signaling of "light–dark" variations in surface reflectance for the reduced, two-dimensional case of R- and G-cone signals; the result is analogous for the three-dimensional case. On the left-hand side are shown the quantum catches that arise in R- and G-cones from a scene lit by either illuminant I (top) or illuminant I' (bottom). Variations in the intensity of the illuminants describe the axes I and I'. These axes also describe variations in the weight on $S_1(\lambda)$ because $S_1(\lambda)$ is spectrally flat. The average quantum catches (W_R W_G) and (W'_R W'_G) arise from the product of the illuminant (I or I') and some amount a of the reflectance $S_1(\lambda)$. Scaling quantum catches in R and G cones each by their average produces the cone signals r and g; although the two axes I and I' that describe the variations f_L in lightness differ in the space of quantum catches (left), they coincide in the space of scaled cone signals (right).

Recovering Relative Reflectance

Scaling cone signals by their averages makes it hard to recover the absolute values of lightness f_L. Figure 5 shows how the average value of reflectance a, which can vary from scene to scene, always comes to lie in a fixed position in the space of scaled cone signals. There is no way to recover completely the description of surface reflectance **p** from scaled cone signals, unless the intensity of the illuminant or some absolute value of reflectance becomes known at some later stage in the visual system. The best that one can do without such information is to find a description of relative reflectance in which **p** is scaled by some unknown number t. This limitation affects any color-constancy scheme that uses reflections to find and discount the chromaticity of the illuminant.

Scaling the three terms f_L, $f_d n_2$, and $f_d n_3$ by some number t does not, however, affect the computation of saturation or hue. The number t disappears in the ratios that are formed to measure saturation and hue.

Recovering Color

Figure 6 shows how light adaptation in individual cone pathways affects the representation of variations in the contributions of the "red–green" and the "yellow–blue" basis functions to reflectance.

Consider again a scene that is lit by either illuminant I or illuminant I'. As shown on the left-hand side of Fig. 6, variations in the contributions of "red–green" and "yellow–blue" basis functions to reflectance describe chromatic axes in the (R, G, B) space of quantum catches. For I these axes are given by varying the contributions of the lights IS_2 and IS_3 to the total light, while for I' the axes are determined by $I'S_2$ and $I'S_3$. The problem is that scaling cone signals, which can make the axes of the two illuminants coincide, cannot make the chromatic axes coincide, as shown on the right-hand side of Fig. 6. The reason for this lies in the relationship between surface reflectance and quantum catches described in formula (9a). Because any reasonable basis functions for illuminants and reflectances are spectrally broadband, the integrals that describe the link $G(\mathbf{m})$ between reflectance and quantum catches are typically nonzero. The three numbers used in scaling cone signals cannot undo what it takes nine numbers to describe.

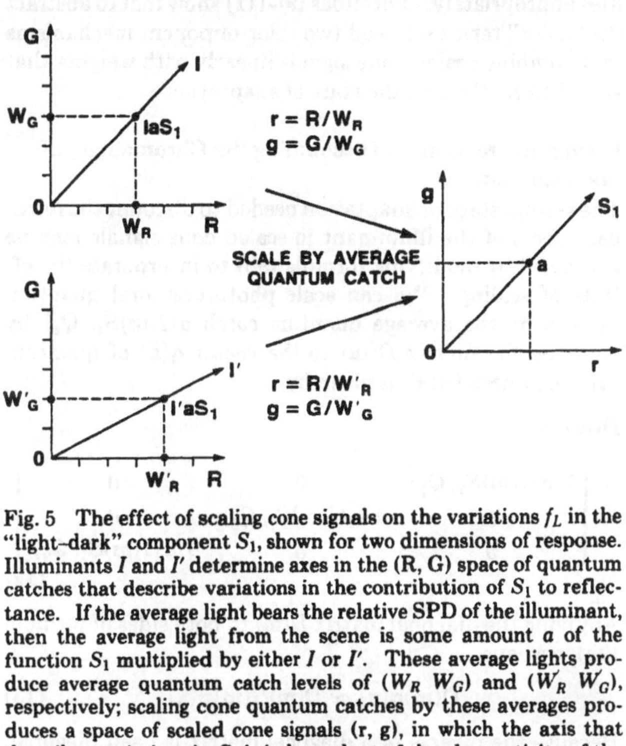

Fig. 5 The effect of scaling cone signals on the variations f_L in the "light–dark" component S_1, shown for two dimensions of response. Illuminants I and I' determine axes in the (R, G) space of quantum catches that describe variations in the contribution of S_1 to reflectance. If the average light bears the relative SPD of the illuminant, then the average light from the scene is some amount a of the function S_1 multiplied by either I or I'. These average lights produce average quantum catch levels of (W_R W_G) and (W'_R W'_G), respectively; scaling cone quantum catches by these averages produces a space of scaled cone signals (r, g), in which the axis that describes variations in S_1 is independent of the chromaticity of the illuminant. The average reflectance value a comes to lie in a fixed position in (r, g) space: without knowledge of a only relative values of reflectance may be recovered from scaled cone signals.

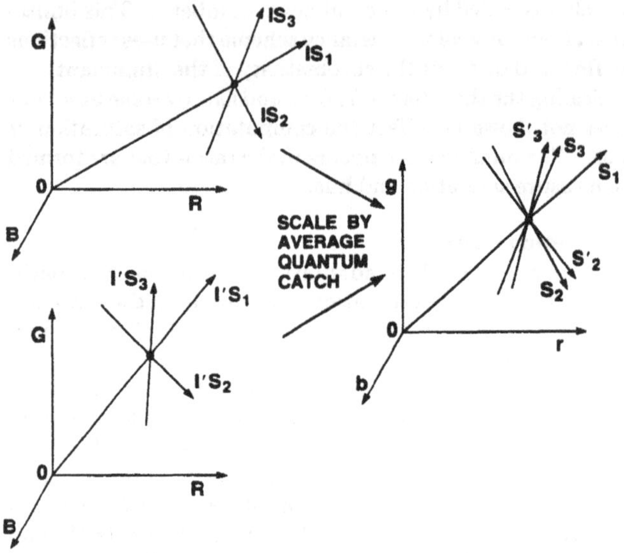

Fig. 6. Inability of simple scaling of cone signals to produce color constancy. Variations in the contributions of S_1, S_2, and S_3 to surface reflectance in a scene lit by illuminants I or I' are shown in the (R, G, B) space of quantum catches (left) and in the (r, g, b) space of scaled cone signals (right). Although scaling by average quantum catches can make the axes IS_1 and $I'S_1$ coincide along axis S_1 in (r, g, b) space, scaling will not make coincident the axes that describe the contributions of S_2 and S_3 to reflectance.

Color-opponent mechanisms that are set to abstract variations in $S_2(\lambda)$ and $S_3(\lambda)$ from scaled cone signals under one illuminant will fail if the illuminant is changed unless there is a second stage of adaptation that can adjust their sensitivities appropriately. Formulas (9)–(11) show that to abstract the "color" terms we need two color-opponent mechanisms that combine scaled cone signals linearly with weights that vary differently with the state of adaptation.

Formal Expression for Discounting the Chromaticity of the Illuminant

The second stage of adaptation needed to discount the residual effects of the illuminant in scaled cone signals may be expressed by modifying formula (9b) to incorporate the effects of scaling. We can scale photoreceptoral quantum catches by the average quantum catch $a\langle I(\mathbf{m})S_1, Q_k\rangle$ by applying the matrix $D(\mathbf{m})$ to the vector $\mathbf{q}(\mathbf{x})$ of quantum catches, where $D(\mathbf{m})$ is given by

$$D(\mathbf{m}) \simeq$$

$$\begin{bmatrix} 1/a\langle I(\mathbf{m})S_1, Q_1\rangle & 0 & 0 \\ 0 & 1/a\langle I(\mathbf{m})S_1, Q_2\rangle & 0 \\ 0 & 0 & 1/a\langle I(\mathbf{m})S_1, Q_3\rangle \end{bmatrix} \tag{12}$$

Applying the diagonal matrix $D(\mathbf{m})$ to both sides of formula (9b), we have

$$D(\mathbf{m})\mathbf{q}(\mathbf{x}) \simeq D(\mathbf{m})G(\mathbf{m})\mathbf{p}(\mathbf{x}). \tag{13}$$

Because the inverse of a diagonal matrix is itself diagonal, the matrix $D(\mathbf{m})$ is not the inverse of the matrix $G(\mathbf{m})$, and so the multiplicative scaling of cone signals (left) cannot completely discount the illuminant.

Formula (13) indicates that to recover the description of reflectance $\mathbf{p}(\mathbf{x})$, the inverse $E(\mathbf{m})$ of the product $D(\mathbf{m})G(\mathbf{m})$ must be applied to scaled cone signals: if

$$E(\mathbf{m}) = [D(\mathbf{m})G(\mathbf{m})]^{-1}, \tag{14}$$

then

$$E(\mathbf{m})D(\mathbf{m})\mathbf{q}(\mathbf{x}) \simeq \mathbf{p}(\mathbf{x}). \tag{15}$$

Because the average value of reflectance a is assumed to be unknown, we take the second stage $E(\mathbf{m})$ to be implemented by an otherwise identical matrix $E'(\mathbf{m})$ with terms that have been scaled by some constant. The two stages $D(\mathbf{m})$ and $E'(\mathbf{m})$ of adaptation thus recover a relative description of reflectance $t\mathbf{p}(\mathbf{x})$.

Second Sites of Adaptation

Two major lines of psychophysical evidence point to the existence of second-stage adaptive mechanisms that act after the multiplicative scaling in the individual cone pathways. The first of these come from experiments on what Wyszecki and Stiles call "asymmetric matching"—matching the appearance of lights viewed under one state of chromatic adaptation to lights viewed under a different state of chromatic adaptation.[11] Color matches made across different states of adaptation cannot be explained by the hypothesis (due to Von Kries[30]) that the sensitivity of the visual system is regulated solely by multiplicative processes in the individual cone mechanisms.[11,31]

The second line of evidence comes from experiments in which Stiles's two-color increment threshold technique has been used to study the spectral sensitivities of the mechanisms that regulate sensitivity to background lights.[28,32–37] These experiments have revealed adaptive mechanisms that appear to be driven by opposed signals from different classes of cone.[32–37]

Evidence from electroretinography points to the existence of second sites of adaptation within the retina, distal to the ganglion cells.[38] Single-unit recordings from the macaque's retina and lateral geniculate nucleus (LGN) reveal two classes of chromatically opponent neurons, whose preferred chromaticities fall broadly along either a "red–green" axis or a "blue–yellow" axis.[39–42] It is therefore natural to suppose that these mechanisms convey the terms $f_d n_2$ and $f_d n_3$, from which stable hue descriptors may be derived.

Adaptation to the Space-Averaged Light

Mechanisms of light adaptation (both in the separate cone pathways and at color-opponent sites) can discount the illuminant only if they adjust sensitivity over a period long enough to ensure that the space-averaged light, accumulated through the normal pattern of eye movements used in viewing a scene, regulates the adjustment.

The time course of light adaptation has been assessed in several experiments, typically by measuring the threshold for detection of a brief flash at various times after a step change in the illuminance of a steady background light. Although a considerable change in sensitivity occurs during the first 200 msec following the change in background,[43,44] a stable state may be reached much more slowly. This is particularly evident in observations that appear to tap the adaptation of second-stage mechanisms. The sensitivity of the Π_1 mechanism, which is believed to be regulated at an opponent site,[34] changes slowly following either an increment[33] or a decrement[32] in background illumination; indeed, some observations[35] suggest that the adaptive state of the second site may stabilize only after several hundred seconds.

Eye, head, and body movements serve to expose small regions of retina to many parts of a scene over a short period: typical fixations in the viewing of a complex scene last from 0.25 to 0.50 sec.[45] Active scanning of a scene may thus ensure that the adaptive state at any point on the retina reflects the space-averaged quantal absorption rate. In this steady state of adaptation, a stimulus falling on a particular retinal region will evoke a response that depends on the degree to which the chromaticity and intensity of the stimulus differ from those to which that region has become adapted.

CENTRAL REPRESENTATION OF THE HUE DESCRIPTOR

We have shown how two stages of adaptation, coupled with eye movements, can discount the illuminant. The output of second-stage chromatic mechanisms in a steady state of adaptation can provide a scaled representation of surface reflectance. We now consider how signals from retinal ganglion cells might be combined by central mechanisms to provide a representation of hue.

Spatiochromatic Opponency

Physiological recordings from single neurons in the primate's retina and LGN show that cells of the slow conducting (parvocellular) type have chromatically opponent receptive fields and are excited by lights that have a limited range of chromaticities.[39–42,46,47] When in a steady state of light adaptation, these neurons combine linearly the signals that arise in the different classes of cone.[46,47]

This linear behavior allows us to represent the chromatic sensitivity of a cell in a three-dimensional chromaticity space by a vector **G** drawn from the "white point" W that represents the steady state of adaptation; see Fig. 7. This vector **G** represents the best stimulus to use in stimulating the cell and is the cell's response gradient.[48] The preferred hue θ_G for a cell is found by projecting its gradient vector onto the chromaticity plane.

Ganglion cells and neurons in the LGN lack an attribute that might be thought important for mechanisms that provide hue constancy. Because the chromatically opponent mechanisms are separated in the center and surround of the receptive field, the responses of these neurons depend on both the spatial and chromatic properties of the stimulus.[47,49] As spatial frequency is raised, the cells respond better to achromatic contrast, so the gradient vectors that represent the chromatic signatures of such cells rotate away from the chromaticity plane. The responses of such units therefore cannot be used directly to estimate hue because they confound variations in lightness with variations in chromaticity in a space-dependent manner. To obtain geometrically stable hue descriptors we need mechanisms that have gradients with fixed directions in the chromaticity plane.

Labeled Lines for Hue

The hue descriptor is represented by the ratio n_3/n_2, which may be thought of as the ratio of signals in the "red–green" and "blue–yellow" opponent pathways. We think it unlikely that any mechanism signals hue by taking this ratio directly because ratios are not well behaved. A more plausible scheme is that hue is represented not by the size of the response of a neuron but by the identification of which neuron responds best. This idea of "labeled lines" leads us to think of a population of neurons tuned to different azimuths; the hue of the stimulus is established by finding the most active neuron.

Figure 8 shows how the linear combination of signals from spatiochromatically opponent units can give rise to chromatic mechanisms with gradient vectors that retain their directions through changes in the spatial properties of the stimulus. The strength of the gradient representation is that gradients are additive, and so a cell that linearly combines the signals from two or more cells will have a gradient that is the same linear combination of these cells' gradients. One eliminates responses to achromatic stimuli by summing responses of appropriate pairs of color-opponent cells that have identical locations and receptive-field space constants. Opposing tendencies to respond to achromatic stimuli are canceled through such addition.

The result of such selective combination is cells that at all spatial frequencies respond best to purely chromatic modulations in a steady state of adaptation. Linear combinations of these units in turn result in mechanisms that respond maximally to any desired hue. Evidence from unpublished physiological experiments by Lennie, Sclar, and Krauskopf suggests that in the striate cortex of the macaque monkey

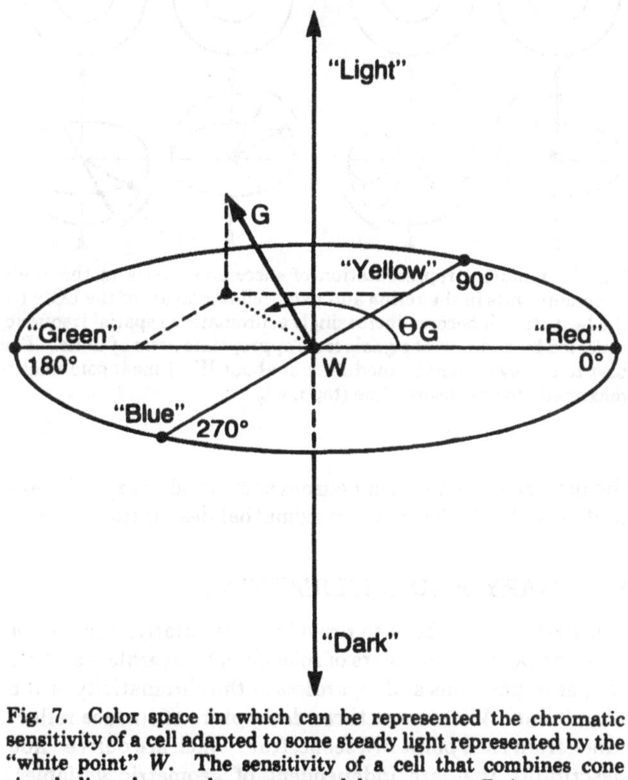

Fig. 7. Color space in which can be represented the chromatic sensitivity of a cell adapted to some steady light represented by the "white point" W. The sensitivity of a cell that combines cone signals linearly may be represented by a vector **G**, its response gradient. The vector **G** represents the most effective choice of chromatic contrasts to use in stimulating the cell. The preferred hue θ_G of the cell may be found by projecting **G** onto the chromaticity plane.

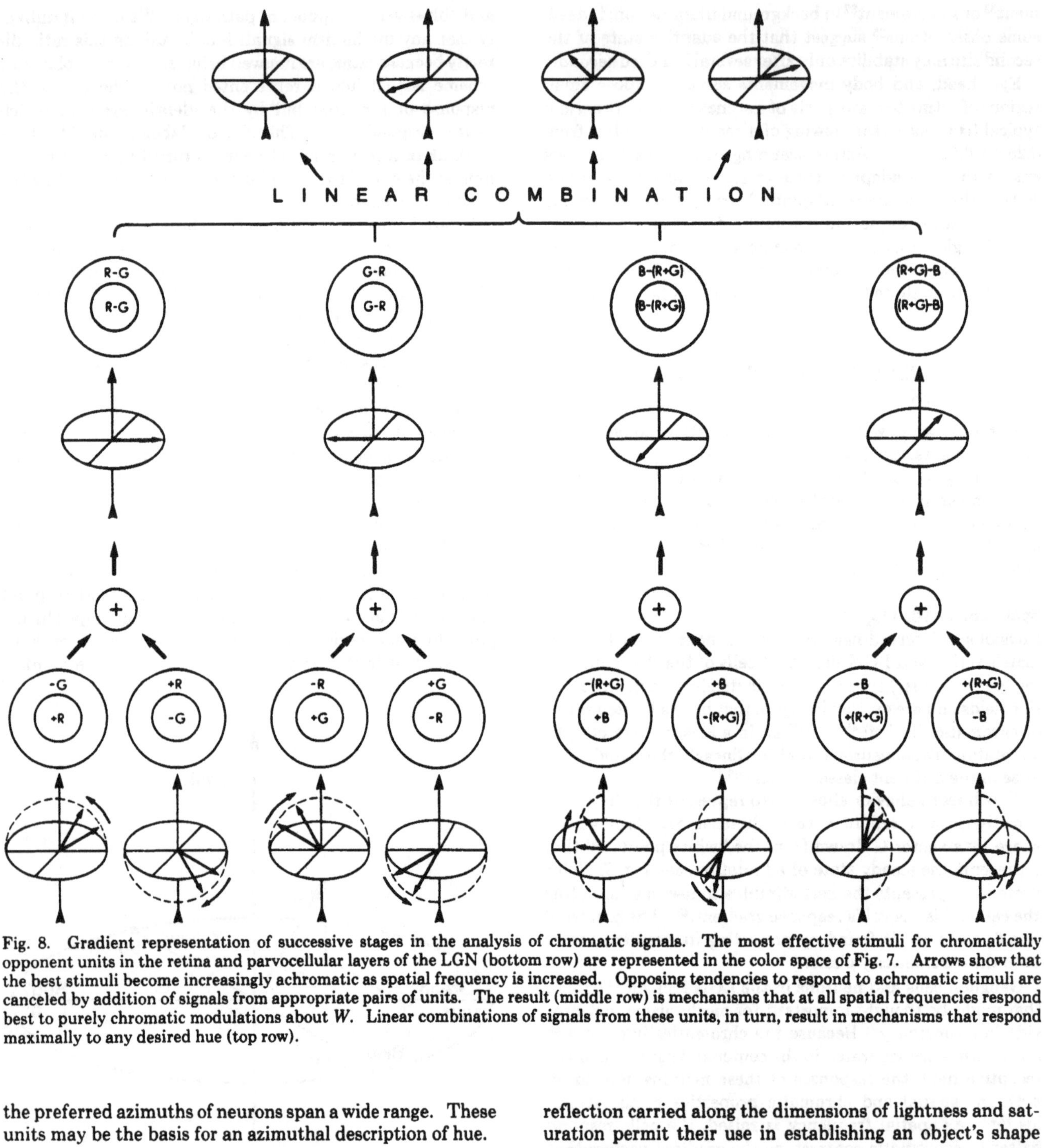

Fig. 8. Gradient representation of successive stages in the analysis of chromatic signals. The most effective stimuli for chromatically opponent units in the retina and parvocellular layers of the LGN (bottom row) are represented in the color space of Fig. 7. Arrows show that the best stimuli become increasingly achromatic as spatial frequency is increased. Opposing tendencies to respond to achromatic stimuli are canceled by addition of signals from appropriate pairs of units. The result (middle row) is mechanisms that at all spatial frequencies respond best to purely chromatic modulations about W. Linear combinations of signals from these units, in turn, result in mechanisms that respond maximally to any desired hue (top row).

the preferred azimuths of neurons span a wide range. These units may be the basis for an azimuthal description of hue.

SUMMARY AND IMPLICATIONS

Our first goal has been to provide a quantitative account of how one can find the colors of solid objects regardless of their shapes or positions and regardless of the chromaticity of the illuminant. We recover three descriptors of surface reflectance from each point in a scene, from which are derived hue descriptors that are independent of geometric variables. Our scheme does not require that the scene be segmented before chromatic descriptors are extracted; indeed, the geometric stability of hue permits its use in segmenting a scene, and the variations in specular and diffuse components of reflection carried along the dimensions of lightness and saturation permit their use in establishing an object's shape and position.

Our formal analysis assumes that the SPD of the illuminant is known, but in any physiologically plausible account of color constancy this knowledge is unlikely to be represented explicitly. Our second goal has therefore been to provide an account of how known mechanisms of the human visual system might discount the illuminant. We have shown that if the space-averaged light from a scene bears the relative SPD of the illuminant, then two stages of light adaptation (a multiplicative change in sensitivity in the independent cone mechanisms followed by an adaptive linear transformation of scaled cone signals at color-opponent sites), when coupled with eye movements that expose them to the average light, yield signals in second-stage mechanisms from which hue descriptors can easily be derived.

The Unknown Illuminant

Second-stage mechanisms that are in a steady state of adaptation provide good estimates of surface reflectance only if the average chromaticity of a scene is the chromaticity of the illuminant. Our scheme suggests that the more a scene departs from this condition, the poorer the estimates of surface reflectance provided by the retina will be.

Buchsbaum[8] suggests that the visual system uses another means to find the illuminant: the space-averaged surface reflectance is known and is not necessarily flat. Buchsbaum's method has difficulty handling a collection of surfaces with unknown properties but may be a useful means of establishing the SPD of the illuminant when the visual environment does not meet the conditions needed for the postulated mechanisms of adaptation to provide color constancy (for example, a room whose surfaces have a strongly predominant color). In this case, knowledge of the surface properties of objects might be used to correct biases in the signals generated by second-stage mechanisms.

Another alternative to the assumption that the average chromaticity is that of the illuminant is to assume, as have Maloney and Wandell,[9] that surface reflectances encountered by a trichromatic visual system are described adequately by only two parameters. In this case no further assumption about surface reflectances need be made, but the penalty is less robust constancy of color appearance.

Physiological Mechanisms

Our account leaves unresolved several questions about the estimates of surface reflectance made by chromatically opponent neurons. We do not know that ganglion cells incorporate (implicitly) the basis functions required to render adequate descriptions of natural surfaces. Furthermore, adaptation at color-opponent sites is required to adjust the sensitivity of a neuron so that it signals a constant reflectance descriptor; the range of sensitivities required suggests that opponent mechanisms must receive inputs from all three classes of cone. Although some chromatically opponent ganglion cells and neurons in the LGN receive inputs from all three classes of cone, when adapted to white light most show no discernible input from B cones.[46,47] However a change in the chromaticity to which a cell is adapted might reveal otherwise hidden inputs from B cones (a possibility for which DeMonasterio *et al.*[42] provide some evidence).

Alternatives to the postulated second stage of adaptation within the retina are that adaptation may occur in cortex, where we know that many color-opponent neurons receive inputs from all three classes of cone, or that the adjustments to scaled cone signals made by our second stage are not actually made by the visual system. At the moment too little is known about adaptation at second sites (especially adaptation brought about by modest chromatic excursions from white) to be sure that the mechanisms so far revealed have the properties that we require.

Physical Limitations

Our model of lights from surfaces assumes several conditions that need not be true of the real world. The illuminant must be spatiochromatically homogeneous, and its SPD must be adequately represented by weights on a particular set of three basis functions that characterize daylight. We therefore expect color constancy to be impaired if the illuminant is heterogeneous (e.g., when objects in a room with a window are illuminated by daylight from an overcast sky and also by a tungsten source) or if its SPD departs markedly from the range represented by the phases of daylight (e.g., when it is composed of one, two, or even three spectrally narrow-band sources). Moreover, we have taken scenes to be shadowless and have assumed negligible interreflectance among surfaces. We expect the colors of shadowed objects to be less constant than those of objects lit directly. Finally, color constancy should be poorer for surfaces whose reflectances cannot be described accurately by the three basis functions built into the visual system or whose reflectances cannot simply be divided into specular and diffuse components (e.g., fluorescent and iridescent surfaces).

ACKNOWLEDGMENTS

This work was supported by National Institutes of Health grants EY-04440 and EY-01319. M. D'Zmura held an IBM graduate fellowship.

REFERENCES

1. D. B. Judd, D. L. MacAdam, and G. Wyszecki, "Spectral distribution of typical daylight as a function of correlated color temperature," J. Opt. Soc. Am. **54**, 1031–1040 (1964).
2. E. R. Dixon, "Spectral distribution of Australian daylight," J. Opt. Soc. Am. **68**, 437–450 (1978).
3. J. Cohen, "Dependency of the spectral reflectance curves of the Munsell color chips," Psychon. Sci. **1**, 369–370 (1964).
4. L. T. Maloney, "Computational approaches to color constancy," Applied Psychology Lab. Tech. Rep. 1985-01 (Stanford University, Stanford, Calif., 1985).
5. P. Sällström, "Colour and physics: some remarks concerning the physical aspects of human colour vision," Institute of Physics Rep. 73-09 (University of Stockholm, Sweden, 1973).
6. M. H. Brill, "A device performing illuminant-invariant assessment of chromatic relations," J. Theor. Biol. **71**, 473–478 (1978).
7. M. H. Brill, "Further features of the illuminant-invariant trichromatic photosensor," J. Theor. Biol. **78**, 305–308 (1979).
8. G. Buchsbaum, "A spatial processor model for object colour perception," J. Franklin Inst. **310**, 1–26 (1980).
9. L. T. Maloney and B. A. Wandell, "Color constancy: a method for recovering surface spectral reflectance," J. Opt. Soc. Am. A **3**, 29–33 (1986).
10. H. Yilmaz, "Color vision and a new approach to general perception," in *Biological Prototypes and Synthetic Systems*, E. E. Bernard and M. R. Kare, eds. (Plenum, New York, 1962).
11. G. Wyszecki and W. S. Stiles, *Color Science. Concepts and Methods, Quantitative Data and Formulae* (Wiley, New York, 1982).
12. K. E. Torrance and E. M. Sparrow, "Theory of off-specular reflection from roughened surfaces," J. Opt. Soc. Am. **57**, 1105–1114 (1967).
13. T. S. Trowbridge and K. P. Reitz, "Average irregularity representation of a rough surface for ray reflection," J. Opt. Soc. Am. **65**, 531–536 (1975).
14. B. T. Phong, "Illumination for computer generated pictures," Commun. ACM **18**, 311–317 (1975).
15. J. N. Lythgoe, *The Ecology of Vision* (Clarendon, Oxford, 1979).
16. K. Nassau, *The Physics and Chemistry of Color; The Fifteen Causes of Color* (Wiley, New York, 1983).
17. S. A. Shafer, "Using color to separate reflection components," Department of Computer Sciences Tech. Rep. 136 (University of Rochester, New York, 1984).
18. V. C. Smith and J. Pokorny, "Spectral sensitivity of the foveal

cone photopigments between 400 and 500 nm," Vision Res. 15, 161–171 (1975).
19. E. H. Land and J. J. McCann, "Lightness and retinex theory," J. Opt. Soc. Am. 61, 1–11 (1971).
20. J. J. McCann, S. P. McKee, and T. H. Taylor, "Quantitative studies in retinex theory: a comparison between theoretical predictions and observer responses to the 'color mondrian' experiments," Vision Res. 16, 445–458 (1976).
21. H. Helson, "Fundamental problems in color vision. I. The principle governing changes in hue, saturation and lightness of non-selective samples in chromatic illumination," J. Exp. Psychol. 23, 439–476 (1938).
22. D. B. Judd, "Hue saturation and lightness of surface colors with chromatic illumination," J. Opt. Soc. Am. 30, 2–32 (1940).
23. H. Helson, D. B. Judd, and M. H. Warren, "Object-color changes from daylight to incandescent filament illumination," Illum. Eng. 47, 221–232 (1952).
24. E. H. Land, "The retinex theory of color vision," Sci. Am. 237, 108–128 (1977).
25. E. H. Land, "Recent advances in retinex theory and some implications for cortical computations: color vision and the natural image," Proc. Natl. Acad. Sci. USA 80, 5163–5169 (1983).
26. E. H. Land, "Recent advances in retinex theory," Vision Res. 26, 7–21 (1986).
27. W. S. Stiles, "Increment thresholds and the mechanisms of colour vision," Doc. Ophthalm. 3, 138–163 (1949).
28. W. S. Stiles, *Mechanisms of Colour Vision* (Academic, New York, 1978).
29. J. M. Valeton and D. van Norren, "Light adaptation of primate cones: an analysis based on extracellular data," Vision Res. 23, 1539–1547 (1983).
30. H. von Helmholtz, *Treatise on Physiological Optics*, J. P. C. Southall, ed. (Dover, New York, 1962), Vol. II.
31. D. Jameson and L. M. Hurvich, "Color adaptation: sensitivity, contrast, after-images," in *Handbook of Sensory Physiology*, D. Jameson and L. M. Hurvich, eds. (Springer-Verlag, New York, 1972), Vol. VII/4.
32. J. D. Mollon and P. G. Polden, "An anomaly in the response of the eye to light of short wavelengths," Phil. Trans. R. Soc. London Ser. B 278, 207–240 (1977).
33. E. J. Augenstein and E. N. Pugh, "The dynamics of the Π_1 colour mechanism: further evidence for two sites of adaptation," J. Physiol. 272, 247–281 (1977).
34. E. N. Pugh and J. D. Mollon, "A theory of the Π_1 and Π_3 colour mechanisms of Stiles," Vision Res. 19, 293–312 (1979).
35. J. M. Loomis, "Transient tritanopia: failure of time-intensity reciprocity in adaptation to long-wavelength light," Vision Res. 20, 837–846 (1980).
36. B. A. Wandell and E. N. Pugh, "A field-additive pathway detects brief-duration, long-wavelength incremental flashes," Vision Res. 20, 613–624 (1980).
37. B. A. Wandell and E. N. Pugh, "Detection of long-duration incremental flashes by a chromatically coded pathway," Vision Res. 20, 625–635 (1980).
38. J. M. Valeton and D. van Norren, "Transient tritanopia at the level of the erg *b*-wave," Vision Res. 19, 689–693 (1979).
39. R. L. DeValois, I. Abramov, and G. H. Jacobs, "Analysis of response patterns of LGN cells," J. Opt. Soc. Am. 56, 966–977 (1966).
40. T. N. Wiesel and D. H. Hubel, "Spatial and chromatic interactions in the lateral geniculate body of the rhesus monkey," J. Neurophysiol. 29, 1115–1156 (1966).
41. F. M. DeMonasterio and P. Gouras, "Functional properties of ganglion cells of the rhesus monkey retina," J. Physiol. 251, 167–195 (1975).
42. F. M. DeMonasterio, P. Gouras, and D. J. Tolhurst, "Concealed colour opponency in ganglion cells of the rhesus monkey retina," J. Physiol. 251, 217–229 (1975).
43. B. H. Crawford, "Visual adaptation in relation to brief conditioning stimuli," Proc. R. Soc. London Ser. B 134, 283–302 (1947).
44. M. M. Hayhoe, N. Benimoff, and D. C. Hood, "The time course of multiplicative and subtractive adaptation processes," Vision Res. (to be published).
45. A. L. Yarbus, *Eye Movements and Vision*, B. Haigh, translator, L. A. Riggs, ed. (Plenum, New York, 1967).
46. C. C. A. M. Gielen, J. A. M. Van Gisbergen, and A. J. H. Vendrik, "Reconstruction of responses of colour-opponent neurones in monkey lateral geniculate nucleus," Biol. Cybern. 44, 211–221 (1982).
47. A. M. Derrington, J. Krauskopf, and P. Lennie, "Chromatic mechanisms in lateral geniculate nucleus of macaque," J. Physiol. 357, 241–265 (1984).
48. B. O'Neill, *Elementary Differential Geometry* (Academic, New York, 1966).
49. C. R. Ingling, Jr. and E. Martinez-Uriegas, "The relationship between spectral sensitivity and spatial sensitivity for the primate r, g *x*-channel," Vision Res. 23, 1495–1500 (1983).

In our paper,[1] Eq. (11) should be corrected to read as

$$H(\mathbf{m})\mathbf{q}(\mathbf{m}) \simeq \mathbf{p}(\mathbf{x}). \tag{11}$$

REFERENCE

1. M. D'Zmura and P. Lennie, "Mechanisms of color constancy," J. Opt. Soc. Am. A 3, 1662–1672 (1986).

Component estimation of surface spectral reflectance

Shoji Tominaga

Faculty of Engineering, Osaka Electro-Communication University, Neyagawa, Osaka, 572 Japan

Brian A. Wandell

Department of Psychology, Stanford University, Stanford, California, 94305

Received February 6, 1989; accepted September 27, 1989

For inhomogeneous materials, the standard reflectance model suggests that under all viewing geometries surface reflectance functions can be described as the sum of a constant function of wavelength (specular) and a diffuse function that is characteristic of the material. As the viewing geometry varies, the relative contribution of these two terms varies. In a previous study [J. Opt. Soc. Am. A 6, 576 (1989)] we described how to use light reflected from inhomogeneous materials, measured in different viewing geometries, to estimate the relative spectral power distribution of the ambient light. Here we show that two restrictions, that (a) surface reflectance functions are all nonnegative and (b) surface reflectance functions are the positive weighted sum of subsurface (diffuse) and interface (specular) components, may be used to estimate the subsurface component of the surface reflectance function. A band of surface spectral reflectances is recovered, as possible solutions for the subsurface estimates.

1. INTRODUCTION

The surface reflectance function of an object varies with the viewing geometry. A simple model of surface reflectance is proposed for inhomogeneous materials (see, e.g., Refs. 1 and 2). Inhomogeneous materials are composed of different component materials, such as a vehicle at the surface layer and embedded pigments at the colorant layer. Plastics and paints are inhomogeneous. The model suggests that to a good approximation the reflectance function can be described as the weighted sum of two functions. One function represents the interface (specular) reflection, and the second function represents the subsurface (diffuse) reflection.

Let $S_I(\lambda)$ and $S_S(\lambda)$ be the surface spectral reflectances for the two components of interface and subsurface reflections, and let $E(\lambda)$ be the spectral power distribution of the incident light. Then the reflected light is

$$Y(\theta, \lambda) = c_I(\theta)S_I(\lambda)E(\lambda) + c_S(\theta)S_S(\lambda)E(\lambda), \tag{1}$$

where $c_I(\theta)$ and $c_S(\theta)$ are the geometric scale factors. The parameters θ vary with the viewing geometry such as the incident angle of light on the surface, the viewing angle, and the phase angle. In a previous paper[2] we evaluated this model and showed how to use light reflected from pairs of surfaces to estimate the spectral power distribution of the illuminant (see also Refs. 3 and 4). From a component analysis of the reflected light, we concluded that the model is correct and that the spectral composition of the specular component of the reflected light is the same as the spectral composition of the incident light. The estimation problem of the illuminant spectrum can then be reduced to finding the common spectral information from the measurements from two or more surfaces. We presented an algorithm to obtain an illuminant estimate without using a reference white standard.

The subsurface reflectance function $S_S(\lambda)$ is characteristic of the surface and therefore is an important clue to the surface's identity. The interface term $S_I(\lambda)$, which is present in all homogeneous materials, does not contain much information about the surface's identity, although it may be useful in determining the geometry of the objects in the image.[5]

The methods in our previous paper provided a unique estimate of the illuminant, but the method did not provide an estimate of the subsurface spectral reflectance term, $S_S(\lambda)$. The estimate of the subsurface reflectance function was restricted to be a function of the form $\alpha + \beta S(\lambda)$, where $S(\lambda)$ is derived from the measurements and α and β are free parameters.

In our previous study we did not take advantage of two important constraints on the surface reflectance functions. First, the subsurface reflectance function $S_S(\lambda)$ must be nonnegative. Second, for all viewing geometries the parameters c_I and c_S must both be nonnegative. These constraints were identified by Lawton and Sylvestre,[6] who also offered an elegant mathematical analysis of how the constraints might be applied. Our methods are also related to recent studies in multispectral image analysis by Kawata *et al.*[7] In Section 2 we state these constraints more precisely.

2. PHYSICAL CONSTRAINTS

The standard reflectance model assumes that the color signals $Y(\theta, \lambda)$ can be expressed as a linear combination of the two component vectors of light reflection,

$$Y(\theta, \lambda) = c_I(\theta)L_I(\lambda) + c_S(\theta)L_S(\lambda), \tag{2}$$

where $L_I(\lambda) = S_I(\lambda)E(\lambda)$ and $L_S(\lambda) = S_S(\lambda)E(\lambda)$. The two vectors $L_I(\lambda)$ and $L_S(\lambda)$ span a two-dimensional space of the possible observations. All the color signals observed from an inhomogeneous surface fall within this two-dimensional subspace.

Using our estimate of the ambient light, we can define the

two-dimensional space of surface reflectances by dividing the two-dimensional space of color signals by $E(\lambda)$:

$$\frac{Y(\theta, \lambda)}{E(\lambda)} = S(\theta, \lambda) = c_I(\theta)S_I(\lambda) + c_S(\theta)S_S(\lambda). \quad (3)$$

Note that this surface reflectance plane defined by $S_I(\lambda)$ and $S_S(\lambda)$ does not change with different illuminants. The surface reflectance plane is inherent in the object surface, unlike the color signal plane defined by $L_I(\lambda)$ and $L_S(\lambda)$. By hypothesis, the interface component $S_I(\lambda)$ of the two dimensions is a constant specular function. The subsurface function $S_S(\lambda)$ must (a) be all nonnegative over the visible wavelength as $S_S(\lambda) \geq 0$ and (b) contribute only nonnegative weights for all of the measured color signals as $c_S(\theta) \geq 0$.

Lawton and Sylvestre treated the general problem of determining the shapes of two functions with such physical constraints from an observed set of additive mixtures of the two functions. Their solution method was based on principal component analysis. Their method produced two bands of estimated functions. In our application, one function, the interface function $S_I(\lambda)$, is fixed. Thus our analysis is a special case of their general two-dimensional analysis. It should be noted that our problem is the estimation of a band of the subsurface reflectance functions rather than the estimation of a unique $S_S(\lambda)$; that is, we can recover only a range of subsurface estimates, not a unique estimate. In Section 3 we introduce the special case of the Lawton–Sylvestre analysis that we call the quarter-circle analysis.

3. QUARTER-CIRCLE ANALYSIS

A. Observation of Surface Reflectances

First, assume that we measure m color signals from an object surface under various geometric conditions $\theta_1, \theta_2, \ldots, \theta_m$. All the measurements are made with a single fixed illuminant. Each color signal that we measure, $Y(\theta_i, \lambda)(i = 1, 2, \ldots, m)$, is sampled at n points $\lambda_1, \lambda_2, \ldots, \lambda_n$ in the visible wavelength region. Next let $\hat{E}(\lambda_j)(j = 1, 2, \ldots, n)$ be an estimate of the spectral power distribution of the illuminant. Then the observations of m surface spectral reflectances are obtained by $S(\theta_i, \lambda_j) = Y(\theta_i, \lambda_j)/\hat{E}(\lambda_j)$. We represent these reflectances by n-dimensional column vectors $\mathbf{s}_i$ ($i = 1, 2, \ldots, m$). These vectors are normalized so that $\|\mathbf{s}_i\|^2 = 1$, where the notation $\|\mathbf{s}\|^2$ is defined as $\mathbf{s}^T\mathbf{s}$.

Since the surface spectral reflectances are two dimensional, the normalized reflectances $\mathbf{s}_i$ can be represented in terms of two orthonormal basis vectors as

$$\mathbf{s}_i = p_{i1}\mathbf{u}_1 + p_{i2}\mathbf{u}_2 \qquad (i = 1, 2, \ldots, m). \quad (4)$$

This orthogonal expansion can be determined by using the singular value decomposition of an $n \times m$ matrix of the observed reflectances (see, e.g., Ref. 2). The vectors $\mathbf{u}_1$ and $\mathbf{u}_2$ are the first two of the n-dimensional left singular vectors. The coefficients p_{i1} and p_{i2} are specified from the singular values σ_j and the elements of the m-dimensional right singular vectors $\mathbf{v}_j$ as $p_{ij} = \sigma_j v_{ij}$ ($j = 1, 2$). The first basis vector $\mathbf{u}_1$ has all nonnegative elements, so that $\mathbf{u}_1 \geq 0$ ($u_{j1} \geq 0, j = 1, 2, \ldots, n$) and $p_{i1} \geq 0$. However, because of the orthogonality condition imposed on eigenvectors, the second $\mathbf{u}_2$ has some negative elements. Because $\mathbf{u}_1$ is orthogonal to $\mathbf{u}_2$, it follows that

$$\begin{aligned} 1 = \|\mathbf{s}_i\|^2 &= \|p_{i1}\mathbf{u}_1 + p_{i2}\mathbf{u}_2\|^2 \\ &= \|p_{i1}\mathbf{u}_1\|^2 + \|p_{i2}\mathbf{u}_2\|^2 \\ &= p_{i1}^2 + p_{i2}^2. \end{aligned} \quad (5)$$

Thus the representation of the normalized surface reflectance measurements will have unit length in the principal component coordinate frame ($\mathbf{u}_1$, $\mathbf{u}_2$).

B. Basis Transformations

The unit-length basis vectors $\mathbf{u}_1$ and $\mathbf{u}_2$ are computed from the normalized observations of surface reflectance functions. These vectors depend on experimental factors such as the choice of measurement angles θ and the number of measurements at each of the measurement angles. These vectors do not characterize the surface properties. Our goal is to find a method of transforming the vectors $\mathbf{u}_1$ and $\mathbf{u}_2$ to a new representation that estimates a range of the subsurface reflectance vector $\mathbf{S}_S$ for $S_S(\lambda)$. This vector $\mathbf{S}_S$ characterizes the object surface and is independent of the experimental parameters. The transformation that we apply to the vectors $\mathbf{u}_1$ and $\mathbf{u}_2$ to estimate $\mathbf{S}_S$ is based on three premises. First, according to theory (see, e.g., Ref. 1) and confirmation by our experimental measurements,[2] the unit vector $\mathbf{S}_I$ in the direction of interface reflectance, $[S_I(\lambda) = 1/\sqrt{n}$ $(\lambda = \lambda_1, \lambda_2, \ldots, \lambda_n)$, $\sum_{i=1}^{n} S_I^2(\lambda_i) = 1]$, is within a linear transformation of the vectors $\mathbf{u}_1$ and $\mathbf{u}_2$. Second, the subsurface reflectance $S_S(\lambda)$ is nonnegative. Third, the surface reflectances measured under different angles θ are positive mixtures of the subsurface and the interface reflectances; that is, all measurements are of the form $c_I(\theta)S_I(\lambda) + c_S(\theta)S_S(\lambda)$, where $c_I(\theta), c_S(\theta) \geq 0$. On the basis of these three premises, we can estimate a band functions that must include the subsurface function $S_S(\lambda)$.

For conceptual simplicity, we begin by transforming the data representation from the principal component coordinates (p_1, p_2) with respect to the basis functions $\mathbf{u}_1$ and $\mathbf{u}_2$ into a new coordinate frame with respect to the coordinate system defined by the vector S_I and a unit-length vector perpendicular to $\mathbf{S}_I$ that we call $\mathbf{S}_I{}^{\perp}$. From our experimental measurements we know that $\mathbf{S}_I$ can be obtained by a linear combination of $\mathbf{u}_1$ and $\mathbf{u}_2$. In other words, $\mathbf{S}_I$ is within the surface reflectance plane spanned by ($\mathbf{u}_1$, $\mathbf{u}_2$). Then we can find the unique unitary transformation $\mathbf{T}$ such that

$$[\mathbf{S}_I, \mathbf{S}_I{}^{\perp}] = [\mathbf{u}_1, \mathbf{u}_2]\mathbf{T}. \quad (6)$$

The unitary transformations $\mathbf{T}$ rotates $\mathbf{u}_1$ into $\mathbf{S}_I$. The second vector $\mathbf{S}_I{}^{\perp}$ is the unique unit-length vector that is orthogonal to $\mathbf{S}_I$ within the surface reflectance plane.

In practice, because of measurement error and small failures of the model, the interface reflectance component $\mathbf{S}_I$ cannot be assumed to be precisely in the plane of the first two principal components ($\mathbf{u}_1$, $\mathbf{u}_2$) of pixel data. Moreover the same situation may occur when more than three objects are used to infer the illuminant spectrum and the interface reflectance $\mathbf{S}_I$. The interface component is then obtained as the projection of a unit vector onto the ($\mathbf{u}_1$, $\mathbf{u}_2$). The unitary matrix $\mathbf{T}$ can be found by solving the least-squares problem that minimized $\|\mathbf{i} - t_1\mathbf{u}_1 - t_2\mathbf{u}_2\|^2$, where $\mathbf{i}$ is a unit vector with entries of $1/\sqrt{n}$. By convention we choose $\mathbf{u}_2$ so that t_2 is positive. (If the original choice of $\mathbf{u}_2$ leads to a negative solution, then simply replace $\mathbf{u}_2$ with $-\mathbf{u}_2$, since this sign is

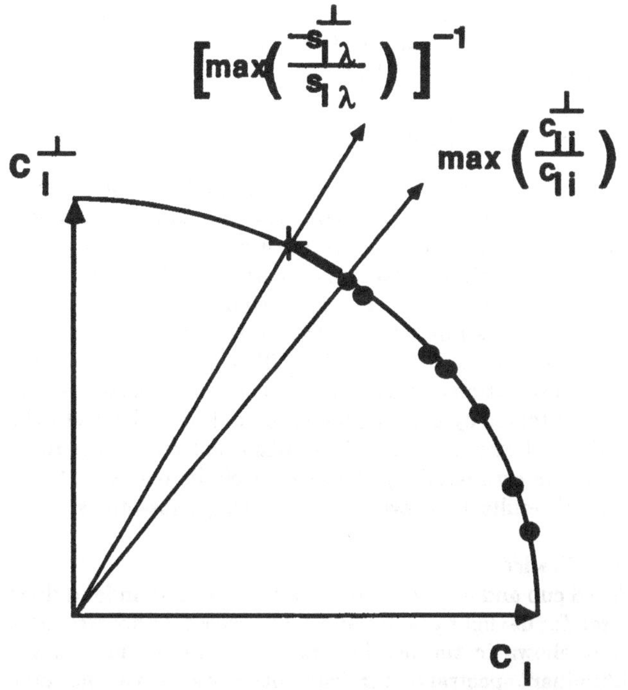

Fig. 1. Pictorial diagram of the $(c_I, c_I^{\perp})$ coordinate system in which the observed surface spectral reflectances are projected onto the quarter of the unit circle within the positive quadrant. Each dot indicates the observation point. A + on the quarter circle indicates the upper limit of the permissible directions for describing the surface reflectance vectors. The possible solution band is shown by the bold arc on the circle.

arbitrary in the construction of the principal components.) The matrix **T** is

$$\mathbf{T} = \frac{1}{(t_1^2 + t_2^2)^{1/2}} \begin{bmatrix} t_1 & t_2 \\ t_2 & -t_1 \end{bmatrix}. \tag{7}$$

An observation that is represented by (p_{i1}, p_{i2}) in the principal component coordinate frame will be represented by $[c_{Ii}, c_{Ii}^{\perp}] = [p_{i1}, p_{i2}]\mathbf{T}$ in the transformed coordinate frame.

There are several simplifying features of this coordinate frame that are illustrated in Fig. 1. First, because the surface reflectances observed have been normalized to unit length, and **T** is a unitary transformation, the coordinates $(c_I, c_I^{\perp})$ satisfy $c_{Ii}^2 + c_{Ii}^{\perp 2} = 1$. Thus the representation of the measurements falls on the unit circle. Next, the value of c_I must exceed 0 because this component describes the contribution from the interface reflectance, which is always nonnegative. Furthermore, the value of $c_I^{\perp}$ is proportional (but not equal) to the subsurface contribution, c_S, which is always positive as well. By our choice of transformation **T**, we have assured ourselves that the constant of proportionality is positive. Thus the observations $(c_{Ii}, c_{Ii}^{\perp})$ must lie on the quarter of the unit circle within the positive quadrant.

The point (1, 0) is consistent with a perfectly specular interface reflection. The point (0, 1) cannot be observed because $\mathbf{S}_I^{\perp}$ has negative elements.

C. Solution Band

The observation of the surface reflectances is totally ordered along the quarter circle. As the contribution of the subsurface component increases, the observation point moves counterclockwise on the circle. The point for the pure subsurface reflectance function lies somewhere on the quarter circle. The only issue that remains, then, is to determine the direction of the vector $\mathbf{S}_S$ representing the subsurface reflectance $S_S(\lambda)$.

Lawton and Sylvestre[6] made the following observations. First, the requirement that both c_S and c_I be positive implies that the observations must fall within a cone defined by the direction of $\mathbf{S}_S$ and the horizontal axis c_I. To understand this point, notice that, if a data point falls counterclockwise with respect to the vector $\mathbf{S}_S$, then the data point would be represented with a negative value of c_I; this is impermissible. It follows that directions that qualify as candidates for $\mathbf{S}_S$ must fall counterclockwise with respect to all the data points. The data point closest to the $c_I^{\perp}$ axis (y axis) defines this limit.

Second, the vector $\mathbf{S}_S$ representing the subsurface reflectance $S_S(\lambda)$ must have all nonnegative elements. Not all vector directions satisfy this requirement. For example, the vector defined by the $c_I^{\perp}$ axis contains negative elements. As we consider unit vectors in the clockwise direction from the $c_I^{\perp}$ axis, there will be a first direction in which all the entries of the reflectance vectors are nonnegative. This direction is a second bound on the region of candidate directions for $\mathbf{S}_S$. The true vector direction representing $\mathbf{S}_S$ must fall between this bound (marked by a + on the quarter circle in Fig. 1) and the data point at the farthest counterclockwise direction. The set of permissible directions for describing the subsurface reflectance vector $\mathbf{S}_S$ is given by this band.

Both of these constraints may be stated by using our notation. Suppose that we represent the vector direction of the subsurface reflectance as (r_1, r_2); that is, the shape of the subsurface reflectance is represented in the form $S_S(\lambda) = r_1 S_I(\lambda) + r_2 S_I^{\perp}(\lambda)$. The requirement that the direction of $\mathbf{S}_S$ be counterclockwise to the last data point can be written as

$$\frac{r_2}{r_1} \geq \max_{1 \leq i \leq m} \left(\frac{c_{Ii}^{\perp}}{c_{Ii}} \right), \tag{8}$$

where the subscript i ranges over the m observations. The second requirement, that all the entries of $\mathbf{S}_S$ be nonnegative, can be expressed as $r_1 S_I(\lambda) + r_2 S_I^{\perp}(\lambda) \geq 0$ for all visible wavelengths $\lambda_1, \lambda_2, \ldots, \lambda_n$. This can be written as

$$\frac{r_1}{r_2} \geq \max_{1 \leq \lambda \leq n} \left(\frac{-s_{I\lambda}^{\perp}}{s_{I\lambda}} \right), \tag{9}$$

where $s_{I\lambda}$ and $s_{I\lambda}^{\perp}$ are the elements of the vectors $\mathbf{S}_I$ and $\mathbf{S}_I^{\perp}$, respectively. Both r_1/r_2 and the maximum value are always positive, so we can rewrite this relation as

$$\frac{r_2}{r_1} \leq \left[\max_{1 \leq \lambda \leq n} \left(\frac{-s_{I\lambda}^{\perp}}{s_{I\lambda}} \right) \right]^{-1}. \tag{10}$$

We can summarize the bounds on the direction of the vector $\mathbf{S}_S$ in a single equation as

$$\max_{1 \leq i \leq m} \left(\frac{c_{Ii}^{\perp}}{c_{Ii}} \right) \leq \frac{r_2}{r_1} \leq \left[\max_{1 \leq \lambda \leq n} \left(\frac{-s_{I\lambda}^{\perp}}{s_{I\lambda}} \right) \right]^{-1}. \tag{11}$$

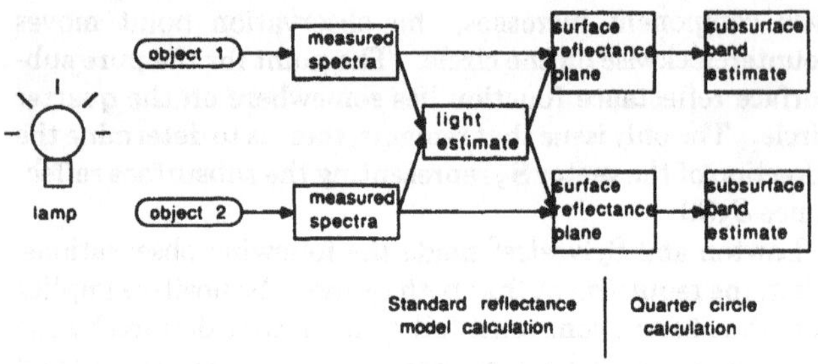

Fig. 2. Schematic representation of the experimental procedure.

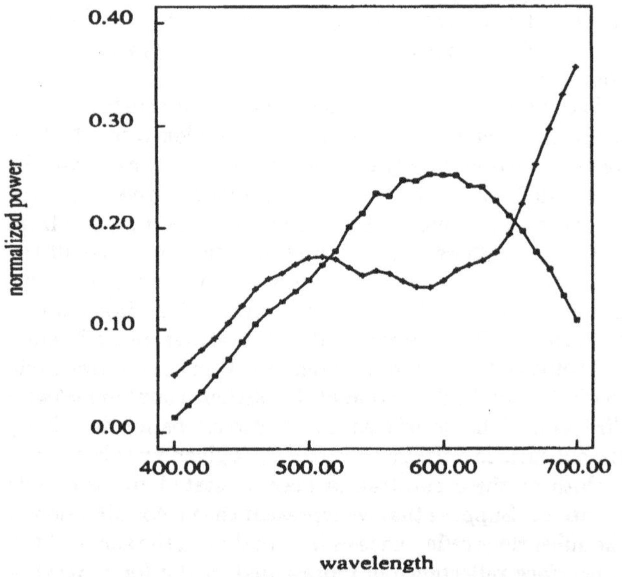

Fig. 3. Estimation results of the illuminant spectral power distributions of two light sources.[2] Crosses represent the estimate of the flood lamp, and filled squares represent that of the tungsten halogen lamp of a slide projector.

Any estimate of $S_S(\lambda)$ with this bounded region satisfies all the nonnegativity constraints and provides an equally satisfactory explanation of the data.

4. EXPERIMENTS

We performed several experiments to evaluate the quarter-circle analysis method of estimating the subsurface reflectance function. The flow chart in Fig. 2 describes the procedure. Two object surfaces are measured with different geometric conditions under a light source. The illuminant spectrum is estimated from the measured color signals by using the method described by Tominaga and Wandell.[2] The observations of the surface spectral reflectances are then obtained by dividing the measured color signals by the estimated ambient light. The subsurface reflectance functions are estimated by the quarter-circle analysis. Some typical results are shown in the following subsections.

A. Plastics

A red cup and a green ashtray were measured under a flood lamp for daylight photograph. The measured spectral data were shown in the previous paper.[2] The estimate of the illuminant spectral power distribution is shown as the curve with crosses in Fig. 3. Figure 4 shows a set of the normalized curves of the surface spectral reflectances of the green ashtray $S(\theta_i, \lambda)$ $(i = 1, 2, \ldots, 8)$. Two principal components are extracted from the singular value decomposition, and the 2 $\times$ 2 transformation matrix is determined on the least-squares fitting to the unit vector with entries of $1/\sqrt{n}$. Figure 5 shows the estimated basis curves of $S_I(\lambda)$ and $S_I{}^{\perp}(\lambda)$. The nearly straight line indicates a constant spectrum of the interface reflectance function $S_I(\lambda)$. Figure 6 shows the projection of the observed reflectances into the weighting coefficient coordinates $(c_I, c_I{}^{\perp})$. Note that the observed points (filled squares) lie on the unit quarter circle. The bounded region for $S_S(\lambda)$ is given as the interval $0.842 \leq r_2/r_1$

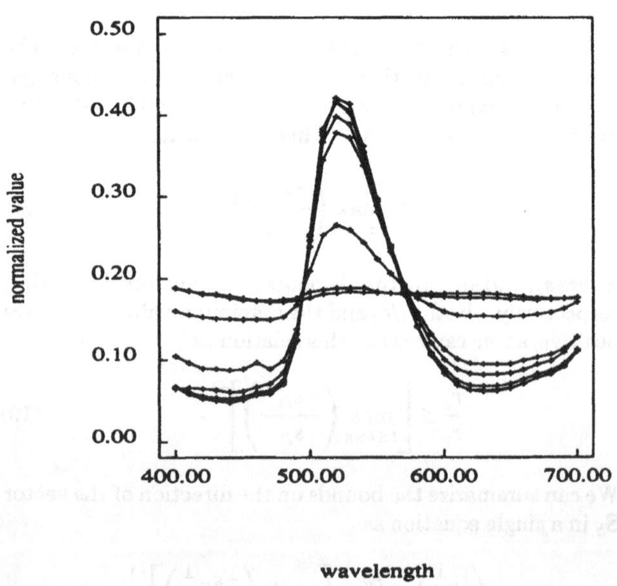

Fig. 4. Normalized curves of the observed surface spectral reflectances of the green plastic ashtray.

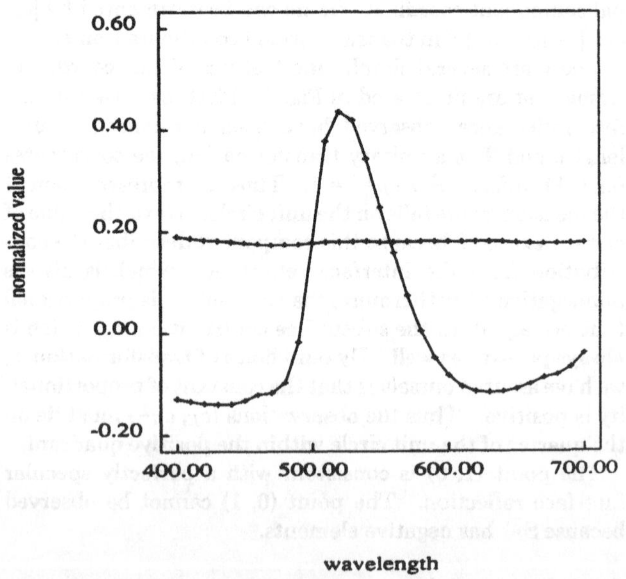

Fig. 5. Basis curves of the observed surface reflectances. The nearly straight line indicates a constant spectrum of the interface reflectance function.

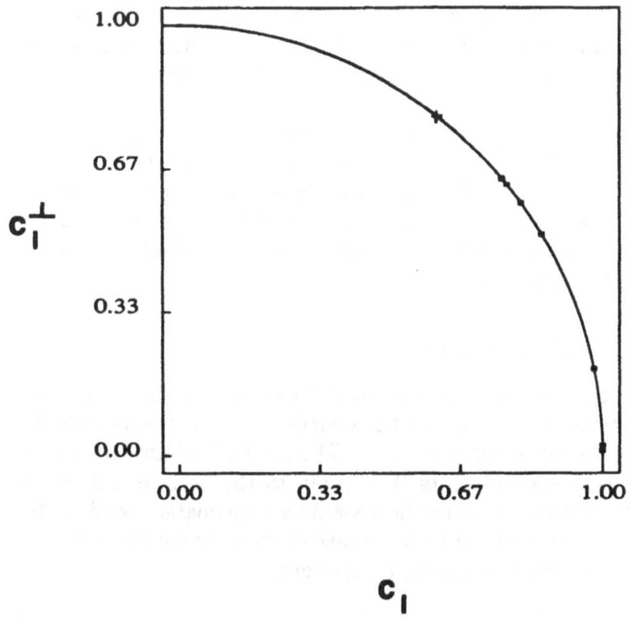

Fig. 6. Projection points of the surface reflectances. The eight points are indicated with filled squares. The upper limit of the solution band is marked with a +.

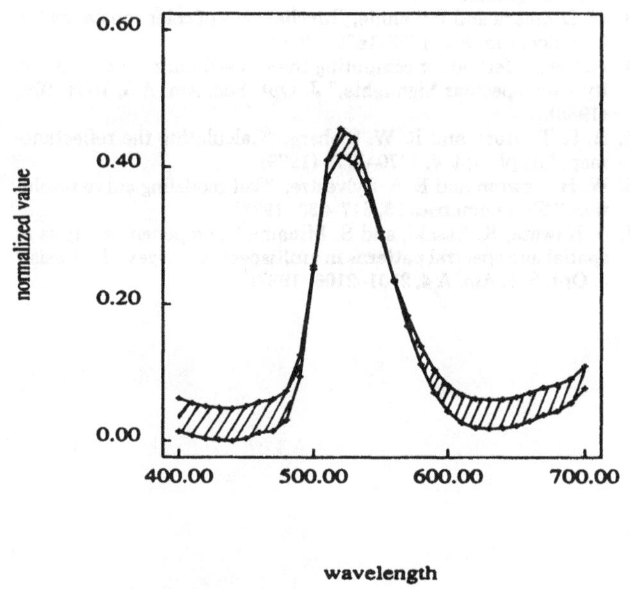

Fig. 7. An estimate of the band of the subsurface reflectance functions of the green plastic ashtray.

≤ 1.293 on the circle. A + indicates the upper limit and corresponds to the limit of physically realizable reflectance curves. An estimate of the band of the subsurface reflectance function is shown in Fig. 7, in which a range of the subsurface estimates is given as the hatched band. The curve that has a point of tangency to the wavelength axis is the theoretically limiting reflectance from nonnegativity. The other curve is the most extreme reflectance of real measurements. The width of the band represents the uncertainty of the estimates.

B. Fruits

An apple and a lemon were measured under the tungsten halogen lamp of a slide projector. The illuminant estimate is shown in a curve with filled squares in Fig. 3 (also refer to the previous paper[2]). Figure 8 shows the normalized curves of the surface spectral reflectances of the lemon. The same computational procedure has been applied. Figure 9 shows an estimated band of the subsurface reflectance function. We have the interval $0.841 \leq r_2/r_1 \leq 0.851$. In this case, the uncertainty of the estimates is small. There is little discrepancy between the theoretically limiting reflectance and the most extreme real reflectance.

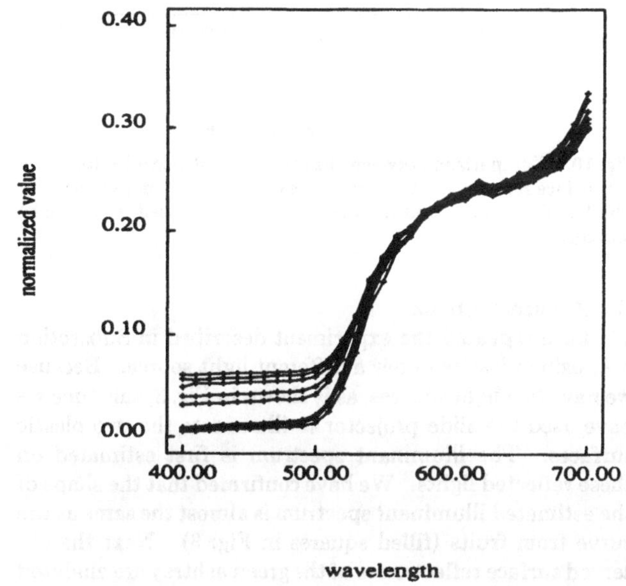

Fig. 8. Normalized curves of the observed surface spectral reflectances of the lemon.

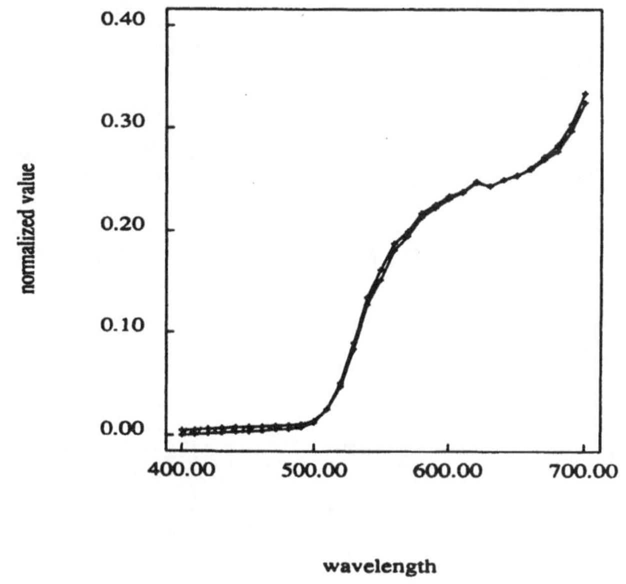

Fig. 9. Estimate of the band of the subsurface reflectance functions of the lemon.

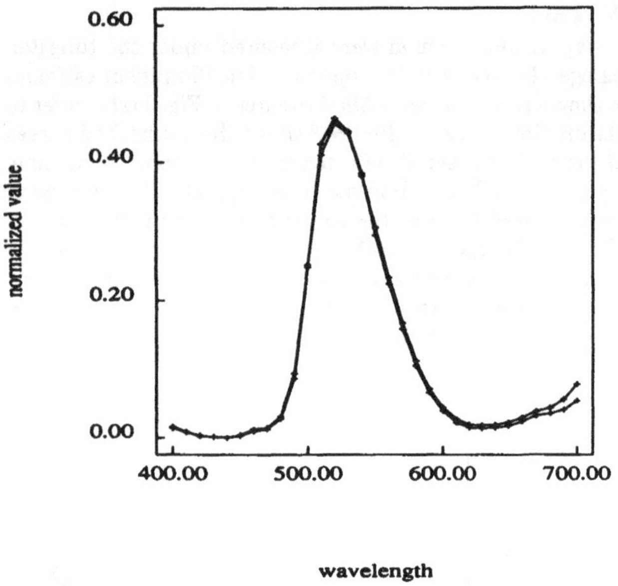

Fig. 10. Comparison between the estimated physical limits of the subsurface reflectance function of the green ashtray in two cases, in which a flood lamp and a slide projector were used as the light sources.

C. Accuracy Check

We have repeated the experiment described in Subsection 4.A, using plastics under a different light source. Because we have two light sources, as is shown in Fig. 3, this time we have used the slide projector to illuminate the two plastic surfaces. The illuminant spectrum is first estimated on these reflected lights. We have confirmed that the shape of the estimated illuminant spectrum is almost the same as the curve from fruits (filled squares in Fig. 3). Next the observed surface reflectances of the green ashtray are analyzed in the same way. The resulting estimate of the subsurface reflectance function is compared with the result in Fig. 7, which was obtained under the flood lamp. Note that the theoretical limits of the subsurface reflectances should be the same in the two experiments with different light sources, although the reflectances of real measurements are different. Figure 10 shows a comparison of the physical limit curve of the subsurface reflectance function between the two experiments. The two curves are coincident to within 2%. These results show the reliability of the whole process of the illuminant spectral estimation and the subsurface reflectance estimation.

5. CONCLUSION

We have described a method for estimating the subsurface (diffuse) component of the surface reflectance function for inhomogeneous materials. The method is based on several physical constraints that apply to the surface reflectance functions. The method yields an estimated band of the spectral reflectance functions as possible solutions for the subsurface reflectance component.

REFERENCES

1. G. J. Klinker, S. A. Shafer, and T. Kanade, "The measurement of highlights in color images," Int. J. Comput. Vision 2, 7–32 (1988).
2. S. Tominaga and B. A. Wandell, "The standard surface reflectance model and illuminant estimation," J. Opt. Soc. Am. A 6, 576–584 (1989).
3. M. D'Zmura and P. Lennie, "Mechanisms of color constancy," J. Opt. Soc. Am. A 3, 1662–1672 (1986).
4. H. Lee, "Method for computing the scene-illuminant chromaticity from specular highlights," J. Opt. Soc. Am. A 3, 1694–1699 (1986).
5. B. K. P. Horn and R. W. Sjoberg, "Calculating the reflectance map," Appl. Opt. 4, 1770–1779 (1979).
6. W. H. Lawton and E. A. Sylvestre, "Self modeling curve resolution," Technometrics 13, 617–633 (1971).
7. S. Kawata, K. Sasaki, and S. Minami, "Component analysis of spatial and spectral patterns in multispectral images. I. Basis," J. Opt. Soc. Am. A 4, 2101–2106 (1987).

A Novel Algorithm for Color Constancy

D.A. FORSYTH*

Robotics Research Group, Department of Engineering Science, Oxford University, England OX1 3PJ

Abstract

Color constancy is the skill by which it is possible to tell the color of an object even under a colored light. I interpret the color of an object as its color under a fixed canonical light, rather than as a surface reflectance function. This leads to an analysis that shows two distinct sets of circumstances under which color constancy is possible. In this framework, color constancy requires estimating the illuminant under which the image was taken. The estimate is then used to choose one of a set of linear maps, which is applied to the image to yield a color descriptor at each point. This set of maps is computed in advance.

The illuminant can be estimated using image measurements alone, because, given a number of weak assumptions detailed in the text, the color of the illuminant is constrained by the colors observed in the image. This constraint arises from the fact that surfaces can reflect no more light than is cast on them. For example, if one observes a patch that excites the red receptor strongly, the illuminant cannot have been deep blue.

Two algorithms are possible using this constraint, corresponding to different assumptions about the world. The first algorithm, **Crule** will work for any surface reflectance. **Crule** corresponds to a form of coefficient rule, but obtains the coefficients by using constraints on illuminant color. The set of illuminants for which **Crule** will be successful depends strongly on the choice of photoreceptors: for narrowband photoreceptors, **Crule** will work in an unrestricted world. The second algorithm, **Mwext**, requires that both surface reflectances and illuminants be chosen from finite dimensional spaces; but under these restrictive conditions it can recover a large number of parameters in the illuminant, and is not an attractive model of human color constancy.

Crule has been tested on real images of Mondriaans, and works well. I show results for **Crule** and for the Retinex algorithm of Land (Land 1971; Land 1983; Land 1985) operating on a number of real images. The experimental work shows that for good constancy, a color constancy system will need to adjust the gain of the receptors it employs in a fashion analagous to adaptation in humans.

1 Introduction

People experience color as a surface property that is largely unaffected by the color of the illuminating light. This phenomenon is known as color constancy. The mechanisms of human color constancy and the circumstances under which these mechanisms work are the subjects of extensive debate. Color constancy is a valuable skill for machine vision, because it allows statements about surface properties that are invariant to viewing conditions. There is a wide and active literature on color constancy in humans and in machines, which is extensively reviewed in (Forsyth 1989b).

*The author acknowledges the support of the Rhodes Trust and of Magdalen College, Oxford.

Apparently the oldest algorithm for color constancy is the coefficient rule, which is normally attributed to von Kries (von Kries 1878; see also West and Brill 1982; Worthey 1985; Worthey and Brill 1986). This algorithm adjusts the gain of each class of photoreceptor (for example, the red, green, and blue channels in a color camera) independently to obtain surface color descriptors. The factors by which the gains are adjusted are called the coefficients. Different authors use different factors: for example, Brill and West (Brill and West 1981) divide the output of each photoreceptor class by its output for a surface patch known to be white, and Land (Land and McCann 1971) chooses coefficients such that the geometric average of photoreceptor responses is constant for each class (Brainard and Wandell 1986).

Adjusting gains is equivalent to regarding the camera output at each point in the image as a vector with one entry for each photoreceptor class, and multiplying this vector by a diagonal matrix, where the diagonal elements are the coefficients, which do not change from point to point.

The coefficient rule can fail when the spectral sensitivities of the photoreceptor classes are not disjoint because in this case the photoreceptor measurements are not independent. Land (Land 1983) has observed that if some linear combination of photoreceptor spectral sensitivities is independent, it is possible to adjust this linear combination instead. To achieve this adjustment, one multiplies the camera output vector by a constant matrix (say **A**) which changes the basis to that in which the outputs are independent, by a diagonal matrix (say Λ), and then by $\mathbf{A}^{-1}$. The diagonal terms in Λ can again be chosen in a number of different ways.

Recent research, for example (Sallstrom 1973; Buchsbaum 1980; Brill and West 1981; Brainard and Wandell 1986; D'Zmura and Lennie 1986; Maloney 1986; Maloney and Wandell 1986; Wandell 1987; Gershon 1988; and Ho et al. 1988), has modeled surface reflectances by a finite dimensional space of functions and illuminants by another finite dimensional space of functions. Surface reflectances and illuminants can then each be represented by a finite coefficient vector, and the imaging process is an algebraic interaction between the surface reflectance coefficients and the illuminant coefficients. The richness and complexity of these interactions means that the coefficient rule is insufficient to model them. This has led to a belief that the coefficient rule is an inappropriate technique for color constancy (Brainard and Wandell 1986). Despite this belief, this paper demonstrates that the coefficient rule has much to recommend it.

Given a surface with a known illuminant, it is possible to measure only a finite number of properties of its surface reflectance—at most, one property for each class of photoreceptor. As a result, it is neither correct nor helpful to see color constancy as a problem of measuring surface reflectance. Rather, color constancy programs attempt to provide surface color descriptors that are unaffected by changes in the illuminant, and are not trivial (e.g., are always fixed at some value for every surface under every light). In this article, I use the appearance of a patch under a canonical illuminant as a descriptor. According to this approach, color constancy involves predicting what an image would have looked like, had it been taken under the canonical illuminant. This requires determining what illuminant was in fact used. The illuminant information in turn specifies a map, taken from a set prepared in advance, which is applied to the image. The output of this map is the predicted appearance of the image under the canonical illuminant—that is, the color descriptors.

First, I use this approach to determine under what circumstances color constancy is possible. Then, to determine the illuminant, I use the observation that only a limited set of photoreceptor responses is possible under any given illuminant, because surfaces can reflect no more light than is cast on them. This leads to a constraint on the illuminant, because observing a photoreceptor response is prima facie evidence that certain illuminants were not employed. For example, one will not observe a strong response from a red photoreceptor under a blue light, because to do so would require that some surface reflect more red light than is actually falling on it. In fact, if one observes a strong response from a red photoreceptor, then the illuminant cannot be blue.

Two algorithms result from this analysis, distinguished by the assumptions necessary for them to work. **Crule** is a simple algorithm that works for an infinite dimensional set of surface reflectances, given sufficiently strongly controlled illuminants. **Crule** forms descriptors using the coefficient rule, with a novel technique for choosing coefficients. A more complex algorithm, **Mwext**, arises in the case of finite dimensional sets of surface reflectances. Restricting surface reflectances to a finite dimensional set means that the world is strongly constrained, so that this algorithm is able to recover a large number of parameters in the illuminant. Although **Crule** requires either a strong constraint on the illuminant or particular photoreceptor properties to succeed, **Mwext** requires unusual properties of both the illuminant and of surface reflectances.

2 Preliminary Assumptions

Many effects can confound color constancy in the real world. Surfaces receive light from other surfaces as well as from the primary source, so that the effective illuminant seen by one surface may differ considerably from that seen by another. Shadows create an ambiguity between one patch, half of which is in shadow, and two patches, one of which is darker than the other. If surfaces are not at a fixed orientation to the light source, it is not always possible to determine whether a patch is dark, or is oriented so that it receives little light.

No present programs can resolve these ambiguities. Very little is known about how they could be resolved. Research on color constancy is commonly confined to considering a Mondriaan world, a world of flat, frontally presented collages of colored paper, where these ambiguities do not occur. In this article, I consider only the Mondriaan world, because, although the problem of color constancy in this world is only marginally relevant to building machines that can see, solving this problem is necessary before it will be possible to build sophisticated color constancy programs.

A number of preliminary assumptions are necessary. These are detailed here.

1. All surfaces are flat, and frontally presented and there are no significant mutual illumination effects. No shadowing occurs.
2. There is a single illuminant, which does not vary over space.
3. All surfaces are Lambertian, and all reflection is diffuse. A surface's albedo varies with wavelength only, and this variation is described by a surface reflectance function. Surfaces do not fluoresce.
4. Color constancy consists of two distinct problems. Firstly, the illuminant must be estimated in some way. Secondly, some statement about the properties of imaged surfaces must be obtained from that estimate and from the responses of the receptors.
5. The product of any surface reflectance function, any illuminant function, and any photoreceptor sensitivity can be integrated with respect to wavelength. Surface reflectance functions are never greater than one, nor less than zero. These are very weak assumptions.

Many of these assumptions are violated in practice. Assumption 1 clearly does not apply to the real world. A body of work exists that attempts to relax assumption 2. This is discussed briefly in section 7. The wings of most butterflies and many birds, for example, violate assumption 3, because their color comes from interference effects. Although surfaces are often assumed Lambertian in computer vision, there is very little information about the validity of this assumption in general—for a clear presentation of the available evidence, see (Brelstaff 1988). Some recent work has used specular components of reflected light to measure the illuminant color (Lee 1986; Klinker et al. 1987); the idea is an old one, and earlier work is described in (Judd 1960). This work encounters difficulties in the dynamic range of most cameras (when the aperature is sufficiently narrow that a specularity does not saturate the camera, it is often difficult to determine the color of objects to more than two or three bits), and with metals. Furthermore, it is not always obvious how surface color can be recovered when one knows the illuminant color.

Further assumptions that constrain properties of the illuminant will become necessary as the analysis proceeds. Although their precise statement requires the analysis of the following section, and is deferred until that section, the meaning of the assumptions is intuitively reasonable and will be stated here.

- Illuminants are "reasonable." For example, it is unreasonable to expect that one could predict the appearance under white light of a surface that one has seen under monochromatic light. Furthermore, it must be possible to describe each member of the set of illuminants that one observes (e.g., with a parametrization).
- The photoreceptors are "reasonable," given the surfaces and the illuminants. For example, it is unreasonable to expect nontrivial color constancy when the illuminants do not excite the photoreceptors.
- Given an illuminant for which two patches are metamers (that is, evoke the same receptor responses), a color constancy algorithm need not predict that they will look different under a different light. Metamers are common in theory, because there are many degrees of freedom in illuminant and surface reflectance. In practice, physical constraints on surface reflectances suggest that metamerism is uncommon (Stiles et al. 1977; Maloney 1986).
- There is a variety of different colored objects in the image. The colors of these objects are not "unreasonably" distributed (that is, the scene does not consist of only shades of some color, for example), and there are "sufficient" different colors. Neither term can be made precise at this stage. This assumption is necessary because the constraints on the illuminant become very weak when there are few colors, and the process of estimating the illuminant is then prone to error.
- For any illuminant, a reasonable measurement of the photoreceptor outputs is possible. This is a property

of the sensor quantization system. Without this assumption, an illuminant could be chosen that is so deeply colored that some photoreceptor classes have poor effective resolution, because they are excited weakly or not at all by the illuminant.

These assumptions are in practice strong. It is not always possible to ensure that images contain a wide variety of surface colors. Metamerism does occur. It is not always possible to control the surface reflectances and illuminants such that color constancy is possible. Further work is necessary to relax all these assumptions.

3 The Color Constancy Equation

In general in what follows, vector quantities are denoted by underlining, and vector components by subscripting.

E	The set of observably distinct illuminant functions, parametrized by some parameter vector $\underline{t}$.
λ	The wavelength parameter.
$e(\lambda; \underline{t})$	The illuminant function corresponding to the parameter $\underline{t}$.
$e(\lambda; \underline{t}^c)$	The canonical illuminant.
S	A set of surface reflectance functions.
$s(\lambda)$	A surface reflectance.
L	The number of receptors.
$\rho_k(\lambda)$	The kth receptor sensitivity, where $k = 0, \ldots, L - 1$.
$\Phi_l(\lambda; \underline{t})$	$\Phi_l(\lambda; \underline{t}) = \rho_l(\lambda)e(\lambda; \underline{t})$, where $l = 0, \ldots, L - 1$. This apparently arbitrary term clarifies the development of the theory.
$\phi_m(\lambda)$	A member of an orthogonal basis for the space spanned by all the $\Phi_l(\lambda; \underline{t})$, where $m = 0, \ldots, L - 1$.
$\underline{\Phi}(\lambda; \underline{t})$	The vector whose components are $\Phi_l(\lambda; \underline{t})$, $l = 0, \ldots, L - 1$.
$\underline{\Psi}(\cdot; \underline{t})$	A function chosen from a set of functions, parametrized by $\underline{t}$ which takes the appearance of a surface under the canonical illuminant to its appearance under the light represented by $\underline{t}$.

Two illuminants are "observably distinct" when, if one images all possible surface reflectances under these illuminants, one obtains two gamuts where are not exactly the same set. The response of the kth photoreceptor, viewing a surface reflectance $s(\lambda)$ with an illuminant $e(\lambda; \underline{t})$ is

$$\int \rho_k(\lambda)e(\lambda; \underline{t})s(\lambda)\, d\lambda$$

where the integral is over some appropriate range of wavelengths. For linear receptors such as CCD cameras, observing a very broad class of surfaces without specular reflection, this equation will correctly represent the output of the kth receptor. The position is more complex with human receptors, not the least because there is reason to believe that the sensitivity of receptors itself varies with the brightness and duration of exposure of a stimulus (see, for example, Bartleson 1977; Wyszecki and Stiles 1980), but at present I consider a situation where receptor sensitivity is a function of wavelength alone. This is a reasonable model for a CCD camera, and any departure from this model is likely to introduce unmanageable difficulties. I assume that the L classes of photoreceptors in the camera system have different, linearly independent spectral sensitivies. If this is not the case, some of the receptors are redundant and can be ignored. It will appear from the analysis below that the choice of receptor spectral sensitivities can strongly affect color constancy. The illuminant parameter is written as a vector, although it is not obvious that it is a vector. Two cases will emerge from the analysis below. In the first case, the set of illuminants is a finite-dimensional vector space. In the second case, it is a smooth manifold.

The color of a surface will be described by its appearance under $e(\lambda; \underline{t}^c)$, the canonical illuminant. Note that many different surface reflectances cause the same receptor responses under the canonical illuminant. Thus, each descriptor refers to an equivalence class of surface reflectances. Color constancy then involves taking an image of a set of surfaces illuminated by some member of E, $e(\lambda; \underline{t})$, and predicting its appearance had it been illuminated by $e(\lambda; \underline{t}^c)$. It is possible to do this by constructing a function $\underline{\Psi}$ that predicts the receptor responses generated by some surface under colored lights, given the lighting parameter $\underline{t}$ and the receptor responses generated by that surface under the canonical illuminant. Assume that the illuminant is known, and is given by $\underline{t}$. We then have

$$\underline{\Psi}\left(\int \underline{\Phi}(\lambda; \underline{t}^c)s(\lambda)\, d\lambda; \underline{t}\right) = \int \underline{\Phi}(\lambda; \underline{t})s(\lambda)\, d\lambda \qquad (1)$$

for all $s(\lambda)$. In this way an operation on surface color is associated with an illuminant.

Given this function, the descriptor for a patch in an image taken under a known illuminant, whose parameter is $\underline{t}$, can be found by computing its preimage in $\underline{\Psi}(\cdot; \underline{t})$. Figure 1 shows a gamut observed under white light. If $\underline{t}$ represented green or blue light, $\underline{\Psi}(\cdot; \underline{t})$ would skew the gamut to those of figures 2 or 3, respectively. If $\underline{\Psi}(\cdot; \underline{t})$ is bijective, then there is a single possible descriptor for that patch. If not, some number of degrees of freedom in the descriptor will remain unconstrained. This case is easily detected in practice, because the gamut of receptor responses will have fewer than the maximum number of degrees of freedom.[1] This means that one or more of the photoreceptors is redundant. In this case if one ignores the redundant photoreceptors, the analysis below passes through (it will emerge that $\underline{\Psi}$ is linear, so that no intersecting singular behavior is possible when it is not bijective). Since there are many different possible conventions for managing the unmeasured degrees of freedom, and because it does not affect the analysis, I do not deal with this case further.

Now consider an orthonormal basis $\{\phi_m(\lambda), 0 \leq m \leq L - 1\}$ for the space spanned by the component functions $\Phi_l(\lambda; \underline{t}^c)$ of $\underline{\Phi}(\lambda; \underline{t}^c)$. Without loss of generality, assume that the L functions $\Phi_l(\lambda; \underline{t}^c)$ are linearly independent. If they are not, either the receptor spectral

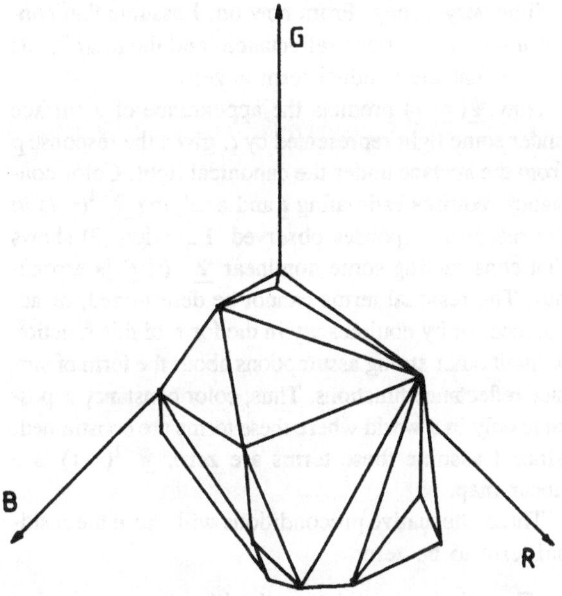

Fig. 1. The convex hull of the gamut obtained by observing 180 color chips under white light. This formed the observed canonical gamut for the experimental work. If $\underline{t}$ represents the green light for the gamut shown in figure 2, $\underline{\Psi}(\cdot; \underline{t})$ would take this set to a superset of that gamut.

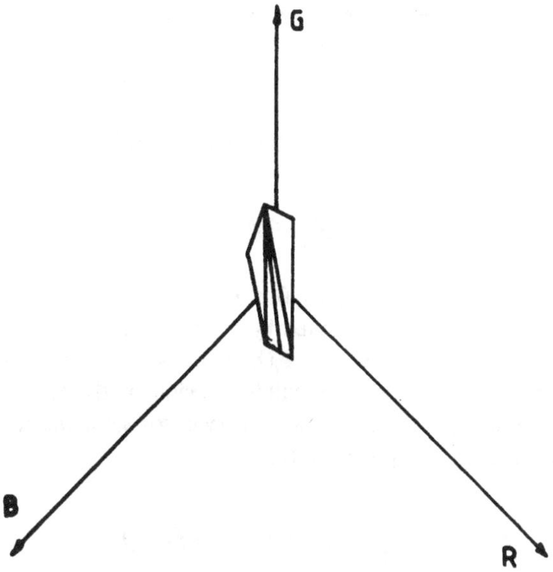

Fig. 2. The convex hull of the gamut observed for a Mondriaan image under green light. Notice that there are significant differences between this gamut, and that of figures 1 and 3. The feasible set consists of maps taking this gamut to a subset of the gamut of figure 1.

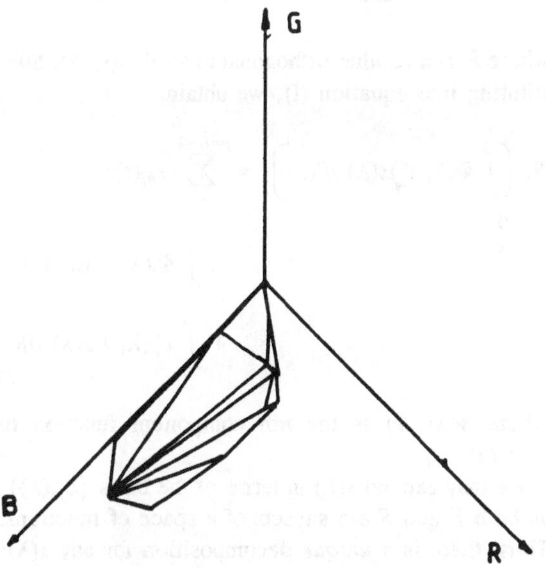

Fig. 3. The convex hull of the gamut observed for a Mondriaan image under blue light. Notice that there are significant differences between this gamut, and those of figures 1 and 2. Crule uses this skewing effect to infer possible illuminants.

sensitivities are not linearly independent or the canonical illuminant has been poorly chosen. It is clear that such an orthonormal basis exists (apply the Gramm-Schmidt algorithm; see, for example, Eisenschitz 1966). Then there is a ***unique*** decomposition of both $\Phi_l(\lambda; \underline{t}')$, and of $\Phi_l(\lambda; \underline{t})$ in terms of this orthonormal basis: say,

$$\Phi_l(\lambda; \underline{t}') = \sum_{j=0}^{j=L-1} a_{lj}\phi_j(\lambda)$$

Since this basis $\{\phi_m(\lambda), 0 \leq m \leq L\}$ was introduced to span the space spanned by the functions $\Phi_l(\lambda; \underline{t}')$, it may be the case that $\Phi_l(\lambda; \underline{t})$ does not belong to $Span\{\phi_m(\lambda)\}$, and there will be a residue in the expansion of $\Phi_l(\lambda; \underline{t})$ on this basis. Hence, for some matrix with l, jth component $r_{lj}(\underline{t})$,

$$\begin{aligned}\Phi_l(\lambda; \underline{t}) &= \sum_{j=0}^{j=L-1} \alpha_{lj}(\underline{t})\phi_j(\lambda) + F_l(\lambda; \underline{t}) \\ &= \sum_{i=0}^{i=L-1}\sum_{j=0}^{j=L-1} r_{lj}(\underline{t})a_{ji}\phi_i(\lambda) + F_l(\lambda; \underline{t}) \\ &= \sum_{i=0}^{i=L-1} r_{li}(\underline{t})\Phi_i(\lambda; \underline{t}') + F_l(\lambda; \underline{t})\end{aligned}$$

where F_l is a residue orthogonal to all the $\phi_m(\lambda)$. Substituting into equation (1), we obtain

$$\begin{aligned}\Psi_n\left(\int \underline{\Phi}(\lambda; \underline{t}')s(\lambda)\, d\lambda; \underline{t}\right) &= \sum_{j=0}^{j=L-1} r_{nj}(\underline{t}) \\ &\quad \times \int \Phi_j(\lambda; \underline{t}')s(\lambda)\, d\lambda \\ &\quad + \int F_n(\lambda; \underline{t})s(\lambda)\, d\lambda\end{aligned} \tag{2}$$

where $\Psi_n(\cdot; \underline{t})$ is the nth component function of $\underline{\Psi}(\cdot; \underline{t})$.

We may expand $s(\lambda)$ in terms of the basis $\{\phi_m(\lambda)\}$, as both E and S are subsets of a space of functions. Thus, there is a ***unique*** decomposition for any $s(\lambda)$:

$$s(\lambda) = \sum_{i=0}^{i=L-1} \sigma_i\phi_i(\lambda) + s^*(\lambda)$$

where $s^*(\lambda)$ is a residue orthogonal to each of the ϕ_m. Equation (2) becomes

$$\begin{aligned}\Psi_n\left(\sum_{j=0}^{j=L-1} a_{ij}\sigma_j; \underline{t}\right) &= \sum_{i=0}^{i=L-1}\sum_{j=0}^{j=L-1} r_{ni}(\underline{t})a_{ij}\sigma_j \\ &\quad + \int F_n(\lambda; \underline{t})s^*(\lambda)\, d\lambda\end{aligned} \tag{3}$$

Equation 3 is fundamental to any analysis of color constancy. I call equation (3) the *color constancy equation*, and refer to the term $\int F_n(\lambda; \underline{t})s^*(\lambda)\, d\lambda$ as the residual term. This term is the only impediment to success on the part of a color constancy algorithm.

In effect, the residual term is that part of the receptor response under the illuminant $\underline{t}$ resulting from properties of the surface reflectance function *that were not observed under the canonical illuminant*. Thus, it is not possible to predict the residual term without a constraint on surface reflectance functions, or knowledge of the illuminant. The simplest case occurs when either illuminants or surface reflectances are constrained so that the residual term is always zero. If the residual term satisfies some relation, color constancy may be possible with appropriate constraints on the surface reflectances and illuminants. This case is of little interest, and is not dealt with here, because the constraints required will be very strong. From now on, I assume that constraints on the surface reflectances and the illuminants ensure that the residual term is zero.

Now $\underline{\Psi}(\underline{p}; \underline{t})$ predicts the appearance of a surface under some light represented by $\underline{t}$, given the response $\underline{p}$ from the surface under the canonical light. Color constancy requires estimating $\underline{t}$ and applying $\underline{\Psi}^{-1}(\cdot; \underline{t})$ to the receptor responses observed. Equation (3) shows that constructing some nonlinear $\underline{\Psi}^{-1}(\cdot; \underline{t})$ is erroneous. The residual terms cannot be determined, or accounted for by nonlinearity in the form of this function without other strong assumptions about the form of surface reflectance functions. Thus, color constancy is possible only in a world where these terms are constrained. Since I assume these terms are zero, $\underline{\Psi}^{-1}(\cdot; \underline{t})$ is a linear map.

Three alternative preconditions will cause the residual term to be zero:

1. There is a constraint on the illuminant, such that the residual terms are zero. In particular, for any illuminant, parametrized by $\underline{t}$, the residual term $F_n(\lambda; \underline{t}) = 0$, for all n. In a world where this is the case, one can build an algorithm that will work for an infinite dimensional set of surface reflectance

functions if t can be estimated. I show that one can in fact estimate $\underline{t}$ and that this constraint naturally implies a version of the coefficient rule.

2. Surface reflectance functions are only L dimensional, and in particular are members of the space spanned by $\{\phi_0(\lambda), \ldots, \phi_{L-1}(\lambda)\}$. In this case a more general mapping of the receptor responses will achieve color constancy. Constancy will be possible for an infinite dimensional space of illuminants in this case, but only a finite dimensional subset of the illuminants causes any response from the surface at all.
3. In the third case, surface reflectances are $M > L$ dimensional, where M is finite. In this case, we assume that the space of surface reflectances is spanned by

$$\{\phi_0(\lambda), \ldots, \phi_{L-1}(\lambda), \phi_L(\lambda), \ldots, \phi_{M-1}(\lambda)\}$$

where $\phi_0(\lambda), \ldots \phi_{L-1}(\lambda)$ are derived from the receptors and the canonical illuminant in the manner indicated. Now if every possible illuminant is a member of a finite dimensional space of functions such that for every $e_i(\lambda)$ in some basis of the space (and hence for any basis) $\int \rho_k(\lambda)e_i(\lambda)\phi_j(\lambda)\,d\lambda = 0$ for $L \leq j < M$ and $0 \leq k < L$, the residual terms will be zero.

In what follows below, I refer to case 1 as a minimal assumption of the first kind, and to cases 2 and 3 being minimal assumptions of the second kind. This is because the distinction between the case 2 and case 3 assumptions is not strong, whereas the distinction between the two kinds of minimal assumption is large. The minimal assumption of the first kind corresponds to assuming an infinite dimensional set of surface reflectances, and a strong constraint on illuminants described in section 5.1. Minimal assumptions of the second kind correspond to requiring that surface reflectance functions be finite dimensional, with a weaker constraint on the illuminants. These assumptions lead to an algorithm for color constancy that differs in its details depending on the assumptions employed. Since the minimal assumptions of the second kind are stronger than those of the first kind, it is possible to recover more information from an image in a world that satisfies the second kind than in one that satisfies the first kind. Unfortunately, it is difficult to be convinced that minimal assumptions of the second kind are appropriate for real pictures. If none of these assumptions (or stronger ones) apply, color constancy is not possible.

In fact, the results in section 6 suggest that it is sufficient to ensure that residual terms are always small with respect to the receptor responses, at least for minimal assumptions of the first kind.

4 Recovering the Illuminant

Assume that in the world in which the color constancy algorithm is going to work, the residual terms are zero. Thus, the illuminants and surface reflectances must be such that under any illuminant, one observes only the degrees of freedom in surface reflectance that one observes under the canonical illuminant. I assume that the world in which a color constancy program works contains only illuminants that have this property. In section 5, I discuss what illuminants this world could contain, but for now assume that the set of such illuminants is nontrivial. Equation 3 becomes

$$\Psi_n\left(\sum_{j=0}^{j=L-1} a_{ij}\sigma_j;\, \underline{t}\right) = \sum_{i=0}^{i=L-1}\sum_{j=0}^{j=L-1} r_{ni}(\underline{t})a_{ij}\sigma_j \qquad (4)$$

This means that the gamut under some illuminant $\underline{t}$ is the image in a linear map of the gamut under the canonical illuminant. As a result, it is possible to use geometrical properties of the gamut to estimate which linear map was applied, and as a result to determine the illuminant.

Intuitively, one observes only a limited number of colors under any given light, because a surface can reflect no more light than falls on it, and cannot reflect less than no light at all. That is, surface reflectance functions are never less than zero, nor greater than one. Thus, they form a bounded, closed, and convex subset of the space of all square integrable functions. Boundedness and closedness are obvious: convexity is true because if $f(\lambda)$ and $g(\lambda)$ are two such functions, $\mu f(\lambda) + (1 - \mu)g(\lambda)$ must lie between zero and one for all λ and for $0 \leq \mu \leq 1$, and so this is a third such function. If one were to image every possible surface reflectance under some illuminant, the resulting gamut would also be bounded, closed, and convex, because it is the image in a linear map of this subset.[2] Define the canonical gamut to be the gamut obtained when one images every possible surface reflectance under the canonical illuminant. Assume that the canonical gamut is known.

There is a linear map, $\underline{\Psi}(\cdot;\, \underline{t})$ corresponding to each illuminant $\underline{t}$, which takes the appearance of a surface under the canonical illuminant to its appearance under the illuminant $\underline{t}$. By assumption, this map is bijective. Given a picture of a set of surfaces, the gamut of this picture is a subset of the image of the canonical gamut in some bijective linear map. Color constancy requires determining what the map was, and applying the inverse map. Consider these inverse maps. By definition, no observation that originated in a point outside the canonical gamut corresponds to a surface reflectance. We can then immediately rule out any inverse map that takes the observed gamut to a proper superset of the canonical gamut, because this implies that some surface reflectance does not lie in the canonical gamut when imaged under the canonical illuminant.

This process leaves a feasible set of linear maps, any of which might be the inverse of the map associated with the illuminant that formed the image. The minimal assumption made corresponds to a further constraint on this feasible set, because it is effectively a constraint on the form of the illuminant. These constraints are explored below, in section 5. The feasible set is the set of linear maps that satisfy both the gamut constraint and the constraints resulting from the minimal assumptions. The feasible set is thus the set of inverse maps corresponding to the illuminants under which the picture could have been taken. An estimator can be used to choose the most likely inverse map within this set.

Thus, an algorithm for performing color constancy has this form:

1. Construct the canonical gamut, by observing as many surfaces as possible under a single given light. This light will be the canonical illuminant, and surface color is defined to mean the color that the surface has when observed under this light. The canonical gamut can be approximated by taking the convex hull of the union of the gamuts obtained by these observations. Call the canonical gamut C.
2. To construct the feasible set for any patch imaged under a constant illuminant:
   - Form the convex hull of the gamut observed. Call this gamut D.
   - Form the set of the L-by-L matrixes, M, of rank L such that $M(D) \subset C$. The particular minimal assumption under which the color constancy algorithm is operating determines a subset of this set, which is the feasible set.
3. Within this feasible set, use some estimator to choose that map most likely to correspond to the illuminant.
4. Apply the chosen map to the receptor responses to obtain color descriptors. The color descriptors are an estimate of the appearance of the surfaces in the original image under the canonical light.

Which minimal assumption is chosen dramatically affects the feasible set. In the next section, I analyze these effects and show how the feasible set can be completely determined.

5 The Structure of the Feasible Set

By assumption, the color constancy algorithm encounters only illuminants with particular properties. In this section, I show which illuminants possess these properties, and demonstrate that the restrictions on illuminant properties result in restrictions on the linear maps associated with the illuminants. For minimal assumptions of the first kind, only a small class of linear maps need be considered, and the algorithm can recover few degrees of freedom in the illuminant. For minimal assumptions of the second kind, the number of degrees of freedom in the illuminant that the algorithm can recover depends on geometric properties of the gamut, but is surprisingly large.

5.1 Minimal Assumptions of the First Kind

Assuming that for all $\underline{t}$, $F_k(\lambda, \underline{t}) = 0$ implies, by equation 2, that for some $r_{lj}(\underline{t})$:

$$\Phi_l(\lambda; \underline{t}) = \sum_{j=0}^{j=L-1} r_{lj}(\underline{t})\Phi_j(\lambda; \underline{t}^c)$$

Recall that $\underline{\Phi}(\lambda; \underline{t})$ is the vector with components $\Phi_l(\lambda; \underline{t}) = \rho_l(\lambda)e(\lambda; \underline{t})$. Define a vector-valued function, $\underline{\rho}(\lambda)$ such that $\underline{\Phi}(\lambda; \underline{t}) = e(\lambda; \underline{t})\underline{\rho}(\lambda)$. Then for a matrix $R(\underline{t})$ having l, jth component $r_{lj}(\underline{t})$, we have

$$\underline{\rho}(\lambda)e(\lambda; \underline{t}) = R(\underline{t})\underline{\rho}(\lambda)e(\lambda; \underline{t}^c)$$

Assume that R has full rank; if it doesn't, one may ignore classes of receptors until the diminished system has full rank, and proceed. If $e(\lambda; \underline{t}) = 0$, then either $e(\lambda; \underline{t}^c) = 0$ or $\underline{\rho}(\lambda) = 0$. This case is of little interest because it means that either the scene is not illuminated, or the surfaces reflect none of the illuminating light. Recall that $e(\lambda; \underline{t})$, and $e(\lambda; \underline{t}^c)$ are scalar-valued functions, and consider

$$\underline{\rho}(\lambda) = R(\underline{t})\,\underline{\rho}(\lambda)\frac{e(\lambda; \underline{t}^c)}{e(\lambda; \underline{t})}$$

Then this requires for all λ that $\underline{\rho}(\lambda)$ is an eigenvector of $R(\underline{t})$. But $R(\underline{t})$ is a function of $\underline{t}$ alone and has at most L distinct eigenvectors, so the components of $\underline{\rho}(\lambda)$ must be chosen such that $\rho(\lambda)$ is for any λ an eigenvector of R. This is possible because $\underline{\rho}(\lambda)$ represents the spectral sensitivity of the sensors, and can be chosen. Since the eigenvectors of $R(\underline{t})$ are fixed by the choice of sensors, the form of $R(\underline{t})$ is restricted. This gives the form of the illuminants for which the algorithm will work.

For $\underline{\rho}(\lambda)$ to be an eigenvector of R for all λ, there must exist functions $k_j(\lambda)$, such that

$$\underline{\rho}(\lambda) = \sum_{j=0}^{j=L-1} k_j(\lambda)\underline{\epsilon}_j$$

where

$$\text{support}\,(k_i(\lambda)) \cap \text{support}\,(k_j(\lambda)) = \emptyset,\ i \neq j$$

and the $\underline{\epsilon}_j$ are arbitrarily chosen, linearly independent vectors. The terms $\underline{\epsilon}_j$, will be the L eigenvectors of $R(\underline{t})$. These are a result of the choice of receptor. If the eigenvectors $\underline{\epsilon}_j$ are not chosen to be linearly independent, then the receptor sensitivities arising will not be linearly independent.

Given that the eigenvectors of R are fixed, there are L free linear parameters for R, that is, its eigenvalues. This means that if minimal assumptions of the first kind apply, the map that achieves constancy consists in form of a change of basis, a diagonal scaling, and another change of basis. In particular, these changes of basis are fixed in advance by the choice of receptors. Furthermore, there exist choices of receptors for which this algorithm cannot work, because they do not have the form given above.

Intuitively, for such a scheme to perform, color constancy requires a strong property of the illuminant. Consider some $\lambda = \lambda_0$; then

$$\underline{\rho}(\lambda_0) = \frac{e(\lambda_0; \underline{t}^c)}{e(\lambda_0; \underline{t})} R(\underline{t})\underline{\rho}(\lambda_0)$$

In particular, as a result of the expansion of $\underline{\rho}$ and the fact that the supports of the coefficient functions are disjoint, there is some $0 \leq i < L$ such that

$$\underline{\rho}(\lambda_0) = k_i(\lambda_0)\underline{\epsilon}_i$$

(no summation over i). Then we have

$$\underline{\rho}(\lambda_0) = \frac{e(\lambda_0; \underline{t}^c)}{e(\lambda_0; \underline{t})} R(\underline{t})k_i(\lambda_0)\underline{\epsilon}_i$$

$$k_i(\lambda_0)\underline{\epsilon}_i = \frac{e(\lambda_0; \underline{t}^c)}{e(\lambda_0; \underline{t})} \mu_i k_i(\lambda_0)\underline{\epsilon}_i$$

where μ_i is the eigenvalue associated with $\underline{\epsilon}_i$. From this equation, one obtains

$$\frac{e(\lambda_0; \underline{t}^c)}{e(\lambda_0; \underline{t})} \mu_i = 1$$

This is true over the support of $k_i(\lambda)$. Thus, minimal assumptions of the first kind require that over the support of each $k_i(\lambda)$, $e(\lambda; \underline{t}^c)/e(\lambda; \underline{t})$ is constant, where these regions are given by the choice of receptor spectral sensitivity. This is a strong requirement on the illuminant.

Thus, minimal assumptions of the first kind mean that the feasible set consists of all those linear maps that satisfy the gamut restriction and whose eigenvectors are fixed. Since the feasible set is defined, the algorithm is specified. I call this algorithm **Crule**, for the fact that it involves a coefficient rule. The common belief that a coefficient rule is insufficiently general to achieve color constancy originates in the requirement that surface reflectances be finite dimensional. Assuming that surface reflectances belong to a finite dimensional set corresponds to a minimal assumption of the second kind, given that the illuminant functions are also constrained. Minimal assumptions of the second kind are treated in section 5.2.

From the point of view of machine vision, these results are encouraging. If one chooses receptors with disjoint, narrow-band sensitivities, the illuminant will be effectively constant over the support of the receptor. Furthermore, because the sensitivities are disjoint, the receptor sensitivities will be eigenvectors of a diagonal matrix, and the receptor gains can be adjusted independently. Such a system could achieve a very high degree of color constancy for all real surfaces under almost all real lights if the receptors responded to a sufficiently narrow band of wavelengths.

The color constancy equation and the above analysis allows insight into the quality of the results reported by McCann, McKee, and Taylor (1976). Failures from a color constancy system are to be expected as a result of the nature of the problem, and the results that they show for the algorithm they use are extremely good.

Recall, however, that McCann, McKee, and Taylor used as an illuminant a weighted sum of narrowband lights, such that each receptor effectively responded to only one light. With this form of illuminant, in all circumstances the map taking receptor responses under one light as compared to those under another such light has only small off-diagonal terms, and that since the lights employed are within a linear map of one another, the residual is always zero (i.e., the minimal assumption of the first kind is always satisfied), so that constancy may be achieved by independent scaling of receptor responses. This experiment does not fully test the Retinex algorithm, because theory shows it should work well in an experiment of this sort. The performance of the algorithm under more general lights remains untested.

Furthermore, McCann, McKee, and Taylor did not note the Retinex algorithms' requirement that some average of surface reflectances remains constant, because they tested it on only one Mondriaan. In section 6, I demonstrate this flaw in the Retinex algorithm, which was originally observed by Brainard and Wandell (1986).

5.2 *Minimal Assumptions of the Second Kind*

When either minimal assumption of the second kind is justified, the arguments employed above no longer apply. Intuitively, the fact that this kind of assumption is stronger than the first kind suggests that more parameters in the illuminant can be recovered than could be with the first kind. In fact, using this approach one can recover a surprisingly large number of parameters in the illuminant.

The gamut of an image of all possible surfaces, taken under some unknown illuminant, is the image in a linear map of the canonical gamut. The number of illuminant parameters that can be recovered is the number of degrees of freedom of this linear map that can be computed simply by observing the gamut. This number depends on geometric properties of the gamut itself. For example, consider the set of vectors in $\mathbf{R}^3$ whose length is less than, or equal to, one. By observing its image in a linear map operating on $\mathbf{R}^3$, one may make assertions about the shear components of the map, but not about the rotation components.

For the purposes of the analysis that follows, no distinction need be drawn between the two cases of minimal assumptions of the second kind. Although in the second case the dimension of the space of surface reflectances is higher, the illuminants have been chosen such that only an L-dimensional subset of the space of surface reflectances ever causes a receptor response. The second kind of assumption implies that the space of surface reflectances is finite dimensional, and the dimension of the subspace whose effects appear in the receptor responses is at most L. If it is less than L, some of the receptors are redundant and may be discarded, so that the case where the dimension of this space is precisely L is the only interesting case.

To formalize these intuitions, define $Gs[\Sigma]$ to be the set of bijective linear maps on $\mathbf{R}^L$ that map a subset Σ of $\mathbf{R}^L$ to itself. $Gs[\Sigma]$ is clearly a subgroup of $GL(L)$, where $GL(L)$ is the group of bijective linear maps operating on $\mathbf{R}^L$. Call $Gs[\Sigma]$ the *linear similarity* group of the set. Denote by $A(\Sigma)$ the image of Σ in the linear map A. Then

$$A \in Gs[\Sigma] \Leftrightarrow A(\Sigma) = \Sigma$$

PROPOSITION. For $\Sigma \subset R^L$, and $F, G \in Gl(L, \mathbf{R})$,

$$F(\Sigma) = G(\Sigma) \Leftrightarrow G = FA$$

for some $A \in Gs[\Sigma]$.

Proof: (See appendix II.)

PROPOSITION. For $\Sigma \subset \mathbf{R}^L F \in Gl(L)$, $Gs[\Sigma]$ and $Gs[F(\Sigma)]$ are isomorphic.

Proof: (See appendix II.)

The important point here is that one cannot tell if members of the linear similarity group of some set have operated on it by observing the image of that set in a linear map. *It cannot "lose linear similarities."*

Now recall that the gamut is the image of the canonical gamut in some bijective linear map. Clearly, for $F \in GL(L)$ and A in $Gs[\Sigma]$, it is impossible to tell the map F from the map $F \circ A$ by observing only the image of Σ in these maps. Furthermore, we can distinguish F and G if and only if there is no $A \in Gs[\Sigma]$ such that $F = G \circ A$. Thus, there is a correspondence between the maps we can distinguish and the collection of cosets defined by the right action of $Gs[\Sigma]$ on $GL(L)$. This collection is a manifold only if $Gs[\Sigma]$ is closed in $GL(L)$ (Boothby 1986, p. 94). Assume that this is the case, as the most likely candidates for $Gs[\Sigma]$ are either a finite discrete group of rotations, or a one-parameter rotation group. Then the collection of cosets can be written as $GL(L)/Gs[\Sigma]$.

The set of maps taking the canonical gamut to a superset of the observed gamut is a subset of $GL(L)$. Call this set $\mathfrak{F}$, and write $\mathfrak{C}$ for the canonical gamut. $\mathfrak{F}$ needs to be collapsed to represent the distinctions in the illuminant that can actually be drawn. The wisest strategy is in constructing the system to determine one representative of each right coset of $Gs[\mathfrak{C}]$ in $GL(L)$ corresponding to a legitimate illuminant (call this a *coset leader*). Then the set of matrixes corresponding to the *observable* constraint set is obtained by replacing each member of $\mathfrak{F}$ by its coset leader. This is likely to be computationally laborious, but can be simplified by a sensible choice of coset leaders.

This discussion, in conjunction with section 4, gives a complete specification for a color constancy algorithm, since the set of linear maps that can be distinguished is determined. I call this algorithm **Mwext**, because it is the natural extension of Maloney and Wandell's work, with physical realizability incorporated. However, the way that this algorithm has been defined means that it can recover more parameters in the illuminant than did Maloney and Wandell, for the same number of receptors.

Now the number of illuminant parameters that can be recovered is exactly the dimension $GL(L)/Gs[\mathfrak{C}]$, which may be rather large. Theorem 9.2 of (Boothby 1986, p. 166) states that for Lie groups (the groups involved are Lie groups) G and H, H a closed subgroup of G, G/H has a unique structure as a differentiable manifold, and the dimension of G/H is $\dim(G) - \dim(H)$. Thus, if the linear similarity group is a discrete group, the dimension of the manifold of observably distinct illuminants is L^2, and if the linear similarity group is a one-parameter subgroup, the dimension is $L^2 - 1$. This is an embarrassment of parameters for a model of the human vision system. People do not recover either 9 or 8 parameters in the illuminant.

This formalism, apart from determining exactly how many parameters are distinguishable in the illuminant, clarifies aspects of Maloney and Wandell's algorithm (Maloney and Wandell 1986; Maloney 1986; Wandell 1987), of which it is in some sense a natural generalization. Maloney and Wandell (1986) formulated a color constancy technique using finite dimensional linear models of illuminant and surface reflectance. The mathematical details of this algorithm are in Maloney's thesis, (Maloney 1984). The description given here is somewhat more abstract than the one Maloney presents, but has the advantage of compactness.

Maloney and Wandell represent surface reflectance functions by $\mathbf{R}^N$, and have M receptors, $M > N$. They use a linear sensor and require that no two unequal surface reflectances produce the same receptor response under any light. Then taking an image corresponds to applying an injective linear map F, $F : \mathbf{R}^N \rightarrow \mathbf{R}^M$. In particular, if $F(S)$ denotes the image of the set S in the map F, the gamut observed is $F(\mathbf{R}^N)$. It is possible to distinguish between maps F that take $\mathbf{R}^N$ to distinct N-dimensional subspaces through the origin in $\mathbf{R}^M$. Maloney (1986) discusses the case where there are two degrees of freedom in surface reflectance, and three receptors. Under these conditions, he can recover either two degrees of freedom in surface reflectance and one in the illuminant or one in surface reflectance and two in the illuminant. The asymmetry is due to a scaling ambiguity between surface lightness and illuminant brightness.

I shall discuss this particular example for simplicity, although with little difficulty both the number of receptors and the number of degrees of freedom in the surface reflectance functions may be changed, as long as there are more receptor bases than surface reflectance bases. Recall that they do not require physical plausibility of surface reflectances, so the set of representations of surface reflectance functions becomes a space, $\Sigma_w = \mathbf{R}^2 \times (0)$.

The linear similarity group of a two dimensional linear subspace of R^3 is the group of matrixes M which, in some basis, have the form

$$\begin{matrix} \cdot & \cdot & \cdot \\ \cdot & \cdot & \cdot \\ 0 & 0 & \cdot \end{matrix}$$

such that $\det(M) \neq 0$. Again, this is easily seen to be a group which, using the notation above, is denoted $Gs[\Sigma_w]$. In particular, the dimension of this (Lie) subgroup considered as a differentiable manifold is 7 (count the dots!). Thus, it is possible to recognize as distinct objects only maps corresponding to distinct points on $GL(3)/Gs[\Sigma_w]$ (this is an object known as a Grassmannian (Boothby 1986), felicitously named after Hermann Grassmann, a pioneer in color vision). To recover a representation of surface reflectance, therefore, there must be only one illuminant corresponding each point in $GL(3)/Gs[\Sigma_w]$. The dimension of $GL(3)$ is 9, and so it is possible to distinguish $9 - 7 = 2$ dimensional manifold of illuminants. Hence, Maloney and Wandell's approach requires a priori constraint on *both* illuminant and surface reflectance.

Mwext can recover more parameters in surface reflectance than Maloney and Wandell did, for the same number of receptors. Maloney and Wandell point out that their approach is incapable of uniquely determining surface lightness. This is as a result of the linear similarity:

$$\begin{matrix} k & 0 & 0 \\ 0 & k & 0 \\ 0 & 0 & k \end{matrix}$$

(where $k \neq 0$) of the plane. Since the canonical gamut does not have this similarity, it is possible for **Mwext** to recover surface lightness. This, however, depends on the assumption that there are many different colors in the image. If the image contains only very dark colors, for example, the brightness of the illuminant will be poorly constrained, and **Crule** will report surface lightness incorrectly.

6 Experimental Results

Very few experimental demonstrations of color constancy programs have been described in the literature. Published tests exist for Maloney and Wandell's (1986) algorithm working on synthetic images only (Maloney 1984). With the exception of results published on one Mondriaan image by McCann, McKee, and Taylor (1976), no published account of the color constancy performance of the Retinex algorithm on real data exists. Brainard and Wandell (1986) tested their model of the Retinex algorithm on synthetic data. Gershon's algorithm (Gershon 1988) has been demonstrated on a single real image. This result is flawed by the fact that there was only a single object on a black background in the image. Brill (1979) implies that he had a color constancy algorithm that worked on real pictures, but I have been unable to find details of this work elsewhere in the literature. Buchsbaum's paper (Buchsbaum 1980) does not disclose whether the results he presents originate in real or in synthetic images.

6.1 Preliminary Information

All images were taken with a monochrome CCD camera with its gain control defeated. Three separate exposures of each image were made using Kodak Wratten filters (no.'s 29, 47B, and 58) for color separation, and a sharp near infrared cut filter, which is essential as a result of the pronounced near infrared sensitivity of CCD cameras. The lenses used were conventional photographic lenses. The CCD camera appears to measure no chromatic aberration, probably because its pixels are large with respect to any fringes. The illuminants used were two 500W photographer's lamps of unknown color temperature, and a warm white appearance. The illuminants were colored by the use of translucent colored plastic filters hung in front of the bulbs. At no stage have the properties of these filters been measured. I refer to the light provided by these lamps as "white light," and to colored illuminants by the conventional color names of the plastic filters used.

Neutral density filters were used to weight the separation filters so that the aperture of the lens did not need to be adjusted between exposures. I chose the weights so that images closely resembled the objects imaged when displayed on the screen of a Sun workstation. Choosing weights on the basis of the overall appearance of the image appears to distribute the errors in the color displayed evenly, making it easier to compare objects with images. A number of other strategies for choosing weights are possible. For example, one might choose the set of weights that led to a bright white patch causing the same response in all three separations, although monitor nonlinearities can cause this approach to skew the color of darker patches.

In fact, the choice of weights is not particularly important in demonstrating color constancy. A color constant algorithm should produce near constant, nontrivial, color descriptors for objects, when presented with well-populated scenes imaged under widely varying lights. It is possible to tell whether an algorithm can do this without ever knowing how the color descriptors produced by the algorithm relate to surface color as people see it. Dealing with an algorithm on such abstract terms is unattractive, however, and a choice of weights that makes the image look recognizable avoids this problem. Color constancy, or any lack of it, is easily recognized from the descriptors.

For this series of experiments, I used Mondriaans made of Color-aid papers (a set of 202 papers, with standard colors, available from the Color Aid corporation[3]) and recorded the colorimetric description (supplied by the manufacturers of the paper, in terms of their own color space) of each patch for comparison and calibration. The patches that comprised each test Mondriaan were chosen from a shaken bag, to provide some randomness in the structure of their gamut. Each Mondriaan contained 60 patches of paper. Figures 1–3 show the gamuts of images of these Mondriaans taken under different colored lights, and confirm the claim that the gamut is skewed by colored lights.

6.2 Implementation Details

Crule is simple to implement. Since the support of the photoreceptor spectral sensitivities is nearly disjoint, the matrixes in the feasible set can be approximated with diagonal matrixes.

A canonical gamut can only be approximated. The gamut of any set of observations of a finite set of surfaces consists of a cloud of points. Since we know that the canonical gamut is convex, we can approximate it by the convex hull of this cloud of points. I refer to this polyhedron as the observed canonical gamut. The observed canonical gamut was formed by imaging 180 of the set of 202 Color-aid papers under white light.

Although the gamut that we observe in an image is just a cloud of points, every point in the convex hull of this gamut corresponds to a real surface reflectance. This is because the canonical gamut is convex, and the observed gamut is just a sampling of the image of the canonical gamut in a linear map. As a result, we can consider the convex hull of this gamut. I shall call this polyhedron the observed gamut.

The feasible set consists of those diagonal maps that take the observed gamut to a subset of the observed canonical gamut. The observed gamut is convex, the observed canonical gamut is convex, and the maps are linear. As a result, any map that takes every vertex of the observed gamut to a point inside the canonical gamut, takes the observed gamut to a subset of the canonical gamut, and therefore lies in the feasible set.

We wish to construct the set of diagonal maps that take every vertex of the observed gamut to a point inside the observed canonical gamut. To do this, for each vertex of the observed gamut, we form the set of maps that take this vertex to a point inside the canonical gamut, and intersect these sets. In turn, the set of maps that take a single vertex p of the observed gamut to a point inside the observed canonical gamut is a convex polyhedron, which I shall refer to as $\mathfrak{M}_p$. For a proof that $\mathfrak{M}_p$ is a convex polyhedron, see appendix I.

The vertices of $\mathfrak{M}_p$ consist of those maps that take a vertex, p, of the observed gamut, to a vertex of the observed canonical gamut (proved in appendix I). Because the maps are diagonal, the vertices of $\mathfrak{M}_p$ (and hence $\mathfrak{M}_p$ itself) are easily computed for any p. The final implementation works as follows:

- The convex hull of the observed gamut is computed.
- For each vertex p of this hull, $\mathfrak{M}_p$ is computed.
- Intersect all the sets computed in this way. This process is relatively simple, because the sets are convex polyhedra in three dimensions. The result is the feasible set.
- Within the feasible set, apply an estimator to choose the map most likely to achieve color constancy. Apply this map to the image, to obtain the color descriptors.

This algorithm is relatively simple to implement, but intersecting the convex hulls requires care as 3D convex hulls are rather difficult to manipulate. Explicitly intersecting the hulls is an unwise technique to use, as it tends to generate clouds of hull points nearly on the same plane, with attendant difficulties of representation and computation. The problem is better approached using cuboid approximations (see Cameron 1989; Woodwark and Quinlan 1984).

The estimator used simply chose the map that gave the gamut with the largest volume. This means that the image of the mapping of the gamut of the original picture will fit inside the gamut obtained under the canonical illuminant (because we have chosen a map in the constraint set), and will be the "largest" such image. The parameters corresponding to this map are easily found by noting that a diagonal linear map takes a volume to that volume multiplied by the trace of the map. Thus, the feasible set is simply searched for the map with the largest trace.

This discussion implicitly assumes that the observed gamut is a good approximation of the gamut that would have been seen if every possible surface reflectance were shown under the illuminant. If the observed gamut does have this property, then the feasible set will be small. If it doesn't, the feasible set can be large, and the estimator is more likely to err. This case occurs when the surface reflectances in the scene are not well distributed—for example, when the scene contains only shades of one or two colors.

6.3 Results

Crule's performance has been tested on three Mondriaans each of sixty chips of colored paper, each viewed under six different lights. The Brainard and Wandell model of the Retinex algorithm (Brainard and Wandell 1986) was implemented for comparison. Both algorithms achieve roughly the same degree of constancy, with the Retinex algorithm perhaps slightly outperforming the **Crule** algorithm, when tested on a single Mondriaan viewed under many different lights. Color figures 1 through 6 show the inputs for each algorithm. Although the images appear widely skewed in color, during the imaging it was possible for human observers to give reasonable color names to the Mondriaan patches. Color figures 7 through 12 show the outputs from **Crule** for the given images. The images are similar in color, suggesting that **Crule** is capable of good color constancy. However, the output for an image taken under red light, shown in color figure 8, shows that **Crule** performed poorly on this image. In color figure 9, the output of **Crule** for red light has been omitted because the algorithm failed as badly. This is a quantization effect, discussed more fully in section 7.

Notice that **Crule** has performed rather more poorly on the third Mondriaan than on the other two. This is because the third Mondriaan has a narrower range of colors in it, so that the illuminant is less well constrained.

Figures 4 through 6 show the receptor responses observed for a set of color chips, selected from the first Mondriaan to give a fair impression, under six different lights. These are widely scattered, demonstrating how strongly illuminant color can skew receptor responses. Figures 7 through 9 show the descriptors computed by **Crule** for the same set of chips, from images taken under six different lights. The descriptors are far less widely scattered, indicating that **Crule** is capable of good color constancy.

Color figures 13 through 18 show the outputs of the Retinex algorithm, on the same images, for comparison. The Retinex algorithm is also susceptible to the quantization effect mentioned above, as color figures 13 and 15 show. On the whole, for the case of a single Mondriaan and a set of different lights, the Retinex algorithm appears to out perform **Crule**. However, the Retinex algorithm depends on a geometric average of surface color remaining constant, and is confounded when, for example, red borders are appended to a Mondriaan to change this average (see color figure 19). Large failures in constancy can be caused in this way, as color figure 20 demonstrates. This figure compares the performance of **Crule** and the Retinex algorithm on an image of the first Mondriaan to their performance when red borders have been attached to the Mondriaan, both viewed under white light. Because the Retinex algorithm is sensitive to the spatial average of surface color, a change in this average skews its surface color descriptors. This causes the slight green cast to the output of the Retinex

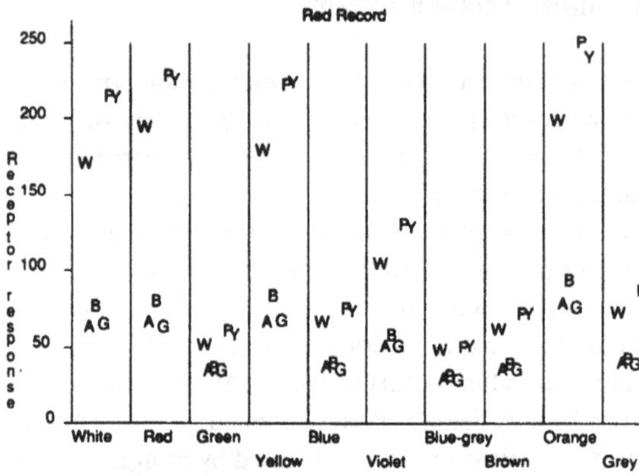

Fig. 4. This and the next five figures demonstrate the performance of **Crule** for 10 chips, selected to give a fair impression. This and the next two figures show the receptor responses for 10 chips, under six different lights. Each column represents a chip and is labeled with a color name for each chip under incandescent light. This figure shows the responses for the camera's red channel. The response under white light is plotted as a "W," under blue-green light as an "A," under blue light as a "B," under green light as a "G," under purple light as a "P," and under yellow light as a "Y." See section 6.1 for the meaning of the color names for the illuminants. Note the wide spread of responses, which will make it difficult to use these values to describe chips. The precise color name, in the Color Aid corporation's scheme, for each chip is as follows. The chip labeled white is white, red is RO-Hue, yellow is Y-T1, blue is BGB-S3, violet is VRV-S1, blue-grey is GYG-S1, brown is OYO-S1, orange is YO-T3, and grey is grey.

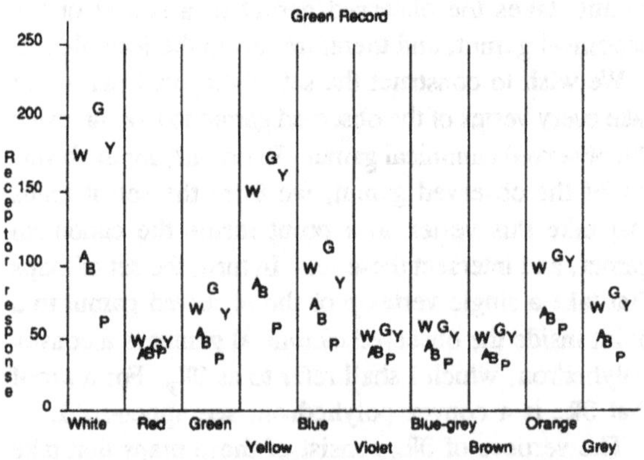

Fig. 5. This figures shows the receptor responses for the camera's green channel, using the same conventions as figure 4.

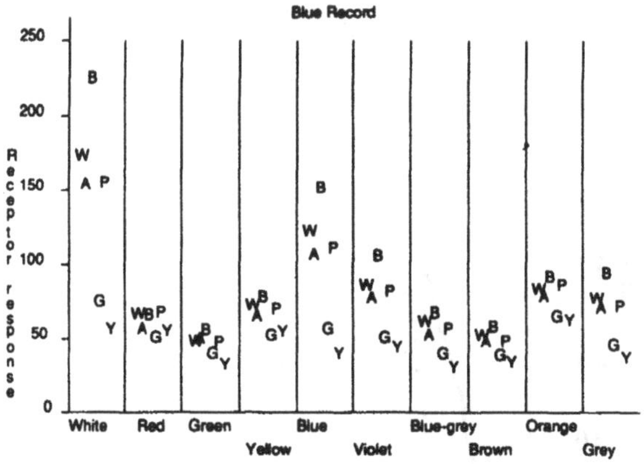

Fig. 6. This figure shows the receptor responses for the camera's blue channel, using the same conventions as figure 4.

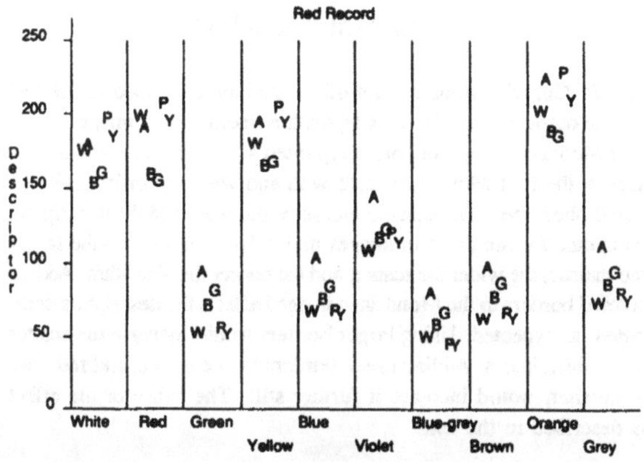

Fig. 7. This figures shows the descriptors in the red channel, output by **Crule**, using the same conventions as figure 4. Notice that the descriptors are clustered, indicating a high degree of color constancy.

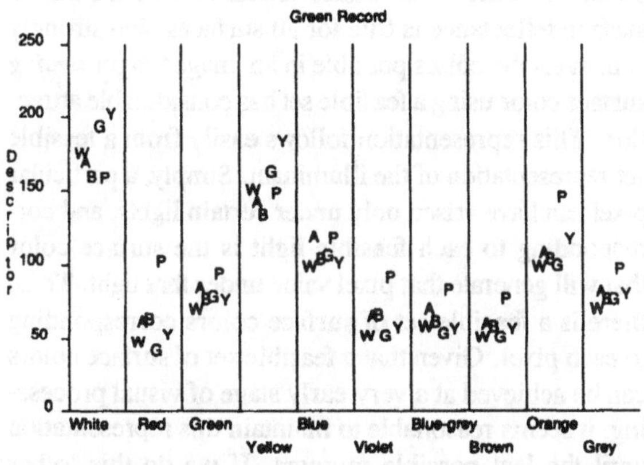

Fig. 8. This figure shows the descriptors in the green channel, output by **Crule**, using the same conventions as figure 4. Again the descriptors are clustered, indicating a high degree of color constancy.

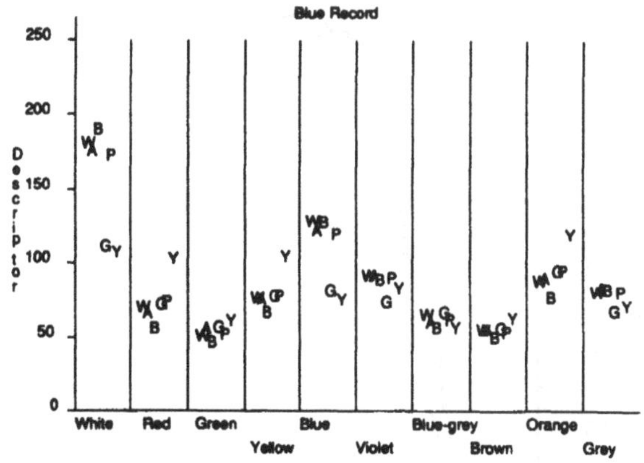

Fig. 9. This figure shows the descriptors in the blue channel, output by **Crule**, using the same conventions as figure 4. Again, the descriptors are clustered, indicating a high degree of color constancy.

for the Mondriaan with red borders in color figure 20. This effect can be made very large by increasing the size of the borders, or by comparing images with different colored borders. Since **Crule** does not depend on the spatial extent of a color, the borders do not affect its descriptors.

A statistical analysis is desirable, because images may conceal misbehavior on the part of the algorithms. There is no standard measure of color constancy. The analysis shown uses the median Euclidean distance of the outputs from the average (over the different lights) output, for each chip. This median is then normalized by the Euclidean magnitude of the outputs. Clearly, this statistic measures clustering, and will be zero for every chip for a perfect color constancy algorithm; and as the algorithm delivers increasingly poor performance, the statistic will increase. Normalization is necessary to prevent an algorithm improving its apparent performance by multiplying its outputs by a small number.

The graphs in figures 10 through 12 show the cumulative distribution for this statistic, for two separate images of sixty chips under six different lights (white, green, blue-green, yellow, and purple). The results for both **Crule** and Retinex under red light were skewed for reasons described in section 7, and were omitted from this analysis. The graphs plot the statistic for descriptors computed from images of the first Mondriaan, and the statistic for descriptors computed for images of the first Mondriaan with and without red borders. The red borders skew the descriptors computed by the Retinex algorithm, and the second plot for the Retinex algorithm reflects this substantially larger scatter in its descriptors. This can

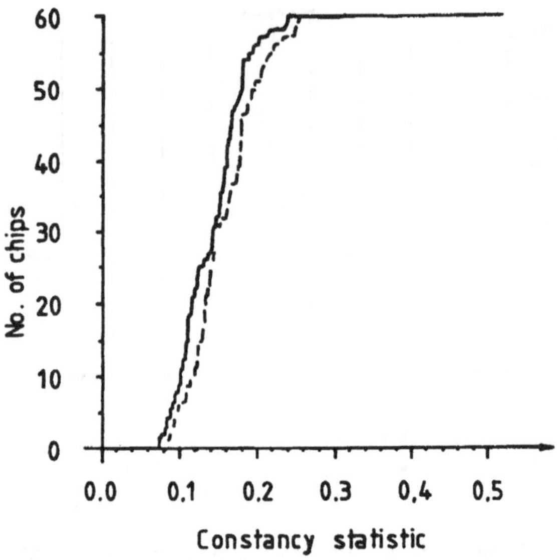

Fig. 10. Cumulative distribution of the statistic described in the text for sixty descriptors computed by **Crule**, operating on images of the first Mondriaan without borders (plotted as a solid line), and on images of the first Mondriaan both with and without borders (plotted as a dashed line). The statistic measures the scatter of the descriptors computed for the chips in images under different lights—the larger the statistic, the wider the scatter, and the poorer the algorithm. Note that adding colored borders to the Mondriaan imaged does not significantly affect the descriptors that **Crule** computes, as expected.

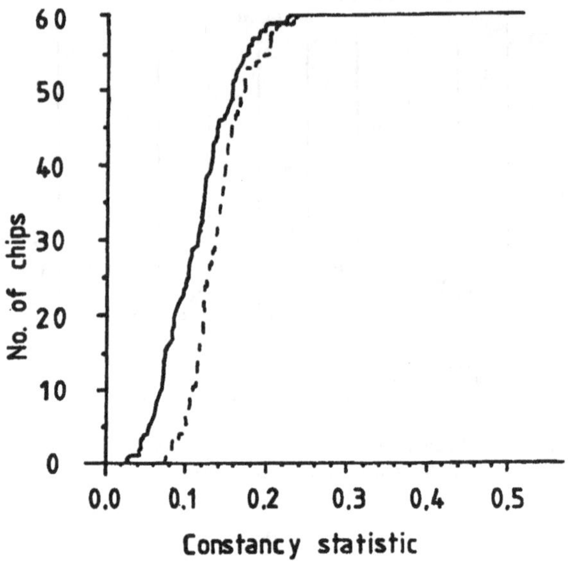

Fig. 11. Cumulative histogram of the statistic described in the text for the outputs of the Retinex algorithm operating on images of the first Mondriaan without borders (plotted as a solid line), and on images of the first Mondriaan both with and without borders (plotted as a dashed line). The statistic measures the scatter of the descriptors computed for the chips in images under different lights—the larger the statistic, the wider the scatter, and the poorer the algorithm. Adding colored borders to the Mondriaan imaged affects the descriptors computed, as expected. Using larger borders would increase the scatter of the descriptors. Adding green borders to one image, and red ones to another, would increase it further still. The cause of the effect is described in the text.

be made very large, by appending large borders, or by testing the Retinex algorithm on one scene with green borders, one with red borders, and one without borders. This effect is a basic property of any algorithm that assumes that some spatio-temporal average of surface color is constant. Other algorithms with this flaw include those of Buchsbaum (1981) and Gershon (1988). This skewing effect can happen easily in real images: for example, one does not expect the average surface color of a view of a forest in spring to be the same as that of the same view, taken in autumn.

Crule does not have this flaw, but if one considers descriptors for only one image under many lights, the Retinex algorithm performs better than **Crule**, because in this case the average surface color is trivially constant. This may be alleviated by computing the feasible set to higher accuracy; at present, the descriptors computed by the new algorithm may have errors as large as 2% of full range simply as a result of the scheme adopted for computing 3D convex hulls, which makes computing the feasible set to high accuracy expensive in space.

7 Discussion

Crule is simple to implement, and works well for real pictures. **Crule** is successful because the constraint on surface reflectance is true for all surfaces, and strongly constrains the colors possible in an image.[4] *Representing* surface color using a feasible set has considerable attraction. This representation follows easily from a feasible set representation of the illuminant. Simply, a particular pixel can have arisen only under certain lights, and corresponding to each feasible light is the surface color that will generate that pixel value under that light. Thus, there is a feasible set of surface colors corresponding to each pixel. Given that a feasible set of surface colors can be achieved at a very early stage of visual processing, it seems reasonable to maintain this representation until the last possible moment. If we do this, other sources of visual information may provide cues that reduce the size of the feasible set—shape and surface

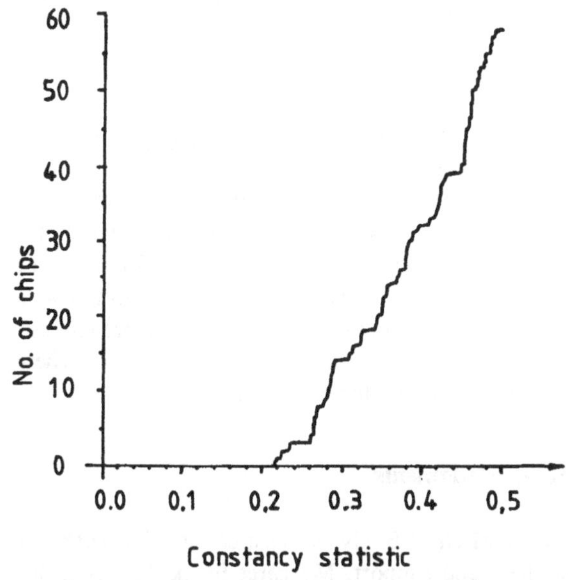

Fig. 12. Cumulative histogram of the statistic described in the text for the first Mondriaan. The statistic measures the scatter of the descriptors for the chips in images under different lights—the larger the statistic, the wider the scatter. In this case, the statistic is computed for the receptor responses, so that one can confirm that both **Crule** and the Retinex algorithm can produce descriptors that are an improvement on the receptor responses alone.

color, for example, are deeply entwined by mutual illumination effects (Gershon et al. 1986; Forsyth and Zisserman 1989; Forsyth and Zisserman 1990; Ho et al. 1989).

Crule will produce color descriptors for two color images, of the kind used by, for example Land (Land 1959a; Land 1959b). In these experiments Land took slides of a series of scenes. One slide was exposed only in the red band, the other only in the green band, so that for each scene he had a red slide and a green slide. The red separation was inserted in a slide projector casting red light, and the green separation in a projector casting white light. The two slide images were shown in register to an audience, which reported a wide range of colors. Naive colorimetric methods would predict that the images would contain only a range of shades of pink. This effect was not in fact first noticed by Land; the history of this effect is well presented in (Judd 1960).

It would be a definite advantage if **Crule** was able to reproduce this property. A simple geometric argument suggests that this is the case. The gamut of a two-color image is on a triangular planar patch in surface receptor space, with one corner at the origin, one corner at the "brightest white" point, and one corner at a "deep red" point. We can reasonably expect to be able to construct a map that takes this into a larger patch within the gamut.

Two color images were constructed from conventional color imges, by mapping the green and blue components to r times the original green component, and the red component to the sum of r times the original green and s times the original red components, where $0 \leq r, s \leq 1$. These receptor responses model the receptor responses expected from the slide experiment.

Crule has been run on such images. The result is shown in color figure 21. No claim is made that this result is quantitatively similar to human performance, only that the algorithm can infer a broad gamut from a two-color image. This is obviously the case, although the images display in neither the original data nor in the photographs quite the breadth of gamut that Land describes (he talks of "lovely color photographs" (Land 1959a, p. 120). Green patches are conspicuous by their absence, as are yellow patches.

Both **Crule** (color figure 8) and the Retinex algorithm (color figures 13 and 15) performed very poorly on images taken under red light. This failure is explained as follows. There is very little information in the input data either in the green or in the blue channels, because to keep the camera from saturating in the red channel, the aperture had to be closed. It would clearly be unfair to attempt to demonstrate the effectiveness of an algorithm for color constancy that employs the coefficient rule, and then to adjust the aperture separately for each color exposure, for that would presuppose a constancy algorithm. Equally, if this is not done, *constancy is impaired* by measurement errors.

The constancy algorithm is connected to its receptors by a quantizing system with constant quantization thresholds. As a result, under red light, the signal on the green channel has a dynamic range of about three bits. The constancy algorithm is thus more ignorant of what the green transducer is measuring, than it is about what the red transducer is measuring, and its constancy is impaired. The human vision system faces a similar problem, in that the operating range of human photoreceptors is very much greater than the signaling capacity of the nerve channels (Barlow and Mollon 1982). In humans, this problem is accounted for by a process known as adaptation, which is described in more detail elsewhere (for example, Bartleson 1977; Beck 1972; Wyszecki and Stiles 1982). The effect of adaptation is to use a spatio-temporal average of the receptor's inputs to adjust its gain so that it can measure effects over a wide dynamic range. The color constancy of a machine vision system could be significantly improved by a similar approach. One might, for example, use a low bandwidth control loop to ensure that the spatio-temporal average of the video signal is constant, by adjusting the integrating time of the CCD camera. By supplying the integrating time information to the color constancy system, one could improve its performance on deeply colored lights.

This work had not addressed the case where the illuminant varies over space. The Retinex algorithm, and algorithms that devolve from it (Horn 1974; Blake 1985; Brelstaff and Blake 1987; Funt and Drew 1988) represent the only existing implemented techniques for dealing with this problem. These algorithms work by declaring a spatial frequency threshold, and asserting that changes below this threshold are due to lighting effects, and changes above it arise from surface color effects. It is unclear how well such a technique will work for other than Mondriaan images. There is every reason to believe that illuminant effects occur at high spatial frequencies (Gilchrist 1983). This suggests that complex processes are necessary to determine whether a change in an image occurred as a result of a change in surface color or a change in lighting, before color constancy programs can work successfully in any but very constrained worlds.

8 Conclusion

By analyzing the circumstances under which color constancy is possible, I have developed a color constancy algorithm, **Crule**, and have demonstrated that it achieves color constancy on real images of Mondriaan's. **Crule** estimates the illuminant in colored pictures using a system of constraints. These contraints derive from physical restrictions on the form of surface reflectance functions.

The performance of **Crule** on real images compares favorably with that of the Retinex algorithm of Land (Land and McCann 1971) when run on single Mondriaan images. However, changing the spatial average of surface reflectance disrupts the color constancy of the Retinex algorithm, but does not affect **Crule**. I have shown that **Crule** can mimic qualitatively Land's (Land 1959a; Land 1959b) work on two-color images. This experimental work lead us to see chromatic adaptation as a measurement strategy employed by humans, and to realize that machines require a similar strategy to perform accurate constancy.

Although this algorithm gives very good results, its success at computing surface color depends very much on the unrealistic assumptions that underly the Mondriaan world. These may fail in the real world: for example, illuminants change quickly spatially (Gilchrist 1983). It is clear that surface color algorithms in vision systems will need to be able to cope with more interesting objects than flat, frontally presented surfaces. There is much scope for further work.

Acknowledgments

I thank Michael Brady, who supervised this work, for his help and support. Margaret Fleck discussed this work at length with me. Andrew Blake, Don Pearson, and Tom Troscianko have made helpful comments. Brian Wandell and Laurence Maloney patiently helped me to understand how and why their techniques worked. An anonymous referee made numerous, careful comments that were an important help in making this paper readable.

Notes

1. This assumes that there is always a wide variety of surface reflectances in the scene.
2. Koenderink (1987) observes that Luther (1921) and Nyberg (1928) were aware of this result.
3. Address: 37 East 18th St., New York, NY 10003.
4. This constraint has been successfully exploited by Kawata et al. (1987) in a system that segments microscope images, based on dye concentrations.

References

H.B. Barlow and J.D. Mollon, *The Senses*. Cambridge University Press: Cambridge, 1982.

C.J. Bartleson, "A review of chromatic adaptation." In F.W. Billmeyer and G. Wyszecki (eds.), *Color 77*. Adam Hilger: Bristol, 1977.

J. Beck, *Surface Color Perception*. Cornell University Press: Ithaca, NY, 1972.

A. Blake, "Boundary conditions for lightness computation in Mondriaan world." In D. Ottoson and S. Zeki (eds.), *Central and Peripheral Mechanisms of Color Vision*. Macmillan: New York, 1985.

W. Boothby, *An Introduction to Differentiable Manifolds and Riemannian Geometry*. Academic Press: San Diego, CA, 1986.

G. Buchsbaum, "A spatial processor model for object color perception," *J. Franklin Inst.* 310: 1–26, 1980.

D.H. Brainard and B.A. Wandell, "Analysis of the Retinex theory of color vision," *J. Opt. Soc. Amer. A* 3: 1651–1661, 1986.

G.J. Brelstaff, "Inferring surface shape from specular reflections." Ph.D. thesis, University of Edinburgh: Edinburgh, 1988.

G. Brelstaff and A. Blake, "Computing lightness," *Pattern Recognition Letters* 5: 129–138, 1987.

M.H. Brill, "A device performing illuminant-invariant assessment of chromatic relations," *J. Theor. Biol.* 71: 473–478, 1978.

M.H. Brill, "Computer simulation of object color recognisers," *J. Opt. Soc. Amer.* 69: 1405, 1979.

M.H. Brill and G. West, "Contributions to the theory of invariance of color under the condition of varying illumination," *J. Math. Biol.* 11: 337–350, 1981.

S.A. Cameron, "Efficient intersection tests for objects defined constructively," *Intern. J. Robotics Res.* 8: 1, 1989.

M. D'Zmura and P. Lennie, "Mechanisms of color constancy," *J. Opt. Soc. Amer. A* 3 (10): 1662–1672, 1986.

R.K. Eisenschitz, *Matrix Algebra for Physicists*. Heinemann: London, 1966.

D.A. Forsyth, "A novel algorithm for colour constancy," *Proc. 2nd Intern. Conf. Comput. Vision*, Tampa, 1988a.

D.A. Forsyth, "Colour constancy and its applications in machine vision." Ph.D. thesis, University of Oxford, Oxford, 1988b.

D.A. Forsyth and A. Zisserman, "Mutual illumination," *Proc. 26th IEEE Conf. Comput. Vision and Pattern Recog.*, Las Vegas, 1989.

D.A. Forsyth and A. Zisserman, "Shape from shading in the light of mutual illumination," *Image and Vision Computing*, 1990 (in press).

B.V. Funt and M. Drew, "Color constancy computation in near-Mondriaan scenes using a finite dimensional linear model," *Proc. IEEE Conf. Comput. Vision and Pattern Recog.* 1988.

R. Gershon, "The use of colour in computational vision." Ph.D. thesis, University of Toronto, 1988.

R. Gershon, A. Jepson, and J. Tsotsos, "Ambient illumination and the determination of material changes," *J. Opt. Soc. Amer. A* 3: 1700–1707, 1986.

A.L. Gilchrist, S. Delman, and A. Jacobsen, "The classification and integration of edges as critical to the perception of reflectance and illumination," *Perception and Psychophysics* 33: 425–436, 1983.

J. Ho, B.V. Funt, and M.S. Drew, "Disambiguation of illumination and surface reflectance from spectral power distribution of color signal: Theory and applications," CSS/LCCR TR 88-18, Centre for Systems Science, Simon Fraser University, 1988.

J. Ho, M.S. Drew, and B.V. Funt, "Color constancy from mutual reflection," CSS/LCCR TR 89-02, Centre for Systems Science, Simon Fraser University, 1989.

B.K.P.H. Horn, "Determining lightness from an image," *Comput. Graph. Image Proc.* 3: 277–299, 1974.

D.B. Judd, "Appraisal of Land's work on two-primary colour projections," *J. Opt. Soc. Amer.* 50: 254–268, 1960.

S. Kawata, S. Sasaki, and S. Minami, "Component analysis of spatial and spectral patterns in multispectral images. I. Basis," *J. Opt. Soc. Amer. A* 4: 2101–2106, 1987.

G.J. Klinker, S.A. Shafer, and T. Kanade, "Using a colour reflection model to separate highlights from object colour," *Proc. 1st Intern. Conf. Comput. Vision*, London, 1987.

J.J. Koenderink, "Color atlas theory," *J. Opt. Soc. Amer. A* 4: 1314–1321, 1987.

J. von Kries, "Beitrag zur Physiologie der Gesichtsempfindung," *Arch. Anat. Physiol.* 2: 505–524, 1878.

E.H. Land, "Color vision and the natural image. Part I," *Proc. Natl. Acad. Sci. U.S.A.* 45: 115–129, 1959a.

E.H. Land, "Color vision and the natural image. Part II," *Proc. Natl. Acad. Sci. U.S.A.* 45: 636–644, 1959b.

E.H. Land, "Color vision and the natural image. Part III: Recent advances in Retinex theory and some implications for cortical computations," *Proc. Natl. Acad. Sci. U.S.A.* 80: 5163–5169, 1983.

E.H. Land, "Recent advances in Retinex theory." In D. Ottoson and S. Zeki (eds.), *Central and Peripheral Mechanisms of Colour Vision*. Macmillan: New York, 1985.

E.H. Land, "Recent advances in Retinex theory," *Vision Research* 26: 7–21, 1986.

E.H. Land and J.J. McCann, "Lightness and Retinex theory," *J. Opt. Soc. Amer.* 61 (1): 1–11, 1971.

H.-C. Lee, "Method for computing the scene-illuminant chromaticity from specular highlights," *J. Opt. Soc. Amer. A* 3: 1694–1699, 1986.

R. Luther, "Aus dem Gebiet der Fabreizmetrik," *Z. Techn. Phys.* 12: 540–558, 1921.

L.T. Maloney, "Computational approaches to color constancy." Ph.D. dissertation, Stanford University, Stanford, CA, 1984.

L.T. Maloney, "Evaluation of linear models of surface spectral reflectance with small numbers of parameters," *J. Opt. Soc. Amer. A* 3 (10): 1673–1683, 1986.

L.T. Maloney and B.A. Wandell, "A computational model of color constancy," *J. Opt. Soc. Amer.* A1: 29–33, 1986.

J.J. McCann, S.P. McKee, and Taylor, "Quantitative studies in Retinex theory," *Vision Research* 16: 445–458, 1976.

N. Nyberg, "Zum Aufbau des Farbkoerpers im Raume aller Lichtempfindungen," *Z. Phys.* 52: 506–419, 1928.

P. Sallstrom, "Colour and physics: some remarks concerning the physical aspects of human colour vision," University of Stockholm: Institute of Physics Report 73-09, 1973.

W.S. Stiles, G. Wyszecki, and N. Ohta, "Counting metameric object-color stimuli using frequency limited spectral reflectance functions," *J. Opt. Soc. Amer.* 67 (6): 779–784, 1977.

B.A. Wandell, "The synthesis and analysis of color images," *IEEE Trans. PAMI* 9 (1): 2–13, 1987.

G. West and M.H. Brill, "Necessary and sufficient conditions for von Kries chromatic adaptation to give color constancy," *J. Math. Biol.* 15: 249–258, 1982.

J.R. Woodwark and K.M. Quinlan, "Reducing the effect of complexity on volume model evaluation," *Comput.-Aided Design J.* 14 (2), 1984.

J.A. Worthey, "Limitations of color constancy," *J. Opt. Soc. Amer. A* 2 (7): 1014–1026, 1985.

J.A. Worthey and M.H. Brill, "Heuristic analysis of von Kries color constancy," *J. Opt. Soc. Amer. A* 3 (10): 1708–1712. 1986.

G. Wyszecki and W.S. Stiles, *Color Science: Concepts, and Methods, Qualitative Data and Formulae*. J. Wiley: New York, 1982.

G. Wyszecki and W.S. Stiles, "High-level trichromatic colour matching and the pigment bleaching hypothesis, *Vision Research* 20: 23–37, 1980.

Appendix I

Proposition. The set of maps that take a single vertex p of the observed gamut to a point inside the observed canonical gamut, $\mathcal{C}_o$, is a convex polyhedron, $\mathcal{M}_p$, whose vertexes are the maps that take p to the vertexes of $\mathcal{C}_o$.

Proof: If $f(p) \in \mathcal{C}_o$ and $g(p) \in \mathcal{C}_o$, then so is $(1 - \mu)f(p) + \mu g(p)$ for $0 \leq \mu \leq 1$, by the convexity of $\mathcal{C}_o$. Thus, $\mathcal{M}_p$ is convex. Clearly the maps that take p to some vertex of $\mathcal{C}_o$ are on the boundary of $\mathcal{M}_p$. Consider the vertexes, v_i, of a face of $\mathcal{C}_o$, and maps $F_i(p)$, with $F_i(p) = v_i$. Clearly F_i lies on the boundary of $\mathcal{M}_p$. Furthermore, if the set $(1 - \mu)v_i + \mu v_j, \forall\mu \in [0, 1]$ is an edge of $\mathcal{C}_o$, then $(1 - \mu)F_i + \mu F_j$ lies on the boundary of $\mathcal{M}_p$. One shows that the boundary of $\mathcal{M}_p$ consists of plane faces, edges, and vertexes in this way. Thus, $\mathcal{M}_p$ is a convex polyhedron.

Appendix II

Proposition. For $\Sigma \subset R^L$, and $F, G \in Gl(L, \mathbf{R})$

$$F(\Sigma) = G(\Sigma) \Leftrightarrow G = FA$$

for some $A \in Gs[\Sigma]$, where $Gs[\Sigma]$ is the linear similarity group of Σ.

Proof: $\Leftarrow$ is clearly true. Now $F, G \in GL(L)$, and so F and G are bijective, by definition. Then

$$G = FA$$

and for some A. In fact,

$$A = F^{-1}G$$

but since $F(\Sigma) = G(\Sigma)$, and F and G are bijective, $F^{-1}G(\Sigma) = \Sigma$. Thus $\Rightarrow$ is true.

Proposition. For $\Sigma \subset \mathbf{R}^L$, $F \in GL(L)$, $Gs[\Sigma]$ and $Gs[F(\Sigma)]$ are isomorphic.

Proof: Consider $f: Gs[\Sigma] \rightarrow Gs[F(\Sigma)]$ corresponding in the natural way to $F: \Sigma \rightarrow F(\Sigma)$. Clearly, for $A \in Gs[\Sigma]$, $f(A) = FAF^{-1} \in Gs[F(\Sigma)]$. We verify easily that the map f is a homomorphism, for $FI_nF^{-1} = I_n$, and $FABF^{-1} = FAF^{-1}FBF^{-1}$. In fact, the map is injective, for $FAF^{-1} = FBF^{-1} \Leftrightarrow A = B$. But we may define a similar injective homomorphism $\bar{f}: Gs[F(\Sigma)] \rightarrow Gs[\Sigma]$ by $\bar{f}(B) = F^{-1}BF$. Hence, the groups are isomorphic.

COLOR FIGURES

See color plate E for Color Figures 1-4.
See color plate F for Color Figures 5-10.
See color plate G for Color Figures 11-16.
See color plate H for Color Figures 17-21.

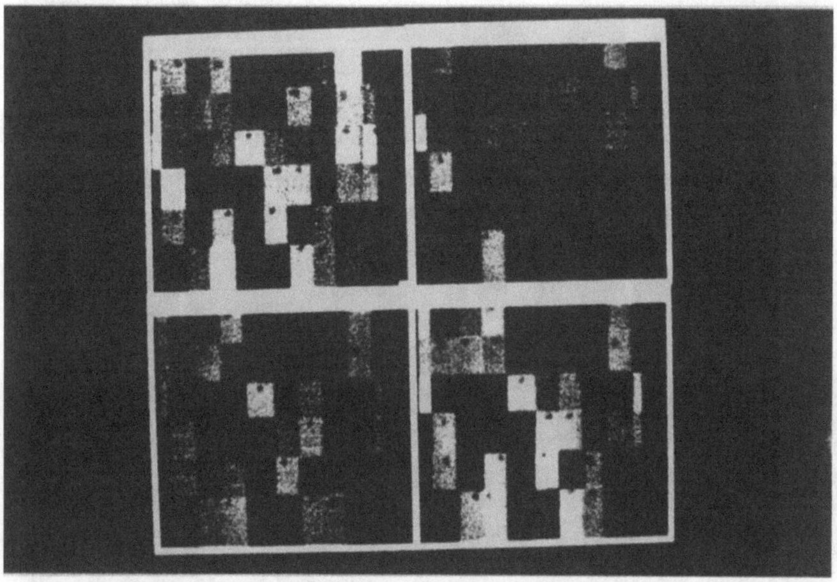

Color Figure 1. Images of the first Mondriaan, imaged under (clockwise, from top left) white, green, purple, and red lights. Note the enormous differences. These images show receptor responses. In each case, a human observer was able to name the colors of the Mondriaan patches.

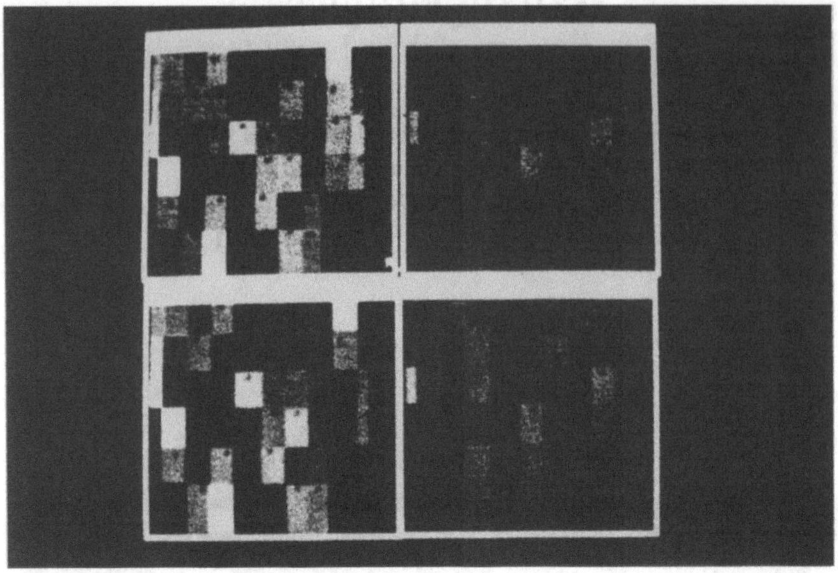

Color Figure 2. Images of the first Mondriaan, imaged under (clockwise, from top left) white, blue-green, blue, and yellow lights.

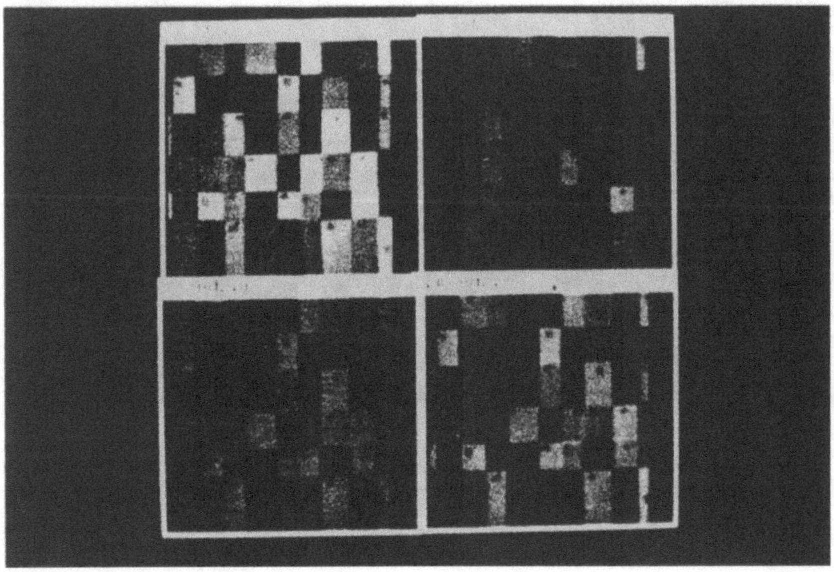

Color Figure 3. Images of the second Mondriaan, imaged under (clockwise, from top left) white, green, purple, and red lights.

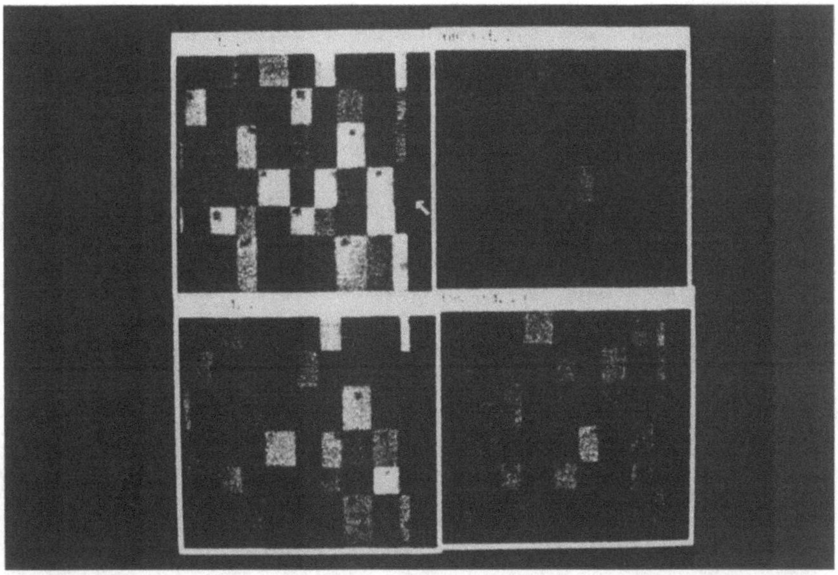

Color Figure 4. Images of the second Mondriaan, imaged under (clockwise, from top left) white, blue-green, blue, and yellow lights.

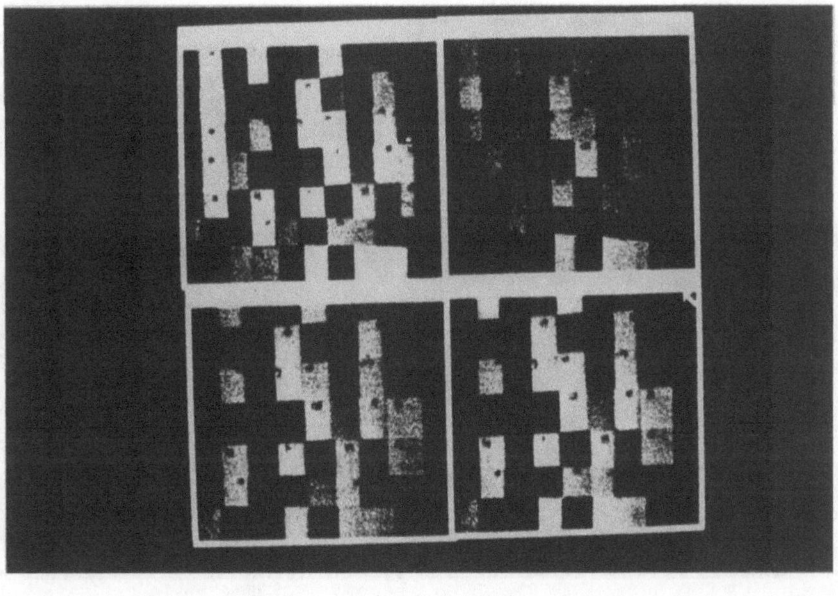

Color Figure 5. Images of third Mondriaan, imaged under (clockwise, from top left) white, green, purple, and red lights.

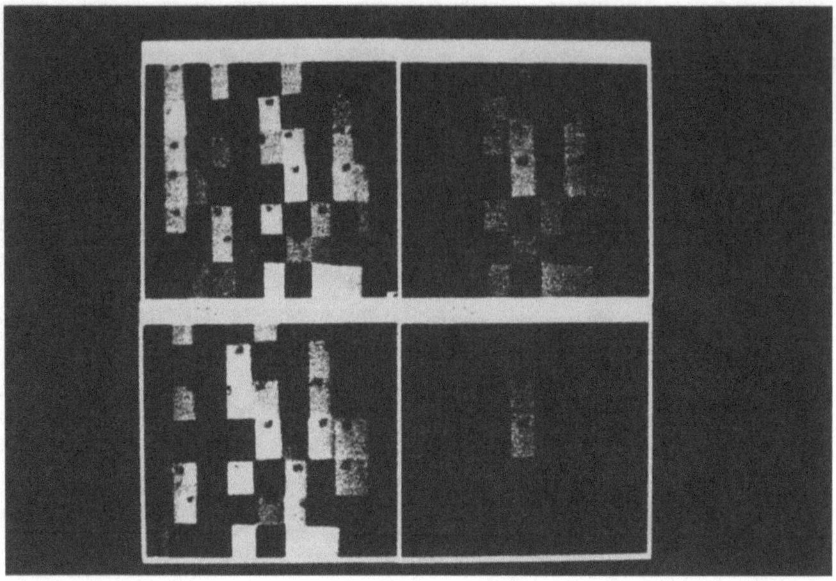

Color Figure 6. Images of third Mondriaan, imaged under (clockwise, from top left) white, blue-green, blue, and yellow lights.

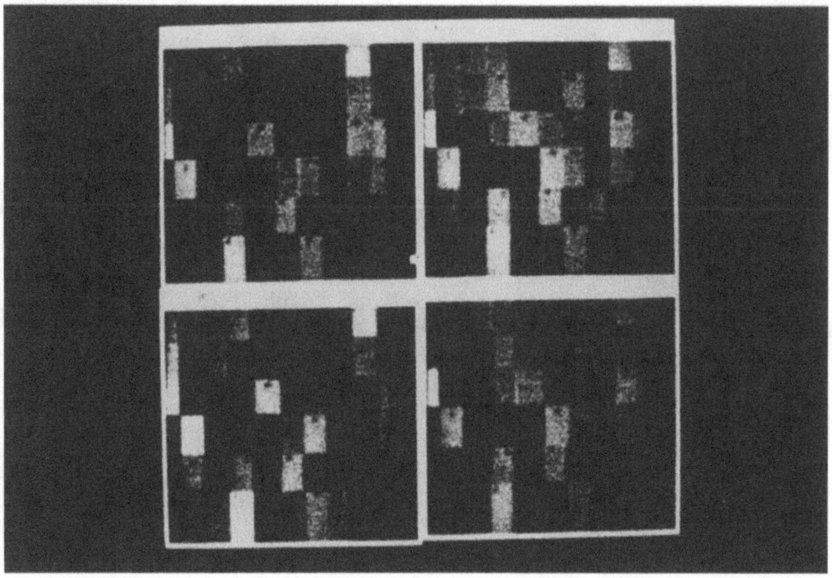

Color Figure 7. Outputs of **Crule** for the first Mondriaan derived from images under (clockwise, from top left) white, blue-green, blue, and yellow lights. **Crule** performs well, because these images look similar.

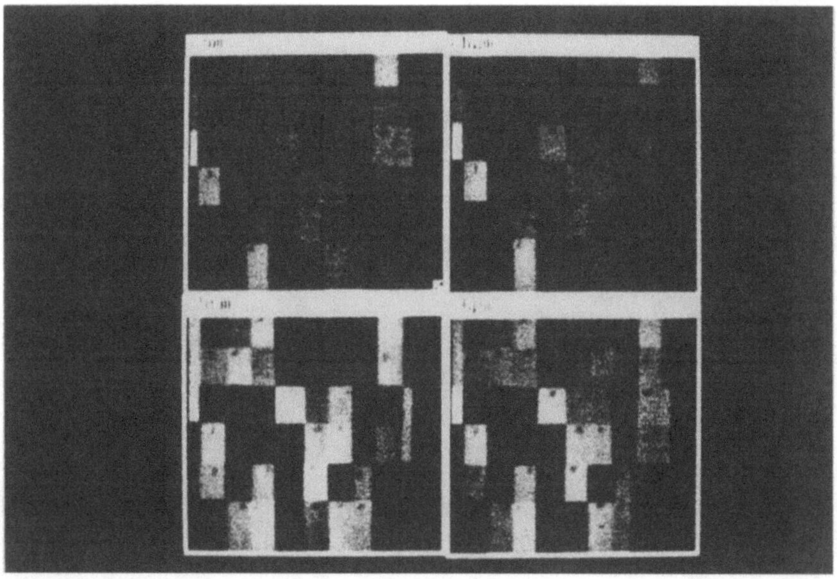

Color Figure 8. Outputs of **Crule** for the first Mondriaan derived from images under (clockwise, from top left) white, green, purple, and red lights. Notice that **Crule** performs poorly on the image taken under red light. This effect is explained in the text.

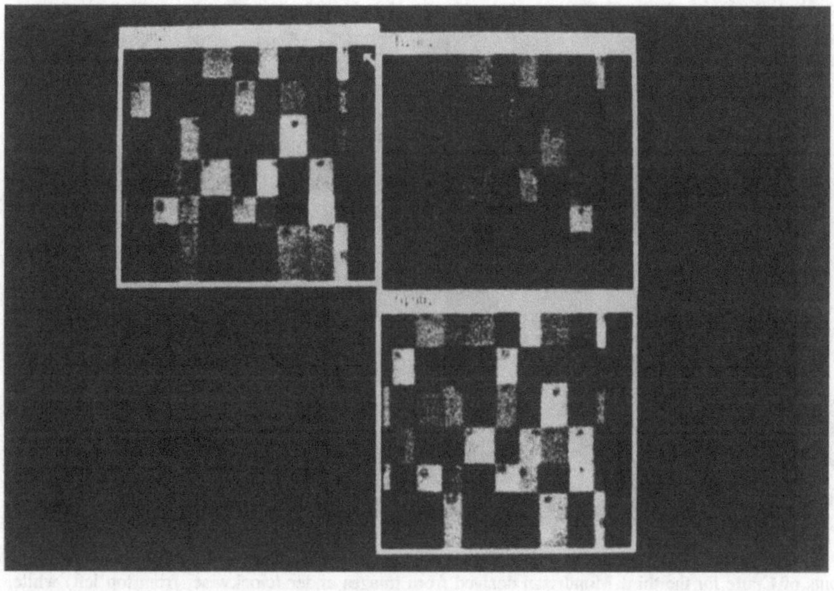

Color Figure 9. Outputs of **Crule** for the second Mondriaan derived from images under (clockwise, from top left) white, green, and purple lights. Again, **Crule** performed poorly on the image taken under red light, which was omitted for this reason.

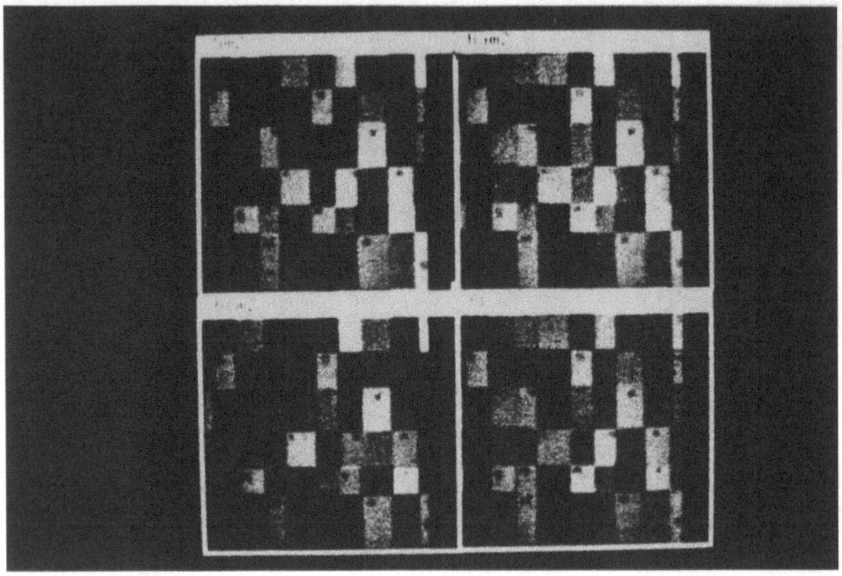

Color Figure 10. Outputs of **Crule** for the second Mondriaan derived from images under (clockwise, from top left) white, blue-green, blue, and yellow lights.

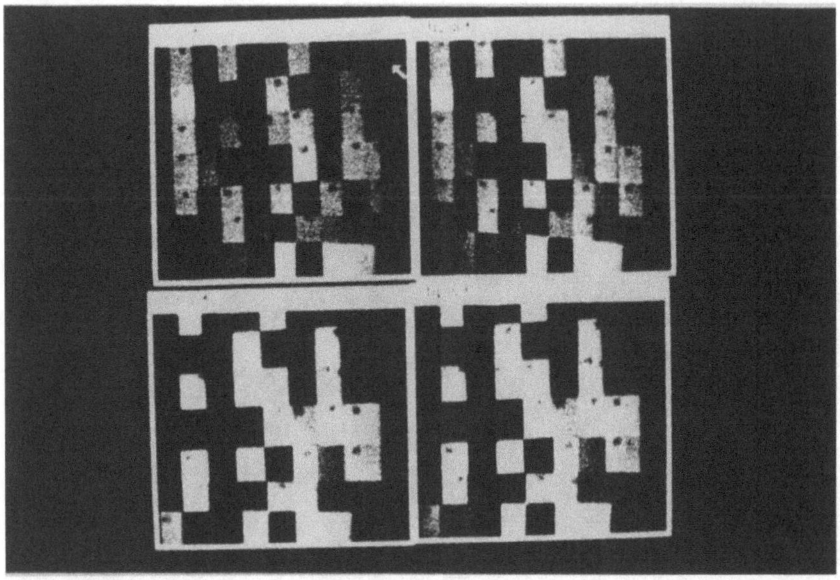

Color Figure 11. Outputs of **Crule** for the third Mondriaan derived from images under (clockwise, from top left) white, green, purple, and red lights. **Crule** is not performing very well on these images, because they do not contain a wide range of colors.

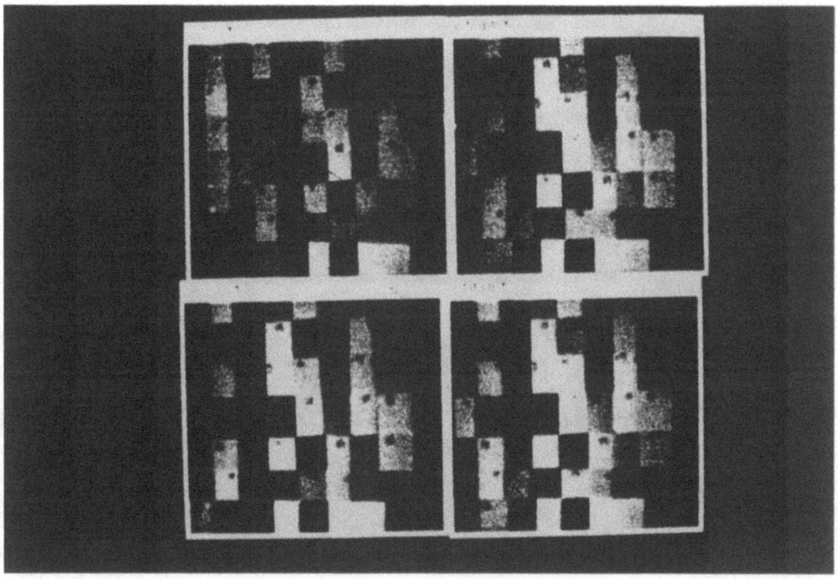

Color Figure 12. Outputs of **Crule** for the third Mondriaan derived from images under (clockwise, from top left) white, blue-green, blue, and yellow lights. **Crule** is not performing very well on these images, because they do not contain a wide range of colors.

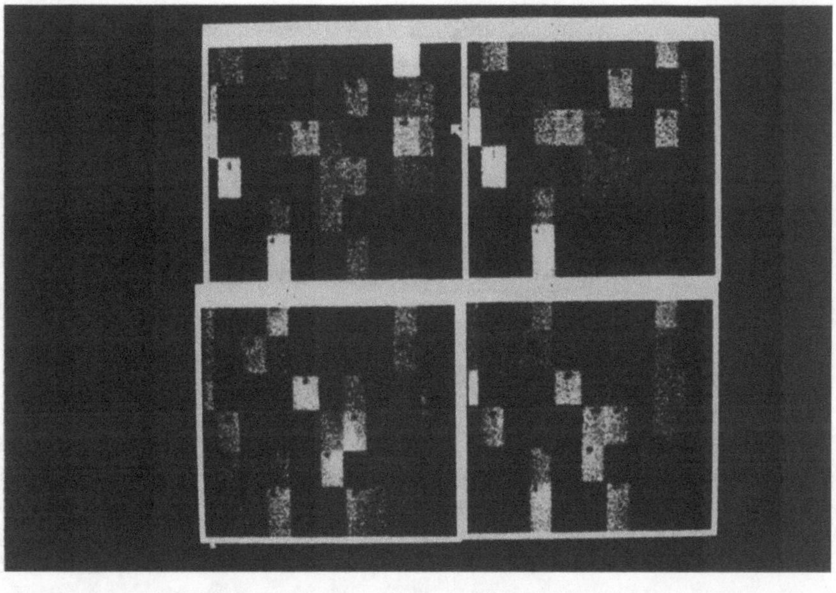

Color Figure 13. Outputs of the Brainard and Wandell (1986) model of the Retinex algorithm for the first Mondriaan derived from images under (clockwise, from top left) white, green, purple, and red lights.

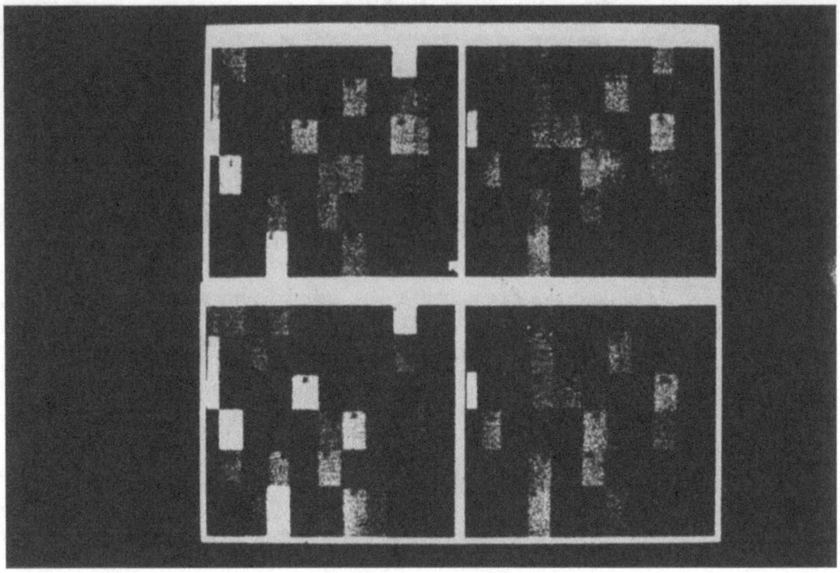

Color Figure 14. Outputs of the Brainard and Wandell (1986) model of the Retinex algorithm for the first Mondriaan derived from images under (clockwise, from top left) white, blue-green, blue, and yellow lights.

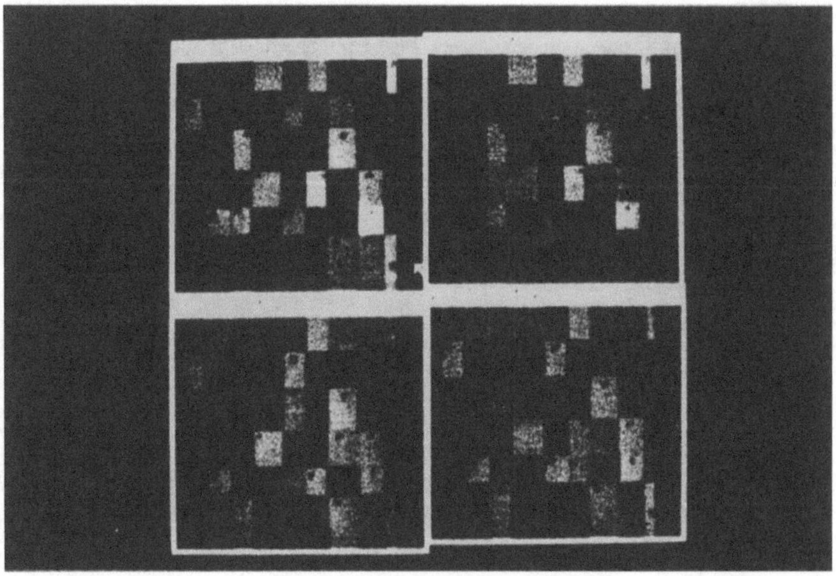

Color Figure 15. Outputs of the Brainard and Wandell (1986) model of the Retinex algorithm for the second Mondriaan derived from images under (clockwise, from top left) white, green, purple, and red lights.

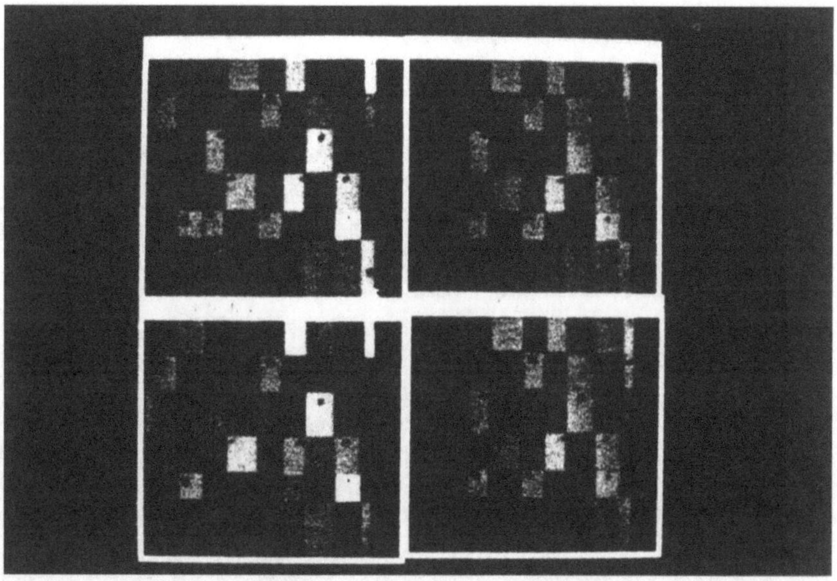

Color Figure 16. Outputs of the Brainard and Wandell (1986) model of the Retinex algorithm for the second Mondriaan derived from images under (clockwise, from top left) white, blue-green, blue, and yellow lights.

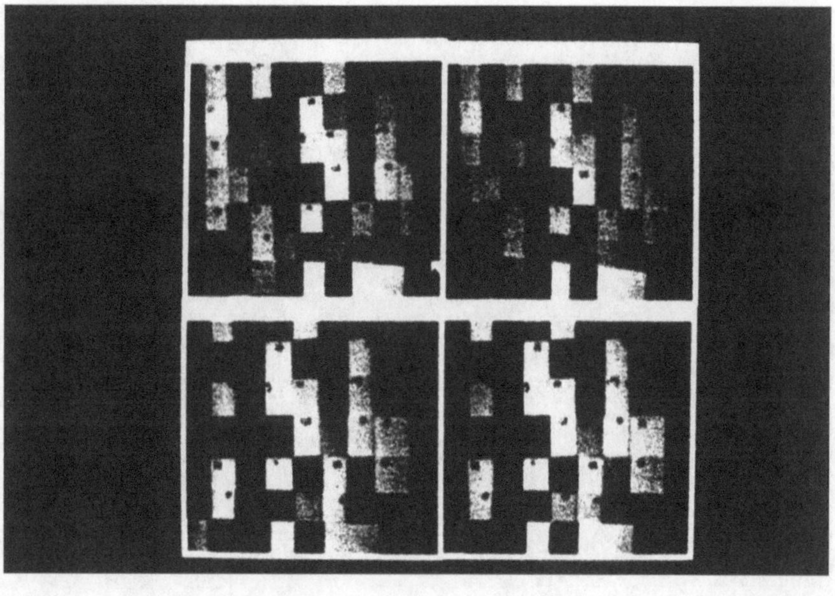

Color Figure 17. Outputs of the Brainard and Wandell (1986) model of the Retinex algorithm for the third Mondriaan derived from images under (clockwise, from top left) white, green, purple, and red lights.

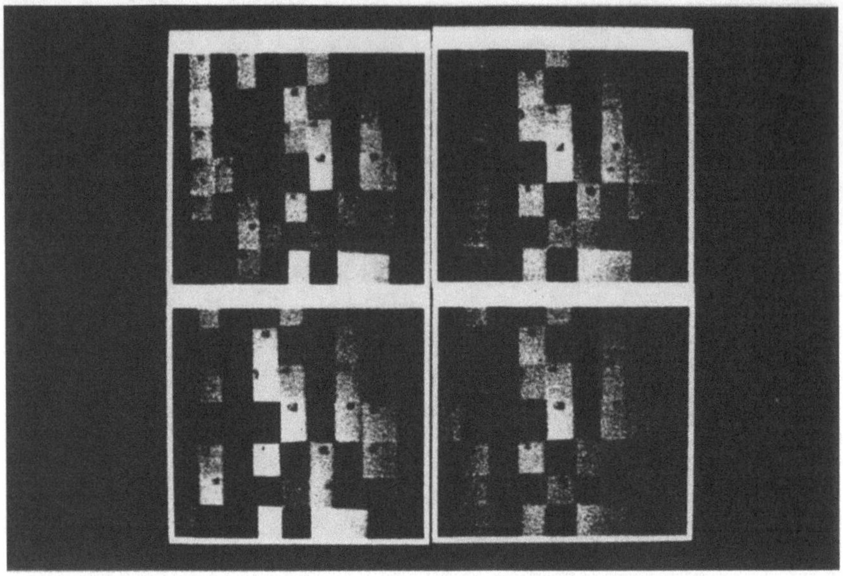

Color Figure 18. Outputs of the Brainard and Wandell (1986) model of the Retinex algorithm for the third Mondriaan derived from images under (clockwise, from top left) white, blue-green, blue, and yellow lights.

Color Figure 19. Photograph of first Mondriaan with red borders attached.

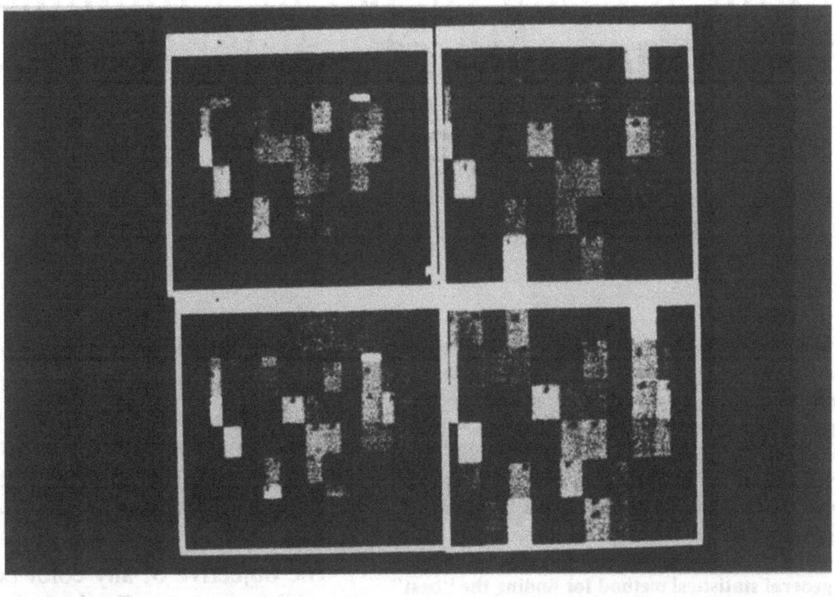

Color Figure 20. **Top: The outputs computed by the Retinex algorithm, for an image of the first Mondriaan under white light (right), and for an image of the first Mondriaan with red borders (left), again under white light. The greenish appearance of the left-hand image is caused by the Retinex algorithm confusing a change in average surface color with a colored illuminant. Bottom: Outputs computed by Crule for an image of the first Mondriaan under white light (right), and for an image of the first Mondriaan with red borders (left), again under white light. Because Crule does not use average surface color, appending borders to the Mondriaan does not affect the color descriptors it computes.**

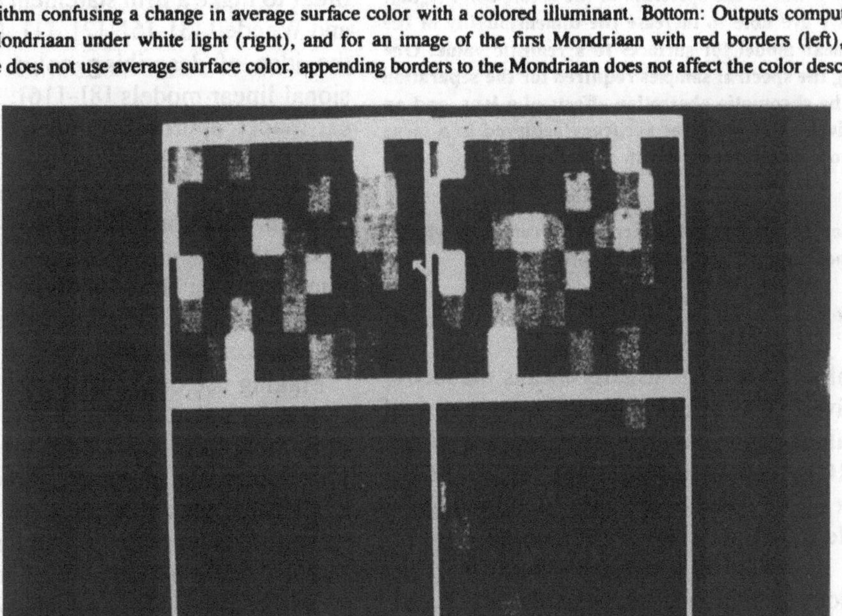

Color Figure 21. **Crule outputs for two separate two color images. The bottom two images are the input images, the top the output images. The image on the left is for a red filter which passed very much more light than that on the right. Details appear in section 7.**

Separating a Color Signal into Illumination and Surface Reflectance Components: Theory and Applications

JIAN HO, BRIAN V. FUNT, AND MARK S. DREW

Abstract—We present a "separation" algorithm for achieving color constancy and theorems concerning its accuracy. The algorithm requires extra information from the optical system, over and above the usual three values mapping human cone responses. However, with this additional information—specifically, a sampling across the visible range of the reflected, color-signal spectrum impinging on the optical sensor—we are able to separate the illumination spectrum from the surface reflectance spectrum contained in the color-signal spectrum which is, of course, the product of these two spectra. At the heart of the separation algorithm is a general statistical method for finding the "best" illumination and reflectance spectra, within a space represented by finite-dimensional linear models of statistically typical spectra, whose product closely corresponds to the spectrum of the actual color signal. Using this method, we are able to increase the dimensionality of the finite-dimensional linear model for surfaces to a realistic value. One method of generating the spectral samples required for the separation algorithm is to use the chromatic aberration effects of a lens, and an example of this is given. We detail the accuracy achieved in a large range of tests and show that agreement with actual surface reflectance is excellent.

Index Terms—Color constancy, color perception, computer vision, finite-dimensional linear model, statistical methods.

I. Introduction

THE term "color constancy" usually means the recovery of perceived surface color from the strengths of three receptor values representing cone responses of the visual system, RGB responses of a color camera or the like, independent of the color of the light illuminating the object. Being able to extract color descriptors that are independent of the illumination is desirable because of the variety of situations in which color is important, but illumination conditions cannot be controlled. Such situations include visually guided robots, automatic terrain classification by remote sensing, and perhaps somewhat less obvious, color reproduction [1], [2].

In color reproduction, the knowledge of surface spectral reflectance helps in setting the intensities of the three color guns in a video display so as to render the scene colors best. Images could even be shot under one ambient illumination and displayed as if they had been shot under a different illumination. Other uses for color constancy information include matching paints or dyes (especially at separate locations), helping a robot outdoors continue to see the grass as green even as a cloud appears and changes the illuminant spectrum, sorting ripe fruit from unripe, and estimating the ambient light spectrum for aid in correct filtering for photography—to name just a few applications.

The objective of any color constancy algorithm is substantial: one must effectively recover some representation of the entire surface reflectance spectrum at each pixel in order to make a firm statement on surface colors. Maloney and Wandell [3]–[5], [2], [1], [6] and Yuille [7] start with the idea of describing color spectra with finite-dimensional linear models [8]–[16]. The finite-dimensional linear-model approach condenses all spectral information into a few numbers by supposing that illumination and reflectance can each be approximated by weighted sums of basis functions. The basis functions for illumination model the principal characteristics of spectra of natural illuminants, such as daylight [17], while those for reflectance model the characteristics of a large set of spectral reflectances [18], [19].

Making use of the relationship between receptor values and basis-function weights, Maloney, Wandell, and Yuille show how to recover both the weights representing surface reflectance and those representing illumination. The method requires solving a set of equations based on the receptor values from several regions of different color. The method's primary limitation is that a system of only three receptor classes is necessarily restricted to being able to model surfaces with only a two-dimensional vector space. But this dimensionality is too small to capture the richness of surface color.

In this paper, our goal is perhaps more modest in that in order to carry out our "separation" algorithm, we demand additional measurements be provided as input, over and above the response values of three photoreceptor classes. In particular, our algorithm requires a sampling of the entire (visible) color-signal spectrum. The product of the illumination spectrum with that of the surface reflectance comprises the color signal, the light reflected from the surface and entering the camera or eye.

From this additional information, though, our separation algorithm extracts the illumination and reflectance with higher dimensionality. It also applies to situations

Manuscript received July 22, 1988; revised May 15, 1990. Recommended for acceptance by W. E. L. Grimson.

The authors are with the School of Computing Science, Simon Fraser University, Vancouver, B.C., Canada V5A 1S6.

IEEE Log Number 9037572.

for which previous algorithms, based on receptor values alone, are incapable of supplying answers.

Given the sampled color signal, the separation algorithm divides it into one component due to the illumination, and a second due to the reflectance. In the result, the dimensionality of the vector space describing surfaces is increased to 3, while the same dimension of 3 for the illumination space and the number of receptor classes is maintained. Therefore, the separation algorithm can better model surface reflectances than previous models. In fact, provided that the basis functions for the surface and illumination vector spaces together form a set satisfying certain requirements given below, our analysis is unrestricted with regard to the dimensionality of each space. As well, we do not require the illumination to be spatially constant.

While it is straightforward to imagine how one might acquire the color-signal spectrum in machine applications, such as satellite imaging or color xerography, it seems on first consideration that such extra information is inaccessible in systems such as the human visual system. This, however, turns out not to be the case. In [20]–[22], it was shown that, in fact, spectral information can be gleaned from the chromatic aberration inherent in lens systems, including human ones. As an example of the power of our method, in Section II, we show how color constancy can be achieved using spectral information derived from chromatic aberration.

Section III contains our main result, which is that by applying statistical methods, the surface reflectance can be obtained from the color signal. In Section IV, we study over 2000 applications of the separation algorithm, and we find that the average error in the reflectance component it extracts from the color signal is only about twice that obtainable by least squares fitting of the reflectance basis functions directly to the actual reflectance. For objects described well by the underlying model, we recover the surface spectral reflectance, and hence achieve color constancy, with great accuracy. In Section V, we outline some of the further directions this work is presently taking.

II. Color Constancy

This section describes the separation algorithm, how it achieves color constancy, and why this leads to an increase in the dimensionality of the surface reflectance vector space. First, finite-dimensional linear models are reviewed.

A. Finite-Dimensional Models for Illumination and Surface Reflectance

Earlier work on using finite-dimensional linear models in color vision [9], [16] showed that many surface spectral reflectance functions are approximated well by a specific finite-dimensional linear model comprising frequency-limited functions. The idea is the following. A set of spectral reflectance basis functions called $S_j(\lambda)$, with j ranging from 1 to n, are selected; then, any surface spectral reflectance is represented as a linear combination of these basis functions. Modeling the surface in this way constitutes a finite-dimensional model for surface spectral reflectance. The basis functions are constrained to being linearly independent. That is, if $\sigma_1 S_1(\lambda) + \cdots + \sigma_n S_n(\lambda) = 0$, all the coefficients σ_j must be zero. Illumination is modeled by m basis functions, with weights ϵ_i.

The surface spectral reflectance corresponding to point x in an image is expressed as

$$S^x(\lambda) = \sum_{j=1}^{n} \sigma_j^x S_j(\lambda).$$

Similarly, the spectral power distribution of ambient light expressed in terms of the basis functions is

$$E^x(\lambda) = \sum_{i=1}^{m} \epsilon_i^x E_i(\lambda).$$

A finite-dimensional model for surface spectral reflectance may not capture the surface characteristics of an object exactly, so we measure the fit by examining the sum-of-squares error formed by the square of the difference between the actual reflectance and the model's approximation of it, integrated over wavelength. Turning the integral into a sum, we consider an error formula

$$\text{error} = \left\{ \sum_{\lambda} (S_{\text{exact}} - S_{\text{approx}})^2 \right\}^{1/2}$$

(cf. [23, p. 177]). The coefficients σ_j of the model should be chosen so as to minimize this error.

Over the visible spectrum (400–700 nm), the surface spectral reflectance curves of natural objects are usually reasonably smooth and continuous. Many experiments on empirical surface spectral reflectances show that the surface reflectances of natural objects are very limited, and most of them can be modeled using only a few basis functions. For example, Cohen [18] analyzed the surface spectral reflectances of 150 Munsell chips and computed the characteristic vectors for these chips. His results show that the first three characteristic vectors account for 99.2% of the variance in the fit. Adding another characteristic vector only yields a slight improvement (about 0.5%) [18] (see also [24]).

Maloney [3] showed that the most surface spectral reflectances of natural objects can be modeled adequately by means of three basis functions. Higher dimensions result in better approximations, yet three basis functions still suffice when the filtering effect of the cone response functions is taken into account [5]. Similarly, three–five basis functions are sufficient for modeling most daylight spectra. In any case, our model is essentially unrestricted in terms of the dimensionality of the vector spaces for illumination and reflectance, although, in practice, most of our computations are carried out with both dimensionalities equal to 3.

B. Spectral Reflectance from Edge Information

As a preliminary example, we consider a simple situation in which two regions of different color abut. With a sharp-edged boundary, this situation allows determination of the color signal $I(\lambda)$ corresponding to the *difference* between the color signals reflected from the two regions.

At each image location, we assume that there are s sensor classes corresponding to human cone responses or camera sensitivities. Let the spectral response curves of the sensors be $R_k(\lambda)$, $k = 1\text{–}s$.

For the *color signal*,

$$I^x(\lambda) \equiv E^x(\lambda)S^x(\lambda),$$

the response ρ_k^x at location x of sensor class k can be expressed as the integral over λ of the product of $R_k(\lambda)$ and the color signal

$$\rho_k^x = \int E^x(\lambda)S^x(\lambda)R_k(\lambda)\,d\lambda.$$

Hence, at each pixel in the image, we have

$$\rho_k^x = \int E^x(\lambda)S^x(\lambda)R_k(\lambda)\,d\lambda = \sum_{i=1}^{m}\sum_{j=1}^{n} \epsilon_i^x\sigma_j^x g_{ijk}$$

where $g_{ijk} \equiv \int E_i^x(\lambda)S_j^x(\lambda)R_k(\lambda)\,d\lambda$.

For now, with Maloney [3], let us assume that the ambient light remains constant over the whole scene. If there are s sensor classes, then the ρ_k^x, $k = 1\text{–}s$ are known at each point, so that there are s equations for the $m + n$ variables ϵ_i and σ_j (representing the illumination and reflectance) within each region, viz.

$$\rho_k^x = \sum_{i=1}^{m}\sum_{j=1}^{n} \epsilon_i\sigma_j^x g_{ijk}, \qquad k = 1\text{–}s,\ \epsilon_i \text{ independent of } x.$$

In an image in which two regions of different color meet, there will be $n + n$ variables for the surface spectral reflectances for the two regions and m variables for the spectral power distribution of the ambient light. Let the two regions be labeled 1 and 2. Then there are $s + s = 2s$ equations

$$\rho_k^X = \sum_{i=1}^{m}\sum_{j=1}^{n} \epsilon_i\sigma_j^X g_{ijk} \tag{1}$$

for $k = 1\text{–}s$ and $X = 1$ or 2.

Now, we have shown elsewhere [20]–[22] that the *difference* $\Delta(\lambda)$ of the color signals in the two regions can be determined from the chromatic aberration occurring at the edge. Essentially, chromatic aberration at a color edge spatially separates the light in such a way that it can be used to determine the complete difference in color signal for all λ.

Hence, from chromatic aberration, we can determine for every λ the set of differences

$$E(\lambda)S^1(\lambda) - E(\lambda)S^2(\lambda) = \Delta(\lambda)$$

with $\Delta(\lambda)$ known.

Substituting the finite-dimensional expansions, we thus have

$$\left\{\sum_{i=1}^{m} \epsilon_i E_i(\lambda)\right\}\left[\sum_{j=1}^{n} (\sigma_j^1 - \sigma_j^2)S_j(\lambda)\right] = I(\lambda). \tag{2}$$

Specific values of $I(\lambda)$ corresponding to specific wavelengths λ can be substituted to obtain as many equations as desired, but only independent equations are actually needed. The dimension of the set of product functions $E_i(\lambda)S_j(\lambda)$ fixes the number of independent equations for the variables ϵ_i and for σ_j^1, σ_j^2—the number of independent equations equaling the dimension of the set. Other methods, such as a statistical approach via a least squares fit, can also be used to derive independent equations for the unknowns ϵ_i and σ_j.

When all the functions $E_i(\lambda)S_j(\lambda)$ are linearly independent, we can certainly solve for all $\epsilon_i(\sigma_j^1 - \sigma_j^2)$ products. Since the unknowns themselves appear as products, however, it is only possible to solve for values of the ϵ_i and σ_j relative to each other. Nevertheless, a constraint such as $\Sigma_{i=1}^m(\epsilon_i)^2 = 1$ or $\|\Sigma_{i=1}^m \epsilon_i E_i(\lambda)\| = 1$ can be used[1] to render the equations solvable, up to an *overall* multiplicative constant, for all the ϵ_i and the differences $\sigma_j^1 - \sigma_j^2$. In this case, then, chromatic aberration alone provides n equations for the variables σ_j^1 and σ_j^2 themselves, as well as providing a complete solution for all the ϵ_i. So we do not yet have enough equations to solve for all the σ_j^1 and σ_j^2 individually.

In addition to $\Delta(\lambda)$, however, we also have available the information provided by (1) involving the sensor responses. Each region yields s equations, so we can make use of our knowledge of the ϵ_i and find all the σ_j^1 for region 1, say, provided they are not too numerous. Specifically, for $n \le s$, we can solve for all the σ_j in both regions.

If, as in the human visual system, there are three sensors at each point (i.e., $s = 3$), then the dimension of the linear model for surfaces can be as large as 3, subject to the linear independence condition holding. We have tested the first three basis vectors provided by Judd *et al.* for ambient illumination [17] coupled with the first three basis vectors provided by Cohen for surface reflectance [18], and we found that they do indeed form a set of nine linearly independent product functions.

Being able to model the surface reflectance space with dimensionality 3 represents a distinct improvement on the work of Maloney *et al.* [3], [1] in that their work permits a dimensionality of only 2 when $s = 3$, as in the human visual system. Such a low dimensionality imposes severe limitations when modeling real surfaces.

The above example shows that injecting additional spectral information can result in substantially better color constancy. The reflectance coefficients σ_j obtained are illumination-independent color descriptors. Of course, the improvement is gained at the price of having to measure the color signal difference which, at least in a laboratory

[1]Here, $\|f\|$ stands for the (Euclidean) norm of f.

setting, can be gleaned from chromatic aberration [20]–[22].

We now turn to the general situation in which the complete color signal, not just the color signal difference, is made available by any measurement method, for example, direct spectrometry.

III. Separating Illumination from Surface Reflectance

In this section, we show that if the basis functions for illumination and those for surface spectral reflectance have certain properties, it is possible to recover both the illumination and the reflectance from the spectral power distribution of the incoming light. We then prove a number of theorems justifying our claim.

A. Exact Models

When both the spectral power distribution of the illumination and the surface spectral reflectance are exactly represented by finite-dimensional models, they both can be derived easily and exactly from the color signal, up to a multiplicative scale factor.

To show this, we start with the product relating illumination, reflectance, and color signal, which is $E(\lambda)S(\lambda) = I(\lambda)$, and substitute the finite-dimensional model expansions:

$$I(\lambda) = \sum_{i=1}^{m} \sum_{j=1}^{n} \epsilon_i \sigma_j E_i(\lambda) S_j(\lambda).$$

If all the *product* functions $E_i(\lambda)S_j(\lambda)$ are linearly independent, then we can choose nm different λ values $\lambda_1, \lambda_2, \cdots, \lambda_{nm}$ such that the matrix M, whose columns consist of $[E_i(\lambda_1)S_j(\lambda_1), E_i(\lambda_2)S_j(\lambda_2), \cdots, E_i(\lambda_{nm})S_j(\lambda_{nm})]^T$, is nonsingular. Hence, when these nm values of λ are substituted into the above equations, they can be interpreted as nm linearly independent equations for the unknowns ϵ_i and σ_j in terms of the known function $I(\lambda)$ sampled at the wavelengths λ. Because of the linear independence property, the solution for the coefficients $(\epsilon_i\sigma_j)$ is unique. If the illumination function $E(\lambda)$ is normalized[2] such that $\epsilon_1 = 1$, then we have unique solutions for all ϵ_i and σ_j. Thus, it is possible to derive the exact illumination and surface spectral reflectance, except for a scaling factor applied to normalize ϵ_1 to 1.

B. Approximate Models

In general, finite-dimensional models will not describe the spectral power distribution of illumination and the surface spectral reflectance exactly, although they should approximate them well. If

$$E(\lambda) \approx \sum_{i=1}^{m} \epsilon_i E_i(\lambda)$$

and

$$S(\lambda) \approx \sum_{j=1}^{n} \sigma_j S_j(\lambda)$$

[2]Wandell [1] and Yuille [7] propose a similar constraint on the ϵ vector.

then we have

$$I(\lambda) \approx \sum_{i=1}^{m} \sum_{j=1}^{n} \epsilon_i \sigma_j E_i(\lambda) S_j(\lambda).$$

Therefore, the previous method for exact models cannot be applied. Instead, we claim that a nonlinear least squares fit will correctly extract the surface reflectance and the illumination. In particular, we select those coefficients ϵ_i and σ_j that minimize the sum of squares

$$\sum_{\lambda} \left[\sum_{i=1}^{m} \sum_{j=1}^{n} \epsilon_i \sigma_j E_i(\lambda) S_j(\lambda) - I(\lambda) \right]^2$$

with the condition $\epsilon_1 \equiv 1$ as above. The solution of this equation constitutes a nonlinear least squares problem for illumination and surface reflectance.

In general, one would not necessarily expect a satisfactory fit to be obtained because of the product structure of the sum. For example, if $f(\lambda)g(\lambda)$ were the result of a least squares fit for some function $I(\lambda)$, then $[f(\lambda)/c(\lambda)][g(\lambda)c(\lambda)]$ would also be a solution for any nonzero function $c(\lambda)$. In addition, there may be an infinite number of solutions to a least squares fit. However, if the set of possible illumination distributions and the set of possible surface spectral reflectances each are modeled well enough individually by a finite-dimensional model, a least squares fit will provide good results. In the next section, we prove explicitly the relationship between the individual fits and the correct separation of illumination from surface reflectance.

C. Theorems

The separation algorithm derives finite-dimensional approximations to the actual illumination and reflectance, and Theorem 2 below expresses how the error in these approximations varies with the minimum error possible using a fixed set of basis functions. Before proceeding to Theorem 2, we first establish an error bound for the best approximation to a product of two functions ES where this best approximation is generated as the product of two approximations f and g to the original functions. We state the error in terms of the error bounds for approximations to E and S taken separately.

Theorem 1: Let F and G be sets of bounded functions and let $E(\lambda)$ and $S(\lambda)$ be two given bounded functions; let all functions have the same domain. Suppose F consists of functions $f(\lambda)$. Let f' be that function which is closest to E. Then f' is that f which minimizes $\| f - E \|$.

Let e_F be this least error:

$$e_F = \min_{f \in F} \{ \| f - E \| \} = \| f' - E \|.$$

Similarly, let the function g' be closest to S, and define the least error for the approximation to S as

$$e_G = \min_{g \in G} \{ \| g - S \| \} = \| g' - S \|.$$

If c is the maximum norm for *all* the functions in sets F and G *and* functions $E(\lambda)$ and $S(\lambda)$, then

$$\min_{f\in F \& g\in G} \| fg - ES \| \leq c(e_F + e_G).$$

The best approximation formed from the best available product functions fg has an error bounded by a bounded multiple of the sum of the individual best approximation errors. In other words, the better each function E and S can be approximated individually, the better the best product function fg, chosen from the set of pairwise products of the available approximating functions, will fit ES.

The Appendix contains all proofs.

Corollary 1: Let $E_i(\lambda)$, $i = 1\text{–}m$, $S_j(\lambda)$, $j = 1\text{–}n$, $E(\lambda)$, and $S(\lambda)$ be given, bounded functions, and let all functions have the same domain. Let $I = [a, b]$ be a bounded, closed interval. Let

$$e_F = \min_{\epsilon_i \in I} \left\{ \left\| \sum_{i=1}^{m} \epsilon_i E_i - E \right\| \right\}$$

and

$$e_G = \min_{\sigma_j \in I} \left\{ \left\| \sum_{j=1}^{n} \sigma_j S_j - S \right\| \right\}.$$

Then there exists some constant c such that

$$\min_{\epsilon_i, \sigma_j \in I} \left\{ \left\| \sum_{i=1}^{m} \epsilon_i E_i \sum_{j=1}^{m} \sigma_j S_j - ES \right\| \right\} \leq c(e_F + e_G).$$

This corollary states that the error in the least squares approximation to the color signal is *proportional* to the errors in the individual fits to the illumination and surface spectral reflectance. If the finite-dimensional linear models closely approximate the illumination and surface spectral reflectance, the error in the least squares fit of the product of the illumination and the surface spectral reflectance will be small.

Now we adopt the constraint $\epsilon_1 \equiv 1$ and show a type of converse of the previous theorem. Namely, we show that the coefficients ϵ_i and σ_j representing the best fit to a given color signal $E(\lambda)S(\lambda)$ and obtained by solving the nonlinear least squares problem for illumination and surface reflectance, in fact, provide a good fit to $E(\lambda)$ and $S(\lambda)$ individually. Again, the error found is proportional to the errors involved in fitting illumination and surface reflectance separately with the finite-dimensional model.

Theorem 2: Let $E_i(\lambda)$, $i = 1\text{–}m$, $S_j(\lambda)$, $j = 1\text{–}n$, $E(\lambda)$, and $S(\lambda)$ be given, bounded functions, and let all functions have the same domain. Let $I = [a, b]$ be a bounded, closed interval. Let

$$e_F = \min_{\epsilon_i \in I} \left\{ \left\| \sum_{i=1}^{m} \epsilon_i E_i - E \right\| \right\}$$

where $\epsilon_1 \equiv 1$ and let

$$e_G = \min_{\sigma_j \in I} \left\{ \left\| \sum_{j=1}^{n} \sigma_j S_j - S \right\| \right\}.$$

Let x_i, $i = 1\text{–}m$ and $x_1 = 1$, and let y_j, $j = 1\text{–}n$ be those particular values of ϵ_i and σ_j such that a best fit is provided for the product function ES:

$$\left\| \sum_{i=1}^{m} x_i E_i \sum_{j=1}^{n} y_j S_j - ES \right\| = \min_{\epsilon_i, \sigma_j \in I} \left\{ \left\| \sum_{i=1}^{m} \epsilon_i E_i \sum_{j=1}^{n} \sigma_j S_j - ES \right\| \right\}$$

with $\epsilon_1 = 1$. If all the product functions $E_i(\lambda)S_j(\lambda)$, $i = 1\text{–}m$ and $j = 1\text{–}n$ are linearly independent, then there exist constants C and D such that

$$\left\| \sum_{i=1}^{m} x_i E_i - E \right\| \leq C(e_F + e_G)$$

and

$$\left\| \sum_{i=1}^{n} y_i S_i - S \right\| \leq D(e_F + e_G).$$

See the Appendix for a proof.

D. Discussion

From Theorem 2, we know that if the basis functions $E_i(\lambda)$ for the spectral power distribution of the illumination and $S_j(\lambda)$ for the surface spectral reflectance have the property that all $E_i(\lambda)S_j(\lambda)$ are linearly independent, a good approximation for the surface spectral reflectance and illumination can be achieved by using a least squares fit.

As we stated in Section II, when $n = 3$ and $m = 3$, the nine illumination–reflectance, basis-function products do form a linearly independent set. In fact, the linear independence condition can be relaxed; as long as the product functions form a set with dimension not less than the number of unknown variables minus 1, then our method will find the correct solution. In the important (3, 3) case, we do find linear independence so that our analysis can be applied directly, and we can be confident that a good least squares fit will result.

Our scheme for determining illumination and surface spectral reflectance consists of finding the most plausible separation of illumination and surface reflectance from their product. In general, functions $f(\lambda)$ and $g(\lambda)$ cannot be derived from their product $f(\lambda)g(\lambda)$, but this separation can be done if the two functions are very different. The difference between the illumination and the surface spectral reflectance is captured by the linear independence of the set of functions $E_i(\lambda)S_j(\lambda)$. Since Judd's basis functions for daylight and Cohen's basis functions for reflectance have the required property, illumination and surface spectral reflectance can be recovered from their product.

E. Implementation

We implemented the separation algorithm using the first three of Cohen's vectors [18] as basis functions for surface spectral reflectance and the first three of Judd's vectors [17] as basis functions for the spectral power distribution of ambient light. If these particular basis functions are used, all $E_i(\lambda)S_j(\lambda)$ products are linearly independent. Color signals are sampled at 10 nm intervals from 400 to 650 nm since published data for sample spectra exist in this range [19].

Given the spectral power distribution of the color signal $I(\lambda)$, the least squares approach consists of minimizing the quantity

$$\sum_{l=0}^{v}\left\{\left[\sum_{i=1}^{m}\epsilon_i E_i(\lambda_l)\right]\left[\sum_{j=1}^{n}\sigma_j S_j(\lambda_l)\right] - I(\lambda_l)\right]^2.$$

This overdetermined set of equations is to be solved for unique solutions ϵ_i, σ_j. By the calculus of variations, the corresponding "normal" equations are as follows.

For $k = 1\text{–}m$,

$$\sum_{l=0}^{v} E_k(\lambda_l)\left[\sum_{j=1}^{n}\sigma_j S_j(\lambda_l)\right]\left\{\left(\sum_{i=1}^{m}\epsilon_i E_i(\lambda_l)\right)\cdot\left(\sum_{j=1}^{n}\sigma_j S_j(\lambda_l)\right) - I(\lambda_l)\right\} = 0.$$

For $k = 1\text{–}n$,

$$\sum_{l=0}^{v} S_k(\lambda_l)\left[\sum_{i=1}^{m}\epsilon_i E_i(\lambda_l)\right]\left\{\left(\sum_{i=1}^{m}\epsilon_i E_i(\lambda_l)\right)\cdot\left(\sum_{j=1}^{n}\sigma_j S_j(\lambda_l)\right) - I(\lambda_l)\right\} = 0.$$

We began by using the mathematical library package *minpack* [25] as a solver for this system of nonlinear equations in the unknowns ϵ_i and σ_j. Since the daylight spectral power distribution is very limited, we can make a very good guess for the ϵ_i, and hence for the σ_j, as initial values for the *minpack* programs. However, the *minpack* program turns out to be very unstable for this problem. That is, for slightly different initial values, it produces different solutions.

As a result, a new algorithm producing stable results was developed based on the fact that the first m equations above are linear in ϵ_i and the last n equations are linear in σ_j. Given an initial value for the ϵ_i, the last n equations will yield the σ_j. Substituting these σ_j into the first m equations generates another set of ϵ_i. Iteration results in convergent, stable solutions.[3]

IV. Applications

For a rigorous appraisal of the above theoretical results, tests were conducted using two types of synthetic color signals. The first kind of color signal is generated by multiplying published spectra for ambient daylight [17] by published surface reflectance functions [19]. The second type consists of random perturbations to the spectra in the space of spectra spanned by the basis functions.

Since the original illumination spectrum is free to not have $\epsilon_1 = 1$, the separation algorithm's solutions σ_j must be scaled before comparing the resulting reflectance function to the original reflectance function. The appropriate scale factor is that which minimizes the least squares error between them.

As a measure of the separation algorithm's accuracy, we compare its results to that of the best fit possible using the basis functions of the underlying finite-dimensional model.

A. Natural Color Signals

We tested the algorithm by combining each of the 370 surface reflectance functions recorded by Krinov [19] with each of the five standard daylight illumination spectra given by Judd *et al.* [17]. These five daylight spectra were reconstituted from the set of illumination basis functions developed by characteristic vector analysis [17] of spectral distributions of daylight. Of the 370 available Krinov reflectance spectra, we used only those 345 that were complete in the visible range.

The results were very good in that for most samplings, the overall error obtained in recovering the surface reflectance was only about 18%, the error being defined as

$$\left\{\sum_{\lambda}(S - S')^2 \Big/ \sum_{\lambda} S^2\right\}^{1/2}$$

where S is an actual spectrum and S' is the estimate. We generally found the above error measure to be equal to a multiple of ~0.5–0.8 times the CIELUV color difference ΔE in three-dimensional color space [27, p. 828]. This overall 18% error must be compared to the direct fit of the underlying finite-dimensional model using three basis functions to the reflectance alone, which had an average error of 10.0%.

Overall, compared to the errors produced by a finite-dimensional model, least squares fit to the $S(\lambda)$ spectra separately, the errors for the separation algorithm's extraction of surface reflectance were at most a multiple of 4.18 times as large, with a minimum multiple of 1.00. The average ratio of errors is only about 1.95 times the error of the direct least squares fit for reflectance, so that the error coefficient D in Theorem 2 is about 2. The median ratio was 1.80 with standard deviation 0.55.

In absolute terms, the largest error for any color signal input was 47.4%; and that case was also one of the most poorly fit by the finite-dimensional model directly, with an error of 26.7%. The ratio of the errors was thus 1.78 in this case.

The best fit had an error of 1.9%, compared to a 1.8% error for the direct least-squares fit to reflectance alone, a ratio of 1.04.

As an example, consider Fig. 1. Here, a color signal has been constructed from the surface spectral reflectance for "river valley with meadows" (general view from a

[3]In fact, convergence can easily be proved since all the equations are continuous (see, e.g., [26]).

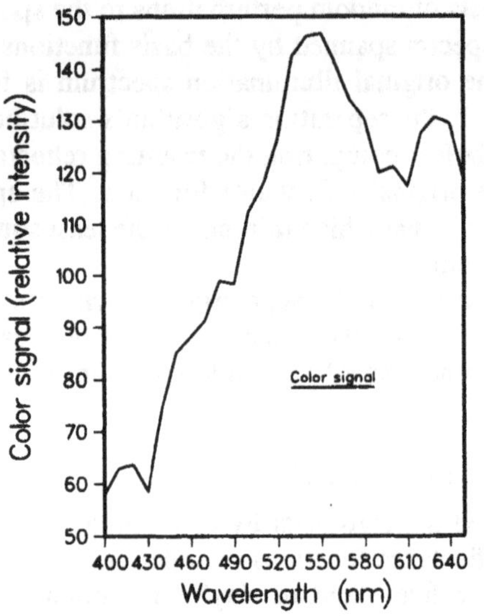

Fig. 1. Color signal formed from product of surface spectral reflectance of river valley with meadows (Krinov spectrum 54) and Judd *et al.* spectrum for 4800 K daylight. This spectrum is a typical input to the separation algorithm.

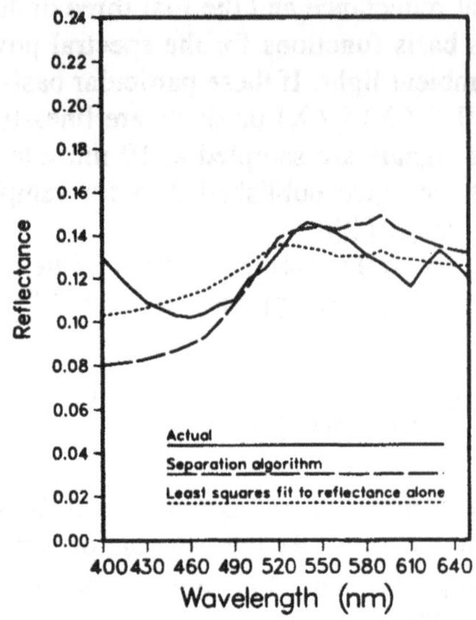

Fig. 2. Original Krinov reflectance spectrum 54 and results of separation algorithm applied to the color signal of Fig. 1—error 15.2%. Also shown are the results of using the finite-dimensional model to match the reflectance spectrum alone, without taking any illumination into consideration—error 8.1%. Ratio of errors: 1.87.

distance of about 3 km; covered with trees and meadows, end of summer) which is Krinov's sample 54, multiplied by Judd's spectrum for 4800 K correlated color temperature daylight.

Fig. 2 shows the original surface reflectance spectrum given by Krinov, along with the separation algorithm's results. The separation error in this case is 15.2%. The figure also shows the result of using the finite-dimensional model to match the reflectance spectrum directly, without taking any illumination into consideration—an error of 8.1%. So the ratio of the error from our method as compared to the ability of the underlying finite-dimensional model to match the reflectance spectrum is only 1.87. Since the direct least squares fit to reflectance alone is as good as one can get using three basis functions, the separation algorithm does very well.

Fig. 3 shows Judd's daylight spectrum, along with the illumination calculated by the separation algorithm. As can be seen, the results are again quite good. The error for the illumination curve derived by the separation algorithm is 12.8%. The worst case was Krinov reflectance 162 illuminated by 4800 K daylight. The results are shown as Fig. 4. The best case was Krinov reflectance 295 illuminated by 7500 K daylight. The results are shown as Fig. 5.

B. Perturbation Spectra

As a further test, we applied the algorithm to a large set of synthetic reflectances derived by taking the sum of the three reflectance basis functions with specific weights and adding random noise in proportions of 1–5%. The separation algorithm's results in recovering these reflectances under a variety of daylight spectra compare favorably to the best fits of the finite-dimensional model to these spectra.

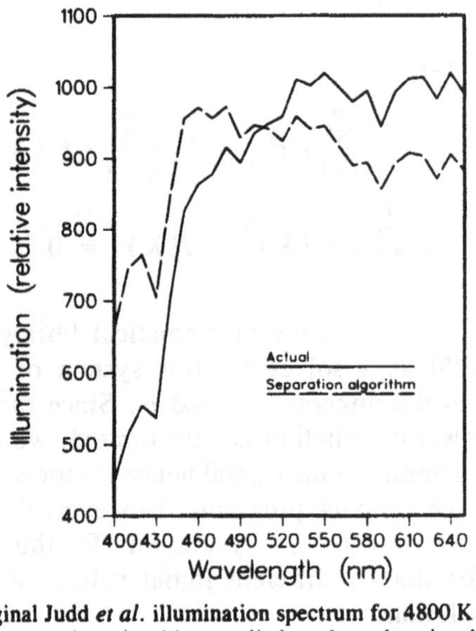

Fig. 3. Original Judd *et al.* illumination spectrum for 4800 K daylight and results of separation algorithm applied to the color signal of Fig. 1—error 12.8%.

The reflectance functions defined by weights (1, 1, 1) and (1, 2, 1.5) combined with each of the five available daylight illumination spectra generate ten cases. For the synthetic reflectances produced in this way without noise added, the basis function matching, of course, is exact. In accordance with Theorem 2, the separation algorithm also extracts the exact reflectance spectrum in this case because ϵ_F and ϵ_G are both zero.

When a perturbative noise contribution is added, the average error of both methods rises. Overall, the ratio of the error of the separation algorithm in recovering the reflectance spectrum to the error involved in directly matching the reflectance spectrum using the reflectance basis functions was about 1.17.

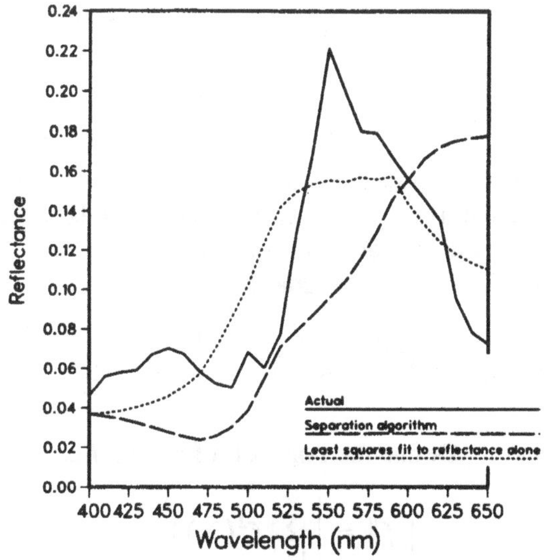

Fig. 4. Worst case: Krinov reflectance 162 illuminated by 4800 K daylight. Error from separation algorithm: 47.4%. Error using finite-dimensional model to fit reflectance alone: 26.7%. Ratio: 1.78.

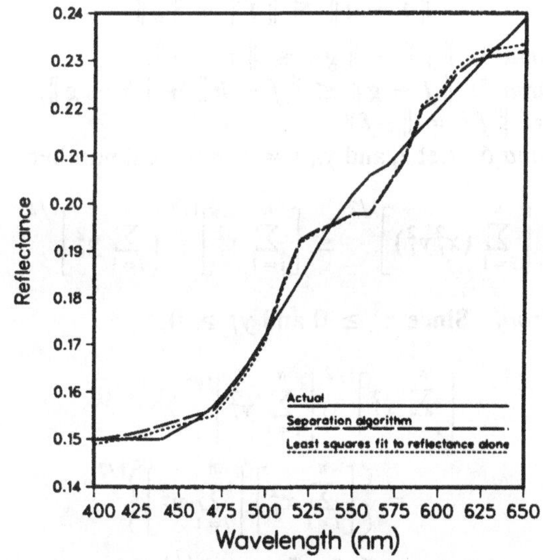

Fig. 5. Best case: Krinov reflectance 295 illuminated by 7500 K daylight. Error from separation algorithm: 1.9%. Error using finite-dimensional model to fit reflectance alone: 1.8%. Ratio: 1.04.

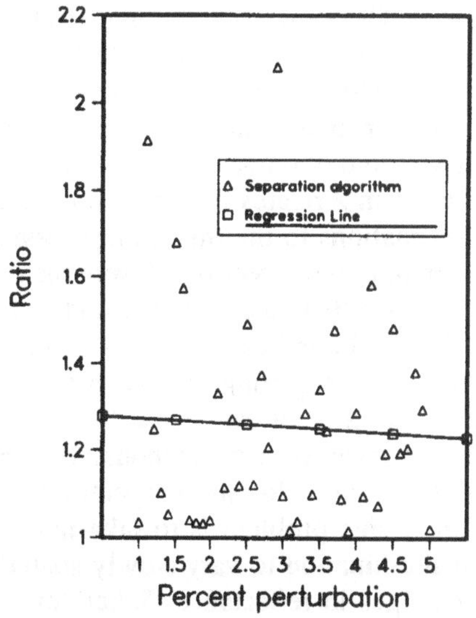

Fig. 6. Perturbed synthetic spectra. Ratio of errors for separation algorithm compared to finite-dimensional model for synthetic spectrum alone, with weightings (1, 1, 1) and combined with 10 000 K daylight, additive perturbation from 1 to 5%. For reflectance separation, average error: 18.3%, maximum error: 35.2%, minimum error: 5.0%; average ratio: 1.25, maximum ratio: 2.08, minimum ratio: 1.02.

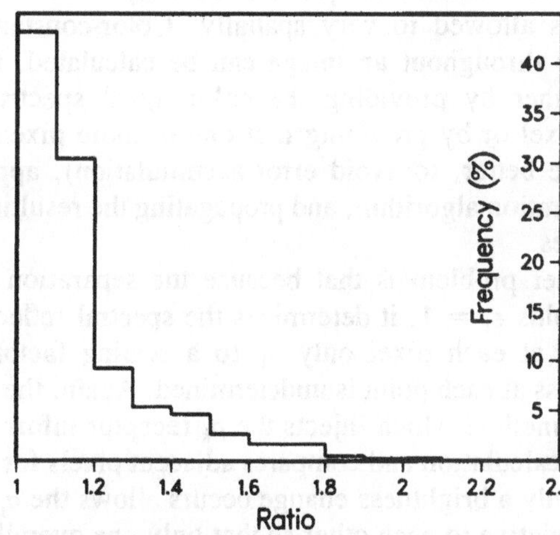

Fig. 7. Histogram of ratio of errors for separation algorithm compared to finite-dimensional model for perturbed synthetic spectra for 410 cases. For reflectance separation, average error: 17.0%, maximum error: 35.9%, minimum error: 4.9%; average ratio: 1.17, maximum ratio: 2.08, minimum ratio: 1.00, median 1.12.

Fig. 6 displays the performance of the separation algorithm in *comparison* to the accuracy of the underlying finite-dimensional model for the color signal of the synthetic reflectance defined by weights (1, 1, 1) combined with 10 000 K daylight. Perturbations from 1 to 5% by 0.1% make up 41 separate sets. From the regression line to the results, one sees that while the errors for both methods rise as the perturbation percentage increases, the ratio of the errors for the two methods stays approximately constant at about 1.25. This means that the separation algorithm does about as well as the finite-dimensional model itself with respect to these synthetic, unphysical spectra over a broad range of perturbations. The maximum ratio was 2.08, while the minimum ratio was 1.02.

A histogram of the test results for 410 color signals (41 perturbations of each reflectance under five illuminations) is shown in Fig. 7. The ratio of the error in the separation algorithm's reflectance estimate compared to direct modeling of the perturbed reflectance with the finite-dimensional model has an average of 1.17, a maximum of 2.08, and a minimum of 1.00. As can be seen from the histogram, most of the ratios lie below the 75% percentile value of 1.2. The standard deviation is 0.183.

V. Conclusion

The separation algorithm divides a color signal into two components, one due to surface reflectance, and another due to the incident illumination. As such, it provides a color-constant descriptor of surface reflectance, although it does require the entire color signal, not just a three-receptor sampling of the color signal.

As shown by Theorem 2 and the perturbation tests, the method works well when compared to the finite-dimensional model on which it is based. Needless to say, linear models of low dimensionality will not approximate surface reflectances and illuminant spectra well in every case, however. Theorem 2 relates the error in the finite-dimensional approximations to the true illumination and reflectance to the error in those recovered from the color signal. The perturbation tests explore the method's sensitivity to noise, and the results indicate that it is relatively stable.

For applications[4] requiring color-constant descriptors, the color signal at a pixel can be easily measured by any of a number of spectrometry methods. It is much more difficult to obtain the color signal spectrum at all pixels. One solution to this problem is to take into account the tendency of illumination to vary slowly spatially in comparison to changes in reflectance. When this is the case, given an accurate reflectance measurement at one pixel, correct reflectances can be propagated to the rest of the image. The propagation method [28] is based on Horn's lightness algorithm [29], and extends it by use of finite-dimensional models to colored Mondrian images in which the spectral content, not just the intensity, of the illumination is allowed to vary spatially. Color-constant descriptors throughout an image can be calculated, therefore, either by providing the color-signal spectrum at every pixel or by providing it at one or more pixels (the more the better, to avoid error accumulation), applying the separation algorithm, and propagating the resulting reflectances.

Another problem is that because the separation algorithm holds $\epsilon_1 = 1$, it determines the spectral reflectance function at each pixel only up to a scaling factor; the brightness at each point is undetermined. Again, the propagation method which injects the ρ_k receptor information into the calculation and compares adjacent pixels for those where only a brightness change occurs allows the σ_j to be scaled relative to each other so that only one overall multiplicative factor is required for the entire image.

Besides traditional spectrometry and chromatic aberration, another way of obtaining enough information about the color-signal spectrum to obtain color constancy is to use mutual reflection [30].

Perhaps the most important finding in the present work is the linear independence of the set of product basis functions $E_i S_j$ formed from the illumination bases of Judd [17] and the reflectance bases of Cohen [18]. This observation seems to capture an underlying physical property of a large sample of illumination and spectral reflectance functions, namely, that they combine into color signals in a way that allows them to be separated back into their original components.

APPENDIX

First, we state some useful results that are needed later (see, e.g., [31]).

[4]B. V. Funt, J. Ho, and M. Drew, "A color constancy device," U.S. patent pending.

Lemma 1: Let x_i and y_i, $i = 1$–n be real numbers. Then

$$\sum_{i=1}^{n} (x_i y_i) \le \left\{ \sum_{i=1}^{n} x_i^2 \right\}^{1/2} \left\{ \sum_{i=1}^{n} y_i^2 \right\}^{1/2}.$$

This is a well-known theorem. It simply states that the cosine of the angle between two vectors is less than or equal to 1 (Cauchy–Schwarz inequality).

Lemma 2: Triangle inequality:

$$\left[\sum_{i=1}^{n} (x_i + y_i)^2 \right]^{1/2} \le \left[\sum_{i=1}^{n} x_i^2 \right]^{1/2} + \left[\sum_{i=1}^{n} y_i^2 \right]^{1/2}.$$

Lemma 3: Define the norm of a function f as

$$\| f \| = \left[\sum_{i=1}^{k} f^2(\lambda_i) \right]^{1/2}$$

where all the λ_i are given distinct values in the domain of the function. Then

$$\| f + g \| \le \| f \| + \| g \|.$$

Lemma 4: $\| f \| - \| g \| \le \| f - g \|$.

Lemma 5: $\| f - g \| \le \| f - h \| + \| h - g \|$.

Fact: $\| f \| = \| -f \|$.

Lemma 6: Let x_i and y_i, $i = 1$–n be real numbers. Then

$$\left[\sum_{i=1}^{n} (x_i^2 y_i^2) \right]^{1/2} \le \left[\sum_{i=1}^{n} x_i^2 \right]^{1/2} \left[\sum_{i=1}^{n} y_i^2 \right]^{1/2}.$$

Proof: Since $x_i^2 \ge 0$ and $y_i^2 \ge 0$,

$$\left[\sum_{i=1}^{n} x_i^2 \right]^{1/2} \left[\sum_{i=1}^{n} y_i^2 \right]^{1/2} = \left\{ \left[\sum_{i=1}^{n} x_i^2 \right] \left[\sum_{i=1}^{n} y_i^2 \right] \right\}^{1/2} = \left\{ \sum_{i=1}^{n} \sum_{j=1}^{n} x_i^2 y_j^2 \right\}^{1/2} \ge \left\{ \sum_{i=1}^{n} x_i^2 y_i^2 \right\}^{1/2}.$$

Lemma 7: $\| fg \| \le \| f \| \times \| g \|$.

With these results, we can address the problem of minimizing the least squares sum.

Proof of Theorem 1: Since c is the bound of the norms for all functions in sets F and G and functions $E(\lambda)$ and $S(\lambda)$, we have $\| f(\lambda) \| \le c$, $\| g(\lambda) \| \le c$, $\| E(\lambda) \| \le c$, and $\| S(\lambda) \| \le c$ for all $f \in F$ and $g \in G$. We have defined f' in F and g' in G as those functions for which $e_F = \| f' - E \|$ and $e_G = \| g' - S \|$ hold. Then letting f and g range over all possibilities, we seek that product fg which best approximates the given product function ES. Therefore, consider the least error of such an approximation:

$$\min_{f \in F \& g \in G} \| fg - ES \| \le \| f'g' - ES \|$$

since the minimum norm is certainly less than or equal to the norm formed from any particular f and g, including f' and g'.

$$= \| f'g' - f'S + f'S - ES \|$$
$$= \| f'(g' - S) + (f' - E)S \|$$
$$\leq \| f'(g' - S) \| + \| (f' - E)S \| \text{ by Lemma 3}$$
$$\leq \| f' \| \times \| g' - S \| + \| f' - E \| \times \| S \| \text{ by Lemma 7}$$
$$\leq ce_G + e_F c,$$

substituting the definitions of e_G and e_F and the fact that c is the bound of all the norms:

$$= c(e_F + e_G).$$

Proof of Corollary 1: Since $E_i(\lambda)$, $i = 1\text{–}m$ are bounded functions,

$$\sum_{i=1}^{m} \epsilon_i E_i(\lambda)$$

is also a bounded function for $\epsilon_i \in I$, $i = 1\text{–}m$. Let

$$F = \left\{ \sum_{i=1}^{m} \epsilon_i E_i(\lambda),\ \epsilon_i \in I, \quad i = 1\text{–}m \right\}.$$

Then F is a set of bounded functions. Similarly,

$$G = \left\{ \sum_{j=1}^{n} \sigma_j S_j(\lambda),\ \sigma_j \in I, \quad j = 1\text{–}n \right\}$$

is also a set of bounded functions, and the result follows directly from Theorem 1.

Proof of Theorem 2: The coefficients x_i and y_j in the statement of the theorem are those found by fitting the product function ES. Let u_i and v_j denote those coefficients providing a best fit to E and S separately. That is, assuming that $E(\lambda)$ and $S(\lambda)$ were known, let u_i, $i = 1\text{–}m$ with $u_1 = 1$ and v_j, $j = 1\text{–}n$ be those values such that

$$e_F = \left\| \sum_{i=1}^{m} u_i E_i - E \right\|$$

and

$$e_G = \left\| \sum_{j=1}^{n} v_j S_j - S \right\|.$$

Define the difference function between the best fit of E alone, approximated independently of $S(\lambda)$, and the actual E as $P(\lambda)$:

$$P(\lambda) = \sum_{i=1}^{m} u_i E_i(\lambda) - E(\lambda).$$

Similarly, for the independent fit to $S(\lambda)$, define

$$Q(\lambda) = \sum_{j=1}^{n} v_j S_j(\lambda) - S(\lambda).$$

Then $\| P \| = e_F$ and $\| Q \| = e_G$.

For general coefficients x_i and y_j, we write

$$f(\lambda) = \sum_{i=1}^{m} x_i E_i(\lambda), \qquad g(\lambda) = \sum_{j=1}^{n} y_j S_j(\lambda),$$

and for the best fit coefficients u_i and v_j, define

$$f'(\lambda) = \sum_{i=1}^{m} u_i E_i(\lambda) \quad \text{and} \quad g'(\lambda) = \sum_{j=1}^{n} v_j S_j(\lambda).$$

Now consider the least squares problem for the product function ES:

$$\| fg - ES \| = \| fg - (f' - P)(g' - Q) \|$$
$$= \| fg - f'g' + f'Q + g'P - PQ \|$$
$$= \| fg - f'g' - (-f'Q - g'P + PQ) \|.$$

Since $\| fg - ES \| \leq c(e_F + e_G)$ for some constant c by Corollary 1, we have

$$\| fg - f'g' - (-f'Q - g'P + PQ) \| \leq c(e_F + e_G).$$

In addition,

$$\sum_{i=1}^{m} \epsilon_i E_i \quad \text{and} \quad \sum_{j=1}^{n} \sigma_j S_j$$

are bounded functions. Therefore, there exists a constant d such that

$$\left\| \sum_{i=1}^{m} \epsilon_i E_i \right\| \leq d$$

and

$$\left\| \sum_{j=1}^{n} \sigma_j S_j \right\| \leq d.$$

As a result of Lemma 4, we know that

$$\| fg - f'g' \| - \| (-f'Q - g'P + PQ) \|$$
$$\leq \| fg - f'g' - (-f'Q - g'P + PQ) \|$$

or

$$\| fg - f'g' \| \leq \| fg - f'g' - (-f'Q - g'P + PQ) \|$$
$$+ \| (-f'Q - g'P + PQ) \|.$$

Note that for the last term on the right-hand side, we have

$$\| (-f'Q - g'P + PQ) \|$$
$$\leq \| -f'Q \| + \| -g'P \| + \| PQ \| \text{ by Lemma 3}$$
$$= \| f'Q \| + \| g'P \| + \| PQ \|.$$

Hence, collecting these results, we find that the difference between the best fit to the product and the product of the best fits to the individual functions satisfies

$$\| fg - f'g' \| \leq c(e_F + e_G) + \| Qf' \|$$
$$+ \| Pg' \| + \| PQ \|$$
$$\leq c(e_F + e_G) + \| Q \| \times \| f' \| + \| P \|$$
$$\times \| g' \| + \| P \| \times \| Q \|$$
$$\leq c(e_F + e_G) + e_G d + e_F d + e_F e_G$$
$$= (c + d)(e_F + e_G) + e_F e_G.$$

That is, substituting the coefficient expansions of f, g, f', and g',

$$\left\| \sum_{i=1}^{m} \sum_{j=1}^{n} (x_i y_i - u_i v_j) E_i S_j \right\| \leq (c + d)(e_F + e_G) + e_F e_G.$$

Let the difference function on the left be called $H(\lambda)$:

$$\sum_{i=1}^{m} \sum_{j=1}^{n} (x_i y_i - u_i v_j) E_i S_j \equiv H(\lambda).$$

Then

$$\|H\| \leq (c + d)(e_F + e_G) + e_F e_G$$

so we have a bound on the difference function.

We can now make use of this bound to develop bounds on the errors involved in modeling $E(\lambda)$ and $S(\lambda)$ separately by using the results x_i, y_i of the nonlinear least squares fit to the product function $E(\lambda)S(\lambda)$. We shall use the bound on H to establish the required properties of x_i and y_j by supposing $H(\lambda)$ to exist and solving for $(y_j - v_j)$ in terms of it. Then the bound on H will translate into a bound on $(y_j - v_j)$.

To solve for the coefficients $(x_i, y_j - u_i v_j)$, consider the vector function

$$V(\lambda) = [E_1(\lambda)S_1(\lambda), E_1(\lambda)S_2(\lambda), \cdots, E_m(\lambda)S_n(\lambda)].$$

The product function $E_i(\lambda)S_j(\lambda)$ forms the $[(i-1)n + j]$th element of the vector function $V(\lambda)$. To see this, substitute $i = 1$ and $j = 1$; then $E_1 S_1$ will be in column $(1 - 1)n + 1 =$ column 1 of $V(\lambda)$. For $i = m$ and $j = n$, we have $(i - 1)n + j = (m - 1)n + n = nm$ so $E_m S_n$ will be in column nm of $V(\lambda)$.

Now choose nm λ values $\lambda_1, \cdots, \lambda_{nm}$ so that the nm vectors $V(\lambda_1), \cdots, V(\lambda_{nm})$ are linearly independent. Such a choice ensures that the matrix $M \equiv [V(\lambda_1), \cdots, V(\lambda_{nm})]^T$ is not singular where $[V_1, \cdots, V_k]^T$ is the column vector transpose of the row vector $[V_1, \cdots, V_k]$. That this choice is possible follows from the linear independence of the $E_i S_j$'s. Substituting the nm values of λ_l, $l =$ 1–nm into the previous equations for $H(\lambda)$, we have nm linearly independent equations, namely,

$$\sum_{i=1}^{m} \sum_{j=1}^{n} (x_i y_j - u_i v_j) E_i(\lambda_l) S_j(\lambda_l) = H(\lambda_l), \quad l = 1\text{–}nm$$

or, written as a matrix equation,

$$MW = A$$

where W is the column vector $[x_1 y_1 - u_1 v_1, x_1 y_2 - u_1 v_2, \cdots, x_m y_n - u_m v_n]^T$ and A is the column vector $[H(\lambda_1), \cdots, H(\lambda_{nm})]^T$.

Let $M^{-1} = (b_{l,q})$ where M^{-1} is the inverse matrix of M. Since the λ_l are predetermined, we can easily form the following number (the matrix norm):

$$b = \sum_{l=1}^{nm} \sum_{q=1}^{nm} |b_{l,q}|,$$

which is a constant.

Now, the system of equations in $H(\lambda_l)$ is solvable, and after solving this system for the coefficients of the functions $E_i S_j$, we have

$$W = M^{-1}A$$

or.

$$x_i y_j - u_i v_j = \sum_{l=1}^{nm} b_{(i-1)n+j,l} H(\lambda_l), \quad i = 1\text{–}m,\ j = 1\text{–}n.$$

Taking the absolute value, we have

$$\begin{aligned} |x_i y_j - u_i v_j| &= \left| \sum_{l=1}^{nm} b_{(i-1)n+j,l} H(\lambda_l) \right| \\ &\leq \sum_{l=1}^{nm} |b_{(i-1)n+j,l}| \times |H(\lambda_l)| \\ &\leq \sum_{l=1}^{nm} |b_{(i-1)n+j,l}| \times \|H\| \\ &\leq b[(c + d)(e_F + e_G) + e_F e_G]. \end{aligned}$$

Now we introduce the restriction that normalization of the ϵ vectors will be determined by $\epsilon_1 = 1$. Since $x_1 = u_1 = 1$, the above implies

$$|y_j - v_j| \leq b[(c + d)(e_F + e_G) + e_F e_G], \quad j = 1\text{–}n.$$

Hence, we finally arrive at a restriction on the fit of $g(\lambda) = \Sigma_{j=1}^{n} y_j S_j(\lambda)$ to the function $S(\lambda)$ where the coefficients y_j are determined from a nonlinear least squares fit to the color signal $E(\lambda)S(\lambda)$:

$$\begin{aligned} \|g - S\| &= \|g - (g' - Q)\| \\ &\leq \|g - g'\| + \|Q\| \\ &= \left\| \sum_{j=1}^{n} (y_j - v_j) S_j \right\| + e_G \\ &\leq \sum_{j=1}^{n} |y_j - v_j| \, \|S_j\| + e_G \\ &\leq \left(\sum_{j=1}^{n} \|S_j\| \right) b[(c + d)(e_F + e_G) \\ &\quad + e_F e_G] + e_G. \end{aligned}$$

From the above inequality, it follows that there exists a constant D such that $\| g - S \| \leq D(e_F + e_G)$. And it is straightforward to show that there exists another constant C for which the inequality $\| f - E \| \leq C(e_F + e_G)$ holds, so that the theorem is proved.

References

[1] B. A. Wandell, "The synthesis and analysis of color images," *IEEE Trans. Pattern Anal. Machine Intell.*, vol. PAMI-9, pp. 2-13, 1987.

[2] ——, "Color rendering of color camera data," *Color Res. Appl.*, vol. 11, pp. S30-S33, 1986.

[3] L. T. Maloney, "Computational approaches to color constancy," Ph.D. dissertation, Appl. Psychol. Lab., Stanford Univ., Stanford, CA, 1985.

[4] L. T. Maloney and B. A. Wandell, "Color constancy: A method for recovering surface spectral reflectance," *J. Opt. Soc. Amer. A*, vol. 3, pp. 29-33, 1986.

[5] L. T. Maloney, "Evaluation of linear models of surface spectral reflectance with small numbers of parameters," *J. Opt. Soc. Amer. A*, vol. 3, pp. 1673-1683, 1986.

[6] D. A. Brainard, B. A. Wandell, and W. B. Cowan, "Black light: How sensors filter spectral variation of the illuminant," *IEEE Trans. Biomed. Eng.*, vol. 36, pp. 140-149, 572, 1989.

[7] A. Yuille, "A method for computing spectral reflectance," *Biol. Cybern.*, vol. 56, pp. 195-201, 1987.

[8] P. Sallstrom, "Colour and physics; Some remarks concerning the physical aspects of human colour vision," Univ. Stockholm, Stockholm, Sweden, Inst. Phys. Rep. 73-09, 1973.

[9] W. S. Stiles, G. Wyszecki, and N. Ohta, "Counting metameric object-color stimuli using frequency-limited spectral reflectance functions," *J. Opt. Soc. Amer.*, vol. 67, pp. 779-784, 1977.

[10] M. H. Brill, "A device performing illuminant-invariant assessment of chromatic relations," *J. Theor. Biol.*, vol. 71, pp. 473-478, 1978.

[11] M. H. Brill and G. West, "Contributions to the theory of invariance of color under the condition of varying illumination," *J. Math. Biol.*, vol. 11, pp. 337-350, 1981.

[12] G. West and M. H. Brill, "Necessary and sufficient conditions for Von Kries chromatic adaptation to give color constancy," *J. Math. Biol.*, vol. 15, pp. 249-258, 1982.

[13] G. Buchsbaum and J. L. Goldstein, "Optimum probabilistic processing in colour perception. II. Colour vision as template matching," *Proc. Roy. Soc. London B*, vol. 205, pp. 249-266, 1979.

[14] G. Buchsbaum, "A spatial processor model for object colour perception," *J. Franklin Inst.*, vol. 310, pp. 1-26, 1980.

[15] G. Buchsbaum and A. Gottschalk, "Trichromacy, opponent colours coding and optimum colour information transmission in the retina," *Proc. Roy. Soc. London B*, vol. 220, pp. 89-113, 1983.

[16] ——, "Chromaticity coordinates of frequency-limited functions," *J. Opt. Soc. Amer. A*, vol. 1, pp. 885-887, 1984.

[17] D. B. Judd, D. L. MacAdam, and G. Wyszecki, "Spectral distribution of typical daylight as a function of correlated color temperature," *J. Opt. Soc. Amer.*, vol. 54, pp. 1031-1040, Aug. 1964.

[18] J. Cohen, "Dependency of the spectral reflectance curves of the Munsell color chips," *Psychon. Sci.*, vol. 1, pp. 369-370, 1964.

[19] E. L. Krinov, "Spectral reflectance properties of natural formations," Nat. Res. Council of Canada, Tech. Transl. TT-439, 1947.

[20] J. Ho, "Chromatic aberration: A new tool for colour constancy," Master's thesis, School Comput. Sci., Simon Fraser Univ., Vancouver, B.C., Canada, 1988.

[21] J. Ho and B. V. Funt, "Color constancy from chromatic aberration," School Comput. Sci., Simon Fraser Univ., Tech. Rep. CSS/LCCR TR 88-18, 1988.

[22] B. V. Funt and J. Ho, "Color from black and white," in *Proc. 2nd IEEE Int. Conf. Comput Vision*, Tarpon Springs, FL, Dec. 1988, pp. 2-8; also, *Int. J. Comput. Vision*, vol. 3, pp. 109-117, 1989.

[23] D. B. Judd and G. Wyszecki, *Color in Business, Science, and Industry*. New York: Wiley, 1975.

[24] J. P. S. Parkkinen, J. Hallikainen, and T. Jaaskelainen, "Characteristic spectra of Munsell colors," *J. Opt. Soc. Amer. A*, vol. 6, pp. 318-322, 1989.

[25] J. J. More, B. S. Garbow, and K. E. Hillstrom, "User guide for MINPACK-1," Argonne Nat. Lab., Argonne, IL, Tech. Rep. ANL-80-74, 1980.

[26] G. Miel, "An updated version of the Kantorovich theorem for Newton's method," *Computing*, vol. 27, pp. 237-244, 1981.

[27] G. Wyszecki, and W. S. Stiles, *Color Science: Concepts and Methods, Quantitative Data and Formulas*, 2nd ed. New York: Wiley, 1982.

[28] B. V. Funt and M. S. Drew, "Color constancy computation in near-mondrian scenes using a finite dimensional linear model," in *Proc. IEEE Comput. Vision Pattern Recognition Conf.*, June 1988, pp. 544-549.

[29] B. K. P. Horn, "Determining lightness from an image," *Comput. Graphics Image Processing*, vol. 3, pp. 277-299, 1974.

[30] B. V. Funt, M. S. Drew, and J. Ho, "Color constancy from mutual reflection," School Comput. Sci., Simon Fraser Univ., Vancouver, B.C., Canada, Tech. Rep. CSS/LCCR TR 89-02, 1989.

[31] J. B. Fraleigh and R. A. Beauregard, *Linear Algebra*. Reading, MA: Addison-Wesley, 1987.

Jian Ho was born in Canton, People's Republic of China, in 1963. He received the B.Sc. degree in computing science and mathematics from Simon Fraser University, Vancouver, B.C., Canada, in 1986. At the time he finished his undergraduate study, he was recognized as the best student in the graduating class (in all disciplines) of the University. He also received the University's most prestigious award, the Governor General of Canada's Gold Medal. He received the M.Sc. degree in computing science from Simon Fraser University in 1988.

He has been working for Bell-Northern Research, Ottawa, Ont., Canada, since 1988. Currently, he is working in the area of distributed real-time control and fault-tolerant systems.

In 1988 one of his joint papers with Dr. B. Funt, "Color from black and white," was selected as the best paper at the Second International Conference on Computer Vision, and won the Marr Prize.

Brian V. Funt received the Ph.D. degree from the University of British Columbia, Vancouver, in 1976.

He was a Visiting Scientist at Stanford University for two years, moved to SUNY Buffalo for another two years, and then joined Simon Fraser University in 1980 where he is currently an Associate Professor. His research focuses on computational approaches to color: color perception, color constancy, and color reproduction.

Dr. Funt is a corecipient of the 1988 Marr Prize awarded at the Second International Conference on Computer Vision.

Mark S. Drew received undergraduate education in engineering science and his subsequent education was in theoretical physics. He received the B.A.Sc. degree from the University of Toronto, Toronto, Ont., Canada, in 1970, and studied the foundations of quantum mechanics for the M.Sc. degree in 1971 in the Mathematics Department there. He studied field theory at the University of British Columbia, Vancouver, B.C., Canada, and received the Ph.D. degree in 1976.

Combining work on energy systems and computer applications, he held an Industrial Postdoctoral Fellowship and subsequently an Industrial Research Fellowship in industry until he joined Simon Fraser University in 1982. His interests lie in the fields of computer vision, computer algorithms for color reproduction, and scientific applications of algebraic programming. Recently, he has also been interested in the application of Lie groups to the solution of systems of nonlinear partial differential equations.

Supervised Color Constancy for Machine Vision

Carol L. Novak and Steven A. Shafer

School of Computer Science, Carnegie Mellon University, Pittsburgh, PA 15213

ABSTRACT

In machine vision, color constancy is the ability to match object colors in images taken under different colors of illumination. This is a difficult problem because the image color will depend upon the spectral reflectance function of the object and the spectral distribution function of the incident light, both of which are generally unknown. Previous methods to solve this problem have represented these functions with a small number of basis functions and used some sort of reference knowledge to calculate the coefficients. Most of these methods have the weakness that the reference property may not actually hold for all images, or that it provides too few constraints to allow an adequate recovery of the functions.

We present here a method for color constancy that uses a color chart of known spectral characteristics to give stronger reference criteria, and with a large number of colors to give enough constraints to calculate the illuminant to the desired degree of accuracy. We call this approach "supervised color constancy" since the process is supervised by a picture of a reference color chart. We demonstrate two methods for computing supervised color constancy, one using least squares estimation, the other using a neural network. We present results for simulated experiments of the calculation of the spectral power distribution of an unknown illuminant.

1. INTRODUCTION

We usually think of color as an invariant property of objects, but in machine vision the measured color of an object varies all the time because it depends upon the lighting environment. For machines, color depends upon the spectral characteristic of light emanating from an object or light source; this varies with every little change in the illumination. It is well known that the human visual system can automatically correct for fairly large variations in the incident light. An object will usually appear to us to be the same color under two different lighting conditions. Since color remains constant for us, it is often considered to be a useful identifying feature, allowing us to quickly distinguish red, ripe fruit from green, unripe fruit. For machines to identify objects by their color, a machine vision system must be able to automatically correct for different colors of incident light.

We can view the problem as first selecting some object and recording its color under an initial light source. Given a picture of the same object taken under any other light source, we would like to calculate a transform that would change the recorded color into the color the object appears to possess in the new picture. A good transform will yield transformed colors very close to the actual color in the picture, so that we can locate the object in the new picture quickly and accurately.

However we do not require that the method must work in any way similar to the way that humans do, nor do we wish to limit ourselves to only doing as well as humans. We would like our method to succeed even in instances where the human fails. In order to achieve this, we will use spectral power distributions of illuminants and spectral reflectance functions of objects because the spectral reflectance curve is an invariant property of the object that is useful for object recognition.

These spectral functions may in general be arbitrary, so their calculation seems formidable. The usual approach is to represent a spectral function in terms of some weighted basis functions and to calculate the coefficients by reference to some *a priori* knowledge about the image in which the object appears. Most often, only one image is used for the calculation; this requires the use of an unreliable reference criterion and provides few constraints for calculating the coefficients. In this paper we propose a new approach that uses an image of a color chart with known properties to estimate the spectral properties of the illumination. The presence of the chart in the image is a reliable referent and yields a large number of constraints used for calculating the coefficients of the illuminant. We call this approach "supervised color constancy" because the calculation is supervised by an image of the reference chart.

We begin the discussion with a mathematical formulation of the image formation problem. We then present two methods for calculation, least squares estimation and a neural network, and compare the results.

1.1. Mathematical Formulation of Color Constancy

In our scenario a color image is taken of a scene illuminated by some unknown type of light. In a typical color picture three measurements are made at each point in the scene, yielding a three dimensional measurement vector M for each point. The goal of color constancy is to automatically calculate corresponding measurements M' of the same scene as it would appear if illuminated by a standard light. This allows a computer to perform color matching.

The light reflected from a surface is equal to the light incident upon that object $L(\lambda)$ (known as the *illuminant*) times the *spectral reflectance* function of that surface $\rho(\lambda)$. The measurement m obtained from a sensor looking at that surface is the light reflected from the surface times the responsivity function of the sensor $s(\lambda)$, integrated over the range of the spectrum for which $s(\lambda) > 0$ (usually the "visible" wavelengths of light)[1]

$$m = \int L(\lambda)\, \rho(\lambda)\, s(\lambda)\, d\lambda \qquad (1\text{–}1)$$

If the surfaces being observed are matte, then there is no appreciable specular reflection, so $\rho(\lambda)$ is an invariant property of the surface. Typically there are three sensors or one sensor with three different filters; in our work we use red, green, and blue filters, although other filters may be chosen. Let $s_R(\lambda)$ be the function defined by multiplying the sensor responsivity with a red filter transmittance function, $s_G(\lambda)$ be the sensor responsivity times a green filter transmittance, and $s_B(\lambda)$ be the sensor times a blue filter. Then for each surface the three measurements are

$$m_R = \int L(\lambda)\, \rho(\lambda)\, s_R(\lambda)\, d\lambda$$
$$m_G = \int L(\lambda)\, \rho(\lambda)\, s_G(\lambda)\, d\lambda$$
$$m_B = \int L(\lambda)\, \rho(\lambda)\, s_B(\lambda)\, d\lambda$$

In order to represent an arbitrary curve in a finite manner, we approximate the illuminant $L(\lambda)$ in terms of some some basis functions $b_1(\lambda)$ ... $b_n(\lambda)$. This way the spectral power distribution (SPD) of the light can be approximated by the sum of appropriately weighted basis functions:

$$L(\lambda) \approx \sum_{j=1}^{n} L_j\, b_j(\lambda)$$

If the basis functions are well chosen and a large enough number are used, then any type of illuminant or reflectance function can be closely approximated in this representation. Using this new representation, the sensor measurements may be rewritten as

$$m_i = \sum_{j=1}^{n} L_j \int b_j(\lambda)\, \rho(\lambda)\, s_i(\lambda)\, d\lambda$$

where the i subscript refers to one of the three filters.

Looking at the same surface with the same camera but under a standard illuminant $L'(\lambda)$ can be modeled as changing the illumination coefficients L_j to L_j' to produce sensor measurements m_i'. If we let $r_{ij} = \int b_j(\lambda)\, \rho(\lambda)\, s_i(\lambda)\, d\lambda$, then the measurements are simply

$$m_i = \sum_{j=1}^{n} L_j r_{ij}$$

$$m_i' = \sum_{j=1}^{n} L_j' r_{ij}$$

If we let M be the vector of the sensor measurements m_i, L be the vector of light coefficients L_j, and R be the matrix with entries r_{ij}, then we have a matrix equation for the measurements under the unknown illuminant

$$M = R\,L$$

and a similar one for the measurements under the standard illuminant

$$M' = R\,L'$$

It should be noted that R is not necessarily square, since there may be more measurements than basis functions.

If the camera measurements M are known and the matrix R has been calculated, the illumination coefficients L may be estimated by least squares. We will show how this can be done. We also suggest how the surface reflectance function $\rho(\lambda)$ can be calculated.

1.2. Previous Work In Color Constancy

Proposed methods for color constancy can be classified by the basis functions b_i they use for performing the transformation, and the by the reference property that goes into determining the correspondence between measurements M' taken under the standard illuminant and measurements M taken under the unknown illumination.

One type of reference property is to assume some kind of statistic holds true for the image. It may be assumed that the different colored objects in the scene are distributed in such a way that the average reflectance value is some known quantity. This quantity may be chosen arbitrarily (typically that the average color of the scene is gray,[2]) or may be based on an averaging of color data collected from a variety of materials that are considered likely to be in the scene.[3] The desired average is compared with the observed average of all areas in the picture; then the color bands are scaled to create a new picture with the desired property. Other researchers have looked for reference properties in an area of the scene rather than the whole scene. An object of known reflectance (typically a white test object) may be intentionally put in the scene, or it may be assumed that a white object is likely to occur in the scene naturally. A slight variation is to assume that the brightest object scene in each color band should average out to white.[4] The appearance of the object or objects yield scale factors to correct for colored incident light. Another method uses known correspondences between certain areas of the scene under unknown illumination and the values they should have under standard illumination; an algorithm is then used to learn the best transformation that can then be applied to the rest of the scene.[5] These methods of achieving color

Other researchers have appealed to the dichromatic reflection model and assumed that highlighted inhomogeneous objects are present in the scene. The highlight areas have the same color as the illumination since the model assumes that the highlights are not colored by the objects. Once the highlights are found, the color of the illumination can be calculated and factored out of the scene.[2, 7, 8] Mathematical constraints can also be employed to limit the possible forms of illumination and surface reflectances; this may be done by using more sensors than degrees of freedom in the SPDs,[9] by including spectral data in addition to or instead of RGB values,[10, 8] or by using qualitative knowledge about the nature of the SPDs (e.g. real-world reflectances are never negative).[11]

Another issue of research in color constancy is the method of representing the SPDs of the illumination and surface reflectances in the calculations. Spectral functions may be approximated by sampling at regular intervals with a spectroradiometer (e.g. every 10 nm within the visible range); by sampling with broad-band physically realizable receptors such as those used in color cameras; by representation with a set of abstract basis functions such as Fourier functions[12] or normalized Legendre polynomials;[13] or by using a set of basis functions derived from a principal components analysis of spectral data taken of natural and man-made objects.[3] There is also some question as to the number of basis functions needed to give an adequate representation of real objects and illuminants. Some researchers have argued that many SPDs corresponding to physically realizable objects are band-limited functions and thus a good fit may be achieved with even a small number of basis functions.[14] However some illuminants, particularly fluorescent lights, have sharp peaks, making them difficult to represent under most schemes. Recently some researchers have argued that 3 basis functions are insufficient even for surface reflectances.[15]

Since most methods generate each constraint from one sensor value, the number of basis functions is limited by the number of sensors. Clearly, to disambiguate as many as colors as possible, a large number of constraints would be desirable. However a practical number of sensors is typically three or four, so we would like to generate more constraints without needing more sensors.

1.3. Supervised Color Constancy

We would like the reference property used to calculate the color constancy transform to be a reliable process rather than a heuristic that works some of the time. One way of getting "ground truth" is to introduce a surface of known reflectance properties into the picture we are interested in, or take a test picture beforehand. If we take a picture of a known white card in some known position under a particular incident lighting, the measured color gives some important information about the color of the light. This method will work identically to the method of assuming that we can find a white surface somewhere in the picture, except that in this case we know that the surface we think is white is in fact white.

We call this approach "supervised" because it involves taking some test data which trains the computer as to what type of light to expect. This type of process is already used in some applications.[16] Illumination changes are common in the real world, and calibrated color standards are easily obtained. At the cost of introducing a known surface into the scene from time to time, we can provide a transform that yields invariant color information as the illumination changes.

In this paper we present a method of supervised color constancy using a color chart instead of a simple white card. We use a chart with *m* color patches with distinct known reflectance functions, so that a three-band color image of the chart will yield $3m$ constraints upon the spectral curve of the incident light., In the general case where there are s sensors, there will be sm constraints. Then we show how these constraints may be solved to give an estimate of the illuminant $L(\lambda)$. Not only is the color chart a reliable reference property, but the large number of constraints it generates allows us to do our calculations in a high dimensional space.

We then present an alternate method that uses a neural network trained to calculate the illuminant vector from the RGB values the camera recorded for the color chart under that illumination. This method has an initial training cost as the network must be shown a number of examples. An advantage over the least squares method is that no measurement of camera spectral sensitivity is needed; such measurements can be difficult to obtain.

2. THE COLOR CHART METHOD FOR SUPERVISED COLOR CONSTANCY

Our method has two stages. In the first stage the illuminant is estimated using the color chart. Once this has been done, the appearance of any object of known reflectance can be reliably predicted. This is useful for situations where a target object (such as a stop sign) must be repeatedly identified under changing light conditions.

In the second stage the reflectance of an unknown object is estimated by using images under several illuminants. In this paper we present the theory used in calculating both stages and experimental results of simulations for the first stage.

2.1. Estimation of the Illuminant Using the Color Chart

In order to estimate the illuminant we assume that the sensitivity functions of the sensors are known, as well as the reflectance functions of the surfaces on the color chart. These functions can usually be obtained from the manufacturer or measured directly with a spectroradiometer. The matrix R is calculated by factoring these together with the basis functions. If three sensors are used that are responsive to red/green/blue, the first three rows of R consist of

$$\begin{matrix} \int b_1(\lambda)s_R(\lambda)\rho_1(\lambda)d\lambda & \cdots & \int b_n(\lambda)s_R(\lambda)\rho_1(\lambda)d\lambda \\ \int b_1(\lambda)s_G(\lambda)\rho_1(\lambda)d\lambda & \cdots & \int b_n(\lambda)s_G(\lambda)\rho_1(\lambda)d\lambda \\ \int b_1(\lambda)s_B(\lambda)\rho_1(\lambda)d\lambda & \cdots & \int b_n(\lambda)s_B(\lambda)\rho_1(\lambda)d\lambda \end{matrix}$$

where $\rho_1(\lambda)$ is the reflectance function of the first surface on the color chart. The next three rows consist of the same calculations for the second surface on the color chart, and so on, so that for m surfaces on the color chart there are $3m$ rows in the matrix. In the general case where there are s sensors there are sm rows in the matrix. Because we use the same color chart, basis functions, and sensors each time, the matrix R and its least squares inverse may be calculated off-line once and then used to quickly estimate the illuminant.

Measurements of the color chart are taken under the unknown illuminant, producing a measurement vector M with $3m$ values. The first three entries in M are the RGB values that the camera measured for the first surface, the second three entries correspond to the second surface, and so on in the same order that the rows of R were calculated. The goal is then to solve the equation

$$M = RL$$

for L. Since there are m surfaces on the color chart, and 3 sensors used to measure the chart, and since we choose to represent the light with n basis functions, M is $3m\times1$, R is $3m\times n$, and L is $n\times1$. Typically n is chosen so that $n<3m$ since it is generally the case that there is some redundancy in the color chart functions. Therefore the least squares method can be used to solve for L, yielding an estimated solution L^*. The estimated illuminant $L^*(\lambda)$

is recovered from L^* by multiplying the coefficients by the appropriate basis functions

$$L^*(\lambda) = \sum_{i=1}^{n} L^*_i b_i(\lambda)$$

The number of illuminant coefficients n may be chosen empirically or through eigenvector analysis. A calculation of the eigenvectors of $R^T R$ gives a set of vectors that span the solution space for L, while the magnitude of each eigenvalue tells how significant the corresponding eigenvector is. Each eigenvalue is divided by the largest one and sorted in decreasing order. A threshold value is chosen based on the signal-to-noise ratio of the sensors. The number of eigenvalues greater than that threshold is the number of basis functions n that we choose for calculating L^*.

By adding more distinct colors, an increasing number of illuminant coefficients can be accurately calculated. This only works up to the point that it becomes too difficult to formulate enough distinct colors that can be differentiated by the camera. Each additional color will add some information to our constraints, but the amount of new information added by each new color will decrease as the number of colors grows. Eventually, that amount will be below the threshold that can be reliably detected by the camera. However, in practice it is easy to obtain 8 or more coefficients which provide a relatively detailed description of any typical illuminant.

2.1.1. The Least Squares Method

The least squares method is a standard technique for finding the best fit curve to a collection of data. The mathematical model we use is

$$m_i = L_1 r_{i,1} + \ldots + L_n r_{i,n} + e$$

where m is the value measured in the image, r_{ij} is the matrix calculated form the combination of the surface, sensor, and basis functions, L_j is the desired illumination description, and e is the residual error.

In order for the least squares calculation to produce a good result, several things must be true:

1) the hypothesized relationship must closely approximate reality
2) there must be more measurements than desired illuminant coefficients
3) the independent variables must be independent

We have already shown that our model is a good approximation of the physics of spectral integration. The number of measurements can be manipulated by simply adjusting the number of colored surfaces on the chart. The third condition depends on the nature of the matrix R, so the basis functions and the colors used in the color chart must be chosen carefully.

2.1.2. Choice of Basis Functions

Although many different families of basis functions may theoretically be used, in practice some care must be taken in choosing among them. A simple choice would be a series of unit impulses ranging over the visible wavelengths. For example the first function could be a curve consisting of an impulse at 400nm and zero everywhere else; the second function could have an impulse at 405nm and a zero everywhere else; and so on. The problem with this choice is that in solving the least squares formulation, the assumption that the independent variables are in fact linearly independent may be violated. Many illuminants are fairly smooth across the spectrum; as a result the value of L_2 would be approximately the average of L_1 and L_3. With such a set of basis functions, variables would tend to be highly correlated with each other. This state of affairs is known as *multicollinearity* and the result is that very small errors in the measurements can cause very large errors in the calculation of the L_i coefficients. Moreover if one variable is over-estimated, the next is likely to be under-estimated, leading to wild oscillations in the calculated values of the coefficients.

A more sound choice is a set of basis functions which describe characteristics which are likely to be independent such as the Fourier components of the curve. We use the Legendre polynomials, suggested by Healey and Binford.[7] These functions are orthogonal and the ith basis function is an ith degree polynomial. By scaling the visible wavelengths of light to range over [-1,+1] and using a suitable number of basis functions, we can easily represent the visible portion of any illuminant with a high degree polynomial. We present results for this method in section 3.4.

2.2. Estimation of the Illuminant Using a Neural Network

Recall that we are trying to solve the equation

$$M = RL$$

for L. Another approach to solving it is to train a neural network to calculate L given M so that it is implicitly calculating the constraints generated by the image of the color chart , or rather their inverse R^{-1}.

Inputs to the network consist of the $3m\times1$ measurement vector M made of the appearance of the color chart under a particular type of light. Outputs from the network are the n coefficients of that light. Since multicollinearity is not an issue, the basis functions can be simple unit samples, so the illumination coefficients are simply the spectral power measured at each wavelength by a spectroradiometer.

The network configuration we have used has a single hidden layer with n hidden units with complete interconnection between the input and hidden layer, between the hidden and output layer, and between the input and output layer (shortcut connections). The weights were updated using continuous back-propagation.[17, 18]

The network is trained using examples of the color chart appearance and the corresponding illumination. Since the constraints generated by the chart reflectances and sensor responses are calculated implicitly by the network, these functions do not have to measured or obtained from the manufacturer. Measuring the spectral response of a camera can be difficult even with the aid of a spectroradiometer. On the other hand, this method does require the spectral measurement of the lights used to train the network and this will require access to a spectroradiometer or to calibrated light sources. However, once the network has been properly trained, the spectroradiometer will not be needed and the method will calculate the spectral power of further lights very quickly. In effect we have constructed a "poor man's spectroradiometer", obtaining a detailed spectral power distribution function of illuminants using only a color chart and a color camera, with a neural network that "learns" the relationships among them. Results for this method are presented in section 3.5.

3. SIMULATED EXPERIMENTS

To find out what kind of results we could theoretically expect from our method, we simulated the picture taking process and applied our method to the simulation data.

3.1. The Simulation Process

In section 1 we introduced Equation (1-1) which says that given the spectral power distribution of the illumination, the spectral responsivity of the sensor, and the spectral reflectance of a surface, we should be able to calculate the value that will be obtained by a sensor pointing at that surface. Accordingly we collected spectral curves of our camera equipment, a color chart, some test objects, and some common types of illumination. The spectral curves were multiplied together and the integrals were calculated. The resulting values were considered the simulated sensor values of a hypothetical camera looking a a known color chart under an unknown illuminant.

Then the calculations described in the previous section were performed to see if the light factored into the sensor values could be recovered. Simulated sensor values were also used to train the neural network, which was then tested on new simulated inputs it had not seen before.

The above calculations were performed on data with simulated noise added. The noisy data was simulated by adding pseudo-random values, weighted by a Gaussian function, to the simulated camera measurements.

3.2. The Spectral Distribution Functions

Although our experiments used simulation and so could have used any kind of spectral curves, we attempted to obtain spectral curves that were fairly close to the kinds of curves associated with the equipment we actually use. Fortunately, spectral data for most camera equipment can be obtained from the manufacturers. We have used the standard MacBeth ColorChecker in our work.[19] The ColorChecker has 24 colored squares composed of Munsell color papers. The spectral curves of these papers were obtained from the manufacturer and used in the simulation. These are considered hypothetical values since there is no guarantee that these curves will be realized on an actual MacBeth chart.

We collected spectral power distribution functions for two classes of hypothetical illuminants. To simulate lights that might be obtained indoors, we used the CIE standard illuminant A with or without hypothetical filters placed in front of it. Illuminant A is supposed to represent light from a full radiator at 2856K; it is similar to what one can expect from incandescent lights. The hypothetical filters were based on filters sold by Edmund Scientific Co.; the spectral curves were obtained from the manufacturer.[20]

In order to capture in our simulation the way outdoor lighting varies from day to day, we used simulated daylight curves generated by the standard CIE method of calculating daylight illuminants with different correlated color temperatures.[21] These curves are are supposed to approximate natural daylight under a wide range of conditions, including direct sunlight, skylight, and daylight under various degrees of cloud coverage. They are calculated from a weighted combination of three spectral curves; the weights are determined by a complex formula that depends upon the correlated color temperature.

3.3. Evaluation Method

To evaluate the success of a color constancy method, researchers need to state what precisely is the goal of the method, and then develop a criterion for measuring how close they have come to that goal. We have examined two related yet different goals and propose a measure of success for each.

One goal is to evaluate the accuracy with which we can predict an object's color under a new type of light. In other words, after estimating the spectral power distribution of the illumination or the spectral reflectance function of an object, how close is the color we predict from these estimates to the true color? Therefore we need a measure of the distance between two colors. In our work in machine vision, we represent these colors with RGB values. For a given color value (R, G, B) and its predicted color value (R_p, G_p, B_p) the error E could be measured by

$$E = (|R-R_p|^n + |G-G_p|^n + |B-B_p|^n)^{1/n}$$

for some value n. In particular, if $n = 1$, then this is the sum of the errors in each band. If $n = 2$, then this is the euclidean distance between the two values. If $n =$ infinity then this is the maximum of the three errors.

The problem with this error measure is that is is biased against bright colors. If we are viewing mostly dark surfaces, the change in their measured values from one light source to another will be small, and so even an ineffective method of color constancy will yield a low error. Therefore we prefer to use a percentage error measure that takes into account how much room there is for error. We will use the error calculation

$$E = \frac{(|R-R_p|^n + |G-G_p|^n + |B-B_p|^n)^{1/n}}{(|R|^n + |G|^n + |B|^n)^{1/n}}$$

and choose $n = 1$. Note that this error measure points out differences in intensity as well as differences in hue and saturation. The CIE color difference formula based on $L^*u^*v^*$ coordinates is often used to give a measure of how far apart colors look to humans.[22] However, since we are concerned with machine vision applications, our formula relates directly to the RGB values measured by the computer system.

Another possible goal for our system would be to measure how well it performs as a spectroradiometer since the immediate output of the least squares calculation or the network is a spectral curve. Since the color appearance predictions are calculated from these spectral curves, a more accurate spectral curve should yield a more accurate color. In this case the error measure of interest is the difference between the predicted spectral curve and the actual one.

A simple difference measure for this criterion is the difference of the spectral power of the two curves at selected wavelengths. These differences can be squared and then added together to compute a total error measure for the predicted curve ΔL.

$$\Delta L = \sum_{\lambda=\lambda_0}^{\lambda_k} (L(\lambda) - L^*(\lambda))^2$$

However, this formulation treats the contribution from all wavelengths equally. If the spectral function is to be used in calculating other quantities, such as color appearance, accuracy at the extremes of the visible spectrum may be much less important than accuracy in the center of the visible spectrum where sensors are much more responsive. Accordingly, we prefer to use the sum of squared differences weighted by the human luminous efficiency function $V(\lambda)$.

$$\Delta_w L = \sum_{\lambda=\lambda_0}^{\lambda_k} V(\lambda)(L(\lambda) - L^*(\lambda))^2$$

We have found that weighted spectral error corresponds more closely than the unweighted error to the color appearance error measure E.

3.4. Results Using Least Squares Method

Our simulation assumed 24 colored squares and 3 color filters which yield 72 constraints upon the illuminant. However, the eigenvector analysis showed that most of that information was redundant in a camera that can only detect 1 part in 100 reliably (which is typical of our equipment). Eigenvector analysis of the matrix R formed with our hypothetical color chart, filters, and camera showed that at most 10 coefficients are significant if the camera is assumed to have a signal-to-noise ratio of 100:1. If the camera had a SNR of 1000:1 then up to 21 coefficients could be reliably calculated.

Figures 1 and 2 show recovery with an empirically derived number of basis functions. That is, we tried the least squares calculation repeatedly, varying the number of basis functions, to find the number that would give the lowest weighted error $\Delta_w L$. The empirical number was 3 for the outdoor lights (these originally came from 3 weighted functions) and 5 for the indoor lights. Surprisingly, the smaller number of basis functions does better than a larger number when noise is introduced into the calculation. The empirically derived numbers also give the smallest color difference E. (The unweighted spectral differences ΔL give a different experimental answer which leads to a large color difference.)

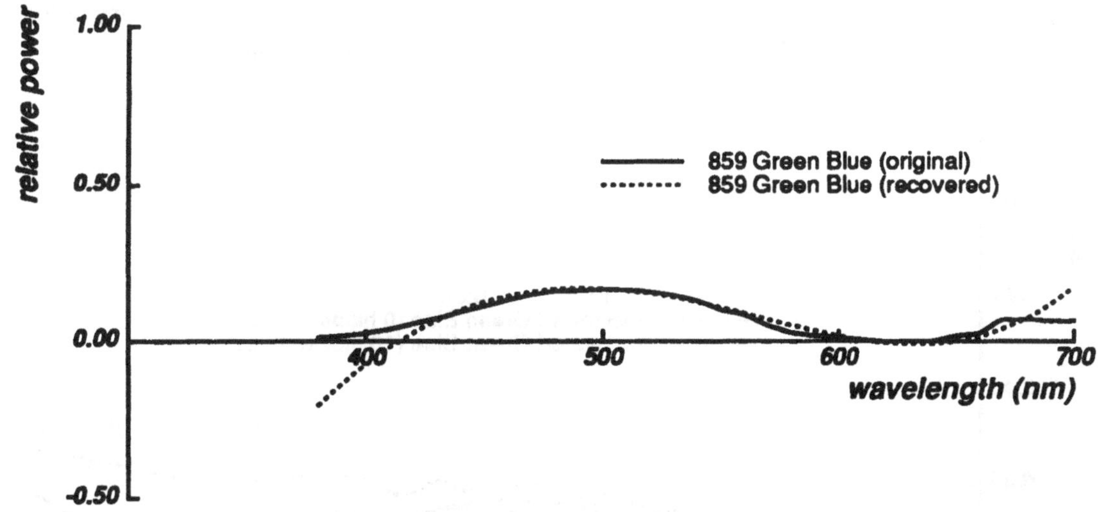

Figure 1: Noisy simulation of indoor light using 5 basis functions

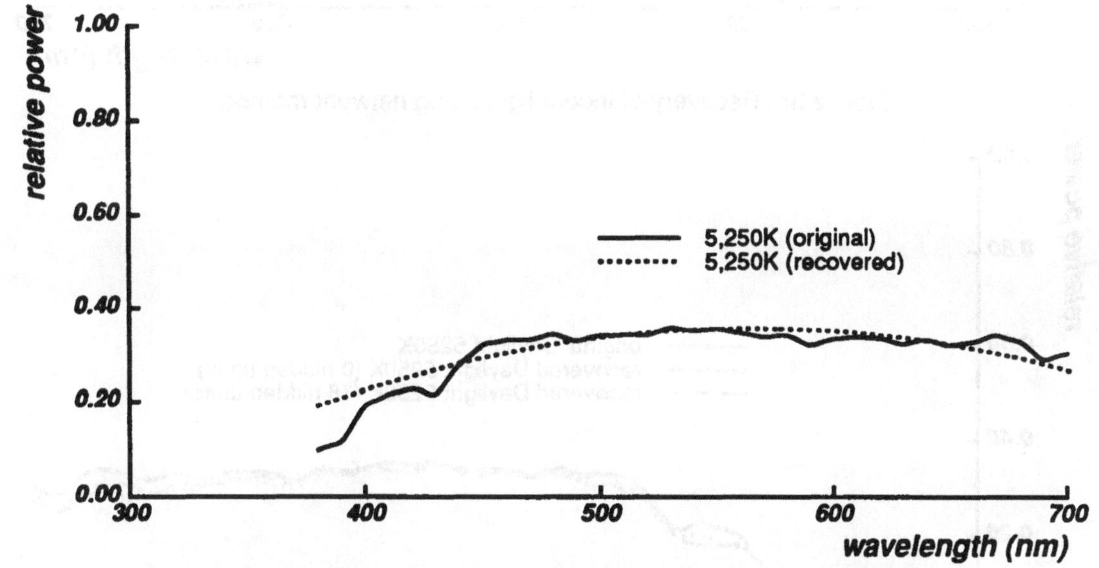

Figure 2: Noisy simulation of outdoor light using 3 basis functions

Our explanation for this less-is-more result is that the larger number of basis functions are magnifying the errors in the simulated camera measurements. It is possible that a different set of basis functions would not suffer so much from this problem. The greatest errors occur at the extremes of the visible range where the differences may be less important for many applications.

3.5. Results Using Neural Network Method

To train the neural network we used 70 outdoor light samples or 47 indoor light samples. The network was then tested on a small number of light samples it had not seen before, although of the same type (indoor or outdoor). The number of hidden units was varied in the different experiments, including some trials with no hidden layer at all. All networks were trained with simulated camera measurements with noise added.

Figures 3 and 4 show the results from the network, comparing the original light curves with the recovered lights. The results are shown for experiments where no hidden layer was used, and where a hidden layer with 16 units was used. For the outdoor lights, the hidden layer did not make much difference in the accuracy of the network in recovering spectral curves. With the indoor lights, the hidden layer gave a big advantage.

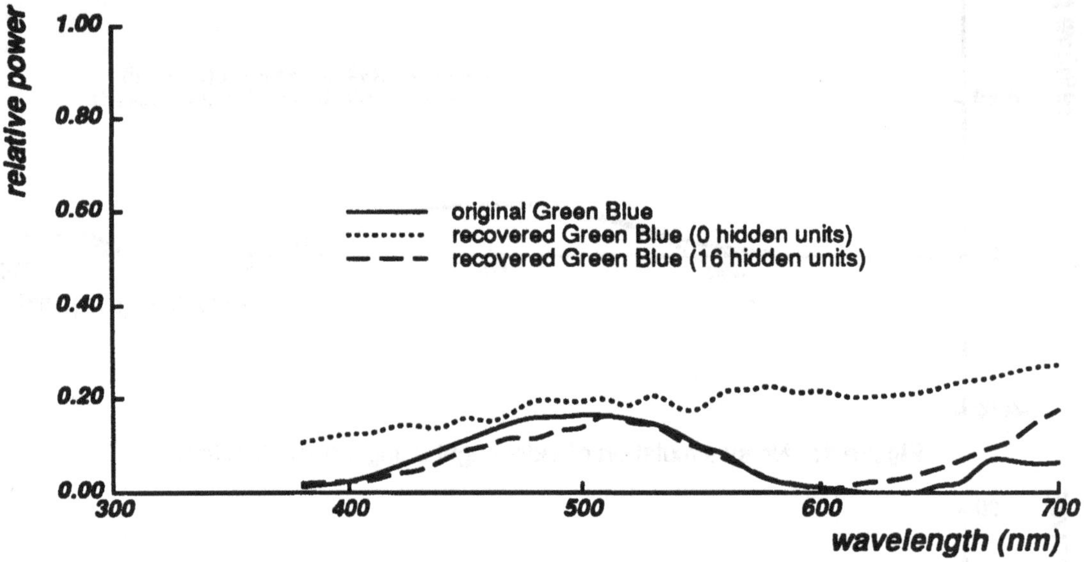

Figure 3: Recovery of indoor light using network method

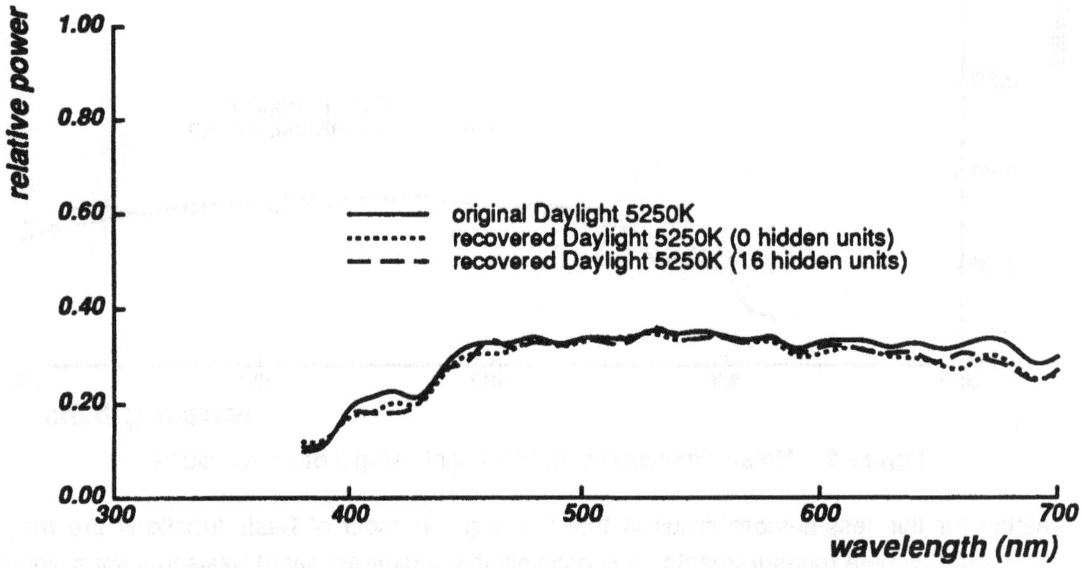

Figure 4: Recovery of outdoor light using network method

3.6. Comparison of Least Squares Method with Neural Network

Figure 5 shows the overall average error in recovering the two types of illuminants as a function of the number of basis functions used in the least squares method. This is shown using the weighted error measurement $\Delta_w L$, averaged for all the illuminants used in the experiment. As can be seen, the lowest error occurs at a different number of basis functions for the two types of lights. After this optimum number of basis functions, adding more functions does not help, and in fact hurts the recovery of the illumination. As is seen in Figure 6, the method comes closest to predicting the correct color when 3 basis functions are used for outdoor lights and 5 for indoor lights. For both types of light, the method can estimate the color appearance with an error of less than 2 percent.

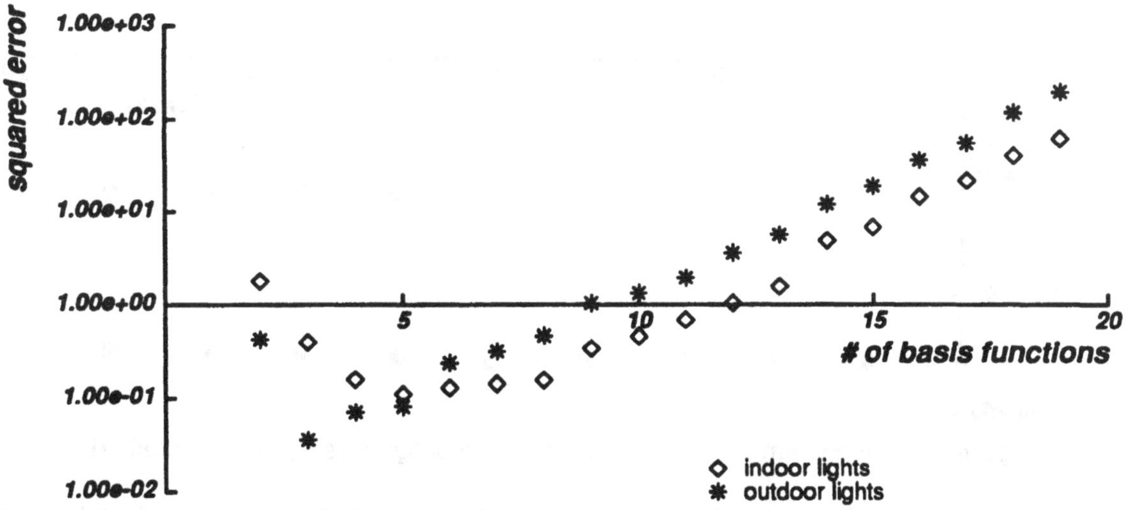

Figure 5: Average error $\Delta_w L$ for indoor and outdoor lights using least squares method

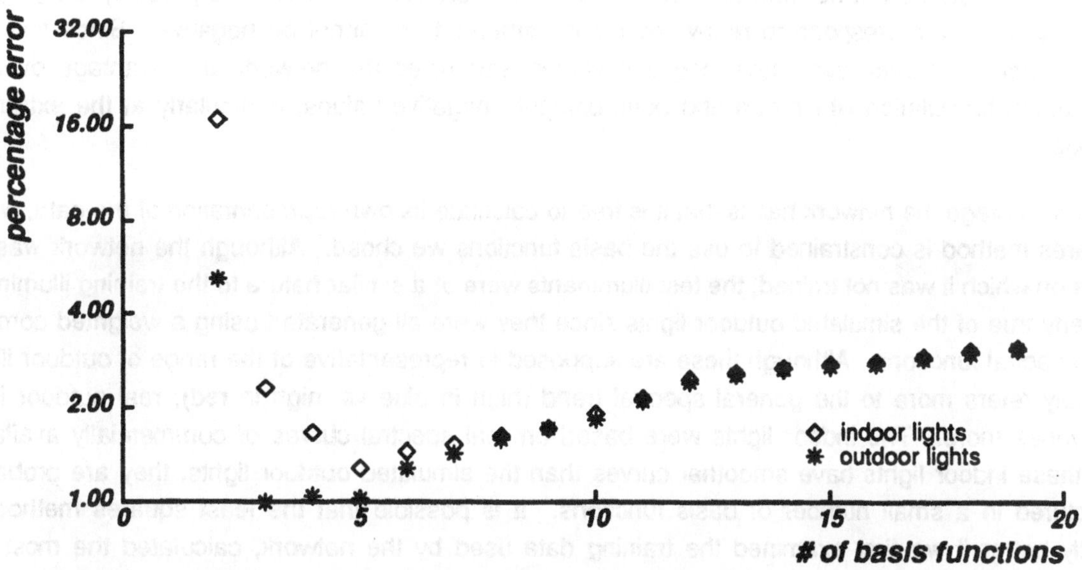

Figure 6: Average error E for indoor and outdoor lights using least squares method

Figure 7 shows the average error in recovering illuminants as a function of the number of hidden units used in the neural network. The error in recovering outdoor lights is very low and the addition of a hidden layer makes very little difference. The network does a better job with the outdoor lights than the least squares method using any number of basis functions.

For the indoor lights, a hidden layer gives a significant improvement. Although we found the lowest error with 16 hidden units, the addition of units beyond 4 gave only a very slight improvement. The best least squares method (with 5 basis functions) performed slightly better than the network method in terms of weighted squared error. If unweighted error is used, then the network performed significantly better than all the least squares methods used.

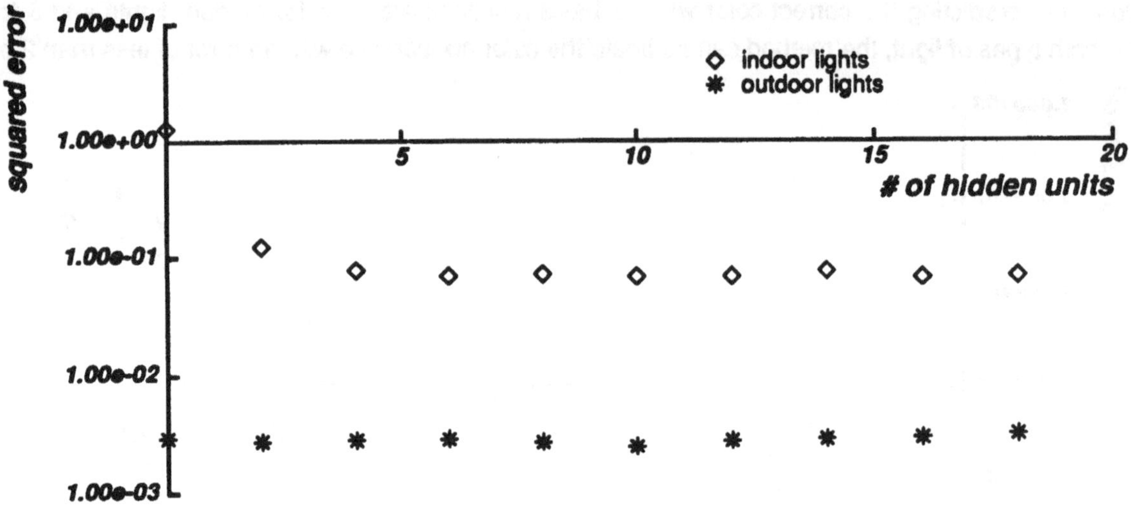

Figure 7: Average error $\Delta_w L$ for indoor and outdoor lights using network method

The neural network method has several advantages over the least squares method, which help explain its success. The outputs from a network are constrained to lie between 0 and 1. This property is highly useful for calculating values that correspond to real-world measurements that cannot be negative. Since spectral power distribution functions cannot ever have negative values, this gives the network an advantage over the least squares method formulation which can and does compute negative values, particularly at the extremes of the visible range.

Another advantage the network has is that it is free to calculate its own representation of the data, whereas the least squares method is constrained to use the basis functions we chose. Although the network was tested on illuminants on which it was not trained, the test illuminants were of a similar nature to the training illuminants. This is particularly true of the simulated outdoor lights since they were all generated using a weighted combination of just three spectral functions. Although these are supposed to representative of the range of outdoor illumination, this probably refers more to the general spectral trend (high in blue vs. high in red); real outdoor illumination probably varies more. The indoor lights were based on real spectral curves of commercially available filters. Although these indoor lights have smoother curves than the simulated outdoor lights, they are probably not so easily captured in a small number of basis functions. It is possible that the least squares method might do significantly better if we first examined the training data used by the network, calculated the most significant eigenvectors of the spectral data, and used those as basis functions rather than the Legendre polynomials.

A disadvantage of the neural network method is of course the time and effort it takes to acquire the training data and the amount of time it takes to train the network adequately. However, once this has been done the network will operate very quickly.

4. FUTURE WORK

In the experiments just presented, we assumed that the reflectance functions of objects of interest were already known to us. Thus, given an accurate estimate of the illumination, we can predict the appearance of an object under that type of illumination by multiplying the estimated illumination by the known sensor functions and object reflectance and integrating over the visible wavelengths. However in general vision scenarios the reflectances of interesting objects in the scene will not be known in advance. Therefore it is necessary to have a way to estimate object reflectances in addition to the illumination.

We propose that surface reflectance functions might be estimated in much the same way that we estimated the illuminant. In the methods described for calculating the illumination, constraints upon the illuminant were generated by observations of the illuminant's effect upon known surfaces on the color chart (or by learning the relationship between spectral power of the light and chart appearance, in the case of the network). With such estimates of the illuminant in hand, constraints upon a surface reflectance function can be generated by observations of the surface illuminated by known or estimated illuminants.

As before, we start with the equation describing the measurement made by the sensor, but this time using the estimated illuminant in the calculation

$$m = \int L^*(\lambda)\, \rho(\lambda)\, s(\lambda)\, d\lambda \qquad (4\text{-}1)$$

This time we choose to represent the unknown surface reflectance function, rather than the illuminant, in terms of some basis functions, so that

$$\rho(\lambda) \approx \sum_{j=1}^{n} \rho_j\, b_j(\lambda)$$

and substituting into Equation (4-1):

$$m_i \approx \sum_{j=1}^{n} \rho_j \int b_j(\lambda)\, L^*(\lambda)\, s_i(\lambda)\, d\lambda$$

where the i subscript refers to one of the three filters. If we group together the known quantities, letting $\Lambda_{ij} = \int b_j(\lambda)\, L^*(\lambda)\, s_i(\lambda)\, d\lambda$ then the equation can be rewritten simply as

$$m_i \approx \sum_{j=1}^{n} \rho_j \Lambda_{ij}$$

We let M be the vector of the sensor measurements m_i under one type of light and let Λ be the matrix with entries of Λ_{ij}, yielding a simple matrix equation

$$M = \Lambda \rho$$

which is then solved for ρ, the vector of reflectance coefficients ρ_j.

If there are only three sensors and one light source then the equation only provides three constraints upon the estimate of the reflectance coefficients ρ. In the previous sections we described how we were able to increase our knowledge about an unknown light source by looking at its influence upon several known surfaces. Now we can turn this around and increase the number of constraints upon an unknown surface reflectance by looking at the influence of several known, or estimated, illuminants. If we have measured the appearance of the object under m illuminants, there will $3m$ rows in the matrix Λ.

The least squares method is used to solve for ρ yielding an estimated solution ρ^{**}. (The double star indicates that the estimate was calculated using data that had also been estimated. If the surface reflection were estimated using known illuminants rather than estimated ones, it could be designated ρ^*.) The estimated surface reflection function $\rho^{**}(\lambda)$ is then recovered by multiplying the coefficients by the appropriate basis functions

$$\rho^{**}(\lambda) \approx \sum_{j=1}^{n} \rho_j^{**}\, b_j(\lambda)$$

5. CONCLUSIONS

In this paper we have presented a mathematical formulation of color constancy; we also classified modern color constancy methods according to the reference criterion chosen and the number of basis functions used. We propose a new approach that we call "supervised color constancy" that uses a color chart of known properties to provide information about illuminants. The presence of the chart in an image provides an extremely reliable reference criterion, and the large number of colored squares provide enough constraints to calculate more basis coefficients than other color constancy methods.

We have shown simulated results for two methods of calculating the illuminant: least squares estimation and neural networks. Both these methods were tested on data with simulated noise to try to give a more realistic scenario. Under these conditions the neural network which was trained on similar illuminants tended to perform better than the least squares estimation technique which used Legendre polynomials as basis functions. The estimation of the illuminant allows us to predict the appearance of objects with known reflectance to a high degree of accuracy.

This paradigm can also be used on objects with unknown reflectances to solve the more general color constancy problem. This is accomplished by using the illumination estimates calculated by the above methods to impose constraints upon the reflectance functions of objects. In this way multiple observations under different illuminants can be accumulated to provide incremental improvements in the estimate of surface colors. Taken together, these methods comprise a powerful new approach for reliable and accurate color recognition in robot vision.

6. ACKNOWLEDGMENTS

Thanks to Barak Pearlmutter for his advice and the use of his neural network simulator and to Dave Touretzky for his advice and encouragement.

This research was supported by the Defense Advanced Research Projects Agency, DoD, through ARPA Order Number 4976, monitored by the Air Force Avionics Laboratory under Contract F33615-87-C-1499; by the Jet Propulsion Laboratory, California Institute of Technology, sponsored by the National Aeronautics and Space Administration under Contract 957989; and by the Air Force Office of Scientific Research under Contract F49620-86-C-0127. Any opinions, findings, and conclusions or recommendations expressed in this publication are those of the authors and do not necessarily reflect the views of the United States Government or the Jet Propulsion Laboratory.

7. REFERENCES

1. C. L. Novak and S. A. Shafer, "Color Vision", in *Encyclopedia of Artificial Intelligence*, John Wiley & Sons, Inc., New York, 1990.

2. M. D'Zmura and P. Lennie, "Mechanisms of color constancy", *J. Opt. Soc. Am.*, Vol. 3, No. 10, October 1986, pp. 1622-1672.

3. R. Gershon, *The Use of Color in Computational Vision*, PhD dissertation, University of Toronto, 1987.

4. E. H. Land, "The Retinex Theory of Color Vision", *Scientific American*, Vol. 237, No. 6, December 1977, pp. 108-128.

5. A. C. Hurlbert and T. A. Poggio, "Synthesizing a Color Algorithm from Examples", *Science*, Vol. 236, January 1988, pp. 482.

6. J. A. Worthey, "Heuristic analysis of von Kries color constancy", *J. Opt. Soc. Am.*, Vol. 3, No. 10, October 1986, pp. 1708-1712.

7. G. Healey and T. O. Binford, "The Role and Use of Color in a General Vision System", *Proceedings of the ARPA Image Understanding Workshop*, DARPA, 1987, pp. 599-613.

8. S. Tominaga and B. A. Wandell, "Standard surface-reflectance model and illuminant estimation", *J. Opt. Soc. Am. A*, Vol. 6, No. 4, April 1989, pp. 576-584.

9. L. T. Maloney and B. A. Wandell, "Color constancy: a method for recovering surface spectral reflectance", *J. Opt. Soc. Am. A*, Vol. 3, No. 1, January 1986, pp. 29-33.

10. J. Ho and B. V. Funt and M. S. Drew, "Disambiguation of Illumination and Surface Reflectance from Spectral Power Distribution of Color Signal: Theory and Applications", CSS/LCCR 88-18, Simon Fraser University, 1988.

11. S. Tominaga and B. A. Wandell, "Component estimation of surface spectral reflectance", *J. Opt. Soc. Am. A*, Vol. 7, No. 2, February 1990, pp. 312-317.

12. B. A. Wandell, "The Synthesis and Analysis of Color Images", *IEEE Transactions on Pattern Analysis and Machine Intelligence*, Vol. PAMI-9, No. 1, January 1987, pp. 2-13.

13. G. Healey and T. O. Binford, "A Color Metric for Computer Vision", *Proceedings of the IEEE Conference on Computer Vision and Pattern Recognition*, IEEE, New York, 1988, pp. 10-17.

14. L. T. Maloney, "Evaluation of linear models of surface spectral reflectance with small numbers of parameters", *J. Opt. Soc. Am.*, Vol. 3, No. 10, October 1986, pp. 1673-1683.

15. J. P. S. Parkkinen and J. Hallikainen and T. Jaaskelainen, "Characteristic spectra of Munsell colors", *J. Opt. Soc. Am. A*, Vol. 6, No. 2, February 1989, pp. 318-322.

16. R. Bajcsy and S. W. Lee and A. Leonardis, "Image Segmentation with Detection of Highlights and Inter-reflections Using Color", MS-CIS 89-39, University of Pennsylvania, June 1989.

17. D. E. Rumelhart and G. E. Hinton and J. L. McClelland, "A general framework for parallel distributed processing", in *Parallel Distributed Processing: Explorations in the microstructures of cognition. Volume 1: Foundations,* D. E. Rumelhart and J. L. McClelland, ed., Bradford Books/MIT Press, Cambridge, MA, 1986.

18. D. E. Rumelhart and G. E. Hinton and R. J. Williams, "Learning internal representations by error propagation", in *Parallel Distributed Processing: Explorations in the microstructures of cognition. Volume 1: Foundations,* D. E. Rumelhart and J. L. McClelland, ed., Bradford Books/MIT Press, Cambridge, MA, 1986.

19. Macbeth, "Macbeth ColorChecker", trademarked product.

20. Edmund Scientific Co., "No. 9081 Spectral Curves", Barrington, NJ, 1980.

21. G. Wyszecki and W. S. Stiles, *Color Science: Concepts and Methods, Quantitative Data and Formulae,* Wiley, New York, 1982.

22. F. Grum and C. J. Bartleson (editors), *Optical Radiation Measurements,* Academic Press, New York, Vol. 2, Colorimetry, 1980.

Color Highlight Analysis

Highlights in color images present an interesting challenge and opportunity for machine vision. For systems not based on appropriate reflection models, highlights are often interpreted erroneously as bright objects or surface markings causing the breakdown of segmentation and correspondence algorithms. On the other hand, algorithms can be designed using color reflection models not only to remove the effects of highlights from images but also to use highlights to recover additional scene information such as illuminant color or surface shape. The papers in this section describe some of the ways in which highlights in color images can be located and exploited.

In the first article, H.C. Lee shows how highlights from two or more surfaces sharing a common illuminant can be used to estimate the chromaticity of the illuminant [1]. The method is based on the dichromatic reflection model under the assumption that the spectral composition of the interface reflection is the same as the spectral composition of the illumination. Using such a model, the locus of points in chromaticity space for a surface will lie on a line connecting the chromaticity of the illuminant and the chromaticity of the body reflection. Thus, the lines corresponding to two or more surfaces will intersect at the chromaticity of the illuminant. This principle is illustrated by an experiment with a synthetic image.

A technique for separating the highlight component from the body reflection component at every pixel in a color image is developed by Klinker, Shafer, and Kanade [2]. The algorithm is based on the dichromatic reflection model and the prediction that the color pixels corresponding to a single dielectric material will form a planar cluster in color space. The resulting intrinsic highlight and body reflection images are easier to analyze separately than the original. In order to apply this method to digital color images obtained using a CCD camera, the authors consider the effects of limited dynamic range, blooming, and camera response nonlinearity. They also correct for effects due to the nonuniform spectral responsivity of CCDs, such as low response for short visible wavelengths and high infrared response. The method is demonstrated successfully on color images of plastic and ceramic cups as well as glossy paper folders.

Healey presents an algorithm for computing the spectral reflectance of a material from a color image [3]. Spectral reflectance is useful for material identification because it is independent of illumination conditions and is largely independent of scene geometry. The method requires only a single highlight and is based on the assumption that highlights have the same spectral composition as the illuminant. Central to the method is a sensor model and the use of basis functions to approximate the reflected color signal from a color image. Experiments with images of plastic cups illustrate the accuracy of the approach.

In [4], H.C. Lee develops an algorithm for estimating illuminant chromaticity that is more general than the technique presented in [1]. As before, this work assumes the dichromatic reflection model with highlights having the same spectral composition as the illuminant. The new algorithm, however, is capable of estimating illuminant chromaticity from the image of a single material. Since the image irradiance due to the matte (body) component of reflection is a much smoother function of surface geometry than the irradiance due to the highlight component, the algorithm uses a smoothness constraint to distinguish the two reflection components and then estimates illuminant chromaticity. Encouraging algorithm performance is demonstrated on synthetic images.

– G.E.H.

Papers Included in This Collection

[1] Lee, H.-C. "Method for Computing the Scene-Illuminant Chromaticity from Specular Highlights." *Journal of the Optical Society of America A* **3**(10): 1694-1699, October 1986.

[2] Klinker, G.J., Shafer, S.A., and Kanade, T. "The Measurement of Highlights in Color Images." *International Journal of Computer Vision* **2**(1): 7-32, 1988.

[3] Healey, G. "Estimating Spectral Reflectance Using Highlights." *Image and Vision Computing* **9**(5): 333-337, October 1991.

[4] Lee, H.-C. "Illuminant Color from Shading." In *Proc. SPIE #1250: Perceiving, Measuring, and Using Color.* Santa Clara, CA SPIE, 1990. Pages 236-244.

Method for computing the scene-illuminant chromaticity from specular highlights

Hsien-Che Lee

Research Laboratories, Eastman Kodak Company, Rochester, New York 14650

Received February 24, 1986; accepted May 20, 1986

The perception of an unchanging surface color under different illuminations requires the computation of the scene-illuminant color either directly or indirectly. A possible source for the computation is the specular highlight of the surface reflection. Some issues related to color constancy are discussed, and a theory for computing the scene-illuminant chromaticity from specular highlight is described. An interesting result of the theory is that in an ideal situation, two surfaces of different colors will be sufficient for the computation.

1. INTRODUCTION

Color is one of the major attributes that we use in identifying an object. Nevertheless, most people have found it difficult to identify their cars under the narrow-band illuminants usually found in supermarket parking lots. We are so used to seeing the world in unchanging colors that an occasional failure of "color constancy" sometimes takes us by surprise. However, if we were to take a closer look at the environment, we might be even more surprised that we can see constant colors at all.

The interesting thing about color perception is that the spectral-energy distribution of the light reflected from an object surface is the product of the spectral energy of the illuminant and the spectral-reflectance function of the surface, and yet, given this product, we seem to be able to see both the illuminant color and the surface color at the same time.

Many models have been proposed to explain how human color perception works.[1–12] The retinex theory[4] provides a complete computational scheme for color perception. However, in a natural scene, shadows, light sources, and curved surfaces often make the retinex theory difficult to apply. Since so many factors are involved in visual perception, it seems likely that more than a single process is responsible for color perception. There have been reports that perception of colors depends on how the subject interprets the spatial arrangement of the scene.[13,14] Although this is difficult to quantify, most people have had a similar experience at one time or another. For example, a bright yellow spot on the basement floor can be mistaken for a shiny object, until it is realized that it is the sunlight projected on the floor through a small opening in the shutter. Then that spot becomes part of the floor and assumes more of the floor color than was previously perceived.

The question of whether the visual computation of color is done before, after, in parallel, independently, or interactively with that of the spatial perception is an important one for theories and models. Since we can see colors in randomly constructed color patches, some process in color perception must be independent of cognitive interpretation. If we formulate the problem of surface-color perception as one of separating the illuminant component and the surface-reflectance component from their product, the situation is a familiar problem in all aspects of visual perception: a mathematically underdetermined problem. The additional constraints for a unique solution must be drawn from the regularity of the physical world.

2. ILLUMINATION: INFERRED OR OBSERVED?

One type of regularity of the physical world has been known for a long time: the specular reflection of most surfaces appears to be white (or the color of the illuminant). In general, many inhomogeneous surfaces reflect light in the following manner (see Fig. 1).

When light illuminates a surface, some light is reflected immediately from the interface between the surface and the air and some penetrates into the surface, is partially absorbed and partially scattered within the surface material, and then is reflected back to the surface and into the air again. The first component, which is reflected immediately at the interface, is usually called the specular component of the total reflected light, and the second component is called the diffuse component since the light coming out of the surface is almost random in direction.[15–19] (The specular component is usually more directional than the diffuse component, unless the surface is very rough.)

For many surfaces of inhomogeneous materials, the specular component has a spectral-energy distribution close to that of the incident light, i.e., the color of the specular component is close to the illuminant color. This is true for painted surfaces, plastic surfaces, plant leaves, etc. but is not true for surfaces made of homogeneous materials such as polished metal. Accordingly to Cook and Torrance's model,[16] this specular (Fresnel) component is a function of the index of refraction of the interface material. Because many types of material serving as binders or as overcoats are some sort of oil, and most oils have very constant indices of refraction for the visible wavelengths,[20] we see specular reflection with color close to that of the incident light (e.g., the specular

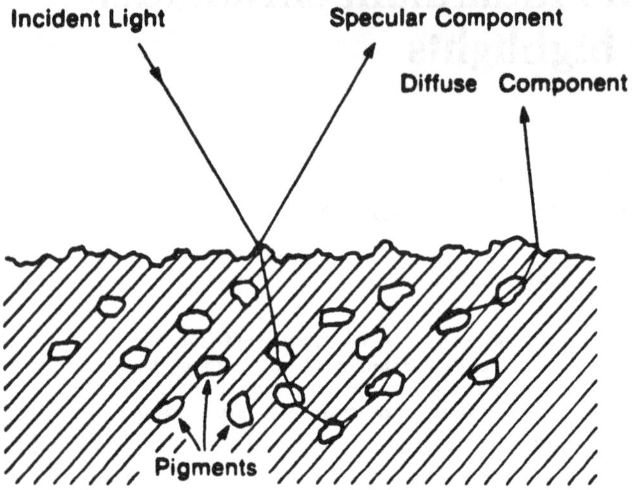

Fig. 1. Reflection from an inhomogeneous surface with two components, specular and diffuse.

reflection from our foreheads). The spectral-energy distribution of the diffuse component is modified by the pigments in the surface material and is what is meant by the actual color of the surface.

Since the quality (color) of the light reflected from the interface is close to that of the illuminant, researchers[1,5] have suggested that it may be this component of reflection that allows the human visual system to perceive the color of the illuminant. If this is true, it may partially support Helmholtz's theory that the impression of illumination serves as a reference level for determining the colors of the surfaces in the field (see p. 4 of Ref. 1). (I say "partially" because, as we will see later, the computation of the scene-illuminant color from specular reflection can be difficult and unreliable for scenes without strong specular reflections, and therefore not every scene can have its illuminant color computed this way.)

3. EVERY SURFACE REFLECTION CASTS A VOTE FOR ITS ILLUMINANT

In the CIE 1931 (x, y) chromaticity diagram,[3] each coordinate position represents a unique color match for the CIE 1931 Standard Colorimetric Observer. If two lights (with chromaticity coordinates $x1, y1$ and $x2, y2$) are combined in various proportions by additive mixture, the chromaticities of the mixtures will lie on the locus formed by a straight line connecting the two points $x1, y1$ and $x2, y2$.

Because the light reflected from a uniform surface is a mixture of the specular (illuminant-color) component and the diffuse (surface-color) component, in the CIE (x, y) chromaticity diagram the coordinates of the colors from different spots of the same surface will fall on the straight line connecting the illuminant color point (x_i, y_i) and the surface color point (x_s, y_s) (see Fig. 2). Perceptually, this means simply that the specular component dilutes the saturation of the color of the diffuse component. Colorimetrically, colors from the same surface are said to have the same dominant wavelength with respect to the illuminant color but different purities. The loci of such points form straight lines that tend to be oriented radially with respect to the chromaticity coordinates of the illuminant. Given two surfaces of different spectral-reflectance functions, there will be two lines pointing to the point representing the illuminant color (unless the two spectral-reflectance functions are metamers or complementary colors to the Standard Observer). Since two lines uniquely determine one point, the chromaticity coordinates of the illuminant can be computed. Therefore, if a given scene contains more than two surfaces

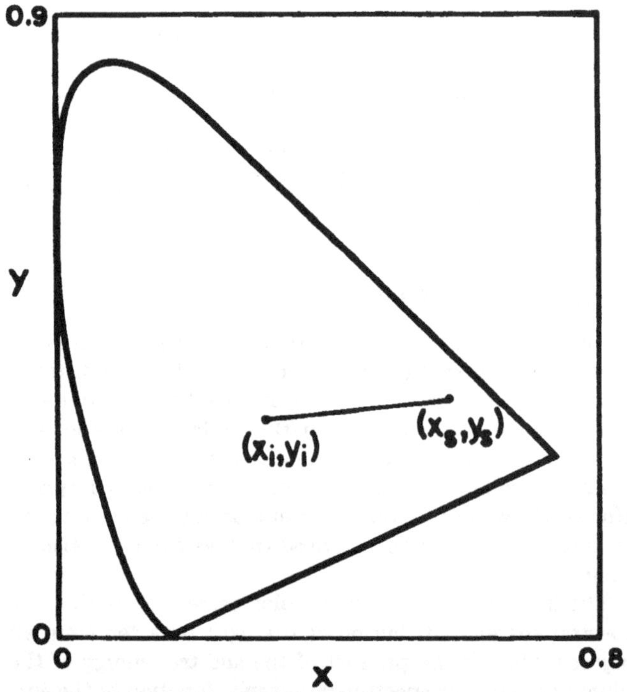

Fig. 2. CIE 1931 x, y chromaticity diagram showing the locus of chromaticities corresponding to colors produced by mixtures of the surface color (x_s, y_s) and the illuminant color (x_i, y_i).

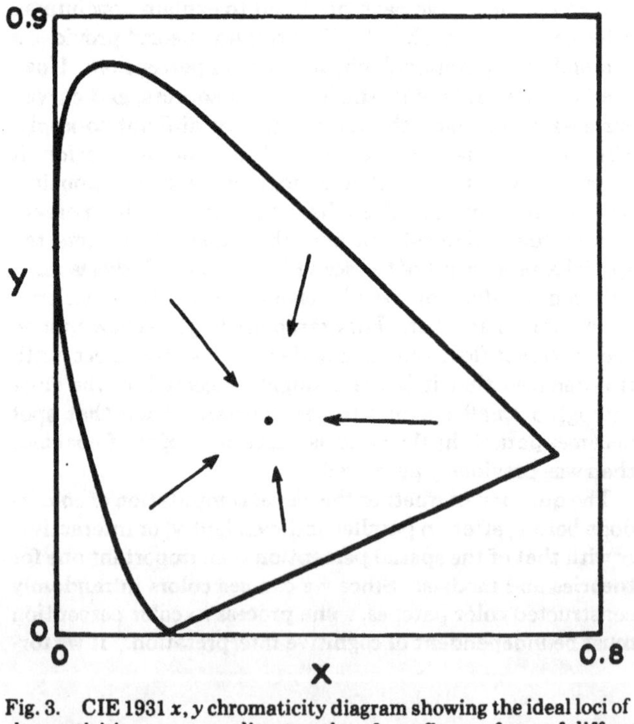

Fig. 3. CIE 1931 x, y chromaticity diagram showing the ideal loci of chromaticities corresponding to colors from five surfaces of different colors.

of different colors, the illuminant chromaticity can, in principle, be determined (see Fig. 3). In practice, if there are more than two surfaces of different colors, there can be more than two straight lines, and they may not all pass through exactly the same point because of the noise, the mutual illumination among the objects, or the presence of more than one illuminant. In this case, computing the illuminant color becomes a voting process. The chromaticity that gets the majority vote can be chosen as the illuminant chromaticity. (This is similar to the Hough transformation[21] for finding a straight line in the parametric space.)

4. THE COMPUTATIONAL MODEL

The computation of the CIE (x, y) coordinates requires that the spectral-response curves of the visual system (or any imaging system) be a linear combination of the CIE color-matching functions $\bar{x}(\lambda)$, $\bar{y}(\lambda)$, and $\bar{z}(\lambda)$. For most imaging systems, this is not practically achievable. Fortunately, the computational model can be developed by using any reasonable spectral-response functions. An imaging system with a set of three sensors, sensitive to light of long, middle, and short wavelengths, and with a bandwidth of about 100 nm, is appropriate for the model developed here. If the bandwidth of each response function is too narrow, or the wavelength ranges are not appropriate, there will be computational difficulty in distinguishing different illuminant chromaticities.

The proper color space and the related equations can be derived from the following reflection model:

$$Er = k \times Lr \times [R \times f(i, v) + S \times h(i, v)], \tag{1}$$

$$Eg = k \times Lg \times [G \times f(i, v) + S \times h(i, v)], \tag{2}$$

$$Eb = k \times Lb \times [B \times f(i, v) + S \times h(i, v)], \tag{3}$$

where Lr, Lg, and Lb are the surface irradiances obtained by integrating the incident radiances over all incoming light directions with respect to the red, green, and blue response functions of the imaging system. R, G, and B are the diffuse reflectance factors, and S is the regular (specular) reflectance factor. Note that the specular component is assumed to be neutral, i.e., all wavelengths are reflected equally. Er, Eg, and Eb are the image irradiances recorded by the system. The functions f and h represent the angular dependence of the reflectance factors. The variables i and v are the incident and viewing angles with respect to the surface normal. It should be noted that no assumptions have been made about the functional forms of f and h, which describe the directional distributions of the two reflection components. However, the reflection model does assume that the spectral composition of the reflected light can be separated from the geometric factors in both the specular and the diffuse components. (This is the same assumption used in Shafer's dichromatic reflection model.[17])

From Eqs. (1)–(3), we have the following relations: assuming R, G, and B are all different, then

$$Eb/Er = A \times (Eg/Er) + C, \tag{4}$$

where

$$A = (Lb/Lg) \times (B - R)/(G - R) \tag{5}$$

and

$$C = (Lb/Lr) \times (G - B)/(G - R). \tag{6}$$

Equation (4) represents a straight line in the $(Eg/Er, Eb/Er)$ color space. If G = R, then it is a vertical line. If R = G = B, then it degenerates into a single point, which is the illuminant chromaticity. To prove that the line of Eq. (4) passes the illuminant chromaticity point, we can substitute Lg/Lr and Lb/Lr for Eg/Er and Eb/Er of Eq. (4), and the equation still holds. Therefore, if there are two or more surfaces of different colors in the scene, then more than two lines can be determined; the point of intersection of these lines is the illuminant chromaticity.

Given an image of a complex scene that contains many object surfaces, the (Er, Eg, Eb)'s measured from the image must be clustered into groups corresponding to different surfaces before Eq. (4) can be applied to each group. The problem of segmenting an image into regions corresponding to surfaces of different colors is a difficult one. A method that avoids this problem is proposed in Section 5 to compute the illuminant chromaticity by using Eq. (4).

5. IMPLEMENTATION

The color-purity change of the reflected light due to the presence of the specular component can be quite small if the surface is not very smooth. To reduce the error in the computation, information should be taken from where the rate of the color-purity change is high. However, without knowing what the reference illuminant chromaticity is, the purity of a color cannot be defined. One simple way to detect color-purity change without being confused by the luminance change is to use ratios between Er, Eg, and Eb. We can define either $(Eg/Er, Eb/Er)$ or $[Er/(Er + Eg + Eb), Eg/(Er + Eg + Eb)]$ as new variables. Changes in one of the variables signal where the chromaticity is changing (either purity change or dominant wavelength change or both). The points at which the rates of chromaticity change are local maxima will be called the color edge points. Once we locate where the chromaticity is changing, the next step is to see whether this change is caused by material change or by the changing amount of specular component. We want to reject chromaticity changes due to material and keep the chromaticity changes due to the specular component. In this sense, we are working on the opposite problem posed by Rubin and Richards.[22]

The criterion we use to distinguish both types of chromaticity change is Eq. (4). If the chromaticity change is due to the specular component, the $(Eg/Er, Eb/Er)$ measurements collected on both sides of the color edge that we have detected should require the same set of coefficients A, C, in Eq. (4). If not, the color edge is due to material change and should be rejected (Fig. 4).

An input three-color image is first transformed into three images $(Eg/Er, Eg, Eb/Er)$. The color edges are detected by first convolving the Eg/Er, Eb/Er images with a Laplacian Gaussian filter and then marking all the zero-crossing points.[23] These zero-crossing points are the detected color edges. The Gaussian filter smooths out the noise, and the Laplacian operator takes the nondirectional second derivatives of the smoothed image. The local gradient maxima will have zero second-derivative values.

To collect data on both sides of the color edge, it is best to

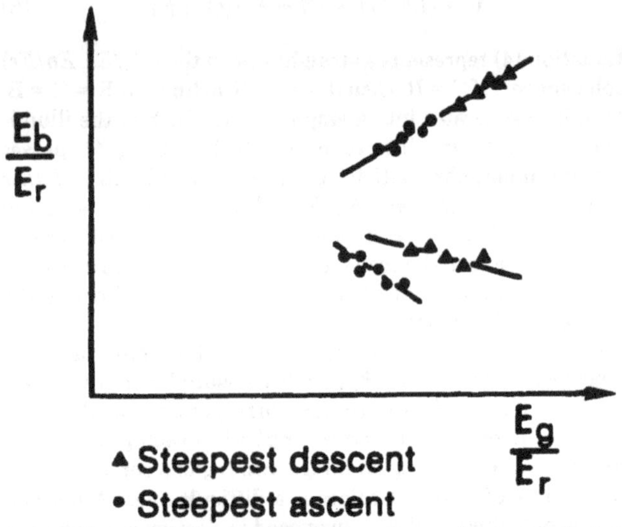

Fig. 4. The criterion for data selection: the two data sets collected on both sides of a color edge point must be on the same line in the color space if they are from the same surface color. The upper group satisfies the criterion but the lower group does not.

follow a path where the rate of the color purity change is the highest, so that the data can be fitted to Eq. (4) more reliably. However, change in purity can not be defined unless the reference illuminant chromaticity is known. Fortunately, for smooth surfaces the diffuse component looks about equally bright from different viewing angles, and the major change in the image irradiance is due mainly to the specular component. High rates of change in image irradiance are accompanied by correspondingly high rates of change in color purity. Therefore we can collect the data along the steepest ascent and steepest descent directions of any one of the image irradiances. In this implementation, we use the green image irradiance.

From each color edge point, the algorithm searches along the steepest ascent and the steepest descent directions to collect $(Eg/Er, Eb/Er)$ values, which are stored in two sets: one set contains data from the steepest ascent search and the other from the steepest descent search. The search will stop either when another zero crossing is found or when the path length is longer than the size of the Gaussian smoothing window. A straight line [Eq. (4)] is fitted through each of the two sets of $(Eg/Er, Eb/Er)$ values so collected. If the fit to either set of data is not good or the two lines differ significantly in slopes, then the two data sets are rejected. Otherwise, the two data sets are merged and a straight line is fitted to them to determine the parameters A and C in Eq. (4). This procedure is done for every color edge point.

After the whole image is processed, a set of (A, C)'s is determined. Because all the lines so determined must pass through one common point, i.e., the illuminant chromaticity point, all the (A, C)'s must fall on a straight line in the (A, C) plane. This can be seen by substituting Lg/Lr and Lb/Lr for Eg/Er and Eb/Er of Eq. (4):

$$Lb/Lr = A \times (Lg/Lr) + C,$$

and therefore,

$$C = -(Lg/Lr) \times A + (Lb/Lr).$$

By fitting a straight line K through all the (A, C) points, the illuminant color $(Lg/Lr, Lb/Lr)$ can be determined as

Lb/Lr = the intercept of line K on the (A, C) plane,

$-Lg/Lr$ = the slope of line K on the (A, C) plane.

An alternative to fitting a straight line K through all the (A, C) points is to use the majority voting method mentioned in Section 3. The $(Eg/Er, Eb/Er)$ coordinate plane is first quantized into discrete accumulator cells. For each (A, C) a straight line can be drawn on the plane. The count of the cell is incremented by one when a straight line passes through its coordinate interval.[21] The cell that gets the majority vote is chosen as the illuminant chromaticity. This accumulator method may be less sensitive to noise but requires more time and memory space.

6. EXPERIMENT

To test the proposed algorithm, we generated a scene (Fig. 5) that consisted of two spheres illuminated by a light source and a uniform background light. The image was 512 × 512, with 8 bits/color (in log irradiance). The parameters used in the reflectance model [Eqs. (1)–(3)] are Lr = 128.0, Lg = 160.0, Lb = 200.0. For the right sphere, R = 0.5, G = 0.4, B = 0.32, S = 0.3. For the left sphere, R = 0.2, G = 0.5, B = 0.3, S = 0.3. For both spheres, the diffuse component was modeled as perfect Lambertian reflection and the specular component as a function of the cosine raised to the twentieth power of the angle between the viewing angle and the specular reflection angle.

The computer-generated sphere scene was a three-color image (Dr, Dg, Db), where Dr, Dg, and Db are logarithms of the image irradiances, Er, Eg, and Eb. The input image was then transformed into three images $(Dg–Dr, Dg, Db–Dr)$. These three images were convolved with Laplacian Gaussian

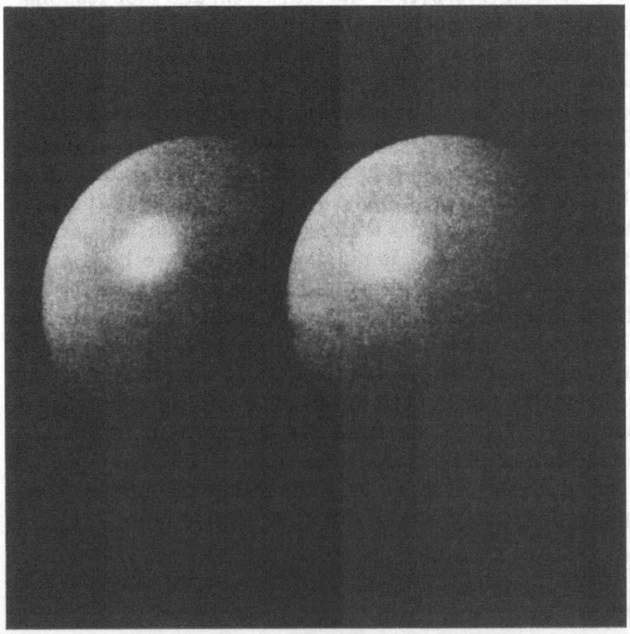

Fig. 5. An image generated according to the reflectance model in Eqs. (1)–(3). The specular component falls off as a power function of the cosine of the off-specular angle. The diffuse component is modeled as Lambertian reflection.

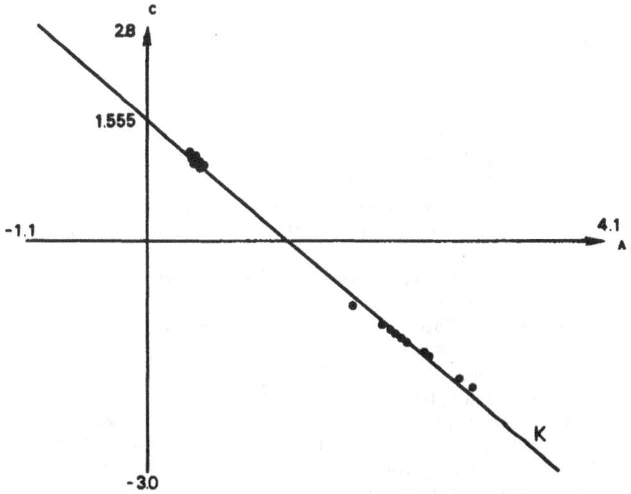

Fig. 6. The (slope, intercept) space representation of line segments that are the least-squares fits to the accepted data. Each point in this space represents a line segment in the chromaticity space. All the points fall approximately on a straight line K. The chromaticity of the illuminant (Lg/Lr, Lb/Lr) can be determined from the slope and the intercept of K.

filters of different sizes. (The standard deviations are 1, 2, 4, and 8 pixels wide.) Only one of the filter shows the detection of the illuminant color. Other filters, owing to the smaller sizes, did not detect edges for the required computation and will not be discussed further. The standard deviation of the useful Gaussian smoothing filter is 8 pixels. The size of the filter is 65 × 65. The filter was scaled by 6400 and the response amplitudes less than 1.0 were set to 0.0. (The convolution was done in real numbers.) The zero crossings were determined by thresholding the horizontal and the vertical gradients of the convolved images. The threshold used was 16.

From each color edge point, the algorithm collects data from both sides of the edge. Straight lines are fitted to the accepted data collected for all the color edge points, and the slopes and intercepts of these lines are determined. When the slopes A and the intercepts C of the line segments are plotted in the (A, C) space, these (A, C) pairs all fall on a straight line K, as shown in Fig. 6. The slope of line K is -1.2424, and the intercept is 1.5555. The illuminant chromaticity used in generating the sphere scene is

$$Lg/Lr = 160.0/128.0 = 1.25,$$
$$Lb/Lr = 200.0/128.0 = 1.5625.$$

Comparing with the slope and the intercept of line K, the computed illuminant color is very close to that used to generate the input sphere scene.

7. DISCUSSION

The problem of color constancy is a very complex one. Although we seem to see constant colors under different illuminations, the color sensations are rarely exactly the same when thoroughly examined. Furthermore, different light sources have different color-rendering properties; therefore, knowledge of the chromaticity of an illuminant is not, in general, sufficient to predict the color appearance of a surface. For example, the color of a tungsten light source can be matched by a mixture of three narrow-band lights. Objects that appear to be one color under the tungsten illuminant can appear to be a very different color under the other illuminant. In this case there is no color constancy, and knowledge of the illuminant chromaticity is of no use. The problem of metamerism is a fundamental limitation of any imaging system that has only a few spectrally different types of sensors. At the moment of image sensing, all differences between the two metameric stimuli are lost. However, for any reasonably broadband illuminants, the perceived colors of object surfaces are usually stable. We propose that estimating the chromaticity of the scene illuminant may be the first logical step in achieving color constancy. In this work, I did not study how knowledge of the illuminant chromaticity could be used to predict the color appearances of a surface under different illuminants, but there are a few empirical models that seem to work moderately well when information about the illuminants is given.[3,7,9,12]

The computational method presented here requires the measurements of at least three image irradiances at each picture point. The response of most imaging sensors to a wide range of light intensity is often a nonlinear function. Since ratios of image irradiances are used in the computation, a nonlinear distortion in the measurements tends to bend a straight line in the chromaticity diagram into a curve. Therefore, the proposed method seems to require careful calibration for measurements. However, it seems that the important information carried by the specular reflection may not be the exact direction along which the illuminant chromaticity must be located. The mere fact that the chromaticity loci are converging to some small area on the chromaticity diagram is useful enough to constrain the solution space to a small region. Furthermore, methods for extending curves to find a common intersection point may be worked out from the characteristics of the sensors.

One special problem related to film as the recording medium is that the integral densities measured on the processed film are not a simple function of the scene irradiance. The major complications are the interimage effect and the nonlinear characteristics of films. These may have to be corrected before the method is applied.

The most difficult problem is that many cases do not have much strong specular reflection, and the signals available can be very weak. This theory of computing the scene-illuminant chromaticity from specular highlights, therefore, cannot entirely account for the perception of illumination. However, it may be one of the visual processes that work to maintain the color constancy.

ACKNOWLEDGMENTS

I would like to thank the following for their comments and suggestions: C. James Bartleson, Edwin J. Breneman, Robert M. Goodwin, and Martin C. Kaplan.

REFERENCES

1. J. Beck, *Surface Color Perception* (Cornell U. Press, Ithaca, N.Y., 1972).
2. H. Helson, *Adaptation-Level Theory* (Harper and Row, New York, 1964).

3. G. Wyszecki and W. S. Stiles, *Color Science*, 2nd ed. (Wiley, New York, 1982).
4. E. H. Land, "The retinex theory of color perception," Sci. Am. **237(6)**, 108–128 (1977).
5. H. R. Flock, "Illumination: inferred or observed?" Percept. Psychophys. **35**, 293 (1984).
6. D. Marr, *Vision* (Freeman, San Francisco, 1982).
7. W. Richards and E. A. Parks, "Model for color conversion," J. Opt. Soc. Am. **61**, 971–976 (1971).
8. C. H. Graham, ed., *Vision and Visual Perception* (Wiley, New York, 1965).
9. C. J. Bartleson, "A review of chromatic adaptation," in *Color 77*, S. W. Billmeyer and G. Wysvecki, eds. (Hilger, Bristol, 1978).
10. R. M. Boynton, *Human Color Vision* (Holt, Rinehart and Winston, New York, 1979).
11. L. M. Hurvich, *Color Vision* (Sinauer Associates Inc., Sunderland, Mass., 1981).
12. R. W. G. Hunt, "A model of color vision for predicting color appearance," Color Res. Appl. **7(2)**, (1982), Part 1.
13. A. Gilchrist, S. Delman, and A. Jacobsen, "The classification and integration of edges as critical to the perception of reflectance and illumination," Percept. Psychophys. **33**, 425–436 (1983).
14. A. Gilchrist, "Perceived lightness depends on perceived spatial arrangement," Science **195**, 185–187 (1977).
15. R. M. Evans, *Eye, Film, and Camera in Color Photography* (Wiley, New York, 1959), p. 72.
16. R. L. Cook and K. E. Torrance, "A reflectance model for computer graphics," Comput. Graphics **15**, 307–316 (1981).
17. S. A. Shafer, "Using color to separate reflection components," Technical Report TR 136 (Computer Science Department, University of Rochester, Rochester, New York, 1984).
18. R. S. Hunter, *The Measurement of Appearance* (Wiley, New York, 1975).
19. W. Wendlandt and H. G. Hecht, *Reflectance Spectroscopy* (Wiley, New York, 1966).
20. R. Kanthack, *Tables of Refractive Indices* (Hilger, London, 1921), Vol. II., App. III.
21. R. O. Duda and P. E. Hart, *Pattern Classification and Scene Analysis* (Wiley, New York, 1973), pp. 335–337.
22. J. M. Rubin and W. A. Richards, "Color vision and image intensity: When are changes material?" AI Memo 631 (Artificial Intelligence Laboratory, Massachusetts Institute of Technology, Cambridge, Mass., 1981).
23. D. Marr and E. Hildreth, "Theory of edge detection," Proc. R. Soc. London Ser. B **207**, 187–217 (1980).

The Measurement of Highlights in Color Images*

GUDRUN J. KLINKER, STEVEN A. SHAFER, AND TAKEO KANADE
Computer Science Department, Carnegie Mellon University, Pittsburgh, PA 15213-3890

Abstract

In this paper, we present an approach to color image understanding that accounts for color variations due to highlights and shading. We demonstrate that the reflected light from every point on a dielectric object, such as plastic, can be described as a linear combination of the object color and the highlight color. The colors of all light rays reflected from one object then form a planar cluster in the color space. The shape of this cluster is determined by the object and highlight colors and by the object shape and illumination geometry. We present a method that exploits the difference between object color and highlight color to separate the color of every pixel into a matte component and a highlight component. This generates two intrinsic images, one showing the scene without highlights, and the other one showing only the highlights. The intrinsic images may be a useful tool for a variety of algorithms in computer vision, such as stereo vision, motion analysis, shape from shading, and shape from highlights. Our method combines the analysis of matte and highlight reflection with a sensor model that accounts for camera limitations. This enables us to successfully run our algorithm on real images taken in a laboratory setting. We show and discuss the results.

1 Introduction

When we look at an image, we can interpret it as a collection of shiny and matte surfaces, smooth and rough, interacting with light, shape, and shadow. However, computer vision has not yet been successful at deriving a similar description of surface and illumination properties from an image. The key reason for this failure has been a lack of models or descriptions rich enough to relate pixels and pixel-aggregates to scene characteristics. In the past, most work with color images has considered object color to be a constant property of an object; and color variation on an object was attributed to noise [28]. However, color variation in real scenes depends to a much larger degree on the optical reflection properties of the scene. This variation causes the perception of object color, highlights, shadows, and shading [14, 31], scene properties that can be determined and used by color vision algorithms.

This paper presents an approach to color image understanding that accounts for color variations due to highlights and shading. We use a reflection model that describes pixel colors as a linear combination of an object color and a highlight color [34]. All color pixels from one object then form a planar cluster in the color space. The cluster shape is determined by the object and highlight colors and by the object shape and illumination geometry. We combine our reflection model with a sensor model that accounts for camera properties, such as a limited dynamic range, blooming, and gamma-correction.

*This material is based upon work supported by the National Science Foundation under Grant DCR-8419990 and by the Defense Advanced Research Projects Agency (DOD), ARPA Order No. 4976, monitored by the Air Force Avionics Laboratory under contract F33615-84-K-1520. Any opinions, findings, and conclusions or recommendations expressed in this publication are those of the authors and do not necessarily reflect the views of the National Science Foundation, the Defense Advanced Research Projects Agency, or the US Government.

By combining these explicit models, we separate color images into two intrinsic images, one showing the scene without highlights, the other one showing only the highlights. The intrinsic images can be a useful tool for a variety of algorithms in computer vision that do not account for highlights in images, such as stereo vision and motion analysis [7, 35]. Our approach can also be used to determine the color of the illumination from highlights in the image and as a preprocessor for methods to determine object shape from shading or highlights [12, 14].

We begin by describing the major influences on the measurement of a color pixel value: the physics of light reflection in the scene, the object shape, and the sensor characteristics of the camera. We then demonstrate how our reflection model can be used for the analysis of color images. We present a method to detect and remove highlights from color images and to determine the illumination color. Finally, we discuss the implications and the future direction of our work.

2 What's in a Color Pixel?

On its path from a light source to the camera, a light ray is altered in many characteristic ways by the objects in the scene. The camera then encodes the measured light in a (color) pixel. It is the goal of image understanding methods to use properties and relationships between pixels to recover a description of the scene. It is essential to the success of such methods that they understand and model the reflection processes in the scene, as well as the sensing characteristics of the camera. We will now describe the processes that occur when light interacts with inhomogeneous, dielectric materials. We will also discuss how the measured pixel values depend on camera characteristics.

2.1 The Physics of Light Reflection from Dielectric Materials

When light interacts with dielectric materials, two kinds of reflection exist: The first takes place when light encounters a change in the refraction index at the interface between two substances; the second occurs within a substance when light is scattered and selectively absorbed [16, 19, 37]. These two kinds of reflection differ in their geometric and photometric properties, and, as we will show, those differences can be exploited for computer vision.

2.1.1 Properties of Inhomogeneous Dielectric Materials. Dielectric materials are nonconducting materials. Most of them, such as paints, plastics, paper, and ceramics, are inhomogeneous. They consist of a medium and some embedded pigments, as shown in figure 1. The medium comprises the bulk of the matter and is generally approximately transparent, while the pigments selectively absorb light and scatter it by reflection and refraction.

When we look at objects that are made out of dielectric material, we usually see the reflected light as composed of two distinct colors that typify the highlight areas and the matte object parts. This is a characteristic property of many dielectrics, and our reflection model capitalizes on this characteristic color change between matte and highlight areas. Our model is applicable to a large set of typical dielectric materials, as we will demonstrate in section 3. It is not applicable to nondielectric (i.e., conducting) materials because those reflect light in only a single color. It is also not applicable to dielectrics that imitate the reflection behavior of metals, e.g., by containing metallic flakes in their material body. We now describe the typical characteristics of dielectrics in more detail.

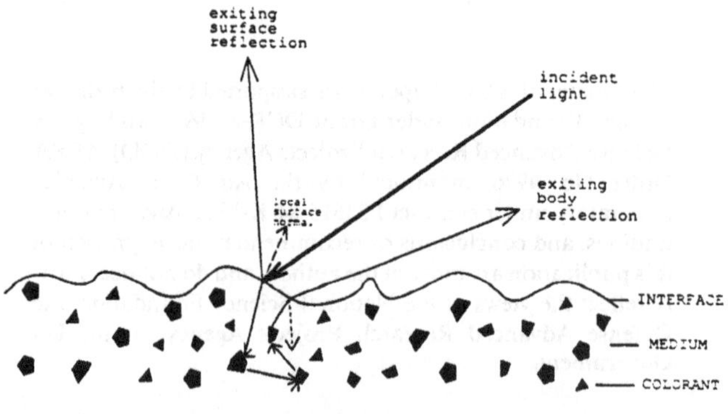

Fig. 1. Light reflection of dielectric materials.

2.1.2 Reflection at the Material Surface. When light hits the surface of a dielectric material, it must first pass through the interface between the surrounding medium (e.g., air) and the material. Since the refraction index of the material is generally different from that of the surrounding medium, some percentage of the incident light is reflected at the surface of the material. We will refer to this light reflection process as *surface reflection.* This process has been referred to in the literature as interface reflection, or Fresnel reflection [13, 20, 21, 34].

If the interface between the surrounding medium and the material is completely smooth, the light is reflected from this interface in only one direction, such that the incident light beam, the exiting light beam, and the surface normal are coplanar and the surface normal bisects the angle between incident and exiting light. This is the direction of perfect mirror reflection. However, the interfaces of many materials have some roughness or texture. The direction of the reflected light depends then on the orientation of the local surface normals, which varies along the object. In such cases, the light is scattered to some degree around the global angle of perfect mirror reflection. Several models have been developed in the physics and computer graphics communities to describe the geometric properties of light reflection from rough surfaces [2, 30, 33, 36]. Issues in modeling this process involve the roughness scale of the surface, compared to the wavelengths of the incident light, and self-shadowing effects on the surface which depend on the viewing direction and the direction of the incident light.

The optical properties of the material also determine the amount and color of the light that is reflected at the material surface. Fresnel's law describes how the reflected light depends on the refraction indexes of the material and the surrounding medium, on the incidence angle, and on the polarization of the light [25]. In general, the refraction index of a material is a function of wavelength, and the Fresnel coefficient varies over the light spectrum. Furthermore, pigment particles may be protruding from the medium at some places [17]. The reflected light may thus be also a function of the position on the object. However, it is common to assume that the pigments are completely embedded in the medium and to approximate the Fresnel coefficient of the medium as a constant over the visible spectrum. This is reasonable since the refraction coefficient of most media changes very little over the visible spectrum. Under these assumptions, the light reflected at the surface has the same color as the illuminating light.

2.1.3 Reflection in the Material Body. For dielectric materials, not all incident light is reflected at the material surface. Some percentage of the light penetrates into the material body. The refracted light beam travels through the medium, hitting pigments from time to time. The material medium is generally approximately transparent, transmitting the refracted light beam without a color or orientation change. When the light beam hits a pigment particle, some of the light is scattered, due to the change in refraction indexes. The rest of the beam penetrates through the particle, where some wavelengths may be partially or entirely absorbed. The remaining light returns afterwards into the surrounding medium [16, 19]. Some part of the beam may travel then to the material surface while the rest stays within the body and hits other pigments. The part of the light beam that arrives at the material surface is there partially refracted and partially reflected, as described above for the surface reflection process. Thus, some fraction of the light exits from the material while the rest is reflected back into the body and undergoes more scattering and absorption.

We refer to the entire reflection process described above as *body reflection.* Its geometric and photometric properties depend on many factors: the transmitting properties of the medium, the scattering and absorption properties of the pigments, and the shape and distribution (including density and orientation) of the pigments. If we assume a random distribution of the pigments, the light exits in random directions from the body. In the extreme, when the exiting light is uniformly distributed, the distribution can be described by Lambert's law. The distribution of the pigments also influences the amount and the color of the reflected light. If the pigments are distributed randomly in the material body, we may expect that, on the average, the same amount and color will be absorbed everywhere in the material before the light exits. In such a case, the light that is reflected from the material body has the same color over the entire surface.

The distribution, size, and scattering properties of the pigments also influence the opacity of the material body. A material is opaque if no part of the refracted light beam is able to penetrate through the entire material. Otherwise, the material is transparent or translucent.

In previous literature of computer vision, the terms *specular* and *diffuse* reflection have frequently been used to refer to surface and body reflection, respectively. However, this is not strictly correct. *Specular reflection* refers to light reflection in the direction of ideal mirror-like reflection, while *diffuse reflection* implies reflection with scattering. This is a purely geometric distinction, while the terms *surface* and *body* reflection refer to different physical processes with distinct spectra. For a very smooth surface, the surface reflection will be specular and the body reflection will be diffuse. However, for typical objects with rough surfaces, the surface reflection will be diffused around the direction of perfect specular reflection, as modeled for example in [36]. Observing this distinction makes it possible to describe many types of material succinctly; for example, metals have only surface reflection, which may be specular or diffuse depending on the roughness.

2.2 The Dichromatic Reflection Model

We now present a mathematical model that approximates the reflection processes of dielectric, inhomogeneous, opaque materials [34]. In order for this model to be a suitable basis for work in color image understanding, it makes several restricting and simplifying assumptions about the properties of surface and body reflection. We begin by assuming that the pigments are distributed randomly in the material body and that they are completely embedded in the medium. The surface then exhibits a single spectrum of body reflection and a single spectrum of surface reflection. We also restrict the illumination conditions of the scene, allowing only one light source and no ambient light or inter-reflection between objects. These assumptions, in concept, restrict the applicability of the model. However, we will show in section 3 that the model still provides a reasonable and very useful approximation to the physics of light reflection for many dielectric materials.

We begin by describing the light, L, which is reflected from an object point as a mixture of the light L_s reflected at the material surface and the light L_b reflected from the material body.

$$L(\lambda,i,e,g) = L_s(\lambda,i,e,g) + L_b(\lambda,i,e,g) \qquad (1)$$

The reflectance geometry is illustrated in figure 2. It defines several photometric angles [14, 33]: the *incidence angle, i*, between the illumination direction, **I**, and the surface normal, **N**, the *exitance angle, e*, between **N** and the viewing direction, **V**, and the *phase angle, g*, between **I** and **V**. The parameter λ represents wavelengths of the light spectrum.

Since we assume that a dielectric, inhomogeneous surface exhibits the same spectrum of surface reflection at any point on the object, we can separate the spectral reflection properties of L_s from its geometric reflection properties [26]. We thus model it as a product of a spectral power distribution, $c_s(\lambda)$, and a geometric scale factor, $m_s(i,e,g)$, which describes the intensity of the reflected light. Similarly, we separate the body reflection component L_b into a spectral power distribution, $c_b(\lambda)$, and a geometric scale factor, $m_b(i,e,g)$. Substituting these terms into equation

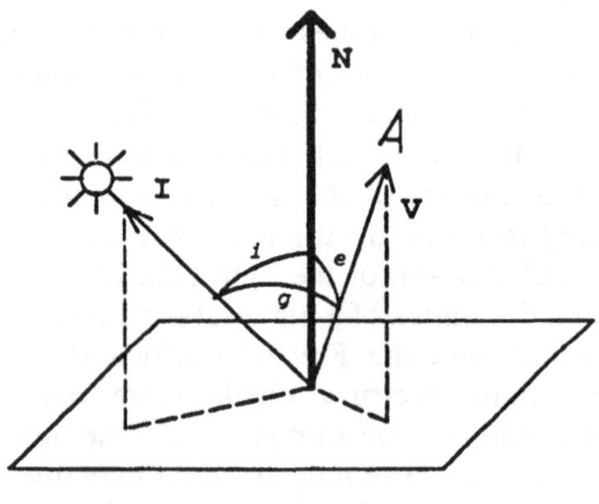

Fig. 2. Photometric angles.

(1), we obtain the *Dichromatic Reflection Model* equation:

$$L(\lambda,i,e,g) = m_s(i,e,g)c_s(\lambda) + m_b(i,e,g)c_b(\lambda) \qquad (2)$$

We thus describe the light that is reflected from an object point as a mixture of two distinct spectral power distributions, $c_s(\lambda)$ and $c_b(\lambda)$, each of which is scaled according to the geometric reflection properties of surface and body reflection. This is shown in figure 3. In the infinite-dimensional vector space of spectral power distributions (each wavelength defines an independent dimension in this vector space [15, 32]), the reflected light can thus be described as a linear combination of the two vectors $c_s(\lambda)$ and $c_b(\lambda)$.

Many reflection models developed in the physics and computer graphics communities [5, 14, 30, 36] are special cases of the model described here [34]. They replace the geometric variables, m_s and m_b, by specific functions that approximate the measured reflection data of a chosen set of typical materials. In our work, we concentrate on the spectral variables in equation (2), $c_s(\lambda)$ and $c_b(\lambda)$, exploiting the color difference between them. We leave the geometrical factors unspecified.

2.3 Object Shape and Color Variation

We will now discuss the relationship between the light mixtures of all points on an object. We study the spectral variation over an entire object by analyzing the histogram of the light mixtures from all object points.

The Dichromatic Reflection Model states that the light spectra of body reflection, $c_b(\lambda)$, and surface reflection, $c_s(\lambda)$, are constant over an entire object, while the geometric scale factors, $m_b(i,e,g)$ and $m_s(i,e,g)$, change with the viewing and illumination angles. Accordingly, $c_b(\lambda)$ and $c_s(\lambda)$ span a *dichromatic plane* in the infinite-dimensional vector space of spectral power distributions, and the light $L(\lambda,i,e,g)$ reflected from any object point lies in this plane. We can assume without loss of generality that $m_s(i,e,g)$ and $m_b(i,e,g)$ are variables in the range from 0 to 1. All light mixtures from an object can then be circumscribed in the dichromatic plane by a parallelogram, as shown in figure 3.

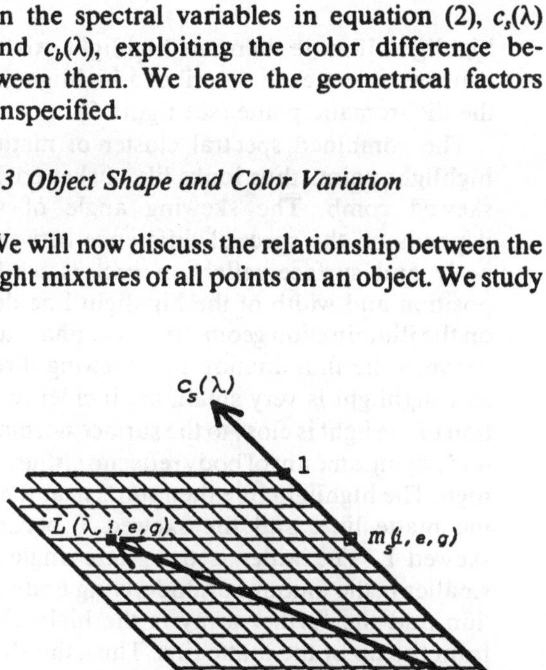

Fig. 3. **A mixture of the lights reflected from the material surface and body.**

An investigation of the geometrical properties of surface and body reflection reveals that the light mixtures form a dense spectral cluster in the dichromatic plane. The shape of this cluster is closely related to the shape of the object. We can use this relationship to determine characteristic features of the spectral clusters. For illustration purposes, we will assume in the following discussion of spectral histograms that body reflection is approximately Lambertian and that surface reflection is describable by a function with a sharp peak around the angle of perfect mirror reflection. Note, however, that this analysis is not limited to a particular geometric reflection model.

Figure 4 shows a sketch of a shiny cylinder. The left part of the figure displays the magnitudes of the body and surface reflection components. The curves show the loci of constant body or surface reflection. The darker curves are the loci of constant surface reflection. Since $m_s(i,e,g)$ decreases sharply around the object point with maximal surface reflection, $m_{b\max}$, these curves are shown only in a small object area. We call the points in this area *highlight points.* The remaining object points are *matte points.* The right part of the figure shows the corresponding spectral histogram in the dichromatic plane. As we will describe below, the object points form two linear clusters in the histogram.

For matte points, the surface reflection component of the reflected light is negligible and all the

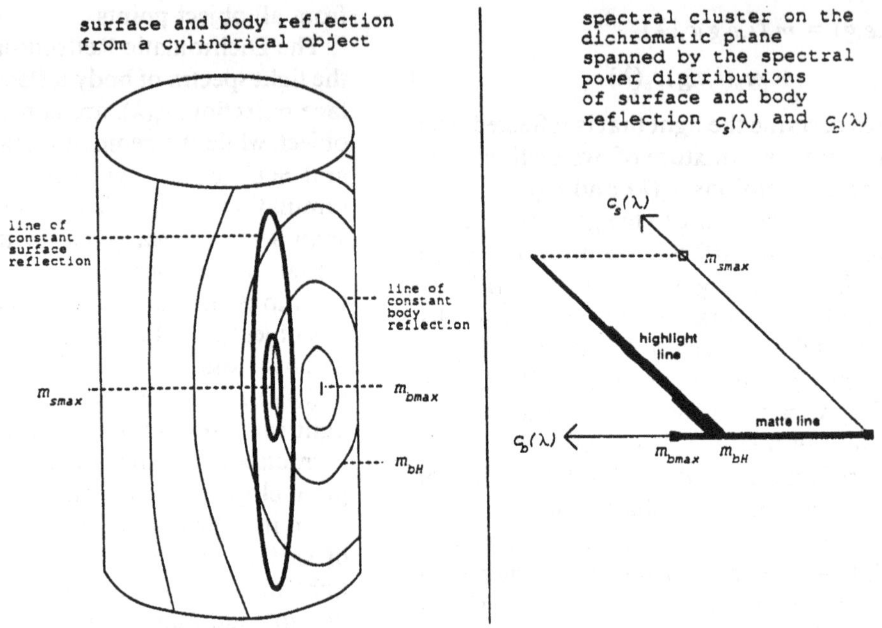

Fig. 4. The shape of the spectral cluster for a cylindrical object.

reflected light comes from body reflection. The observed light at such points thus depends only on $c_b(\lambda)$, scaled by $m_b(i,e,g)$ according to the geometrical relationship between the local surface normal of the object and the viewing and illumination directions. Consequently, the matte points form a *matte line* in the dichromatic plane in the direction of the body reflection vector, $c_b(\lambda)$, as shown in the right part of figure 4.

Highlight points exhibit both body reflection and surface reflection. However, since $m_s(i,e,g)$ is much more sensitive to a small change in the photometric angles than $m_b(i,e,g)$, the body reflection component is generally approximately constant in a highlight area, as displayed by the curve with label m_{bH} in figure 4. Accordingly, the second term of the dichromatic reflection model equation (2) has a constant value, $m_{bH}c_b(\lambda)$, and all spectral variation within the highlight comes from varying amounts of $m_s(i,e,g)$. The highlight points thus form a *highlight line* in the dichromatic plane in the direction of the surface reflection vector, $c_s(\lambda)$. The line departs from the matte line at position $m_{bH}c_b(\lambda)$, as shown in figure 4. More precisely, the highlight cluster looks like a slim, skewed wedge because of the small variation of the body reflection component over the highlight. If more than one highlight exists on a concave object, each describes a highlight line in the dichromatic plane (see figure 5).

The combined spectral cluster of matte and highlight points thus looks like a skewed T or a skewed comb. The skewing angle of the T depends on the spectral difference between the body and surface reflection vectors while the position and width of the highlight line depend on the illumination geometry: If the phase angle g between the illumination and viewing direction at a highlight is very small, the incidence direction of the light is close to the surface normal. The underlying amount of body reflection, thus, is very high. The highlight line then starts near the tip of the matte line, and the skewed T becomes a skewed L. The wider is the phase angle g, the smaller is the amount of underlying body reflection and the further away is the highlight line from the tip of the matte line. Thus, the shape of the histogram is related to the geometry of the scene.

2.4 *Taking Color Images with Real Cameras: A Sensor Model*

So far, we have described light reflection in a theoretical, physical model. However, the obser-

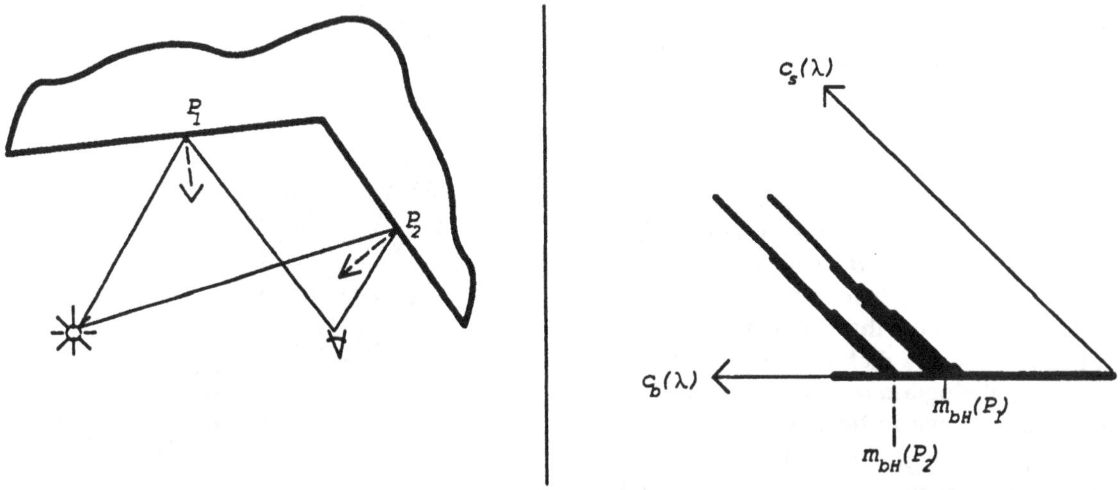

Fig. 5. Cluster shapes for concave objects.

ved pixels values are also influenced by the characteristics of the recording camera. This section covers the influence of such camera characteristics on the image data. Since some of these influences disturb the light reflection properties stated in the dichromatic reflection model, we need to provide methods that restore the physical properties of light reflection. Where this is impossible, our image analysis algorithms must be able to detect or tolerate the inaccuracies in the image data.

We will first discuss the implications of representing a continuous light spectrum as a finite-dimensional vector. Next, we present problems associated with the limited dynamic range of cameras. Finally, we analyze the effects of the changes in the camera's responsivity as a function of both wavelength and intensity. Since we use a CCD-camera (SONY AVC-D1) to obtain our images, we concentrate on the issues occurring with such cameras.

2.4.1 Spectral Integration. The Dichromatic Reflection Model describes light reflection using the physically precise representation: the continuous light spectrum. It is, however, impossible to measure and process light information in this way. The human eye uses, as well as light sensing devices, a finite set of samples to describe the spectrum. Such sample measurements are obtained by filtering the light spectrum and integrating over the filtered spectrum. We call this process *spectral integration.* It integrates the amount of incoming light, $L(\lambda,i,e,g)$, weighted by the spectral transmittance, $\tau_f(\lambda)$, of the respective filter and the spectral responsivity of the camera, $s(\lambda)$, over all wavelengths λ:

$$C_f = \int L(\lambda,i,e,g)\tau_f(\lambda)s(\lambda)\,d\lambda \tag{3}$$

We use a red, green, and blue color filter (Wratten filters with numbers 25, 58, and 47), thus reducing the infinite-dimensional vector space to a three-dimensional space. The spectrum of an incoming light beam at pixel position (x,y) is represented by a triple $C(x,y) = [R,G,B]$, where x and y are determined by i, e, and g and by the position of the object relative to the camera.

Spectral integration from the infinite-dimensional vector space of spectral power distributions to a three-dimensional color space is a linear transformation [10, 32]. Thus, if a light beam $L(\lambda)$ is a mixture of two lights, $L_1(\lambda)$ and $L_2(\lambda)$, in the infinite color space, then the color vector $\mathbf{C}$, resulting from spectral integration on $L(\lambda)$ is the same linear combination of the color vectors $\mathbf{C}_1$ and $\mathbf{C}_2$ that result from spectral integration on $L_1(\lambda)$ and $L_2(\lambda)$. For this reason, and because the photometric angles i, e, and g depend on x and y, the linear relationship between reflected light and the colors of surface and body reflection, as stated in equation (2), is maintained under spectral integration. We thus obtain the dichromatic reflection model for the three-Dimensional Color Space:

$$\mathbf{C}(x,y) = m_s(i,e,g)\mathbf{C}_s + m_b(i,e,g)\mathbf{C}_b \qquad (4)$$

The color pixel value $\mathbf{C}(x,y)$ is thus a linear combination of the two color vectors, $\mathbf{C}_s = [R_s,G_s,B_s]$ and $\mathbf{C}_b = [R_b,G_b,B_b]$, which describe the colors of surface and body reflection on an object in the scene. Within the three-dimensional color space, $\mathbf{C}_s$ and $\mathbf{C}_b$ span a dichromatic plane which contains a parallelogram in which the color clusters lie. This is shown in figure 6.

Note that, in principle, spectral integration introduces a nonuniqueness, called *metamerism* [18, 19], in the resulting finite representation: the infinite-dimensional color space is partitioned into a set of equivalence classes, each of which contains an infinite number of spectral curves that are all represented by the same color. However, this is of no concern to us since the finite color space maintains the linear relationships between the reflected light mixtures from one object.

2.4.2 Limited Dynamic Range. Real cameras have only a limited dynamic range to sense the brightness of the incoming light. This restricts our analysis of light reflection, as measured in real color images, to a *color cube,* as shown in figure 7. Its walls denote the upper and lower limits in the dynamic range of the respective color bands.

If the incoming light is too bright at some pixel position, thus exceeding the dynamic range of the camera, the camera cannot sense and represent it

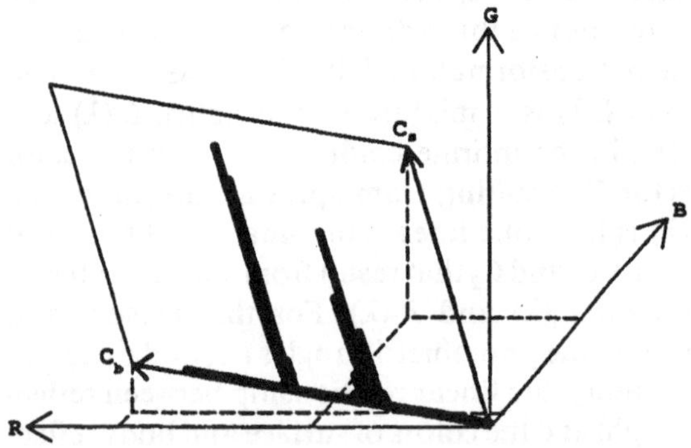

Fig. 6. Color cluster in a three-dimensional color space.

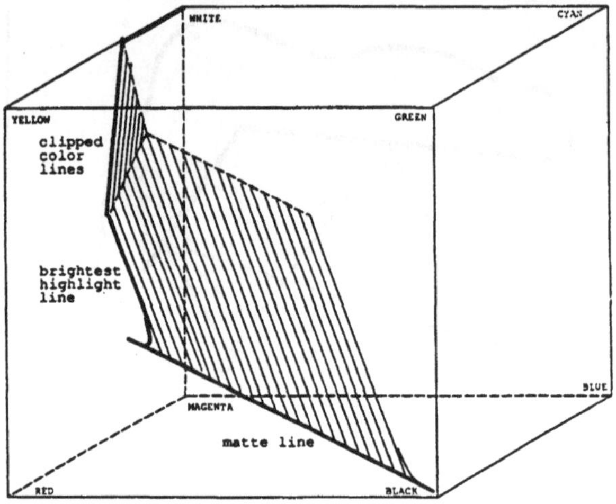

Fig. 7. Color cluster in the color cube, with color clipping.

adequately. This effect may occur in one, two, or all three color bands, depending on the color of the light. We refer to this effect as *color clipping.* It can be a problem for measuring the color of highlights or bright objects. In the color histograms, clipping causes the clusters to bend along the walls and edges of the color cube (see figure 7). In a very bright area, the dynamic range of the camera may be exceeded in all three color bands, resulting in white pixels even though the color of the light incident at the camera may not be white. Such *clipped color pixels* do not follow the characteristics of the dichromatic reflection model and must thus be distinguished from matte pixels and highlight pixels.

If a CCD-camera is used to obtain the images, too much incident light at a pixel may completely saturate the sensor element at this position [4]. This may cause *blooming* in the camera [3] which means that more charges are generated at a pixel than the relevant sensor element can hold. The excess carriers then spread out to the adjacent pixels and change their values proportional to the magnitude of the overload. In case of very bright spots in the scene, blooming may travel quite far from the sensor element that was originally overloaded, thus causing extended white areas in an image. We call pixels which have increased color values because of blooming *bloomed color pixels.* Their color values may be increased arbitrarily in one, two or all three color bands.

Both color clipping and color blooming are a

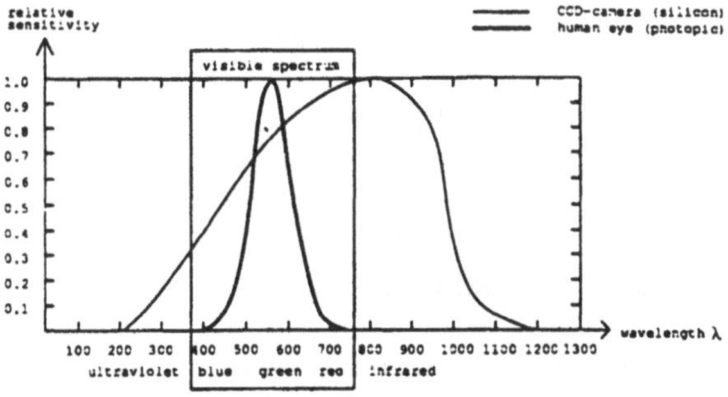

Fig. 8. Spectral responsivity of CCD-cameras and the human eye.

problem for image sensing with most currently available cameras; especially, for the sensing of scenes with highlights. They interact badly with another effect that is common in CCD-cameras: spatial averaging between neighboring pixels, which is commonly built into CCD-cameras to reduce the variance in the sensing characteristics in the array. Because of such spatial averaging, color clipping and blooming is "smoothed" over an image area. As a consequence, we cannot detect these effects by merely thresholding the color image at maximal intensity: because of the averaging, the effect is spread to neighboring pixels, increasing their values and decreasing the values of the clipped pixels. For this reason, we suspect color values in the upper 10% of the intensity scale in our images to be potentially influenced by color clipping or blooming.

2.4.3 Color Balancing and Spectral Linearization. We now describe how the measured color pixel values are influenced by the responsivity of CCD-cameras to varying wavelengths and incident flux. This analysis applies to black-and-white CCD-cameras.

A typical curve of the spectral responsivity of CCD-cameras is shown in figure 8 [4]. Because of the ramp-like spectral responsivity of CCD-cameras within the visible spectrum, the blue band registers much less contrast than the red band and thus lowers the signal-to-noise ratio. In order to provide an equal scaling on the three color axes in the color space, we need to rescale the pixel data separately in the color bands. We refer to this procedure as *color balancing.* A simple multiplication of the color data by constant factors, however, does not improve the dynamic range in the color bands and thus does not improve the quality of our color data; it only introduces gaps into the color distribution in the color histogram. Instead, we balance the color bands by controlling the camera aperture during the picture-taking process, using aperatures for green and blue exposures under tungsten light that are ½ and 1½ *f*-stops (respectively) higher than the aperture used for the red exposure. The upper left quarter of color Figure 1* shows the image of a Macbeth Color Checker color chart that we obtained using color balancing by aperture control. The lower left quarter shows an image of the same color chart taken without color balancing. The unbalanced image looks more orange, since the blue and green color bands do not contribute appropriately to the pixel colors. For this reason, its grey scale in the lowest row of the chart is brown. The grey scale of the color balanced image, on the other hand, is approximately grey. The pictures shown hereafter have been color balanced by aperture control.

Note that CCD-cameras are very sensitive to infrared light. Since most color filters in the visible spectrum are nearly transparent in the infrared, camera responsivity to the infrared range can add a significant amount of measured intensity to the color values. This may "wash out" the colors in the image significantly. To overcome this, we always use a total IR suppressor (Corion FR-400) in front of our camera.

The color pixels also depend on the camera response to incident light flux. Most modern sensors have a linear response to flux [4]. However, since the luminous output of display devices is related by a power law to the driving voltage, $L = kV^{\gamma}$, it is common practice in the television industry to produce a signal proportional to $Y^{1/\gamma}$ in the camera, with 2.2 as a typical value for γ. This is known as *gamma-correction.* As a result, the camera output is related by this inverse power law to the incident flux. We have measured this relationship for our camera in the Calibrated Imaging Laboratory at Carnegie Mellon University by comparing the known reflectance values of the grey scale of the Macbeth color chart with the measured, averaged intensity values in the corres-

*All color figures are at the end of the paper, pages 27–32.

ponding image areas in the three color bands. The resulting diagrams are shown in the right quarters of color figure 1. The upper diagram displays the measurements from a color image that was taken with color balancing by aperture control. The horizontal axis represents the known reflectance values of the grey blocks. The vertical axis represents the measured intensities in the three color bands. The points in the diagram lie approximately on the typical curve for gamma-correction. The lower diagram displays the same measurements from the unbalanced color image. It demonstrates the decreasing spectral responsivity of our camera to short wavelengths.

Color figure 2 demonstrates how the gamma-correction of our camera disturbs the linear properties of the Dichromatic Reflection Model. The upper left quarter shows the image of an orange cup. The cup does not reflect a significant amount of surface reflection into the camera. Thus, all pixels in the outlined image area are matte pixels and should form a straight matte line in the color space. The upper right quarter of color figure 2 demonstrates that this is not the case: gamma-correction introduces curvature into the color space, distorting the linear properties of the dichromatic reflection model.

To linearize the color data, we fit interpolating cubic splines to the measured responsivity data, separately in each color band, as suggested by LeClerc [22]. To model very bright light reflection, as can occur in highlights, we rescale the linearization function and extrapolate it from the brightest measurement outwards by a square root curve. The rescaling is necessary because linearizing very bright spots in the original image produces values beyond the representable range of pixel values, due to the slope of the fitted curve. We then use these functions to generate a look-up table relating the measured intensities in every color band to the incident flux. The upper right box of color figure 3 shows the cubic splines that were fitted to the responsivity data of color figure 1. The lower left box of color figure 3 displays the linearized image of the color chart. When compared with an unlinearized image (upper left box), the colors of the linearized image look more saturated. The image is also darker, due to the rescaling of the intensities. The lower half of color figure 2 shows that when our linearization method is applied to the image of the orange cup, the resulting color histogram is approximately linear, pointing in the direction of the body reflection vector. All pictures shown hereafter are linearized.

3 Analysis of Real Color Images

After discussing in section 2 what influences the pixel values in a color image, we will now show how the described properties can be used for color image understanding. We first demonstrate that the Dichromatic Reflection Model accounts well for color variation found in real color images. We will then present an algorithm that uses the described properties to generate intrinsic reflection images and discuss the results. Such images show only the body reflection component or only the surface reflection component of every pixel. Since the intrinsic reflection images have simpler relationships to the illumination geometry than the original image, we expect that this algorithm will be a suitable preprocessor for other image understanding methods, such as stereo vision, shape from shading, and shape from highlights.

3.1 Color Clusters from Real Color Images

There exist some inherent assumptions in the Dichromatic Reflection Model. We will now demonstrate that real color data from dielectric, inhomogeneous materials under varying illumination colors follows the theory of this model. We have taken a series of color images in the Calibrated Imaging Laboratory of plastic cups, glossy paper folders, and a ceramic cup under white, yellow, and pink light. Black curtains on the walls and a black ceiling were used to eliminate indirect illumination in the visible spectrum of the scene.

Color figure 4 shows the image of a green, a yellow, and an orange plastic cup, illuminated by yellow light. Color figure 5 shows the color histogram that results from projecting the color pixels from the outlined areas of color figure 4 into the color cube. The color histogram consists of several linear clusters. There are three linear clusters which start from the origin of the color cube at the black corner. These clusters contain the matte pixels of the three cups. They lie along the orange, yellow, and green matte lines of the three cups. The matte clusters are connected to linear highlight clusters which lie along the respective highlight lines of the cups. For the yellow cup, the matte and highlight clusters are nearly collinear because the material and illumination color are nearly the same. Note that the highlight clusters from all three cups are approximately parallel to each other. Furthermore, they are parallel to the color vector describing the yellow illumination because the refraction indexes of the cups are fairly constant over the visible spectrum. We thus can use the direction of the highlight clusters to determine the illumination color. This is useful for color constancy algorithms [6, 23, 24] which try to remove the influence of the illumination color from the body reflection component, thus normalizing the image to a standard white illumination.

Further linear clusters in the color histogram contain clipped color pixels. These are displayed in color figure 5 in their complementary color (blue). We identify such color clipping through a high threshold on the measured intensities in the three color bands. The clipped color pixels come from the centers of the highlights where the reflected light exceeds the dynamic range of the camera. Accordingly the centers of the highlights contain white pixels, whereas the pixels closer to the highlight boundaries exhibit yellow surface reflection.

Color figures 6 and 7 demonstrate how blooming influences the color distribution in the color histogram. The figures show an image and color histogram of an orange cup under yellow light. The highlight center on the cup is so bright that charges spread out to adjacent pixels, increasing their values. Because the cup is orange and the illumination is yellow, blooming occurs mainly in the red color band. Bloomed color pixels thus have an increased red color value, while the green and blue values are correct. As a result, the highlight cluster is spread out along the red axis of the color cube. The same effect is present—to a smaller degree—in the green highlight cluster in color figure 5. Since blooming cannot always be completely avoided in pictures with highlights, we need to account for it in our color histograms. We assume that blooming occurs only at a few pixels in the image and that it increases the color pixel values arbitrarily. It is then unlikely that such color pixel values occur often in the image. We thus suspect that colors with a low frequency count in the color histogram are caused by blooming or noise. Color figure 7 displays such color pixels in white.

Color figure 8 shows that the dichromatic reflection model applies to a variety of dielectric, inhomogeneous materials. The figure displays the images and the color histograms of a ceramic cup and two folders made out of glossy paper. The folders are slightly bent such that they have highlights in only a small object area. The corresponding color histograms consist of matte and highlight clusters, as described above, thus supporting the theory of the Dichromatic Reflection Model.

3.2 Detecting and Removing Highlights

We will now show how the Dichromatic Reflection Model can be used to detect and remove highlights in color images. The Dichromatic Reflection Model describes the color of every pixel in a color image as a mixture of two colors, the color of the body reflection component of the viewed object, $\mathbf{C}_b$, and the color of the surface reflection component, $\mathbf{C}_s$. We have developed an algorithm that exploits the T-shape of a color cluster to reverse this color mixing process. It fits a matte and a highlight line to the cluster and decomposes the color pixels into their constituent matte and highlight components.

Currently, our algorithm does not have the capability to distinguish between several color clusters in the color cube, since some of its operations (such as plane fitting) are applied globally to all entries in the color cube. If there are several objects in the scene, our algorithm requires that the image be previously segmented into image areas each of which corresponds to a single color cluster. It then applies its analysis of color variations separately to each segmented area. Most current color image segmentation methods [27, 28] cannot provide the required segmentation since they separate the highlights from the surrounding matte object areas and thus cut the color clusters into pieces. Gershon has reported encouraging first results in merging matte and highlight areas in a postprocessing step after an initial segmentation [9]. We are currently developing a segmentation algorithm that exploits the Dichromatic Reflection Model for its analysis of local color variations in an image, generating local hypotheses about dominant linear color clusters and dichromatic planes. The result is reported in [38]. For the work described here we have segmented the images by hand.

3.2.1 Determining the Body and Surface Reflection Vectors of an Image Area. We will now describe our method to determine the body and surface reflection vectors of the segmented image areas. The algorithm projects the color pixels from one chosen image area into the color cube, ignoring clipped or bloomed pixels as described in section 3.1. According to the Dichromatic Reflection Model, we may expect that the color pixels lie in a dichromatic plane in the color cube. Our algorithm thus fits a plane to the cluster, using principal component analysis to determine the eigenvalues and eigenvectors of the covariance matrix of the cluster [1, 11]. We sort the eigenvalues by decreasing size. The third eigenvector then determines the direction of minimal color variation in the cluster. The physically relevant color variation occurs in the plane that is perpendicular to the third eigenvector. We thus project all color pixels perpendicularly into this dichromatic plane.

Our analysis of the relationship between object shape and cluster shape in section 2.3. predicts that the color cluster forms a skewed T in the dichromatic plane and that the orientations of the linear sections of the T determine the body and surface reflection vectors of the cluster. Determining the corresponding linear matte and highlight subclusters in the cube is not trivial because neither a classification of the pixels into matte and highlight pixels nor the orientation and position of the subclusters are known. Our algorithm determines the two subclusters in a two-stage process.

The first stage is intended to render an initial, rough estimate of the matte and highlight line. We fit a convex hull of line segments to the contour of the cluster, using a recursive line splitting method [29]. To do so, we choose the line L between the brightest and darkest points in the cluster as the starting line. We divide the plane into buckets that have the form of slim stripes running perpendicularly to L (see Figure 9). We compute the color distance of all pixels from L and store the maximal distance wtihin every stripe s as a function $d(s)$. This function approximates the contour of the color cluster. Its shape, however, is very susceptible to noise. In order to have it reflect only the large-scale color variations, we smooth $d(s)$, averaging its values over a window. We then approximate the smoothed contour by a convex polygon. For this, we determine the stripe s_i with maximal distance $d(s_i)$ from L. If $d(s_i)$ exceeds a given threshold, L is split into two segments, and the distances of the color pixels from the new line segments are recomputed. Both segments are then split recursively into smaller segments until all distance measurements are below the threshold. The choice of the threshold determines how significant an orientation change of the contour needs to be before it influences the shape of the polygon. For all results shown in this paper, we have required $d(s_i)$ to exceed 20.

Our program next classifies the line segments. We know, from the discussion of the cluster shape in section 2.3., that matte pixels depend only on the body reflection component. We thus expect that the matte pixels form a matte line that approaches the origin of the color cube. Accordingly, we assume that the line segment closest to the black corner of the cube approximates the matte

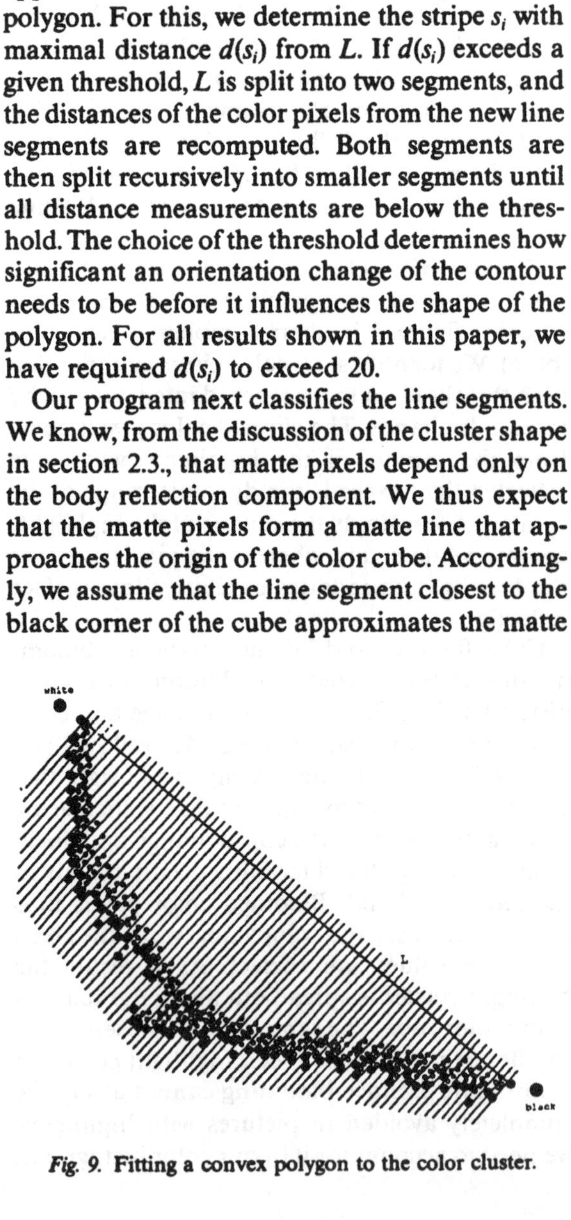

Fig. 9. Fitting a convex polygon to the color cluster.

line. Section 2.3. further states that the highlight line starts from the matte line. Accordingly, our algorithm classifies the next line segment connected to the matte line as the highlight line. If there are further line segments connected to the highlight line, we assume that they are related to clipped color pixels that were not detected by the heuristics of section 3.1. All further line segments are thus classified as clipped color lines.

The line fitting method described so far provides us with an initial estimate for the matte and highlight lines. But for several reasons, it cannot be expected to generate very accurate results. Most importantly, the fitted convex polygon sweeps across the concavity at the corner between the matte and highlight line. As a result, the estimated highlight line will generally describe an orientation with too little slope. Secondly, our initial plane fitting step was applied blindly to all pixels in the color cluster. If some clipped color pixels escaped the discarding heuristics, they may have biased the orientation of the dichromatic plane towards the white corner of the color cube. Since the fitted lines lie in the plane, they are then also biased toward white. For these reasons, we consider the matte and highlight lines to be only an initial approximation to the linear parts of the color cluster. Their orientations and positions will be improved in the second stage of the line fitting process.

In the second stage, we use the preliminary result of the first stage to coarsely classify the color pixels into matte pixels, highlight pixels, and clipped pixels, depending on the line they are closest to (see figure 10). We then refit the matte and highlight lines separately to the matte pixels and to the highlight pixels, using principal component analysis. The first eigenvectors describe the directions of the largest color variation in the two clusters. They determine our new estimates for the matte and the highlight line. The corresponding body and surface reflection vectors are parallel to the matte and highlight lines, but start at the origin of the color space. In the above-described steps, we assume that the classification assigns correct labels to most of the pixels in the cluster. We know, however, that some highlight pixels may have been misclassified as matte pixels because the initial estimate of the highlight line may have been wrong, as discussed above.

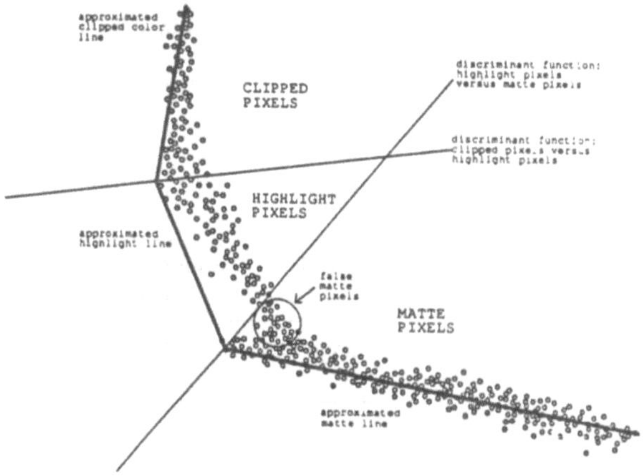

Fig. 10. Classifying the color pixels.

Some highlight pixels near the starting point of the highlight line (i.e., in the concavity) may be closer to the initial matte line than to the initial highlight line (see figure 10). We expect that this will generally not be a problem because the number of such false matte pixels is generally very small, when compared to the number of correct matte pixels. The orientation of the matte line is thus not influenced significantly by false matte pixels. This argument relies on the fact that highlights are generally much smaller than the surrounding matte object area. For flat objects, this may not be true. We may then have to alternate several times between reclassifying the pixels and refitting the lines.

Color figure 9 shows the matte and highlight lines that were fitted to the color histogram of the three cups under yellow light. It displays a matte line and a highlight line (M0, H0) for the green cup, a matte and a highlight line (M1, H1) for the yellow cup, and a matte line, a highlight line and a clipped color line (M2, H2, C2) for the orange cup. A list of the start and end points of all lines is given in Table 1. Table 2 contains the corresponding body and surface reflection vectors, normalized to unit length. As expected, the body reflection vector of the green cup is dominated by its green component while the body reflection vectors of the yellow and orange cup are increasingly dominated by their red component. Currently, we do not have the facilities to evaluate these results on a quantitative basis. The surface reflection vectors of all three cups are approximately the same, thus supporting the observation

Table 1. Matte and highlight lines of the three plastic cups under yellow light.

Fitted Lines: Plastic Cups under Yellow Light			
	matte line	highlight line	clipped line
green cup	from (5, 1, 1) to (23, 56, 9)	from (24, 54, 9) to (177,192, 52)	—
yellow cup	from (2, 0, 2) to (189,106, 19)	from (185,108, 21) to (255,175, 41)	—
orange cup	from (1, 0, 2) to (159, 36, 15)	from (155, 41, 17) to (220,101, 34)	from (220,101, 34) to (238,191, 54)

Table 2. Body and surface reflection vectors of the three plastic cups under yellow light.

Reflection Vectors: Plastic Cups under Yellow Light		
	body reflection vector	surface reflection vector
green cup	(0.31,0.94,0.14)	(0.73,0.66,0.20)
yellow cup	(0.87,0.49,0.07)	(0.71,0.68,0.20)
orange cup	(0.97,0.22,0.07)	(0.72,0.67,0.19)
illumination vector		(0.73,0.66,0.19)

that the three highlight clusters are parallel. To obtain an estimate of the illumination color, we have taken a picture of a grey card (Munsell N 8.5/ with 68.4% reflectivity) under yellow light. The card reflects the illuminating light without changing its spectral curve significantly. We can thus use the averaged color pixel values of the image as an indication of the color of the light source. The resulting, normalized estimate for our yellow illumination is also shown in table 2. It is very close to the fitted surface reflection vectors of the three cups. This suggests that our method for determining the body and surface reflection vectors of an image area is also a useful estimator of the illumination color.

3.2.2 Generating Intrinsic Reflection Images. Once we have determined the body and surface reflection vectors, $\mathbf{C}_b$ and $\mathbf{C}_s$, of an image area, we can decompose the color pixels into their constituent body and surface reflection components. Equation (4) in section 2.4.1. describes the color of every pixel as linear combination of these two color vectors:

$$\mathbf{C}(x,y) = m_s\mathbf{C}_s + m_b\mathbf{C}_b$$

$\mathbf{C}_b$, $\mathbf{C}_s$ and their cross product, $\mathbf{C}_b \times \mathbf{C}_s$, define a new (not necessarily orthogonal) coordinate system in the color cube. This coordinate system describes every color in the cube in terms of their amounts of body reflection, surface reflection, and noise ε, as given by the color distance from the dichromatic plane (see figure 11). There exists an affine transformation, and thus a linear transformation matrix T, which transforms any color vector $\mathbf{c} = [R,G,B]^T$ from the initial coordinate system into a vector $\mathbf{d} = [m_b,m_s,\varepsilon]^T$ in the new coordinate system

$$\mathbf{d} = T\mathbf{c} \qquad (5)$$

After computing T from $\mathbf{C}_b$ and $\mathbf{C}_s$, we can thus transform every color pixel in the image area into its constituent body and surface reflection components, m_b and m_s. By selecting the m_b component of every transformed pixel, we generate the *body reflection image* of the image area. By selecting the m_s component, we generate the corresponding *surface reflection image.*

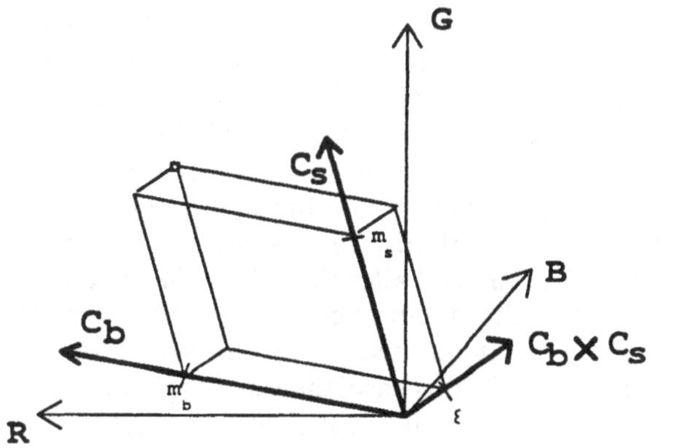

Fig. 11. Decomposing a color pixel into its constituent body and surface reflection components.

Color figures 10 and 11 show the intrinsic body and surface reflection images that are thus generated from Color Figure 4. The colors of clipped and bloomed pixels have not been separated into their reflection components. They are displayed in black, as are the pixels outside the marked image areas in color figure 4. We expect that flagging clipped and bloomed pixels in a separate intrinsic *error image* will provide useful information to other image-understanding algorithms like shape from shading. It helps them to base their analysis on only undistorted image data. If necessary, they may then extrapolate the derived shape information into areas with clipped or bloomed pixels. Our method of detecting and discarding distorted color pixels may also enable us to automatically control the camera aperture so that we can obtain color images with minimal clipping and blooming.

The intrinsic body reflection image in color figure 10 demonstrates the success of our algorithm as a method for removing highlights in undistorted image areas. It also shows that the algorithm detects very well the areas in which the pixel values are distorted. The salt-and-pepper noise on the yellow cup is due to the approximate collinearity of its body and surface reflection vectors. Small amounts of noise in its color pixels have a large effect on the computed amounts of underlying body reflection. The intrinsic surface reflection image in color figure 11 shows that our algorithm is able to reliably detect highlights in color images. It also determines the illumination color, as reflected from the surfaces of the cups. Note that this approach to the detection of highlights in color images provides quantitative information on the amount of surface reflection that is present at every pixel, as opposed to a boolean decision. Our highlight images may thus be useful input for methods that determine object shapes from highlights [12].

3.2.3 Restoring the Colors of Clipped and Bloomed Pixels. In addition to flagging clipped and bloomed color pixels, we try to restore their correct color values. For this operation, we distinguish between bloomed matte pixels and bloomed highlight pixels, depending on the reflection line they are closest to. To restore the color values of clipped or bloomed pixels, we exploit the observation that in many cases, clipping and blooming occurs only in one or two color bands. The pixels may thus have correct data in the other color bands. We assume that the smallest of the three values of a color pixel stems from a color band without clipping and blooming. We then replace the clipped or bloomed pixels by pixels on the matte or highlight line that have the same value in the undistorted band.

Color figures 12 and 13 show the intrinsic images of color figure 4 after clipped and bloomed pixels were restored. In the intrinsic body reflection image, the pixels were restored reasonably well. However, there are visible intensity changes between some distorted and undistorted pixels. There is a limit to how much can be done. Since all clipped pixels and all bloomed highlight pixels are restored to lie exactly on the highlight line, they are all projected into the same body reflection component. We thus lose all shading information for these pixels, which is normally encoded in the distance of a highlight pixel from the highlight line. This may cause a perceptible intensity change between distorted and undistorted pixels in the body reflection image. We expect this data to be used only in combination with the corresponding intrinsic error image. It may serve as additional, approximate information to infer the object shape in areas with distorted color data.

The intrinsic surface reflection image, on the other hand, has gained significantly from our restoring procedure. The clipped and bloomed pixels fill the centers of the highlights vividly with the proper surface reflection color.

3.3 Results and Discussion

We have applied our algorithm to several images of plastic cups, glossy paper folders, and a ceramic cup under white, yellow, and pink illumination. Some results have already been presented and discussed in previous sections. We will now show further results.

Color figures 14 through 17 demonstrate that our algorithm is able to detect and remove highlights well on a variety of dielectric materials under varying illumination colors. In all cases, the body reflection images have been restored very well. When judging by eyeball inspection, the body reflection images of the paper folders and of the ceramic cup do not seem to contain any indication of a highlight. We will evaluate this statement quantitatively in the future by testing whether shape from shading methods determine the correct object shape from these images. In the images of the plastic cups, the highlights could not be perfectly removed at clipped and bloomed pixels for reasons that were discussed in section 3.2.3.

The surface reflection images locate very well the highlights in the images. Note the gradual change of the amount of surface reflection on the paper folders. Since the surface reflection image captures this gradual change, it may provide the means for a quantitative approach to determining object shape from highlights.

Tables 3 through 6 show the normalized reflection vectors of the objects in all images. Our algorithm performed reasonably well in detecting the white illumination color from the highlights of the paper folders and the ceramic cup and the pink illumination from the highlights of the cups. We are currently investigating what caused the larger variations in the surface reflection vectors of the cups under white light (table 6). There may be several contributing factors. Most importantly, color clipping can alter the orientation of a surface reflection vector significantly. This is most evident for the surface reflection vector of the yellow cup. In this case, the starting point of the highlight cluster is located at very high color values, especially in the red component of the color pixels. This does not leave a lot of room for the highlight cluster before it is bent at the red wall of the cube. Furthermore, our clipping threshold does not capture all clipped pixels on this object; and it provides a cutoff plane which is parallel to the wall of the cube, but at a skewed angle to the highlight cluster. As a result, disproportionately many color pixels are close to the clipping threshold and on the inner side of the skewed T, and the angle between the fitted highlight line and the normal to the wall is too large.

Another reason for wrong orientations of the surface reflection vectors may be that the classifications of pixels into matte and highlight pixels is imprecise (see section 3.2.1, figure 10). This can result in selecting a bad starting point for the highlight line. The determination of the starting point also depends on the selection of the threshold for recursive line splitting and on the width of the averaging window for smoothing the distance function $d(s)$.

There seems to be a general trend in table 6 of overestimating the blue component and underestimating the red component of the illumination

Table 3. Body and surface reflection vectors of the paper folders under white light.

Reflection Vectors: Paper Folders under White Light		
	body reflection vector	surface reflection vector
blue folder	(0.11,0.22,0.97)	(0.58,0.53,0.62)
red folder	(0.98,0.16,0.14)	(0.45,0.63,0.64)
illumination vector		(0.58,0.57,0.58)

Table 4. Body and surface reflection vectors of the stripes on the ceramic cup under white light.

Reflection Vectors: Ceramic Cup under White Light		
	body reflection vector	surface reflection vector
gree stripe	(0.53,0.76,0.37)	(0.50,0.53,0.68)
blue stripe	(0.10,0.19,0.98)	(0.56,0.56,0.62)
illumination vector		(0.58,0.57,0.58)

Table 5. Body and surface reflection vectors of the three plastic cups under pink light.

Reflection Vectors: Plastic Cups under Pink Light		
	body reflection vector	surface reflection vector
green cup	(0.38,0.91,0.17)	(0.83,0.51,0.23)
yellow cup	(0.93,0.36,0.05)	(0.72,0.60,0.32)
orange cup	(0.99,0.16,0.05)	(0.78,0.53,0.32)
illumination vector		(0.71,0.59,0.38)

Table 6. Body and surface reflection vectors of the three plastic cups under white light.

Reflection Vectors: Plastic Cups under White Light		
	body reflection vector	surface reflection vector
green cup	(0.21,0.92,0.32)	(0.43,0.47,0.77)
yellow cup	(0.82,0.56,0.15)	(0.26,0.46,0.85)
orange cup	(0.96,0.25,0.14)	(0.43,0.59,0.68)
illumination vector		(0.58,0.57,0.58)

vector. There may be measurement errors involved, concerning the color balancing and spectral linearization parameters. They might have been caused, for example, by a shading gradient on the color chart.

All of these hypotheses are subject to further investigations, as are ways of avoiding such problems or improving the results. One way to improve the orientations of the surface reflection vectors is given by the observation that the vectors should be the same for all highlights. This constraint can be useful, if there exist several approximately correct surface reflection vectors and only a few bad ones. We may then determine a single, good surface reflection vector that best fits a weighted combination of all highlight clusters.

Our examples show that we can detect and remove highlights reasonably well in the given color images. We will now discuss what the limitations of our approach are. The main feature of our algorithm is that it exploits the color difference between surface reflection and body reflection. There needs to be a certain angle between the orientations of the body and surface reflection vectors of an object, in order for our program to detect them correctly. How big this angle needs to be depends on the lengths of the color clusters and on the selected thresholds for recursive line splitting and smoothing. These, in turn, depend on the expected amount of noise in the color clusters. Our example of the yellow cup under yellow light in section 3.2.1. has demonstrated that the algorithm was able to distinguish the surface reflection vector correctly from the body reflection vector when the angle betwen them was only 15 degrees. In this example, however, both color clusters were fairly long. In general, we may have problems with very dark or very bright objects (when either of the two color clusters is very short) and with objects with unsaturated colors, such as pastels.

We cannot apply our algorithm to metals because they do not have a significant body reflection component. This means that the color pixels from a metal form only a highlight cluster in the color space; no matte cluster exists. We thus do not need to separate its surface reflection component from its body reflection component. Problems may arise when a scene contains both metallic and dielectric objects and we need to determine whether a single linear cluster in the color cube should be interpreted as a highlight cluster, generated by a metal, or as a matte cluster, generated by a dielectric object without a highlight. To answer this question, we have to evaluate geometric information. One viable approach might be to use shape from shading to generate the shape of a hypothetical dielectric object and to compare it with the shape of a hypothetical metal, as generated by a shape from highlights method.

Another important requirement in our algorithm is that the subclusters form lines. There may be several reasons why this may not be the case for the matte clusters of all objects in real scenes. Firstly, the matte color clusters may be curved because the illumination color changes in the scene. The most important factor here is inter-reflection between objects. If inter-reflection exists between two objects, *A* and *B*, the illumination at some points on object *B* is a mixture of the direct illumination and the reflected light from parts of *A*. Consequently, the light reflected from those points on *B* depends on the color of *A*. The color cluster then contains entries that spread toward the color of object *A*.

The matte cluster may also be nonlinear because the reflection properties of the object vary, either gradually or abruptly. This happens if there are color texture patterns on the object. The ceramic cup in color figure 15 is an example of an object that has abrupt color changes. The resulting color clusters are widely separated in the color cube. Thus, if we can somewhat determine object areas belonging to a single color cluster, our current algorithm is able to detect and remove highlights in each of these areas. If the color changes are more gradual, as in a rainbow, or if the object areas with constant reflection properties are too small, the color pixels do not form distinctive, separate color clusters in the color cube; and our method breaks down.

4 Conclusions and Further Work

In this paper, we have demonstrated that it is possible to analyze real color images by using a color reflection model. Our model accounts for highlight reflection and matte shading, as well as for some characteristics of cameras. By developing a physical description of color variation in color images, we have developed a method to separate highlight reflection from matte object reflection. The resulting intrinsic reflection images have a simpler relationship to the illumination geometry than the original image and may thus improve the results of many other computer vision algorithms, such as motion analysis, stereo vision, and shape from shading or highlights [7, 12, 14, 35]. Since the surface reflection component of dielectric materials generally has the same color as the illumination, we can also determine the illumination color from the intrinsic surface reflection image, information that is needed by color constancy algorithms [6, 23, 24].

The key points leading to the success of this work are our modeling of highlights as a linear combination of both body and surface reflection and our modeling of the camera properties. With few exceptions [6, 8, 23, 31], previous work on highlight detection in images has assumed that the color of highlight pixels is completely unrelated to the object color. This assumption would result in two unconnected clusters in the color space: one line or ellipsoid representing the object color and one point or sphere representing the highlight color. Our model and our color histograms demonstrate that, in real scenes, there exists a transition area on the objects from purely matte areas to the spot that is generally considered to be the highlight. This transition area determines the characteristic shapes of the color clusters, which is the information that we use to detect and remove highlights. This view of highlights should open the way for quantitative shape-from-gloss analysis, as opposed to binary methods based on thresholding intensity.

By modeling the camera properties, we are able to obtain color images of high quality (through color balancing and spectral linearization), in which most pixels maintain the linear properties of light reflection, as described in the Dichromatic Reflection Model. We can also detect most distorted color pixels in an image and thus generate an intrinsic error image which then guides out algorithm to separate only undistorted color pixels into their reflection components. We expect that the intrinsic error image will be similarly useful in guiding other computer vision algorithms, such as shape from shading. It may also enable us to automatically control the camera aperture so that we can obtain color images with mineral clipping and blooming.

Our approach to color image analysis may influence research in other areas of color computer vision. The color histograms demonstrate that all color pixels from one object form a single cluster in the color space. A color cluster thus groups pixels into areas of constant material properties, independently of influences from the illumination geometry such as shading or highlights. This finding can be used for color image segmentation to distinguish material changes from shading or highlight boundaries. Furthermore, the shape of the color clusters reveals some information about the illumination geometry and the object shapes. We are currently investigating these implications of the model to improve color-image-understanding methods, and we are also working on extending our model to account for other material types, such as metals, and more complex illumination conditions including ambient light, light sources in several colors, shadow casting, and interreflection between objects.

Although the current method has been applied only in a laboratory setting, its initial success shows the value of modeling the physical nature of the visual environment. Our work and the work of others in this area may lead to methods that will free computer vision from its current dependence on statistical signal-based methods for image segmentation.

Acknowledgments

We would like to thank Ruth Johnston-Feller for her comments and suggestions. Richard Szeliski and Chuck Thorpe provided valuable comments on previous drafts of the article. We are also grateful to the reviewers. They helped to improve the paper significantly.

References

1. D.H. Ballard and C.M. Brown. *Computer Vision.* Prentice Hall: Englewood Cliffs, NJ 07632, 1982.
2. P. Beckmann and A. Spizzichino. *The Scattering of Electromagnetic Waves from Rough Surfaces.* MacMillan: New York, 1963, pp. 1–33, 70–96.
3. J.D.E. Beynon and D.R. Lamb (eds.). *Charge-coupled devices and their application.* McGraw-Hill: London, 1980.
4. W. Budde. *Optical Radiation Measurements.* Volume 4: *Physical Detectors of Optical Radiation.* Academic Press: New York, 1983.
5. R.L. Cook and K.E. Torrance. "A reflectance model for computer graphics." *ACM Trans. Graphics* 1(1):7–24, January 1982. Also published in COMPUTER GRAPHICS 15(3), SIGGRAPH'81.
6. M. D'Zmura and P. Lennie. "Mechanisms of color constancy." *J. Opt. Soc. Amer. A* (*JOSA-A*), 3(10):1662–1672, October 1986.
7. L. Dreschler and H.-H. Nagel. "Volumetric model and 3D trajectory of a moving car derived from monocular TV frame Sequences of a Street Scene." *Comput. Graphics Image Process.*, 20:199–228, 1982.
8. R. Gershon. "The use of color in computational vision." PhD thesis, Department of Computer Science, Univ. of Toronto, 1987.
9. R. Gershon, A.D. Jepson, and J.K. Tsotsos. "Highlight identification using chromatic information." In *Proc. first Int. Conf. Computer Vision (ICCV), London,* Computer Society Press (IEEE), June 8–11, 1987, pp. 161–171.
10. H. Grassmann. "On the theory of compound colors." *Phil. Mag.*, April 1854.
11. H.H. Harman. *Modern Factor Analysis,* 2nd ed. University of Chicago Press: Chicago and London, 1967.
12. G. Healey and T.O. Binford. "Local shape from specularity." In J.M. Brady and A. Rosenfeld (eds.). *Proc. First Int. Conf. Computer Vision (ICCV),* Computer Society Press

(IEEE), London, June 8–11, 1987, pp. 151–161. Also appeared in DARPA-Image Understanding Workshop. Los Angeles, CA, February 1987, pp. 874–887.

13. G. Healey and T.O. Binford, "The role and use of color in a general vision system," In L.S. Bauman (ed.), *DARPA-Image Understanding (IUS) workshop*, Los Angeles, CA, February 1987, pp. 599–613.
14. B.K.P. Horn, "Understanding image intensities," *Artifical Intelligence* 8(11):201–231, 1977.
15. B.K.P. Horn, "Exact reproduction of colored images," *Computer Vision, Graphics, and Image Processing (CVGIP)* 26:135–167, 1984.
16. R.S. Hunter, *The Measurement of Appearance*. Wiley: New York, 1975.
17. R.M. Johnston, "Geometric metamerism," *Color Engineering* 5(3):42, May–June, 1967.
18. R.M. Johnston, "Color theory," *Pigment Handbook*. Wiley: New York, 1974, pp. 229–287, ch. III-D-b.
19. D.B. Judd and G. Wyszecki, *Color in Business, Science and Industry*. Wiley, New York, 1975.
20. G.J. Klinker, S.A. Shafer, and T. Kanade, "Measurement of gloss from color images." In *Inter-Society Color Council (ISCC) 87 Conf. on Appearance*, Williamsburg, VA, February 8–11, 1987, pp. 9–12.
21. G.J. Klinker, S.A. Shafer, and T. Kanade, "Using a color reflection model to separate highlights from object color," In J.M. Brady and A. Rosenfeld (ed.), *Proc. First Int. Conf. Computer Vision (ICCV)*, Computer Society Press (IEEE), London, June 8–11, 1987, pp. 145–150. Also appeared in DARPA-Image Understanding Workshop, Los Angeles, CA, February 1987, pp. 614–619.
22. Y. LeClerc, "A Method for Spectral Linearization." Private communication, 1986.
23. H.-C. Lee, "Method for computing the scene-illuminant chromaticity from specular highlights," *J. Opt. Soc. Amer. A (JOSA-A)* 3(10):1694–1699, October 1986.
24. L.T. Maloney and B.A. Wandell, "Color constancy: a method for recovering surface spectral reflectance," *J. Opt. Soc. Amer. A (JOSA-A)* 3(1):29–33, January 1986.
25. P. Moon, "A table of Fresnel reflection," *J. Math. Phys.* 19(1), 1940.
26. F.E. Nicodemus, J.C. Richmond, J.J. Hsia, I.W. Ginsberg, and T. Limperis, *Geometrical Considerations and Nomenclature for Reflectance*, Tech. Report NBS Monograph 160, National Bureau of Standards, October 1977.
27. R. Ohlander, K. Price, and D.R. Reddy, "Picture segmentation using a recursive region splitting method," *Computer Graphics and Image Processing* 8:313–333, 1978.
28. Y. Ohta, T. Kanade, and T. Sakai, "Color information for region segmentation," *Computer Graphics and Image Processing* 13:222–231, 1980.
29. T. Pavlidis, *Structural Pattern Recognition*. Springer Verlag: Berlin, Heidelberg, New York, 1977.
30. B.T. Phong. "Illumination for computer generated pictures," *Communications of the ACM* 18:311–317, 1975.
31. J.M. Rubin and W.A. Richards, "Color vision and image intensities: When are changes material?" *Biological Cybernetics* 45:215–226, 1982.
32. S.A. Shafer, "Describing light mixtures through linear algebra," *J. Opt. Soc. Amer. (JOSA)* 72(2):299–300, February 1982.
33. S.A. Shafer, "Optical phenomena in computer vision," *Canad. Soc.* Computational Studies of Intelligence (CSCSI-84), Ontario, May 1984. Also available as Tech. Report TR-135, Computer Science Department, Univ. Rochester, March 1984.
34. S.A. Shafer, "Using color to separate reflection components," *COLOR res. appl.* 10(4):210–218, Winter 1985. Also availble as Tech. Report TR-136, Computer Science Department, Univ. of Rochester, NY, April 1984.
35. C.E. Thorpe, "FIDO: Vision and navigation for a robot rover." PhD thesis, Computer Science Department, Carnegie-Mellon University, December 1984. Available as Tech. Report CMU-CS-84-168.
36. K.E. Torrance and E.M. Sparrow, "Theory of off-specular reflection from roughened surfaces," *J. Opt. Soc. Amer.* 57:1105–1114, September 1967.
37. S.J. Williamson and H.Z. Cummins, *Light and Color in Nature and Art*. Wiley, New York, 1983.
38. G.J. Klinker, S.A. Shafer, and T. Kanade, "Image segmentation and reflection analysis through color." In SPIE's 1988 Techn. Symp. on Ophics, Electro-Ophics, and Sensors, April 4–8, 1988, Orlando, Florida. (Also in DARPA-Image Understanding Workshop, Boston, MA, April 1988.)

COLOR FIGURES

See color plate I for Color Figures 1-6.
See color plate J for Color Figures 7-12.
See color plate K for Color Figures 13-17.

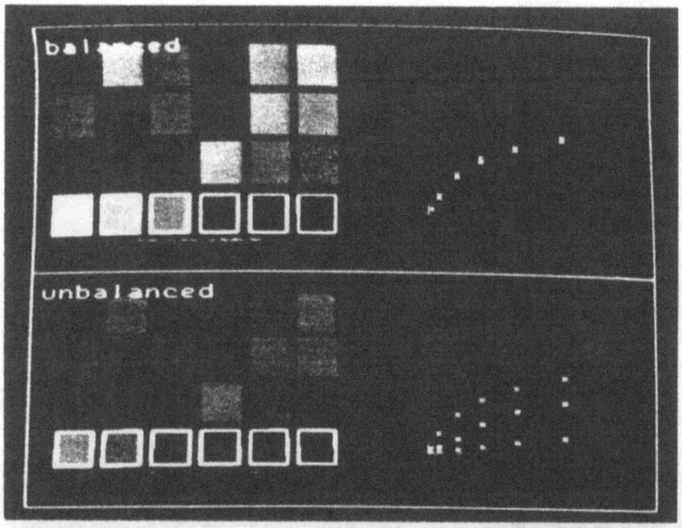

Color Figure 1. The effect of color balancing and gamma correction on color images.

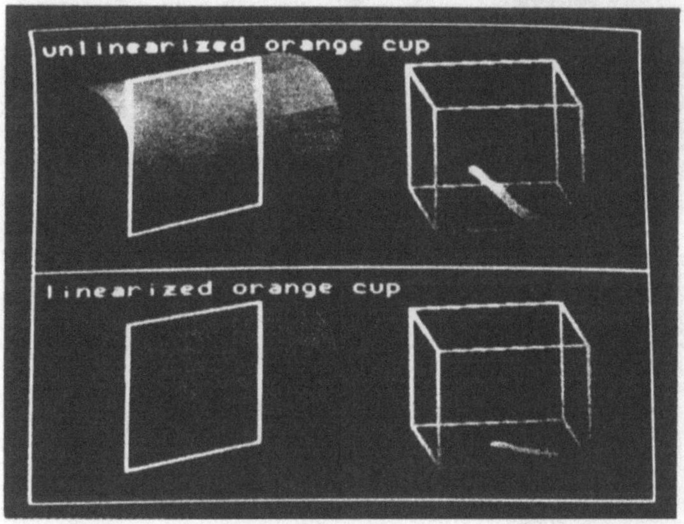

Color Figure 2. Curvature in the color space, caused by gamma-corrected cameras.

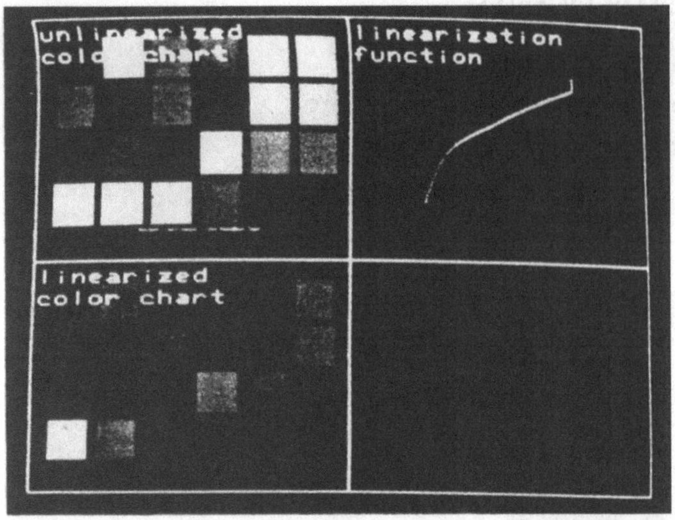

Color Figure 3. Spectral linearization with cubic spline functions.

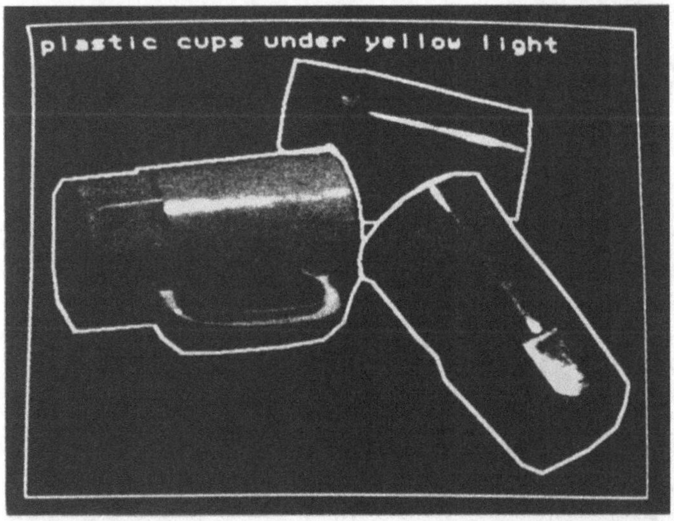

Color Figure 4. Three plastic cups under yellow light.

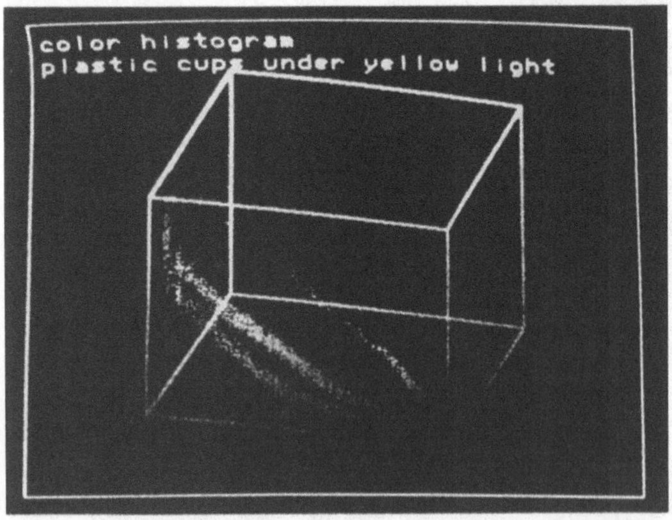

Color Figure 5. Color histogram of three plastic cups under yellow light.

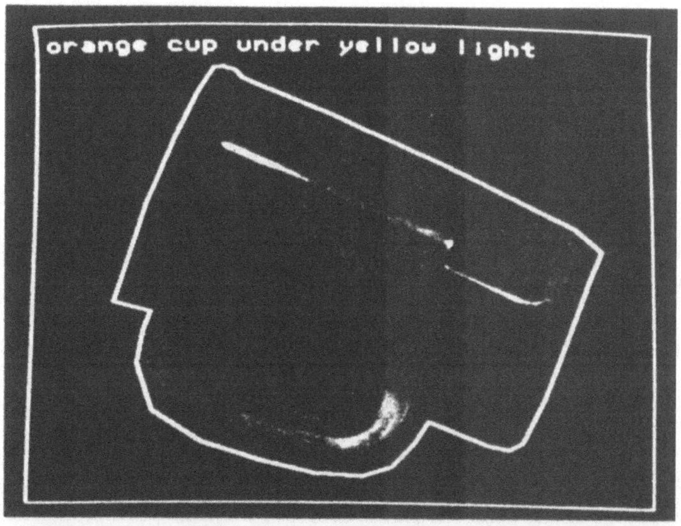

Color Figure 6. Orange cup under yellow light.

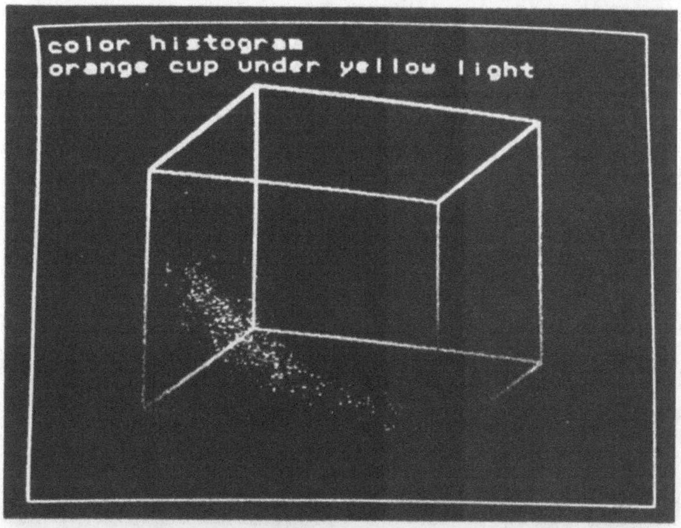

Color Figure 7. Color histogram of the orange cup under yellow light.

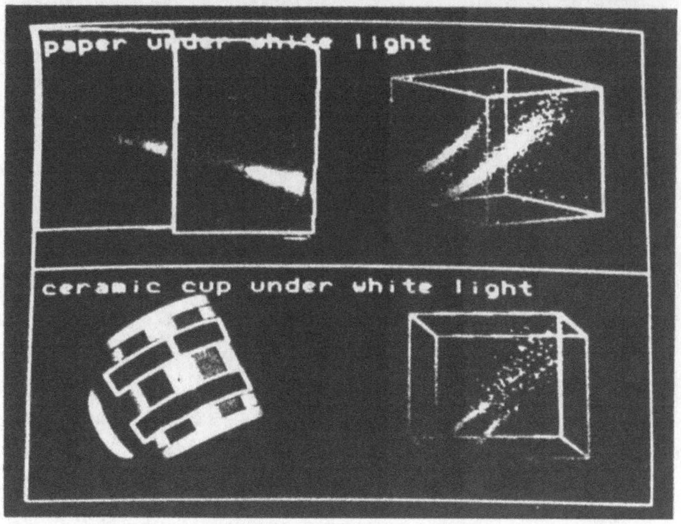

Color Figure 8. Images and color clusters of glossy paper and a ceramic cup under white light.

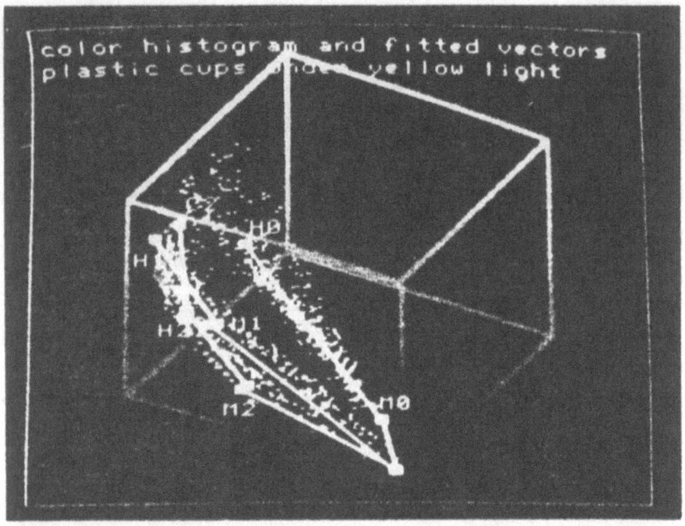

Color Figure 9. Color histogram and fitted vectors of the three plastic cups under yellow light.

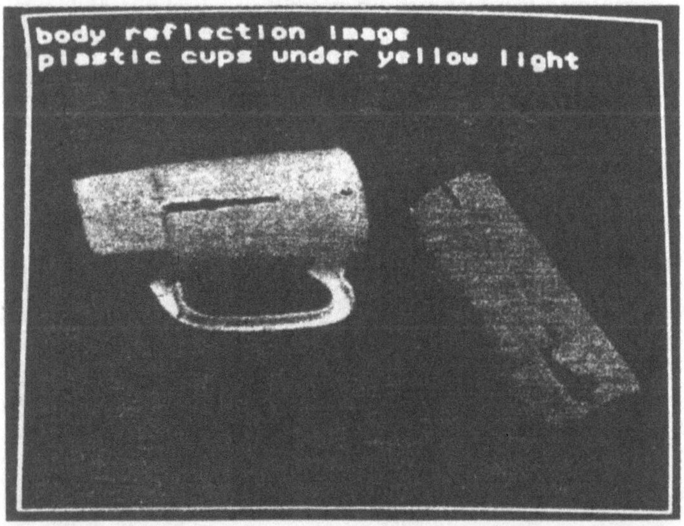

Color Figure 10. Intrinsic body reflection image of the three plastic cups under yellow light; clipped pixels and bloomed pixels are shown in black.

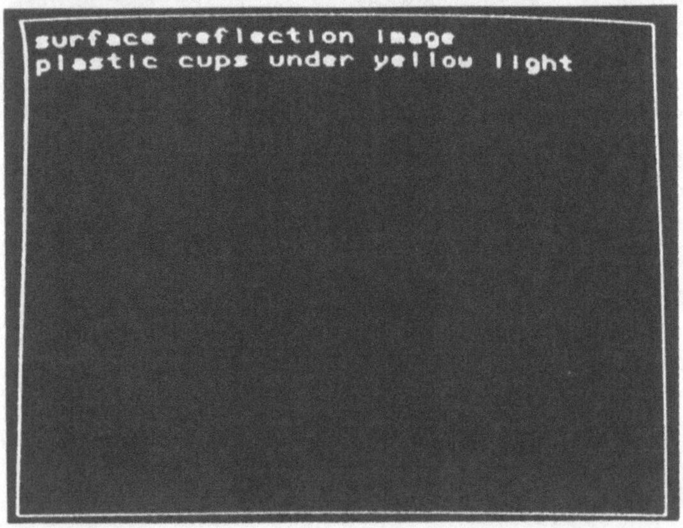

Color Figure 11. Intrinsic surface reflection image of the three plastic cups under yellow light; clipped and bloomed pixels are shown in black.

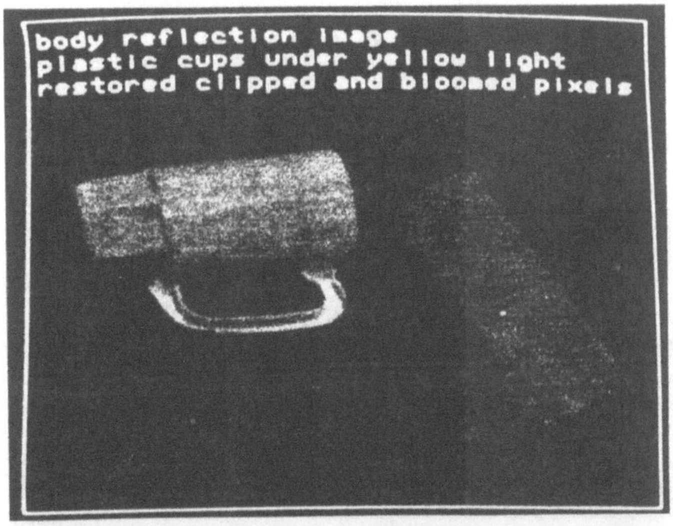

Color Figure 12. Intrinsic body reflection image of the three plastic cups under yellow light; with restored clipped and bloomed pixels.

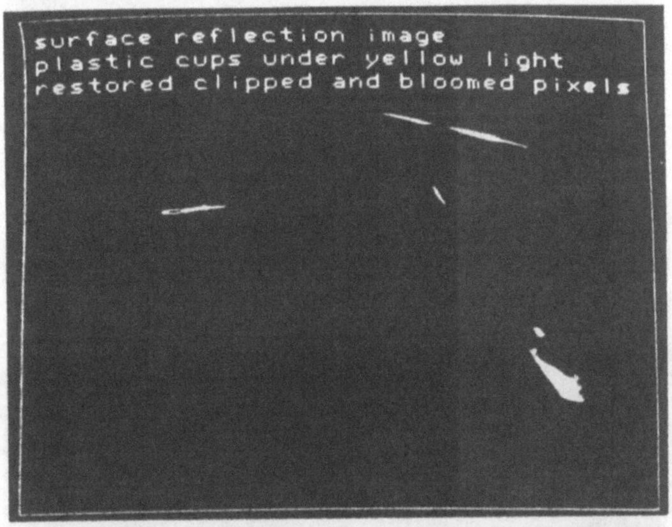

Color Figure 13. Intrinsic surface reflection image of the three plastic cups under yellow light; with restored clipped and bloomed pixels.

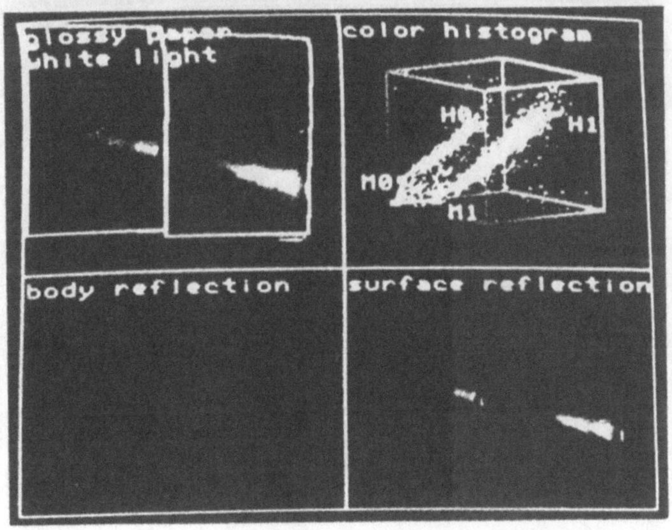

Color Figure 14. Intrinsic reflection images of the glossy paper folders under white light.

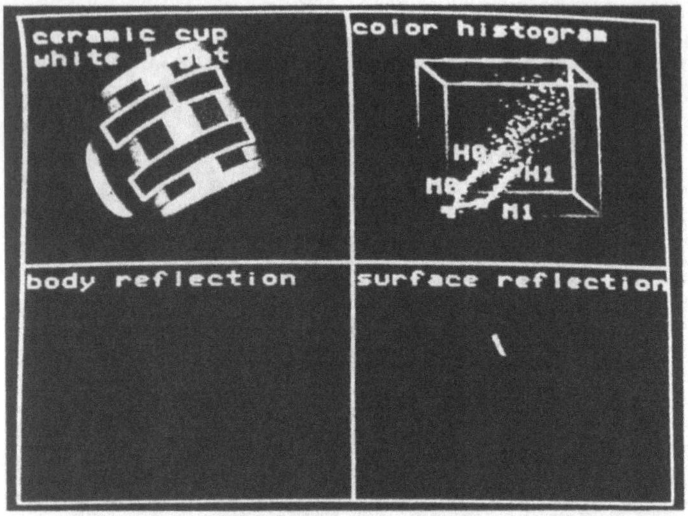

Color Figure 15. Intrinsic reflection images of the ceramic cup under white light.

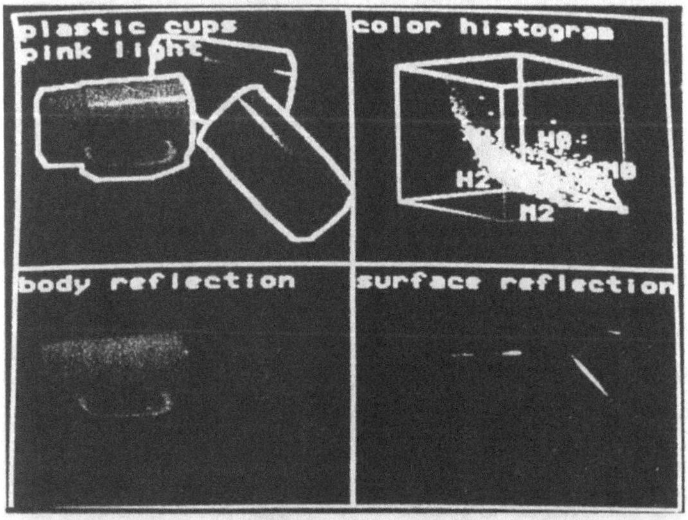

Color Figure 16. Intrinsic reflection images of the three plastic cups under pink light.

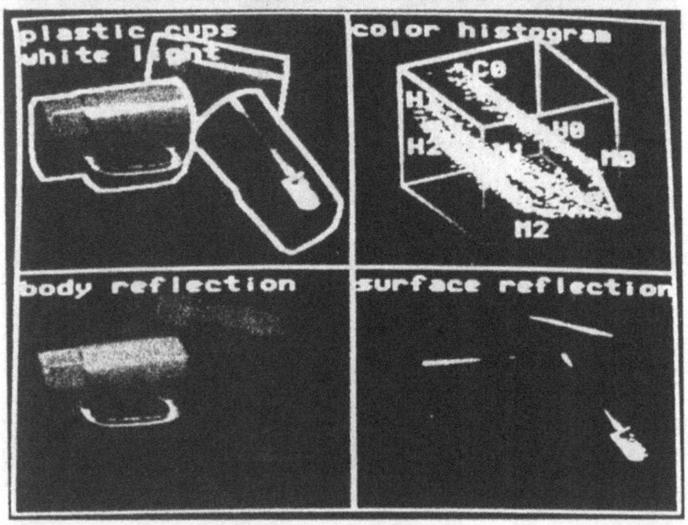

Color Figure 17. Intrinsic reflection images of the three plastic cups under white light.

Estimating spectral reflectance using highlights

Glenn Healey

A new algorithm is presented for estimating the spectral reflectance of a surface from a colour image using highlights. The technique is based on detailed reflectance models and is more general than existing techniques. An estimate of spectral reflectance is useful for surface recognition and characterization because it is intrinsic to the material, i.e. it does not depend on illumination conditions. Experimental results obtained by applying the algorithm to colour images are presented.

Keywords: image processing, spectral reflectance, surface recognition

INTRODUCTION

A property of an object which is valuable for recognition is the object's spectral reflectance. Spectral reflectance is the fraction of incident light that an object reflects as a function of wavelength. Unfortunately, spectral reflectance is not determined by the spectral distribution of the light reflected by an object. It is this reflected light which is directly sensed by a vision system. The light which is reflected by an object is the product of the spectral distribution of the incident light and the spectral reflectance of the object material. To recover the spectral reflectance of an object, some mechanism must be available to factor out the effects of the incident light. Many experiments have shown that the human vision system is capable of making this computation. This ability of humans to see objects as having a constant colour despite varying illumination conditions is called colour constancy. Many theories have been advanced to explain colour constancy[1-5].

It is shown elsewhere[6] that the reflectance R of an opaque sample of matter with one relevant interface can be approximated accurately by:

$$R(g, \lambda) = R_S(g, \lambda) + R_B(g, \lambda) = \begin{cases} M_S(g)\, C_S(\lambda) & \text{Metal} \\ M_S(g)\, C_S(\lambda) + M_B(g)\, C_B(\lambda) & \text{I.D.} \end{cases} \quad (1)$$

Department of Electrical and Computer Engineering, University of California, Irvine, CA 92717, USA
Paper received: 29 September 1989; Revised paper received: 7 June 1990

where g indicates dependence on the photometric angles and λ indicates dependence on wavelength. The subscript S denotes terms associated with surface (interface) reflection and the subscript B indicates terms associated with body reflection. I.D. indicates an inhomogeneous dielectric material. Equation (1) has been shown to be accurate for a wide class of materials. It is unable, however, to approximate accurately the behaviour of surface reflection from red metals for a small range of angles near glancing incidence[6].

Surface reflection is the process that is responsible for highlights or specularities on objects. The fraction of the incident light that is not reflected from the surface penetrates into the body of the material. For inhomogeneous dielectrics this light is scattered by colourant particles in the body of the material and some fraction of this light can exit back through the surface in a range of directions. Body reflection refers to this process of light scattering by colourant particles in inhomogeneous dielectrics.

In this paper, a method for estimating the functions $C_S(\lambda)$ and $C_B(\lambda)$ of (1) to within a multiplicative constant is derived. Estimating these functions is particularly useful for material recognition, because these functions do not depend on illumination conditions or scene geometry.

APPROXIMATING THE COLOUR SIGNAL

Sensing elements in a camera have a certain quantum efficiency which depends on how light interacts with the material of the sensor. The quantum efficiency of the sensor material is defined as the ratio of generated electron flux to incident photon flux. In general, quantum efficiency depends on wavelength.

If we let $q(\lambda)$ denote the quantum efficiency of the sensor material and $E_I(\lambda)$ denote the spectral irradiance for a one pixel area on the sensor plane, then the sensor value measured at that pixel will be:

$$V = \int_\lambda q(\lambda)\, E_I(\lambda)\, d\lambda \quad (2)$$

where λ ranges over the entire electromagnetic spectrum.

It is useful to be able to recover more information about the variation of E_I with wavelength than is available in the single measurement V. This information can be obtained by using several sensors with differing wavelength responses. Colour filters are often used for this purpose. Suppose n filters are used. Let $t_i(\lambda)$ be the spectral transmission of the ith filter $(0 \leq i \leq n-1)$. The effective spectral response of a filter-camera combination is then $f_i(\lambda) = t_i(\lambda) q(\lambda)$. Therefore, at each pixel a vector S is measured with components:

$$s_i = \int_\lambda f_i(\lambda)\, E_I(\lambda)\, d\lambda \quad (0 \leq i \leq n-1) \tag{3}$$

where λ ranges over the entire electromagnetic spectrum.

In this work, we will be primarily interested in the behaviour of $E_I(\lambda)$ in the visible region of the spectrum. Since many imaging sensors, e.g. CCD cameras, respond to light in the near infrared, it is often necessary to use an IR cutoff filter to block infrared radiation. In this work, the functions $f_i(\lambda)$ are assumed to be zero outside of the visible wavelength range. The visible wavelength range will be denoted by $[\lambda_1, \lambda_2]$.

The spectral irradiance at the sensor plane is often referred to as the colour signal. A method is described elsewhere[7] for recovering a continuous approximation to the colour signal from a colour sensor vector S measured at one pixel in the image. This method will be used by the reflectance estimation algorithm below. Similar approximations have been used for different purposes in the past[8,9]. A thorough discussion of the use of basis functions for colour constancy is given by Forsyth[2]. Below, the approximation technique is briefly summarized.

Consider approximating $E_I(\lambda)$ as a linear combination of n basis functions $b_j(\lambda)$:

$$E_I(\lambda) \approx \sum_{0 \leq j \leq n-1} a_j b_j(\lambda) \tag{4}$$

Let

$$k_{ij} = \int_{\lambda_1}^{\lambda_2} f_i(\lambda)\, b_j(\lambda)\, d\lambda \tag{5}$$

and let K be the $n \times n$ matrix such that $K(i, j) = k_{ij}$.

We approximate $E_I(\lambda)$ as in Forsyth[2] by:

$$E_I(\lambda) \approx B^T(\lambda) A \tag{6}$$

where:

$$A = K^{-1} S, \tag{7}$$

A is the vector $[a_0, a_1, \ldots, a_{n-1}]^T$, $B(\lambda)$ is the vector $[b_0(\lambda), b_1(\lambda), \ldots, b_{n-1}(\lambda)]^T$, and S is the vector with components given by (3). An analysis of the properties of this approximation is given elsewhere[10].

ESTIMATING REFLECTANCE

It has been frequently observed (e.g. see Healey[6]) that the spectral power distribution of highlights from inhomogeneous dielectrics is approximately the same as the spectral power distribution of the illumination. This section derives a technique for estimating the spectral reflectance of a material from a colour image based on this fact. This technique uses the result of a highlight detection process (e.g. Brelstaff and Blake[11], Klinker *et al.*[12]). Lee[13] has proposed an approach based on the same physical principle for computing the scene illuminant chromaticity from highlights. The method of this paper is more general in that only one highlight per illuminant (instead of two) is required. Also, the method of this paper is capable of generating estimates of spectral reflectance directly. To the author's knowledge, the experimental results of this paper[14] represent the first time an approach based on highlights has been used to estimate spectral reflectance from colour images.

The index of refraction for an inhomogeneous dielectric material is nearly constant across the visible spectrum[15]. From the Fresnel equations[16], the specular reflectance of an inhomogeneous dielectric material is a nearly constant function of wavelength for fixed geometry. Moreover the specular reflectance for fixed wavelength is nearly constant with respect to geometry for almost all incidence angles[6]. Therefore, for inhomogeneous dielectric materials the surface component of the reflectance can be regarded as constant with respect to both geometry and wavelength.

Klinker *et al.*,[12] describe a method for locating highlights on inhomogeneous dielectrics in colour images. This method locates highlight regions, associated background regions of pure body reflection, and the boundary between the two regions. From this input, it is possible to recover an approximation to spectral reflectance up to a multiplicative constant. Using the reflectance model of (1), the function $E_I(\lambda)$ for fixed geometry is given by:

$$E_I(\lambda) = [R_S(\lambda) + R_B(\lambda)]\, L(\lambda) \tag{8}$$

where $L(\lambda)$ is the spectral power distribution of the illumination. On the pure body reflection side of a highlight boundary, $R_S(\lambda) = 0$ giving:

$$E'_I(\lambda) = R_B(\lambda)\, L(\lambda) \tag{9}$$

where $R_B(\lambda)$ can be considered to be the same as in (8) since $R_B(\lambda)$ changes slowly with respect to geometry. We can solve for $R_S(\lambda)\, L(\lambda)$ using:

$$R_S(\lambda)\, L(\lambda) = E_I(\lambda) - E'_I(\lambda) \tag{10}$$

But $R_S(\lambda)$ is constant with respect to λ and geometry. Let $R_S(\lambda) = k$. Therefore, using (10) we can compute $L(\lambda)$ up to the constant k by:

$$kL(\lambda) = E_I(\lambda) - E'_I(\lambda) \tag{11}$$

From (9), the body reflectance $R_B(\lambda)$ can now be computed up to the constant $1/k$ by:

$$\frac{1}{k} R_B(\lambda) = \frac{E'_I(\lambda)}{kL(\lambda)} \tag{12}$$

For a metal with the same illumination $L(\lambda)$, we have:

$$E_I(\lambda) = R_S(\lambda)\, L(\lambda) \tag{13}$$

because $R_B(\lambda) = 0$ in (1). Thus we can compute

$R_S(\lambda)$ to within $1/k$ by:

$$\frac{1}{k}\; R_S(\lambda) = \frac{E_I(\lambda)}{kL(\lambda)} \tag{14}$$

Therefore, using (12) and (14) we can compute $R_B(\lambda)$ and $R_S(\lambda)$ for any surface illuminated by $L(\lambda)$ up to a multiplicative constant. From (1), this immediately gives us $C_B(\lambda)$ and $C_S(\lambda)$ up to a multiplicative constant. Examples of the performance of this method on real images are given below.

It should not be surprising that there is a fundamental ambiguity in the computed spectral reflectance corresponding to the constant k. Since:

$$E_I(\lambda) = L(\lambda)\, R(\lambda), \tag{15}$$

we see that an arbitrary constant t can be introduced into both $L(\lambda)$ and $R(\lambda)$ such that the resultant $E_I(\lambda)$ will be indistinguishable from $E_I(\lambda)$ in (15):

$$E_I(\lambda) = \left(\frac{L(\lambda)}{t}\right)(tR(\lambda)) \tag{16}$$

Thus without using additional assumptions, we cannot expect to determine $R(\lambda)$ better than to within a multiplicative constant.

EXPERIMENTS

A laboratory setup has been used to test the above algorithm. Colour images are digitized using a CCD camera and gelatin filters. The camera is equipped with an infrared cutoff filter. Figure 1 illustrates the spectral characteristics of the sensors. The large amplitude curve indicates the spectral sensitivity of the camera. The four smaller curves are the spectral transmission functions of the filters multiplied by the spectral sensitivity of the camera. These four smaller curves correspond to the functions $f_0(\lambda)$, $f_1(\lambda)$, $f_2(\lambda)$ and $f_3(\lambda)$ given above. These curves have been taken from manufacturer's specifications.

Figures 2–11 show the performance of the reflectance estimation algorithm on colour images of plastic cups illuminated by a lamp with colour temperature 4800°K which is intended to simulate

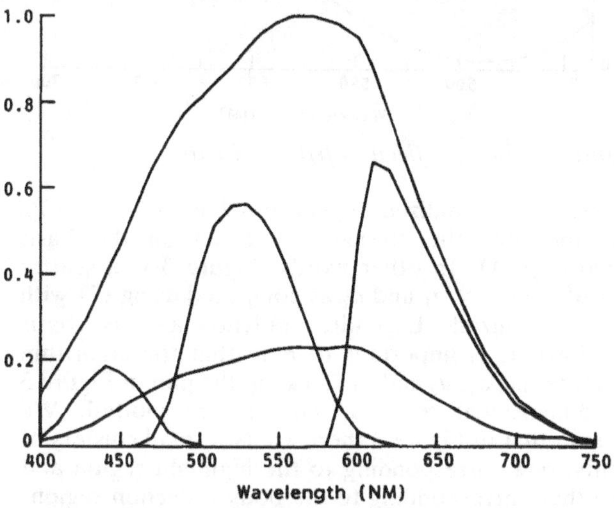

Figure 1. Sensor response functions

Figure 2. Image of a blue cup

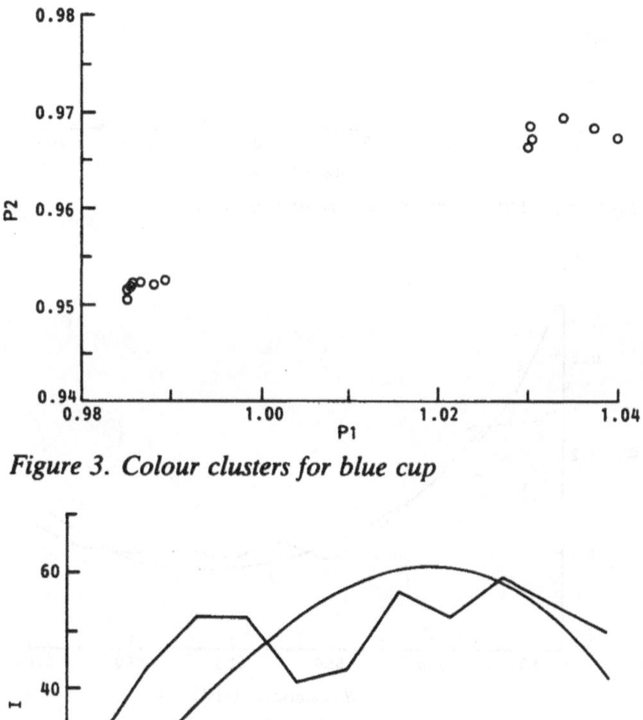

Figure 3. Colour clusters for blue cup

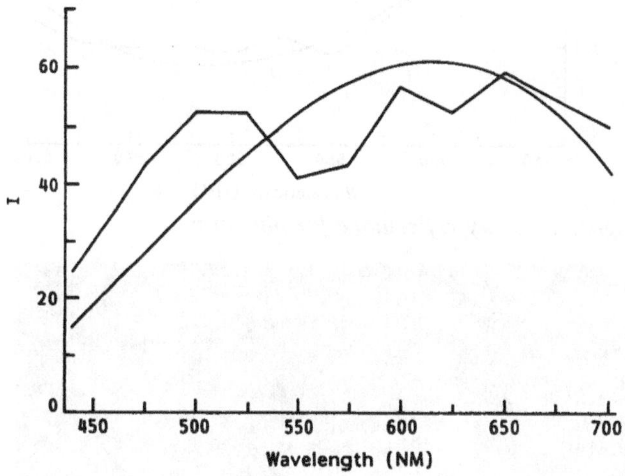

Figure 4. Surface reflection from blue cup

daylight. Neutral density filters are used to prevent highlights from saturating the camera. Inhomogeneous dielectric highlights are located using the method described by Klinker *et al.*[12].

The above method is used to recover an estimate of the function $E_I(\lambda)$ for both the surface reflected light (highlights) and the body reflected light. The normalized Legendre polynomials[10] are used as the basis functions $B(\lambda)$. In Figures 4, 6, 9 and 11 two curves are plotted. The smooth curves represent the computed approximations on the polynomial basis. For comparison, the jagged curves in these figures indicate reference data obtained using a spectroradiometer. In each case, the computed approximation is normalized to allow comparison with the radiometer data.

Figure 2 is an image of a blue cup. Figure 3 shows the result of applying the colour signal recovery method to both the surface and body reflection regions of the image. The P1 axis in Figure 3 indicates the computed strength of $E_I(\lambda)$ on the basis function $p_1(\lambda)$ (the degree one normalized Legendre polynomial). The P2 axis indicates the strength of $E_I(\lambda)$ on the basis function $p_2(\lambda)$. In other words, Figure 3 is a scatter plot of values of a_1 and a_2 as computed using (7) with the normalized Legendre polynomials as basis functions. It is important to note that the algorithm computes a_0, a_1, a_2 and a_3. To keep the plot of Figure 3 two-dimensional, only a_1 and a_2 are plotted. We observe that in Figure 3 there are two small clusters of points, one corresponding to the highlight region and the other corresponding to the body reflection region. The algorithm computes average pixel values for the

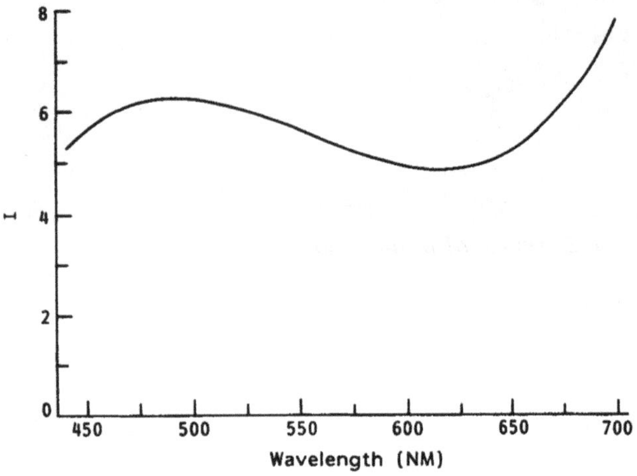

Figure 5. Body reflection from blue cup

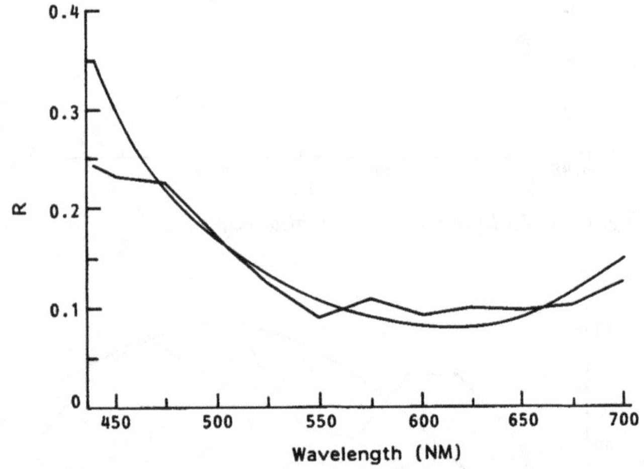

Figure 6. Body reflectance for blue cup

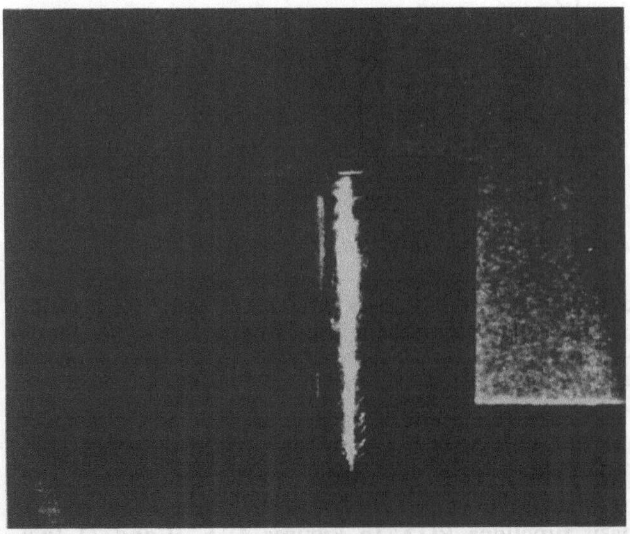

Figure 7. Image of a red cup

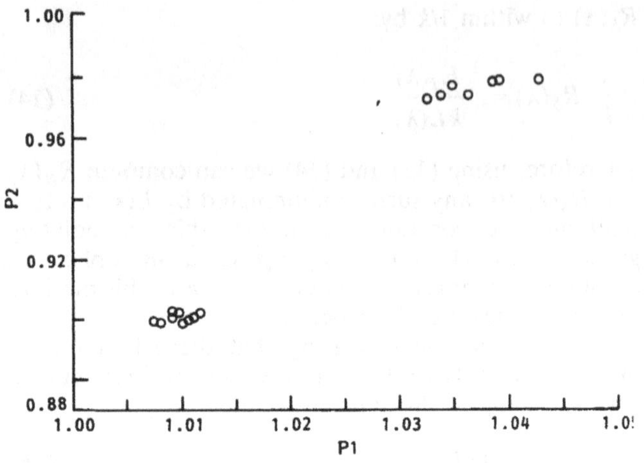

Figure 8. Colour clusters for red cup

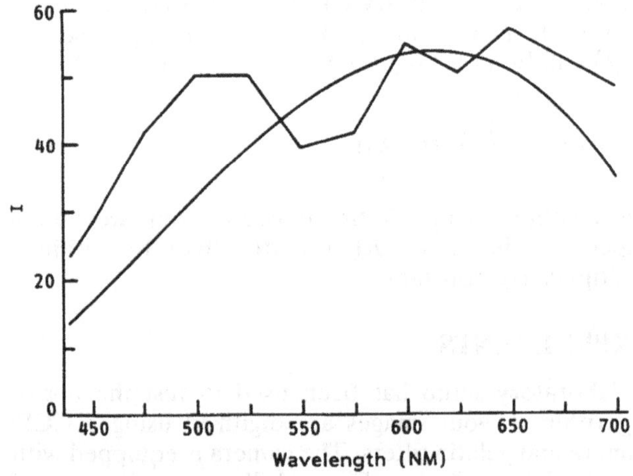

Figure 9. Surface reflection from red cup

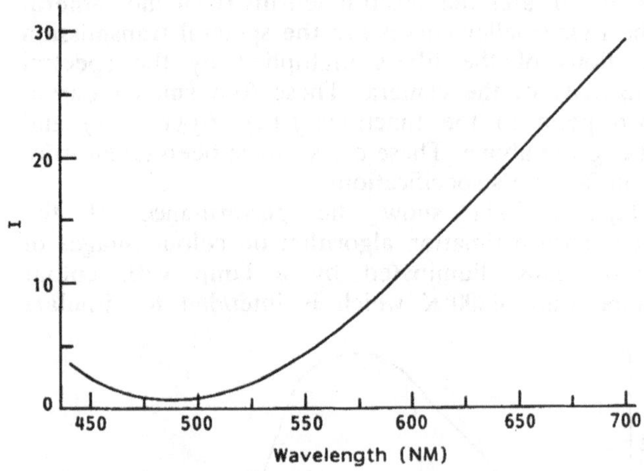

Figure 10. Body reflection from red cup

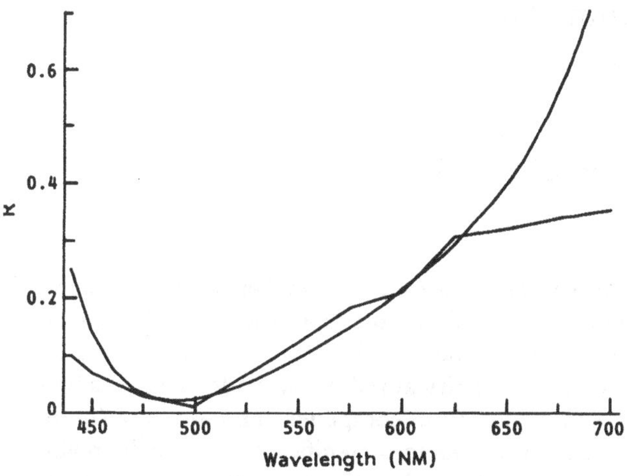

Figure 11. Body reflectance for red cup

highlight cluster and for the body reflection cluster to generate input data for the reflectance estimation algorithm. The smooth curve in Figure 4 is the computed approximation to $E_I(\lambda)$ for the surface reflection from the blue cup. The jagged curve is spectral radiometer data for the spectral power distribution of the light source being used. Figure 5 is the computed approximation to $E_I(\lambda)$ for the body reflection from the blue cup. Note that it would be difficult to infer that the cup is blue simply by inspecting the spectral distribution of the reflected light $E_I(\lambda)$ in Figure 5. To determine the colour of the cup (as distinct from the colour of the light reflected from the cup), we must compute the surface spectral reflectance. This is done using the method given above. Figure 6 shows the spectral reflectance computed for the blue cup. Figures 7–11 show the performance of the algorithms on a colour image of a red plastic cup.

We observe that for both images, the technique recovers reasonable approximations to the physical data. The assumption that the colour of surface reflected light is approximately the same as the colour of the illumination holds well for these plastic cups. As expected, the third order polynomials are not expressive enough to capture the small wiggles or sharp changes in the actual physical signals. The general shapes of the approximating polynomials, however, agree well with the radiometer data. These approximations are certainly sufficient to conclude, for example, that the cup of Figure 2 is made of an inhomogeneous material that is bluish and that the cup of Figure 7 is made of an inhomogeneous material that is reddish.

We would expect to be able to compute more accurate approximations by measuring the spectral responses of the sensors directly instead of using manufacturer's specifications. Another way to improve the computed approximations is to take more care in selecting basis functions $B(\lambda)$. One undesirable property of normalized Legendre polynomials is that they all take on maximum values at the endpoints of their range. This tends to amplify the effects of noise on the computed approximations in these parts of the spectrum[10]. This is evident to a certain extent in these results and is particularly prominent in Figure 11 where there is a large error near the red end of the visible spectrum. Recent work has investigated the use of other basis functions which may produce approximations displaying more uniform noise amplification characteristics[7].

ACKNOWLEDGEMENTS

I thank the referees for providing many useful suggestions. This work was done while the author was at the Robotics Lab of the Computer Science Department at Stanford University. This work has been supported by an NSF graduate fellowship, AFOSR contract F33615-85-C-5106, and ARPA contract N000-39-84-C-0211.

REFERENCES

1 **Beck, J** *Surface Colour Perception* Cornell University Press, NY, USA (1972)
2 **Forsyth, D A** 'A novel approach to colour constancy', *2nd Int. Conf. Comput. Vision*, Tarpon Springs, FL, USA (December 1988) pp 9–18
3 **Helson, H** *Adaptation-Level Theory*, Harper and Row, NY, USA (1964)
4 **Land, E** 'The retinex theory of colour perception', *Scientific American*, Vol 237 No 6 (1977) pp 108–128
5 **Maloney, L and Wandell, B** 'Colour constancy: a method for recovering surface spectral reflectance', *J. Opt. Soc. Am. A*, Vol 3 (October 1986) 1673–1683
6 **Healey, G** 'Using colour for geometry insensitive segmentation', *J. Opt. Soc. Am. A*, Vol 6 (June 1989) pp 920–937
7 **Healey, G** 'Colour sensing and discrimination by computer', *IEEE Trans. Syst. Man & Cybern.* (1989)
8 **Horn, B K P** 'Exact reproduction of coloured images', *Comput. Vision, Graph. & Image Process*, Vol 26 (1984) pp 135–167
9 **Wandell, B** 'The synthesis and analysis of colour images', *IEEE Trans. PAMI.*, Vol 9 No 1 (January 1987)
10 **Healey, G and Binford, T O** 'A colour metric for computer vision', *Proc. IEEE Conf. Comput. Vision & Pattern Recogn.*, Ann Arbor, MI, USA (1988)
11 **Brelstaff, G and Blake, A** 'Detecting specular reflections using lambertian constraints', *Proc. 2nd Int. Conf. Comput. Vision*, Tarpon Springs, FL, USA (December 1988) pp 297–302
12 **Klinker, G, Shafer, S and Kanade, T** 'Colour image analysis with an intrinsic reflection model', *Proc. 2nd Int. Conf. Comput. Vision*, Tampa, FL, USA (1988) pp 292–296
13 **Lee, H C** 'Method for computing the scene-illuminant chromaticity from specular high-lights', *J. Opt. Soc. Am. A.* Vol 3 No 10 (October 1986) pp 1694–1699
14 **Healey, G and Binford, T O** 'The role and use of colour in a general vision system', *Proc. DARPA IU Workshop*, USC, CA, USA (1987)
15 **Gray, D (ed.)**, *American Institute of Physics Handbook*, McGraw-Hill, NY, USA (1972)
16 **Born, M and Wolf, E** *Principles of Optics*, Pergamon Press, NY, USA (1959)

Illuminant color from shading

Hsien-Che Lee

Imaging Science Laboratory
Eastman Kodak Company, Rochester, NY 14650

ABSTRACT

The image of a uniform wall illuminated by a spotlight often gives a strong impression of the illuminant color. How can it be possible to know if it is a white wall illuminated by yellow light or a yellow wall illuminated by white light? If the wall is a Lambertian reflector, it would not be possible to tell the difference. However, in the real world, some amount of specular reflection is often present. An empirical reflection model describes light reflection from an inhomogeneous surface as a mixture of a specular (interface) component and a diffuse (body) component. Since the spatial scale over which the interface reflection changes significantly is much smaller than that of the body reflection, it can be shown that one can effectively exploit the scale difference to find a unique solution, which is often quite accurate. The method can also be generalized to compute the illuminant chromaticity for a nonuniform smooth surface.

1. INTRODUCTION

When we walk through many rooms inside a building with different lightings in various locations, we are always aware of the change of the illuminant color. Although there are few quantitative measurements available, our experiences tell us that the human visual system seems to be able to perceive qualitatively the scene illuminant quite well. Even when we have difficulty judging the "true" color of a piece of fabric under certain indoor lighting (an illustration of the breakdown of color constancy), we seldom fail to tell the color of the illuminant even without looking at it directly. This effortless perception of scene illumination does not depend on stereo or motion, as our experience in watching projection slides of still natural scenes can tell us. It does seem to depend on the presence of gradual shadings on the object surfaces. It is well known that color patches of uniform chromaticity and luminance do not give the perception of illumination. This is called the aperture (or film) mode of color perception. In order to give an impression of illumination, the surface shading has to have gradual variations. This is called the surface mode of color perception, that is, the perceived color seems to belong to the object surface [1].

Computing the scene illuminant color from a given color image is not a simple problem. The difficulty is that the recorded color image-irradiances are functions of the illuminant, the surface shape, and the surface reflectances. Without knowing any two of them, there are infinite possible solutions (a mathematically ill-posed problem [2]). Recent work suggests that the specular (interface reflection) component of surface reflection can be used to compute the illuminant chromaticity. This could be done either by finding the converging point of the chromaticity loci of different surfaces in the image (the chromaticity convergence method) [3] [4], or by separating the two reflection components based on certain signatures in the (R,G,B) color space [5] [6].

The general light reflection model of a uniform surface for a three color imaging system used by the methods mentioned above is as follows (see [7] for details). Assume that the incident radiance on a surface can be written as:

$$L(\theta_i, \phi_i, \lambda) = c(\lambda) L_0(\theta_i, \phi_i) \tag{1}$$

with $c(\lambda)$ normalized to one at its maximum, and θ_i and ϕ_i are the incident angles. Let $R_r(\lambda)$, $R_g(\lambda)$, and $R_b(\lambda)$ be the spectral responsivity functions of the color imaging system, and

$$\begin{aligned} L_r &= \int c(\lambda) R_r(\lambda) d\lambda \\ L_g &= \int c(\lambda) R_g(\lambda) d\lambda \\ L_b &= \int c(\lambda) R_b(\lambda) d\lambda; \end{aligned} \tag{2}$$

then

$$\begin{aligned} E_r(x,y) &= kL_r(\rho_r f(x,y) + \rho_s h(x,y)) \\ E_g(x,y) &= kL_g(\rho_g f(x,y) + \rho_s h(x,y)) \\ E_b(x,y) &= kL_b(\rho_b f(x,y) + \rho_s h(x,y)), \end{aligned} \tag{3}$$

where x and y are the spatial coordinates on the image plane, and r, g, and b indicate the conventional red, green, and blue channels or any other combinations of them. E_r, E_g, and E_b are the red, green, and blue image irradiances, ρ_r, ρ_g, and ρ_b are the diffuse (body) reflectance factors, ρ_s represents the specular (interface) reflectance factor, and $f(x,y)$ and $h(x,y)$ represent the factors dependent on the imaging geometry and surface shapes. This reflection model makes a strong assumption that the reflectance factor of the interface (specular) reflection component is independent of the imaging geometry as well as the spectral responsivity of each color channel. We will call this reflection model the neutral interface reflection (NIR) model. This assumption, of course, is not strictly true in practice (especially for homogeneous material, as discussed in detail by Cook and Torrance [8]), but there are experimental data showing that it is a reasonable approximation for many types of surface material [7]. Although the NIR model is a special case of the dichromatic reflection model, described by Shafer [9], which does not assume that the interface reflection is neutral, the NIR model is often used in real applications, because of its good approximation and simplicity.

2. THE CHROMATICITY DIAGRAM

In general, one cannot recover the absolute magnitudes of both the light-source intensity and the reflectance factor at the same time from the image irradiance signal alone, because if one raises the light-source intensity by a factor of 2 and reduces the reflectance factor by a factor of 2, the image irradiance will remain the same. Therefore, it is useful to define quantities that would specify the color of the light, independent of its intensity. For this purpose, the chromaticity of a given beam of light irradiating the image plane is defined by the following ratios [10]:

$$\begin{aligned} u &= E_r/(E_r + E_g + E_b), \\ v &= E_g/(E_r + E_g + E_b), \\ w &= E_b/(E_r + E_g + E_b). \end{aligned} \tag{4}$$

It should be pointed out that the standard notations for CIE chromaticity coordinates are x, y, and z. To avoid confusion with the spatial coordinates, we will use u, v, and w for the chromaticity coordinates. Since $u + v + w = 1$, one needs only u and v to specify the chromaticity coordinates of the light. It is also easy to see that they are independent of the light intensity. One useful property of the chromaticity coordinates is that if one additively mixes two lights to produce the third light, then the chromaticity coordinates of the third light are a linear combination of the chromaticity coordinates of the first two lights. Let (u_1, v_1) and (u_2, v_2) be the chromaticity coordinates of the first two lights, and let the "total irradiance" $E = E_r + E_g + E_b$ of the first light be E_1 and that of the second light be E_2; then it can be easily shown that the chromaticity coordinates of the mixed light are:

$$\begin{aligned} u_3 &= \left(\frac{E_1}{E_1 + E_2}\right)u_1 + \left(\frac{E_2}{E_1 + E_2}\right)u_2, \\ v_3 &= \left(\frac{E_1}{E_1 + E_2}\right)v_1 + \left(\frac{E_2}{E_1 + E_2}\right)v_2. \end{aligned} \tag{5}$$

The important consequence of this property for the NIR model is that *since the light reflected from a uniform surface is the additive mixture of the interface reflection component and the body reflection component, its chromaticity locus will fall on a straight-line segment connecting the chromaticity of the interface reflection and the chromaticity of the body reflection.* Therefore, each uniform surface will show up in the chromaticity space as a short line segment. If there are two uniform surfaces of different colors, the intersection point is the chromaticity locus of the illuminant. This is the essence of the chromaticity-convergence method for determining the illuminant color.

The theory behind the chromaticity-convergence method requires at least two surfaces of different colors be present in the scene in order to compute the illuminant chromaticity. The two surfaces have to reflect the illuminating light as mixtures of the specular and the diffuse components in various proportions.

However, if one looks at a uniform wall illuminated by a spotlight, one has a strong impression not only of the shading variations, but also of the color of the light source. How can it be possible for us to perceive that it is a white wall illuminated by yellow light, but not a yellow wall illuminated by white light? Since there is only one surface in the image, the theory of the chromaticity-convergence method tells us that the light-source chromaticity is constrained to be along a straight line in the chromaticity diagram, having infinitely many possible solutions. Some thought will lead us to the following solution: The chromaticity-convergence method needs two surfaces because it does not assume any knowledge about surface shape, lighting geometry, or physical characteristics other than those explicitly expressed in the NIR model. For example, the specular reflection is more angle dependent than the diffuse reflection. If our visual system is aware of the difference, it may be able to compute the solution uniquely. As we will show in the next section, this is indeed possible.

3. ILLUMINANT COLOR FROM SHADING: THE SMOOTHNESS CONSTRAINT

If we inspect the image irradiance equations 3 for a uniform surface, we will discover that by subtracting a properly scaled green irradiance signal from the red irradiance signal we can completely cancel spatially either the interface-reflection component or the body-reflection component. For example,

$$E_r(x,y) - (\frac{L_r}{L_g})E_g(x,y) = kL_r(\rho_r - \rho_g)f(x,y) \tag{6}$$

$$E_r(x,y) - (\frac{L_r\rho_r}{L_g\rho_g})E_g(x,y) = kL_r\rho_s(1 - \frac{\rho_r}{\rho_g})h(x,y). \tag{7}$$

The important implication is that if we can find the proper factors, we can completely recover the shapes of the two reflection components, and the proper factors happen to determine the chromaticity of the illuminant. It is also true that the difference signal need not be from the red and the green signals. Any other combination of the three image-irradiance signals will also serve well. If we want the factor to come out as the chromaticity of the illuminant, we can define:

$$\begin{aligned} E(x,y) &= E_r(x,y) + E_g(x,y) + E_b(x,y) \\ &= kL(\rho f(x,y) + \rho_s h(x,y)), \end{aligned} \tag{8}$$

where

$$\begin{aligned} L &= L_r + L_g + L_b, \\ \rho &= (L_r\rho_r + L_g\rho_g + L_b\rho_b)/L. \end{aligned} \tag{9}$$

Now, the difference signals become:

$$\begin{aligned} E_r(x,y) - (\frac{L_r}{L})E(x,y) &= kL_r(\rho_r - \rho)f(x,y) \\ E_g(x,y) - (\frac{L_g}{L})E(x,y) &= kL_g(\rho_g - \rho)f(x,y) \\ E_b(x,y) - (\frac{L_b}{L})E(x,y) &= kL_b(\rho_b - \rho)f(x,y), \end{aligned} \tag{10}$$

and the factors L_r/L, L_g/L, and L_b/L are precisely the chromaticity coordinates of the illuminant. Similar expressions can be written for the $h(x,y)$ component:

$$\begin{aligned} E_r(x,y) - (\frac{L_r\rho_r}{L\rho})E(x,y) &= kL_r\rho_s(1 - \frac{\rho_r}{\rho})h(x,y) \\ E_g(x,y) - (\frac{L_g\rho_g}{L\rho})E(x,y) &= kL_g\rho_s(1 - \frac{\rho_g}{\rho})h(x,y) \\ E_b(x,y) - (\frac{L_b\rho_b}{L\rho})E(x,y) &= kL_b\rho_s(1 - \frac{\rho_b}{\rho})h(x,y). \end{aligned} \tag{11}$$

Now we know that we can recover the shape of one of the reflection components by computing $E_r - s_r E$, if we can select s_r "properly." The remaining question is how to define what we mean by "properly." If we have some prior knowledge about the functions $f(x,y)$ or $h(x,y)$, then we can continue changing the factor s_r until the resulting difference signal $E_r - s_r E$ behaves the way we know it should. The knowledge need not be precise or even quantitative. It serves only as a criterion for selecting the right s_r. One obvious choice for the needed constraint is to require that the reflection component $f(x,y)$ be the smoothest function among all the possible alternatives in the family of functions generated by $E_r - s_r E$. The choice does have a good physical basis, because when the two reflection components, each having its own shape, are mixed together, the combined function is not as smooth as the smoother component alone. This is true for the following two reasons:

1. The spatial scale at which the irradiance change occurs is, in general, much smaller for the specular (interface) component than for the diffuse (body) component. The diffuse component as modeled by a Lambertian surface varies as a function of cosine, while the specular component [11] is typically modeled as the cosine function raised to the 20th (or more) power. Therefore, of the two reflection components, the diffuse component $f(x,y)$ is almost always the smoother one.

2. The peak of the diffuse component rarely occurs at the same place as that of the specular component, because the former is at the place where the surface normal is pointing to to the light source, whereas the latter occurs at the place where the surface normal is approximately the bisector of the angle between the source vector and the viewing vector. Any slight mixture of the specular component $h(x,y)$ with the diffuse component $f(x,y)$ will create an extra "bump" in the irradiance signal. Therefore, if the factor s_r is not selected properly, the difference signal $E_r - s_r E$ will not be as smooth as when it is.

One possible violation of the first condition is when the light source is very close to the illuminated surface, and therefore creates a sharply changing $f(x,y)$ because of the inverse-square fall-off of the light intensity, and because of the large change of the incident angle within a short distance. Also, a surface with large curvatures has a similar effect. Another possibility is a microscopically (in terms of the resolution of the imaging system) very rough surface, making the specular component very nondirectional. Violation of the second condition can happen under the single-light-source assumption only when the light source is located on the optical axis of the imaging system, a physically unlikely situation. With multiple light sources and mutual illumination among surfaces, the second condition can be violated, but not frequently. In most practical cases, the two types of violation are rare, and the smoothness constraint should give a fairly reasonable answer.

It is important to understand that the smoothness constraint here refers to the assumption that the body reflection function, $f(x,y)$, is much smoother than the interface reflection, $h(x,y)$. It does not imply that the underlying surface has to be macroscopically smooth. Because the relations between the surface shape and the shape of the image irradiance can be very complicated, smoothness in the shape of the image-irradiance signal does not imply smoothness in the surface shape, and vice versa.

It seems that equations 10 and 11 would give two "good" solutions for selecting the "proper" s_r. However, as a consequence of the smoothness constraint, the solution in equation 10 is chosen most of the time, because the body reflection component $f(x,y)$ is almost always smoother than the interface reflection component $h(x,y)$.

The next question is how to define smoothness. For computational reasons, we choose the Laplacian operator for its symmetry. It is very likely that another choice of measure of smoothness could give equally good or better results. We can now put the problem in the following mathematical form: select s_r and s_g such that the following are minimized:

$$\int [\nabla^2(E_r(x,y) - s_r E(x,y))]^2 dx dy,$$

$$\int [\nabla^2(E_g(x,y) - s_g E(x,y))]^2 dx dy. \tag{12}$$

The solution turns out to be quite simple:

$$s_r = \frac{\int \nabla^2 E_r(x,y) \nabla^2 E(x,y) dx dy}{\int \nabla^2 E(x,y) \nabla^2 E(x,y) dx dy}$$

$$s_g = \frac{\int \nabla^2 E_g(x,y) \nabla^2 E(x,y) dx dy}{\int \nabla^2 E(x,y) \nabla^2 E(x,y) dx dy}. \tag{13}$$

The solution given by 13 has the following nice property. As we have seen in Section 2, the chromaticity of the light reflected from a uniform surface falls on a straight line pointing toward the illuminant chromaticity. Let the straight line equation be: $au + bv = c$, i.e.,

$$a(\frac{E_r}{E}) + b(\frac{E_g}{E}) = c, \tag{14}$$

or

$$a(E_r) + b(E_g) = c(E). \tag{15}$$

It can be shown that the solution given by 13 is also on the same line, and therefore it satisfies the (only) constraint that can be derived from physics without making assumptions about the surface shape and imaging geometry. That means the solution is guaranteed to be a feasible solution. This is true because the solution in 13 is still a linear combination of the irradiance signals. The proof is as follows:

$$\begin{aligned} a(s_r) + b(s_g) &= a(\frac{\int \nabla^2 E_r \nabla^2 E dx dy}{\int \nabla^2 E \nabla^2 E dx dy}) + b(\frac{\int \nabla^2 E_g \nabla^2 E dx dy}{\int \nabla^2 E \nabla^2 E dx dy}) \\ &= \frac{\int \nabla^2 (aE_r + bE_g) \nabla^2 E dx dy}{\int \nabla^2 E \nabla^2 E dx dy} \\ &= \frac{\int \nabla^2 (cE) \nabla^2 E dx dy}{\int \nabla^2 E \nabla^2 E dx dy} \\ &= c. \end{aligned}$$

If we expand the solution 13 in terms of the function $f(x,y)$ and $h(x,y)$, we get the following expressions:

$$\begin{aligned} s_r &= (\frac{L_r}{L}) \frac{\int (\rho_s^2 (\nabla^2 h)^2 + \rho_r \rho (\nabla^2 f)^2 + \rho_s(\rho_r + \rho) \nabla^2 h \nabla^2 f) dx dy}{\int (\rho_s^2 (\nabla^2 h)^2 + \rho^2 (\nabla^2 f)^2 + 2\rho_s \rho \nabla^2 h \nabla^2 f) dx dy}, \\ s_g &= (\frac{L_g}{L}) \frac{\int (\rho_s^2 (\nabla^2 h)^2 + \rho_g \rho (\nabla^2 f)^2 + \rho_s(\rho_g + \rho) \nabla^2 h \nabla^2 f) dx dy}{\int (\rho_s^2 (\nabla^2 h)^2 + \rho^2 (\nabla^2 f)^2 + 2\rho_s \rho \nabla^2 h \nabla^2 f) dx dy}. \end{aligned} \tag{16}$$

They show that the chromaticity estimation is accurate only when the term $\rho_s^2(\nabla^2 h(x,y))^2$ is much larger than the sum of the other terms, and this suggests methods to improve the accuracy of the estimate. For example, we can give more weight to the regions of large irradiance because the specular-reflection region is often brighter than the neighboring regions. Another way is to compute the rate of chromaticity change at each pixel and use it as a weighting function, because only when the specular reflection is significant in magnitude relative to the diffuse reflection would it cause a noticeable chromaticity change.

4. ESTIMATING THE ILLUMINANT CHROMATICITY

4.1 From the shading of a uniform surface

To test the estimator on two-dimensional images, I generated images of a sphere whose center is 395cm away ($z = -395$) from the image plane ($z = 0$). The radius of the sphere is 120cm. The z-axis is the optical axis. The focal point is at (x=0, y=0, z=5cm). The point light-source is located at x = 300cm, y = 300cm, and z = 400cm. The specular reflectance-factor is 2.0, and it is modeled as the cosine raised to the 40th power. In this experiment, the illuminant chromaticity is changed from image to image, and the estimator is applied to each image to estimate the illuminant color. As we discussed in the last section, there are several ways to improve the estimation results by selectively weighting each pixel differently, i.e., by selecting s_r and s_g such that the following are minimized:

$$\begin{aligned} &\int (W(x,y))[\nabla^2(E_r(x,y) - s_r E(x,y))]^2 dx dy, \\ &\int (W(x,y))[\nabla^2(E_g(x,y) - s_g E(x,y))]^2 dx dy. \end{aligned} \tag{17}$$

The solution becomes:

$$s_r = \frac{\int W(x,y)\nabla^2 E_r(x,y)\nabla^2 E(x,y)dxdy}{\int W(x,y)\nabla^2 E(x,y)\nabla^2 E(x,y)dxdy}$$

$$s_g = \frac{\int W(x,y)\nabla^2 E_g(x,y)\nabla^2 E(x,y)dxdy}{\int W(x,y)\nabla^2 E(x,y)\nabla^2 E(x,y)dxdy} \quad (18)$$

It can easily be shown that the modified solution still satisfies the chromaticity collinearity constraint. So the improvement incurs some cost for computation, but does not sacrifice the original nice property. Since the highlight is always brighter than its surrounds, a simple weighting scheme is to make $W(x,y)$ a function of the image irradiance. The following table shows how the estimator with a simple weighting function $W(x,y) = (E(x,y))^4$ performs on the set of sphere images.

image	ρ_r	ρ_g	ρ_b	true illuminant (u,v)	estimated illuminant (u,v)
w000	1.0	0.4	0.6	0.5027 0.3995	0.5037 0.3985
w001	1.0	0.4	0.6	0.2631 0.4245	0.2638 0.4238
w002	0.4	0.6	1.0	0.5027 0.3995	0.5022 0.3996
w003	1.0	0.4	0.6	0.1566 0.3586	0.1570 0.3581

4.2 From the shading of a nonuniform surface

Extension of the above illuminant estimator to a nonuniform surface is conceptually straightforward. If we examine equations 6 and 7, we find that the derivation of the estimator based on the smoothness constraint still applies to the case of a nonuniform surface, provided that we can detect and exclude, from our computation of Laplacians, the edge pixels between regions of different body relectance factors. Although it is well known that edge detection itself is difficult and sensitive to noise, the situation here is not as bad, because we can afford to exclude the false edges caused by noise from our computation without losing much of the essential information for illuminant color estimation.

If the nonuniform surface consists of two differently colored uniform regions i and j, then

$$\begin{aligned} s_r &= \frac{\int_{i,j} \nabla^2 E_r \nabla^2 E dxdy}{\int_{i,j} \nabla^2 E \nabla^2 E dxdy} \\ &= (\alpha_i)\frac{\int_i \nabla^2 E_r \nabla^2 E dxdy}{\int_i \nabla^2 E \nabla^2 E dxdy} + (\alpha_j)\frac{\int_j \nabla^2 E_r \nabla^2 E dxdy}{\int_j \nabla^2 E \nabla^2 E dxdy}, \end{aligned} \quad (19)$$

where,

$$\alpha_i = \frac{\int_i \nabla^2 E \nabla^2 E dxdy}{\int_{i,j} \nabla^2 E \nabla^2 E dxdy}$$

$$\alpha_j = \frac{\int_j \nabla^2 E \nabla^2 E dxdy}{\int_{i,j} \nabla^2 E \nabla^2 E dxdy}.$$

The illuminant estimator extended this way therefore has a very good physical interpretation: it is the weighted average of the estimates made by each uniform region on the nonuniform surface. This is true even if the uniformly colored regions are not connected together, because the relation between f and h should hold in each region. For example, on a tiled surface, all the areas covered by the same-colored tiles are combined together to make one estimate, which is in turn combined with estimates from other colors. Therefore, in principle, it does not matter how many regions any color is broken into. Of course, in practice, the boundary pixels have to be excluded, and therefore there is additional loss of information for every breakup.

To generate images of nonuniform surface, random rectangles of different sizes and colors are mapped to the surface of a sphere. The center of the sphere is placed again at 395cm away from the image plane. The position of the light source and the distribution of the colors of the random rectangles are changed to generate images with different shading and color distributions. A simple gradient edge detector with nonmaximum suppression is applied to the individual red, green, and blue records. Any pixels detected as edges in one of the color records are declared as color edges. To ensure that all region boundaries are excluded properly, pixels that are less than two pixels away from the detected color edges are also excluded from the Laplacian computation. The results, after applying the simple weighting function $W(x,y) = (E(x,y))^4$, are as follows:

image	true illuminant (u,v)	estimated illuminant (u,v)	gray-world estimate (u,v)
r000	0.5027 0.3995	0.5032 0.3992	0.4849 0.4176
r001	0.2631 0.4245	0.2634 0.4252	0.2552 0.4192
r002	0.5027 0.3995	0.5001 0.4019	0.4956 0.4009
r003	0.1566 0.3586	0.1571 0.3597	0.1495 0.3718

image	true illuminant (u,v)	estimated illuminant (u,v)	gray-world estimate (u,v)
f000	0.5027 0.3995	0.5040 0.3986	0.6700 0.2707
f001	0.2631 0.4245	0.2651 0.4219	0.4550 0.2815
f002	0.5027 0.3995	0.5043 0.3978	0.7096 0.2218
f003	0.1566 0.3586	0.1589 0.3580	0.2836 0.3250

image	true illuminant (u,v)	estimated illuminant (u,v)	gray-world estimate (u,v)
s000	0.5027 0.3995	0.5040 0.3986	0.6761 0.2622
s001	0.2631 0.4245	0.2658 0.4231	0.4435 0.3141
s002	0.5027 0.3995	0.5045 0.3981	0.6918 0.2453
s003	0.1566 0.3586	0.1584 0.3589	0.2830 0.3122

If we move the light source to x = 300cm, y = 300cm, and z = 40000cm, we have the following results:

image	true illuminant (u,v)	estimated illuminant (u,v)	gray-world estimate (u,v)
g000	0.5027 0.3995	0.5039 0.3987	0.6799 0.2576
g001	0.2631 0.4245	0.2659 0.4219	0.4380 0.3294
g002	0.5027 0.3995	0.5046 0.3977	0.6830 0.2566
g003	0.1566 0.3586	0.1581 0.3591	0.2840 0.3059

As can be seen, all the estimates are consistently very accurate. They are better than the gray-world estimates even when the images are generated by gray-world statistics (images r000 to r003).

It should be pointed out that the Laplacian signal of the interface reflection component $h(x,y)$ is relatively small compared with that of a reflectance edge. If any reflectance-edge pixel is not excluded from the estimation, the result can be very wrong. For example, the gradient edge detector sometimes computes the location of the true edge incorrectly. If we have not had excluded pixels that are neighbors of the detected edges, the estimation results will be useless for some images, as can be seen in the following table. The columns marked as 1-pixel and 2-pixel are the estimation results from excluding pixels, which are less than or equal to one- and two-pixel distance away from the detected edges.

image	true (u,v)	edge alone (u,v)	1-pixel (u,v)	2-pixel (u,v)
f000	0.5027 0.3995	0.5044 0.3982	0.5041 0.3985	0.5040 0.3986
f001	0.2631 0.4245	0.4059 0.3412	0.2653 0.4216	0.2651 0.4219
f002	0.5027 0.3995	0.5487 0.3622	0.5044 0.3977	0.5043 0.3978
f003	0.1566 0.3586	0.1621 0.3537	0.1598 0.3569	0.1589 0.3580

5. CONCLUDING REMARKS

Light-reflection components usually have different spatial scales. The interface (specular) reflection component typically has a much finer scale than the body (diffuse) component. Because of this scale difference, one can effectively separate them by imposing a smoothness constraint to extract the shape of the diffuse component. In doing so, an estimate of the illuminant color can be obtained. This illuminant estimator happens to satisfy the chromaticity collinearity constraint automatically, which makes it particularly attractive for applications.

When it is generalized to a nonuniform surface, it essentially computes the weighted average of the individual estimates made by each of the uniformly colored regions, provided that one can eliminate the edge pixels.

The estimator has been tested on computer-generated images with reasonably good performance. Unlike the gray-world estimator, it is quite immune to heavy bias in the reflectance statistics of the scene. However, in order to use it for a nonuniform surface, explicit exclusion of edges is necessary. Fortunately, one need not worry about false edges because they can be excluded from the estimator without much loss of information. On the other hand, one has to worry about undetected edges because their inclusion in computing the estimator can introduce a large error in the estimation.

The estimation accuracy can be greatly improved by weighting each pixel with a function of its irradiance. Other types of weighting function and definition of smoothness can be used, but are not pursued in this work. One final question one might ask is: Can this estimator be applied to images with many surfaces ? The answer is yes if all the surfaces are spatially near each other in depth. For surfaces far away from each other, one has to distinguish the scale difference due to depth from the scale difference between the reflection components. Compared with the chromaticity-convergence method, the illuminant-color estimator developed here is simpler to implement. It works well for a single surface, and does not explicitly search for straight-line signatures in the chromaticity space. Furthermore, the computation is more local and therefore potentially can take better care of images with multiple light sources. However, the chromaticity-convergence method does not use the difference-in-spatial-scale assumption, and would not have any problem when several surfaces are at different depths. Combination of the two methods to compute a robust estimation of the scene-illuminant chromaticity is a subject of future research.

6. ACKNOWLEDGMENT

This work was done at the Artificial Intelligence Laboratory of the Massachusetts Institute of Technology, while I was on leave from the Imaging Science Laboratory/Kodak. I would like to thank Professor Tomaso Poggio of MIT, Mr. Terence Lund, and Mr. William Hutchinson of Kodak for making this possible. Anya Hurlbert, Tomaso Poggio, and James Little provided many good suggestions and comments on an earlier, more complete version of this paper, which was published as MIT AI Memo No. 1068, August 1988.

References

[1] *The Science of Color*, Committee on Colorimetry, Optical Society of America, Optical Society of America, Washington, D.C., 1963, 145-152.

[2] T. Poggio and C. Koch, "Ill-posed problems in early vision: from computational theory to analog networks, *Proceedings of Royal Society*, London, B, 226, 1985, pp. 303-323.

[3] H.-C. Lee, "Method for computing the scene-illuminant chromaticity from specular highlights," *Journal of the Optical Society of America*, A, 3, 10, October 1986, pp.1694-1699.

[4] M. D'Zmura and P. Lennie, "Mechanisms of color constancy", *Journal of the Optical Society of America*, A, 3, 10, October 1986, pp.1662-1672.

[5] R. Gershon, A.D. Jepson, and J.K. Tsotsos, "The use of color in highlight identification," *Proceedings of the 10th International Joint Conference on Artificial Intelligence*, Milan, Italy, August 1987, pp. 752-754.

[6] G.J. Klinker, S.A. Shafer, and T. Kanade, "Image segmentation and reflection analysis through color," *Proceedings of Image Understanding Workshop*, Cambridge, Massachusetts, April 1988, pp. 838-853.

[7] H.-C. Lee, E.J. Breneman, and C.P. Schulte, "Modeling light reflection for computer color vision," To appear in IEEE PAMI.

[8] R.L. Cook and K.E. Torrance, "A reflectance model for computer graphics," *ACM Transactions on Graphics*, 1, 1, January 1982, pp. 7-24.

Color Interreflection

Interreflection arises in nearly every scene. We sometimes model illumination as though each surface receives light from some identifiable (point or diffuse) light source, but in reality that model is not accurate. Each surface also receives light that is reflected off of other objects in the scene. This phenomenon is *interreflection*. In order to carry out accurate analysis of images, we need to recognize and account for interreflection. This has given rise to two branches of literature in physics-based vision – interreflection intensity analysis, which is presented in the section on "Shape Recovery from Interreflection" in the Shape Recovery volume of this series; and interreflection color analysis, which is described here.

In interreflection color analysis, we generally simplify the model by considering just two mutually reflecting surfaces at a time. Color interreflection is actually a very rich phenomenon because it is determined by the geometry of the surfaces and their arrangement as well as the colors of the illumination and of each of the two surfaces. So, the analysis of color interreflection can have several goals:

- to detect where interreflection occurs and "remove" its effects from the scene, allowing simpler analysis of the resulting data;
- to determine the geometry of the surfaces;
- to determine the spectral reflectance (color) of the surfaces.

Color interreflection can be modeled by extension of the dichromatic reflection model [1][2]. The dichromatic model states that each surface reflection has a "surface" component $\rho_s(\lambda)$ and a "body" component $\rho_b(\lambda)$. Under interreflection, if light bounces from surface A to surface B, then the illumination on B is determined by both the direct lighting $I(\lambda)$ and the light produced by both components of A's reflection, $I(\lambda)\rho_{sA}(\lambda)$ and $I(\lambda)\rho_{bA}(\lambda)$. These three colors all illuminate B, so the reflected light from B has six components:

- $I(\lambda)\rho_{sB}(\lambda)$ – light from the light source directly to B, reflecting by surface reflection from B ("surface reflection").
- $I(\lambda)\rho_{bB}(\lambda)$ – light from the light source to B, reflecting by body reflection from B ("body reflection").
- $I(\lambda)\rho_{sA}(\lambda)\rho_{sB}(\lambda)$ – light from the light source to A, reflecting by surface reflection from A onto B, then reflecting by surface reflection from B ("surface/surface reflection").
- $I(\lambda)\rho_{sA}(\lambda)\rho_{bB}(\lambda)$ – light source to surface reflection from A to body reflection from B ("surface/body reflection").
- $I(\lambda)\rho_{bA}(\lambda)\rho_{sB}(\lambda)$ – light source to body reflection from A to surface reflection from B ("body/surface reflection").
- $I(\lambda)\rho_{bA}(\lambda)\rho_{bB}(\lambda)$ – light source to body reflection from A to body reflection from B ("body/body reflection").

The first two components are the components of the Dichromatic Reflection Model for surface B; the last four components are the components of interreflection from A onto B. Each of these six components has an identifiable geometry-dependent behavior. Paper [2] in this collection is a conceptual paper that begins with a discussion of the Dichromatic Reflection Model, summarizes results (included elsewhere in this volume) of applying the dichromatic model to images, and develops the above discussion of six reflection components under interreflection. The paper presents no results of analyzing interreflection; it only points out the color-space ramifications of the above model. Thus, it is a proposal of a model but not a presentation of results.

The most striking point in that discussion comes from the observation that surface reflection is generally about the same color as direct illumination; thus, the above six interreflection components only display four distinct colors:

- color of the illumination: surface and surface/surface reflection;
- color of B: body and surface/body reflection;
- color of A: body/surface reflection;
- color of A × B: body/body reflection.

Although it is located in the "Color Image Segmentation" section of this volume, [1] is also very important in the literature of color interreflection analysis. In that paper, there is no attempt to model or analyze interreflection processes quantitatively. However, the paper does develop and utilize a color-space model of the appearance of interreflection in the image. In particular, hue segmentation is used to identify regions that are suspected to exhibit interreflection. As can be seen from the above list of four colors, hue segmentation of region B in the image can be expected to distinguish the latter two color components from the former two. However, it cannot distinguish surface/surface from surface reflection, nor surface/body from body reflection. Thus, the method is limited to the case that surface A has negligible surface reflection (i.e., is very matte).

No one has yet presented a comprehensive method for analyzing all the components of interreflection. Papers [3] and [4] in this collection do, however, present fairly detailed analysis of the how the colors of B and A × B can be used to estimate the underlying surface color $\rho_{bB}(\lambda)$. This analysis is similar to methods for color constancy, using interreflection as a reference condition instead of "true white." Because these methods do not treat colors other than B and A × B, they assume only body and body/body reflection. Thus, they assume that both A and B are completely matte. Note that interreflection in this situation will provide information only if A and B have different body colors, i.e., if they are distinct surfaces with different colors.

Paper [3] begins with a finite-dimensional reformulation of the color analysis problem, using a small number of spectral basis functions to model illumination, reflection, and sensing. The interreflection scenario then gives rise to a number of equations with a number of unknowns. By extracting an overall reflectance scale factor, the problem is reduced so that the number of color filters required is related simply to the number of dimensions in the approximations of illumination and reflection. In particular, with three color filters as in color TV images, both illumination and reflection can be recoved with three dimensions each.

The recovery method is based on the solution of a set of nine nonlinear equations; the authors reduce the calculation to a linear solution in a multi-stage algorithm. The paper also shows that the magnitude of the body/body reflection term can be given by "configuration factors" as in the *radiosity* method of heat-transfer calculations (also used in computer graphics).

In paper [4], a more appealing computational method is presented to determine body color and body/body reflection color based on the simultaneous analysis of both surfaces A and B. For surface B, the colors in the image will be a linear combination of the colors of B and A × B; thus, they will lie on a plane in the color space. (Note: This is *not* the same plane as that predicted by the dichromatic reflection model.) Likewise, the colors of surface A are a linear combination of colors A and A × B. Both of these planes contain the color vector A × B, so their intersection in color space yields color A × B directly.

– S.A.S.

Reference

[1] Bajcsy, R., Lee, S.W., and Leonardis, A. "Color Image Segmentation with Detection of Highlights and Local Illumination Induced by Inter-reflections." In *Proc. International Conference on Pattern Recognition*. Atlantic City, NJ. IEEE, 1990. Reprinted in the "Color Image Segmentation" section of this volume.

Papers Included in This Collection

[2] Shafer, S.A., Kanade, T., Klinker, G.J., and Novak, C.L. "Physics-Based Models for Early Vision by Machine." In *Proc. SPIE #1250 Perceiving, Measuring, and Using Color*. Santa Clara CA. SPIE, 1990. Pages 222-235.

[3] Funt, B.V., Drew, M.S., and Ho, J. "Color Constancy from Mutual Reflection." *Intl. J. of Computer Vision* **6**(1): 5-24, 1991.

[4] Funt, B.V., and Drew, M.S. "Color Space Analysis of Mutual Illumination." Simon Fraser University CSS/LCCR Report TR 91-03, 1991. To appear in *IEEE Trans. on Pattern Analysis and Machine Intelligence.*

Physics-Based Models for Early Vision by Machine

Steven A. Shafer, Takeo Kanade, Gudrun J. Klinker*, Carol L. Novak

Robotics Institute, Carnegie Mellon University, Pittsburgh PA 15213

*Cambridge Research Laboratory, Digital Equipment Corp., Cambridge MA 02139

Abstract: *Research in early (low-level) vision, both for machines and humans, has traditionally been based on the study of idealized images or image patches such as step edges, gratings, flat fields, and Mondrians. Real images, however, exhibit much richer and more complex structure, whose nature is determined by the physical and geometric properties of illumination, reflection, and imaging. By understanding these physical relationships, a new kind of early vision analysis is made possible. In this paper, we describe a progression of models of imaging physics that present a much more complex and realistic set of image relationships than are commonly assumed in early vision research. We begin with the Dichromatic Reflection Model, which describes how highlights and color are related in images of dielectrics such as plastic and painted surfaces. This gives rise to a mathematical relationship in color space to separate highlights from object color. Perceptions of shape, surface roughness/texture, and illumination color are readily derived from this analysis. We next show how this can be extended to images of several objects, by deriving local color variation relationships from the basic model. The resulting method for color image analysis has been successfully applied in machine vision experiments in our laboratory. Yet another extension is to account for inter-reflection among multiple objects. We have derived a simple model of color inter-reflection that accounts for the basic phenomena, and report on this model and how we are applying it. In general, the concept of illumination for vision should account for the entire "illumination environment", rather than being restricted to a single light source. This work shows that the basic physical relationships give rise to very structured image properties, which can be a more valid basis for early vision than the traditional idealized image patterns.*

1. Introduction

Vision systems are faced with the task of analyzing optical image data to produce a description of the objects in the world, their properties, and their relationships to each other. Researchers in vision have been successful at analyzing very specific elements of visual perception such as edge detection, color matching, or binocular stereopsis, but there are many gaps in the understanding. One of the largest of these gaps is the unknown relationship between elementary "2D" image features on the retina or sensor and the perception of "3D" scene properties such as surface color, texture, and shape. While this is true of research both in analyzing human vision and in building machine vision systems, this paper addresses primarily machine vision.

The reason for the difficulties in the traditional research in this area is illustrated in Figure 1(a). Traditional research has proceeded by developing "feature detectors" for image features such as edges, color differences, and so on, with no consideration whatsoever for the structure of optical energy in the

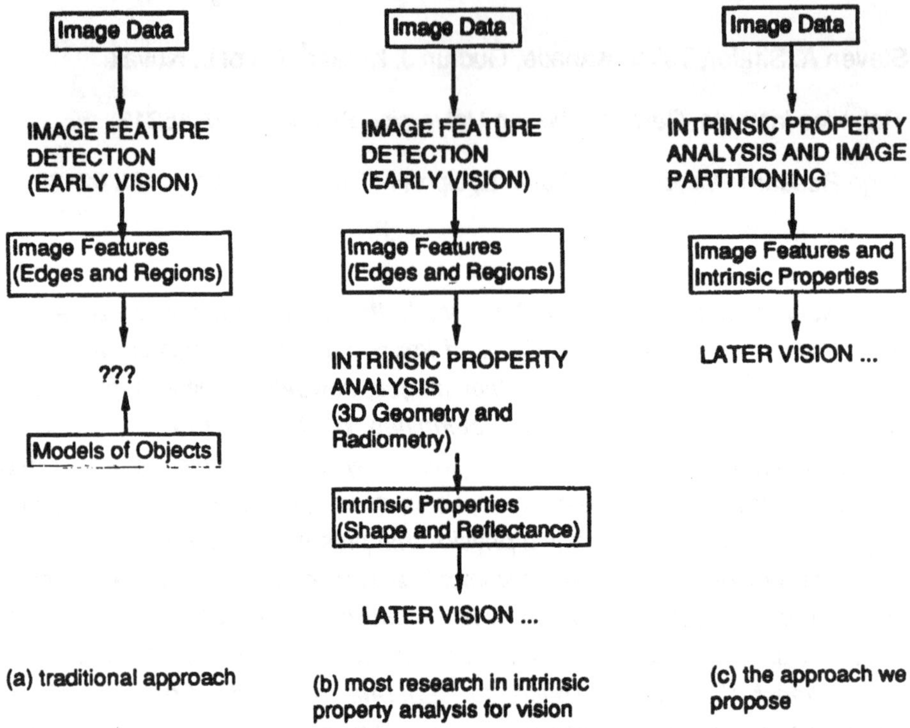

Figure 1: Relationship Between 2D and 3D Visual Features

real 3D world. The presumption has been that these image features are only to provide a starting point for later analysis, so they need not correspond very perfectly to meaningful 3D entities. Detecting these features is called "early vision". In reality, phenomena such as shadows, highlights, uneven lighting, surface roughness and texture, and so on cause many tiny variations in the image data. As a result, "edge images" and "region segmentations" in machine vision are typically very noisy, full of both false alarms and missed features. Thus, this early processing creates many problems for later visual analysis.

(While the authors are not scholars in the area of human vision, we sometimes observe a similar situation in that literature: There are numerous studies of how spatial frequencies, color, and other image properties are sensed, but relatively few ideas about how those features are related to the perception of 3D shape and surface properties. Visual fields used for experimentation are typically sine-gratings, split-color fields, and Mondrians, rather than complex images of real objects with varying shape, material type and texture, etc.)

In the 1970s, machine vision researchers proposed to address this problem by postulating a level of representation of world features based on generic laws of physics and geometry, but still independent of any specific object models (Figure 1(b)) [14, 19]. This level of features, sometimes called "intrinsic properties", includes 3D surface orientation, 3D edge orientation and angles, surface material type, surface roughness and texture, etc. Research in this area, while sporadic for many years, is now quite strong within the community. However, this research still has not produced robust vision systems that work on complex images. The reason is that underlying all of this research is still the idea that physics and geometry are not applied to the raw image data, but rather that an early vision stage has *preceded* the intrinsic property analysis to yield a meaningful set of edge features, or a meaningful partitioning of the image into uniform regions for later analysis [4]. For example, shape-from-shading generally assumes that the image has already been partitioned into regions over which a single reflectance function can be applied [7]; skewed symmetry and other 3D geometric analysis assumes that surface boundary edges have already been detected [9]. Thus, analysis of intrinsic properties has been viewed as a second stage of vision that follows after image feature detection.

That view presents grave difficulties when the image features do not have an immediate interpretation in terms of intrinsic scene properties, which is true in nearly all real images of complex scenes. For this reason, we propose a new way to think of early vision and intrinsic property analysis, as shown in Figure 1(c): We propose that the raw image data itself be analyzed in terms of intrinsic models; and that this be done concurrently with feature detection, as an integral part of the same process. Thus, rather than partitioning the image on the basis of color similarity and assuming that these regions correspond to surfaces, we propose to begin by hypothesizing object material types and color features in the 3D world, and use the boundaries of these hypothesized scene features to define the partitioning of the image into regions. In this paper, we illustrate this approach for a very simple case: analysis of images of plastic objects with highlights. We first present a basic theory and show how an integral analysis-and-segmentation system can be based on that theory. Of course, analysis of more complex scenes demands that many such theories be combined somehow. While we don't yet know exactly how that can be done, we will present our efforts to date in developing a simple case of multiple interacting theories: analysis of interreflection among multiple objects. We believe that this overall approach of early vision analysis using intrinsic models may have significant advantages over the traditional approaches to early vision based on edges and region segmentation.

2. The Dichromatic Reflection Model

For several years, we have been studying the analysis of one particular intrinsic property for vision: the intensity of the highlight at each point in an image of plastic objects [18, 11, 12, 10]. This is an interesting property because highlights have been utterly confounding for traditional approaches to early vision by machine, yet it is based on a very simple color model. This property is also interesting because if it can be successfully analyzed in early vision, it paves the way for the analysis of other more subtle properties such as surface roughness.

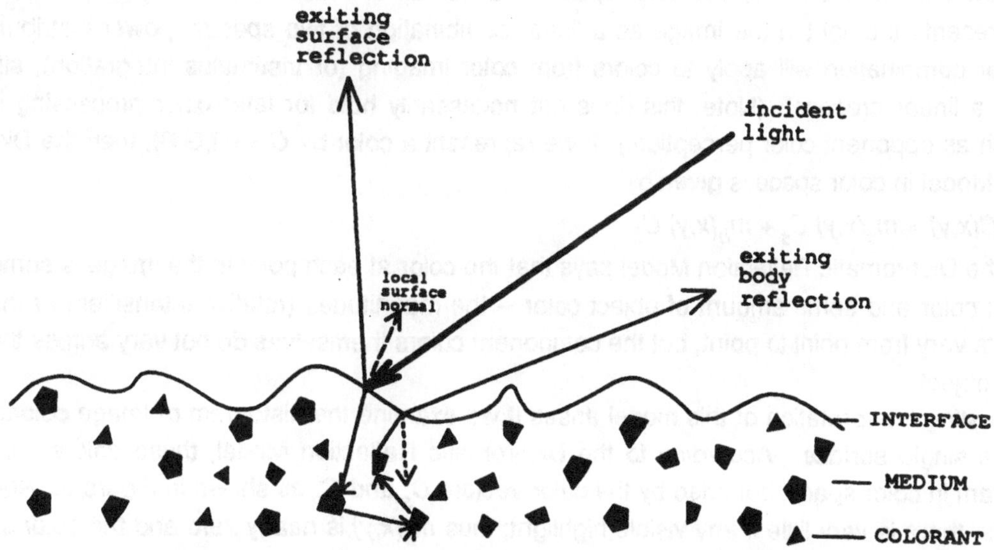

Figure 2: Reflection from an Inhomogeneous Dielectric

Figure 2 sketches the primary reflection processes in an inhomogeneous dielectric material such as plastic or paint [18]. When light from the illumination source strikes the surface, it strikes the interface with the transparent medium. Some of the light is reflected immediately, according to Fresnel's laws. This light, which we call *surface reflection*, has a spectral distribution that is typically about the same as the illuminant (depending on the variation of the index of refraction over the visible spectrum). It is also highly directional -- if the surface is smooth, the surface reflection will very specular; if the surface is rough, it will be somewhat diffused. Surface reflection is the process that creates highlights.

The light that is not reflected at the interface penetrates into the bulk of the material, and there it begins to scatter off of the pigment or other colorant particles in the material; eventually, it may be absorbed, transmitted through the material, or reflected back across the interface into the air. This last process is what we call *body reflection*. Light from the body reflection has a spectral distribution determined by the object colorant as well as the illuminant, and it is fairly random in direction (typically approximately Lambertian). Body reflection is responsible for "object color", the characteristic color of a specific object.

In an image of such material, for example ordinary colored plastic, the light reflected from each point on the surface is composed of both types of reflection, added together. If we symbolize the light energy incident on the sensor as $L(x,y,\lambda)$, the reflected energy from surface reflection as $L_s(x,y,\lambda)$, and the energy from body reflection as $L_b(x,y,\lambda)$, we have:

$$L(x,y,\lambda) = L_s(x,y,\lambda) + L_b(x,y,\lambda)$$

Here, x and y represent coordinates in the image and the corresponding points on the object's surface; λ is the wavelength of light. In most circumstances, the spectral distribitions of the surface and body reflection are approximately constant at all points on the surface of a single object; thus, L_s and L_b can be factored into *magnitudes m* and *spectral reflectances S* as follows:

$$L(x,y,\lambda) = m_s(x,y)\ S_s(\lambda) + m_b(x,y)\ S_b(\lambda)$$

This is what we call the *Dichromatic Reflection Model.*

For application to vision, the Dichromatic Reflection Model can be projected from the domain of spectral power distributions into the color space of a human or machine vision system. The equation above represents the light in the image as a linear combination of two spectral power distributions; this same linear combination will apply to colors from color imaging (or tristimulus integration), since color sensing is a linear process. (Note: this does not necessarily hold for later color processing in human vision, such as opponent color perception.) If we represent a color by $C = [R,G,B]$, then the Dichromatic Reflection Model in color space is given by:

$$C(x,y) = m_s(x,y)\ C_s + m_b(x,y)\ C_b$$

In words, the Dichromatic Reflection Model says that the color at each point in the image is some amount of highlight color and some amount of object color -- the magnitudes (relative intensities) of these color components vary from point to point, but the component colors themselves do not vary across the surface of a single object.

An interesting interpretation of this model arises if we examine the histogram of image colors from the points on a single surface. According to the Dichromatic Reflection Model, these colors must form a parallelogram in color space, bounded by the color vectors C_s and C_b as shown in Figure 3. Note that for most points, there is very little if any visible highlight; thus $m_s(x,y)$ is nearly zero and the color simply lies somewhere along the vector C_b. We call these "matte points". All the points with significant amounts of highlight come from a small area on the object surface and thus all have nearly the same amount of body

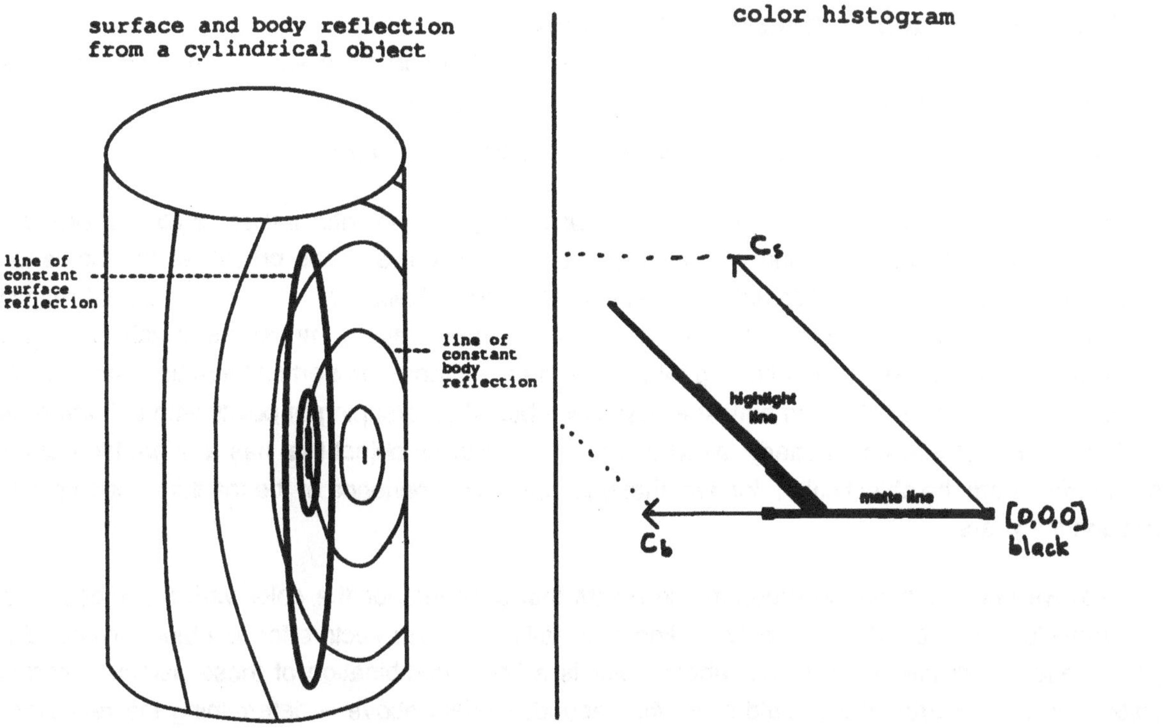

Figure 3: Color Space Histogram Forms a Parallelogram

reflection $m_b(x,y)$; they form a sort of spike in the color space whose direction is parallel to the illuminant color C_s.

Using this model, the color at any individual point in the image can be decomposed into its two reflection components by projecting it onto the two bounding color vectors of the object corresponding to that point. The magnitudes of these projections tell the relative intensity of the surface reflection and body reflection at that point in the image (Figure 8). This kind of analysis is very useful for vision because it allows us to decompose the image into an image of body reflection with no highlights, useful for shape estimation, and another image of just highlights, which reveals a great deal about surface roughness and structure.

The algorithm for this analysis can be summarized as follows [11]:

1. Identify a region of the image that corresponds to a single surface.
2. Suppose that this surface is an inhomogeneous dielectric material.
3. Histogram the image colors within this region.
4. In the color histogram, fit the vectors for C_s, the surface reflection color, and C_b, the body reflection color.
5. At each point in the region, project its color $C(x,y)$ onto these two reflection color vectors to determine $m_s(x,y)$ and $m_b(x,y)$, the magnitudes of surface and body reflection at that point.

We have tried this analysis on several images and obtained results as shown in Figure 8, which is described in more detail below.

3. A New Approach to Image Segmentation

The algorithm just presented has two chief difficulties (aside from the presumption of material type):

- How can we identify the extent of the image region that corresponds to one single surface?
- How can we fit the reflection color vectors to the color histogram data?

The first issue, how to determine the extent of a single meaningful image region, is precisely the problem alluded to in the Introduction. As presented so far, our analysis presumes that somehow, the image has already been partitioned into meaningful regions *before* the intrinsic model of reflection is applied to separate the reflection components. This would be virtually impossible to achieve in practice: On what basis should such a partitioning take place, before intrinsic properties have been determined? In the past, color segmentation has almost always been based on grouping of points with uniform or nearly-uniform color; but that would utterly fail when presented with an object that has a bright highlight whose color differs from the object color. Indeed, that has been the experience of the machine vision community through the years.

What we need is a way to group image points that accounts for the color variation modeled by the Dichromatic Reflection Model. If only we knew the reflection color vectors for an object, we could simply group together all the image points whose color is a linear combination of those particular component colors. In other words, if we could solve the second problem above -- determining the reflection color vectors -- then we could solve the first problem -- measuring the extent of the image region corresponding to that object. In fact, all we really need is a way to examine a small neighborhood of the image and make a good guess about the reflection color vectors. If we could do that, then we could make such a hypothesis for each neighborhood and then group together those neighborhoods with consistent hypothesized reflection colors.

We have developed a method for segmentation that does just such an analysis [12, 10]. The method is based on the observation that neighborhoods in different parts of the object -- the matte areas, the highlight areas, the edge -- will have different numbers of non-zero color components. Thus, the dimensionality of the color histogram for each neighborhood differs as shown in Figure 4. For points on the matte area of a surface, the color at every point in the neighborhood is a scaled version of the body reflection color for the surface, C_b, because there is no appreciable highlight. If the neighborhood is entirely within such a matte region, all the colors will lie on the line containing C_b, and the histogram will be one-dimensional. If a neighborhood contained some or all points from the highlight region, the colors would have appreciable amounts of both object and highlight colors, and the distribution in color space would be two-dimensional. Points along the edge of the object may have neighborhoods that overlap other nearby objects, giving more different color components than are consistent with the Dichromatic Reflection Model -- the histograms will generally be three-dimensional (or more if more sensor bands are used than three). Thus, the dimensionality of the color variation within each local neighborhood of the image forms a powerful feature for grouping image regions.

Real image data is not so perfect, but rather is subject to noise. To deal with this, we estimate the dimensionality of a set of colors in color space by performing an eigenvector analysis (K-L transform): we use the number of eigenvalues significantly above the noise level (e.g. 4σ) as our estimate of the dimensionality of the distribution. We can assign each neighborhood to a class based on the dimensionality of its color distribution: 0=point, 1=linear, 2=planar, 3=volume. The idea of the algorithm for segmentation is then to classify each neighborhood, and group together those that have consistent

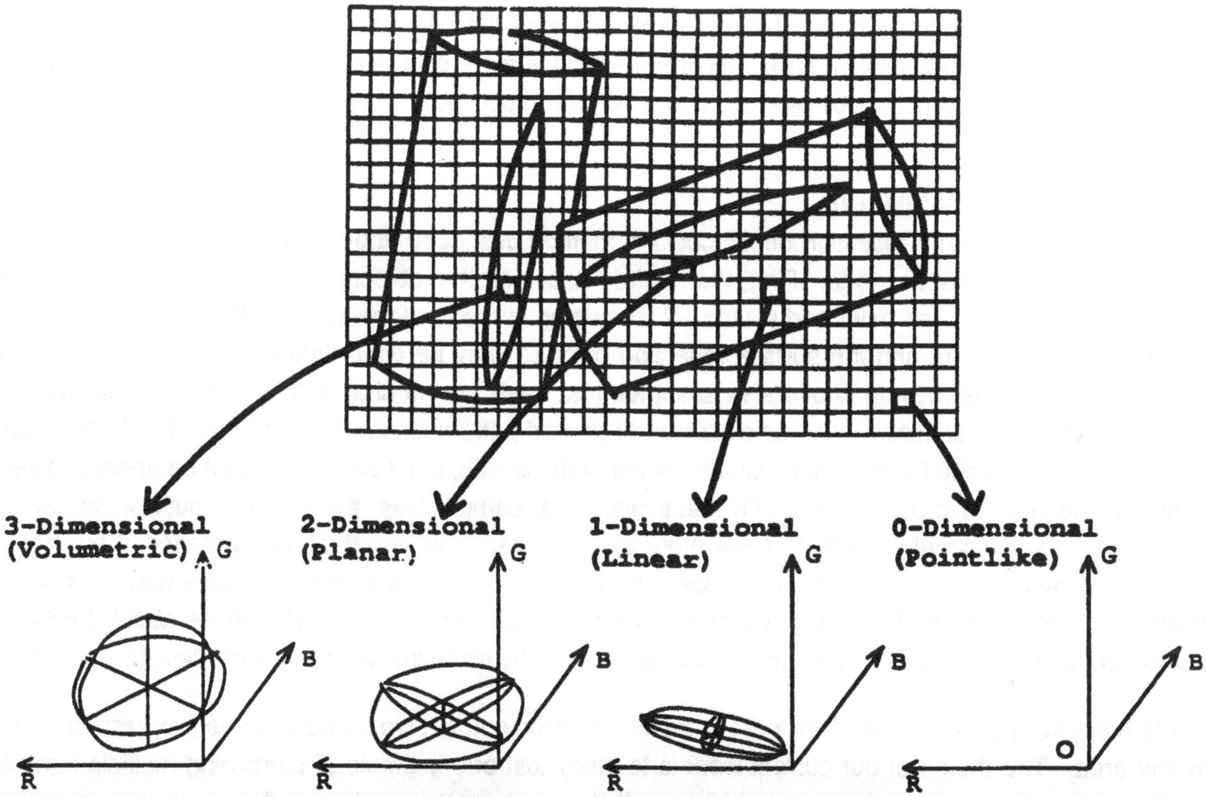

Figure 4: Dimensionality of Neighborhood Color Histograms

classifications. Note that "consistent" does not necessarily mean "the same classification". For example, two adjacent neighborhoods may both be linear, but come from different objects and thus lie on different lines in color space. Consistency must include a check that the "line" in the histogram is the same "line" for all neighborhoods to be grouped into a single region. Similarly, an orange neighborhood from a matte area on an orange object will be "linear" along an orange line in color space; an adjacent neighborhood from the highlight on that object will be "planar" on an orange-and-white plane. Such hypotheses are perfectly consistent for grouping. Thus, the criterion for grouping is consistency with a single set of reflection color vectors C_s and C_b according to the Dichromatic Reflection Model. In other words, image regions are defined as sets of points that are consistent with the same instantiation of the Dichromatic Reflection Model. Note how different this is than the traditional segmentation criterion: grouping points with identical or nearly identical color in the image [16, 5].

Figure 8 shows the results of applying this method to a color image of several objects. The upper left corner shows the original image; the upper right shows the regions that result from segmentation; the lower left and right show the body and surface reflection components, respectively Note that our method hypothesizes reflection color vectors as part of the segmentation process, thus the projection into reflection components is a simple by-product of segmenting the image. The distinction between image segmentation and intrinsic property analysis has been completely eliminated, as these are now integral activities. Thus, this approach fulfils our model described in the Introduction, shown in Figure 1(c): it is a method for finding image features based on the analysis of intrinsic properties of the world. Furthermore, this analysis is based on a local image property: dimensionality of the local color distribution. Such a grouping criterion would be very difficult to conceive by thinking only of traditional image features such as edges, color difference metrics, or spatial frequency analysis.

For comparison, Figure 9 shows the results of analyzing the same image using a traditional method (histogram-based region splitting) [16]. Our new method is more accurate and useful, produces valuable reflection component data as a by-product, and is actually much faster [10].

4. Modeling Interreflection

The segmentation and reflection decomposition method described above is promising, yet it does not address a very realistic scenario. First, the method assumes that objects are made of inhomogeneous dielectric material such as paint and plastic. If something is made of metal, there is effectively no "body reflection" (i.e. $m_b = 0$), and the surface reflection color C_s may be quite different from the color of the illuminant [6, 3]. The method above also assumes that there are no shadows, a single light source, no noticeable colorant variations (such as a painted or printed pattern, or wood grain), etc. The Dichromatic Reflection Model is fine for the limited situations that it deals with, but it cannot by itself be applied to very complex scenes. In order to deal with these additional complexities, the method must somehow be combined with additional models of these other phenomena. This is not simply a matter of adding an equation or two; rather, the entire control flow and mechanism of proposing hypotheses may need to be changed. For example, it may be necessary to propose several different hypotheses about the same portion of the image, and allow the various interacting models settle into a single "most likely" hypothesis.

We have not yet determined how best to attack this problem, but we do have some progress to report in this area. The theme of our current work is to study just one example of combining multiple intrinsic models: interreflection among multiple objects. If the segmentation in Figure 8 (upper right corner) is examined closely, it can be seen that several areas are mis-labeled where the hue of one object is significantly changed by light reflected off of a nearby object. We are currently in the process of studying this phenomenon because it appears to be quite noticeable in our images, it is utterly beyond the ability of the above segmentation method to deal with, and yet it is fairly simple to model and understand.

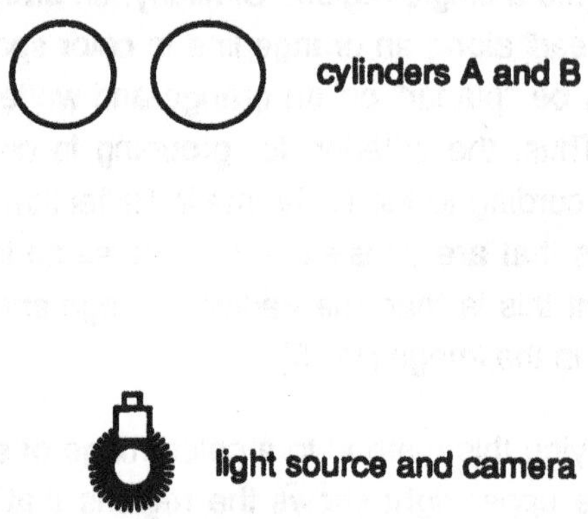

Figure 5: Interreflection Between Two Cylinders

Our model of interreflection is an extension of the Dichromatic Reflection Model [8]. Suppose we have two nearby objects A and B, as shown in Figure 5 -- we use cylinders for simplicity of the example. We know from the Dichromatic Reflection Model that the light we see reflected from A has two terms: surface reflection and body reflection. These are modeled relative to the incident illumination. In the past, we have implicitly assumed that the illumination comes from a single light source. But, when B is nearby, there is also illumination on A produced by light reflected from B. Thus, each of the reflection components of A must be further modeled as the sum of reflection from direct illumination, and reflection from light that was already reflected by B. How can we characterize the light reflected off of B onto A? It will have the usual surface and body reflection terms of the Dichromatic Reflection Model. Thus, in this situation, we can model the reflected light from A as a sum of six terms:

1. *S:* Light source -> surface reflection from A
2. *B:* Light source -> body reflection from A
3. *S-S:* Light source--> surface reflection from B -> surface reflection from A
4. *S-B:* Light source -> surface reflection from B -> body reflection from A
5. *B-S:* Light source -> body reflection from B -> surface reflection from A
6. *B-B:* Light source -> body reflection from B -> body reflection from A

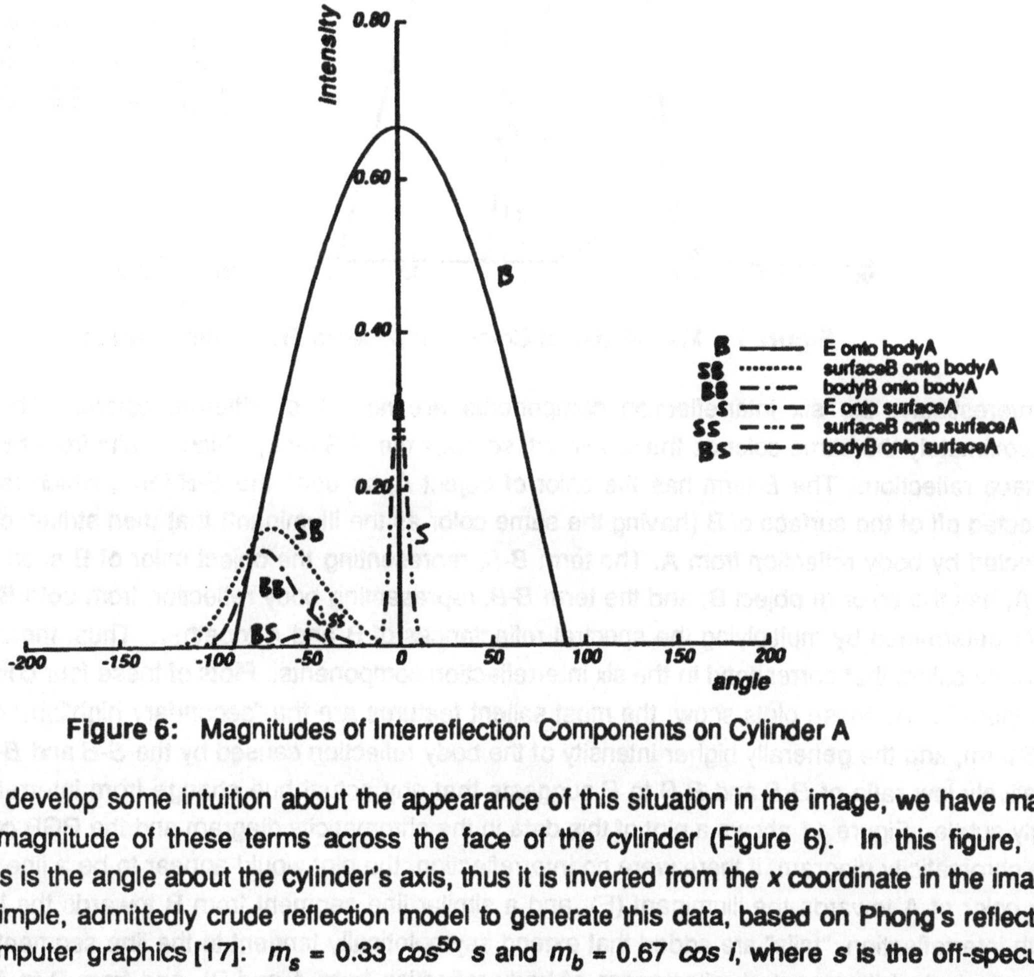

Figure 6: Magnitudes of Interreflection Components on Cylinder A

In order to develop some intuition about the appearance of this situation in the image, we have made plots of the magnitude of these terms across the face of the cylinder (Figure 6). In this figure, the horizontal axis is the angle about the cylinder's axis, thus it is inverted from the *x* coordinate in the image. We used a simple, admittedly crude reflection model to generate this data, based on Phong's reflection model for computer graphics [17]: $m_s = 0.33 \cos^{50} s$ and $m_b = 0.67 \cos i$, where *s* is the off-specular angle (between the reflected ray and the direction of specular reflection) and *i* is the angle of incidence (between the incident ray and the surface normal). This corresponds to surfaces A and B that are both somewhat matte, with noticeable highlights that are not too sharp. The situation is shown in Figure 10, a computer-generated image. The top band of this image shows two cylinders with no inter-reflection; the large middle band shows the result when modeling a single bounce of interreflection (as described here); the bottom band shows the result of modeling two bounces of interreflection, i.e. A to B to A and B to A to B, by natural extension of the above model. As can be seen, there is a very noticeable difference between the model with no interreflection and one bounce, but the effect of modeling additional bounces of light is relatively small.

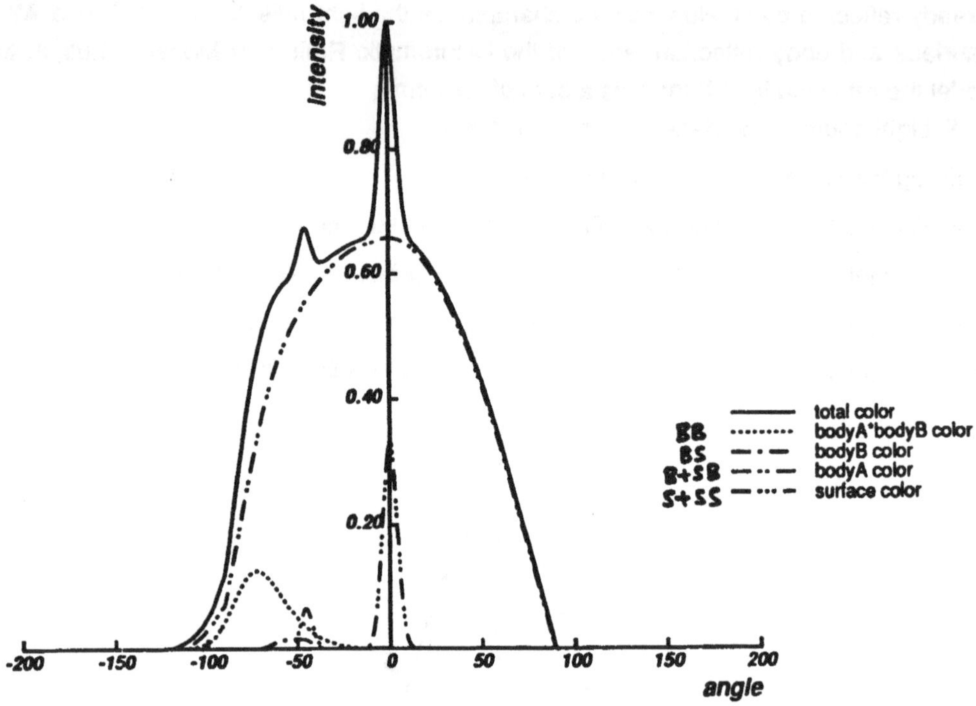

Figure 7: Magnitudes of Color Components From Interreflection

Interestingly, the six interreflection components are not all of different colors. The *S* term has approximately the same color as the illuminant; so does the *S-S* term, which results from two instances of surface reflection. The *B* term has the color of object A; so does the *S-B* term, which represents light reflected off of the surface of B (having the same color as the illuminant) that then strikes object A and is reflected by body reflection from A. The term *B-S*, representing the object color of B seen as a highlight on A, has the color of object B; and the term *B-B*, representing body reflection from both B and A, has a color determined by multiplying the spectral reflectances of B and A together. Thus, there are only four different colors that correspond to the six interreflection components. Plots of these four colors are shown in Figure 7. As these plots show, the most salient features are the "secondary highlight" caused by the *S-S* term, and the generally higher intensity of the body reflection caused by the *S-B* and *B-B* terms. The relatively low ratio of *B-S* and *B-B* to *B* suggests that any actual hue change from interreflection will be fairly subtle. Figure 11 shows a plot of this data in the chromaticity diagram and the RGB color space. In the chromaticity diagram, if there were no interreflection, the plot would appear to be a line segment from the color of A towards the illuminant (E), and a similar line segment from B towards the illuminant [13]. With interreflection, "tails" are added that extend asymptotically tangent to the line segment from A to AB (the product of the spectral reflectances of body reflection from A and B), and from B to AB; these tails are interrupted by the spikes that correspond to the secondary highlights from the *S-S* terms. In the RGB color space, shown by a color cube, the loci of the plots have straight "skewed-T" shapes (as predicted by the Dichromatic Reflection Model [4, 11]), with additional curves trailing off as predicted by our interreflection model.

Now that we have developed this model, our next task is to confirm that it is an accurate model of real interreflection. Figure 12 shows a real photograph of two drinking cups taken in our lab, using the precision color imaging methods described elsewhere in these proceedings [15]. The graph shown above the photograph depicts the overall intensity of the image along a single scan line of the image (marked in black). It appears to conform well to the prediction we made as shown in Figure 7. So, we are encouraged that our model is valid -- still, we have to verify that it models each of the color components accurately. To do this, we are analyzing data from spectroradiometer scans of the same scene, taken at closely spaced intervals along the same row. This data, having a higher dimensionality than RGB color image data, should allow us to identify the spectral signature of each color component and measure its magnitude along the row.

Our original goal in studying this topic, developing a deeper understanding of how to combine multiple intrinsic models and maintain multiple hypotheses about the intrinsic interpretation of the image, has not yet been fulfilled. Our model so far appears useful, but we have not yet reached the point of proposing a specific method for image segmentation that embodies these ideas. Noteworthy efforts by others in this area include the work of Brill [2], which unfortunately requires more than three sensor bands (in this case, five); and the work of Bajcsy, Lee, and Leonardis [1], which uses an initial hue-thresholding segmentation to identify object regions before interreflection analysis is performed. For the reasons outlined in this paper, we are still hopeful to develop a method that integrates the interreflection model into the segmentation process itself, as we believe this will result in a more robust and informative analysis of the image.

Ultimately, this model must be combined with other intrinsic models of surface properties. For example, metals exhibit strong surface reflection but no body reflection [6]; thus, if A were metal, the *S*, *B-S*, and *S-S* terms would be strong and the other terms weak. A rough surface produces more diffuse surface reflection [20]. If A were rough, the *S*, *B-S*, and *S-S* terms would be geometrically diffused; whereas, if B were rough, then the *S-B* and *S-S* terms would be diffused. With extensions such as these, the model we have presented can also deal with scenes containing mixed material types, mirror-like reflection, and other related visual phenomena.

5. Conclusions

This paper is a half-technical, half-philosophical presentation that surveys our work over the last five years as well as our motivations. The general philosophy of our work is to develop a new way to think about vision based on geometry and optical physics, rather than relying on heuristic methods that are simpler but don't work very well. A whole segment of the machine vision community now subscribes to this general approach of "physics-based vision"; yet, within that school of thought, we stand at a fairly extreme position. We hope, and we currently believe, that the application of integrated physical models of many scene phenomena can take place directly on the raw image data, before it is subjected to the torments of edge detection, region segmentation, and other such heuristic, image-based analyses. Indeed, we propose that edges and regions should be defined by the extent of coverage of each hypothesized intrinsic property, thus they ought to be a *result* of the intrinsic analysis rather than the traditional view, which is that edges and regions are computed *before* the intrinsic property analysis and form the input to that analysis.

The general methodology we present is as follows:

1. Examine local properties, and let each neighborhood suggest a hypothesized instantiation of an intrinsic model.
2. Group together those neighborhoods with consistent hypotheses. This grouping forms a hypothesis at a higher level.

3. The boundaries of this grouping determine the segmentation of the image; the parameters used to instantiate the intrinsic model form a proposed description of the scene.

Currently, this methodology has been demonstrated for a single intrinsic model, the Dichromatic Reflection Model, and the results have been encouraging. Our efforts now are to understand how to extend this approach to deal with multiple intrinsic models, which might require very fundamental changes to the methodology such as the explicit maintenance of multiple conflicting hypotheses. As our initial study case, we have developed a model for interreflection and are now validating that model by comparing its predictions with actual image data.

More generally, we wish to suggest that it is very useful to think about the 3D scene from the earliest stages of vision. This may give rise to image operators, such as our measure of the dimensionality of the local color distribution, that are very different from the traditional edge-finding and region-grouping operators, and that have significant advantages for visual analysis.

Acknowledgements

Thanks to Reg Willson, John Krumm, Jim Moody, and others who have helped to create the imaging facilities in the Calibrated Imaging Laboratory that have made this research possible.

This research was supported by the Defense Advanced Research Projects Agency, DoD, through ARPA Order Number 4976, monitored by the Air Force Avionics Laboratory under Contract F33615-87-C-1499; by the Jet Propulsion Laboratory, California Institute of Technology, sponsored by the National Aeronautics and Space Administration under Contract 957989; and by the Air Force Office of Scientific Research under Contract F49620-86-C-0127. Any opinions, findings, conclusions or recommendations expressed in this publication are those of the authors and do not necessarily reflect the views of the United States Government or the Jet Propulsion Laboratory.

6. Bibliography

1. Bajcsy, R., Lee, S.W., and Leonardis, A. Image Segmentation With Detection of Highlights and Inter-Reflections Using Color. GRASP LAB 182 MS-CIS-89-39, U. Penn. Dept. of Computer and Info. Science, June, 1989.

2. Brill, M. H. Object-based segmentation and color recognition in multispectral images. SPIE-SPSE Meeting, January, 1989. Paper 1076-11.

3. Cook, R.L. and Torrance, K.E. "A Reflectance Model for Computer Graphics". *ACM Trans. on Graphics 1*, 1 (January 1982), 7-24.

4. Gershon, R. *The Use of Color in Computational Vision*. Ph.D. Th., Dept. of Computer Science, U. Toronto, 1987.

5. Hanson, A.R. and Riseman, E.M. Segmentation of Natural Scenes. In *Computer Vision Systems*, A.R. Hanson and E.M. Riseman, Ed., Academic Press, 1978, pp. 129-164.

6. Healey, G. and Binford, T.O. The Role and Use of Color in a General Vision System. Proc. DARPA Image Understanding Workshop, February, 1987, pp. 599-613.

7. Horn, B.K.P. "Understanding Image Intensities". *Artificial Intelligence 8*, 11 (1977), 201-231.

8. Kanade, T. and Shafer, S. Image Understanding Research at Carnegie Mellon. Proc. DARPA Image Understanding Workshop, May, 1989, pp. 32-48. Note: Interreflection was described in the presentation at the workshop, though it does not appear in the paper in the written proceedings..

9. Kanade, T. and Kender, J.R. Mapping 'mage Properties into Shape Constraints: Skewed Symmetry, Affine-Transformable Patterns, and the Shape-from-Texture Paradigm. In *Human and Machine Vision*, Academic Press, 1983.

10. Klinker, G.J. *A Physical Approach to Color Image Understanding*. Ph.D. Th., Dept. of Computer Science, Carnegie Mellon University, May 1988.

11. Klinker, G.J., Shafer, S.A., and Kanade, T. "The Measurement of Highlights in Color Images". *Intl. J. of Computer Vision 2*, 1 (June 1988), 7-32.

12. Klinker, G.J., Shafer, S.A., and Kanade, T. "A Physical Approach to Color Image Understanding". *Intl. J. of Computer Vision* (1990). In press..

13. Lee, H.-C. "Method for computing the scene-illuminant chromaticity from specular highlights". *J. of the Optical Society of America A 3*, 10 (October 1986), 1694-1699.

14. Marr, D. Representing Visual Information. In *Computer Vision Systems*, A.R. Hanson and E.M. Riseman, Ed., Academic Press, 1978, pp. 61-80.

15. Novak, C.L., Shafer, S.A., and Willson, R.G. Obtaining Accurate Color Images for Machine Vision. This proceedings, 1990.

16. Ohlander, R., Price, K., and Reddy, D.R. "Picture Segmentation Using a Recursive Region Splitting Method". *Computer Graphics and Image Processing 8* (1978), 313-333.

17. Phong, B.T. "Illumination for Computer Generated Pictures". *Communications of the ACM 18* (1975), 311-317.

18. Shafer, S.A. "Using Color to Separate Reflection Components". *Color Research and Application 10*, 4 (Winter 1985), 210-218.

19. Tenenbaum, J.M., Fischler, M.A., and Barrow, H.G. "Scene Modeling: A Structural Basis for Image Description". *Computer Graphics and Image Processing 12* (1980), 407-425.

20. Torrance, K.E. and Sparrow, E.M. "Theory for Off-Specular Reflection From Roughened Surfaces". *J. of the Optical Society of America 57* (September 1967), 1105-1114.

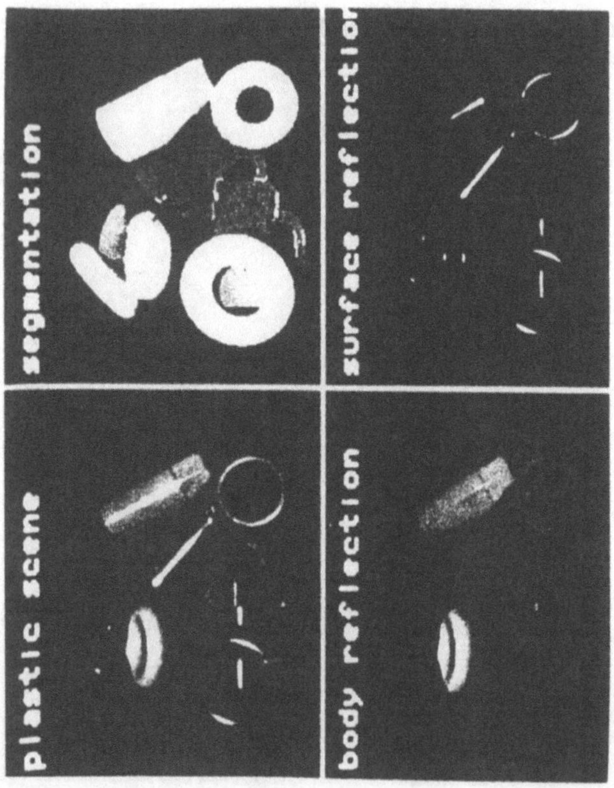

Figure 8: Image Analysis Using the Dichromatic Reflection Model

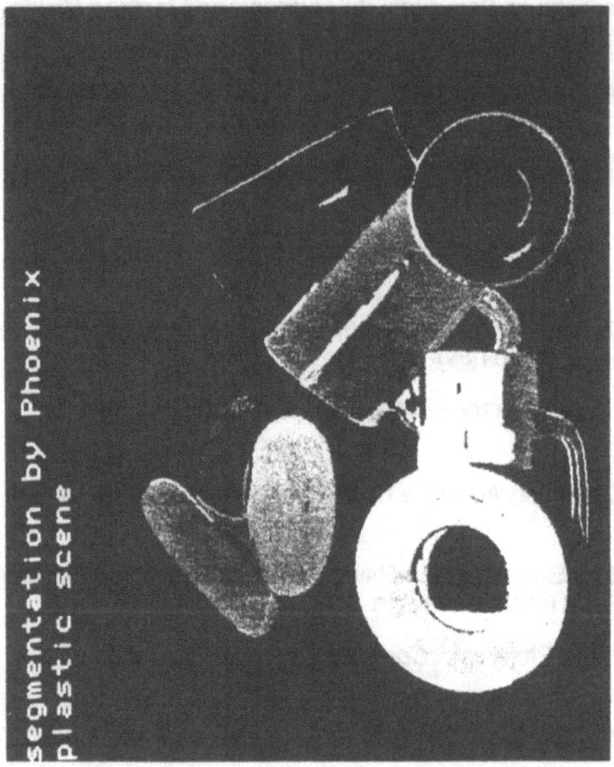

Figure 9: Segmentation by an Older, Traditional Method

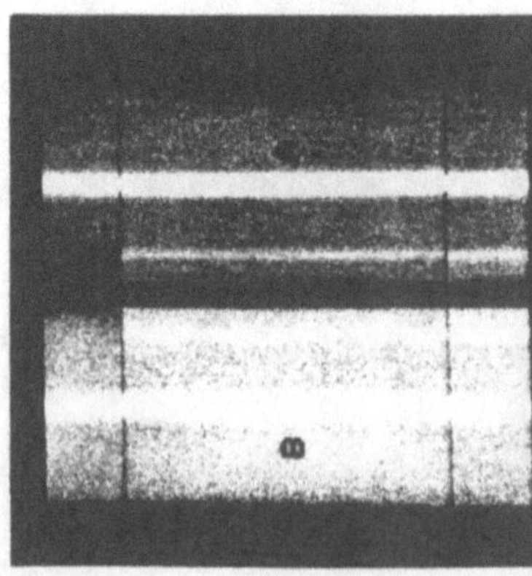

Figure 10: Two Cylinders Showing Interreflection (Synthetic Image)

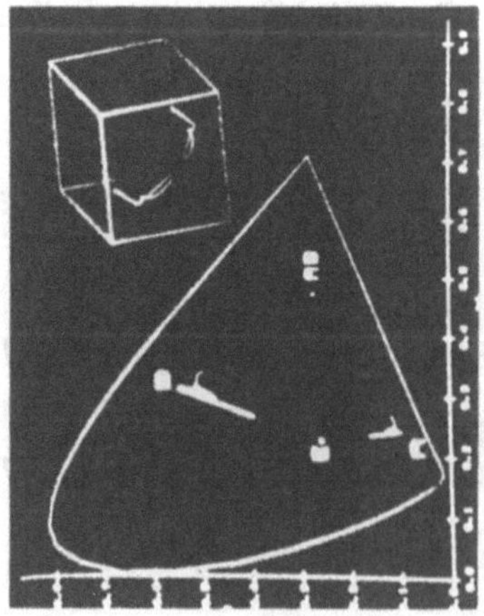

Figure 11: Colors of Interreflection in Chromaticity Diagram and Color Space

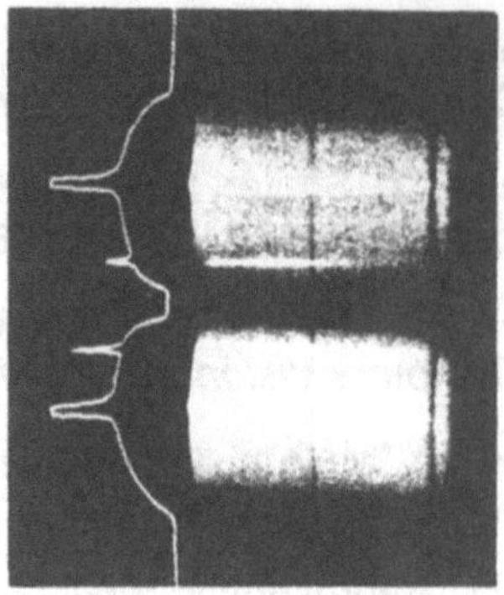

Figure 12: Interreflection in a Real Image

See color plate L for Figures 8-12.

Color Constancy from Mutual Reflection

BRIAN V. FUNT, MARK S. DREW AND JIAN HO
School of Computing Science, Simon Fraser University, Vancouver, British Columbia, Canada V5A 1S6

Abstract
Mutual reflection occurs when light reflected from one surface illuminates a second surface. In this situation, the color of one or both surfaces can be modified by a *color-bleeding* effect. In this article we examine how sensor values (e.g., RGB values) are modified in the mutual reflection region and show that a good approximation of the surface spectral reflectance function for each surface can be recovered by using the extra information from mutual reflection. Thus color constancy results from an examination of mutual reflection. Use is made of finite dimensional linear models for ambient illumination and for surface spectral reflectance. If m and n are the number of basis functions required to model illumination and surface spectral reflectance respectively, then we find that the number of different sensor classes p must satisfy the condition $p \geq (2n + m)/3$. If we use three basis functions to model illumination and three basis functions to model surface spectral reflectance, then only three classes of sensors are required to carry out the algorithm. Results are presented showing a small increase in error over the error inherent in the underlying finite dimension models.

1 Introduction

Two illuminated surfaces of different reflectance can appear to have their colors bleed into one another in regions where light reflected from one surface falls onto the other surface. This is a mutual reflection effect. Figure 1 shows two surfaces that form an edge, and the resulting interreflection effect. In this article we examine how the lights reflected separately from each surface combine to form RGB values in the mutual reflection region. Using vector models for the illumination spectral power distribution and for surface reflectance functions we show that by looking at the extra information coming from measurements of the interreflection RGB values—in addition to sensor measurements from each surface separately—it is possible to recover the surface reflectance properties of both surfaces. In this way we use mutual reflection to obtain "color constancy."

The term *color constancy* refers to the ability of humans and some animals to perceive object colors as approximately constant, independent of changing illumination (see, e.g., Beck 1972). There is evidence that several simultaneous mechanisms contribute to color constancy in humans (Blackwell & Buchsbaum 1988). Nevertheless, we have found in (Ho, et al., 1990), that it is the case that an algorithm can be built to recover reflectance from the color signal, in which illumination and surface characteristics are confounded, without regard to colored surrounds, selective adaptation, memory of colors, etc., provided that the complete color signal is known. Most recent attempts to accomplish this end have made use of finite dimensional linear models in which illumination and surface spectral reflectance are approximated by a weighted sum of a few basis functions of wavelength (Brainard et al. 1989; Brill & West 1986; Buchsbaum 1980; D'Zmura & Lennie 1986; Gershon et al. 1987; Maloney 1985; Maloney & Wandell 1986; Wandell 1987; Yuille 1987).

In the innovative work of Maloney and Wandell 1985; 1986; 1987, several special conditions have to be true for reflectance to be recoverable from the color signal entering the camera. Some of these conditions are that the illumination must be constant over a given segment of the image, and that a sufficiency of different color information must be available over the image. More importantly, in this method the number of sensor classes (e.g., 3 for RGB sensors) must be at least as great as the number of basis functions modeling the surface spectral reflectance, plus 1. This means that for a usable

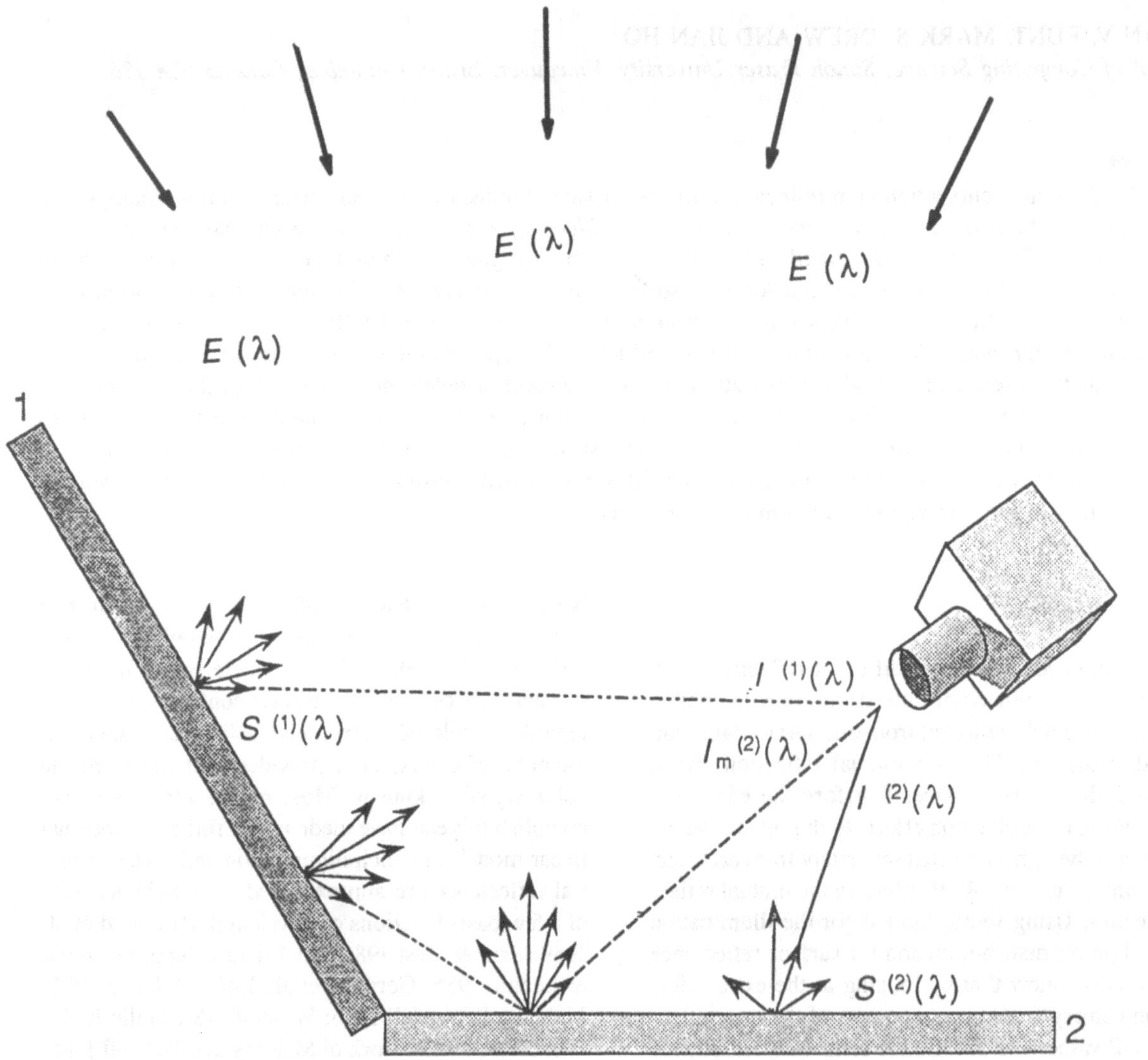

Fig. 1. Isotropic illumination with spectral power distribution $E(\lambda)$ impinges on surfaces 1 and 2, which have surface reflectance functions $S^{(1)}(\lambda)$ and $S^{(2)}(\lambda)$ respectively. Far from the edge, surface 1 reflects color signal $I^{(1)}(\lambda)$ toward the camera; and, similarly, surface 2 reflects $I^{(2)}(\lambda)$. Close to the edge, $E(\lambda)$ is augmented by light reflected from the other surface. The color signals reflected from the surfaces near the edge are $I^{(1)}_m(\lambda)$ and $I^{(2)}_m(\lambda)$.

dimensionality for the set of basis functions modeling reflectance, say 3 or better (Maloney 1986), one must somehow develop a "fourth sensor class" to provide enough information to allow a solution that disentangles illumination and surface spectral reflectance. Alternatively, one can continue using the usual 3 sensor classes, but at the expense of having only a dimensionality of 2 with which to model surface reflectance.

Maloney and Wandell's method essentially involves writing down a set of equations relating the sensor values ρ_k, $k = 1 \ldots p$, to the finite dimensional linear model weights, ϵ_i, $i = 1 \ldots m$ for illumination, and σ_j, $j = 1 \ldots n$ for surfaces, and then coupling the sets of equations at different pixels by using the assumption of constant illumination. Taking into account enough pixels, which must have sufficiently different sensor values, brings the number of equations equal to the overall number of unknowns, allowing a solution for ϵ_i and σ_j.

In Ho et al. (1990) it was shown that the need for additional sensor values can essentially be fulfilled by requiring that a finer sampling of the color signal spectrum be introduced as additional input. In that work, it was shown that from detailed knowledge of the spectral power distribution of the color signal, a complete disambiguation of the reflectance and illumination from the color signal is possible, and in fact is possible independently at each pixel and without the need for constant illumination or any special requirement on the richness of color variation over image segments. This method (Ho et al. 1990) relies on a statistical determination of the finite dimensional model weights using a least-squares method, and has the sole requirement that a special mathematical relationship holds among the basis vectors, viz. that the set of functions generated by forming products of the basis functions for illumination and surface reflectance forms a linearly independent set. The main benefit of this method is that one can increase the dimensionality of the surface reflectance space without having to increase the number of ρ_k values beyond the 3 that correspond to the cones in the human visual system.

In Funt & Ho (1989) and Gershon et al. (1986, 1987) it was shown that the required spectral information for the color signal is derivable from the chromatic aberration effect at edges between adjacent color regions. In fact, what is found via chromatic aberration is the *difference* between color signals from each side of the edge. Coupled with the method of Ho et al. (1990) and given this spectral information plus RGB sensor values from each side of the edge, the values of ϵ_i and σ_j can be determined and color constancy achieved.

In this article, we take a different approach, although the basic motivation is the same—we seek more information than just the ρ_k values in order to circumvent the "fourth sensor problem." Here, we consider the mutual reflection at a 3-dimensional edge, and the resulting bleeding of colors at the 2-dimensional image edge, without regard to chromatic aberration. Such interreflections generally occur near the edges of objects and the boundaries where objects occlude one another (Horn 1986).

We can take measurements of ρ_k values from pixels where mutual reflection effects are present and compare them to values where such effects must be absent or small. By using such measurements on each side of an image edge, relatively far from the edge and also relatively close to the edge, we capture sufficient information to determine ϵ_i and σ_j separately on each side of the edge, and hence determine the true reflectance for each color patch. This is more than enough information to provide color constancy. In Gershon et al. (1986) it was shown that mutual reflection can help to distinguish shadow boundaries from material changes. Here we show that mutual reflection has another positive aspect—it is an effect that can be exploited to achieve color constancy. As well, once constant color descriptors are found, the mutual reflection component can be effectively removed from the image. Such an image filter could be useful for shape-from-shading, etc. (Forsyth & Zisserman 1989, 1990).

The only constraint on the formulation turns out to be that the number p of sensor classes must be greater than or equal to $(2n + m)/3$, where m and n are the number of basis functions required to model illumination and surface spectral reflectance respectively. This constraint is satisfied by values $p = m = n = 3$ and hence we can keep to only 3 sensor classes while still allowing $n = 3$ for describing reflectances. These values of p, m, and n are shown to recover both surface spectral reflectance and illumination spectral curves well within acceptable errors.

Here, we address only a small part of a large problem. We assume a preliminary edge-finding and segmentation algorithm that would identify those edges where mutual reflection might be present. Given such identification of appropriate sites our analysis concentrates on extracting any available reflectance information from mutual reflection effects.

In section 2 we show how ρ_k measurements on each side of a color edge, and inside the mutual reflection region, are related to the values of ϵ_i and σ_j for each side of the edge separately, assuming a finite dimensional linear model for describing natural lighting and reflectances. The resulting model is quite complex, but some reasonable assumptions reduce the equations to solvable form. We show that a reciprocity relation exists between the mutual reflection measurements on each side of an edge, so that only ρ_k values in the mutual reflection zone on a single side of the edge are required.

In section 3 an algorithm is presented for solving for the reflectance weights σ_j on each side of the edge separately, as well as the illumination weights ϵ_i, which are assumed to not change across an edge. We develop a uniqueness proof that guarantees convergence of the algorithm.

In section 4 we set out the geometrical significance of the mutual reflection configuration factor used in the analysis.

In section 5 we examine the accuracy of the solution method by carrying out simulations. The results are seen to be very good in most cases, with accuracies of the resulting reflectance spectra not degraded much from the accuracy of the underlying finite dimensional model.

2 Finite-dimensional Models Applied to Mutual Reflection

2.1 Mutual Reflection Equations

We assume that illumination and surface spectral reflectance are each modeled to an acceptable degree of accuracy by finite dimensional linear models. Depending on the dimensionality of the set of basis functions, not all illumination spectra and surface reflectances will be well modeled; generally, the higher the dimensionality of the set of basis functions the better will be the approximation (see, e.g., Maloney 1986). However, we shall show below using simulations that whether or not the accuracy of the underlying finite dimensional model is good, the accuracy of the solution for color constancy from mutual reflection measurements is not greatly reduced.

For reflectance basis functions we use those determined by Cohen via a statistical principal component analysis of 150 Munsell chips randomly selected from a total of 433 chips (Cohen 1964); see also Parkkinen et al. (1989). For illumination, a similar analysis was carried out by Judd et al. (1964) for 622 typical daylight samples. We demonstrate below that it is possible to use $n = m = 3$ basis functions for surfaces and illumination and still have only $p = 3$ sensor classes.

Most naturally occurring daylights and many surface reflectances can be approximated reasonably well using these basis functions, so that by determining a good approximation of the weights ϵ_i and σ_j, we are in fact deriving an approximation of the entire illumination or reflectance spectrum. Hence what we require from a solution is actually more stringent than simply *color* constancy, in that we seek to recover more information than just chromaticity values. And, in fact, the measure of accuracy we shall use is the error sum of squares over the entire visible spectrum of an approximate reflectance spectrum compared with an actual one.

Consider the situation depicted in figure 1. *Isotropic* illumination $E(\lambda)$, assumed constant across an edge, impinges on two surfaces, 1 and 2. Assume surface 1 has spectral reflectance function $S^{(1)}(\lambda)$ and surface 2 has reflectance $S^{(2)}(\lambda)$, all surfaces being ideal diffuse (i.e., Lambertian) reflectors. Then depending on the geometry, interreflection may be present as shown in figure 1. In fact, the two surfaces need not meet at an edge for a mutual reflection effect to come into play; and we do not require the surfaces to be flat. We show ray geometry in figure 1 to clarify the interreflection process. However, ray geometry is actually unnecessary since perfect diffuse reflectors obliterate directional information: reflections are really in all directions into the hemisphere above each surface. We assume here that illumination is uniform on each surface and hence exclude shadow situations.

Expanding the illumination function in terms of m basis functions, the spectral power distribution can be approximated by the sum

$$E(\lambda) = \sum_{i=1}^{m} \epsilon_i E_i(\lambda)$$

with weights ϵ_i. Similarly, for surface 1 the reflectance can be written

$$S^{(1)}(\lambda) = \sum_{j=1}^{n} \sigma_j^{(1)} S_j(\lambda)$$

where reflectance is modeled by n basis functions with n weights $\sigma_j^{(1)}$. Surface 2 has reflectance

$$S^{(2)}(\lambda) = \sum_{j=1}^{n} \sigma_j^{(2)} S_j(\lambda)$$

The color signal $I(\lambda)$ is the light reflected by the surface, and is given by the product of E and S. On the surface 1 side, at a point relatively far from the edge, the color signal is simply

$$I^{(1)}(\lambda) = E(\lambda)\ S^{(1)}(\lambda)$$

Similarly, on the surface 2 side we have

$$I^{(2)}(\lambda) = E(\lambda)\ S^{(2)}(\lambda)$$

Substituting the basis function decomposition, we have

$$I^{(1)}(\lambda) = \sum_{i=1}^{m}\sum_{j=1}^{n} \epsilon_i \sigma_j^{(1)} E_i(\lambda) S_j(\lambda)$$

$$I^{(2)}(\lambda) = \sum_{i=1}^{m}\sum_{j=1}^{n} \epsilon_i \sigma_j^{(2)} E_i(\lambda) S_j(\lambda)$$

Now consider the mutual reflection region in which part of $I^{(1)}(\lambda)$, the light reflected from surface 1, impinges onto surface 2. Similarly, part of $I^{(2)}(\lambda)$ impinges onto surface 1. The amount of $I^{(1)}(\lambda)$ that is intercepted by surface 2 is dependent on both the configuration of the surfaces and on the shape of each surface, since not every reflected ray is intercepted (consider a small triangle of surface 1 forming an edge with a large plane of surface 2). In computer graphics diffuse interreflections are modeled using *configuration factors* that give the fraction of light from one surface that reaches another surface (Goral et al. 1984). We discuss these in section 4.

Since there is no a priori reason to assume that one surface is more reflective than the other (although for clarity in figure 1 only a single interreflection is shown), we consider the fraction α_{12} of $I(\lambda)$ from 1 that strikes 2, and the similarly defined α_{21}. Denote by $I_m^{(2)}$ the color signal from a spot on surface 2 in the mutual reflection region, so that it includes a contribution from light reflected by surface 1. It consists of two parts. The first part is $I^{(2)}(\lambda)$, the reflection of $E(\lambda)$ from surface 2. This we assume is the same as from a spot relatively far from the edge, outside the mutual reflection region. The second part is due to the light reflected from surface 1. Since the latter is presumed to come from a spot on surface 1 that is near the edge, the second part is a contribution from $I_m^{(1)}$ (cf., Goral 1984). Hence the color signals coming from each side of the edge in the mutual reflection region are

$$\begin{aligned} I_m^{(1)}(\lambda) &= I^{(1)}(\lambda) + \alpha_{21} I_m^{(2)}(\lambda) S^{(1)}(\lambda) \\ I_m^{(2)}(\lambda) &= I^{(2)}(\lambda) + \alpha_{12} I_m^{(1)}(\lambda) S^{(2)}(\lambda) \end{aligned} \tag{1}$$

where α_{12} is not assumed to be the same as α_{21}, and we do not include dependency of the surface spectral reflectance function shape on angle of incidence or angle of line of sight (although it could change by an overall multiplicative constant).

Solving the above set of simultaneous equations, we can write each $I_m(\lambda)$ in terms of $I^{(1)}(\lambda)$ and $I^{(2)}(\lambda)$ from each surface separately and the geometrical factors α:

$$\begin{aligned} I_m^{(1)}(\lambda) &= \frac{I^{(1)}(\lambda) + \alpha_{21} I^{(2)}(\lambda) S^{(1)}(\lambda)}{1 - \alpha_{12}\alpha_{21} S^{(1)}(\lambda) S^{(2)}(\lambda)} \\ I_m^{(2)}(\lambda) &= \frac{I^{(2)}(\lambda) + \alpha_{12} I^{(1)}(\lambda) S^{(2)}(\lambda)}{1 - \alpha_{12}\alpha_{21} S^{(1)}(\lambda) S^{(2)}(\lambda)} \end{aligned} \tag{2}$$

2.2 Simplified Equations

The above set of equations would be quite difficult to solve numerically. However, since both α and $S(\lambda)$ are less than 1, in most cases not much accuracy is lost by rewriting the equations to first order in $\alpha \cdot S$. This amounts to ignoring more than one bounce of the color signal between the surfaces. With typical values of S 0.01–0.3 and typical values of α of ≤ 0.5 this approximation might be off by at most a few percent. Of course, for highly reflective materials (0.3 represents the 92 percentile in the Krinov catalog (Krinov 1947) and for geometric configurations with high values of α, the approximation would be farther from the full model.

To first order, we have

$$\begin{aligned} I_m^{(1)}(\lambda) &= I^{(1)}(\lambda) + \alpha_{21} I^{(2)}(\lambda) S^{(1)}(\lambda) \\ I_m^{(2)}(\lambda) &= I^{(2)}(\lambda) + \alpha_{12} I^{(1)}(\lambda) S^{(2)}(\lambda) \end{aligned} \tag{1a}$$

as the set to solve.

To see how these equations can be further reduced, denote the terms involving α by

$$\begin{aligned} \bar{I}_m^{(1)}(\lambda) &\equiv I_m^{(1)}(\lambda) - I^{(1)}(\lambda) \\ \bar{I}_m^{(2)}(\lambda) &\equiv I_m^{(2)}(\lambda) - I^{(2)}(\lambda) \end{aligned}$$

That is,

$$\bar{I}_m^{(1)}(\lambda) = \alpha_{21} E(\lambda) S^{(2)}(\lambda) S^{(1)}(\lambda)$$

and also

$$\bar{I}_m^{(2)}(\lambda) = \alpha_{12} E(\lambda) S^{(1)}(\lambda) S^{(2)}(\lambda)$$

Therefore, we have a *reciprocity relation*

$$\bar{I}_m^{(1)}(\lambda) = \frac{\alpha_{21}}{\alpha_{12}} \bar{I}_m^{(2)}(\lambda)$$

Since all the signals $I(\lambda)$ are in principle measurable, this relation states that one α can be derived from the other. To obtain a solution that takes into account noise and errors in measurement, samples could be measured at N wavelengths so that the above relation would then represent N equations in 1 unknown. A statistical best estimate results from the least-squares solution for the ratio of the α's. Therefore, the number of unknowns to be solved for is reduced by one. Below, we show how it is possible to use only one of the pair of equations for the $I_m(\lambda)$'s to complete a sufficient set to solve for the remaining unknowns.

To simplify further, consider reducing all the equations involved to equations relating *sensor values* by multiplying by sensor sensitivity functions $R_k(\lambda)$ for $k = 1\ldots p$ sensor classes and integrating over wavelength (e.g., $p = 3$ for RGB signals). Then following Maloney & Wandell (1986) and defining the precalculated tensor

$$g_{ijk} \equiv \int E_i(\lambda)S_j(\lambda)R_k(\lambda)\, d\lambda$$

the equations for $I^{(1)}(\lambda)$ and $I^{(2)}(\lambda)$ become the following:

$$\begin{aligned}\rho_k^{(1)} &\equiv \int I^{(1)}(\lambda)R_k(\lambda)\, d\lambda \\ &= \sum_{i=1}^{m}\sum_{j=1}^{n} \epsilon_i \sigma_j^{(1)} g_{ijk}\end{aligned} \tag{3}$$

and

$$\begin{aligned}\rho_k^{(2)} &\equiv \int I^{(2)}(\lambda)R_k(\lambda)\, d\lambda \\ &= \sum_{i=1}^{m}\sum_{j=1}^{n} \epsilon_i \sigma_j^{(2)} g_{ijk}\end{aligned} \tag{4}$$

with $k = 1\ldots p$.

For the mutual reflection color signal, we first substitute the basis function expansions yielding

$$I_m^{(2)}(\lambda) = \sum_{i=1}^{m}\sum_{j=1}^{n} \epsilon_i \sigma_j^{(2)} E_i(\lambda)S_j(\lambda) + \alpha \sum_{i=1}^{m}\sum_{j=1}^{n}\sum_{j'=1}^{n} \epsilon_i \sigma_j^{(1)} \sigma_{j'}^{(2)} E_i(\lambda)S_j(\lambda)S_{j'}(\lambda)$$

where we have abbreviated α_{12} simply as α. Now multiplying by $R_k(\lambda)$ and integrating, we find the final equation in our set of equations to solve,

$$\begin{aligned}\rho_k^{m} &\equiv \int I^{(2)}(\lambda)R_k(\lambda)\, d\lambda + \alpha \int I^{(1)}(\lambda)S^{(2)}(\lambda)R_k(\lambda)\, d\lambda \\ &= \rho_k^{(2)} + \alpha \sum_{i=1}^{m}\sum_{j=1}^{n}\sum_{j'=1}^{n} \epsilon_i \sigma_j^{(1)} \sigma_{j'}^{(2)} h_{ijj'k}\end{aligned} \tag{5}$$

for $k = 1\ldots p$, where

$$h_{ijj'k} \equiv \int E_i(\lambda)S_j(\lambda)S_{j'}(\lambda)R_k(\lambda)\, d\lambda$$

and we have abbreviated $\rho_k^{m(2)}$ as ρ_k^{m}.

From equation (5) we can see that it is indeed possible to restrict our attention to measurements of $\rho_k^{(1)}$, $\rho_k^{(2)}$, and $\rho_k^{m} \equiv \rho_k^{m(2)}$ on *one* side of the edge, since from the similar equation for $\rho_k^{m(1)}$ we can see that if a solution is found for α_{12} then α_{21} can be determined via a least-square solution involving the $\rho_k^{m(1)}$, $\rho_k^{m(2)}$, $\rho_k^{(1)}$, and $\rho_k^{(2)}$. So that if a solution for ϵ_i, $\sigma_j^{(1)}$, $\sigma_j^{(2)}$, and α_{12} can be found via measurements on one side of an edge in the image then a solution for α_{21} is also available, given measurements $\rho_k^{m(1)}$ in the mutual reflection region on the other side.

Using measurements $\rho_k^{(1)}$, $\rho_k^{(2)}$, and ρ_k^{m}, we are faced with solving the set of equations (3), (4), and (5) for the ϵ_i, for the reflectance weights of both surfaces $\sigma_j^{(1)}$ and $\sigma_j^{(2)}$, and also for the geometrical factor α.

2.3 Assumptions

It is important to underline the assumptions inherent in the pair of equations (1) and equations (3), (4), and (5).

a. We assume isotropic diffuse illumination, i.e., a sky model as opposed to a sun model or a sun-sky combination.
b. The pair of equations (1) encompasses only two surfaces. However, the set of equations we actually solve, (3), (4), and (5), can be generalized to multiple surfaces.
c. We assume that the surfaces are convex—they do not see themselves.
d. We assume that each surface has a surface reflectance function that is independent of position, in contradistinction to the more complex model of Koenderink & van Doorn (1983).
e. In equation (1) we assume that I_m is the same everywhere in the mutual reflection zone. This amounts to assuming that the mutual reflection is between two semi-infinite convex surfaces. With two such surfaces, any point on either surface sees the same fraction of the hemisphere above it filled by the opposite surface (see Horn 1977). Each point, therefore, receives an equal amount of illumination due to mutual reflection from the opposite surface. This is a strong assumption; however, we show below that we can later eliminate it.
f. We also assume that there exists at least one point on each surface where the illumination resulting from mutual reflection from the opposite surface is negligible.

These assumptions mean in effect that we have adopted a two-zone, one-bounce model of mutual reflection. Although this reflects how our initial thinking about the problem developed, a two-zone model is rather restrictive. However, our equations—(1a) in particular—can be reinterpreted in a much more general way. They can be viewed as stating that for any point on side 1 where mutual reflection is present, the factor α_{21} represents the sum of all the contributions of light reflected from side 2 impinging upon it. Thus the factor α can be allowed to vary as a function of position; we can abandon assumption (e) and move to a one-bounce, infinite-zone model of mutual reflection.

The suitability of this interpretation is borne out in section 5, where we test the model on several simulated concave edges including ones derived from a full model of diffuse interreflection.

We are forced to keep assumption (f); however, a simple method of finding pixels where mutual reflection is negligible would be to examine the derivatives of ratios of RGB values. From figure 10 in section 5 we see that the derivatives of ratios can be expected to vanish at pixels where no mutual reflection is present.

Our simulation tests show that the model can be expected to work if one deals with surface reflectance functions for natural materials, but we do not expect the method to work when a one-bounce model is inappropriate. This would be the case for highly reflective surfaces or for geometric configurations in which mutual reflection plays a dominant role, such as inside highly concave enclosures.

2.4 Bound on Dimensionalities

In order that a solution be possible, there must be at least as many equations as unknowns. Counting components in the set $\{\epsilon_i, \sigma_j^{(1)}, \sigma_j^{(2)}\}$ we have $m + 2n$ unknowns, and α adds one more. Equations (3), (4), and (5) each provide p equations, for a total of $3p$ equations. Hence a solution is possible only if the condition $3p \geq m + 2n + 1$ holds. Therefore we cannot use $p = m = n = 3$ unless we impose a further constraint. Since we must only expect solutions to yield illumination and surface reflectance spectra up to a multiplicative constant, because the ϵ_i and σ_j occur in products in the color signal, we must make one further requirement on one of the ϵ_i or σ_j; we choose to set $\epsilon_1 \equiv 1$. That is, for surfaces we determine reflectance but not brightness.

Now our condition reads $p \geq (m + 2n)/3$ and we have a nonlinear set of equations for ϵ_i, $\sigma_j^{(1)}$, $\sigma_j^{(2)}$, and α that is at least sufficient for a solution. Since the set of equations is also clearly independent, in that one cannot write any equation in terms of the others, we would be confident of having fulfilled necessary and sufficient conditions for a solution if the set of equations were linear; however, because they are nonlinear we must proceed with caution.

3 Implementation and Results

We reduce the solution of the nonlienar set of equations (3), (4), and (5) to the solution of *linear* ones by breaking up the solution into a multi-stage algorithm. Starting with some reasonable initialization for ϵ_i, we iteratively solve in turn for $\sigma_j^{(1)}$, then $\sigma_j^{(2)}$, and finally ϵ_i. The condition $\epsilon_1 \equiv 1$ at each step of the iteration also forces α to converge.

In detail, with respect to equations (3), (4), and (5), we use the following algorithm:

Initialize $\epsilon_1 = \epsilon_2 = \epsilon_3 = 1$

Step 1: Use (3) to solve for $\sigma_j^{(1)}$ in terms of ϵ_i, since (3) is linear in these unknowns

Step 2: Use (5) to solve for $\alpha \cdot \sigma_j^{(2)}$ in terms of ϵ_i and $\sigma_j^{(1)}$.

Step 3: Use (4) to solve for $\epsilon_j \div \alpha$ in terms of $\alpha \cdot \sigma_j^{(2)}$.

Step 4: Now set $\epsilon_1 \equiv 1$: i.e., set $\epsilon_1 \div \alpha \leftarrow 1/\alpha$. Therefore, set $\alpha \leftarrow 1/\epsilon_1$ and $\epsilon_i \leftarrow \epsilon_i/\epsilon_1$.

Iterate until all values in the set $\{\epsilon_i, \sigma_j^{(1)}, \sigma_j^{(2)}, \alpha\}$ change less than a prescribed tolerance.

Since the equations are nonlinear, the order of the above steps is important. We found by trial and error that the steps in the order described lead in a stable fashion to the desired solution. The algorithm produced stable, unique results regardless of how the initial values for ϵ_i were chosen over a wide range. However, for the first initial value we imposed the constraint $\epsilon_1 \equiv 1$ from the outset. We also found that when the underlying finite-dimensional model did an extremely poor job of describing both surfaces, the algorithm was slow to converge and did not do as well in recovering surface reflectance. We explore this situation in section 5 by looking at very noisy, poorly modeled surfaces.

An important consideration is that the algorithm should obtain a unique solution. Uniqueness and speed of convergence can be established by converting the problem to be solved into the form of a fixed-point problem (see, Burden et al. 1981). Choosing a particular algorithm corresponds to adopting a particular fixed-point version of a problem. There are often several possible options, and not all of them converge. We can investigate the convergence properties of our algorithm by explicitly converting it to the fixed-point form to which it is equivalent. Then the well-known contraction mapping theorem can be used to test uniqueness as well as give rates of convergence.

Following Steps 1, 2, and 3, we algebraically solve for functions

$$\vec{\sigma}^{(1)} = \vec{\sigma}^{(1)}(\vec{\epsilon}, \vec{\rho}^{(1)}) \tag{6}$$

$$\vec{\sigma}^{(2)} = \vec{\sigma}^{(2)}(\alpha, \vec{\epsilon}, \vec{\rho}^{(1)}, \vec{\mu}) \tag{7}$$

where

$$\vec{\mu} \equiv \vec{\rho}^{\,m(2)} - \vec{\rho}^{(2)}$$

and finally

$$\vec{\epsilon} = \vec{e}(\alpha, \vec{\epsilon}, \vec{\rho}^{(1)}, \vec{\mu}, \vec{\rho}^{(2)}) \tag{8}$$

where we denote by $\vec{e}$ the solution of equation (4) for $\vec{\epsilon}$. Now since we set $\epsilon_1 \equiv 1$, we can solve the ϵ_1 component of equation (8) for α, so that α is given in terms of ϵ_2, ϵ_3, and the camera RGB values $\vec{\rho}$, $\vec{\mu}$ only. Substituting this value of α into the remaining equations for ϵ_2 and ϵ_3 gives a set of two equations in fixed-point form:

$$\begin{aligned} \epsilon_2 &= e_2(\epsilon_2, \epsilon_3, RGB) \\ \epsilon_3 &= e_3(\epsilon_2, \epsilon_3, RGB) \end{aligned} \tag{9}$$

where we have denoted the set of observed camera RGB values simply as *RGB*.

To prove uniqueness we must show (a) that when the values of ϵ_2, ϵ_3 are allowed to range over a reasonable domain *D*, then the right-hand sides of equations (9) will also be confined to *D*; and (b) that the absolute values of all possible first partial derivatives of the two right-hand sides with respect to ϵ_2, ϵ_3 are bounded by a constant $K/2$ with $K < 1$. When conditions (a) and (b) are satisfied, the vector function $[e_2(\epsilon_2, \epsilon_3, RBG), e_3(\epsilon_2, \epsilon_3, RGB)]$ is guaranteed to intersect with the vector function $[e_2, e_3] \equiv [\epsilon_2, \epsilon_3]$ at a single point. These conditions also prove convergence of the sequence provided by the algorithm, with the rate of convergence being controlled by the bound of the derivatives.

What this uniqueness check amounts to is carrying out the *first* iteration of our algorithm over a large domain *D* of *initial values* of ϵ_2, ϵ_3 and examining the resulting solution values e_2, e_3 after a first pass. If the values of e_2, e_3 are within the search space of initial values ϵ_2, ϵ_3 and if the partial derivatives of e_2, e_3 with respect to both ϵ_2 and ϵ_3 are everywhere sufficiently small, then the algorithm is guaranteed to converge to a unique solution.

Since the first-pass estimates e_2, e_3 of ϵ_2, ϵ_3 are generated using the particular observed sensor values $\vec{\rho}^{(1)}$, $\vec{\mu}$, $\vec{\rho}^{(2)}$ corresponding to the pixels under examination, we cannot make a blanket statement of uniqueness for our algorithm. There may be cases in which one cannot prove the theorem guaranteeing unique convergence, even when such cases do in fact converge correctly.

We implemented the uniqueness check as a preprocessing algorithm for the equation-solving algorithm. For a particular input set of sensor values, it is straightforward to search a wide domain of initial values of ϵ_2, ϵ_3 and take partial derivatives of the first-pass estimates e_2, e_3. We found in our simulations (section 5) that uniqueness was guaranteed except in those cases where the finite dimensional model itself represented the surfaces very poorly. Thus the preprocessing step is a useful filter for screening out any pixels for which the algorithm may not converge, such pixels generally corresponding to surface reflectances poorly captured by the finite dimensional model.

In the equation-solving algorithm itself, it is important to note the way in which α appears. Since it always appears in combination with another variable until the final normalizing step, the only place where the above algorithm's accuracy of solution for α and for all the variables is affected by the particular value of α is in forming $I_m^{(2)}(\lambda)$ in equation (2). As a result, the accuracy of solutions depends only very weakly on the particular value of α. We found (see simulations in section 5) that a change of α from 0.01 to 0.5 resulted in only a 6.7% change in the accuracy of the solution for α, with similar results for the other variables.

So long as the mutual reflection effect is indeed present, the solution proceeds to find α. No matter what the actual value of α the percent error in α will be nearly constant for a particular illumination and set of reflectances for the two surfaces. As well, the accuracy of the reflectances recovered depends only weakly on the particular value of α dictated by geometry. Of course, this near-independence with respect to α is true mathematically but noise will break this situation down somewhat. We investigate below the effect of noise on the solution. As well, as $\alpha \rightarrow 0$ we expect the method to fail; this would correspond to the case of negligible interreflection.

In section 5 we apply the algorithm to several cases involving synthesized color signals composed from naturally occurring lights and reflectances. In the next section we investigate the geometrical significance of the factor α.

4 Configuration Factors

One may ask whether in general there will exist points on a surface where we can effectively say that no mutual reflection effect exists. We can answer this question by examining the structure of the parameters α_{12} and α_{21}. The total irradiance onto surface B from surface A as a fraction of the total radiance emanating from surface B into the hemisphere above it is termed the *form-factor* in illumination engineering. Here we are interested instead in how much light reflected from one surface is intercepted at a *particular* spot on the second surface. A catalog of such *configuration factors* (total irradiance as a function of position on surface B incident from all points on surface A) has been assembled by Siegel and Howell (1981). In general, configuration factors can be calculated for any geometry and are independent of wavelength for ideal diffuse reflectors.

For illustrative purposes, consider for simplicity a relatively long (effectively semi-infinite) planar edge (figure 2). Let the fraction of light from the small area ΔA_1 intercepted by the upper surface A_2, divided by the total amount of light from ΔA_1 into the hemisphere above it, be F_{12}. Calculation of the quantity F_{12} will effectively yield the fraction of light F_{21} from A_2 intercepted by ΔA_1 as well, because of the reciprocity relation for configuration factors (Siegel & Howell 1981).

$$A_2F_{21} = \Delta A_1 F_{12}$$

If $I(\lambda)$ is measured in W/m² of unprojected area then the power in a wavelength interval received over A_2 from light $I^{(1)}(\lambda)$ originating from ΔA_1 is given by

$$P^{(2)}(\lambda) = I^{(1)}(\lambda)\ \Delta A_1 F_{12}$$

And the power received at ΔA_1 from $I^{(2)}(\lambda)$ radiated from all of A_2 is

$$P^{(1)}(\lambda) = I^{(2)}(\lambda) A_2 F_{21}$$

assuming that $I^{(2)}(\lambda)$ is uniform over A_2.

In general, one finds that the factor F_{12} is given by the Fredholm area integral (Siegel & Howell 1981)

$$F_{12} = \int_{A_2} \frac{\cos\theta_1 \cos\theta_2}{\pi r^2}\, dA_2$$

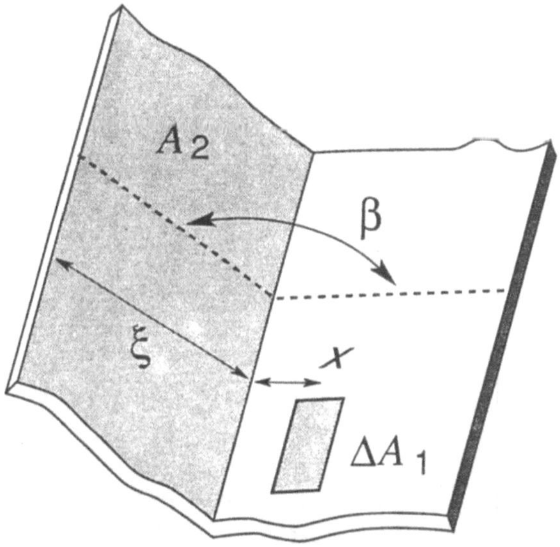

Fig. 2. A long edge formed from flat surfaces. The opening angle is β. ζ is the width of surface 2, with area A_2. The small surface area ΔA_1 is at distance x from the edge.

where r is the length of the vector from ΔA_1 to a point on A_2 and θ_1, θ_2 are the angles made between that vector and the normals to ΔA_1 and A_2.

Given this result, it is in fact straightforward to generalize to the case in which the second surface A_2 is curved, as in figure 3. This figure shows a general curved surface A_2 generated by a line moving parallel to itself and parallel to the plane of ΔA_1; calculating F_{12} for reflection onto A_2, the result is (Siegel & Howell 1981).

$$F_{12} = \frac{1}{2}(\sin\phi_1 - \sin\phi_2)$$

where ϕ_1, ϕ_2 are the angles from ΔA_1 to the top and bottom edges of A_2 (see figure 3). For figure 2, we have $\phi_2 = \pi/2$; from the geometry of figure 2 one finds that

$$\sin\phi_1 = \frac{x - \zeta\cos\beta}{(x^2 + \zeta^2 - 2x\zeta\cos\beta)^{1/2}}$$

Using the reciprocity relations for configuration factors given above and defining the dimensionless distance from the vertex as $X \equiv x/\zeta$, we find (Siegel & Howell 1981) that

$$F_{21} = \frac{\Delta A_1}{2A_2}\left[1 + \frac{\cos\beta - X}{(X^2 + 1 - 2X\cos\beta)^{1/2}}\right]$$

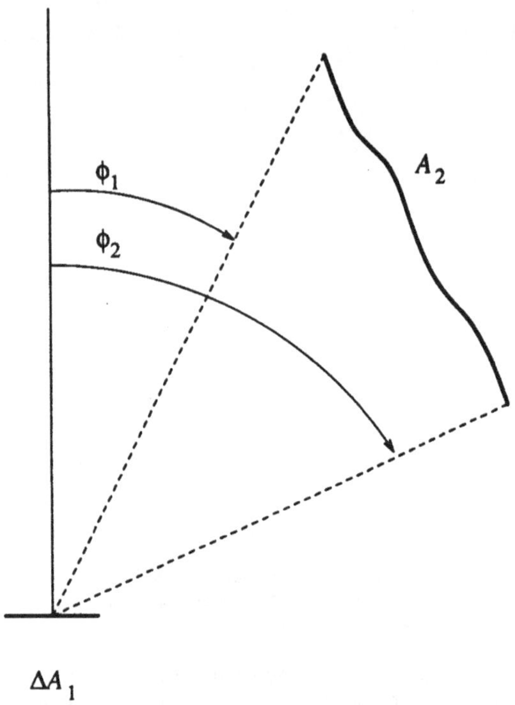

Fig. 3. Angles from ΔA_1 to top and bottom edges of a curved surface A_2.

To relate F_{21} to α_{21}, we must convert power received by ΔA_1 to power received per unit area via division by ΔA_1. Then

$$\alpha_{21}\, I^{(2)}(\lambda) = \frac{I^{(2)}(\lambda)\, A_2 F_{21}}{\Delta A_1}$$

Whereas the algorithm in section 3 extracts α_{21} from pixel RGB information, here we are constructing α_{21} from geometry—(the "forward" problem (Nayar et al. 1990)—and must decide how much of A_2 contributes to mutual reflection onto ΔA_1. For flat surfaces, it is simplest to take all of A_2 into account.

From the identification above, one has

$$\alpha_{21} = \frac{1}{2}\left[1 + \frac{\cos\beta - X}{(X^2 + 1 - 2X\cos\beta)^{1/2}}\right]$$

This factor is plotted in figure 4 for representative opening angles β. One can see that as the position X from the vertex increases, the factor α_{21} decreases monotonically, with limit 0. Typical values, for $\beta = 120°$, *are 0.107 for* $X = 0.6$ and 0.028 for $X = 2.0$. Configuration factors between less simple surfaces may be arrived at directly from the definition or by using Stokes theorem to convert the double integral into a line integral around a contour (Sparrow & Cess 1978).

It is useful to note that in the limit $\zeta \to \infty$, $X \to 0$, and $\alpha_{21} \to (1 + \cos\beta)/2$. Therefore, for semi-infinite planes I_m is independent of position, in agreement with Horn (1977). In that case the α_{21} in (10), generated by integrating over an infinity of one-bounce contributions, goes over into our initial one-bounce model (la). In the next section we test the model derived by reinterpreting the equations to mean that α sums up all the one-bounce contributions from one surface to another, where the surfaces are infinite and need not be planar.

For flat surfaces, figure 4 shows that there are areas where α is relatively small provided one surface is sufficiently wider than the other. For real surfaces, particularly convex ones, there will certainly be many sites classifiable as being far from edges or occlusions.

The form of α_{21} above shows a typical configuration factor that characterizes a mutual reflection effect from a surface A_2 to a position of interest on A_1. This value of α_{21} could be used in equations (2). However, the factor F_{12} above is not related to the number α_{12} discussed in section 2; F_{12} determines how much light from a small spot on A_1 would impinge onto A_2, and is therefore much smaller than F_{21}. In section 2, α_{12} characterizes how much light from A_1 arrives at a small region of A_2, and therefore might be of the same order as α_{21}.

5 Simulations

We test the method by synthesizing color signals from naturally occurring illumination spectra and surface spectral reflectance functions. By passing these synthetic color signals through bandpass filters representing film or other sensor sensitivities, we determine sensor values ρ_k, which then become the input values for our algorithm, along with the proportion α of the light from one surface reaching the second surface. The algorithm is then used to solve for an estimate of the surface spectral reflectance on each side of the edge, the illumination spectrum, and the mutual reflection factor α. In section 5.1 we construct the full mutual reflection input signal equation (2) and test the algorithm by solving the approximate model equations (3), (4), and (5). For simplicity we assume a geometry leading to $\alpha_{12} = \alpha_{21}$ to generate the input values of ρ^m from equation (2). This tests the model itself, but not the assumptions behind it. In section 5.2 we construct a complete infinite-bounce, infinite-zone diffuse mutual reflection edge composed of planar sides and apply our one-bounce model to it. This tests the model's assumptions.

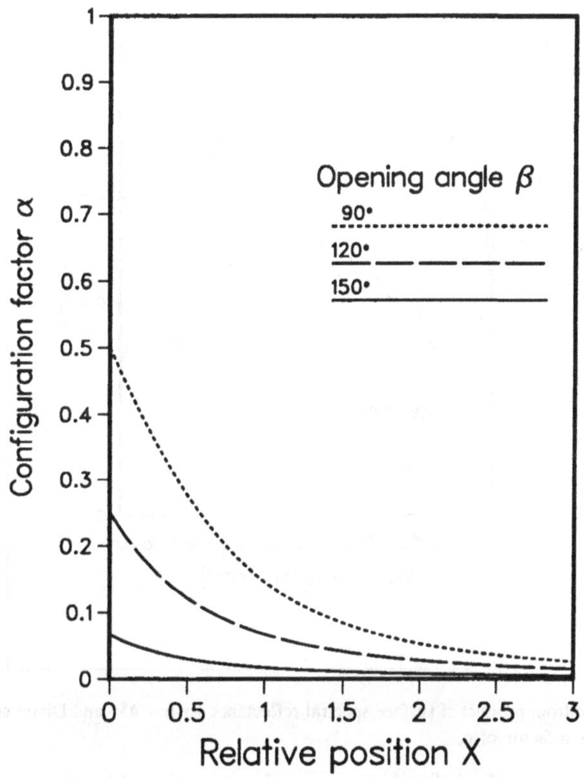

Fig. 4. Configuration factor α as a function of $X = x/\zeta$ for several opening angles β.

5.1 Two-zone Edge

In this section, we use equations (2) to construct an edge. For surfaces we choose any pair of reflectances from the collection of 370 surface spectral reflectance measurements catalogued by Krinov (1947). From Ho et al. (1990) we know how the finite dimensional linear model of reflectance performs for these functions and we start by choosing a pair of reflectances that are typical in terms of the accuracy of the underlying model. We choose Krinov #53 ("heather, dense growth before flowering") and Krinov #54 ("river valley with meadows") as typical curves. For these reflectances, the underlying model with 3 basis functions taken from Cohen gives errors of 11.04% and 8.11%, respectively.

Here, we define an error statistic that measures departures from the correct curve over the entire function of wavelength. It is essentially a rescaled Standard Error divided by the sample root mean square in order to make it a fraction rather than an absolute figure. Error is defined as

$$\left\{ \frac{\Sigma(y - y')^2}{\Sigma y^2} \right\}^{1/2}$$

where y is an actual value and y' is the estimate.

For illumination, we can use one of the standard daylights tabulated by Judd et al. (1964) or any other tabulated daylight, for example, those measured by Dixon (1978). Dixon included ultraviolet wavelengths

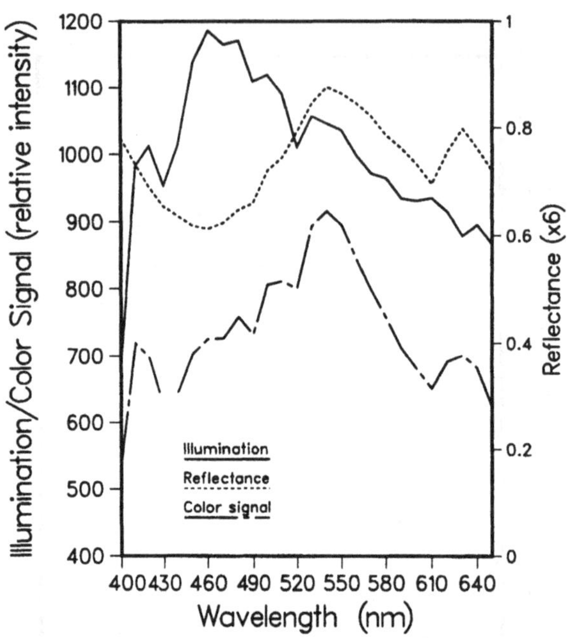

Fig. 5. Typical case: color signal formed from product of surface spectral reflectance Krinov #54 and Dixon spectrum for Australian daylight. The color signal has been scaled up by a factor of 6.

in developing a principal-component analysis for Australian daylight, but here we truncate to the visible, 400 nm–650 nm. The spectrum for the mean vector for daylight in Bendigo, Australia, is shown in figure 5 along with Krinov reflectance curve #54 and the resulting color signal composed of the product of the two spectra. For the domain D of initial value ϵ_2, ϵ_3's to use in the uniqueness check preprocessing step we took the largest absolute values of ϵ_2 and ϵ_3 for all of Judd's curves and then doubled that region. We suppose for simplicity that the shape and geometry of the two surfaces leads to geometrical factors of $\alpha_{21} = \alpha_{12} \equiv \alpha = 0.05$ for mutual reflection.

The algorithm of section 3 applied to the two color signals generated from this illumination spectrum and Krinov reflectances #53 and #54 can be tested by forming sensor values $\rho_k^{(1)}$, $\rho_k^{(2)}$, and ρ_k^{m}. To develop these signals, we used sensitivity functions (see figure 6) corresponding to Kodak filters #25 (red), #58 (green), and #47B (blue). The algorithm results are quite good: the value of α found by the solution method is 0.0495, so that the ratio of the estimated α to the actual one is 0.990. As pointed out above, using the approximate model algorithm the accuracy for this ratio remains nearly constant no matter what the actual value of α is.

To develop an error figure for the spectral reflectance function, it is important to recall that the algorithm gives a solution only up to an overall multiplicative constant. Therefore, the shape of the reflectance curve is derived, but not its absolute scale. To make a fair comparison, we scale by an appropriate factor and then calculate the error statistic. A best match of the derived curve and the actual one is found by solving for the multiplicative factor by minimizing the squared residuals after multiplying by the unknown factor. The normal equation for this minimization problem yields the factor.

As shown in figure 7, the spectral curve for reflectance 2 is quite close to that for Krinov #54; the overall error for the curve found by our algorithm is 9.71%.

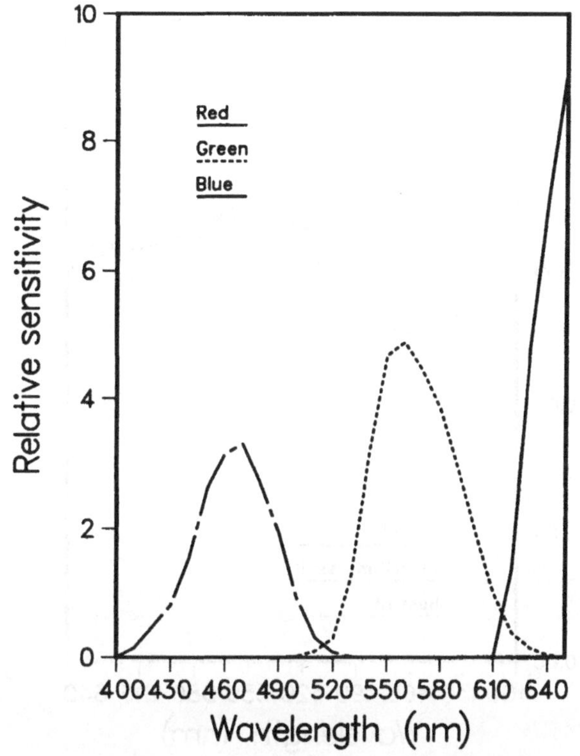

Fig. 6. Kodak filters #25 (red), #58 (green), and #47B (blue).

Since the error of the best fit using the underlying finite dimensional model is 8.11%, the ratio of our model error to the finite dimensional model error is only 1.197. The fit for surface 1, which has reflectance Krinov #53, is similar. We show in figure 8 the results of the algorithm in recovering the illumination compared to the actual spectrum. Again, the increase in error is quite small—we obtain an error of 4.10% as compared to the finite dimensional model error of 3.97%.

In table 1, we show the results of running the algorithm on sensor values derived from the same reflectances as above for all five of Judd's standard daylights, corresponding to five different correlated color temperatures, as well as the Dixon spectrum. As can be seen, the results are similar for all illuminations. As well, uniqueness was validated by the preprocessing step in these cases.

A more stringent test is carried out by utilizing atypical reflectance curves that are not well modeled by the basis functions. For this test we use Krinov #162 ("grass, young, green") along with Krinov #54. A best fit using the finite dimensional model itself has an error of 26.65% for Krinov #162. The results for the algorithm are again reasonably good, as seen in table 2.

To take into account noise, we add a random perturbation to each ρ-vector separately; each component independently receives additive noise equal to a percentage of the vector magnitude. For this test we use the Dixon spectrum coupled to reflectances Krinov #53 and Krinov #54. Running the algorithm repeatedly, we

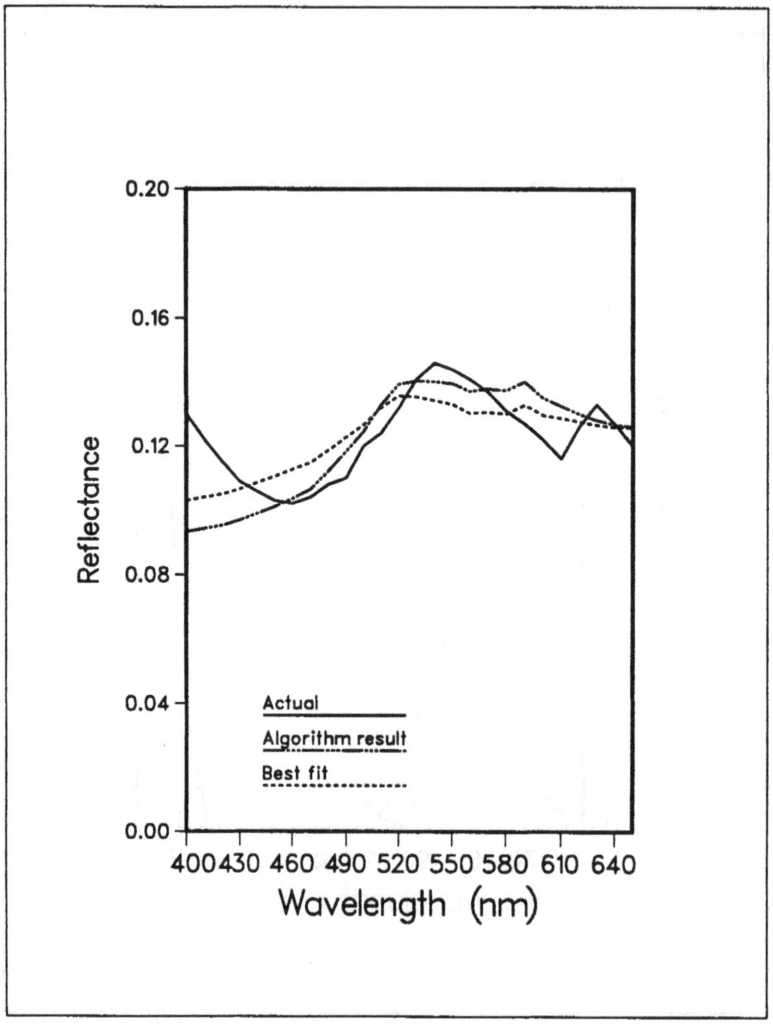

Fig. 7. Comparison of the best fit given by dimension-3 finite dimensional model with surface spectral curve resulting from the algorithm. The best-fit error (here, the reflectance curve is Krinov #54) has error = 8.11%, whereas the algorithm result has error = 9.71%, a factor of 1.197 worse than the best fit. When the geometrical mutual reflection factor α is 0.05, the algorithm gives the value 0.04949, so that the ratio of the estimate to the actual α is 0.9898.

arrive at average results, shown in table 3. Here, the 0% noise case is shown for comparison and all other figures are averages. For a few very poorly fit noisy signals at the higher noise level, the algorithm is less stable and a greater tolerance must be used for halting iteration. For each of these cases the uniqueness check also showed that the results were unreliable. So for a large noise component the results are not good. Nevertheless, the results are reasonably good for low noise, and in that regime the stability of the algorithm is very good. The number of iterations through the algorithm was typically 6 to 8.

If use is made in our model of broad-band sensor functions such as those in the human visual system, instead of bandpass filters common in video and film, then the results of the algorithm are found to be somewhat more accurate since a broader sampling of more frequencies results in a better representation of the entire color signal spectrum in terms of the finite dimensional model weights.

5.2 *Infinite-zone Edge*

In Drew and Funt (1990), we simulated the full physics (infinite-zone, infinite-bounce) of the interreflection at

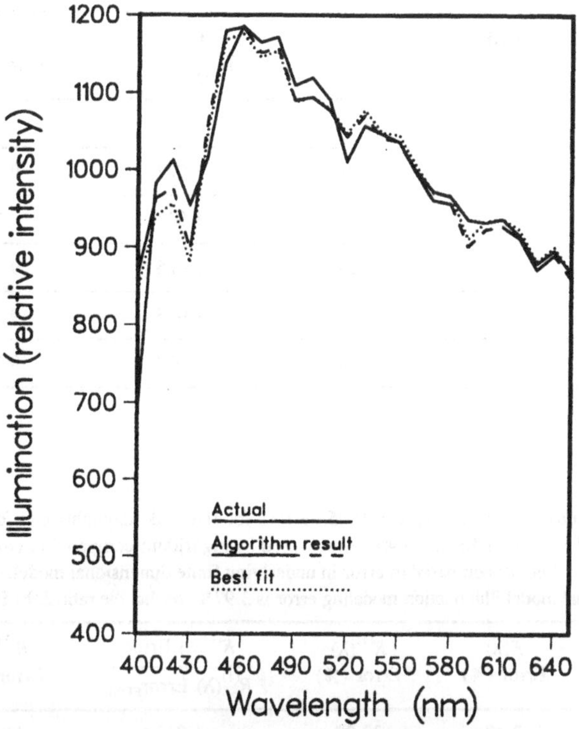

Fig. 8. The original Dixon illumination spectrum for Australian daylight and results of the mutual reflection algorithm applied to the color signal of figure 5—error 4.10%.

an edge under diffuse illumination by employing the radiosity method (Goral et al. 1984). However, unlike the usual approach in computer graphics, we carried out the radiosity calculation separately on each wavelength instead of on the usual three color bands and therefore could not use standard computer graphics radiosity tools to synthesize test images. In generating a correct color distribution at the edge it is crucial that one apply the filter functions to generate RGB values only after calculating the entire color signal spectrum for each pixel.

We used this method on the edge of figure 2, using Krinov spectra #54 and #248 and incident illumination D65. Figure 9 shows the variation in RGB on surface 2 that occurs with increasing distance from the edge. For comparison, the (constant) RGB values resulting from $S^{(2)}(\lambda)E(\lambda)$ in the absence of interreflection are also plotted. The ratios R/B and G/B graphed in figure 10 show that not only the intensities, but the colors too change with distance from the edge. We can see from figure 10 that derivatives of ratios approach zero as one gets farther from the edge.

Running our mutual reflection analysis algorithm on the calculated RGB image data, we find that reflectances on each side are recovered quite accurately. Except at pixels very close to the edge, the algorithm obtains a reflectance $S^{(2)}(\lambda)$ that is virtually independent of position. The error with respect to the actual reflectance spectrum is 4.37%, which translates into a color difference ΔE of 3.54 units in CIELUV uniform color space (see Wyszecki and Stiles 1982). For surface 2, the best least-squares approximation using the finite

Table 1. Errors using typical reflectances: Krinov #53 and #54, $\alpha = 0.5$, and Judd's North American daylights of five different correlated color temperatures as well as Dixon's Australian daylight. The finite dimensional model best curves have errors of 11.04% for Krinov #53 and 8.11% for Krinov #54. Columns show: (1) α recovered by algorithm; (2) ratio of the recovered α to the actual one; (3) error for illumination spectrum found by algorithm; (4) absolute error of reflectance curve found by algorithm for surface 1; (5) error of reflectance curve for surface 1 compared to error in underlying finite dimensional model; (6, 7) similarly for reflectance curve for surface 2. For the Dixon case, the finite dimensional model illumination modeling error is 3.97%, so that the ratio $E(\lambda)$ Error / $E(\lambda)$ Error_{FDM} is 1.032.

Daylight	α	$\frac{\alpha_{EST}}{\alpha_{ACT}}$	$E(\lambda)$ Error (%)	$R^{(1)}(\lambda)$ Error (%)	$\frac{R^{(1)}(\lambda)\ \text{Error}}{R^{(1)}(\lambda)\ \text{Error}_{FDM}}$	$R^{(2)}(\lambda)$ Error (%)	$\frac{R^{(2)}(\lambda)\ \text{Error}}{R^{(2)}(\lambda)\ \text{Error}_{FDM}}$
4800K	0.04971	0.9943	0.30	11.35	1.028	9.81	1.209
5500K	0.04971	0.9942	0.29	11.34	1.027	9.79	1.206
6500K	0.04973	0.9945	0.31	11.34	1.027	9.75	1.202
7500K	0.04972	0.9944	0.31	11.32	1.026	9.73	1.200
10000K	0.04972	0.9944	0.31	11.33	1.026	9.70	1.196
Dixon	0.04949	0.9898	4.10	11.34	1.027	9.71	1.197

Table 2. Poorly fit case: reflectances Krinov #162 and #54, $\alpha = 0.05$—errors for five Judd daylights and for Dixon daylight (best model fit has error 26.65% for Krinov #162 and 8.11% for Krinov #54); α recovered by algorithm; compared to correct α; illumination spectrum error found by algorithm; reflectance curve 1 error; compared to error in underlying finite dimensional model; similarly for reflectance curve 2. For the Dixon case, the finite dimensional model illumination modeling error is 3.97%, so that the ratio $E(\lambda)$ Error / $E(\lambda)$ Error_{FDM} is 1.428.

Daylight	α	$\frac{\alpha_{EST}}{\alpha_{ACT}}$	$E(\lambda)$ Error (%)	$R^{(1)}(\lambda)$ Error (%)	$\frac{R^{(1)}(\lambda)\ \text{Error}}{R^{(1)}(\lambda)\ \text{Error}_{FDM}}$	$R^{(2)}(\lambda)$ Error (%)	$\frac{R^{(2)}(\lambda)\ \text{Error}}{R^{(2)}(\lambda)\ \text{Error}_{FDM}}$
4800K	0.04988	0.9975	3.47	32.29	1.212	11.13	1.371
5500K	0.04987	0.9974	3.41	32.11	1.205	11.04	1.361
6500K	0.04988	0.9976	3.37	31.93	1.198	10.95	1.349
7500K	0.04987	0.9974	3.19	31.83	1.194	10.89	1.342
10000K	0.04987	0.9974	2.86	31.68	1.189	10.80	1.331
Dixon	0.04964	0.9929	5.66	31.95	1.199	10.90	1.343

Table 3. Noise: reflectances Krinov #53 and #54, Dixon spectrum, $\alpha = 0.05$—finite dimensional model best fits have $E(\lambda)$ Error_{FDM} = 3.97%, $R^{(1)}(\lambda)$ Error_{FDM} = 11.04%, and $R^{(2)}(\lambda)$ Error_{FDM} = 8.11%. Columns show: α recovered by algorithm; compared to correct α; illumination spectrum error found by algorithm; reflectance curve 1 error; reflectance curve 2 error.

Noise (%)	α	$\frac{\alpha_{EST}}{\alpha_{ACT}}$	$E(\lambda)$ Error (%)	$R^{(1)}(\lambda)$ Error (%)	$R^{(2)}(\lambda)$ Error (%)
0	0.04983	0.9967	4.13	11.33	9.75
5	0.04385	0.8771	17.13	16.80	15.05
10	0.03912	0.7823	38.50	29.030	28.34

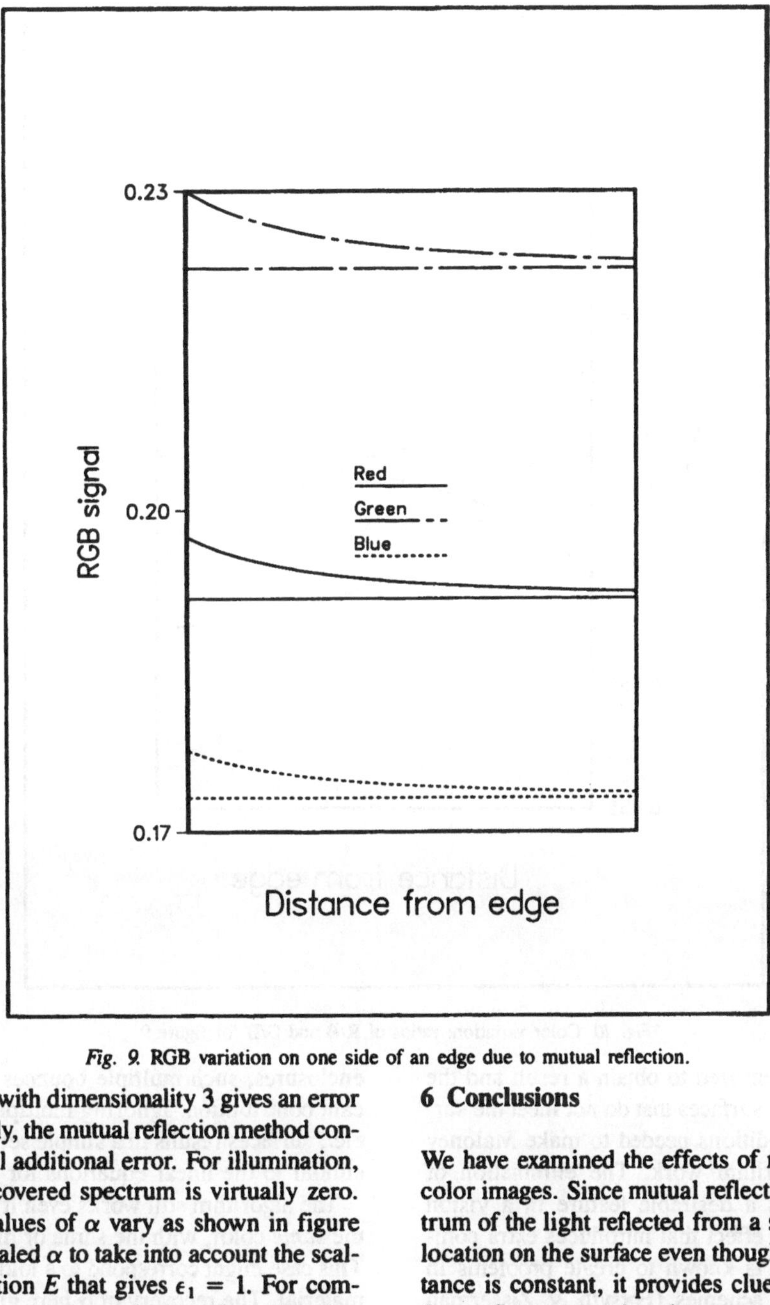

Fig. 9. RGB variation on one side of an edge due to mutual reflection.

dimensional model with dimensionality 3 gives an error of $\Delta E = 3.25$; clearly, the mutual reflection method contributes only a small additional error. For illumination, the error of the recovered spectrum is virtually zero.

The recovered values of α vary as shown in figure 11. Here we have scaled α to take into account the scaling of the illumination E that gives $\epsilon_1 = 1$. For comparison we also show the theoretical configuration factor α as determined by equation (10). We display that factor offset so that it goes to zero and not to the small value it takes on at the pixel farthest from the edge. As one would expect since the recovered α maps the contributions from all bounces into a one-bounce factor, it is marginally larger than the theoretical α generated using equation (10) that models an infinite-zone, one-bounce situation.

6 Conclusions

We have examined the effects of mutual reflection in color images. Since mutual reflection causes the spectrum of the light reflected from a surface to vary with location on the surface even though the surface reflectance is constant, it provides clues to the actual surface reflectance and to the geometry of the situation. Tests show that the algorithm is quite robust.

The method has the potential for eliminating mutual reflection effects and in most cases recovers a good approximation of the complete spectral information. Thus the method yields more than enough information to supply color descriptors for the object surfaces that are independent of ambient illumination; therefore, color constancy is obtained. Sensor values from only three

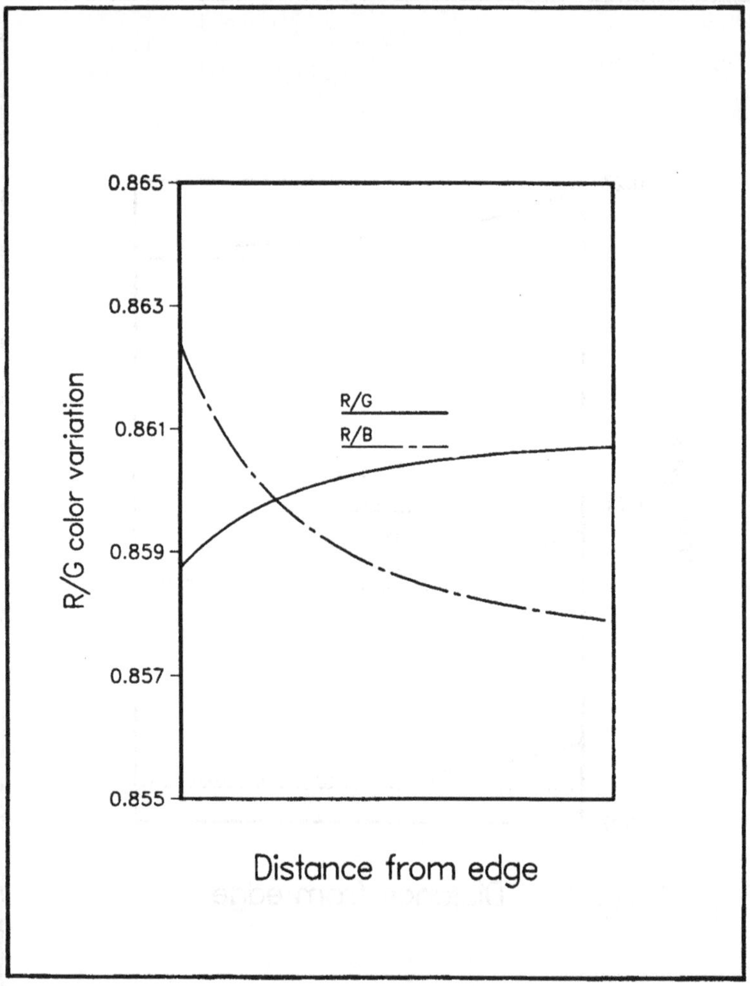

Fig. 10. Color variation: ratios of R/B and G/B for figure 9.

pixels need to be measured to obtain a result and the method will work on surfaces that do not meet the surface complexity conditions needed to make Maloney and Wandell's algorithm work. The elimination of mutual reflection is a desirable feature in a vision system since it is an effect that introduces extra complexities; that is, it is known to create problems in shape-from-shading schemes (Forsyth & Zisserman 1989, 1990).

Clearly, the interreflection model put forward here, that uses basis function decompositions for an accurate whole-spectrum model of reflection, can be extended to the modeling of multisurface enclosures for graphics applications (cf. Goral et al. 1984). Multiple interreflections could be incorporated at the expense of having a more difficult mathematical system to solve; for enclosures, such multiple bounces may add a significant contribution. Ignoring multiple bounces for 3, 4, etc., surfaces results in a simple set of linear equations similar to the linear equations for $I_m(\lambda)$ in section 2.

The algorithm still works even if both surfaces have the *same* color, with the same or differing brightness. This case might correspond to a folded piece of uniform material. The recovery of α here gives shape information near the fold. The method works in this one-color case because the reflected light is a different color from the illumination, and the extra measurements of sensor values supply enough information to disentangle the two.

The results show that, in situations in which mutual reflection is present, measurements of the sensor values taken separately from the two different interreflecting

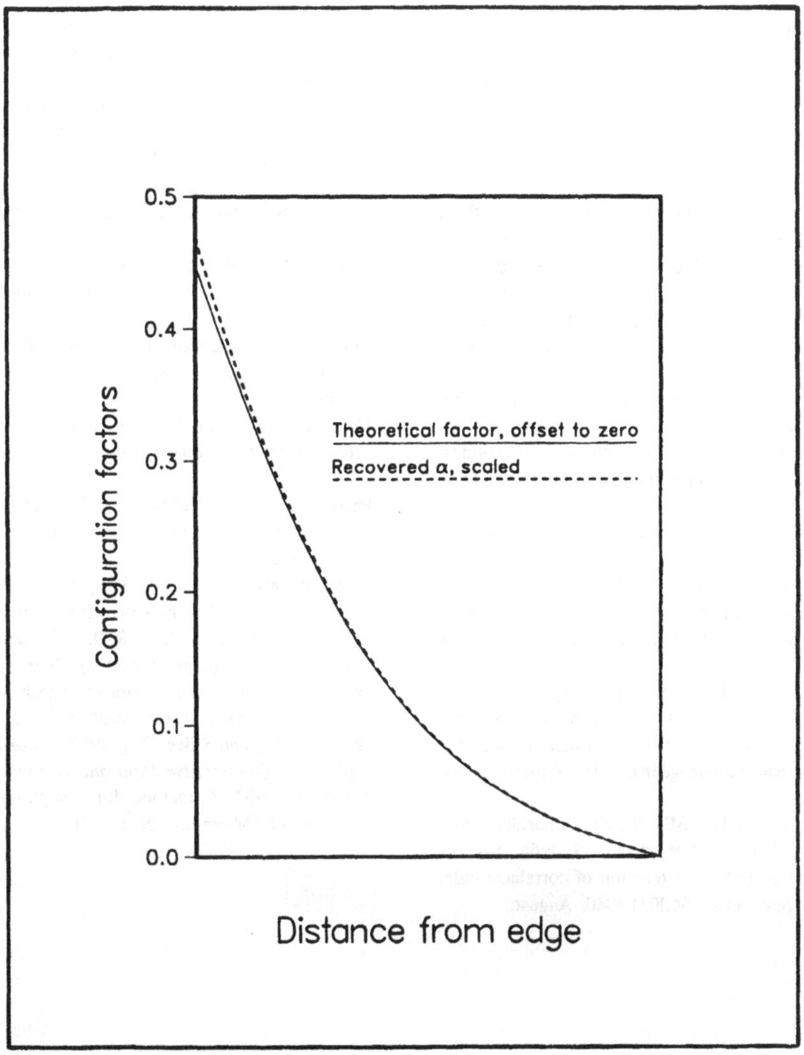

Fig. 11. Variation of one-bounce, two-zone α factor, where the sensor RGB inputs are derived from a full multi-zone model. Theoretical one-bounce, infinite-zone factor is shown for comparison.

surfaces, as well as from within the mutual reflection zone, can convey sufficient information to allow complete recovery of the two surface spectral reflectance functions as well as of ambient illumination. We also recover the factor α, which summarizes information about the relative geometry of the two surfaces.

7 Acknowledgments

M.S. Drew is indebted to the Centre for Systems Science at Simon Fraser University for partial support; B.V. Funt thanks both the CSS and the Natural Sciences and Engineering Research Council of Canada for their support.

References

Beck, J. 1972. *Surface Color Perception*, Cornell Univ. Press: Ithaca, N.Y.

Blackwell, K.T., and Buchsbaum, G. 1988. Quantitative studies of color constancy. *J. Opt. Soc. Amer. A.*, 5:1772–1780.

Brainard, D.A., Wandell, B.A., and Cowan, W.B. 1989. Black light: How sensors filter spectral variation of the illuminant. *IEEE Trans. Biomed. Eng.* 36:140–149, 572.

Brill, M.H., and West, G. 1986. Chromatic adaptation and color constancy: A possible dichotomy. *Col. Res. Appl.* 11:196–204.

Buchsbaum, G. 1980. A spatial processor model for object colour perception. *J. Franklin Inst.* 310:1-26.

Burden, R.L., Faires, J.D. and Reynolds, A.C. 1981. *Numerical Analysis.* Prindle, Weber & Schmidt: Boston.

Cohen, J. 1964. Dependency of the spectral reflectance curves of the Munsell color chips. *Psychon. Sci. Amer. A*, 6:318–322.

Dixon, E.R. 1978. Spectral distribution of Australian daylight. *J. Opt. Soc. Amer.* 68:437–450.

Drew, M.S., and Funt, B.V. 1990. Calculating surface reflectance using a single-bounce model of mutual reflection. *Proc. Intern. Conf. Comput. Vision*, Osaka, December 4–7.

D'Zmura, M. and Lennie, P. Mechanisms of color constancy. *J. Opt. Soc. Amer.* (10):1662–1672.

Forsyth, D., and Zisserman, A. 1989. Mutual illumination. *Proc. conf. Comput. Vision Patt. Recog.*, San Diego, pp. 466–473.

Forsyth, D., and Zisserman, A. 1990. Shape from shading in the light of mutual illumination. *Image Vision Comput.* 8:42–49.

Funt, B.V., and Ho, J. 1988. Color from black and white. *Proc. 2nd Intern. Conf. Comput. Vision*, pp. 2–8, Dec. 5–8, Tarpon Springs, FL.

Funt, B.V., and Ho, J. 1989. Color from black and white. *Intern. J. Comput. Vision* 3(2).

Gershon, R., Jepson, A.D, and Tsotsos, J.K. 1986. Ambient illumination and the determination of material changes. *J. Opt. Soc. Amer.* 3:1700–1707.

Gershon, R., Jepson, A.D. and Tsotsos, J.K. 1987. From [R,G,B] to surface reflectance: Computing color constant descriptors in images. *Proc. 10th IJCAI*, Milan, pp. 755–758.

Goral, C.M., Torrance, K.E., Greenberg, D.P. and Battaile, B. 1984. Modeling the interaction of light between diffuse surfaces. *Computer Graphics* 18:213–222.

Ho, J., Funt, B.V., and Drew, M.S. 1990. Separating a color signal into illumination and surface reflectance components: Theory and applications. *IEEE Trans. Patt. Anal. Mach. Intell.* 12:966–977.

Horn, B.K.P. 1977. Understanding image intensities. *Artificial Intelligence* 8:201–231.

Horn, B.K.P.. 1986. *Robot Vision*. MIT Press: Cambridge, MA.

Judd, D.B., MacAdam, D.L., and Wyszecki, G. 1964. Spectral distribution of typical daylight as a function of correlated color temperature. *J. Opt. Soc. Amer.* 54:1031–1040, August.

Krinov, E.L. 1947. Spectral reflectance properties of natural formations. *Technical Translation* TT–439, National Research Council of Canada.

Koenderink, J.J., and van Doorn, A.J. 1983. Geometrical modes as a general method to treat diffuse interreflections in radiometry. *J. Opt. Soc. Amer.* 73:843–850.

Maloney, L.T. 1985. Computational approaches to color constancy, Ph.D. dissertation, Stanford University, Applied Psychology Laboratory.

Maloney, L.T. 1986. Evaluation of linear models of surface spectral reflectance with small numbers of parameters. *J. Opt. Soc. Amer.* A, 3:1673–1683.

Maloney, L.T., and Wandell, B.A. 1986. Color constancy: a method for recovering surface spectral reflectance. *J. Opt. Soc. Amer.* A, 3:29–33.

Nayar, S.K., Ikeuchi, K., and Kanade, T. 1990. Shape from interreflections. *Proc. Intern. Conf. Comput. Vision*, Osaka, December 4–7.

Parkkinen, J.P.S., Hallikainen, J., and Jaaskelainen, T. 1989. Characteristic spectra of Munsell colors. *J. Opt. Soc. Amer.* A, 6:318–322.

Siegel, R., and Howell, J.R. 1981. *Thermal Radiation Heat Transfer*. Hemisphere Publishing Corp: New York.

Sparrow, E.M., and Cess, R.D. 1978. *Thermal Radiation Heat Transfer*. Hemisphere Publishing Corp.: New York.

Wandell, B.A. 1987. The synthesis and analysis of color images. *IEEE Trans. Patt. Anal. Mach. Intell.* 9:2–13.

Wyszecki, G., and Stiles, W.S. 1982. *Color Science: Concepts and Methods, Quantitative Data and Formulas*. Wiley: New York.

Yuille, A. 1987. A method for computing spectral reflectance. *Biological Cybernetics* 56:195–201.

Color Space Analysis of Mutual Illumination

Brian V. Funt and Mark S. Drew
School of Computing Science
Simon Fraser University
Vancouver, B.C.
Canada V5A 1S6
(604) 291-3126
e-mail funt@cs.sfu.ca

Abstract

Mutual illumination occurs when light reflected from one surface impinges on a second one. The resulting additional illumination incident on the second surface affects both the color and intensity of the light reflected from it. As a consequence, the image of a surface in the presence of mutual illumination differs from what it otherwise would have been in the absence of mutual illumination. Unaccounted for mutual illumination can easily confuse methods that rely on intensity or color such as shape-from-shading or color-based image segmentation. In this paper, we introduce an algorithm that removes mutual illumination effects from images. The domain is that of convex surfaces of uniform color and diffuse reflectance where for each surface the interreflection occurs mainly from one other surface. The algorithm is based on a singular value decomposition of the colors coming from each surface. Geometrical information about where on the surface the colors emanate from is not required. The RGB triples from a single convex surface fall in a plane; intersecting the planes generated from two interreflecting surfaces results in a unique interreflection color. Each pixel can then be factored into its interreflection and no-interreflection components so that a complete no-interreflection image is produced.

1 Introduction

The interreflection of light in color images can substantially alter colors and intensity levels of pixels corresponding to points receiving light, not just from a light source, but by reflection from other surfaces in the scene as well. This interreflected light is often called mutual illumination and its effects are well known [11, 5, 6, 16, 2, 7].

The mutual illumination effect can be slight when there is a lot of illumination directly from a light source [2, 7, 3]; however, in enclosed spaces such as a room, the infinite exchange of radiation between the walls provides a large fraction of the light in the room [12]. In computer graphics, it has been necessary to model these interreflections using the radiosity method [9] in order to obtain realistic images.

For computer vision, the one-bounce model of mutual illumination proposed in [2, 7, 3] provides a good approximation in many situations. The one-bounce model simply makes use of the fact that the intensity of the interreflection diminishes substantially with each bounce.

Even when mutual illumination is small as a percentage of the total illumination, it may profoundly alter the results of computer vision algorithms [5, 6]. In part this is because mutual illumination changes image intensities in a consistent, non-random way so that errors it introduces into, say, a shape-from-intensity solution accumulate rather than cancel out. For example, imagine the region near a concave fold in a surface of a uniformly colored object. Intensities of points near the fold are all increased (but by differing amounts) due to interreflection and this increase will lead most shape-from-intensity algorithms into ascribing an incorrect set of direction cosines to the surface there. Nayar [16] observes that the concavity will always appear as being shallower than it actually is.

Nayar makes an interesting attempt to overcome this problem by using an iterative scheme in which first an estimate of the shape is calculated from the intensity data; then with this shape estimate as input, the radiosity method is applied to estimate a corrected, no-interreflection image intensity distribution. These steps are iterated until convergence. The method is fundamentally flawed, however, because it does not take into account the color of the interreflected light.

Near a concave edge formed by two surfaces of different color, interreflection causes a significant component of the light reflected from either side to have a color which differs from that of either of the original surface colors. The spectrum of the interreflected light is formed as the product of the illumination spectrum and the spectra of the two surface reflectance functions. In fact, the surface reflectance functions multiply together as many times as the light is interreflected, with the total mutual illumination contribution equaling the sum of an infinite number of interreflections. The resulting spectrum is then filtered by the sensor response functions to obtain an RGB triple representing the color.

Instead of calculating the energy leaving each surface facet in terms of the complete spectrum, or in three bands, Nayar *et al.* assume that the reflectance can be summed up by a single number, the albedo. The difficulty with Nayar's approach can be seen by considering a concave edge that is red on one side and blue on the other. In the extreme case, the spectral reflectance function of the red will be zero at the blue end of the spectrum and vice-versa. The product of the red and blue spectra will therefore be zero everywhere. As a result, while mutual illumination is to be expected based on the edge geometry, none in fact occurs; and the radiosity step of Nayar's algorithm, which describes surface reflectance by a single albedo value, will over-estimate the interreflection and lead to an incorrect calculation of the surface shape. The problem persists even for less extreme cases, such as an edge which is red on both sides, because the product of the red spectrum with itself does not equal itself.

Color and mutual illumination are inextricably interrelated. Unless we are to restrict ourselves to surfaces that are varying shades of grey, color will be fundamental to the analysis of mutual illumination. Conversely, mutual illumination can be important to the analysis of color. In previous work, [2, 7, 3] we analyzed mutual illumination to obtain color constancy (i.e. illumination-independent color descriptors). In fact, it is possible to reconstruct good approximations to the *entire* surface spectral reflectance functions for two surfaces participating in mutual illumination.

This is *strong* color constancy in that the entire spectrum is established [10]. Weaker versions of color constancy have also been advocated. For example, Forsyth [4] has suggested that RGB values obtained under a canonical illuminant may be viewed as color-constant descriptors.

In the present paper, we describe an algorithm that produces an image which is free of mutual illumination effects. Since these effects include induced color shifts, their elimination is also a form of color constancy, but in a weaker sense than either of the above definitions. Our algorithm finds the color of each object independent of mutual illumination, where "color" means the RGB values produced under the light source in use, not a canonical source.

The domain is restricted to interreflections between two matte, convex surfaces, each of uniform color, with illumination that can vary spatially in its intensity but not in its spectral composition. The algorithm generates two images. One is a no-interreflection image in which the colors and shading on each surface are as they would have been if the alternate surface had not been in the scene. The second is an image of the mutual illumination distribution, the magnitude of which represents a type of "configuration factor" between the objects [18]. Here, we use the term "interreflection factor" to describe this field because it includes the effect of multiple interreflections within a one-bounce model; whereas, a true configuration factor represents only a single bounce. This interreflection factor encapsulates a good deal of information about the scene geometry including orientation, occlusion, and shape. Potentially, it could be used to extract some of these features.

A one-bounce model of mutual illumination predicts that the RGB triples measured from each of an interreflecting pair of surfaces should consist of linear combinations of a *no-bounce* color and a *one-bounce* color. A no-bounce color means the only-reflected-once (i.e. no interreflections involved) color that results from the reflection of the source illuminant off one of the surfaces. A one-bounce color results from the reflection, by the second surface, of the first surface's no-bounce color. Although in general there are two no-bounce colors, one for each surface, there is only one

one-bounce color because of symmetry.

To the extent that the one-bounce model holds and the domain restrictions apply (i.e. no self-reflection or specularities), the linear combination of two components means that there will be a plane in RGB color space containing all the RGB triples of a surface. A reliable method for determining the plane is to carry out a Singular Value Decomposition (SVD) of the RGB values [8].

We would like to find the RGB vectors of the one-bounce and no-bounce colors. The two principal SVD eigenvectors define the plane, but unfortunately they do not represent these colors. Nonetheless, in principle once a surface's color-space plane is known, bands within which these colors must lie can be determined by the method originated by Lawton and Sylvestre [15] and further elucidated by Kawata et al. [13]. This method treats each surface independently.

We can, however, make use of the fact that the two color planes are related via their common one-bounce color. For example, the spectrum of light from a red surface reflected secondarily by a blue surface is the same as that of light from a blue surface reflected secondarily by a red one; hence, we can expect both the color planes of the two surfaces to contain the *same* one-bounce, red-blue, interreflection color. Intersecting these planes will yield the one-bounce color.

The idea of intersecting color-space planes has been used before, but in the analysis of specularities. In the analysis of interreflections, we will follow the SVD analysis used by Tominaga and Wandell [20] in their treatment of surface and body reflectance from multiple surfaces under a simple model of specularity. They make use of the fact that it is possible to derive the spectrum of the illuminant, within the dichromatic reflection model [17, 14], by crossing the planes of color signals (entire spectra) from each of several surfaces [21, 22] (and see also [19]). Then the SVD analysis for each surface can be combined, via the Lawton and Sylvestre method, to determine a band of spectra within which the body reflection function must lie.

Section 2 sets out the color space analysis of mutual illumination: how the color planes are found and how the one-bounce color is determined. The no-bounce color lies within a band defined by the one-bounce color along with the constraint that all colors must have entirely non-negative components. Given the no-bounce color of a surface, the shading and the geometric interreflection factor are easily calculated at each pixel.

We make use of 3-component vectors, not entire spectra, so that *ab initio* it is not clear how well the SVD method will work. Therefore it is useful to start with a simple, simulated image. Since the physics of a planar edge under diffuse illumination is perfectly understood [18], in Section 3 we examine the simulated image of a two-color, semi-infinite planar edge (cf. [2]). We find that the Lawton and Sylvestre method runs into problems because the vectors have so few components (i.e., 3) and present a modified method for this case.

Section 4 details the result of applying the algorithm to a real image as well as a method for automatically evaluating the confidence with which the method has identified the correct surface colors, shading fields, one-bounce color, and interreflection factor fields.

2 Mutual Illumination and Color Space Analysis

2.1 One-bounce model of mutual illumination

Consider a semi-infinite edge formed by two differently colored materials, as shown in Fig. 1. Mutual illumination can occur between non-touching surfaces as well, as shall be shown in Section 4, but the edge in Fig. 1 will be studied in Section 3 and might be considered fundamental to the problem.

Suppose the edge is illuminated with spectral power distribution $E(\lambda)$. Denote the surface spectral reflectance function for surface A by $S^A(\lambda)$ and that for surface B by $S^B(\lambda)$. Let us assume that all surfaces are Lambertian, or at least that we exclude specularities, and call the RGB values resulting from the direct illumination impinging on both surfaces, plus the results of all interreflection, ρ_k, $k = 1..3$. Naturally, the values of ρ_k vary over the surfaces A and B and especially while moving away from the edge.

To exactly model the ρ_k requires evaluating an integral equation representing the infinite-bounce interreflection at each wavelength and then applying the color filters. The physics of the situation is given in [18]. To simplify the analysis, we adopt a one-bounce approximation of the radiant exchange between the two surfaces [2, 7, 3].

The one-bounce model describes the color signal $C^A(x, \lambda)$ from a pixel at location x on surface A as:

$$C^A(x,\lambda) \simeq \alpha^A(x)\, E(\lambda)S^A(\lambda) + \beta^{BA}(x)S^A(\lambda)E(\lambda)S^B(\lambda)\,. \tag{1}$$

A similar equation can be written for surface B.

Equation (1) states that C is comprised of the light reflected from surface A with no interreflection — the no-bounce color signal — plus the light reflected from surface B and then reflected again from surface A. Of course, this once-reflected light will reflect again off B, but we truncate an infinite series of bounces and call the second term the one-bounce contribution.

Due to changes in surface orientation and position, the intensity of the direct illumination E will generally vary spatially, although its spectrum is assumed to remain constant. The factor $\alpha^A(x)$ represents this intensity variation or *shading* at point x.

For reflection from surface B to surface A, the proportionality factor $\beta^{BA}(x)$ represents the relative contribution of interreflection to the color signal. It encapsulates into a single number all the possible circumstances that can affect the magnitude of the one-bounce contribution, including the local surface orientation at x, the overall shape of both surfaces, the shading on both surfaces, and the possibility that some points on surface B are occluded from the vantage point of pixel x. For image synthesis, calculating $\beta^{BA}(x)$ poses a computationally intensive problem (cf. [9]); for image analysis, on the other hand, *deriving* the proportionality factors $\alpha(x)$ and $\beta(x)$ given an RGB image can be done quite efficiently.

To derive the colors and proportionality factors, we must consider the nature of the color space.

On sensors of sensitivity $R_k(\lambda)$, a color signal C creates sensor response $\vec{\rho}$ according to

$$\rho_k = \int C(\lambda)\, R_k(\lambda)\, d\lambda \tag{2}$$

Application of equation (2) to the $C(\lambda)$ in equation (1) leads to vectors $\vec{\rho}$ on surfaces A and B given by

$$\begin{aligned} \rho_k^A(x) &\simeq \alpha^A(x)\, \rho_k^{A\_nobounce} + \beta^{BA}(x)\rho_k^{onebounce} \\ \rho_k^B(x) &\simeq \alpha^B(x)\, \rho_k^{B\_nobounce} + \beta^{AB}(x)\rho_k^{onebounce} \end{aligned} \tag{3}$$

where

$$\begin{aligned} \rho_k^{A\_nobounce} &= \int E(\lambda)S^A(\lambda)R_k(\lambda)d\lambda \\ \rho_k^{B\_nobounce} &= \int E(\lambda)S^B(\lambda)R_k(\lambda)d\lambda \end{aligned}$$

and

$$\rho_k^{onebounce} = \int E(\lambda)S^A(\lambda)S^B(\lambda)R_k(\lambda)d\lambda$$

All $\vec{\rho}$ from surface A are formed by a linear combination of $\vec{\rho}^{A\_nobounce}$ and $\vec{\rho}^{onebounce}$. Similarly, on surface B the $\vec{\rho}$ are made up of $\vec{\rho}^{B\_nobounce}$ and the *same* $\vec{\rho}^{onebounce}$. To the extent that equation (1) correctly approximates the color signal from surface A, all the $\vec{\rho}$ arising from surface A will lie in a plane in $\vec{\rho}$-space. In Brill's terminology [1], the $\vec{\rho}$ form a rank-2 image field.

2.2 SVD color space analysis

Since the two color-space planes formed by all the ρ_k^A and all the ρ_k^B contain the common color $\vec{\rho}^{onebounce}$ it lies in their intersection and so is easy to compute. Singular value decomposition (SVD) analysis provides a reliable method of finding the planes. The SVD operates on the matrix **R** of all the data from surface A. **R** is $N \times 3$, where N is the number of pixels. The SVD analysis generates additional matrices that decompose **R** as:

$$\mathbf{R} = \mathbf{U}\,\mathbf{\Lambda}\,\mathbf{V}$$

where **U** is an $N \times 3$ matrix, $\mathbf{\Lambda}$ is a 3×3 diagonal matrix of eigenvalues, and **V** is a 3×3 matrix of eigenvectors in 3-dimensional RGB color space. If the third eigenvalue in $\mathbf{\Lambda}$ is much smaller than the first two, then most of the pixels lie in the plane defined by the principal component vectors $\vec{v}_1$, $\vec{v}_2$ formed from the columns of **V**.

2.3 One-bounce color from crossing planes

By analogy to the similar situation that arises with specularities [21, 22, 19], we proceed to cross the color space planes generated from the two surfaces in order to find the one-bounce color $\vec{\rho}^{onebounce}$. In [19] the planes were embedded in high-dimensional spaces of vectors representing entire spectra, so the SVD method was used in calculating their intersection. In the present situation, however, the planes are simply embedded in 3-space, so that the crossed-planes color $\vec{\rho}^{onebounce}$ is unique and can be found simply by algebra.

If the first two principal component vectors on surfaces A and B are $\vec{v}_1^A, \vec{v}_2^A, \vec{v}_1^B, \vec{v}_2^B$ then solving the following vector equation in the 3 unknowns a_2, b_1, b_2 yields the intersection vector:

$$\vec{v}_1^A + a_2 \vec{v}_2^A - b_1 \vec{v}_1^B - b_2 \vec{v}_2^B = \vec{0} \qquad (4)$$

Therefore

$$\vec{\rho}^{onebounce} = \vec{v}_1^A + a_2 \vec{v}_2^A \qquad (5)$$

For surface A, the one-bounce model predicts that any $\vec{\rho}$ is composed of the above color and the no-bounce color for surface A, $\vec{\rho}^{A\_nobounce}$. As will be addressed in the next section, the problem remains of how to find the no-bounce color of each side.

2.4 Quarter-circle analysis

Within the one-bounce model of mutual reflection, each $\vec{\rho}$ is composed of a linear combination of $\vec{\rho}^{onebounce}$ and $\vec{\rho}^{nobounce}$. From a measured set of $\vec{\rho}$'s and the fact that $\vec{\rho}^{onebounce}$ has already been calculated, we wish to derive $\vec{\rho}^{nobounce}$. For this purpose, the most convenient phrasing of the Lawton and Sylvestre [15] analysis is that given by Tominaga and Wandell [20], which they called 'quarter-circle' analysis. Following this method, we first project all the measured $\vec{\rho}$ on surface A onto the $\vec{v}_1^A, \vec{v}_2^A$ plane and then rotate each $\vec{\rho}$ into a new coordinate system with axes $\vec{a_1} \equiv \vec{\rho}^{onebounce}$ and $\vec{a_2}$ perpendicular to $\vec{a_1}$ in that plane.

Described in this new coordinate system and normalized to unit length, each measured $\vec{\rho}$ vector must lie within the first quadrant of a unit circle (see Fig. 2). Each $\vec{\rho}$, therefore, has positive components a_1, a_2 along the $\vec{a_1}, \vec{a_2}$ directions. The task is to establish limits on the direction of $\vec{\rho}^{nobounce}$.

The first constraint on $\vec{\rho}^{nobounce}$ stems from the fact that of the measured $\vec{\rho}$'s, the one closest to the $\vec{a_2}$ axis represents the point on the surface where the mutual illumination is least. The $\vec{\rho}$ nearest to $\vec{a_2}$ has the largest ratio of its components, a_2/a_1 (histogramming the ratios avoids picking an outlier). This $\vec{\rho}$ gives a lower bound on a_2/a_1 for $\vec{\rho}^{nobounce}$:

$$a_2/a_1 \geq \max_{data}(a_2/a_1)$$

An upper bound follows from the condition that $\vec{\rho}^{\,nobounce}$ cannot have negative components. Since $\vec{a_1}$ has positive components and $\vec{a_2}$ is perpendicular to it, it follows that $\vec{a_2}$ has some negative components in RGB color space. As we move away from $\vec{a_2}$ toward $\vec{a_1}$, we reach a point where every component of the $\vec{\rho}$ vector is nonnegative. That point is the upper bound for a_2/a_1.

The limiting vector contains just enough of $\vec{a_1}$ added to $\vec{a_2}$ that the component corresponding to the most negative component of $\vec{a_2}$ becomes precisely *zero*. The condition is

$$a_1 \vec{a_1} + a_2 \vec{a_2} \geq \vec{0}$$

so that

$$a_2/a_1 \leq [\max_{k=1..3}(-a_{2k}/a_{1k})]^{-1}$$

Once the band of possible $\vec{\rho}^{\,nobounce}$ vectors has been established, the original image $\vec{\rho}$ data is easily decomposed into its components in the $\vec{\rho}^{\,onebounce}$ and $\vec{\rho}^{\,nobounce}$ directions. Of course, the previous analysis has established only that $\vec{\rho}^{\,nobounce}$ lies within a band, so the components are only restricted to be within a band as well.

Given a particular choice of $\vec{\rho}^{\,nobounce}$ from the band of possible $\vec{\rho}^{\,nobounce}$ vectors that forms an angle θ with $\vec{\rho}^{\,onebounce}$, the components of any $\vec{\rho}$ in these one-bounce and no-bounce directions are given by

$$\vec{\rho} = b_c \vec{\rho}^{\,nobounce} + b_1 \vec{\rho}^{\,onebounce} , \tag{6}$$

with

$$b_c = a_2/\sin\theta \,,\; b_1 = a_1 - a2/\tan\theta \,,$$

$$a_1 = \vec{\rho} \cdot \vec{\rho}^{\,onebounce} \,,\; a_2 = \vec{\rho} \cdot \vec{a}_2 \,,\; \vec{a}_2 \perp \left\{ \vec{a}_1 \equiv \vec{\rho}^{\,onebounce} \right\} .$$

The coefficient b_c is proportional to the shading, while b_1 is the relative magnitude of the light from the other surface.

There is a pathological case caused by white surfaces in which the SVD analysis will break down. For example, suppose a red surface interreflects with a nearby theroretical, white or grey surface (i.e. $S(\lambda) =$ constant). In that case, the no-bounce color on the red side will equal the one-bounce, interreflection color; and as a result, the second and third eigenvectors will arise from a combination of the second and higher bounces, not the first bounce.

The other obvious situation in which the SVD analysis cannot apply is when there is no interreflection.

In the next Section we apply our method to a simulated image to test the accuracy of the one-bounce model and determine the width of the band of possible no-bounce colors.

3 Simulated Image Test

3.1 Determining the one-bounce and no-bounce colors

We constructed a multispectral image of a theoretical edge composed of two semi-infinite planar surfaces illuminated by diffuse light (see Fig. 1) using the infinite-bounce method described in [2]. We used the reflectances of pink and green Pantone papers and the spectrum of standard illuminant A [23]. From the multispectral image, $\vec{\rho}$ vectors are calculated using the spectral transmission curves of the common color-separation filters, Kodak #25 (red), #58 (green) and #47B (blue).

To challenge the algorithm, we chose a fairly small opening angle $\gamma = 45°$ and made side B twice as long as side A. This means that even at the pixel located at the maximum width of side A, with $x = x_{max}$, we expect a comparatively high mutual illumination contribution. At the maximum width pixel on side B, though, the mutual illumination effect will be negligible.

The first question is how well does the one-bounce model fit this data? An SVD analysis on each side of the edge separately finds eigenvalues for side A of $\Lambda = (22.6, 0.141, 0.00296)$ and for side B of $\Lambda = (22.6, 0.454, 0.030)$. Since the first two eigenvalues outweigh the third in each case, the vectors $\vec{\rho}$ on each side are well modeled by a plane in RGB space. The $\vec{\rho}^{\,onebounce}$ direction is found by crossing these planes, and the normalized result is $\vec{\rho}^{\,onebounce} = (0.613, 0.625, 0.484)$

Does this color match the actual one-bounce color of the simulated edge? In RGB space, the angle between the derived $\vec{\rho}^{\,onebounce}$ and the correct one is 4.4°. As an alternate error measure, we calculate the rms error between the actual and derived one-bounce colors by forming

$$\mathcal{E} \equiv \left[\left(\vec{\rho}^{\,actual} - \vec{\rho}^{\,approx}\right)^2 / \left(\vec{\rho}^{\,actual}\right)^2\right]^{1/2} \times 100\% \tag{7}$$

The result is 7.7%, which seems reasonable.

Of course, we are more interested in whether or not the two no-bounce colors are correct and what choice of $\vec{\rho}^{\,nobounce}$ best represents the band of possible $\vec{\rho}^{\,nobounce}$ vectors. The ratios a_2/a_1 for side A range from 0.497 to 0.525. Therefore, the lower bound on the ratio is 0.525. The non-negativity constraint imposes an upper bound of 0.881 on the same ratio. Using the lower bound to calculate the corresponding no-bounce color for side A results in $\vec{\rho}^{\,nobounce} = (0.900, 0.400, 0.173)$. This compares favorably to the actual $\vec{\rho}^{\,nobounce} = (0.914, 0.376, 0.151)$, with an angular difference of 2.05° or $\mathcal{E} = 3.57\%$.

Using the upper bound to calculate the no-bounce color results in $\vec{\rho}^{\,nobounce} = (0.968, 0.251, 0.000)$ which is off by 13.7°. Clearly, this error is high. Even the average of the two estimates leads to an error of 4.82°. On side B the situation is even more lopsided: the lower-bound, no-bounce color has a small error— 0.435°— whereas the upper-bound vector has a large error— 27.9°.

Although the non-negativity constraint is theoretically well founded, it actually is producing a very poor limit on the band of no-bounce vectors because it requires that one out of the three RGB components be zero. In Tominaga et. al.[20], the same theoretical limit works better because there are many more samples taken across the spectrum. In that case, only one small section of the spectrum is restricted to zero in the limiting vector.

Based on the observation that the correct vector lies much closer to the band limit imposed by the lower bound constraint than that imposed by the upper bound constraint, we abandon the non-negativity part of the quarter-circle analysis as ineffective and shift to estimating a_2/a_1 entirely on the basis of the data point found to be farthest from the crossed-planes color. In essence, this is equivalent to assuming that each surface has at least one point where mutual illumination is negligible.

On a CRT display, the recovered no-bounce colors are indistinguishable by eye from the correct colors. The one-bounce color is an intermediate color between the pink and green of the Pantone papers.

3.2 Shading and Interreflection Factors

Once $\vec{\rho}^{\,onebounce}$ and $\vec{\rho}^{\,nobounce}$ are known, equation (6) will generate coefficients $b_c(x)$ and $b_1(x)$ for every image location x. Using $\vec{\rho}^{\,nobounce}$ for side A, the image $b_c(x)\,\vec{\rho}^{\,nobounce}$ predicts what side A would have looked liked had side B not been in the scene. The image $b_1(x)\vec{\rho}^{\,onebounce}$ represents the mutual illumination incident at each point. Since all the parameters of the simulated edge are known, we can assess how well the algorithm performs by comparing these images to corresponding theoretical ones.

Since the simulated edge was illuminated by diffuse light, we expect that in the absence of mutual illumination there will be no spatial variation in the shading anywhere on either surface. Thus $b_c(x)$ should be constant everywhere and equal to 1 if the normalization is chosen correctly. Fig. 3 shows that this expectation is borne out, with the normalization accomplished by dividing by the exact $\|\vec{\rho}^{\,nobounce}\|$. Normalization is necessary because the derived $\|\vec{\rho}^{\,nobounce}\|$ is always a unit vector; its true magnitude is not found.

The mutual illumination image $b_1(x)\,\vec{\rho}^{\,onebounce}$ should approximately equal the mutual illumination effects of a single bounce since the idea was to capture infinite-bounce effects within a one-bounce model and second and higher bounces are known to be relatively small. If the one-bounce assumption holds, the interreflection factor $b_1(x)$ should equal— up to a single, overall normalization constant— the configuration factor (i.e., the proportion of the light leaving the other surface that arrives at x [2, 7]).

Fig. 4 compares b_1 to the theoretical configuration factor for the simulated scene's geometry. The results are normalized relative to the actual one-bounce color by dividing by the known $\|\vec{\rho}^{\,onebounce}\|$ Results are good for side B, but not for side A, although b_1 still gives the correct shape for the configuration factor there.

The difference between side A and side B is that for side B the mutual illumination effect becomes negligible far from the edge, whereas this is not the case for side A. Since the lower bound estimate of the no-bounce color identifies the pixel with the least amount of one-bounce color as *identical* to the no-bounce color, the results are most satisfactory when there is indeed a spot with very little mutual illumination. Had we used the actual vectors $\vec{\rho}^{nobounce}$ in calculating b_1, results for both sides would have been excellent.

If there is one point where there is very little incident mutual illumination, then there probably will be a region of neighboring points having little or no mutual illumination, because mutual illumination effects tend to vary slowly spatially. This observation points to a verification procedure that can be used to establish whether or not the algorithm has been applied to an image on which it can succeed. Namely, the algorithm's results are reliable when a significant number of pixels in the original image have the same color as the derived no-bounce color. The converse is not the case.

In the next Section we apply the algorithm to a real laboratory image.

4 Real Image Test

4.1 Experimental Setup

We used colored Pantone papers to make the convex surfaces. To the eye, these papers appear relatively free of specular highlights.

Fig. 5 shows an approximately planar strip of pale blue paper (Pantone #290) making an angle of $\sim 45°$ with another strip (#360) colored grass green. The blue strip was nearly vertical and the green strip was inclined, with the camera placed directly above the scene. In the image, the blue strip is at the top and the green strip is at the bottom. We placed black cardboard behind the scene.

Mutual illumination induces a color shift that is surprisingly hard to discern by eye, although it is clearly evident in the numerical image data. Since the color image is not that informative, Figure 5 presents a black and white photograph of the color image.

A tungsten light at a distance of ~ 1 m and at an angle of $\sim 45°$ to the horizontal illuminated the scene. Color images were taken using a CCD Camera with infrared cutoff and color separation filters (Kodak Wratten filters #25, #58 and #47B). The camera responds linearly with intensity, but with unequal sensitivity across the spectrum. For the method described in this paper, sensitivity calibration is not required.

4.2 Results of Color Space Analysis

We first segmented the image into three regions: background, blue strip, green strip. SVD analysis of the green strip produces eigenvalues $\Lambda = (3835.2, 43.0, 16.2)$. For the blue strip, $\Lambda = (5482.9, 72.7, 19.9)$. Clearly, the data does not fit a plane as well for this real image as it did for the simulated image. Factors that may be contributing to the third dimension are noise, stray light, and small amounts of specularity.

The intersection of the two planes defines the green-to-blue (or equivalently blue-to-green), one-bounce color, $\vec{\rho}^{onebounce} = (0.573,0.653,0.496)$. Since the green Pantone paper is more saturated than the blue, this color appears a bit more green than blue.

Fig. 6 histograms the ratio a_2/a_1 for pixels from the green strip. The maximum value for the ratio is 0.154 and it is not an outlier. This ratio defines no-bounce green of $\vec{\rho}^{nobounce} = (0.673, 0.633, 0.383)$.

For comparison, the actual green measured under the same illumination but at a location where there was no mutual illumination from the blue strip was $\vec{\rho}^{nobounce} = (0.662, 0.636, 0.397)$. This measured no-interreflection color differs by 1.05° from that produced by the algorithm.

To assess the algorithm's results without knowing the actual reflectances of the two colored strips, we can reconstruct an image based on the one-bounce analysis and compare it to the actual image, where the reconstructed image is obtained from $b_1(x)$ and $b_c(x)$ via equation (6). For the green strip, the errors $\mathcal{E}$ between actual data and reconstructed one-bounce model values have median value of only 0.27%. For the blue strip, this value is 0.25%. Clearly, the one-bounce model performs well.

To further assess the results, we apply the verification procedure described above. In particular, the test finds that many pixels in the original image are close in color to the derived $\vec{\rho}^{nobounce}$, which indicates that there is a high likelihood that some location in the image had an insignificant amount of mutual illumination impinging on it and therefore that $\vec{\rho}^{nobounce}$ is correct. Fig. 7 shows a histogram of differences $\mathcal{E}$ between the derived no-bounce color on the green side and the RGB values of the original image. Many pixels fall close to the calculated green. The average difference from the no-bounce color is 3.4% and the percentage of pixels within 3% of the no-bounce color is 39.1%.

For the blue strip, Fig. 8 shows the distribution of a_2/a_1. The maximum value of the ratio is 0.232. Composing a no-bounce color from this ratio we find $\vec{\rho}^{onebounce} = (0.735, 0.576, 0.357)$. Fig. 9 shows the distribution of $\mathcal{E}$ from this color for the blue strip. The average difference from the no-bounce color is 3.5% and the percentage of pixels within 3% of the no-bounce color is 35.4%. The actual blue measured at a location without mutual illumination, but under the same direct illumination from the light bulb, was $\vec{\rho}^{nobounce} = (0.724, 0.5834, 0.367)$; the vector produced by the algorithm was 1.0° away from this color.

4.3 Output images

The algorithm decomposes the original image into two component images. Fig. 10 shows the original scene with mutual illumination removed (i.e., $b_c(x)\,\vec{\rho}^{nobounce}$). Fig. 11 shows the image of the mutual illumination component (i.e. $b_1(x)\,\vec{\rho}^{onebounce}$). Since this effect is small, the image has been multiplied by the factor 3 for display. As should be expected, the image of mutual illumination is brightest near the edge, where the interreflection effect is strongest.

Fig. 12 shows which parts of the original image are similar in color to the calculated no-bounce color by masking the no-bounce color with a 1 when the percentage difference $\mathcal{E}$ between the measured pixel values and the no-bounce color is below a threshold of 3%. In the resulting image, there are connected regions close to the calculated color, so we can have confidence in the algorithm's results. These regions are away from the edge, where the effect of mutual illumination diminishes.

5 Conclusion

Under a one-bounce model of interreflection, the RGB vectors for a region of uniform color lie in a plane. Intersection of two such planes from two interreflecting regions yields the color of the interreflected light. This one-bounce color provides one of the two axes on which the original data can be projected. By projecting the data onto this axis, an image of the mutual illumination field is produced. Quarter-circle analysis constrains the second axis— the no-bounce color— to a band, but in fact, only one limit of the band turns out to be useful. Projection of the original image onto the axis defined by this limit results in a no-interreflection image (i.e., an image of each surface *as if* the other surface had not been present).

The method generalizes to interreflections caused by self-reflection. For example, an image of the concave surface inside a cup will have a second illumination component due to the mutual illumination from the cup to itself, and the color of the interreflected light will not in general be the same as the cup itself. In situations generating three components (e.g. a main reflection component, self-reflection and interreflection; or main reflection, interreflection and specularity) the full Lawton-Sylvestre-Kawata method, in principle, should apply, but we have not tested it.

The color space method is fast, simple and includes a reliability check. Results from tests on real and synthesized images show that it works well. Any shape-from-shading method can be applied directly to the no-interreflection image. The mutual illumination image approximates an image of configuration factors encoding other potentially useful geometrical information about the scene.

6 Acknowledgements

M.S. Drew is indebted to the Centre for Systems Science at Simon Fraser University for partial support; B.V. Funt thanks both the CSS and the Natural Sciences and Engineering Research Council of Canada for their support.

References

[1] M.H. Brill. Image segmentation by object color: a unifying framework and connection to color constancy. *J. Opt. Soc. Am. A*, 7:2041–2047, 1990.

[2] M.S. Drew and B.V. Funt. Calculating surface reflectance using a single-bounce model of mutual reflection. In *Proceedings: International Conference on Computer Vision, Osaka, Dec.4-7/90*, pages 393–399. IEEE, 1990.

[3] M.S. Drew and B.V. Funt. Variational approach to mutual illumination in color images. Technical Report CSS/LCCR TR 90-21, Simon Fraser University School of Computing Science, 1990.

[4] D. Forsyth. A novel algorithm for color constancy. *Int. J. Comput. Vision*, 5:5–36, 1990.

[5] D. Forsyth and A. Zisserman. Mutual illumination. In *Proceedings: Computer Vision and Pattern Recognition*, pages 466–473. IEEE, 1989.

[6] D. Forsyth and A. Zisserman. Shape from shading in the light of mutual illumination. *Image and Vision Computing*, 8:42–49, 1990.

[7] B.V. Funt, M.S. Drew, and J. Ho. Color constancy from mutual reflection. *Int. J. Computer Vision*, 6:5–24, 1991.

[8] G.H. Golub and C.F. van Loan. *Matrix Computations*. John Hopkins U. Press, 1983.

[9] C.M. Goral, K.E. Torrance, D.P. Greenberg, and B. Battaile. Modeling the interaction of light between diffuse surfaces. *Computer Graphics*, 18:213–222, 1984.

[10] J. Ho, B.V. Funt, and M.S. Drew. Separating a color signal into illumination and surface reflectance components: Theory and applications. *IEEE Trans. Patt. Anal. and Mach. Intell.*, 12:966–977, 1990.

[11] B.K.P. Horn. Understanding image intensities. *Artif. Intell.*, 8:201–231, 1977.

[12] J.E. Kaufman. *IES Lighting Handbook*. Illuminating Engineering Society, 4th edition, 1966.

[13] S. Kawata, K. Sasaki, and S. Minami. Component analysis of spatial and spectral patterns in multispectral images. i. basis. *J. Opt. Soc. Am. A*, 4:2101–2106, 1987.

[14] G.J. Klinker, S.A. Shafer, and T. Kanade. The measurement of highlights in color images. *Int. J. Comput. Vision*, 2:7–32, 1988.

[15] W.H. Lawton and E.A. Sylvestre. Self modeling curve resolution. *Technometrics*, 13:617–633, 1971.

[16] S.K. Nayar, K. Ikeuchi, and T. Kanade. Shape from interreflections. In *Proceedings: International Conference on Computer Vision, Osaka, Dec.4-7/90*, pages 2–11. IEEE, 1990.

[17] S.A. Shafer. Using color to separate reflection components. *Color Res. Appl.*, 10:210–218, 1985.

[18] R. Siegel and J.R. Howell. *Thermal Radiation Heat Transfer*. Hemisphere Pub. Corp., 1981.

[19] S. Tominaga and B.A. Wandell. Standard surface-reflectance model and illuminanat estimation. *J. Opt. Soc. Am. A*, 6:576–584, 1989.

[20] S. Tominaga and B.A. Wandell. Component estimation of surface spectral reflectance. *J. Opt. Soc. Am. A*, 7:312–317, 1990.

[21] F. Tong and B.V. Funt. Specularity removal for shape from shading. In *Proceedings: Vision Interface 1988*, pages 98–103, Edmonton, Alberta, Canada, 1988.

[22] F. Tong and B.V. Funt. Removing specularities from color images for shape from shading. In A. Krzyzak, editor, *Computer Vision and Shape Recognition*. World Scientific Publishing, 1989.

[23] G. Wyszecki and W.S. Stiles. *Color Science: Concepts and Methods, Quantitative Data and Formulas*. Wiley, New York, 2nd edition, 1982.

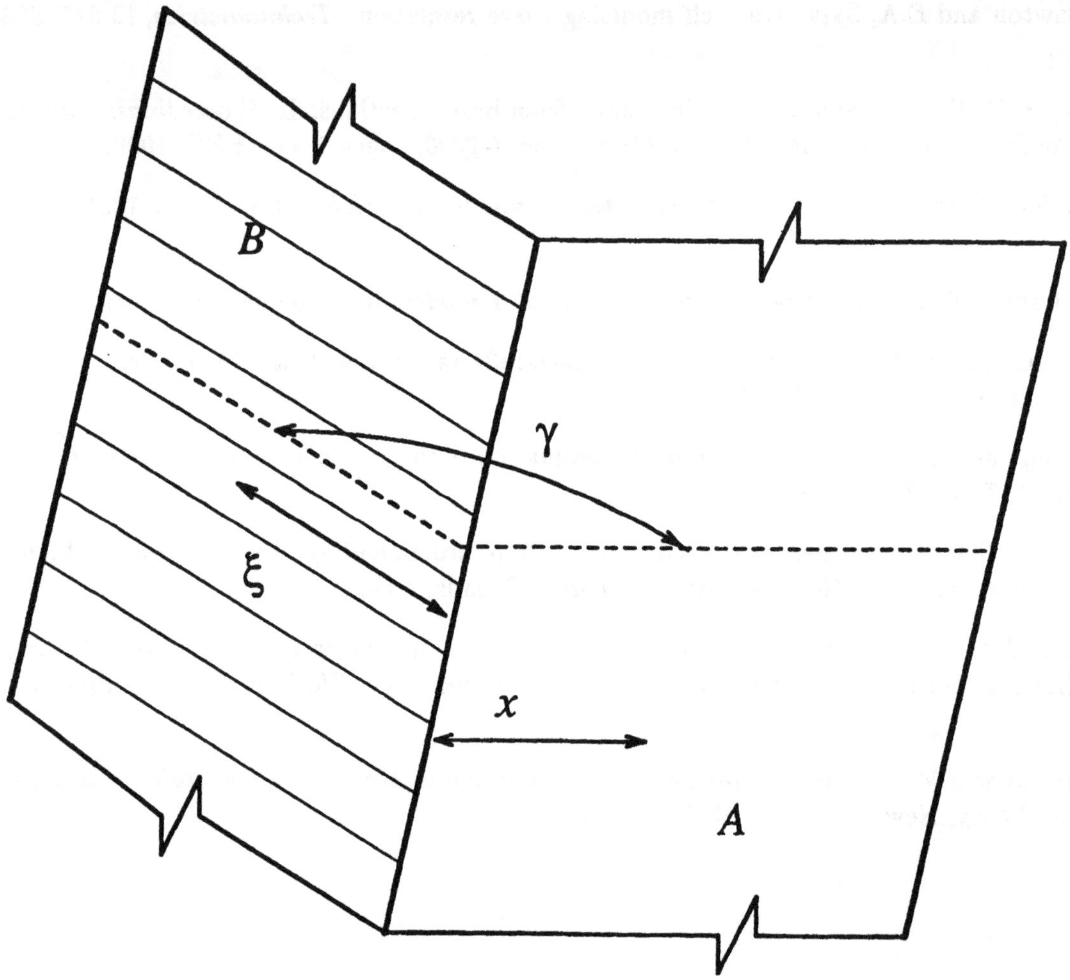

Figure 1. Two semi-infinite planes that join with opening angle γ. In Section 3, we use $\gamma = 45°$, creating an edge that appears much more closed.

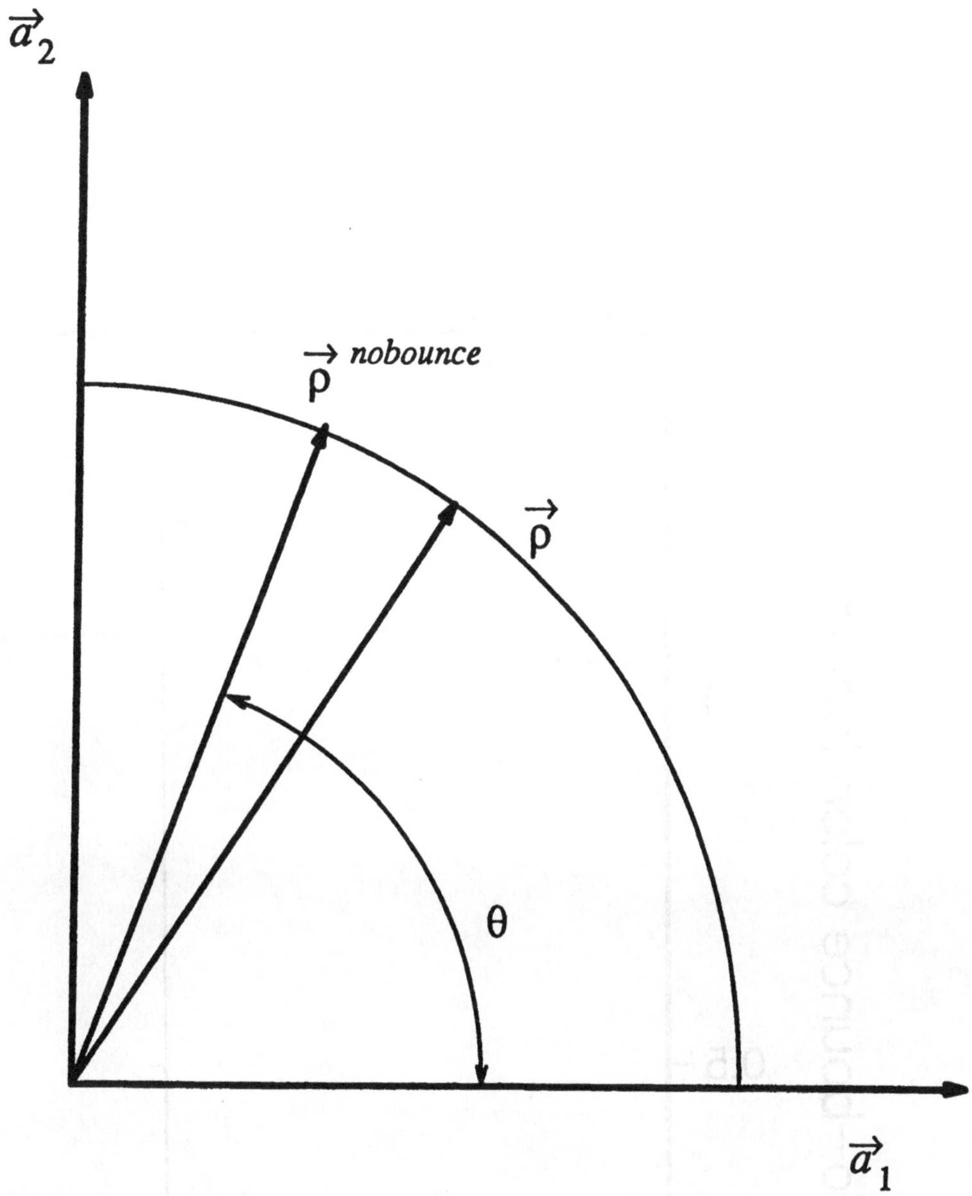

Figure 2. Projecting normalized RGB vectors onto the cross-planes direction and a perpendicular to it in the object's color plane yields vectors all on a quarter-circle.

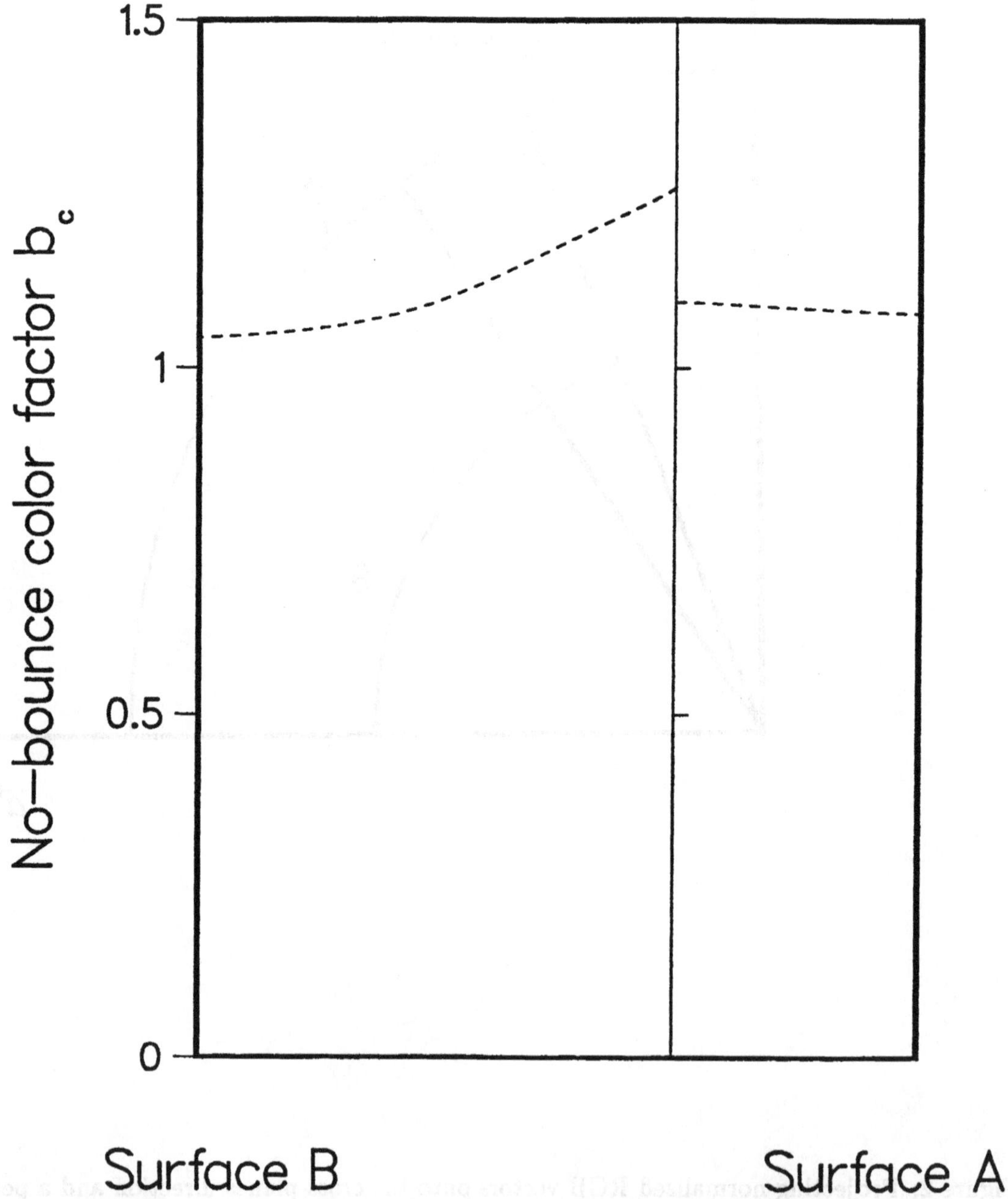

Figure 3. Shading factors b_c recovered by the algorithm are shown for the planar edge of Fig. 1, with the width of surface B equal to twice the width of surface A and $\gamma = 45°$.

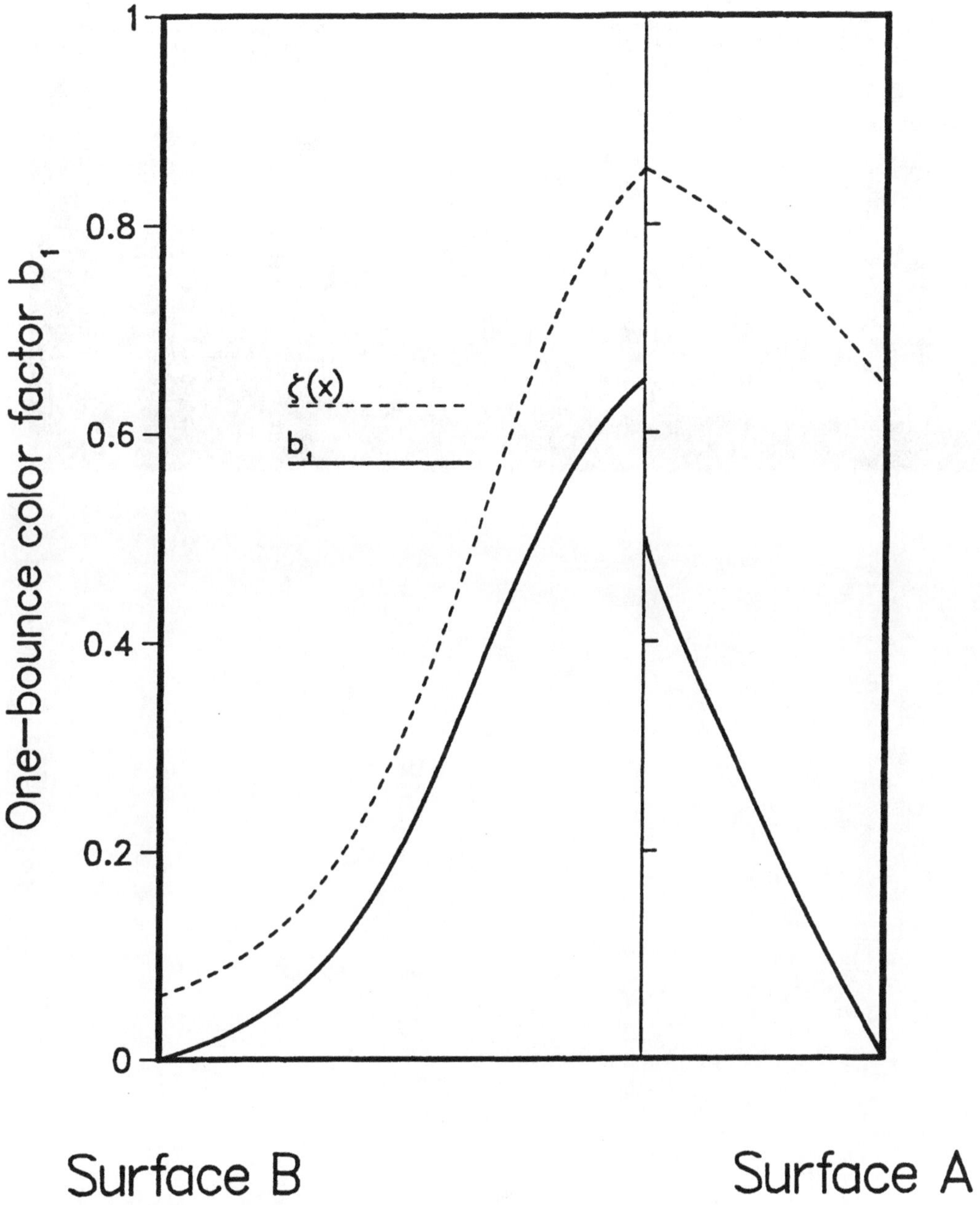

Figure 4. Interreflection factors b_1 recovered by the algorithm for the planar edge (___). For comparison is the theoretical factor ζ for a single bounce (— —).

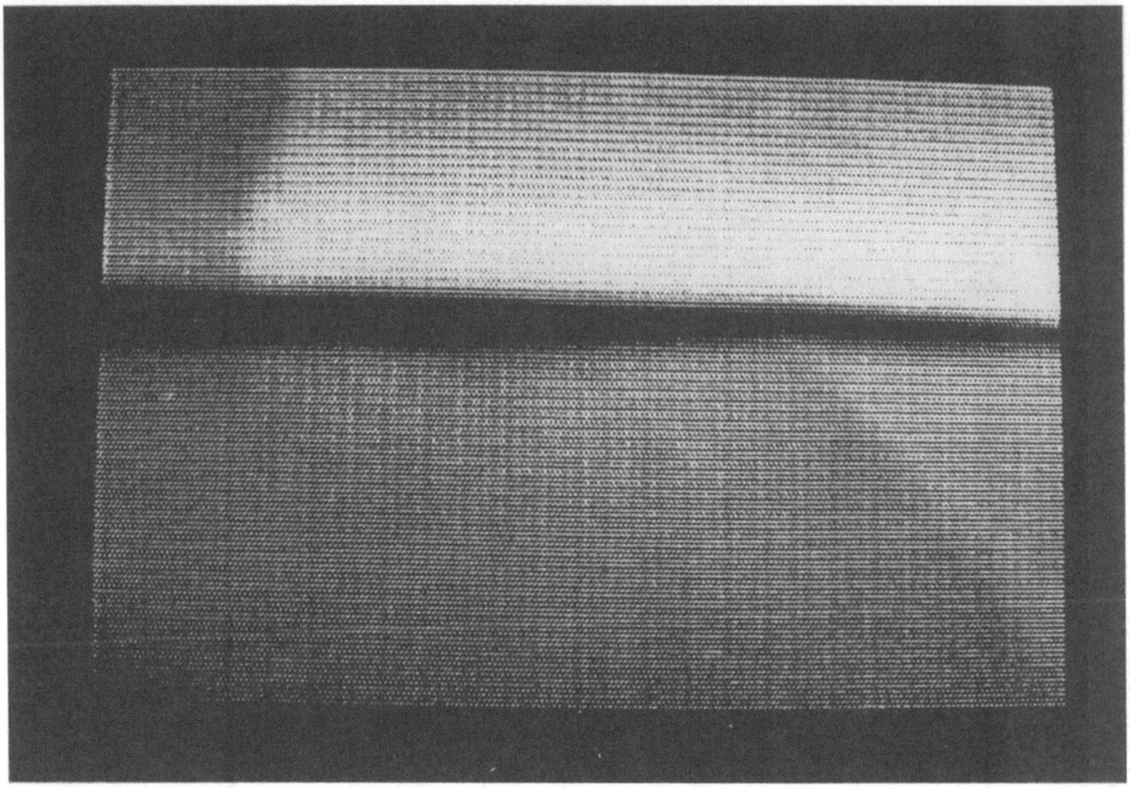

Figure 5. Image of a blue strip and a green strip. Since the color shift due to mutual illumination is not easily discernible in any case, the image is only shown in black and white.

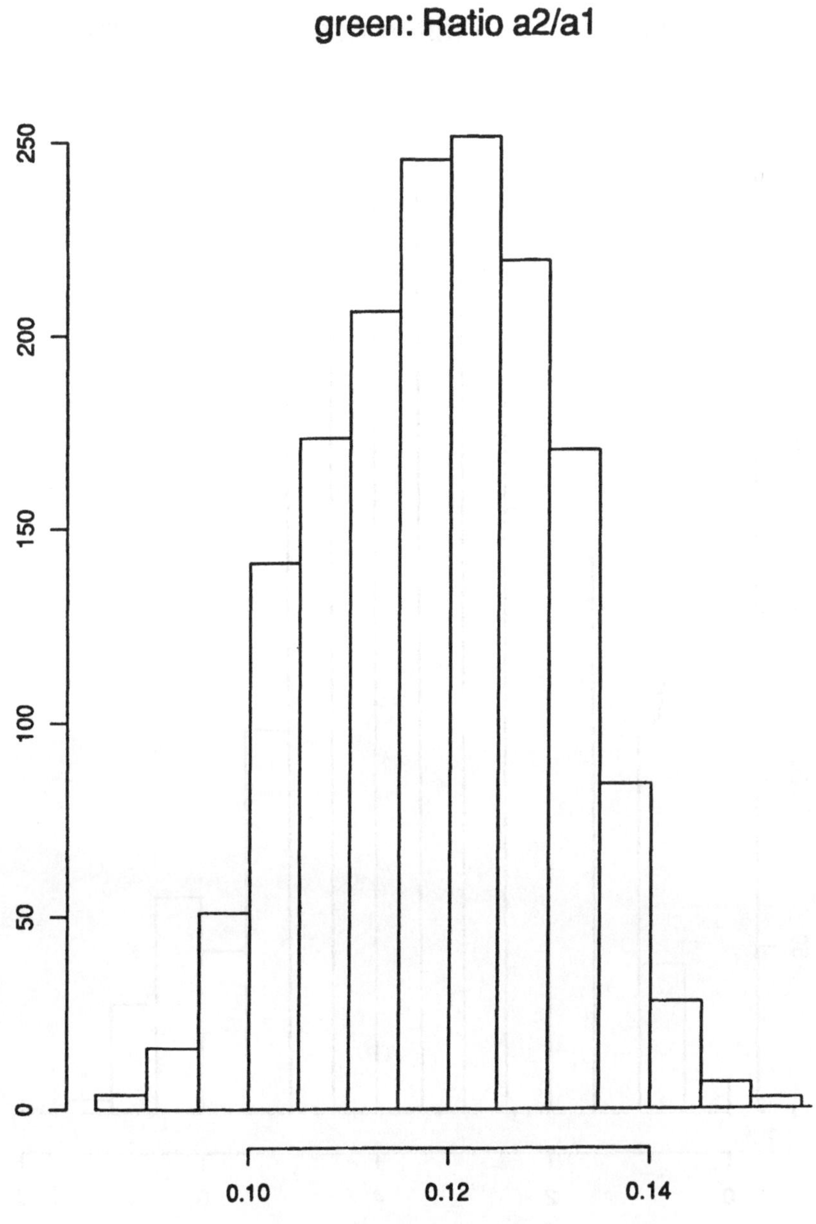

Figure 6. Histogram of ratio a_2/a_1 for the green strip's color space plane.

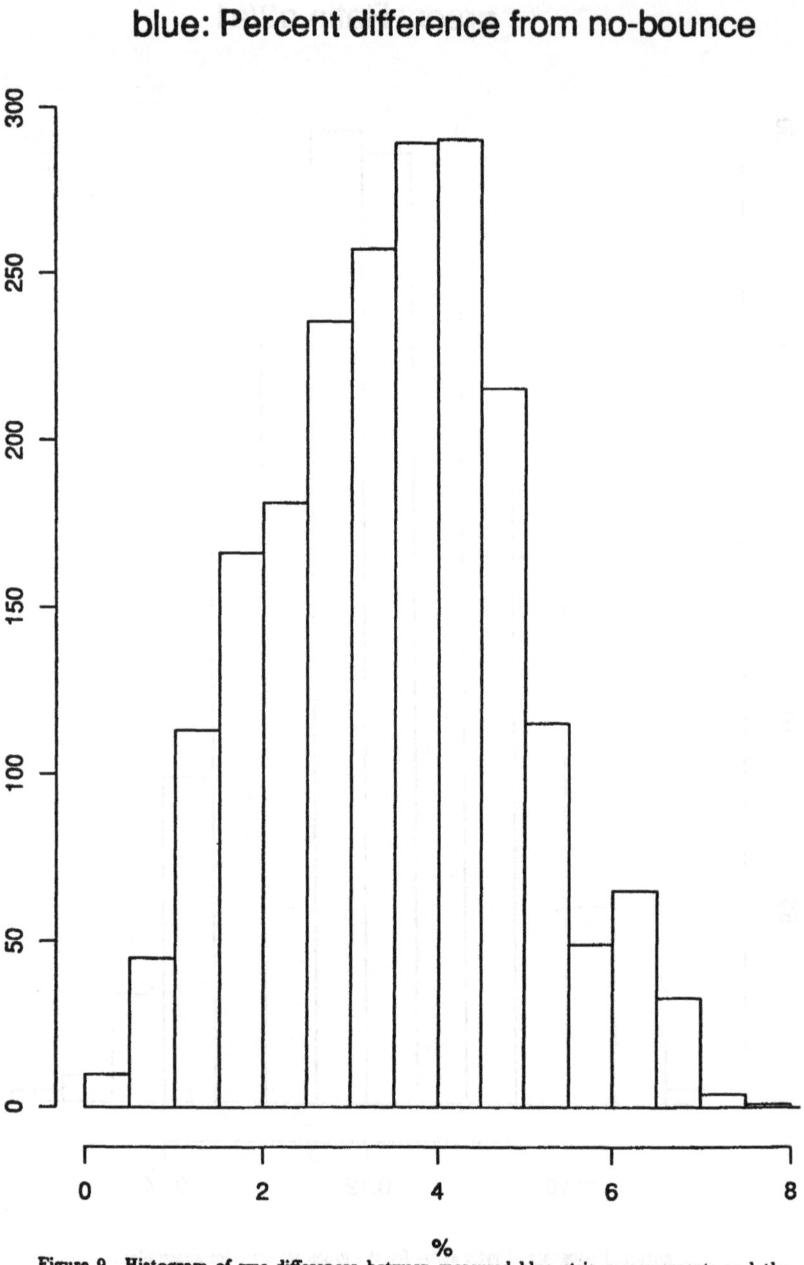

Figure 9. Histogram of rms differences between measured blue strip measurements and the calculated one-bounce color.

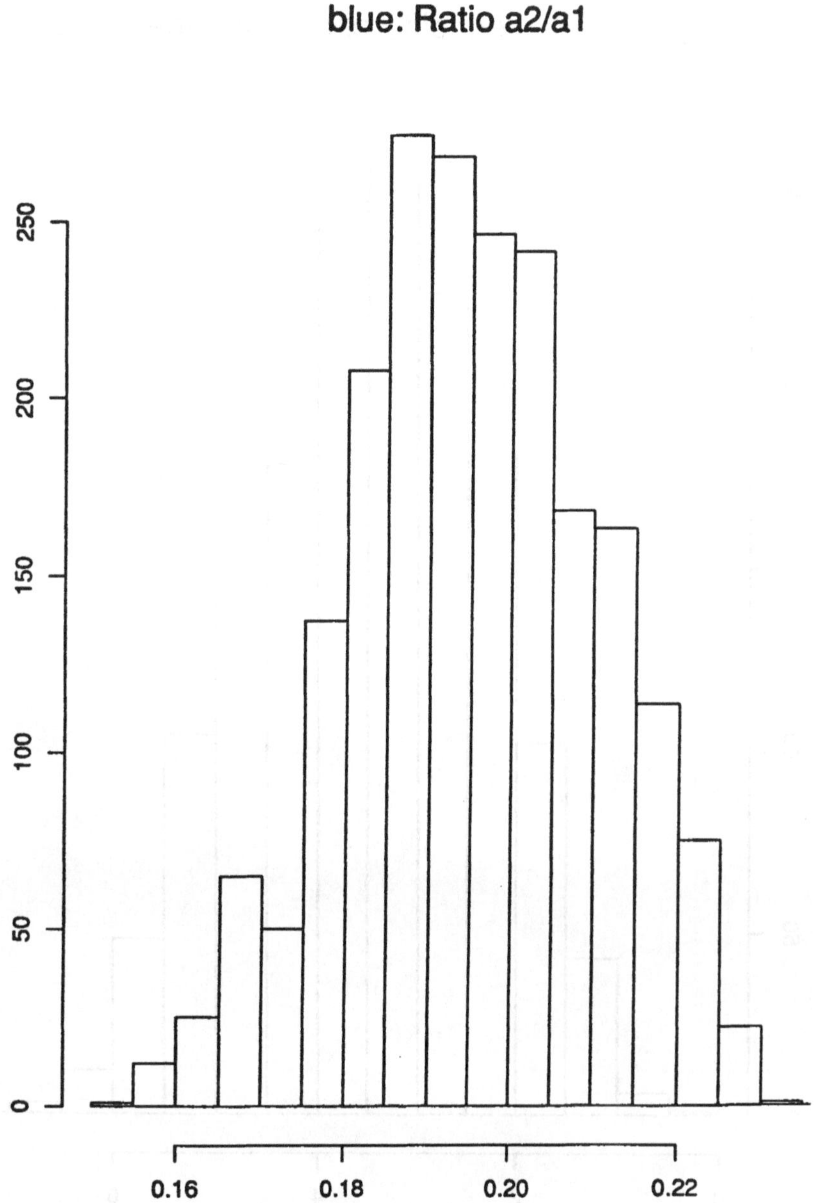

Figure 8. Histogram of ratio a_2/a_1 for the blue strip's color space plane.

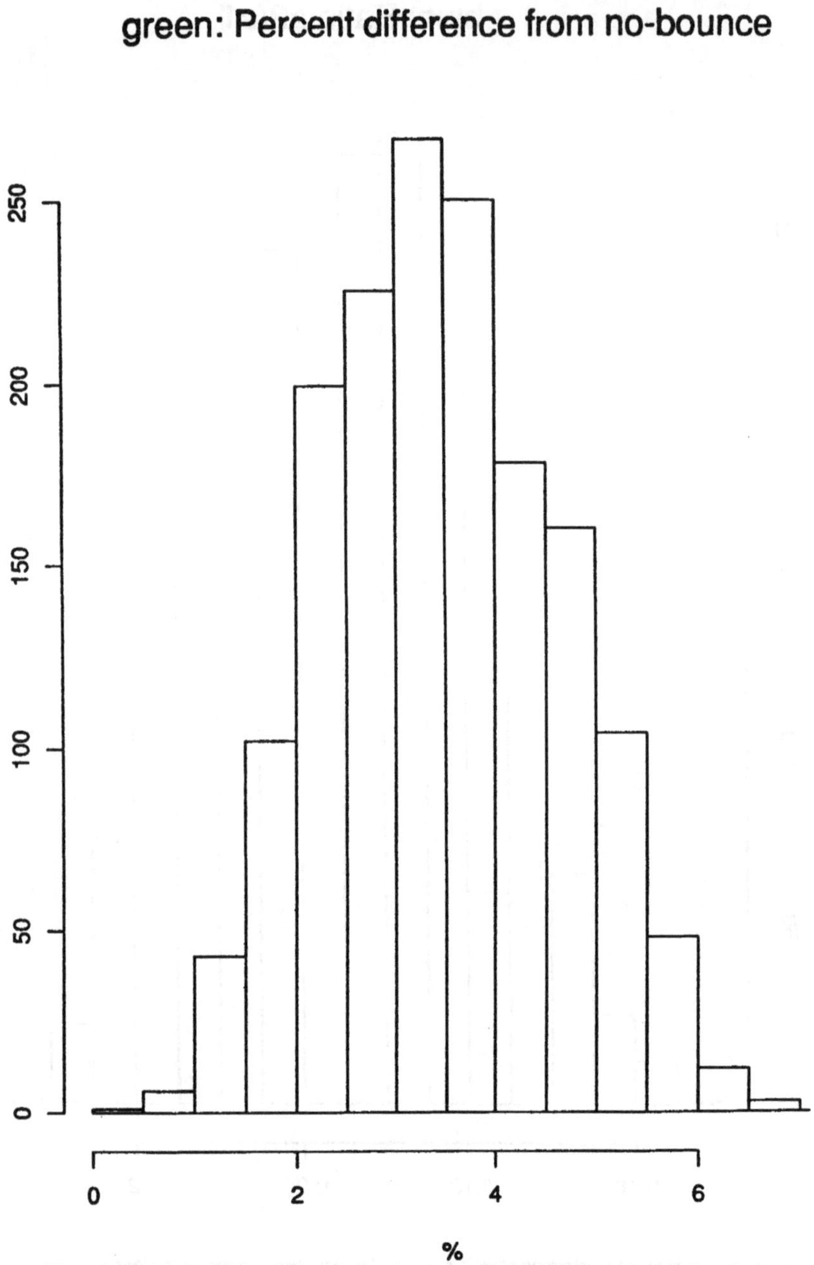

Figure 7. Histogram of rms differences between measured green strip pixels and the calculated one-bounce color.

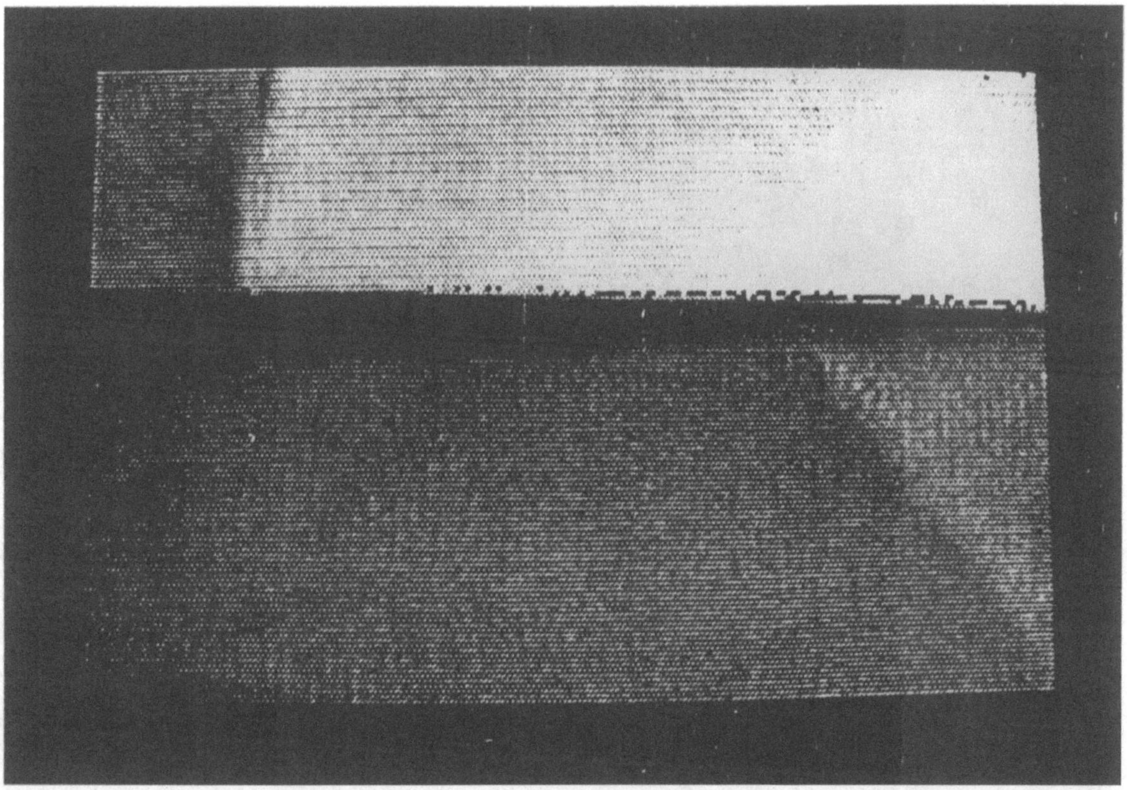

Figure 10. Black and white version of the no-bounce image representing the original image with interreflections removed. The shading is as it would have been in the absence of mutual illumination.

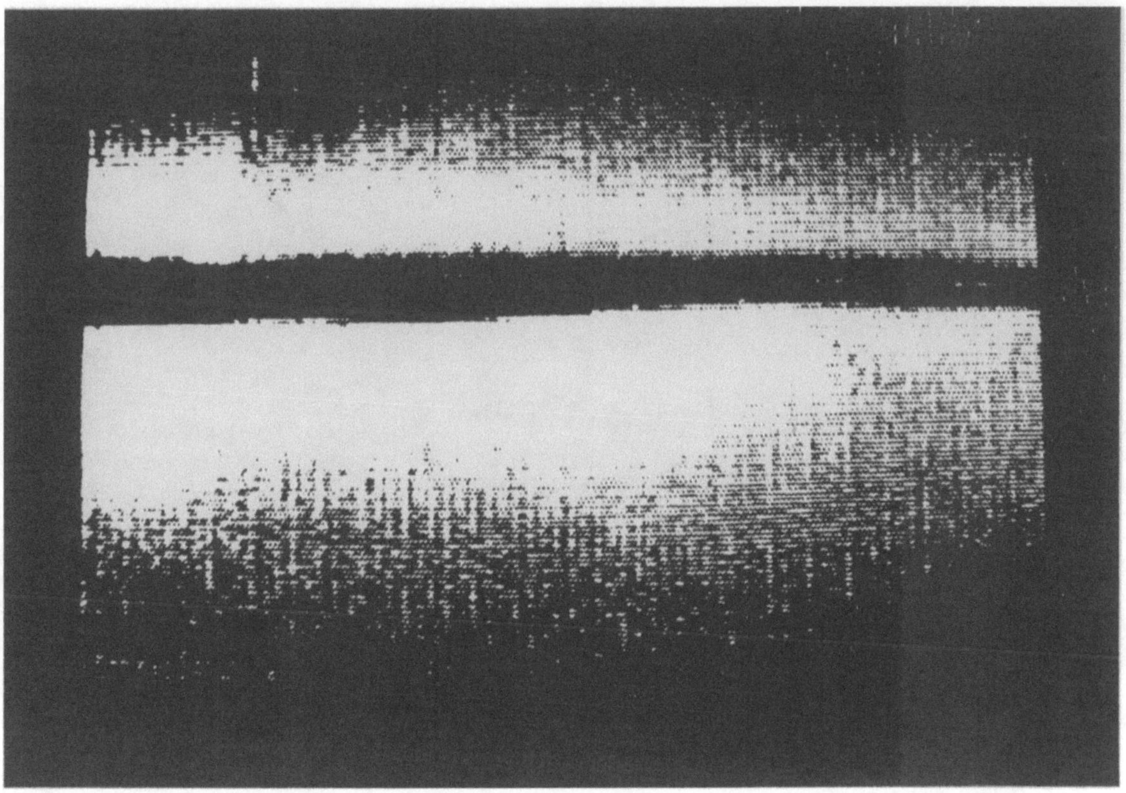

Figure 11. Image of the one-bounce color (i.e., the interreflection factors).

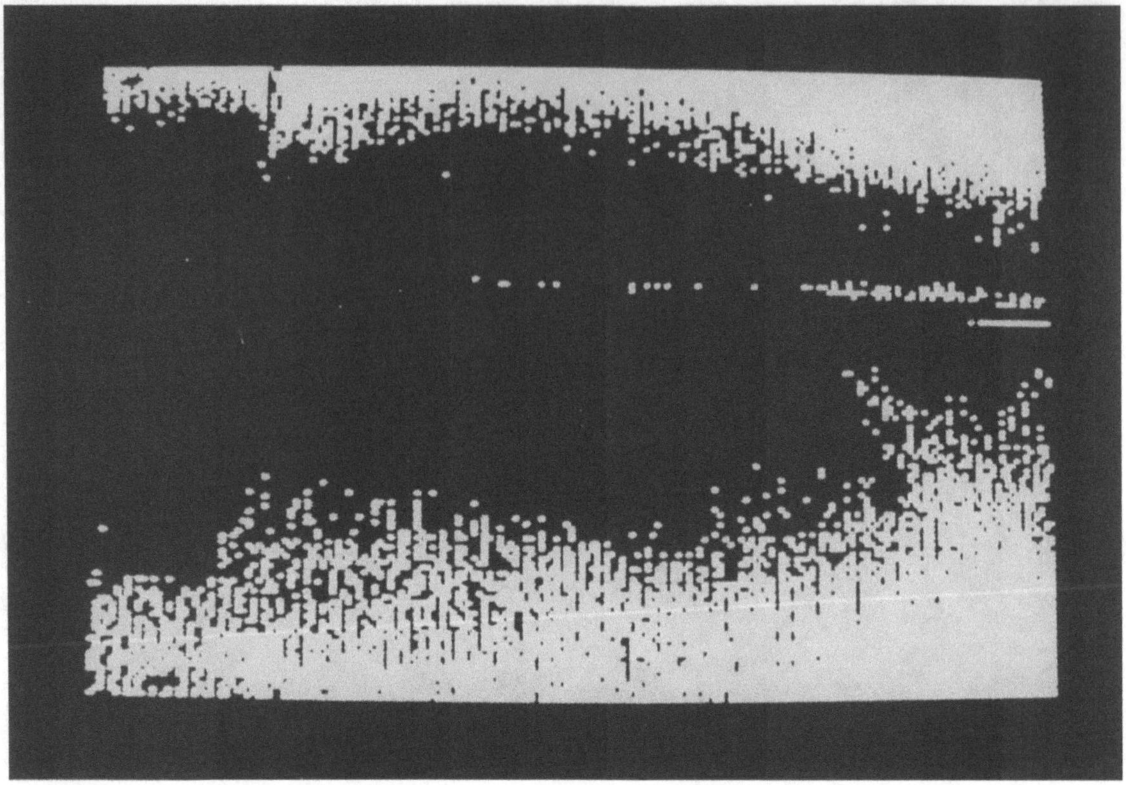

Figure 12. Image of no-bounce colors, shown in black and white, masked with 1 or 0 depending on whether the true color is close to the calculated no-bounce color.

Author Index

Volume abbreviations: **C** Color, **R** Radiometry, **S** Shape Recovery

Ahuja, N. . . . . **R** 361
ASTM . . . . **R** 146

Bajcsy, R. . . . . **C** 199
Barrow, H.G.. . . . . **R** 1
Beyer, H.A. . . . . **R** 245, 266
Binford, T.O.. . . . . **S** 243
Birk, J. . . . . **S** 156
Blake, A. . . . . **S** 277, 287
Boult, T.E. . . . . **R** 167
Brainard, D.H. . . . . **C** 208
Brelstaff, G. . . . . **S** 277
Breneman, E.J. . . . . **C** 79
Brill, M.H.. . . . . **C** 109, 116
Bruckstein, A. . . . . **S** 171

Chellappa, R. . . . . **S** 39, 62
Coleman, E.N. . . . . **S** 180
Cook, R.L.. . . . . **R** 42
Cushing, K. . . . . **R** 256

Das, S. . . . . **R** 361
deFigueiredo, R.J.P. . . . . **S** 213
Drew, M.S.. . . . . **C** 272, 365, 385
Dupuis, P. . . . . **S** 17
D'Zmura, M.. . . . . **C** 224

Ferrie, F.P.. . . . . **S** 102
Fischler, M.A. . . . . **R** 1
Forsyth, D. . . . . **C** 241; **S** 313
Frankot, R.T.. . . . . **S** 62
Funt, B.V. . . . . **C** 272, 365, 385

Gershon, R. . . . . **C** 101
Giblin, P. . . . . **S** 287
Ginsberg, I.W. . . . . **R** 94
Goshtasby, A.. . . . . **R** 274
Grimson, W.E.L. . . . . **S** 434

Haralick, R.M. . . . . **R** 394
Havelock, D.I. . . . . **R** 305
Healey, G.E. . . **C** 52, 129, 166, 335; **R** 279; **S** 243
Ho, J.. . . . . **C** 272, 365
Horn, B.K.P. . . . . . . **R** 16, 84; **S** 5, 389
Howard, W. . . . . **R** 256
Hsia, J.J. . . . . **R** 94

Ikeuchi, K. **R** 60; **S** 200, 268, 322, 467, 484, 493

Jain, R.. . . . . **S** 180
Jepson, A.D.. . . . . **C** 101

Kamgar-Parsi, Behrooz and Behzad . **R** 293
Kanade, T. . . **C** 134, 309, 351; **R** 60; **S** 200, 322, 346, 493
Kelley, R.B. . . . . **S** 156
Kender, J.R.. . . . . **S** 378
Klinker, G.J.. . . . . **C** 134, 309, 351
Koenderink, J.J. . . . . **S** 305
Kondepudy, R. . . . . **R** 279
Koshikawa, K. . . . . **R** 190
Krotkov, E. . . . . **R** 371
Krumm, J. . . . . **R** 316

Lee, C.-H.. . . . . **S** 292
Lee, H.-C.. . . . . **C** 79, 303, 340
Lee, S.W. . . . . **C** 199
Lennie, P. . . . . **C** 224
Leonardis, A. . . . . **C** 199
Levine, M.D.. . . . . **S** 102
Limperis, T. . . . . **R** 94

Maloney, L.T. . . . . **C** 87, 219
McClellan, J.R.. . . . . **R** 256
Murase, H. . . . . **R** 213

Nayar, S.K. . . . . . **R** 60, 347; **S** 200, 322
Nicodemus, F.E. . . . . **R** 94
Novak, C.L. . . . . **C** 1, 13, 284, 351

Oliensis, J. . . . . **S** 17
Onn, R.. . . . . **S** 171

Palmer, A.. . . . . **R** 256
Park, J.-S.. . . . . **S** 298
Pentland, A.P. . . . . . . **R** 331; **S** 29, 422
Poggio, T. . . . . **S** 408

Ray, R. . . . . **S** 156
Richmond, J.C.. . . . . **R** 94

Sato, K. . . . . **S** 484
Schulte, C.P. . . . . **C** 79
Schunk, B.G. . . . . **S** 389
Shafer, S.A. . **C** 1, 13, 43, 134, 284, 309, 351; **R** 316, 386; **S** 346
Shapiro, L.G. . . . . **R** 394
Sjoberg, R.W. . . . . **R** 84
Smith, E.M. . . . . **S** 378
Smith, G.B. . . . . **S** 459

Sparrow, E.M. . . . . R 32
Subbarao, M. . . . . R 340

Tagare, H.D. . . . . S 213
Tarabanis, K. . . . . R 399
Tenenbaum, J.M.. . . . . R 1
Thrift, P. . . . . S 292
Tominaga, S. . . . . C 70, 235
Torrance, K.E.. . . . . R 32, 42
Tou, J.T. . . . . S 298
Tsai, R.Y. . . . . R 221, 399
Tsotsos, J.K.. . . . . C 101

van Doorn, A.J. . . . . S 305
Verri, A. . . . . S 408

Wandell, B.A. . . . C 28, 70, 208, 219, 235
Willson, R.G. . . . . C 13; R 386
Wittels, N. . . . . R 256
Wolff, L.B. . . . . R 167, 193, 200; S 233
Woodham, R.J.. . . . . . . S 115, 121, 417

Yi, S.. . . . . R 394

Zheng, Q.. . . . . S 39
Zisserman, A. . . . . S 287, 313

Jones and Bartlett Books in Computer Science and Related Areas

Barnsley, M., *The Fractal Transform*
ISBN 0-86720-218-1

Bernstein, A.J., and Lewis, P.M., *Concurrency in Programming and Database Systems*
ISBN 0-86720-205-X

Birmingham, W.P., Gupta, A.P., and Siewiorek, D.P., *Automating the Design of Computer Systems: The MICON Project*
ISBN 0-86720-241-6

Chandy, K.M., and Taylor, S., *An Introduction to Parallel Programming*
ISBN 0-86720-208-4

Flynn, A., and Jones, J., *Mobile Robots: Inspiration to Implementation*
ISBN 0-86720-223-8

Geometry Center, University of Minnesota, *Not Knot* (VHS video)
ISBN 0-86720-240-8

Healey, G., *et al.* (eds.), *Physics-Based Vision: Principles and Practice, COLOR*
ISBN 0- 86720-295-5

Iterated Systems, Inc., *Floppy Book: A P.OEM PC Book*
ISBN 0-86720-222-X

Iterated Systems, Inc., *SNAPSHOTS: True-Color Photo Images Using the Fractal Formatter*
ISBN 0-86720-299-8

Klinker, G.J., *A Physical Approach to Color Image Understanding*
ISBN 0-86720-237-8

Lee, E. S., *Algorithms and Data Structures in Computer Engineering*
ISBN 0-86720-219-X

Myers, B.A. (ed.), *Languages for Developing User Interfaces*
ISBN 0-86720-450-8

Parke, F.I., and Waters, K., *Computer Facial Animation*
ISBN 0-86720-243-2

Whitman, S., *Multiprocessor Methods for Computer Graphics Rendering*
ISBN 0-86720-229-7

Wolff, L., *et al.* (eds.), *Physics-Based Vision: Principles and Practice, RADIOMETRY*
ISBN 0-86720-294-7

Wolff, L., *et al.* (eds.), *Physics-Based Vision: Principles and Practice, SHAPE RECOVERY*
ISBN 0-86720-296-3